# Attribute Sampling

# Attribute Sampling

## TABLES AND EXPLANATIONS

Tables for Determining Confidence Limits
and Sample Size Based on Close Approximations
of the Binomial Distribution

## Herman Burstein

*Associate Professor of Economics*
*New College of Hofstra University, Hempstead, New York*

**McGRAW-HILL BOOK COMPANY**

*New York      St. Louis      San Francisco      Düsseldorf      Johannesburg*
*Kuala Lumpur      London      Mexico      Montreal      New Delhi*
*Panama      Rio de Janeiro      Singapore      Sydney      Toronto*

ATTRIBUTE SAMPLING

# Contents

v

## PART THREE

## PART FOUR

# Acknowledgments

This books rests principally on three articles (References [1], [2], and [3] in Appendix F) by Prof. T. W. Anderson (Department of Statistics, Stanford University) and myself. I am profoundly indebted to Prof. Anderson for the groundwork that has made this book possible, as well as for much other guidance during our association.

Articles [1] and [2] are largely based on work for S. D. Leidesdorf & Co., Certified Public Accountants, New York City. This work was done over a period of several years, when Prof. Anderson (at that time Chairman of the Department of Mathematical Statistics at Columbia University) served as the firm's Statistical Consultant, and I as its Economist. Many individuals in the firm were of assistance, and at least the following deserve specific mention: Saul Beldock and Harold Cohan for nurturing the research; Robert Gevirtzman, Edgar Mack, Stephen Reiss, and David Shulman for writing computer programs, making computations, preparing and checking tables, and other phases of our research.

To Dr. J. Arthur Greenwood (New York University) I am deeply indebted for a variety of suggestions by which the above articles and consequently this book benefited. A key role is played by approximation formulas he supplied for confidence limits of the parameter of a Poisson distribution.

Research leading to this book continued after I joined the faculty of New College of Hofstra University in 1967. Here I had the valuable assistance of several students: Jeanne Ferrante and Don Bieniewicz, who skillfully developed computer programs for the principal tables; Jennifer Calagan and Raymond Moy, who painstakingly checked all text and mathematics and made helpful suggestions; and Seymour Liebman, who performed a variety of calculations.

In additional ways this book owes much to Hofstra University—for extensive use of the facilities of the Computer Center to prepare tables, and for grants to cover expenses associated with my research. In these respects I want to thank in particular Dr. Nathan Goldfarb and Michael Goldberg of the Computer Center and Dean David Christman of New College.

Finally, I want to thank my daughter Daury Ann, who helped check a number of the examples and tables in the book.

I hope that the many persons who contributed to this book but are not specifically mentioned above will accept my thanks.

*Herman Burstein*

# Attribute Sampling

# PART ONE

# Introduction

A statistical attribute sample consists of n items drawn at random from a population of N items, with c items in the sample observed to have a designated characteristic. Each c item is an event. Based on the sample proportion $c/n$, a conclusion is reached about p, the proportion of events in the population.

It is unlikely that $c/n$ is exactly the same as p. But based on probability theory one may state a confidence interval presumably containing p, and a degree of confidence that p lies within this interval. The end points of the interval are called *confidence limits*: $\bar{p}$ is the upper limit, and $\underline{p}$ is the lower limit. The degree of confidence is called the *confidence level* $\gamma$, which may take the form of either $\gamma_1$ or $\gamma_2$: $\gamma_1$ pertains to a one-sided limit, that is, to a statement about only $\bar{p}$ or only $\underline{p}$; $\gamma_2$ pertains to two-sided limits, that is to a statement about both $\bar{p}$ and $\underline{p}$.

To illustrate, assume that from a very large population of adults a random sample n is taken, with an event c defined as a person who did not graduate from high school, and that $n = 100$ and $c = 20$.

3

The estimate of the proportion p of nongraduates in the population is c/n = .20. If n = 100, c = 20, and N is infinite or very large, one may state he is 95% confident ($\gamma_2$ = 95%) that p lies between two-sided confidence limits of .292 ($\overline{p}$) and .127 ($\underline{p}$). For increased confidence in the statement about p, the interval is widened. Thus one may be 99% confident ($\gamma_2$ = 99%) that p lies between .321 and .108. Alternatively, a statement can be made about just one of the confidence limits, for example, that one is 95% confident ($\gamma_1$ = 95%) p is at or below .277 ($\overline{p}$). Or one could be 95% confident ($\gamma_1$ = 95%) that p is at or above .137 ($\underline{p}$).

The meaning of confidence level may be explained as follows, using the 95% level for illustration. Assume that we are to take a vast number of random samples from a population with the proportion p, and that we are to calculate confidence limits for each sample. Based on the laws of probability, we can formulate a procedure for calculating confidence limits such that 95% of the samples result in p being within the calculated limits, and 5% of the samples result in p being outside these limits. For any sample where this procedure is used, we say that the confidence level is 95% for the confidence limits obtained. (Similarly, we can formulate a procedure such that 99% of the samples result in p being within the calculated limits, or 90% of the samples, or 80%, etc.)

One of the two basic purposes of this book is to permit easy and accurate calculation of confidence limits based on any attribute sample, regardless of size of n and c. If both upper and lower confidence limits are wanted, these can be calculated at four of the most widely used confidence levels ($\gamma_2$): 80%, 90%, 95%, and 99%; calculations can also be made for confidence levels of 60% and 98%. If only one confidence limit is desired, upper or lower, calculations can be made for six confidence levels ($\gamma_1$): 80%, 90%, 95%, 97.5%, 99%, and 99.5%. (For a given confidence limit at $\gamma_1$, $\gamma_2$ = $2\gamma_1$ − 100%.)

If the population is infinite or, in practical terms, finite but vastly larger than the sample, the binomial distribution is the probability model which provides exact confidence limits at stated confidence levels. However, the binomial distribution entails laborious calculations. There are tables and charts of the binomial distribution, of binomial confidence limits, and of approximations to these. But none of the hitherto available materials are completely satisfactory since they may require extensive interpolation for the desired accuracy of the confidence limits, their range of n and/or c may be insufficient, and the approximations may be of questionable accuracy.

The normal and Poisson distributions, particularly the former, have been used as approximations to the binomial distribution. But the normal distribution tends to fail as an accurate approximation when n grows small or c/n approaches 0 or 1. The Poisson distribution tends to fail as an accurate approximation when n grows small or c/n approaches .5.

The tables, formulas, and procedures in this book permit determination of binomial confidence limits with a relative accuracy of at least .999 for a sample of any size and for any ratio c/n; exact confidence limits are supplied for n of 20 or less. Relative accuracy of .999 signifies that the difference between the approximate confidence limit ($\overline{p}'$ or $\underline{p}'$) and the exact confidence limit ($\overline{p}$ or $\underline{p}$) is no more than .001 of the exact confidence limit. With a desk calculator one can usually compute $\overline{p}'$ or $\underline{p}'$ in a matter of seconds if c is within the scope of Tables 1–4 and in a few minutes if c is outside the scope of these tables and the formulas in Table 7 are required.[1] By hand, the computations are still quite manageable.

When the population is finite (and not vastly larger than the sample), the correct probability model for the confidence limits of an attribute sample is the *hypergeometric distribution*. This is more laborious than the binomial distribution, and tables, charts, and approximations are all the less available. However, if the finite population correction (FPC) is applied to binomial confidence limits, the results are good approximations to hypergeometric confidence limits. Also, the FPC is relatively simple to calculate. Therefore, whether or not the population is infinite (or vast), the binomial distribution remains a serviceable model for determining confidence limits from an attribute sample. This becomes all the more true because of practical considerations: The change produced by the FPC (increase in the lower confidence limit or decrease in the upper confidence limit) tends to border on insignificance when the sample size is less than 5% of the population size. Hence a rule of thumb is to apply the FPC only when sample size is 5% or more of population size; some consider it worth applying only when the ratio is 10% or more.

As an alternative to the FPC, it may be possible to sample with replacement, as discussed in Chapter 4 (pages 81–82) However, as

---

[1] These tables appear in Part Three.

indicated there, such an approach is often unattractive and/or un-feasible. Hence this book assumes that sampling is without replace-ment, with the FPC to be applied when the sample is an appreciable part (5% or more) of the population.

The second basic purpose of this book is to provide substantially accurate means of determining the size of an attribute sample, based on the binomial distribution. At this point it is appropriate to note an important distinction between two types of attribute sampling: proportion sampling and acceptance sampling.

If the primary purpose of the sample is to use c/n as an estimate of p, accompanied by confidence limits $\bar{p}$ and $\underline{p}$, we may call this an *estimate of a proportion,* or *proportion sampling.* Attribute sampling of this type is often employed. One may seek to discover through a sample what proportion of a store's customers own their homes, what proportion of a manufacturer's shipments arrive late, what proportion of the electorate plan to vote for a given candidate, what proportion of precision-made products are reliable (free from defects), what proportion of a company's purchase vouchers pass an audit, etc. The size of the sample is based on an anticipated value of c/n, a specified confidence level, and a specified error margin, namely, the distance between c/n and the confidence limits. If the sample is of appropriate size and if c/n equals its anticipated value, confidence limits based on the binomial distribution will yield an error margin equal to the specified one.

On the other hand, primary interest may focus not on p but on the confidence limits $\bar{p}$ and $\underline{p}$. In such cases attribute sampling takes the form called *acceptance sampling.* To illustrate, a manufacturer pur-chasing bolts as a supply part in lots of 10,000 may seek assurance through a sample that the proportion of error (defective) items in the lot (population) is not so high as to jeopardize the quality of his finished product. If the number of error items c in the sample does not exceed a critical value $\dot{c}$, called the *acceptance number,* he ac-cepts the lot. If the number of error items exceeds $\dot{c}$, he rejects the lot.

Thus the manufacturer might specify that he wishes to take only 1% risk of accepting a production lot with an error rate as high as .04. That is, he specifies an unsatisfactory rate $p_2$ (.04), and a risk $\beta$ (1%) of accepting a lot with such a rate. He further runs the risk, which may be called $\alpha$, of rejecting a satisfactory lot. Therefore he also has to specify a satisfactory rate $p_1$ and risk $\alpha$ of rejecting a lot with such a rate. For example, he might define the satisfactory rate as .01 and specify risk $\alpha$ as 10%; that is, if a lot has an error rate as low as .01, the probability of rejecting the lot—of finding more than $\dot{c}$ items in a sample—should be only 10%. We call $\dot{c}$ + 1, or $\dot{c}_1$, the *rejection number*.

A sample of 324 items, with an acceptance number of 5 and a rejection number of 6, essentially meets the foregoing specifications with respect to risks $\alpha$ and $\beta$ and satisfactory and unsatisfactory rates $p_1$ and $p_2$. If the sample contains five error items or less, the lot is accepted; if it contains six or more, it is rejected.

In acceptance sampling, risks $\alpha$ and $\beta$ are both customarily specified; together with specifications of satisfactory and unsatisfactory error rates, they serve as the basis for calculating sample size and acceptance number. However, in some situations, such as auditing, only risk $\beta$ and unsatisfactory rate $p_2$ are specified, accompanied by an anticipated value of c/n. This may be called a *quasi-form* of acceptance sampling. Although risk $\alpha$ nevertheless exists, it is ignored apparently on the ground that in an application such as auditing it is essentially risk $\beta$ that is crucial.[1] In similar fashion, a quasi-form of acceptance sampling may specify only risk $\alpha$ and satisfactory rate $p_1$, along with the anticipated value of c/n.

This book permits calculation of sample size and acceptance number for either acceptance sampling proper or the quasi-forms. Only single sampling plans are dealt with here; double, sequential, and other forms of acceptance sampling are outside the scope of this book.

In the case of proportion sampling, sample size can be determined for the confidence levels listed on page 4. In the case of acceptance sampling, sample size can be determined for the following risk levels (which are the complements of confidence levels for one-sided confidence limits): 0.5%, 1%, 2.5%, 5%, 10%, and 20%. Determination of sample size assumes that the population is either infinite or finite but much larger than the sample. If the population is not substantially larger than the sample, the FPC can be applied to reduce the sample size. As stated previously in connection with determination of

---

[1] In auditing, *rejection* has a different meaning than in manufacturing. The auditor cannot throw out or send back a population of invoices because the sample contains an excessive number of error items. Rejection means that he is not satisfied with the situation and will further deal with it by means of appropriate audit procedures.

confidence limits, an alternative to the FPC is to sample with replacement (also see Chapter Four, pages 81–82); but since such a course is generally undesirable, this book assumes that sample size is predicated on sampling without replacement.

Fourteen tables (including formulas) for determining confidence limits and sample size are presented in Part Three of this book. Part Two explains, largely through examples, how the tables are used. Chapter One of Part Two deals with confidence limits, Chapter Two with sample size for proportion sampling, and Chapter Three with sample size for acceptance sampling. These chapters assume the population is either infinite or finite but much (at least 10 and preferably 20 times) larger than the sample size, so that the FPC is generally unnecessary for practical purposes.

Chapter Four of Part Two withdraws the assumption of an infinite or relatively large population and explains the FPC, which is desirable when the sample is an appreciable part of a finite population. The FPC brings the confidence limits closer to $c/n$ and reduces sample size. Failure to apply the FPC may be said to have "conservative" results: the distance between the confidence limits and $c/n$ is somewhat exaggerated, and sample size is overstated.

Part Four contains six appendixes, largely devoted to discussing the derivation and accuracy of the tables. For the fullest understanding of the tables, these discussions are recommended reading. However, they may be skipped by the reader interested only in practical application and lacking the necessary statistical background. Appendixes A, B, C, and D respectively deal with confidence limits, sample size for proportion sampling, sample size for acceptance sampling, and the FPC. Thus Appendixes A–D correspond to Chapters One–Four, respectively. Appendix E presents a list of references cited in Appendixes A–D, and Appendix F is a glossary of symbols employed in this book.

# PART TWO

# CHAPTER ONE

# Use of the Tables to Obtain Binomial Confidence Limits

A problem illustrating the use of binomial confidence limits is as follows. In a random sample of 100 eligible voters in a large city, 20 indicate a preference for candidate Jones. Sample size (100 in our illustration) is designated by n; the number of sample events, namely, items having a specified characteristic (20 in our illustration), is designated by c. The sample proportion c/n is .20 and provides an estimate of the population proportion p; i.e., it is estimated that for all voters in the city the proportion favoring candidate Jones is .20. What are the upper and lower binomial confidence limits for p, respectively $\bar{p}$ and $\underline{p}$, at a confidence level (designated by $\gamma_2$) of 95%?

The above problem may be stated in more succinct form, and this form will be used in the examples presented shortly:

$$n \quad = \quad 100$$
$$c \quad = \quad \phantom{0}20$$
$$c/n \quad = \quad .20$$

Obtain $\bar{p}$ and $\underline{p}$ at $\gamma_2 \; = \; 95\%$.

Example 1-1 (page 17) reveals these limits to be .292 and .127; that is, one can be 95% confident the population proportion is between .292 and .127.

Tables 1-7 in Part Three of this book permit determination of binomial confidence limits as required in the above and similar problems. Use of these tables is explained in this chapter by 26 examples, with pertinent notes. First, however, we shall describe the scope of the tables and provide a list of the examples.

## SCOPE OF THE TABLES

Tables 1-7 permit one to obtain binomial confidence limits with relative accuracy of at least .999 for any sample size n, for any proportion c/n, for confidence levels $\gamma_2$ of 60%, 80%, 90%, 95%, 98%, and 99% if two-sided confidence limits (both $\overline{p}$ and $\underline{p}$) are desired, and for confidence levels $\gamma_1$ of 80%, 90%, 95%, 97.5%, 99%, and 99.5% if one-sided confidence limits (only $\overline{p}$ or only $\underline{p}$) are desired. Relative accuracy of at least .999 signifies that the difference between an exact confidence limit and an approximation of this limit is .001 or less of the exact limit. The approximations of $\overline{p}$ and $\underline{p}$ are termed $\overline{p}'$ and $\underline{p}'$, respectively.

Tables 1 and 2 are primary; for sample sizes of 20 or more and c/n of .5 or less, they permit calculation of $\overline{p}'$ and $\underline{p}'$, respectively. Tables 3 and 4 are secondary; they may also be used for calculating $\overline{p}'$ and $\underline{p}'$ but are more limited in scope. They are for optional use when n is 40 or more, c/n is .1 or less, and $\overline{p}'$ is desired, and also when n is 16 or more, c/n is .125 or less, and $\underline{p}'$ is desired. The advantage of using Tables 3 and 4 is that they entail simpler approximation formulas than Tables 1 and 2. Tables 5 and 6 provide exact values of $\overline{p}$ and $\underline{p}$ for sample sizes of 20 or less and c/n of .5 or less.

Confidence limits based on Table 1 generally, but not always, have greater relative accuracy than those based on Table 3 when c/n is within the scope of both tables. Similarly, Table 2 generally affords greater accuracy than Table 4 when c/n is within the scope of both. Although somewhat less accurate than Tables 1 and 2, Tables 3 and 4 nevertheless maintain at least .999 accuracy, barring a few trivial exceptions for Table 3 specifically described in Appendix A (page 426). These exceptions involve the two highest confidence levels, c = 4, and n in the range of 40-41 or 91-95. Relative accuracy then drops below .999 but not below .99876. If this is a matter of concern, one should use Table 1 instead of Table 3 for the few indicated confidence levels and values of c and n.

Tables 1–4 supply formulas for approximating confidence limits and provide approximation constants that, along with n, are entered into these formulas to obtain the confidence limits. For each confidence level and each value of c, a set of approximation constants is given. Constants are given for c from 0 to 1,000, which will suffice in a great many applications. If c exceeds 1,000, the constants can be calculated from extension formulas in Table 7.

Table 7 provides formulas and brief tables that, excepting values called $\overline{m}$ and $\underline{m}$ for c below 50, permit approximation constants to be calculated for any value of c. For c below 160 the calculated constants may differ slightly from the tabular constants in Tables 1–4, leading to confidence limits somewhat reduced in accuracy but nevertheless maintaining relative accuracy of at least .999. Compared with Tables 1–4, Table 7 has the advantage of very compact presentation of data required to calculate confidence limits. On the other hand, Tables 1–4 save the labor of calculating approximation constants for c below 1,000, and, as already mentioned, they tend to provide slightly more accurate limits for c below 160.[1]

Altogether, Tables 1–4 and 7 permit $\overline{p}$ and $\underline{p}$ to be approximated by three different methods. All such approximations are termed $\overline{p}'$ and $\underline{p}'$, denoting that the approximations have relative accuracy of at least .999, although results may differ slightly according to method. To distinguish among methods, we may refer to the following:

1. Approximations based on the Long Formulas, using formulas and tabular constants from Tables 1 and 2.
2. Approximations based on the Short Formulas, using formulas and tabular constants from Tables 3 and 4.
3. Approximations based on calculated constants, using formulas from Tables 1–4 together with constants calculated from formulas in Table 7.

Tables 1–7 apply to c/n of .5 or less. However, they are equally applicable to c/n over .5. Then it is necessary to use the complements of c/n and c, namely r/n and r, where r = n − c. To illustrate, if

---

[1] For c above 160, the values of $\overline{m}$ and $\overline{m}$ in Tables 1–4 are approximations rather than exact values. These approximations have relative accuracy of at least .9999 and therefore permit $\overline{p}$ and $\underline{p}$ to be approximated with at least .999 accuracy.

n = 100, c = 80, and c/n = .80, then r = 20 and r/n = .20. The tables are entered on the basis of r and r/n to find the confidence limits. The limits based on r/n are designated $\overline{p}_r$ and $\underline{p}_r$ in contrast with the designation of $\overline{p}$ and $\underline{p}$ as the limits based on c/n. The limits based on r/n are converted into limits based on c/n by again obtaining complements: $\overline{p} = 1 - \underline{p}_r$, and $\underline{p} = 1 - \overline{p}_r$. The procedure is illustrated in Examples 1-9 to 1-12 and 1-24 to 1-26. (In the rare event that c/n is slightly above .5, c is below 100, and the approximation constants are obtained on the basis of Table 7, the lower confidence limit $\underline{p}'$ may have relative accuracy minutely below .999, namely, between .998 and .999.)

Confidence limits based on Tables 1–7 presume that the population N is infinite or vastly larger than the sample. If such is not the case, a finite population correction (FPC) can be applied, as explained in Chapter Four. The FPC narrows the interval between c/n and each confidence limit; that is, it reduces $\overline{p}$ and increases $\underline{p}$. Unless the sample size is a significant fraction of the population, say, 5% or more, the effect upon the confidence limits is minimal and usually capable of being disregarded.

The scope of Tables 1–7 may be summarized as follows.

**SCOPE OF TABLES 1-7**

| (n) Sample size | (c/n) Proportion of events (c) in sample* | Table numbers | | | |
|---|---|---|---|---|---|
| | | Confidence limits based on tabular constants for c from 0 to 1,000 | | Confidence limits based on calculated constants for any value of c† | |
| | | Upper limit | Lower limit | Upper limit | Lower limit |
| 20 or more | .5 or less | 1 | 2 | 7–A, 7–E | 7–B, 7–E |
| 40 or more | .1 or less | 3 | | 7–C, 7–E | |
| 16 or more | .125 or less | | 4 | | 7–D, 7–E |
| 20 or less | .5 or less | 5 | 6 | | |

*If c/n exceeds .5, calculate r/n = 1 − c/n and enter the tables on the basis of r/n instead.

†The formulas in Table 7-E for approximating $\overline{m}$ and $\underline{m}$ do not apply to c under 50.

## EXAMPLES

The 26 examples in this chapter seek to explain the use of Tables 1–7 and to enable the user to find an example or combination of examples matching his problem. Answers to example problems are enclosed in rectangles. The sample sizes n and sample proportions c/n chosen for these examples aim at simplicity and permitting several comparisons: between two-sided and one-sided confidence limits when the confidence level is the same; between confidence limits based on tabular constants and those based on calculated constants; between approximate and exact confidence limits; and between confidence limits for two proportions an equal distance below and above .5 (for example, limits for .20 compared with limits for .80).

Throughout the examples the confidence level, designated in general terms as $\gamma$, remains the same—95%. If two-sided confidence limits are sought (both the upper and lower limits), the confidence level is specifically designated $\gamma_2$ and the tables and formulas are entered on the basis of $\gamma_2$. If one-sided limits are sought (only the upper limit or only the lower limit), the confidence level is specifically designated $\gamma_1$ and the tables and formulas are entered on the basis of $\gamma_1$. It is important to recognize that at a given confidence level, say 95%, the upper confidence limit varies according to whether one makes a confidence statement only about the upper limit or about both the upper and lower limits; the upper limit drops when a statement is made only about that limit. The lower limit varies in mirror fashion; it rises when a statement is made only about that limit.

In essence the confidence limit $\bar{p}'$ consists of a crude approximation $\bar{m}/n$, which is multiplied by an adjustment factor to yield a refined approximation of $\bar{p}$. Similarly, in essence the confidence limit $\underline{p}'$ consists of a crude approximation $\underline{m}/n$ multiplied by an adjustment factor. The values $\bar{m}$ and $\underline{m}$ are Poisson confidence limits (for m, the parameter of a Poisson distribution, based on an observed value of c). The relationship of $\bar{m}$ and $\underline{m}$ to c corresponds to the relationship of $\bar{p}$ and $\underline{p}$ to c/n; hence $\bar{m}$ and $\underline{m}$ provide a means of estimating $\bar{p}$ and $\underline{p}$. For each value of c at a given confidence level $\gamma$, Tables 1–4 and the formulas in Table 7 supply $\bar{m}$, $\underline{m}$, and a set of constants for the adjustment factor.

## LIST OF EXAMPLES: DETERMINATION OF BINOMIAL CONFIDENCE LIMITS
(Based on Confidence Level $\gamma$ = 95%)

| Example number | (n) sample size | (c) Number of events in sample | (c/n) Sample proportion | Specified confidence limits: upper ($\bar{p}$), lower ($\underline{p}$) | Table entered on basis of $\gamma_1$ or $\gamma_2$ | Table number |
|---|---|---|---|---|---|---|
| Examples Using Tabular Constants (Tables 1–4) | | | | | | |
| 1-1 | 100 | 20 | .20 | $\bar{p}$ and $\underline{p}$ | $\gamma_2$ | 1, 2 |
| 1-2 | 20 | 4 | .20 | only $\bar{p}$ | $\gamma_1$ | 1 |
| 1-3 | 100 | 10 | .10 | $\bar{p}$ and $\underline{p}$ | $\gamma_2$ | 3, 4 |
| 1-4 | 100 | 12 | .12 | $\bar{p}$ and $\underline{p}$ | $\gamma_2$ | 1, 4 |
| 1-5 | 100 | 20 | .20 | only $\bar{p}$ | $\gamma_1$ | 1 |
| 1-6 | 100 | 20 | .20 | only $\underline{p}$ | $\gamma_1$ | 2 |
| 1-7 | 100 | 10 | .10 | only $\bar{\bar{p}}$ | $\gamma_1$ | 3 |
| 1-8 | 100 | 10 | .10 | only $\underline{p}$ | $\gamma_1$ | 4 |
| 1-9 | 100 | 80 | .80 | $\bar{p}$ and $\underline{p}$ | $\gamma_2$ | 2, 1 |
| 1-10 | 100 | 90 | .90 | $\bar{p}$ and $\underline{p}$ | $\gamma_2$ | 4, 3 |
| 1-11 | 100 | 80 | .80 | only $\bar{p}$ | $\gamma_1$ | 2 |
| 1-12 | 100 | 80 | .80 | only $\underline{p}$ | $\gamma_1$ | 1 |
| Examples Using Calculated Constants (Formulas in Table 7) | | | | | | |
| 1-13 | 100 | 20 | .20 | $\bar{p}$ and $\underline{p}$ | $\gamma_2$ | 7-A, 7-B |
| 1-14 | 100 | 10 | .10 | $\bar{p}$ and $\underline{p}$ | $\gamma_2$ | 7-C, 7-D |
| 1-15 | 6,000 | 1,200 | .20 | $\bar{p}$ and $\underline{p}$ | $\gamma_2$ | 7-A, 7-B, 7-E |
| 1-16 | 12,000 | 1,200 | .10 | $\bar{p}$ and $\underline{p}$ | $\gamma_2$ | 7-C, 7-D, 7-E |
| 1-17 | 6,000 | 1,200 | .20 | only $\bar{p}$ | $\gamma_1$ | 7-A, 7-E |
| 1-18 | 6,000 | 1,200 | .20 | only $\underline{\underline{p}}$ | $\gamma_1$ | 7-B, 7-E |
| 1-19 | 12,000 | 1,200 | .10 | only $\bar{p}$ | $\gamma_1$ | 7-C, 7-E |
| 1-20 | 12,000 | 1,200 | .10 | only $\underline{p}$ | $\gamma_1$ | 7-D, 7-E |
| Examples Using Actual Confidence Limits (Tables 5–6) | | | | | | |
| 1-21 | 20 | 4 | .20 | $\bar{p}$ and $\underline{p}$ | $\gamma_2$ | 5, 6 |
| 1-22 | 20 | 4 | .20 | only $\bar{p}$ | $\gamma_1$ | 5 |
| 1-23 | 20 | 4 | .20 | only $\underline{p}$ | $\gamma_1$ | 6 |
| 1-24 | 20 | 16 | .80 | $\bar{p}$ and $\underline{p}$ | $\gamma_2$ | 6, 5 |
| 1-25 | 20 | 16 | .80 | only $\bar{p}$ | $\gamma_1$ | 6 |
| 1-26 | 20 | 16 | .80 | only $\underline{p}$ | $\gamma_1$ | 5 |

*EXAMPLE 1-1* _____

## Problem

n   = 100
c   =   20
c/n =   .20

Obtain $\bar{p}$ and $\underline{p}$ at $\gamma_2$ = 95%.

## Instructions

1. In Table 1, Section 3 ($\gamma_2$ = 95%), for c = 20 find the corresponding constants $\bar{m}$, $\bar{a}$, and $\bar{b}$. Enter these three constants and n into the formula for $\bar{p}'$ on the title page of Table 1.

2. In Table 2, Section 3 ($\gamma_2$ = 95%), for c = 20 find the corresponding constants $\underline{m}$, $\underline{a}$, and $\underline{b}$. Enter these three constants and n into the formula for $\underline{p}'$ on the title page of Table 2.

## Solution

1. $\bar{p}' = \dfrac{\bar{m}}{n} \times \dfrac{n - \bar{a}}{n - \bar{b}} = \dfrac{30.888}{100} \times \dfrac{100 - 7.675}{100 - 2.292}$

   $= .30888 \times .94491 = \boxed{.2919}$

2. $\underline{p}' = \dfrac{\underline{m}}{n} \times \dfrac{n - \underline{a}}{n - \underline{b}} = \dfrac{12.217}{100} \times \dfrac{100 - 4.840}{100 - 8.232}$

   $= .12217 \times 1.0370 = \boxed{.1267}$

## Notes

In Table 1, $\bar{b}$ may be negative when c is small. Then the numerical value of $\bar{b}$ is added to n instead of subtracted from n. (See Example 1-2.)

The exact value of $\bar{p}$ is .2918, compared with $\bar{p}'$ of .2919. The relative error of $\bar{p}'$ is (.2919 − .2918)/.2918, 1 part in 2,918, which is less than .001. Accordingly the relative accuracy of $\bar{p}'$ is greater than .999.

The exact value of $\underline{p}$ to four significant figures is .1267; $\underline{p}'$ has the same value.

*EXAMPLE 1-2* _____

### Problem

n   =  20
c   =   4
c/n = .20
Obtain $\bar{p}$ at $\gamma_1$ = 95%.

### Instructions

In Table 1, Section 4 ($\gamma_1$ = 95%), for c = 4 find the corresponding constants $\bar{m}$, $\bar{a}$, and $\bar{b}$. Enter these three constants and n into the formula for $\bar{p}'$ on the title page of Table 1.

### Solution

$$\bar{p}' = \frac{\bar{m}}{n} \times \frac{n - \bar{a}}{n - \bar{b}} = \frac{9.1535}{20} \times \frac{20 - 1.778}{20 + 0.796}$$

$$= .45768 \times .87623 = \boxed{.4010}$$

### Notes

In actual use, if n is 20 (or less), one would obtain the exact upper confidence limit from Table 5 instead of approximating it by means of Table 1. However, the chief purpose of the present example is to show that sometimes (when c is small) $\bar{b}$ is negative and therefore is added to n instead of subtracted from n.

The reader can compare the approximation $\bar{p}'$ in this example with the exact $\bar{p}$ in Example 1-22, taken from Table 5. The two values are the same to four significant figures.


*EXAMPLE 1-3* _____

### Problem

n   = 100
c   =  10
c/n = .10
Obtain $\bar{p}$ and $\underline{p}$ at $\gamma_2$ = 95%.

## Instructions

1. In Table 3, Section 3 ($\gamma_2$ = 95%), for c = 10 find the corresponding constants $\overline{m}$ and $\overline{k}$. Enter these two constants and n into the formula for $\overline{p}'$ on the title page of Table 3.

2. In Table 4, Section 3 ($\gamma_2$ = 95%), for c = 20 find the corresponding constants $\underline{m}$ and $\underline{k}$. Enter these two constants and n into the formula for $\underline{p}'$ on the title page of Table 4.

## Solution

1. $\overline{p}' = \dfrac{\overline{m}}{n + \overline{k}} = \dfrac{18.390}{100 + 4.33} = \boxed{.1763}$

2. $\underline{p}' = \dfrac{\underline{m}}{n - \underline{k}} = \dfrac{4.7954}{100 - 2.10} = \boxed{.04898}$

## Notes

The exact value of $\overline{p}$ is .1762, compared with $\overline{p}'$ of .1763. The relative error of $\overline{p}'$ is (.1763 − .1762)/.1762, 1 part in 1,762, which is less than .001. Accordingly the relative accuracy of $\overline{p}'$ is greater than .999.

The exact value of $\underline{p}$ is .04900, compared with $\underline{p}'$ of .04898. The relative error of $\underline{p}'$ is (.04900 − .04898)/.04900, 2 parts in 4,900, which is less than .001. Accordingly the relative accuracy of $\underline{p}'$ is greater than .999.

For the present problem, Tables 1 and 2 can also be used to obtain the confidence limits. However, Tables 3 and 4 are used here instead because they entail the simpler Short Formulas and because the value of c/n is sufficiently small (.10 or less) to permit their use with assured relative accuracy of .999 for the confidence limits. Example 1-4 illustrates the situation where c/n is such that Table 1 is required to obtain the upper confidence limit with assured relative accuracy of .999, while Table 4 may still be used to calculate the lower confidence limit with assured relative accuracy of .999.

*EXAMPLE 1-4* _____

### Problem

n   = 100
c   =  12
c/n =  .12

Obtain $\overline{p}$ and $\underline{p}$ at $\gamma_2$ = 95%.

### Instructions

1. In Table 1, Section 3 ($\gamma_2$ = 95%), for c = 12 find the corresponding constants $\overline{m}$, $\overline{a}$, and $\overline{b}$. Enter these three constants and n into the formula for $\overline{p}'$ on the title page of Table 1.

2. In Table 4, Section 3 ($\gamma_2$ = 95%), for c = 12 find the corresponding constants $\underline{m}$ and $\underline{k}$. Enter these two constants and n into the formula for $\underline{p}'$ on the title page of Table 4.

### Solution

1. $\overline{p}' = \dfrac{\overline{m}}{n} \times \dfrac{n - \overline{a}}{n - \overline{b}} = \dfrac{20.962}{100} \times \dfrac{100 - 4.918}{100 - 0.485}$

   $= .20962 \times .95545 = \boxed{.2003}$

2. $\underline{p}' = \dfrac{\underline{m}}{n - \underline{k}} = \dfrac{6.2006}{100 - 2.40} = \boxed{.06353}$

### Notes

The exact value of $\overline{p}$ is .2002, compared with $\overline{p}'$ of .2003. The relative error of $\overline{p}'$ is (.2003 − .2002)/.2002, 1 part in 2,002, which is less than .001. Accordingly the relative accuracy of $\overline{p}'$ is greater than .999.

The exact value of $\underline{p}$ is .06357, compared with $\underline{p}'$ of .06353. The relative error of $\underline{p}'$ is (.06357 − .06353)/.06357, 4 parts in 6,357, which is less than .001. Accordingly the relative accuracy of $\underline{p}'$ is greater than .999.

If desired, one can calculate $\underline{p}'$ by means of Table 2, as in Example 1-1. Then $\underline{p}'$ is .06357—the same as exact $\underline{p}$ to four significant figures. This illustrates the tendency of Table 2 to generally (but not always) provide greater relative accuracy than Table 4 when c/n is within the scope of both tables. Similarly, Table 1 generally provides greater relative accuracy than Table 3 when c/n is within the scope of both.

## EXAMPLE 1-5 _____

### Problem

n    = 100
c    =   20
c/n  =   .20
Obtain $\bar{p}$ at $\gamma_1$ = 95%.

### Instructions

In Table 1, Section 4 ($\gamma_1$ = 95%), for c = 20 find the corresponding constants $\bar{m}$, $\bar{a}$, and $\bar{b}$. Enter these three constants and n into the formula for $\bar{p}'$ on the title page of Table 1.

### Solution

$$\bar{p}' = \frac{\bar{m}}{n} \times \frac{n - \bar{a}}{n - \bar{b}} = \frac{29.062}{100} \times \frac{100 - 7.348}{100 - 2.869}$$

$$= .29062 \times .95389 = \boxed{.2772}$$

### Notes

The above approximation is the same as the exact value of $\bar{p}$ to four significant figures.

The figure of .2772 when a statement is made only about the upper confidence limit compares with a figure of .2919 (Example 1-1) when a statement is made at the same confidence level about both the upper and lower limits.

## EXAMPLE 1-6 _____

### Problem

n    = 100
c    =   20
c/n  =   .20
Obtain $\underline{p}$ at $\gamma_1$ = 95%.

### Instructions

In Table 2, Section 4 ($\gamma_1$ = 95%), for c = 20 find the corresponding constants $\underline{m}$, $\underline{a}$, and $\underline{b}$. Enter these three constants and n into the formula for $\underline{p}'$ on the title page of Table 2.

*Solution*

$$\underline{p}' = \frac{m}{n} \times \frac{n - a}{n - b} = \frac{13.255}{100} \times \frac{100 - 5.102}{100 - 7.975}$$

$$= .13255 \times 1.0312 = \boxed{.1367}$$

*Notes*

The above approximation is the same as the exact value of $\underline{p}$ to four significant figures.

The figure of .1367 when a statement is made only about the lower confidence limit compares with a figure of .1267 (Example 1-1) when a statement is made at the same confidence level about both the upper and lower limits.

*EXAMPLE 1-7* _____

*Problem*

n   = 100
c   =   10
c/n =   .10
Obtain $\overline{p}$ at $\gamma_1$ = 95%.

*Instructions*

In Table 3, Section 4 ($\gamma_1$ = 95%), for c = 10 find the corresponding constants $\overline{m}$ and $\overline{k}$. Enter these two constants and n into the formula for $\overline{p}'$ on the title page of Table 3.

*Solution*

$$\overline{p}' = \frac{\overline{m}}{n + \overline{k}} = \frac{16.962}{100 + 3.58} = \boxed{.1638}$$

*Notes*

The exact value of $\overline{p}$ is .1637, compared with $\overline{p}'$ of .1638. The relative error of $\overline{p}'$ is (.1638 − .1637)/.1637, 1 part in 1,637, which is less than .001. Accordingly the relative accuracy of $\overline{p}'$ is greater than .999.

The figure of .1638 when a statement is made only about the upper confidence limit compares with a figure of .1763 (Example 1-3) when a statement is made at the same confidence level about both the upper and lower limits.

Table 1 could be used instead of Table 3, resulting in $\overline{p}' = .1637$ and illustrating the general tendency of Table 1 to provide greater relative accuracy than Table 3 when c/n is within the scope of both.

*EXAMPLE 1-8* _____

### Problem

n   = 100
c   =   10
c/n =   .10
Obtain $\underline{p}$ at $\gamma_1$ = 95%.

### Instructions

In Table 4, Section 4 ($\gamma_1$ = 95%), for c = 10 find the corresponding constants $\underline{m}$ and $\underline{k}$. Enter these two constants and n into the formula for $\underline{p}'$ on the title page of Table 4.

### Solution

$$\underline{p}' = \frac{\underline{m}}{n - \underline{k}} = \frac{5.4254}{100 - 1.79} = \boxed{.05524}$$

### Notes

The exact value of $\underline{p}$ is .05526, compared with $\underline{p}'$ of .05524. The relative error of $\underline{p}'$ is $(.05526 - .05524)/.05526$, 2 parts in 5,526, which is less than .001. Accordingly the relative accuracy of $\underline{p}'$ is greater than .999.

The figure of .05524 when a statement is made only about the lower confidence limit compares with a figure of .04898 (Example 1-3) when a statement is made at the same confidence level about both the upper and lower limits.

Table 2 could be used instead of Table 4, resulting in $\underline{p}' = .05527$ and illustrating the general tendency of Table 2 to provide greater relative accuracy than Table 4 when c/n is within the scope of both.

*EXAMPLE 1-9* _____

**Problem**

n   = 100
c   =  80
c/n = .80
Obtain $\bar{p}$ and $\underline{p}$ at $\gamma_2$ = 95%.

**Instructions**

1. When c/n exceeds .5, calculate r = n − c, and r/n. Find the confidence limits $\bar{p}_r$ and $\underline{p}_r$ based on r/n. Do so by substituting r and r/n for c and c/n to find an appropriate example and to use the tables and formulas.

2. Calculate $\bar{p}$ = 1 − $\underline{p}_r$.

3. Calculate $\underline{p}$ = 1 − $\bar{p}_r$.

**Solution**

1. r = n − c = 100 − 80 = 20; r/n = 20/100 = .20. Substituting r for c and r/n for c/n, the problem is the counterpart of that in Example 1-1. Further instructions are therefore the same as in Example 1-1, leading to $\bar{p}_r'$ = .2919 and $\underline{p}_r'$ = .1267. (The calculations are the same as in Example 1-1 and therefore need not be repeated here.)

2. $\bar{p}'$ = 1 − $\underline{p}_r'$ = 1 − .1267 = $\boxed{.8733}$

3. $\underline{p}'$ = 1 − $\bar{p}_r'$ = 1 − .2919 = $\boxed{.7081}$

*EXAMPLE 1-10* _____

**Problem**

n   = 100
c   =  90
c/n =  .90
Obtain $\bar{p}$ and $\underline{p}$ at $\gamma_2$ = 95%.

*Instructions*

1. When c/n exceeds .5, calculate r = n − c, and r/n. Find the confidence limits $\overline{p}_r$ and $\underline{p}_r$ based on r/n. Do so by substituting r and r/n for c and c/n to find an appropriate example and to use the tables and formulas.

2. Calculate $\overline{p} = 1 - \underline{p}_r$.

3. Calculate $\underline{p} = 1 - \overline{p}_r$.

*Solution*

1. r = n − c = 100 − 90 = 10; r/n = 10/100 = .10. Substituting r for c and r/n for c/n, the problem is the counterpart of that in Example 1-3. Further instructions are therefore the same as in Example 1-3, leading to $\overline{p}'_r = .1763$ and $\underline{p}'_r = .04898$. (The calculations are the same as in Example 1-3 and therefore need not be repeated here.)

2. $\overline{p}' = 1 - \underline{p}'_r = 1 - .04898 = \boxed{.95102}$

3. $\underline{p}' = 1 - \overline{p}'_r = 1 - .1763 = \boxed{.8237}$

---

*EXAMPLE 1-11* _____

*Problem*

n   = 100
c   =   80
c/n =  .80
Obtain $\overline{p}$ at $\gamma_1$ = 95%.

*Instructions*

1. When c/n exceeds .5, calculate r = n − c, and r/n. Find the confidence limit $\underline{p}_r$ based on r/n. Do so by substituting r and r/n for c and c/n to find an appropriate example and to use the tables and formulas.

2. Calculate $\overline{p} = 1 - \underline{p}_r$.

*Solution*

1. r = n − c = 100 − 80 = 20; r/n = 20/100 = .20. Substituting r for c and r/n for c/n, the problem is the counterpart of that in Example

1-6. Further instructions are therefore the same as in Example 1-6, leading to $p'_r$ = .1367. (The calculations are the same as in Example 1-6 and therefore need not be repeated here.)

2. $\overline{p}'$ = 1 − $\underline{p}'_r$ = 1 − .1367 = $\boxed{.8633}$

---

*EXAMPLE 1-12* _____

### Problem

n   = 100
c   =   80
c/n =   .80

Obtain $\underline{p}$ at $\gamma_1$ = 95%.

### Instructions

1. When c/n exceeds .5, calculate r = n − c, and r/n. Find the confidence limit $\overline{p}_r$ based on r/n. Do so by substituting r and r/n for c and c/n to find an appropriate example and to use the tables and formulas.

2. Calculate $\underline{p}$ = 1 − $\overline{p}_r$.

### Solution

1. r = n − c = 100 − 80 = 20; r/n = 20/100 = .20. Substituting r for c and r/n for c/n, the problem is the counterpart of that in Example 1-5. Further instructions are therefore the same as in Example 1-5, leading to $\overline{p}'_r$ = .2772. (The calculations are the same as in Example 1-5 and therefore need not be repeated here.)

2. $\underline{p}'$ = 1 − $\overline{p}'_r$ = 1 − .2772 = $\boxed{.7228}$

---

*EXAMPLE 1-13* _____

### Problem

n   = 100
c   =   20
c/n =   .20

Obtain $\overline{p}$ and $\underline{p}$ at $\gamma_2$ = 95%.

## Instructions

1. For c = 20 and $\gamma_2$ = 95%, obtain $\bar{m}$ and $\underline{m}$ from Tables 1 and 2 (or Tables 3 and 4). Alternatively, Table 7–E can be used to calculate $\bar{m}'$ and $\underline{m}'$ when c ⩾ 50.

2. Based on $\bar{m}$, $\underline{m}$, and c, and on the formulas and values of f and g in Tables 7–A and 7–B, calculate $\bar{a}'$, $\bar{b}'$, $\underline{a}'$, and $\underline{b}'$.

3. Enter the constants $\bar{m}$, $\underline{m}$, $\bar{a}'$, $\bar{b}'$, $\underline{a}'$, and $\underline{b}'$, together with n, into the formulas for $\bar{p}'$ and $\underline{p}'$ on the title pages of Tables 1 and 2.

## Solution

1. $\bar{m}$ = 30.888;  $\underline{m}$ = 12.217    (Tables 1 and 2)

2. $\bar{a}' = \dfrac{\bar{m}}{4} - f = \dfrac{30.888}{4} - .60 = 7.1220$    (Table 7-A)

$\bar{b}' = \dfrac{2c - \bar{m}}{4} - f = \dfrac{40 - 30.888}{4} - .60 = 1.6780$

(Table 7–A)

$\underline{a}' = \dfrac{\underline{m}}{3} + g = \dfrac{12.217}{3} + .65 = 4.7223$    (Table 7–B)

$\underline{b}' = \dfrac{3c - 3 - \underline{m}}{6} + g = \dfrac{60 - 3 - 12.217}{6} + .65 = 8.1138$

(Table 7–B)

3. $\bar{p}' = \dfrac{\bar{m}}{n} \times \dfrac{n - \bar{a}'}{n - \bar{b}'} = \dfrac{30.888}{100} \times \dfrac{100 - 7.1220}{100 - 1.6780}$

$= .30888 \times .94463 = \boxed{.2918}$    (Table 1)

$\underline{p}' = \dfrac{\underline{m}}{n} \times \dfrac{n - \underline{a}'}{n - \underline{b}'} = \dfrac{12.217}{100} \times \dfrac{100 - 4.7223}{100 - 8.1138}$

$= .12217 \times 1.03691 = \boxed{.1267}$    (Table 2)

## Note

The values of $\bar{p}'$ = .2918 and $\underline{p}'$ = .1267 in this example, using calculated approximation constants (except for $\bar{m}$ and $\underline{m}$), compare with $\bar{p}'$ = .2919 and $\underline{p}'$ = .1267 in Example 1-1, using tabular approximation constants.

*EXAMPLE 1-14* _____

### Problem

n    = 100
c    =    10
c/n =    .10

Obtain $\bar{p}$ and $\underline{p}$ at $\gamma_2$ = 95%.

### Instructions

1. For c = 10 and $\gamma_2$ = 95%, obtain $\bar{m}$ and $\underline{m}$ from Tables 1 and 2 (or Tables 3 and 4). Alternatively, Table 7–E can be used to calculate $\bar{m}'$ and $\underline{m}'$ when c ⩾ 50.

2. Based on $\bar{m}$, $\underline{m}$, and c, and on the formulas and values of d in Tables 7–C and 7–D, calculate $\bar{k}'$ and $\underline{k}'$.

3. Enter the constants $\bar{m}$, $\underline{m}$, $\bar{k}'$, and $\underline{k}'$, together with n, into the formulas for $\bar{p}'$ and $\underline{p}'$ on the title pages of Tables 3 and 4.

### Solution

1. $\bar{m}$ = 18.390;   $\underline{m}$ = 4.7954     (Tables 1 and 2)

2. $\bar{k}' = \dfrac{\bar{m} - c}{2} + d = \dfrac{18.390 - 10}{2} + .126 = 4.3210$

   (Table 7–C)

   $\underline{k}' = \dfrac{c - 1 - \underline{m}}{2} = \dfrac{10 - 1 - 4.7954}{2} = 2.1023$

   (Table 7–D)

3. $\bar{p}' = \dfrac{\bar{m}}{n + \bar{k}'} = \dfrac{18.390}{100 + 4.3210} = \boxed{.1763}$     (Table 3)

   $\underline{p}' = \dfrac{\underline{m}}{n - \underline{k}'} = \dfrac{4.7954}{100 - 2.1023} = \boxed{.04898}$     (Table 4)

### Note

The values of $\bar{p}'$ and $\underline{p}'$ in this example, using calculated approximation constants (except for $\bar{m}$ and $\underline{m}$), are the same as in Example 1-3, using tabular approximation constants.

*EXAMPLE 1-15* _____

### Problem

n    = 6,000
c    = 1,200
c/n =    .20

Obtain $\bar{p}$ and $\underline{p}$ at $\gamma_2$ = 95%.

### Instructions

1. For c = 1,200 and $\gamma_2$ = 95%, calculate $\bar{m}'$ and $\underline{m}'$ from the formulas in Table 7—E.

2. Based on $\bar{m}'$, $\underline{m}'$, and c and on the formulas in Tables 7—A and 7—B, calculate $\bar{a}'$, $\bar{b}'$, $\underline{a}'$, and $\underline{b}'$.

3. Enter the constants $\bar{m}'$, $\underline{m}'$, $\bar{a}'$, $\bar{b}'$, $\underline{a}'$, and $\underline{b}'$, together with n, into the formulas for $\bar{p}'$ and $\underline{p}'$ on the title pages of Tables 1 and 2.

### Solution

1. $\bar{m}'$ = c + A$\sqrt{c + 1}$ + B = 1,200 + 1.9600 $\sqrt{1,201}$
   + 1.9472 = 1,269.87     (Table 7–E)

   $\underline{m}'$ = c − 1 − A$\sqrt{c}$ + B = 1,199 − 1.9600 $\sqrt{1,200}$
   + 1.9472 = 1,133.05     (Table 7–E)

2. $\bar{a}'$ = $\dfrac{\bar{m}'}{4}$ = $\dfrac{1,269.87}{4}$ = 317.47     (Table 7–A)

   $\bar{b}'$ = $\dfrac{2c - \bar{m}'}{4}$ = $\dfrac{2,400 - 1,269.87}{4}$ = 282.53     (Table 7–A)

   $\underline{a}'$ = $\dfrac{\underline{m}'}{3}$ = $\dfrac{1,133.05}{3}$ = 377.68     (Table 7–B)

   $\underline{b}'$ = $\dfrac{3c - 3 - \underline{m}'}{6}$ = $\dfrac{3,600 - 3 - 1,133.05}{6}$

   = 410.66     (Table 7–B)

3. $\overline{p}' = \dfrac{\overline{m}'}{n} \times \dfrac{n - \overline{a}'}{n - \overline{b}'} = \dfrac{1{,}269.87}{6{,}000} \times \dfrac{6{,}000 - 317.47}{6{,}000 - 282.53}$

$\quad = .21165 \times .99389 = \boxed{.2104} \quad$ (Table 1)

$\underline{p}' = \dfrac{\underline{m}'}{n} \times \dfrac{n - \underline{a}'}{n - \underline{b}'} = \dfrac{1{,}133.05}{6{,}000} \times \dfrac{6{,}000 - 377.68}{6{,}000 - 410.66}$

$\quad = .18884 \times 1.00590 = \boxed{.1900} \quad$ (Table 2)

## EXAMPLE 1-16 _____

### Problem

n   = 12,000
c   =   1,200
c/n =      .10

Obtain $\overline{p}$ and $\underline{p}$ at $\gamma_2 = 95\%$.

### Instructions

1. For c = 1,200 and $\gamma_2$ = 95%, calculate $\overline{m}'$ and $\underline{m}'$ from the formulas in Table 7—E.

2. Based on $\overline{m}'$, $\underline{m}'$, and c and on the formulas in Tables 7—C and 7—D, calculate $\overline{k}'$ and $\underline{k}'$.

3. Enter the constants $\overline{m}'$, $\underline{m}'$, $\overline{k}'$, and $\underline{k}'$, together with n, into the formulas for $\overline{p}'$ and $\underline{p}'$ on the title pages of Tables 3 and 4.

### Solution

1. $\overline{m}' = c + A\sqrt{c + 1} + B = 1{,}200 + 1.9600 \sqrt{1{,}201}$
   $\quad + 1.9472 = 1{,}269.87 \quad$ (Table 7-E)

   $\underline{m}' = c - 1 - A\sqrt{c} + B = 1{,}199 - 1.9600 \sqrt{1{,}200}$
   $\quad + 1.9472 = 1{,}133.05 \quad$ (Table 7-E)

2. $\overline{k}' = \dfrac{\overline{m}' - c}{2} = \dfrac{1{,}269.87 - 1{,}200}{2} = 34.935 \quad$ (Table 7-C)

   $\underline{k}' = \dfrac{c - 1 - \underline{m}'}{2} = \dfrac{1{,}200 - 1 - 1{,}133.05}{2} = 32.975$

   $\quad$ (Table 7-D)

3. $\overline{p}' = \dfrac{\overline{m}'}{n + \overline{k}'} = \dfrac{1{,}269.87}{12{,}000 + 34.935} = \boxed{.1055}$ (Table 3)

$\underline{p}' = \dfrac{\underline{m}'}{n - \underline{k}'} = \dfrac{1{,}133.05}{12{,}000 - 32.975} = \boxed{.09468}$ (Table 4)

*Note*

Using the formula of Table 1, one also obtains $\overline{p}' = .1055$. Using the formula of Table 2, one obtains $\underline{p}' = .09469$ instead of .09468 as in the present example. However, the very slight difference between .09469 and .09468 does not prevent $\underline{p}'$ from having accuracy of at least .999.

*EXAMPLE 1-17* _____

*Problem*

n = 6,000
c = 1,200
c/n = .20
Obtain $\overline{p}'$ at $\gamma_1$ = 95%.

*Instructions*

1. For c = 1,200 and $\gamma_1$ = 95%, calculate $\overline{m}'$ from the formula in Table 7–E.

2. Based on $\overline{m}'$ and c, and on the formulas in Table 7–A, calculate $\overline{a}'$ and $\overline{b}'$.

3. Enter the constants $\overline{m}'$, $\overline{a}'$, and $\overline{b}'$, together with n, into the formula for $\overline{p}'$ on the title page of Table 1.

*Solution*

1. $\overline{m}' = c + A\sqrt{c + 1} + B = 1{,}200 + 1.6449 \sqrt{1{,}201}$
   $+ 1.5685 = 1{,}258.57$ (Table 7–E)

2. $\overline{a}' = \dfrac{\overline{m}'}{4} = \dfrac{1{,}258.57}{4} = 314.64$ (Table 7–A)

   $\overline{b}' = \dfrac{2c - \overline{m}'}{4} = \dfrac{2{,}400 - 1{,}258.57}{4} = 285.36$ (Table 7–A)

3. $\overline{p}' = \dfrac{\overline{m}'}{n} \times \dfrac{n - \overline{a}'}{n - \overline{b}'} = \dfrac{1,258.57}{6,000} \times \dfrac{6,000 - 314.64}{6,000 - 285.36}$

$\qquad = .20976 \times .99488 = \boxed{.2087} \qquad \text{(Table 1)}$

## EXAMPLE 1-18

### Problem

$n \quad = \; 6,000$
$c \quad = \; 1,200$
$c/n = \quad .20$
Obtain $\underline{p}$ at $\gamma_1 = 95\%$.

### Instructions

1. For $c = 1,200$ and $\gamma_1 = 95\%$, calculate $\underline{m}'$ from the formula in Table 7—E.

2. Based on $\underline{m}'$ and c, and on the formula in Table 7—B, calculate $\underline{a}'$ and $\underline{b}'$.

3. Enter the constants $\underline{m}'$, $\underline{a}'$, and $\underline{b}'$, together with n, into the formula for $\underline{p}'$ on the title page of Table 2.

### Solution

1. $\underline{m}' = c - 1 - A\sqrt{c} + B = 1,200 - 1 - 1.6449 \sqrt{1,200}$
$\qquad + 1.5685 = 1,143.59 \qquad \text{(Table 7-E)}$

2. $\underline{a}' = \dfrac{m'}{3} = \dfrac{1,143.59}{3} = 381.20 \qquad \text{(Table 7-B)}$

$\quad \underline{b}' = \dfrac{3c - 3 - \underline{m}'}{6} = \dfrac{3,600 - 3 - 1,143.59}{6}$

$\qquad = 408.90 \qquad \text{(Table 7-B)}$

3. $\underline{p}' = \dfrac{m'}{n} \times \dfrac{n - \underline{a}'}{n - \underline{b}'} = \dfrac{1,143.59}{6,000} \times \dfrac{6,000 - 381.20}{6,000 - 408.90}$

$\qquad = .19060 \times 1.00495 = \boxed{.1915} \qquad \text{(Table 2)}$

*EXAMPLE 1-19* _____

### Problem

n   = 12,000
c   =   1,200
c/n =     .10
Obtain $\bar{p}$ at $\gamma_1$ = 95%.

### Instructions

1. For c = 1,200 and $\gamma_1$ = 95%, calculate $\bar{m}'$ from the formula in Table 7–E.

2. Based on $\bar{m}'$ and c, and on the formula in Table 7–C, calculate $\bar{k}'$.

3. Enter the constants $\bar{m}'$ and $\bar{k}'$, together with n, into the formula for $\bar{p}'$ on the title page of Table 3.

### Solution

1. $\bar{m}' = c + A\sqrt{c + 1} + B = 1{,}200 + 1.6449\sqrt{1{,}201}$
   $+ 1.5685 = 1{,}258.57$     (Table 7–E)

2. $\bar{k}' = \dfrac{\bar{m}' - c}{2} = \dfrac{1{,}258.57 - 1{,}200}{2} = 29.285$     (Table 7–C)

3. $\bar{p}' = \dfrac{\bar{m}'}{n + \bar{k}'} = \dfrac{1{,}258.57}{12{,}000 + 29.285} = \boxed{.1046}$     (Table 3)

### Note

Using the formula of Table 1, one also obtains $\bar{p}'$ = .1046.

*EXAMPLE 1-20* _____

### Problem

n   = 12,000
c   =   1,200
c/n =     .10
Obtain $\underline{p}$ at $\gamma_1$ = 95%.

### Instructions

1. For c = 1,200 and $\gamma_1$ = 95%, calculate $\underline{m}'$ from the formula in Table 7—E.

2. Based on $\underline{m}'$ and c, and on the formula in Table 7—D, calculate $\underline{k}'$.

3. Enter the constants $\underline{m}'$ and $\underline{k}'$, together with n, into the formula for $\underline{p}'$ on the title page of Table 4.

### Solution

1. $\underline{m}' = c - 1 - A\sqrt{c} + B = 1,200 - 1 - 1.6449\sqrt{1,200}$
   $+ 1.5685 = 1,143.59$    (Table 7-E)

2. $\underline{k}' = \dfrac{c - 1 - \underline{m}'}{2} = \dfrac{1,200 - 1 - 1,143.59}{2} = 27.705$

   (Table 7-D)

3. $\underline{p}' = \dfrac{\underline{m}'}{n - \underline{k}'} = \dfrac{1,143.59}{12,000 - 27.705} = \boxed{.09552}$    (Table 4)

### Note

Using the formula of Table 2, one obtains $\underline{p}'$ = .09553 instead of .09552 as in the present example. However, the very slight difference between .09553 and .09552 does not prevent $\underline{p}'$ from having accuracy of at least .999.

### EXAMPLE 1-21 _____

### Problem

n   = 20
c   =  4
c/n = .20

Obtain $\overline{p}$ and $\underline{p}$ at $\gamma_2$ = 95%.

### Instructions and Solution

1. Use Table 5 for $\overline{p}$. Enter the line n = 20, c = 4, and the column
   $\gamma_2$ = 95% to find $\overline{p}$ = $\boxed{.43661}$ .

2. Use Table 6 for $\underline{p}$. Enter the line n $=$ 20, c $=$ 4, and the column $\gamma_2$ $=$ 95% to find $\underline{p}$ $=$ $\boxed{.057334}$ .


*EXAMPLE 1-22* ────────────────────────────────────────

**Problem**

n   $=$   20
c   $=$   4
c/n $=$ .20
Obtain $\overline{p}$ at $\gamma_1$ $=$ 95%.

**Instructions and Solution**

Use Table 5. Enter the line n $=$ 20, c $=$ 4, and the column $\gamma_1$ $=$ 95% to find $\overline{p}$ $=$ $\boxed{.40103}$ .


*EXAMPLE 1-23* ────────────────────────────────────────

**Problem**

n   $=$   20
c   $=$   4
c/n $=$ .20
Obtain $\underline{p}$ at $\gamma_1$ $=$ 95%.

**Instructions and Solution**

Use Table 6. Enter the line n $=$ 20, c $=$ 4, and the column $\gamma_1$ $=$ 95% to find $\underline{p}$ $=$ $\boxed{.071354}$ .

*EXAMPLE 1-24* _____

### Problem

n   =   20
c   =   16
c/n  =  .80

Obtain $\overline{p}$ and $\underline{p}$ at $\gamma_2$ = 95%.

### Instructions

1. When c/n exceeds .5, calculate r = n − c, and r/n. Find the confidence limits $\overline{p}_r$ and $\underline{p}_r$ based on r/n. Do so by substituting r and r/n for c and c/n to find an appropriate example and to use the tables.

2. Calculate $\overline{p}$ = 1 − $\underline{p}_r$.

3. Calculate $\underline{p}$ = 1 − $\overline{p}_r$.

### Solution

1. r = n − c = 20 − 16 = 4; r/n = 4/20 = .20. Substituting r for c and r/n for c/n, the problem is the counterpart of that in Example 1-21. Further instructions are therefore the same as in Example 1-21, leading to $\overline{p}_r$ = .43661 and $\overline{p}_r$ = .057334. (The procedure is the same as in Example 1-21 and therefore need not be repeated here.)

2. $\overline{p}$ = 1 − $\underline{p}_r$ = 1 − .057334 = $\boxed{.942666}$

3. $\underline{p}$ = 1 − $\overline{p}_r$ = 1 − .43661 = $\boxed{.56339}$

*EXAMPLE 1-25* _____

### Problem

n   =   20
c   =   16
c/n  =  .80

Obtain $\overline{p}$ at $\gamma_1$ = 95%.

*Instructions*

1. When c/n exceeds .5, calculate r = n − c, and r/n. Find the confidence limit $\underline{p}_r$ based on r/n. Do so by substituting r and r/n for c
   and c/n to find an appropriate example and to use the tables.

2. Calculate $\overline{p} = 1 - \underline{p}_r$ .

*Solution*

1. r = n − c = 20 − 16 = 4; r/n = 4/20 = .20. Substituting r for c and
   r/n for c/n, the problem is the counterpart of that in Example
   1-23. Further instructions are therefore the same as in Example
   1-23, leading to $\underline{p}_r$ = .071354. (The procedure is the same as in
   Example 1-23 and therefore need not be repeated here.)

2. $\overline{p} = 1 - \underline{p}_r = 1 - .071354 =$   $\boxed{.928646}$

*EXAMPLE 1-26* _____

*Problem*

n   =   20
c   =   16
c/n = .80
Obtain $\underline{p}$ at $\gamma_1$ = 95%.

*Instructions*

1. When c/n exceeds .5, calculate r = n − c, and r/n. Find the confidence limit $\overline{p}_r$ based on r/n. Do so by substituting r and r/n for c
   and c/n to find an appropriate example and to use the tables.

2. Calculate $\underline{p} = 1 - \overline{p}_r$ .

*Solution*

1. r = n − c = 20 − 16 = 4; r/n = 4/20 = .20. Substituting r for c and
   r/n for c/n, the problem is the counterpart of that in Example
   1-22. Further instructions are therefore the same as in Example
   1-22, leading to $\overline{p}_r$ = .40103. (The procedure is the same as in
   Example 1-22 and therefore need not be repeated here.)

2. $\underline{p} = 1 - \overline{p}_r = 1 - .40103 =$   $\boxed{.59897}$

# CHAPTER TWO

# Use of the Tables
# to Determine Sample Size
# for Proportion Sampling

Determination of sample size depends upon our anticipations as to the results of the sample. Therefore we shall discuss these anticipated results before dealing with the problem of sample size.

Results of the sample may be described as follows: (1) Given a sample of size n, we find the number of items c in the sample having a designated characteristic and we use the proportion c/n as an estimate of the population proportion p. (2) We obtain the confidence limits $\bar{p}$ and $\underline{p}$ at confidence level $\gamma$ to indicate the range within which we believe p lies. Alternatively, we can obtain only $\bar{p}$ or only $\underline{p}$ at confidence level $\gamma$, as discussed later. To facilitate the present discussion, we assume that both $\bar{p}$ and $\underline{p}$ are obtained. We use $\gamma_2$ to specifically designate the confidence level when both limits are desired and use $\gamma_1$ when only one limit is desired; $\gamma$ is a general term.

An illustration of proportion sampling is as follows. In a random sample of 200 voters in a large city, 20 prefer candidate Jones. The sample proportion is c/n = 20/200 = .10. Hence it is estimated .10 is the proportion of all voters in the city who prefer Jones. At a confidence level $\gamma_2$ of 95%, confidence limits $\bar{p}$ and $\underline{p}$ of .150 and .062

are obtained. Hence, at a confidence level of 95%, it is indicated that the population proportion p lies between .150 and .062.

The interval between c/n and each confidence limit is the error margin, which for convenience we designate by the general term e. Actually there are two error margins: $\bar{e}$, the interval between $\bar{p}$ and c/n; and $\underline{e}$, the interval between c/n and $\underline{p}$. When binomial confidence limits are obtained, $\bar{e}$ equals or exceeds $\underline{e}$; the two are equal only when c/n = .50. In our illustration, $\bar{e}$ = .050 (namely, .150 − .100), while $\underline{e}$ = .038 (namely, .100 − .062). Summing up, for a given sample size n, a given sample proportion c/n, and a given confidence level $\gamma$, we can find the error margin e (more precisely, the error margins $\bar{e}$ and $\underline{e}$).

Turning matters around, we can determine sample size by anticipating the values of c/n, e, and $\gamma$. That is, we must specify: (1) $\hat{p}$, the anticipated value of c/n; (2) $\hat{e}$, the anticipated (acceptable) value of e; and (3) $\gamma$, the confidence level for e. The *Appropriate Sample Size* may then be defined as one which meets the following requirement: If the sample proportion c/n proves equal to $\hat{p}$, e will equal $\hat{e}$ (insofar as possible in using integers for n and c).

A problem illustrating the determination of sample size for an estimate of a proportion is as follows. It is desired to take a random sample to estimate the proportion of eligible voters in a large city who prefer candidate Jones and to obtain the upper and lower confidence limits for this estimate. Based on prior information (such as earlier polls), it is anticipated that the sample proportion will be about .10. The specified (acceptable) error margin at the 95% confidence level is .05. What is the Appropriate Sample Size? In other words, what should be the value of n so that if c/n proves to be .10, the error margin will be .05?

This problem may be stated in more succinct form, which will be used in the examples presented shortly:

$$\hat{p} = .10$$
$$\hat{e} = .05 \text{ at } \gamma_2 = 95\%$$

Both $\bar{p}$ and $\underline{p}$ are desired.
Obtain the Appropriate Sample Size n.

As shown in Example 2-1 (page 46), the answer is n = 200. If c/n proves to be .10, the error margin will be .05. This meets the Appropriate Sample Size requirement that e = $\hat{e}$ if c/n = $\hat{p}$. The reader will recognize that the present illustration is the mirror image of the one given earlier in this chapter.

When both $\overline{p}$ and $\underline{p}$ are desired, the specification $\hat{e}$ signifies that $\overline{e}$ shall equal $\hat{e}$. In other words, we are specifying the anticipated *upper* error margin $\hat{\overline{e}}$, without direct regard to the anticipated lower error margin $\hat{\underline{e}}$. The reason is as follows: As pointed out earlier, $\overline{e} \geqslant \underline{e}$; a sample size procedure that limits $\overline{e}$ to the specified value $\hat{e}$ will inherently limit $\underline{e}$ to the same or a smaller value. Therefore $\hat{\overline{e}}$ is the determining factor in sample size when both $\overline{p}$ and $\underline{p}$ are desired. Accordingly, the specification $\hat{e}$ then denotes $\hat{\overline{e}}$. If only $\overline{p}$ is desired, the specification $\hat{e}$ again denotes $\hat{\overline{e}}$. On the other hand, if only $\underline{p}$ is desired, the specification $\hat{e}$ denotes $\hat{\underline{e}}$.

For a specified $\gamma$ and $\hat{e}$, sample size essentially tends to increase with $\hat{p}$ until $\hat{p}$ = .50. Therefore a conservative approach, which helps make the sample size sufficiently large so that e will not exceed $\hat{e}$, is to consider $\hat{p}$ as the *maximum* anticipated sample proportion, but not over .50. Thus in the preceding illustration, earlier polls might have shown that about .08 of the voters prefer Jones; but since samples vary in their results and the next sample, it is believed, might show a proportion as high as .10, the figure of .10 is used as the specification $\hat{p}$. The most conservative approach is to use $\hat{p}$ = .50, inasmuch as sample size then tends to reach a maximum for a specified $\gamma$ and $\hat{e}$. To the extent that doubt surrounds $\hat{p}$, this approach becomes increasingly desirable. When no information on $\hat{p}$ is available, this approach is generally employed.

Tables 8–12 in Part Three of this book permit determination of sample size for an estimate of a proportion. Use of these tables is explained in this chapter by 19 examples, with pertinent notes. First, however, we shall describe the scope of the tables and provide a list of examples.

## SCOPE OF TABLES 8–12

Tables 8–12 provide formulas and data that permit us to determine sample size for any anticipated sample proportion $\hat{p}$, for any anticipated error margin $\hat{e}$ (provided, of course, that $\hat{p} + \hat{e}$ does not exceed 1 or that $\hat{p} - \hat{e}$ is not less than 0), for confidence levels $\gamma_2$ of 60%, 80%, 90%, 95%, 98%, and 99% if two-sided confidence limits (both $\overline{p}$ and $\underline{p}$) are desired, and for confidence levels $\gamma_1$ of 80%, 90%,

95%, 97.5%, 99%, and 99.5% if one-sided confidence limits (only $\overline{p}$ or only $\underline{p}$) are desired.

Together with the procedures set forth in the examples, Tables 8–12 provide high accuracy; that is, they permit one to obtain exactly or very nearly the Appropriate Sample Size (consistent with the use of integers for n and c). This accuracy is obtained with the aid of a correction factor included in the procedures. Without the correction factor, the tables still provide fairly good accuracy overall. In discussing the individual tables, we shall comment on the accuracy of each before application of the correction factor. To the extent that the specification $\hat{p}$ is a rough guess, and inasmuch as the tables are conservative in the sense of tending more to overstate than understate sample size, the user may decide to dispense with the correction factor (which is quite simple and does not involve much computation).

Tables 8, 10, and 11 are employed where the specification $\hat{e}$ denotes the anticipated upper error margin $\hat{\overline{e}}$, namely where one wishes to obtain both $\overline{p}$ and $\underline{p}$, or only $\overline{p}$. Tables 9–11 are employed where the specification $\hat{e}$ denotes the anticipated lower error margin $\hat{\underline{e}}$, namely, where only $\underline{p}$ is to be obtained. Table 12 is employed where $\hat{p} = 0$ and thus only $\overline{p}$ is to be obtained.

For $\hat{\overline{e}}$, Table 8 applies to problems where $\hat{p} \leqslant .25$ (but not 0) and Table 10 to problems where $\hat{p} > .25$ (but not over .5). The procedure for determining sample size is called the Poisson Procedure in the case of Table 8 and the Modified Normal Procedure in the case of Table 10. The reason for two separate procedures is to keep determination of sample size reasonably simple and at the same time reasonably accurate before application of the correction factor. The less the error in sample size before use of the correction factor, the more nearly does the result after such use meet the requirement for Appropriate Sample Size (and the more readily can one dispense with the correction factor). For samples between 100 and 100,000 and for the confidence levels in the tables, the Poisson Procedure before correction leads to sample-size error between approximately 0% and 4%, while the Modified Normal Procedure before correction leads to sample-size error between approximately −3% and +5%. For samples between 20 and 100, the Poisson Procedure leads to sample-size error between approximately 4% and 10%, while the Modified Normal Procedure leads to sample-size error between approximately −11% and +5%.

For $\underline{\hat{e}}$, Table 9 applies to problems where $\hat{p} \leqslant .25$ and Table 10 to problems where $\hat{p} > .25$. The procedure for determining sample size is called the Poisson Procedure in the case of Table 9 and the Modified Normal Procedure in the case of Table 10. For samples between 100 and 100,000, the Poisson Procedure leads to sample-size error between approximately 0% and 3%, while the Modified Normal Procedure leads to sample-size error between approximately −10% and +3%. For samples between 20 and 100, the Poisson Procedure leads to sample-size error of about 0%, while the Modified Normal Procedure leads to sample-size error between approximately −25% and +11%.

The Poisson Procedure entails finding the anticipated value of c. Dividing c by $\hat{p}$ then yields n. Tables 8 and 9 provide values of c based on the specifications $\hat{p}$, $\hat{e}$, and $\gamma$. Alternatively, c can be obtained by formula from Table 11 (except in a very few cases described in Table 11).

As previously indicated in parenthetical remarks, because n and c are integers a sample may not exactly meet the specifications for $\hat{p}$ and $\hat{e}$; that is, c/n may not be able to prove exactly equal to $\hat{p}$, and e may not be able to prove exactly equal to $\hat{e}$. As sample size decreases, these inequalities tend to become larger, and one must be increasingly prepared to accept what is in effect a change in specifications. To illustrate, using a somewhat extreme situation, assume we specify $\hat{p} = .40$, $\hat{e} = .25$, and $\gamma_2 = 95\%$, and both $\bar{p}$ and $\underline{p}$ are desired. The sample size that most nearly meets these specifications is 18. But for n = 18 the value of c/n closest to .40 is 7/18 = .388; and for n = 18 and c/n = .388, e = .255. Hence, by adopting n = 18, we are, in effect, changing $\hat{p}$ to .388 and $\hat{e}$ to .255.

Tables 8–12 apply to $\hat{p}$ of .5 or less. However, they are equally applicable to $\hat{p}$ over .5. Then it is necessary to find the complement of $\hat{p}$—namely, $\hat{q}$—and substitute $\hat{q}$ for $\hat{p}$, as in Examples 2-5 to 2-7, 2-11 to 2-13, and 2-19. To illustrate, if $\hat{p} = .6$, $\hat{q} = 1 - \hat{p} = .4$, and one proceeds on the basis of $\hat{q} = .4$.

Sample size based on Tables 8–12 presumes that the population N is infinite or vastly larger than the sample. If such is not the case, a finite population correction (FPC) can be applied, as explained in Chapter Four, which reduces sample size. Unless the sample size is a significant part of the population, say, 5% or more, the reduction in sample size is minimal and often capable of being disregarded.

The scope of Tables 8–12 may be summarized as follows.

**SCOPE OF TABLES 8-12**

| $(\hat{p})$ Anticipated sample proportion* | Desired confidence limits: upper $(\overline{p})$, lower $(\underline{p})$ | Table numbers | |
|---|---|---|---|
| | | Poisson Procedure | Modified Normal Procedure |
| .25 or less† | $\overline{p}$ and $\underline{p}$ | 8 or 11 | |
| .25 or less† | Only $\overline{p}$ | 8 or 11 | |
| .25 or less† | Only $\underline{p}$ | 9 or 11 | |
| Over .25 | $\overline{p}$ and $\underline{p}$ | | 10 |
| Over .25 | Only $\overline{p}$ | | 10 |
| Over .25 | Only $\underline{p}$ | | 10 |
| 0 | Only $\overline{p}$ | 12 | |

*If $\hat{p}$ exceeds .5, calculate $\hat{q} = 1 - \hat{p}$ and substitute $\hat{q}$ for $\hat{p}$.
†But not 0.

## EXAMPLES

The 19 examples that follow seek to explain the use of Tables 8–12 and to enable the user to find an example or combination of examples matching his problem. Choice of values for the anticipated sample proportion $\hat{p}$ and the anticipated error margin $\hat{e}$ aims at simplicity and at permitting several comparisons: between the Poisson Procedure and the Modified Normal Procedure; between sample size when two-sided confidence limits (both $\overline{p}$ and $\underline{p}$) are desired and sample size when one-sided limits (only $\overline{p}$ or only $\underline{p}$) are desired; between sample sizes for only $\overline{p}$ and for only $\underline{p}$; and between sample sizes for two values of $\hat{p}$ an equal distance below and above .5 (for example, when $\hat{p} = .10$ and when $\hat{p} = .90$).

Throughout the examples the confidence level, designated in general terms as $\gamma$, remains the same: 95%. If both $\overline{p}$ and $\underline{p}$ are desired, the confidence level is specifically designated $\gamma_2$ and the tables and formulas are entered on the basis of $\gamma_2$. If only $\overline{p}$ or only $\underline{p}$ is sought, the confidence level is specifically designated $\gamma_1$ and the tables and formulas are entered on the basis of $\gamma_1$. It is important to recognize that at a given confidence level, say, 95%, the sample size varies according to whether both confidence limits are desired or only one; sample size is less when it is desired to obtain only one limit rather than both. (Furthermore, for a specified $\hat{p}$ and $\hat{e}$, sample size is less when only $\underline{p}$ is desired than when only $\overline{p}$ is desired, provided $\hat{p} \leqslant .5$.)

The sample-size procedure employed in connection with Tables 8–12 is essentially as follows. (1) We obtain a tentative sample size n with the aid of the tables. (2) We test whether n meets the requirement for appropriate sample size; that is, assuming $c/n = \hat{p}$, we find e and compare e with $\hat{e}$. (3) If the difference between e and $\hat{e}$ is appreciable, we obtain a correction factor, derived from e and $\hat{e}$, and use this correction factor to obtain a final result, n'.

**LIST OF EXAMPLES: DETERMINATION OF SAMPLE SIZE FOR PROPORTION SAMPLING**
(Based on Confidence Level $\gamma$ = 95%)

| Example number | $(\hat{p})$ Anticipated sample proportion | $(\hat{e})$ Anticipated error margin | Desired confidence limits: upper $(\bar{p})$, lower $(\underline{p})$ | Table entered on basis of $\gamma_1$ or $\gamma_2$ | Table number |
|---|---|---|---|---|---|
| | | Examples Using the Poisson Procedure | | | |
| 2-1 | .10 | .05 | $\bar{p}$ and $\underline{p}$ | $\gamma_2$ | 8 |
| 2-2 | .25 | .05 | $\bar{p}$ and $\underline{p}$ | $\gamma_2$ | 8 |
| 2-3 | .10 | .05 | only $\bar{p}$ | $\gamma_1$ | 8 |
| 2-4 | .10 | .05 | only $\underline{p}$ | $\gamma_1$ | 9 |
| 2-5 | .90 | .05 | $\bar{p}$ and $\underline{p}$ | $\gamma_2$ | 8 |
| 2-6 | .90 | .05 | only $\bar{p}$ | $\gamma_1$ | 9 |
| 2-7 | .90 | .05 | only $\underline{p}$ | $\gamma_1$ | 8 |
| | | Examples Using the Modified Normal Procedure | | | |
| 2-8 | .40 | .05 | $\bar{p}$ and $\underline{p}$ | $\gamma_2$ | 10 |
| 2-9 | .40 | .05 | only $\bar{p}$ | $\gamma_1$ | 10 |
| 2-10 | .40 | .05 | only $\underline{p}$ | $\gamma_1$ | 10 |
| 2-11 | .60 | .05 | $\bar{p}$ and $\underline{p}$ | $\gamma_2$ | 10 |
| 2-12 | .60 | .05 | only $\bar{p}$ | $\gamma_1$ | 10 |
| 2-13 | .60 | .05 | only $\underline{p}$ | $\gamma_1$ | 10 |
| 2-14 | .40 | .25 | $\bar{p}$ and $\underline{p}$ | $\gamma_2$ | 10 |
| | | Examples Using Alternative Form of the Poisson Procedure | | | |
| 2-15 | .10 | .05 | $\bar{p}$ and $\underline{p}$ | $\gamma_2$ | 11 |
| 2-16 | .10 | .05 | only $\bar{p}$ | $\gamma_1$ | 11 |
| 2-17 | .10 | .05 | only $\underline{p}$ | $\gamma_1$ | 11 |
| | | Examples Where $\hat{p}$ = 0 or 1 | | | |
| 2-18 | 0 | .05 | only $\bar{p}$ | $\gamma_1$ | 12 |
| 2-19 | 1 | .05 | only $\underline{p}$ | $\gamma_1$ | 12 |

*EXAMPLE 2-1* _____

## Problem

$\hat{p} = .10$

$\hat{e} = .05$ at $\gamma_2 = 95\%$

Both $\bar{p}$ and $\underline{p}$ are desired.

Obtain the Appropriate Sample Size n.

## Instructions

1. For $\hat{p} \leqslant .25$ and where both $\bar{p}$ and $\underline{p}$ are desired, use the Poisson Procedure based on Table 8.

   Calculate $\hat{\bar{p}} = \hat{p} + \hat{e}$ .

   Calculate $\hat{Q} = \dfrac{\hat{\bar{p}}}{\hat{p}} \times \dfrac{2 - \hat{p}}{2 - \hat{\bar{p}}}$ .

2. In Table 8 and the column for $\gamma_2 = 95\%$, find Q nearest to $\hat{Q}$. Find the value of c corresponding to Q. If necessary, interpolate linearly between values of Q bounding $\hat{Q}$ to find c.

   The interpolation formula for Table 8 is

$$c = c_a + \frac{(c_b - c_a)(Q_a - \hat{Q})}{Q_a - Q_b} \qquad \text{c is an integer}$$

   $Q_a$ is the bounding value larger than $\hat{Q}$, $Q_b$ is the bounding value smaller than $\hat{Q}$, $c_a$ is the value corresponding to $Q_a$, and $c_b$ is the value corresponding to $Q_b$.

3. Calculate $n = \dfrac{c}{\hat{p}}$ .

4. Test n for Appropriate Sample Size. Using the tables for confidence limits, obtain $\bar{p}$ based on c, n, and $\gamma_2$. Calculate $\bar{e} = \bar{p} - c/n$. If $\bar{e}$ differs appreciably from $\hat{e}$, calculate $n' = n \times (\bar{e}/\hat{e})^2$.

## Solution

1. $\hat{\bar{p}} = \hat{p} + \hat{e} = .10 + .05 = .15$

$$\hat{Q} = \frac{\hat{\bar{p}}}{\hat{p}} \times \frac{2 - \hat{p}}{2 - \hat{\bar{p}}} = \frac{.15}{.10} \times \frac{1.90}{1.85} = 1.5 \times 1.0270 = 1.541$$

2. For $\hat{Q} = 1.541$ and $\gamma_2 = 95\%$, the nearest $Q = 1.544$ and the corresponding $c = 20$ in Table 8.

3. $n = \dfrac{c}{\hat{p}} = \dfrac{20}{.10} = \boxed{200}$

4. Test, using Table 3:
   For $c = 20$ and $\gamma_2 = 95\%$, $\overline{m} = 30.888$ and $\overline{k} = 5.59$.

$$\overline{p}' = \frac{\overline{m}}{n + \overline{k}} = \frac{30.888}{205.59} = .1502$$

$\overline{e} = \overline{p} - c/n = .1502 - .10 = .0502$. Inasmuch as $\overline{e}$ is very close to $\hat{e}$, $n = 200$ is indicated as the Appropriate Sample Size. That is, the correction factor $(\overline{e}/\hat{e})^2$ is considered unnecessary.

*Note*

As $\hat{p}$ decreases, the Poisson Procedure becomes increasingly accurate as a means of obtaining the Appropriate Sample Size, thereby reducing or eliminating the role of the correction factor. However, this role is illustrated in Example 2-2 and others.

*EXAMPLE 2-2* _____

*Problem*

$\hat{p} = .25$
$\hat{e} = .05$ at $\gamma_2 = 95\%$
Both $\overline{p}$ and $\underline{p}$ are desired.
Obtain the Appropriate Sample Size n.

*Instructions*

1. For $\hat{p} \leqslant .25$ and where both $\overline{p}$ and $\underline{p}$ are desired, use the Poisson Procedure based on Table 8.

   Calculate $\hat{\overline{p}} = \hat{p} + \hat{e}$.

   Calculate $\hat{Q} = \dfrac{\hat{\overline{p}}}{\hat{p}} \times \dfrac{2 - \hat{p}}{2 - \hat{\overline{p}}}$.

2. In Table 8 and the column for $\gamma_2 = 95\%$, find $Q$ nearest to $\hat{Q}$. Find the value of $c$ corresponding to $Q$. If necessary, interpolate

linearly between values of Q bounding $\hat{Q}$ to find integer c, using the interpolation formula in Instruction 2 of Example 2-1.

3. Calculate $n = \dfrac{c}{\hat{p}}$.

4. Test n for Appropriate Sample Size. Using the tables for confidence limits, obtain $\bar{p}$ based on c, n, and $\gamma_2$. Calculate $\bar{e} = \bar{p} - c/n$. If $\bar{e}$ differs appreciably from $\hat{e}$, calculate $n' = n \times (\bar{e}/\hat{e})^2$.

*Solution*

1. $\hat{p} = \hat{p} + \hat{e} = .25 + .05 = .30$

$$\hat{Q} = \frac{\hat{p}}{\hat{p}} \times \frac{2 - \hat{p}}{2 - \hat{p}} = \frac{.30}{.25} \times \frac{1.75}{1.70} = 1.2 \times 1.0294 = 1.235$$

2. For $\hat{Q} = 1.235$ and $\gamma_2 = 95\%$, the bounding values of Q in Table 8 are 1.245 and 1.229, with corresponding c = 80 and 90, respectively. That is,

$$Q_a = 1.245 \qquad c_a = 80$$
$$Q_b = 1.229 \qquad c_b = 90$$
$$\hat{Q} = 1.235$$

Using the interpolation formula in Instruction 2 of Example 2-1,

$$c = 80 + \frac{(90 - 80)\,(1.245 - 1.235)}{(1.245 - 1.229)}$$

$$= 80 + \frac{10 \times .01}{.016} = 86 \text{ (to the nearest integer)}$$

3. $n = \dfrac{c}{\hat{p}} = \dfrac{86}{.25} = 344$

4. Test, using Table 1:
   For c = 86 and $\gamma_2 = 95\%$, $\bar{m} = 106.21$, $\bar{a} = 29.58$, and $\bar{b} = 19.60$.

$$\bar{p}' = \frac{\bar{m}}{n} \times \frac{n - \bar{a}}{n - \bar{b}} = \frac{106.21}{344} \times \frac{344 - 29.58}{344 - 19.60}$$

$$= .30875 \times .96924 = .2993$$

$$\bar{e} = \bar{p} - c/n = .2993 - .25 = .0493$$

$$n' = n \times \left(\frac{\overline{e}}{\hat{e}}\right)^2 = 344 \times \left(\frac{.0493}{.0500}\right)^2 = 344 \times .9722$$

$$= \boxed{334}$$

### Note

If desired we can test n = 334. First calculate $c = n\hat{p} = 84$ (to the nearest integer). Using Table 1 and repeating the procedure of step 4, we find that for n = 334 and $c = 84$, $\overline{e} = .0501$, which is very close to the specification $\hat{e} = .05$. However, a sample size of 334 cannot exactly meet the specification for $\hat{p}$ (inasmuch as .25 × 334 = 83.5). To find n which can exactly yield $c/n = \hat{p} = .25$, we calculate $n = c/\hat{p} = 84/.25 = 336$. Using Table 1 and repeating the procedure of step 4, we find that for n = 336 and $c = 84$, $\overline{e} = .0499$, which is as close as before to the specification $\hat{e} = .05$. Thus the sample size is slightly more precisely determined to be 336.

---

### *EXAMPLE 2-3*

#### Problem

$\hat{p} = .10$

$\hat{e} = .05$ at $\gamma_1 = 95\%$

Only $\overline{p}$ is desired.

Obtain the Appropriate Sample Size n.

#### Instructions

1. For $\hat{p} \leqslant .25$ and where only $\overline{p}$ is desired, use the Poisson Procedure based on Table 8.

   Calculate $\hat{\overline{p}} = \hat{p} + \hat{e}$.

   Calculate $\hat{Q} = \dfrac{\hat{\overline{p}}}{\hat{p}} \times \dfrac{2 - \hat{p}}{2 - \hat{\overline{p}}}$.

2. In Table 8 and the column for $\gamma_1 = 95\%$, find Q nearest to $\hat{Q}$. Find the value of c corresponding to Q. If necessary, interpolate linearly between values of Q bounding $\hat{Q}$ to find integer c, using the interpolation formula in Instruction 2 of Example 2-1.

3. Calculate $n = \dfrac{c}{\hat{p}}$ .

4. Test n for Appropriate Sample Size. Using the tables for confidence limits, obtain $\bar{p}$ based on c, n, and $\gamma_1$. Calculate $\bar{e} = \bar{p} - c/n$. If $\bar{e}$ differs appreciably from $\hat{e}$, calculate $n' = n \times (\bar{e}/\hat{e})^2$.

*Solution*

1. $\hat{\bar{p}} = \hat{p} + \hat{e} = .10 + .05 = .15$

$$\hat{Q} = \frac{\hat{\bar{p}}}{\hat{p}} \times \frac{2 - \hat{p}}{2 - \hat{\bar{p}}} = \frac{.15}{.10} \times \frac{1.90}{1.85} = 1.5 \times 1.0270 = 1.541 .$$

2. For $\hat{Q} = 1.541$ and $\gamma_1 = 95\%$, the nearest $Q = 1.540$ and the corresponding c = 15 in Table 8.

3. $n = \dfrac{c}{\hat{p}} = \dfrac{15}{.10} = \boxed{150}$

4. Test, using Table 3:
   For c = 15 and $\gamma_1$ = 95%, $\bar{m}$ = 23.097 and $\bar{k}$ = 4.15.

$$\bar{p}' = \frac{\bar{m}}{n + \bar{k}} = \frac{23.097}{154.15} = .1498$$

$\bar{e} = \bar{p} - c/n = .1498 - .10 = .0498$. Inasmuch as $\bar{e}$ is very close to $\hat{e}$, n = 150 is indicated as the Appropriate Sample Size. That is, the correction factor $(\bar{e}/\hat{e})^2$ is considered unnecessary.

---

## EXAMPLE 2-4

*Problem*

$\hat{p} = .10$

$\hat{e} = .05$ at $\gamma_1 = 95\%$

Only p is desired.

Obtain the Appropriate Sample Size n.

## Instructions

1. For $\hat{p} \leqslant .25$ and where only $\underline{p}$ is desired, use the Poisson Procedure based on Table 9.

   Calculate $\underline{\hat{p}} = \hat{p} - \hat{e}$.

   Calculate $\hat{R} = \dfrac{\underline{\hat{p}}}{\hat{p}} \times \dfrac{2 - \hat{p}}{2 - \underline{\hat{p}}}$.

2. In Table 9 and the column for $\gamma_1 = 95\%$, find R nearest to $\hat{R}$. Find the value of c corresponding to R. If necessary, interpolate linearly between values of R bounding $\hat{R}$ to find integer c.

   The interpolation formula for Table 9 is

   $$c = c_a + \frac{(c_b - c_a)(\hat{R} - R_a)}{(R_b - R_a)} \qquad \text{c is an integer}$$

   $R_a$ is the bounding value smaller than $\hat{R}$, $R_b$ is the bounding value larger than $\hat{R}$, $c_a$ is the value corresponding to $R_a$, and $c_b$ is the value corresponding to $R_b$.

3. Calculate $n = \dfrac{c}{\underline{\hat{p}}}$.

4. Test n for Appropriate Sample Size. Using the tables for confidence limits, obtain $\underline{p}$ based on c, n, and $\gamma_1$. Calculate $\underline{e} = c/n - \underline{p}$. If $\underline{e}$ differs appreciably from $\hat{e}$, calculate $n' = n \times (\underline{e}/\hat{e})^2$.

## Solution

1. $\underline{p} = \hat{p} - \hat{e} = .10 - .05 = .05$

   $$\hat{R} = \frac{\underline{\hat{p}}}{\hat{p}} \times \frac{2 - \hat{p}}{2 - \underline{\hat{p}}} = \frac{.05}{.10} \times \frac{1.90}{1.95} = .5 \times .9744 = .487$$

2. For $\hat{R} = .487$ and $\gamma_1 = 95\%$, the nearest R $= .498$ and the corresponding c $= 8$ in Table 9.

3. $n = \dfrac{c}{\underline{\hat{p}}} = \dfrac{8}{.10} = \boxed{80}$

4. Test, using Table 4:
   For c $= 8$ and $\gamma_1 = 95\%$, $\underline{m} = 3.9808$ and $\underline{k} = 1.510$.

$$\underline{p}' = \frac{m}{n - \underline{k}} = \frac{3.9808}{78.490} = .05072$$

$\underline{e} = c/n - \underline{p} = .10 - .05072 = .0493$. Inasmuch as $\underline{e}$ is very close to $\hat{e}$, $n = 80$ is indicated as the Appropriate Sample Size. That is, the correction factor $(\underline{e}/\hat{e})^2$ is considered unnecessary.

## Note

Using Table 3, trial and error indicates that for $c = 8$, the closest approach to $\hat{e} = .05$ occurs for $n = 79$; then $\underline{e} = .0499$. However, $c/n$ is then $.1013$, which departs from the specification $\hat{p} = .10$. On balance, $n = 80$ well satisfies the specifications for both $\hat{p}$ and $\hat{e}$.

### *EXAMPLE 2-5* _____

### Problem

$\hat{p} = .90$
$\hat{e} = .05$ at $\gamma_2 = 95\%$

Both $\bar{p}$ and $\underline{p}$ are desired.
Obtain the Appropriate Sample Size n.

### Instructions

When $\hat{p}$ exceeds $.5$, calculate $\hat{q} = 1 - \hat{p}$. Find the sample size for $\hat{q}$. Do so by substituting $\hat{q}$ for $\hat{p}$ to find an appropriate example and to use the tables and formulas.

### Solution

$\hat{q} = 1 - .90 = .10$. Substituting $\hat{q}$ for $\hat{p}$, the problem is the counterpart of that in Example 2-1 (or Example 2-2). Further instructions are therefore the same as in Example 2-1 (or Example 2-2), leading to $n = \boxed{200}$. (The calculations are the same as in Example 2-1 and therefore need not be repeated here.)

## EXAMPLE 2-6 _____

### Problem

$\hat{p}$ = .90
$\hat{e}$ = .05 at $\gamma_1$ = 95%
Only $\overline{p}$ is desired.
Obtain the Appropriate Sample Size n.

### Instructions

When $\hat{p}$ exceeds .5, calculate $\hat{q}$ = 1 − $\hat{p}$. Find the sample size for $\hat{q}$. Do so by substituting $\hat{q}$ for $\hat{p}$ to find an appropriate example and to use the tables and formulas.

When only $\overline{p}$ is desired for $\hat{p}$, this connotes that only $\underline{p}$ is desired for $\hat{q}$. Therefore the example, tables, and formulas to be used are those pertaining to $\underline{p}$.

### Solution

$\hat{q}$ = 1 − .90 = .10. Substituting $\hat{q}$ for $\hat{p}$, the problem is the counterpart of that in Example 2-4. Further instructions are therefore the same as in Example 2-4, leading to n = $\boxed{80}$. (The calculations are the same as in Example 2-4 and therefore need not be repeated here.)

## EXAMPLE 2-7 _____

### Problem

$\hat{p}$ = .90
$\hat{e}$ = .05 at $\gamma_1$ = 95%
Only $\underline{p}$ is desired.
Obtain the Appropriate Sample Size n.

### Instructions

When $\hat{p}$ exceeds .5, calculate $\hat{q}$ = 1 − $\hat{p}$. Find the sample size for $\hat{q}$. Do so by substituting $\hat{q}$ for $\hat{p}$ to find an appropriate example and to use the tables and formulas.

When only $\underline{p}$ is desired for $\hat{p}$, this connotes that only $\overline{p}$ is desired for $\hat{q}$. Therefore the example, tables, and formulas to be used are those pertaining to $\overline{p}$.

## Solution

$\hat{q} = 1 - .90 = .10$. Substituting $\hat{q}$ for $\hat{p}$, the problem is the counterpart of that in Example 2-3. Further instructions are therefore the same as in Example 2-3, leading to $n = \boxed{150}$. (The calculations are the same as in Example 2-3 and therefore need not be repeated here.)

## EXAMPLE 2-8 _____

### Problem

$\hat{p} = .40$

$\hat{e} = .05$ at $\gamma_2 = 95\%$

Both $\bar{p}$ and $\underline{p}$ are desired.

Obtain the Appropriate Sample Size n.

### Instructions

1. For $\hat{p} > .25$, use the Modified Normal Procedure based on Table 10.

   Calculate $\hat{\bar{p}} = \hat{p} + \hat{e}$.

   Calculate $\hat{\bar{q}} = 1 - \hat{\bar{p}}$.

2. In Table 10 find the value of $A^2$ corresponding to $\gamma_2 = 95\%$.

3. Calculate $n = \dfrac{A^2 \hat{\bar{p}} \hat{\bar{q}} + \hat{e}}{\hat{e}^2}$.

4. Test n for Appropriate Sample Size. Calculate $c = n\hat{p}$ (an integer). Using the tables for confidence limits, obtain $\bar{p}$ based on c, n, and $\gamma_2$. Calculate $\bar{e} = \bar{p} - c/n$. If $\bar{e}$ differs appreciably from $\hat{e}$, calculate $n' = n \times (\bar{e}/\hat{e})^2$.

### Solution

1. $\hat{\bar{p}} = \hat{p} + \hat{e} = .40 + .05 = .45$

   $\hat{\bar{q}} = 1 - \hat{\bar{p}} = 1 - .45 = .55$

2. For $\gamma_2 = 95\%$, $A^2 = 3.8415$ in Table 10.

3. $n = \dfrac{A^2 \hat{p}\hat{q} + \hat{e}}{\hat{e}^2} = \dfrac{(3.8415 \times .45 \times .55) + .05}{(.05)^2} = \boxed{400}$

4. Test, using Table 1:

   $c = n\hat{p} = 400 \times .40 = 160$

   For $c = 160$ and $\gamma_2 = 95\%$, $\overline{m} = 186.80$, $\overline{a} = 46.70$, and $\overline{b} = 33.30$.

   $\overline{p}' = \dfrac{\overline{m}}{n} \times \dfrac{n - \overline{a}}{n - \overline{b}} = \dfrac{186.80}{400} \times \dfrac{353.30}{366.70} = .46700 \times .96346$

   $= .4499$

   $\overline{e} = \overline{p} - c/n = .4499 - .40 = .0499$. Inasmuch as $\overline{e}$ is very close to $\hat{e}$, $n = 400$ is indicated as the Appropriate Sample Size. That is, the correction factor $(\overline{e}/\hat{e})^2$ is considered unnecessary.

*Note*

If we apply the correction factor $(\overline{e}/\hat{e})^2$, we obtain $400 \times (499/500)^2 = 398$. Testing $n = 398$ on the basis of Table 1 leads to $\overline{e} = .0501$, compared with the desired $\hat{e} = .0500$. This suggests that an intermediate figure, namely, $n = 399$, is the closest approach to Appropriate Sample Size.

The error in using $n = 400$ instead of 399 may be justified on several counts: (1) it is small in absolute; (2) it is in the conservative direction—sample size too large instead of too small; (3) it is minor in view of $\hat{p}$ being a guess rather than an accurate prediction of $c/n$; and (4) the procedures in this book generally (except for a limited number of exact confidence limits in Tables 5 and 6) involve close approximations rather than exact values.

*EXAMPLE 2-9* _____

*Problem*

$\hat{p} = .40$

$\hat{e} = .05$ at $\gamma_1 = 95\%$

Only $\overline{p}$ is desired.

Obtain the Appropriate Sample Size n.

## Instructions

1. For $\hat{p} > .25$, use the Modified Normal Procedure based on Table 10.

    Calculate $\hat{\hat{p}} = \hat{p} + \hat{e}$.

    Calculate $\hat{\hat{q}} = 1 - \hat{\hat{p}}$.

2. In Table 10 find the value of $A^2$ corresponding to $\gamma_1 = 95\%$.

3. Calculate $n = \dfrac{A^2 \hat{\hat{p}} \hat{\hat{q}} + \hat{e}}{\hat{e}^2}$.

4. Test n for Appropriate Sample Size. Calculate $c = n\hat{p}$ (an integer). Using the tables for confidence limits, obtain $\bar{p}$ based on c, n, and $\gamma_1$. Calculate $\bar{e} = \bar{p} - c/n$. If $\bar{e}$ differs appreciably from $\hat{e}$, calculate $n' = n \times (\bar{e}/\hat{e})^2$.

## Solution

1. $\hat{\hat{p}} = \hat{p} + \hat{e} = .40 + .05 = .45$

    $\hat{\hat{q}} = 1 - \hat{\hat{p}} = 1 - .45 = .55$

2. For $\gamma_1 = 95\%$, $A^2 = 2.7055$ in Table 10.

3. $n = \dfrac{A^2 \hat{\hat{p}} \hat{\hat{q}} + \hat{e}}{\hat{e}^2} = \dfrac{(2.7055 \times .45 \times .55) + .05}{(.05)^2} = \boxed{288}$

4. Test, using Table 1:

    $c = n\hat{p} = 288 \times .40 = 115$

    For $c = 115$ and $\gamma_1 = 95\%$, $\bar{m} = 134.27$, $\bar{a} = 33.57$, and $\bar{b} = 23.93$.

    $\bar{p}' = \dfrac{\bar{m}}{n} \times \dfrac{n - \bar{a}}{n - \bar{b}} = \dfrac{134.27}{288} \times \dfrac{254.43}{264.07} = .46622 \times .96349$

    $= .4492$

$\bar{e} = \bar{p} - c/n = .4492 - .3993 = .0499$. Inasmuch as $\bar{e}$ is very close to $\hat{e}$, $n = 288$ is indicated as the Appropriate Sample Size. That is, the correction factor $(\bar{e}/\hat{e})^2$ is considered unnecessary.

## Note

If we apply the correction factor $(\bar{e}/\hat{e})^2$, we obtain $288 \times (499/500)^2 = 287$. Testing $n = 287$ on the basis of Table 1 leads to $\bar{e} = .0500$. This confirms that $n = 287$ is the closest approach to Appropriate Sample Size.

The error in using $n = 288$ instead of $n = 287$ may be justified on several counts: (1) it is small in absolute; (2) it is in the conservative direction—sample size too large instead of too small; (3) it is minor in view of $\hat{p}$ being a guess rather than an accurate prediction of $c/n$; and (4) the procedures in this book generally (except for a limited number of exact confidence limits in Tables 5 and 6) involve close approximations rather than exact values.

*EXAMPLE 2-10* _____

### Problem

$\hat{p} = .40$

$\hat{e} = .05$ at $\gamma_1 = 95\%$

Only $\underline{p}$ is desired.
Obtain the Appropriate Sample Size n.

### Instructions

1. For $\hat{p} > .25$, use the Modified Normal Procedure based on Table 10.

   Calculate $\underline{\hat{p}} = \hat{p} - \hat{e}$.

   Calculate $\underline{\hat{q}} = 1 - \underline{\hat{p}}$.

2. In Table 10 find the value of $A^2$ corresponding to $\gamma_1 = 95\%$.

3. Calculate $n = \dfrac{A^2 \, \underline{\hat{p}} \, \underline{\hat{q}} + \hat{e}}{\hat{e}^2}$.

4. Test n for Appropriate Sample Size. Calculate $c = n\hat{p}$ (an integer). Using the tables for confidence limits, obtain $\underline{p}$ based on c, n, and $\gamma_1$. Calculate $\underline{e} = c/n - \underline{p}$. If $\underline{e}$ differs appreciably from $\hat{e}$, calculate $n' = n \times (\underline{e}/\hat{e})^2$.

*Solution*

1. $\underline{\hat{p}}$ = $\hat{p} - \hat{e}$ = .40 − .05 = .35

   $\underline{\hat{q}}$ = 1 − $\underline{\hat{p}}$ = 1 − .35 = .65

2. For $\gamma_1$ = 95%, $A^2$ = 2.7055 in Table 10.

3. n = $\dfrac{A^2 \underline{\hat{p}}\, \underline{\hat{q}} + \hat{e}}{\hat{e}^2}$ = $\dfrac{(2.7055 \times .35 \times .65) + .05}{(.05)^2}$ = $\boxed{266}$

4. Test, using Table 2:

   c = $n\hat{p}$ = 266 × .40 = 106

   For c = 106 and $\gamma_1$ = 95%, $\underline{m}$ = 89.653, $\underline{a}$ = 29.88, and $\underline{b}$ = 37.56.

   $\underline{p'}$ = $\dfrac{m}{n}$ × $\dfrac{n - a}{n - b}$ = $\dfrac{89.653}{266}$ × $\dfrac{236.12}{228.44}$ = .33704 × 1.03362

   = .3484

   $\underline{e}$ = c/n − $\underline{p}$ = .3985 − .3484 = .0501. Inasmuch as $\underline{e}$ is very close to $\hat{e}$, n = 266 is indicated as the Appropriate Sample Size. That is, the correction factor $(\underline{e}/\hat{e})^2$ is considered unnecessary.

*Note*

If we apply the correction factor $(\underline{e}/\hat{e})^2$, we obtain 266 × (501/500)$^2$ = 267. Testing n = 267 on the basis of Table 2 leads to $\underline{e}$ = .0500. This confirms that n = 267 is the closest approach to Appropriate Sample Size.

The error in using n = 266 instead of n = 267 may be justified on several counts: (1) it is small in absolute; (2) it is minor in view of $\hat{p}$ being a guess rather than an accurate prediction of c/n; and (3) the procedures in this book generally (except for a limited number of exact confidence limits in Tables 5 and 6) involve close approximations rather than exact values. On the other hand, in contrast with Examples 2-8 and 2-9, the error is in the nonconservative direction— sample size too small instead of too large.

## EXAMPLE 2-11 _____

### Problem

$\hat{p}$ = .60
$\hat{e}$ = .05 at $\gamma_2$ = 95%
Both $\bar{p}$ and $\underline{p}$ are desired.
Obtain the Appropriate Sample Size n.

### Instructions

When $\hat{p}$ exceeds .5, calculate $\hat{q}$ = 1 − $\hat{p}$. Find the sample size for $\hat{q}$. Do so by substituting $\hat{q}$ for $\hat{p}$ to find an appropriate example and to use the tables and formulas.

### Solution

$\hat{q}$ = 1 − .60 = .40. Substituting $\hat{q}$ for $\hat{p}$, the problem is the counterpart of that in Example 2-8. Further instructions are therefore the same as in Example 2-8, leading to n = $\boxed{400}$. (The calculations are the same as in Example 2-8 and therefore need not be repeated here.)

## EXAMPLE 2-12 _____

### Problem

$\hat{p}$ = .60
$\hat{e}$ = .05 at $\gamma_1$ = 95%
Only $\bar{p}$ is desired.
Obtain the Appropriate Sample Size n.

### Instructions

When $\hat{p}$ exceeds .5, calculate $\hat{q}$ = 1 − $\hat{p}$. Find the sample size for $\hat{q}$. Do so by substituting $\hat{q}$ for $\hat{p}$ to find an appropriate example and to use the tables and formulas.

When only $\bar{p}$ is desired for $\hat{p}$, this connotes that only $\underline{p}$ is desired for $\hat{q}$. Therefore the example, tables, and formulas to be used are those pertaining to $\underline{p}$.

*Solution*

$\hat{q}$ = 1 − .60 = .40. Substituting $\hat{q}$ for $\hat{p}$, the problem is the counter-part of that in Example 2-10. Further instructions are therefore the same as in Example 2-10, leading to n = $\boxed{266}$ . (The calculations are the same as in Example 2-10, and therefore need not be repeated here.)

*EXAMPLE 2-13* _____

*Problem*

$\hat{p}$  = .60
$\hat{e}$  = .05 at $\gamma_1$ = 95%

Only $\underline{p}$ is desired.
Obtain the Appropriate Sample Size n.

*Instructions*

When $\hat{p}$ exceeds .5, calculate $\hat{q}$ = 1 − $\hat{p}$. Find the sample size for $\hat{q}$. Do so by substituting $\hat{q}$ for $\hat{p}$ to find an appropriate example and to use the tables and formulas.

When only $\underline{p}$ is desired for $\hat{p}$, this connotes that only $\overline{p}$ is desired for $\hat{q}$. Therefore the example, tables, and formulas to be used are those pertaining to $\overline{p}$.

*Solution*

q = 1 − .60 = .40. Substituting $\hat{q}$ for $\hat{p}$, the problem is the counter-part of that in Example 2-9. Further instructions are therefore the same as in Example 2-9, leading to n = $\boxed{288}$ . (The calculations are the same as in Example 2-9 and therefore need not be repeated here.)

*EXAMPLE 2-14* _____

*Problem*

$\hat{p} = .40$

$\hat{e} = .25$ at $\gamma_2 = 95\%$

Both $\overline{p}$ and $p$ are desired.
Obtain the Appropriate Sample Size n.

*Instructions*

1. For $\hat{p} > .25$, use the Modified Normal Procedure based on Table 10.

   Calculate $\hat{\overline{p}} = \hat{p} + \hat{e}$ .

   Calculate $\hat{\overline{q}} = 1 - \hat{\overline{p}}$ .

2. In Table 10 find the value of $A^2$ corresponding to $\gamma_2 = 95\%$.

3. Calculate $n = \dfrac{A^2 \hat{\overline{p}} \hat{\overline{q}} + \hat{e}}{\hat{e}^2}$ .

4. Test n for Appropriate Sample Size. Calculate $c = n\hat{p}$ (an integer). Using the tables for confidence limits, obtain $\overline{p}$ based on c, n, and $\gamma_2$. Calculate $\overline{e} = \overline{p} - c/n$. If $\overline{e}$ differs appreciably from $\hat{e}$, calculate $n' = n \times (\overline{e}/\hat{e})^2$.

*Solution*

1. $\hat{\overline{p}} = \hat{p} + \hat{e} = .40 + .25 = .65$

   $\hat{\overline{q}} = 1 - \hat{\overline{p}} = 1 - .65 = .35$

2. For $\gamma_2 = 95\%$, $A^2 = 3.8415$ in Table 10.

3. $n = \dfrac{A^2 \hat{\overline{p}} \hat{\overline{q}} + \hat{e}}{\hat{e}^2} = \dfrac{(3.8415 \times .65 \times .35) + .25}{(.25)^2} = 18$

4. Test, using Table 5:

   $c = n\hat{p} = 18 \times .40 = 7$

   For n = 18, c = 7, and $\gamma_2 = 95\%$, $\overline{p} = .64255$ in Table 5.

   $\overline{e} = \overline{p} - c/n = .64255 - .38889 = .25366$

$$n' = n \times \left(\frac{\overline{e}}{\hat{e}}\right)^2 = 18 \times \left(\frac{.25366}{.25000}\right)^2 = \boxed{19}$$

*Note*

This example illustrates the problem of very small sample sizes, which (owing to use of integers) have difficulty in accurately meeting the specifications $\hat{p}$ and $\hat{e}$. Therefore judgment must be used in selecting the sample size that on the whole best meets specifications, as follows.

To test n = 19, we calculate c = $n\hat{p}$ = 8. For n = 19 and c = 8, Table 5 yields $\overline{p}$ = .66500; hence $\overline{e}$ = .66500 − .42105 = .24395. Comparison between n = 19 and n = 18 with respect to deviation from specifications ($\hat{p}$ = .40, $\hat{e}$ = .25) shows:

|  | c/n | $\overline{e}$ | Absolute deviation | |
|---|---|---|---|---|
|  |  |  | $\hat{p} - c/n$ | $\hat{e} - \overline{e}$ |
| n = 19, c = 8 | .42105 | .24395 | .02105 | .00605 |
| n = 18, c = 7 | .38889 | .25366 | .01111 | .00366 |

On the whole, it seems that n = 19 more satisfactorily meets specifications. Although it involves slightly greater absolute deviations from specifications than does n = 18, the deviations for n = 19 are in the conservative direction; that is, sample size of 19 permits c/n to be larger than $\hat{p}$ and at the same time permits $\overline{e}$ to be less than $\hat{e}$.

*EXAMPLE 2-15* _____

*Problem*

$\hat{p}$ = .10
$\hat{e}$ = .05 at $\gamma_2$ = 95%

Both $\overline{p}$ and $\underline{p}$ are desired.
Obtain the Appropriate Sample Size n.

*Instructions*

1. For $\hat{p} \leqslant .25$, one may use an alternative form of the Poisson Procedure based on Table 11.

   Calculate $\hat{\overline{p}}$ = $\hat{p}$ + $\hat{e}$.

Calculate $\hat{Q} = \dfrac{\hat{\hat{p}}}{\hat{p}} \times \dfrac{2 - \hat{p}}{2 - \hat{\hat{p}}}$.

In Table 11 find the values of A, $A^2$, and B corresponding to $\gamma_2$ = 95%.

2. Calculate $c = \left[ \dfrac{A + \sqrt{A^2 + 4(\hat{Q} - 1)(\hat{Q} + B - 1)}}{2(\hat{Q} - 1)} \right]^2 - 1$;

$\qquad\qquad\qquad\qquad\qquad\qquad\qquad\qquad$ c is an integer.

3. Calculate $n = \dfrac{c}{\hat{p}}$.

4. Test n for Appropriate Sample Size. Using the tables for confidence limits, obtain $\overline{p}$ based on c, n, and $\gamma_2$. Calculate $\overline{e} = \overline{p} - c/n$. If $\overline{e}$ differs appreciably from $\hat{e}$, calculate $n' = n \times (\overline{e}/\hat{e})^2$.

*Solution*

1. $\hat{\hat{p}} = \hat{p} + \hat{e} = .10 + .05 = .15$

$\hat{Q} = \dfrac{\hat{\hat{p}}}{\hat{p}} \times \dfrac{2 - \hat{p}}{2 - \hat{\hat{p}}} = \dfrac{.15}{.10} \times \dfrac{1.90}{1.85} = 1.5 \times 1.0270 = 1.541$

For $\gamma_2$ = 95%, A = 1.9600, $A^2$ = 3.8415, and B = 1.9472 in Table 11.

2. $c = \left[ \dfrac{1.9600 + \sqrt{3.8415 + (4 \times .541 \times 2.488)}}{2 \times .541} \right]^2 - 1$

$\quad = \left( \dfrac{1.9600 + 3.037}{1.082} \right)^2 - 1 = 4.62^2 - 1 = 20$

3. $n = \dfrac{c}{\hat{p}} = \dfrac{20}{.10} = \boxed{200}$

4. Test, using Table 3:

For c = 20 and $\gamma_2$ = 95%, $\overline{m}$ = 30.888 and $\overline{k}$ = 5.59.

$$\overline{p}' = \frac{\overline{m}}{n + \overline{k}} = \frac{30.888}{205.59} = .1502$$

$\overline{e} = \overline{p} - c/n = .1502 - .10 = .0502$. Inasmuch as $\overline{e}$ is very close to $\hat{e}$, $n = 200$ is indicated as the Appropriate Sample Size. That is, the correction factor $(\overline{e}/\hat{e})^2$ is considered unnecessary.

### Note

The problem and results are the same here as in Example 2-1. The only difference is in the procedure for finding c. Here c is obtained by formula, whereas in Example 2-1 c is obtained from Table 8.

*EXAMPLE 2-16* _____

### Problem

$\hat{p} = .10$
$\hat{e} = .05$ at $\gamma_1 = 95\%$
Only $\overline{p}$ is desired.
Obtain the Appropriate Sample Size n.

### Instructions

1. For $\hat{p} \leqslant .25$, one may use an alternative form of the Poisson Procedure based on Table 11.

   Calculate $\hat{\overline{p}} = \hat{p} + \hat{e}$.

   Calculate $\hat{Q} = \dfrac{\hat{\overline{p}}}{\hat{p}} \times \dfrac{2 - \hat{p}}{2 - \hat{\overline{p}}}$.

   In Table 11 find the values of A, $A^2$, and B corresponding to $\gamma_1 = 95\%$.

2. Calculate $c = \left[ \dfrac{A + \sqrt{A^2 + 4(\hat{Q} - 1)(\hat{Q} + B - 1)}}{2(\hat{Q} - 1)} \right]^2 - 1$;

   c is an integer.

3. Calculate $n = \dfrac{c}{\hat{p}}$.

4. **Test n for Appropriate Sample Size.** Using the tables for confidence limits, obtain $\bar{p}$ based on c, n, and $\gamma_1$. Calculate $\bar{e} = \bar{p} - c/n$. If $\bar{e}$ differs appreciably from $\hat{e}$, calculate $n' = n \times (\bar{e}/\hat{e})^2$.

### Solution

1. $\hat{\bar{p}} = \hat{p} + \hat{e} = .10 + .05 = .15$

$$\hat{Q} = \frac{\hat{\bar{p}}}{\hat{p}} \times \frac{2 - \hat{p}}{2 - \hat{\bar{p}}} = \frac{.15}{.10} \times \frac{1.90}{1.85} = 1.5 \times 1.0270 = 1.541$$

For $\gamma_1 = 95\%$, $A = 1.6449$, $A^2 = 2.7055$, and $B = 1.5685$ in Table 11.

2. $c = \left[ \dfrac{1.6449 + \sqrt{2.7055 + (4 \times .541 \times 2.110)}}{2 \times .541} \right]^2 - 1$

$$= \left( \frac{1.6449 + 2.6966}{1.082} \right)^2 - 1 = 4.01^2 - 1 = 15$$

3. $n = \dfrac{c}{\hat{p}} = \dfrac{15}{.10} = \boxed{150}$

4. Test, using Table 3:

For $c = 15$ and $\gamma_1 = 95\%$, $\bar{m} = 23.097$ and $\bar{k} = 4.15$.

$$\bar{p}' = \frac{\bar{m}}{n + \bar{k}} = \frac{23.097}{154.15} = .1498$$

$\bar{e} = \bar{p} - c/n = .1498 - .10 = .0498$. Inasmuch as $\bar{e}$ is very close to $\hat{e}$, n = 150 is indicated as the Appropriate Sample Size. That is, the correction factor $(\bar{e}/\hat{e})^2$ is considered unnecessary.

### Note

The problem and results are the same here as in Example 2-3. The only difference is in the procedure for finding c. Here c is obtained by formula, whereas in Example 2-3 c is obtained from Table 8.

## EXAMPLE 2-17 _____

### Problem

$\hat{p} = .10$
$\hat{e} = .05$ at $\gamma_1 = 95\%$

Only p is desired.
Obtain the Appropriate Sample Size n.

### Instructions

1. For $\hat{p} \leqslant .25$, one may use an alternative form of the Poisson Procedure based on Table 11.

   Calculate $\underline{\hat{p}} = \hat{p} - \hat{e}$.

   Calculate $\hat{R} = \dfrac{\underline{\hat{p}}}{\hat{p}} \times \dfrac{2 - \hat{p}}{2 - \underline{\hat{p}}}$.

   In Table 11 find the values of A, $A^2$, and B corresponding to
   $\gamma_1 = 95\%$.

2. Calculate $c = \left[ \dfrac{A + \sqrt{A^2 - 4(B - 1)(1 - \hat{R})}}{2(1 - \hat{R})} \right]^2$ ; c is an in-

   teger. If the term inside the radical is negative (which can happen
   in a limited number of cases—when $\gamma_1 = 99.5\%$ and c should be
   1, 2, or 3, and when $\gamma_1 = 99\%$ and c should be 1 or 2), do not
   use this formula; instead use Table 9 to find c.

3. Calculate $n = \dfrac{c}{\underline{\hat{p}}}$.

4. Test n for Appropriate Sample Size. Using the tables for
   confidence limits, obtain p based on c, n, and $\gamma_1$. Calculate
   $\underline{e} = c/n - \underline{p}$. If $\underline{e}$ differs appreciably from $\hat{e}$, calculate $n' = n \times (\underline{e}/\hat{e})^2$.

### Solution

1. $\underline{\hat{p}} = \hat{p} - \hat{e} = .10 - .05 = .05$

   $\hat{R} = \dfrac{\underline{\hat{p}}}{\hat{p}} \times \dfrac{2 - \hat{p}}{2 - \underline{\hat{p}}} = \dfrac{.05}{.10} \times \dfrac{1.90}{1.95} = .5 \times .9744 = .487$

For $\gamma_1$ = 95%, A = 1.6449, $A^2$ = 2.7055, and B = 1.5685 in Table 11.

$$2. \ c = \left[ \frac{1.6449 + \sqrt{2.7055 - (4 \times .5685 \times .513)}}{2 \times .513} \right]^2$$

$$= \left( \frac{1.6449 + 1.241}{1.026} \right)^2 = 2.81^2 = 8$$

$$3. \ n = \frac{c}{\hat{p}} = \frac{8}{.10} = \boxed{80}$$

4. Test, using Table 4:

For c = 8 and $\gamma_1$ = 95%, $\underline{m}$ = 3.9808, and $\underline{k}$ = 1.510.

$$\underline{p}' = \frac{m}{n - \underline{k}} = \frac{3.9808}{78.490} = .05072$$

$\underline{e}$ = c/n $-$ $\underline{p}$ = .10 $-$ .05072 = .0493. Inasmuch as $\underline{e}$ is very close to $\hat{e}$, n = 80 is indicated as the Appropriate Sample Size. That is, the correction factor $(\underline{e}/\hat{e})^2$ is considered unnecessary.

### Note

The problem and results are the same here as in Example 2-4. The only difference is in the procedure for finding c. Here c is obtained by formula, whereas in Example 2-4 c is obtained from Table 9.

---

### EXAMPLE 2-18 _____

### Problem

$\hat{p}$ = 0
$\hat{e}$ = .05 at $\gamma_1$ = 95%

Obtain the Appropriate Sample Size n.

### Instructions

1. When $\hat{p}$ = 0, this denotes that c = 0 and that only the upper confidence limit $\overline{p}$ is desired at confidence level $\gamma_1$. (The lower confidence limit is of course zero.)

In Table 12 find the value of $\overline{m}$ for c = 0 and $\gamma_1$ = 95%.

2. Calculate n = $\dfrac{\overline{m}}{\hat{e}} - \dfrac{\overline{m}}{2}$.

### Solution

1. For c = 0 and $\gamma_1$ = 95%, $\overline{m}$ = 2.9957 in Table 12.

2. n = $\dfrac{\overline{m}}{\hat{e}} - \dfrac{\overline{m}}{2} = \dfrac{2.9957}{.05} - \dfrac{2.9957}{2}$ = 59.91 − 1.50 = $\boxed{58}$

### EXAMPLE 2-19

### Problem

$\hat{p}$ = 1
$\hat{e}$ = .05 at $\gamma_1$ = 95%
Obtain the Appropriate Sample Size n.

### Instructions

When $\hat{p}$ = 1, this denotes that only the lower confidence limit $\underline{p}$ is desired at confidence level $\gamma_1$. (The upper confidence limit is of course 1.)

When $\hat{p}$ exceeds .5, the sample size procedure is based on $\hat{q}$ = 1 − $\hat{p}$. In the present case $\hat{q}$ = 0. The procedure is therefore the same as in Example 2-18.

### Solution

Substituting $\hat{q}$ = 0 for $\hat{p}$ = 1, the instructions of Example 2-18 lead to n = $\boxed{58}$. (The calculations are the same as in Example 2-18 and therefore need not be repeated here.)

# CHAPTER THREE

# Use of the Tables to Determine Sample Size for Acceptance Sampling

We deal first with acceptance sampling proper and then with what may be called *quasi-forms* of acceptance sampling.

## ACCEPTANCE SAMPLING

Acceptance sampling customarily deals with an error (or defective) rate, that is, with a lot (population) where some of the items contain a designated type of error (defect). Based on the number of error items c in the sample n, the lot is either accepted as having a satisfactorily low error rate or rejected as having an unsatisfactorily high rate. The lot is accepted if the number of error items in the sample is $\dot{c}$ or less and rejected if the number is $\dot{c}_1$ or more: $\dot{c}_1 = \dot{c} + 1$. The terms $\dot{c}$ and $\dot{c}_1$ are the acceptance number and rejection number, respectively. Collectively, they may be called *decision numbers.*

Sample size, acceptance number $\dot{c}$, and rejection number $\dot{c}_1$ are designed to meet the following specifications, as nearly as integers for $\dot{c}$ and n permit: (1) risk $\beta$ of accepting a lot with unsatisfactory rate $p_2$ and (2) risk $\alpha$ of rejecting a lot with satisfactory rate $p_1$. To illustrate, one might specify 1% risk of accepting a lot with an error rate as high as .05 and 5% risk of rejecting a lot with an error rate as low as .01.

If the actual number of error items c proves to be the same as $\dot{c}$, which means acceptance of the lot, this is tantamount to concluding one is 99% confident (complement of 1% risk) that the lot rate is not more than .05. If the number of error items proves to be $\dot{c}_1$, which means rejection of the lot, this is tantamount to concluding one is 95% confident (complement of 5% risk) that the lot rate is at least .01. Fewer error items than $\dot{c}$ permit one to state either that he is more than 99% confident the lot rate does not exceed .05 or that he is 99% confident the lot rate does not exceed a figure below .05. More error items than $\dot{c}_1$ permit one to state either that he is more than 95% confident the lot rate is at least .01 or that he is 95% confident the lot rate is at least a figure greater than .01.

Depending upon the user's viewpoint as to the relative importance of accepting the unsatisfactory or rejecting the satisfactory, the risks associated with $p_1$ and $p_2$ may change. Thus one might specify 5% (rather than 1%) risk of accepting a rate of .05 and 1% (rather than 5%) risk of rejecting a rate of .01. Or one might specify the same risk, say, 5%, of accepting a rate of .05 or rejecting a rate of .01. In the case of a producer checking the quality of an outgoing product and more concerned with the risk (and cost) of rejecting a good lot than with the risk of accepting a bad lot, emphasis might be placed on keeping risk $\alpha$ small; hence risk $\alpha$ is called *producer's risk*. In the case of a consumer testing the quality of an incoming product or material and more concerned with the risk (and cost) of accepting a bad lot than with the risk of rejecting a good one, emphasis might be placed on keeping risk $\beta$ small; hence risk $\beta$ is called *consumer's risk*.

Returning to the first example, which specifies 1% risk of accepting an unsatisfactory rate of .05 and 5% risk of rejecting a satisfactory rate of .01, the problem may be stated succinctly as follows:

$$
\begin{aligned}
p_1 &= .01 \\
p_2 &= .05 \\
\alpha &= 5\% \\
\beta &= 1\%
\end{aligned}
$$

As shown in Example 3-1 (page 76), the answer to the above problem is n = 258, $\dot{c}$ = 5, and $\dot{c}_1$ = 6. For n = 258 and a lot rate of .05, there is 1% probability of obtaining 5 or less error items; for a lot rate of .01, there is about 5% probability of obtaining 6 or more error items. Correspondingly, there is 1% risk of accepting a population with an error rate as high as .05 and about 5% risk of rejecting a population with a rate as low as .01 for the indicated sample size and acceptance number.

Inasmuch as rates $p_1$ and $p_2$ are relatively far apart, the above sampling plan draws rather a broad dividing line between satisfactory and unsatisfactory lots. True, lots with a rate of .01 or less will be accepted most (95% or more) of the time, while lots with rates of .05 or more will be rejected most (99% or more) of the time. But lots with intermediate rates, above .01 and below .05, are less clearly divided between satisfactory and unsatisfactory. A sharper division between satisfactory and unsatisfactory lots is achieved by bringing $p_1$ and $p_2$ closer together, for example, by specifying them as .02 and .05 (instead of .01 and .05). But this entails a substantial increase in sample size. If we continue to specify risks of 5% and 1% for $p_1$ and $p_2$, respectively, n becomes 580 (instead of 258), with $\dot{c} = 17$ and $\dot{c}_1 = 18$.

## QUASI-FORMS OF ACCEPTANCE SAMPLING

Sometimes a quasi-form of acceptance sampling is employed where only risk $\beta$ of accepting rate $p_2$ is specified. Determination of n and $\dot{c}$ requires the user to further specify the largest expected error rate $\hat{p}$ in the sample. To illustrate, an auditor might expect that the sample rate will be, at most, about .02 and might specify 1% risk of accepting a lot with a rate of .05. This problem may be stated succinctly as follows:

$$p_2 = .05$$
$$\hat{p} = .02$$
$$\beta = 1\%$$

As shown in Example 3-2 (page 77), the answer to the above problem is n = 258 and $\dot{c}$ = 5.

Another way of viewing the above problem is that the auditor, assuming he finds an error rate of .02 in the sample, wishes to be 99% confident (complement of 1% risk) that the error rate in the lot is not over .05. If he finds exactly 5 error items in the sample of 258 (a rate of very nearly .02), he may conclude he is 99% confident that the population rate does not exceed .05. If the number of error items in the sample is below 5, he may state he is more than 99% confident the lot rate is not over .05; alternatively, he may state he is 99% confident that the upper confidence limit for the lot rate is a figure below .05. If he finds 6 or more error items in the sample, he lacks the desired assurance about the quality of the lot and, in effect, "rejects" it by taking some appropriate course of action.

In a similar fashion, a quasi-form of acceptance sampling may be employed where only risk $\alpha$ of rejecting rate $p_1$ is specified. To determine n and $\dot{c}_1$, it is further necessary to specify the lowest anticipated sample rate $\hat{p}$.

To illustrate, one might expect the sample error rate to be at least .10 and might wish to be 99% confident that the lot rate is at least .06 before embarking on a proposed course of action. This problem is equivalent to stating that $\hat{p} = .10$ and specifying 1% risk (complement of 99% confidence) of rejecting a lot rate of .06. The problem may be stated succinctly as follows:

$$\begin{aligned} p_1 &= .06 \\ \hat{p} &= .10 \\ \alpha &= 1\% \end{aligned}$$

As shown in Example 3-4, the answer is n = 239 and $\dot{c}_1 = 24$. If 24 or more error items occur in the sample, the lot is rejected in the sense that its error rate is considered unsatisfactorily high. Equivalently, one may be 99% or more confident that the lot rate is at least .06. If less than 24 error items occur in the sample, the lot is considered satisfactory; that is, one cannot be 99% confident the lot rate is at least .06, the confidence limit goes below .06, and the proposed course of action is no longer statistically justified.

A similar problem may occur where it is considered desirable, rather than undesirable, that there be some small proportion of events (items having a designated characteristic) at or above a specified level. To illustrate, a manufacturer might require 99% confidence that at least .06 of his items have a stated characteristic; at the same time he may be operating in a manner so that the anticipated sample rate is at least .10. Hence the problem is again one where $\hat{p} = .10$ and 1% risk of "rejecting" a rate of .06 is specified. And again n = 239, with $\dot{c}_1 = 24$. However, since $\dot{c}_1$ items in the sample signifies matters are satisfactory rather than unsatisfactory, it is no longer appropriate to term $\dot{c}_1$ a rejection number. It is now more appropriate to call $\dot{c}_1$ a *decision number* associated with rate $p_1$.

## SCOPE OF TABLES 8, 9, 11, 12, and 13

Where acceptance sampling proper is employed, involving specifications both as to $p_1$ and $p_2$ and risks $\alpha$ and $\beta$, Table 13 is used. Where the quasi-form of acceptance sampling that involves rate $p_2$ and $\beta$ is employed, Table 8 is used; Table 12 is used for the special case $\hat{p} = 0$. Where the quasi-form of acceptance sampling that involves rate $p_1$ and rate $\alpha$ is employed, Table 9 is used. Tables 8 and 9 serve a dual role, also being used to determine sample size for an estimate of a proportion. Table 13 is used only for acceptance sampling. Hence the range of decision numbers in Tables 8 and 9 exceeds the needs of acceptance sampling and also exceeds the range of decision numbers in Table 13.

Decision numbers $\dot{c}$ and $\dot{c}_1$ can be obtained from formulas instead of from Tables 8, 9, and 13. In the case of the quasi-forms of acceptance sampling, $\dot{c}$ and $\dot{c}_1$ can be obtained from relatively simple formulas, presented in Table 11. The formula for $\dot{c}$ in this table does not apply in the special case $\hat{p} = 0$, but then $\dot{c}$ is, of course, zero, as in Example 3-3. In the case of acceptance sampling proper, the formula for $\dot{c}$ is complex and appears only in Appendix C (page 448). Use of formulas for $\dot{c}$ and $\dot{c}_1$ is further discussed in the Examples and in the notes to Examples 3-1, 3-2, and 3-4.

Table 13 is intended for use with rates $p_1$ and $p_2$ of .25 or less; it has good accuracy within this range of rates. Similarly, Tables 8, 9, and 11 are intended for use with rates $\hat{p}$ of .25 or less (but not zero). As previously stated, Table 12 is for the special case where $\hat{p} = 0$ (and where $p_2$ and $\beta$ are specified). The tables permit sample size to be calculated for risks $\alpha$ and $\beta$ of 0.5%, 1%, 2.5%, 5%, 10%, and 20%. Table 13 presents all combinations of these risks—a total of 36 combinations.

Preceding discussion has indicated that the confidence level $\gamma_1$ is the complement of risk levels $\alpha$ and $\beta$. Inasmuch as $\gamma_1$ is stated as a percentage, strictly speaking $\gamma_1 = 100\% - \alpha$ and $\gamma_1 = 100\% - \beta$. For simplicity, however, it should be noted that we write $\gamma_1 = 1 - \alpha$ and $\gamma_1 = 1 - \beta$. To illustrate, if we seek to enter Table 1 to obtain $\bar{m}$ for risk $\beta = 1\%$ and follow the instruction $\gamma_1 = 1 - \beta$, we enter Table 1 for a confidence level of $\gamma_1 = 99\%$.

Sample size based on Tables 8, 9, 11, 12, and 13 presumes that the lot is much (at least 10 and preferably 20 times) larger than the sample. If such is not the case, a finite population correction (FPC) can be applied, as explained in Chapter 4, which reduces sample size. If the sample size is large relative to the lot size and is not reduced by applying the FPC, the result is to diminish risks $\alpha$ and $\beta$ for specified $p_1$ and $p_2$ or, alternatively, to increase rate $p_1$ and decrease rate $p_2$ for the specified $\alpha$ and $\beta$. (The increased rate $p_1$ is equal to the lower confidence limit based on n, decision number $\dot{c}_1$, confidence level $\gamma_1 = 1 - \alpha$, and the FPC. The decreased rate $p_2$ is equal to the upper confidence limit based on n, decision number $\dot{c}$, confidence level $\gamma_1 = 1 - \beta$, and the FPC.)

The scope of the tables may be summarized as follows.

**SCOPE OF TABLES 8, 9, 11, 12, and 13**

| Risk | Error rate | Table number |
|------|------------|--------------|
| $\alpha$ and $\beta$ | $p_1$ and $p_2$ | 13 |
| Only $\beta$ | Only $p_2$ | 8 or 11; 12 if $\hat{p} = 0$ |
| Only $\alpha$ | Only $p_1$ | 9 or 11 |

## EXAMPLES

Four examples of acceptance sampling are given in this chapter. Example 3-1 applies to acceptance sampling proper, where both rates $p_1$ and $p_2$ and risks $\alpha$ and $\beta$ are specified. Example 3-2 applies to the quasi-form, where only rate $p_2$ and risk $\beta$ are specified along with anticipated sample rate $\hat{p}$ greater than zero; Example 3-3 applies where $\hat{p} = 0$. Example 3-4 applies to the quasi-form, where only rate $p_1$ and risk $\alpha$ are specified along with anticipated sample rate $\hat{p}$.

In the case of Example 3-2, $\dot{c}$ can be determined from a relatively simple formula in Table 11 instead of from Table 8; this formula is based on a quantity $\hat{Q}$ (explained in Example 3-2) and on risk $\beta$. To use the formula in Table 11 for acceptance sampling, it is merely necessary to substitute $\dot{c}$ for c. Whether $\dot{c}$ is determined from Table 11 or Table 8, the procedure for obtaining n is the same as in Example 3-2. Calculation of c (or $\dot{c}$) by formula is illustrated in Example 2-16 (dealing with sample size for an estimate of a proportion).

In the case of Example 3-4, $\dot{c}_1$ can be determined from a relatively simple formula in Table 11 instead of from Table 9; this formula is based on a quantity $\hat{R}$ (explained in Example 3-4) and on risk $\alpha$. To use the formula in Table 11 for acceptance sampling, it is merely necessary to substitute $\dot{c}_1$ for c. Whether $\dot{c}_1$ is determined from Table 11 or Table 9, the procedure for obtaining n is the same as in Example 3-4. Calculation of c (or $\dot{c}_1$) by formula is illustrated in Example 2-17 (dealing with sample size for an estimate of a proportion); in a very few cases, described in Table 11, the formula is not applicable.

In the case of Example 3-1, acceptance number $\dot{c}$ can also be obtained by formula, but in this case a complex one. For the interested reader, the formula appears in Appendix C (page 448). The formula is based on a quantity $\hat{S}$ (explained in Example 3-1) and on risks $\alpha$ and $\beta$. Whether $\dot{c}$ is determined from Table 13 or the formula in Appendix C, the procedure for obtaining n is the same as in Example 3-1.

Because the procedures in this chapter result in substantially accurate sample size, the examples do not include steps for testing and adjusting n. If it is desired to test n obtained by the procedures of Tables 8, 11, 12, and 13, involving $p_2$ and $\beta$, calculate $\overline{p}$ based on $\dot{c}$, n, and $\gamma_1 = 1 - \beta$. If $\overline{p} = p_2$, sample size is correct. Otherwise adjust n by trial and error so that $\overline{p} = p_2$ (as nearly as integers for $\dot{c}$ and n permit). Similarly, to test n obtained by the procedures of Tables 9 and 11, involving $p_1$ and $\alpha$, calculate $\underline{p}$ based on $\dot{c}_1$, n, and $\gamma_1 = 1 - \alpha$, and adjust n by trial and error so that $\underline{p} = p_1$.

**LIST OF EXAMPLES: DETERMINATION OF SAMPLE SIZE FOR ACCEPTANCE SAMPLING**

| Example number | Specified risks | Specified rates | | |
|---|---|---|---|---|
| | | $p_1$ and $p_2$ | $p_2$ and $\hat{p}$ | $p_1$ and $\hat{p}$ |
| 3-1 | $\alpha$ and $\beta$ | .01, .05 | | |
| 3-2 | Only $\beta$ | | .05, .02 | |
| 3-3 | Only $\beta$ | | .05, 0 | |
| 3-4 | Only $\alpha$ | | | .06, .10 |

## EXAMPLE 3-1

### Problem

$p_1$ = .01
$p_2$ = .05
$\alpha$ = 5%
$\beta$ = 1%

Obtain the Appropriate Sample Size n and acceptance number $\dot{c}$.

### Instructions

1. Calculate $\hat{S} = \dfrac{p_2}{p_1} \times \dfrac{2 - p_1}{2 - p_2}$.

2. In Table 13, for $\beta$ = 1% and for $\alpha$ = 5%, find S nearest to $\hat{S}$. Find the value of $\dot{c}$ corresponding to $\hat{S}$; $\dot{c}$ is the acceptance number.

   If necessary, interpolate linearly between values of S bounding $\hat{S}$ to find integer c. The interpolation formula is the same as given in Example 2-1, Instruction 2, except that $\hat{S}$, S, and $\dot{c}$, respectively, are substituted for $\hat{Q}$, Q, and c.

3. Find $\overline{m}$ corresponding to $\dot{c}$ in Table 13 for $\beta$ = 1%. If exact $\dot{c}$ does not appear in Table 13, use Table 1 or 7—E to find $\overline{m}$ corresponding to $\dot{c}$ for $\gamma_1$ = 1 − $\beta$; read c as $\dot{c}$.

4. Calculate n = $\dfrac{\overline{m}}{p_2} - \dfrac{\overline{m} - \dot{c}}{2}$.

### Solution

1. $\hat{S} = \dfrac{p_2}{p_1} \times \dfrac{2 - p_1}{2 - p_2} = \dfrac{.05}{.01} \times \dfrac{1.99}{1.95} = 5 \times 1.0205 = 5.103$

2. For $\beta$ = 1%, $\alpha$ = 5%, and $\hat{S}$ = 5.103, the nearest S = 5.017 and the corresponding $\boxed{\dot{c} = 5}$ in Table 13.

3. For $\dot{c}$ = 5, $\overline{m}$ = 13.109 in Table 13, Section 2, for $\beta$ = 1% (or in Table 1 for $\gamma_1$ = 1 − $\beta$ = 99%)

4. n = $\dfrac{\overline{m}}{p_2} - \dfrac{\overline{m} - \dot{c}}{2} = \dfrac{13.109}{.05} - \dfrac{8.109}{2} = 262.18 - 4.05$

   = $\boxed{258}$

*Note*

Instead of using Instruction 2 and Table 13, ċ may be obtained from a complex formula presented in Appendix C (page 448). This formula is based on Ŝ from Instruction 1 and on risks $\alpha$ and $\beta$.

*EXAMPLE 3-2* _____

*Problem*

$p_2$ = .05
$\hat{p}$ = .02
$\beta$ = 1%

Obtain the Appropriate Sample Size n and acceptance number ċ.

*Instructions*

1. Calculate $\hat{Q} = \dfrac{p_2}{\hat{p}} \times \dfrac{2 - \hat{p}}{2 - p_2}$.

2. In Table 8, for $\beta$ = 1%, find Q nearest to $\hat{Q}$. Find the value of ċ corresponding to Q; ċ is the acceptance number.

   If necessary, interpolate linearly between values of Q bounding $\hat{Q}$ to find integer ċ. The interpolation formula is the same as given in Example 2-1, Instruction 2, except that ċ is substituted for c.

3. In Table 1 or 7–E, for $\gamma_1$ = 1 − $\beta$, find $\overline{m}$ corresponding to ċ; read c as ċ.

4. Calculate $n = \dfrac{\overline{m}}{p_2} - \dfrac{\overline{m} - \dot{c}}{2}$.

*Solution*

1. $\hat{Q} = \dfrac{p_2}{\hat{p}} \times \dfrac{2 - \hat{p}}{2 - p_2} = \dfrac{.05}{.02} \times \dfrac{1.98}{1.95} = 2.5 \times 1.0154 = 2.539$

2. For $\hat{Q}$ = 2.539 and $\beta$ = 1%, the nearest Q = 2.622 and the corresponding $\boxed{\dot{c} = 5}$ in Table 8.

3. For ċ = 5 and $\gamma_1$ = 1 − $\beta$ = 99%, $\overline{m}$ = 13.109 in Table 1.

4. $n = \dfrac{\overline{m}}{p_2} - \dfrac{\overline{m} - \dot{c}}{2} = \dfrac{13.109}{.05} - \dfrac{8.109}{2} = 262.18 - 4.05$

   $= \boxed{258}$

### Note

Instead of using Instruction 2 and Table 8, $\dot{c}$ may be obtained from the formula in Table 11, based on $\hat{Q}$ from Instruction 1 and on risk $\beta$. Example 2-16 illustrates the use of this formula (in this case, for an estimate of a proportion).

*EXAMPLE 3-3* _____

### Problem

$p_2 = .05$
$\hat{p} = 0$
$\beta = 1\%$

Obtain the Appropriate Sample Size n and acceptance number $\dot{c}$.

### Instructions

1. For $\hat{p} = 0$, $\dot{c} = 0$.

2. In Table 12, for risk $\beta = 1\%$, find $\overline{m}$ corresponding to $\dot{c} = 0$.

3. Calculate $n = \dfrac{\overline{m}}{p_2} - \dfrac{\overline{m}}{2}$.

### Solution

1. $\boxed{\dot{c} = 0}$

2. For risk $\beta = 1\%$ and $\dot{c} = 0$, $\overline{m} = 4.6052$ in Table 12.

3. $n = \dfrac{\overline{m}}{p_2} - \dfrac{\overline{m}}{2} = \dfrac{4.6052}{.05} - \dfrac{4.6052}{2} = 92.10 - 2.30 = \boxed{90}$

## EXAMPLE 3-4

### Problem

$p_1 = .06$
$\hat{p} = .10$
$\alpha = 1\%$

Obtain the Appropriate Sample Size n and rejection number $\dot{c}_1$.

### Instructions

1. Calculate $\hat{R} = \dfrac{p_1}{\hat{p}} \times \dfrac{2 - \hat{p}}{2 - p_1}$.

2. In Table 9, for $\alpha = 1\%$, find R nearest to $\hat{R}$. Find the value of $\dot{c}_1$ corresponding to R; $\dot{c}_1$ is the rejection number.

   If necessary, interpolate linearly between values of R bounding $\hat{R}$ to find integer $\dot{c}_1$. The interpolation formula is the same as given in Example 2-4, Instruction 2, except that $\dot{c}_1$ is substituted for c.

3. In Table 2 or 7–E, for $\gamma_1 = 1 - \alpha$, find $\underline{m}_1$ corresponding to $\dot{c}_1$; read $\underline{m}$ as $\underline{m}_1$ and c as $\dot{c}_1$.

4. Calculate $n = \dfrac{m_1}{p_1} + \dfrac{\dot{c}_1 - m_1 - 1}{2}$.

### Solution

1. $\hat{R} = \dfrac{p_1}{\hat{p}} \times \dfrac{2 - \hat{p}}{2 - p_1} = \dfrac{.06}{.10} \times \dfrac{1.90}{1.94} = .6 \times .9794 = .588$

2. For $\hat{R} = .588$ and $\alpha = 1\%$, the nearest R = .587 and the corresponding $\boxed{\dot{c}_1 = 24}$ in Table 9.

3. For $\dot{c}_1 = 24$ and $\gamma_1 = 1 - \alpha = 99\%$, $\underline{m}_1 = 14.089$ in Table 2 (reading $\underline{m}$ as $\underline{m}_1$ and c as $\dot{c}_1$).

4. $n = \dfrac{m_1}{p_1} + \dfrac{\dot{c}_1 - m_1 - 1}{2} = \dfrac{14.089}{.06} + \dfrac{8.911}{2}$

   $= 234.82 + 4.46 = \boxed{239}$

*Note*

Instead of using Instruction 2 and Table 9, $\dot{c}_1$ may be obtained from the formula in Table 11, based on $\hat{R}$ from Instruction 1 and on risk $\alpha$. Example 2-17 illustrates the use of this formula (in this case, for an estimate of a proportion); in a very few cases, described in Table 11, the formula is not applicable.

# CHAPTER FOUR

# The Finite Population Correction

The procedures thus far presented for determining confidence limits and sample size assume the sample is selected without replacement[1] and are based on probabilities given by the binomial distribution for a random sample drawn from an infinite population. However, these procedures remain essentially accurate for a finite population provided the sample size n is quite small relative to the population size N, say 5% or less.

If binomial probabilities are used when the population is finite and sampling is without replacement, an element of conservatism is introduced. That is, confidence limits are overstated (upper limit too high, lower limit too low), and sample size is overstated. When n/N is 5% or less, the degree of overstatement is ordinarily negligible.

As n/N increases, such overstatement increases and can less easily be ignored. The problem can be avoided by sampling with replacement

---

[1] Customarily, once an item is selected to appear in the sample, it is not returned to the population and therefore cannot be selected a second time. This is sampling without replacement. Under sampling with replacement, each item selected to appear in the sample is immediately returned to the population and therefore has the chance of being selected again. This means that a particular item in the population might count as two or more items in the sample.

when the population is finite. Then the binomial probabilities apply. However, sampling with replacement is often unattractive: Counting a particular item in the population as more than one item in the samples does not actually add information about the population. Also, it may not be feasible to sample with replacement (for example, if sample items are selected at systematic intervals). Finally, it may be necessary to obtain confidence limits for a sample that has already been drawn without replacement from a finite population. All in all, sampling with replacement tends to be an unsatisfactory way of dealing with a finite population and n/N relatively large (say over 5%).

For a finite population and sampling without replacement, confidence limits and sample sizes are correctly based on probabilities given by the hypergeometric distribution. However, this distribution entails laborious computations. Instead, one may closely approximate results based on the hypergeometric distribution by applying a finite population correction (FPC) to results based on the binomial distribution. The effect of applying the FPC to binomial confidence limits is to reduce the upper limit and increase the lower limit. The effect of applying the FPC to binomial sample size is to reduce the sample size.

## FPC FOR CONFIDENCE LIMITS

For a sample of size n and containing c events (items having a designated characteristic), the upper and lower binomial confidence limits are $\overline{p}$ and $\underline{p}$. After the FPC is applied to them, they are denoted by $\overline{p}_F$ and $\underline{p}_F$, respectively. Then

$$\overline{p}_F = \frac{c}{n} + \left(\overline{p} - \frac{c}{n}\right) \sqrt{\frac{N - n}{N - 1}}$$

$$\underline{p}_F = \frac{c}{n} - \left(\frac{c}{n} - \underline{p}\right) \sqrt{\frac{N - n}{N - 1}}$$

Example 4-1 illustrates the calculation of $\overline{p}_F$ and $\underline{p}_F$.[1]

---

[1] The following simpler, though less accurate, formulas may be used if n/N does not exceed .5. Their accuracy diminishes as n/N increases, but is nevertheless sufficient for many practical purposes. See the note to Example 4-1.

$$\overline{p}_F \simeq \overline{p} - \frac{n\overline{p} - c}{2N}$$

$$\underline{p}_F \simeq \underline{p} + \frac{c - n\underline{p}}{2N}$$

## FPC FOR SAMPLE SIZE

Prior to application of the FPC, the sample size is denoted by n and after application, by $n_F$. Then

$$n_F = \frac{n \times N}{n + N - 1}$$

Use of this formula is illustrated in Examples 4-2 to 4-6. These examples cover sampling for an estimate of a proportion and acceptance sampling. Examples 4-3 to 4-6 deal with somewhat special situations where $n_F$ tends to be inaccurate, as discussed below.

In the case of acceptance sampling, it should be noted that as sample size is reduced, the decision (acceptance and rejection) numbers are proportionately reduced, as nearly as use of integers permits.

To the extent that the FPC produces a substantial change in sample size and that c and c/n are expected to be quite small, $n_F$ tends to be somewhat inaccurate. When sample size is determined with reference to the upper error margin $\overline{e}$ (proportion sampling) or to the unsatisfactory error rate $p_2$ (acceptance sampling), $n_F$ tends to be too small. When sample size is determined with reference to the lower error margin $\underline{e}$ or to the satisfactory error rate $p_1$, $n_F$ tends to be too large.

Accordingly it may be desirable to adjust $n_F$ up or down, resulting in $n_F'$. The method of doing so is somewhat different in the case of proportion sampling than in acceptance sampling. Therefore the following discussion deals separately with these two types of sampling.

### Proportion Sampling

To begin it is necessary to review the sample-size procedure presented in Chapter 2. We assume that we desire both $\overline{p}$ and $\underline{p}$ or only $\overline{p}$, so that the upper error margin $\overline{e}$ is of interest. Using the methods of Chapter 2 we calculate n on the basis of specifications $\hat{p}$ (anticipated value of c/n), $\hat{e}$ (error margin), and $\gamma$ (confidence level). Then we test n as follows. Assuming $c/n = \hat{p}$, we calculate $c = n\hat{p}$ to the nearest integer. Using the methods of Chapter 1, we calculate $\overline{p}$ at confidence level $\gamma$ on the basis of the calculated n and c. We calculate $\overline{e} = \overline{p} - c/n$. If $\overline{e} = \hat{e}$, sample size is considered accurate. If the difference between $\overline{e}$ and $\hat{e}$ is appreciable, sample size can be adjusted on the basis of $\hat{e}/\overline{e}$, yielding n'.

If we are dealing with $n_F$ instead of n (and with $\overline{p}_F$ instead of $\overline{p}$), then in similar fashion we calculate $\overline{e}_F$. And the test of sample-size accuracy is whether $\overline{e}_F = \hat{e}$. The formula for $n_F$ tends to understate sample size in the sense that the resulting $\overline{e}_F$ exceeds $\hat{e}$. We can adjust $n_F$ on the basis of $\hat{e}/\overline{e}_F$, resulting in a larger sample size:

$$n_F' = \frac{n_F \times N}{n_F + (N - n_F)\,(\hat{e}/\overline{e}_F)^2}$$

Example 4-3 illustrates this type of problem.

In parallel manner, when only $\underline{p}$ is desired, the test of sample-size accuracy is whether $\underline{e} = \hat{e}$. And if we are dealing with $n_F$, the test of sample-size accuracy is whether $\underline{e}_F = \hat{e}$. Now, however, the formula for $n_F$ tends to overstate sample size in the sense that the resulting $\underline{e}_F$ is less than $\hat{e}$. We can adjust $n_F$ on the basis of $\hat{e}/\underline{e}_F$, resulting in a smaller sample size:

$$n_F' = \frac{n_F \times N}{n_F + (N - n_F)\,(\hat{e}/\underline{e}_F)^2}$$

Example 4-4 illustrates this type of problem.

**Acceptance Sampling**

We have an analogous test for sample-size accuracy in the case of acceptance sampling. Now, however, the test is in terms of confidence limits rather than error margins. When sample size is determined with reference to error rate $p_2$, the test is whether $\overline{p}$ (calculated on the basis of n, acceptance number $\hat{c}$, and confidence level $\gamma_1 = 1 - \beta$, where $\beta$ is the risk associated with $p_2$) equals $p_2$. If we are dealing with $n_F$ instead of n (and with $\overline{p}_F$ instead of $\overline{p}$), the test similarly is whether $\overline{p}_F = p_2$. The formula for $n_F$ tends to understate sample size in the sense that the resulting $\overline{p}_F$ exceeds $p_2$. We can obtain a larger sample size $n_F'$ such that $\overline{p}_F = p_2$. However, a straightforward, simple formula for doing so is not available. Instead we follow a trial-and-error procedure, using interpolation to shorten the process. This type of problem is illustrated in Example 4-5.

When sample size is determined with reference to error rate $p_1$, the test of accuracy is whether $\underline{p}$ (calculated on the basis of n, rejection number $\hat{c}_1$, and $\gamma_1 = 1 - \alpha$, where $\alpha$ is the risk associated with $p_1$) equals $p_1$. If we are dealing with $n_F$, the test of sample-size accuracy is whether $\underline{p}_F = p_1$. The formula for $n_F$ tends to overstate sample size in the sense that the resulting $\underline{p}_F$ exceeds $p_1$. However, this tendency may be obscured owing to the fact that $\dot{c}_1$ and n must be integers. An adjusted sample size $n_F'$ is obtained by trial and error, using interpolation, so that $\underline{p}_F = p_1$. This type of problem is illustrated in Example 4-6.

## TABLE OF SQUARE ROOTS (Table 14)

The FPC procedures require the square roots of fractional values. For the convenience of the user, Table 14 provides square roots of values from .001 to .999.

## EXAMPLES

The six examples in this chapter, already described in preceding discussion, may be summarized as follows.

**LIST OF EXAMPLES: THE FINITE POPULATION CORRECTION**

| Example number | FPC applied to | (n/N) Relationship of sample size to population size* |
|---|---|---|
| 4-1 | Confidence limits $\overline{p}$ and $\underline{p}$ (Same procedures are used for only $\overline{p}$ and only $\underline{p}$.) | 100/1,000 |
| 4-2 | Sample size for proportion sampling, with both $\overline{p}$ and $\underline{p}$ desired. (Same procedure is used if only $\overline{p}$ is desired.) | 200/2,000 |
| 4-3 | Sample size for proportion sampling, with both $\overline{p}$ and $\underline{p}$ desired. (Same procedure is used if only $\overline{p}$ is desired.) | 200/500 |
| 4-4 | Sample size for proportion sampling, with only $\underline{p}$ desired. | 80/500 |
| 4-5 | Sample size for acceptance sampling, with both $p_1$ and $p_2$ and both risk $\alpha$ and risk $\beta$ specified. (Same procedure is used for quasi-form of acceptance sampling with only $p_2$ and risk $\beta$ specified.) | 258/1,000 |
| 4-6 | Sample size for quasi-form of acceptance sampling with only $p_1$ and risk $\alpha$ specified. | 239/1,000 |

*In the case of confidence limits, the sample size n is presumably after application of the FPC. In the case of sample size, n is prior to application of the FPC.

*EXAMPLE 4-1* _____

**Problem**

$N \quad = \quad 1,000$
$n \quad = \quad \quad 100$
$c \quad = \quad \quad \ 20$
$c/n \ = \quad \quad .20$

Obtain $\overline{p}$ and $\underline{p}$ at $\gamma_2 \ = \ 95\%$.

**Instructions**

1. Based on n, c, and $\gamma_2$, obtain $\overline{p}$ and $\underline{p}$ as instructed in Chapter 1.
2. Calculate

$$\overline{p}_F \ = \ \frac{c}{n} \ + \ \left( \overline{p} \ - \ \frac{c}{n} \right) \sqrt{\frac{N \ - \ n}{N \ - \ 1}}$$

$$\underline{p}_F \ = \ \frac{c}{n} \ - \ \left( \frac{c}{n} \ - \ \underline{p} \right) \sqrt{\frac{N \ - \ n}{N \ - \ 1}}$$

**Solution**

1. Based on Example 1-1 in Chapter 1,

$\overline{p} \quad = \ .2919$

$\underline{p} \quad = \ .1267$

2. $\overline{p}_F \ = \ .20 \ + \ (.2919 \ - \ .20) \ \sqrt{\dfrac{1,000 \ - \ 100}{1,000 \ - \ 1}}$

$\quad \quad = \ .20 \ + \ (.0919 \ \times \ .9492)$

$\quad \quad = \ .20 \ + \ .0872 \ = \ \boxed{.2872}$

$\underline{p}_F \ = \ .20 \ - \ (.20 \ - \ .1267) \ \sqrt{\dfrac{1,000 \ - \ 100}{1,000 \ - \ 1}}$

$\quad \quad = \ .20 \ - \ (.0733 \ \times \ .9492)$

$\quad \quad = \ .20 \ - \ .0696 \ = \ \boxed{.1304}$

*Note*    Using the simpler formulas in the footnote on page 82, we obtain $\overline{p}_F \ \simeq \ .2873$ and $\underline{p}_F \ \simeq \ .1304$. The substantial accuracy obtained in the present example is due to the fact that n/N is relatively small, namely .1. The simpler formulas become increasingly inaccurate as n/N increases, and should not be used when n/N exceeds .5.

*EXAMPLE 4-2*

## Problem

$N = 2,000$
$\hat{p} = .10$
$\hat{e} = .05$ at $\gamma_2 = 95\%$
Both $\overline{p}$ and $p$ are desired.
Obtain the Appropriate Sample Size $n_F$ for an estimate of a proportion.

## Instructions

1. Based on $\hat{p}, \hat{e}$, and $\gamma_2$, obtain n as instructed in Chapter 2.

2. Calculate $n_F = \dfrac{n \times N}{n + N - 1}$.

3. Test $n_F$:
   Calculate $c_F = n_F \hat{p}$.

   Based on $c_F$, $n_F$, and $\gamma_2$, obtain $\overline{p}$ as instructed in Chapter 1.

   Calculate $\overline{e} = \overline{p} - \dfrac{c_F}{n_F}$.

   Calculate $\overline{e}_F = \overline{e} \sqrt{\dfrac{N - n_F}{N - 1}}$.

   If $\overline{e}_F$ differs appreciably from $\hat{e}$, calculate

   $$n_F' = \frac{n_F \times N}{n_F + (N - n_F)(\hat{e}/\overline{e}_F)^2}$$

## Solution

1. Based on Example 2-1 in Chapter 2, $n = 200$.

2. $n_F = \dfrac{n \times N}{n + N - 1} = \dfrac{200 \times 2,000}{200 + 2,000 - 1}$

   $= \dfrac{400,000}{2,199} = \boxed{182}$

3. Test:
   $c_F = n_F \hat{p} = 182 \times .10 = 18$

Based on $c_F = 18$, $n_F = 182$, and $\gamma_2 = 95\%$, and on Example

1-3 in Chapter 1, $\overline{p} = \dfrac{28.448}{187.36} = .1518$.

$$\overline{e} = \overline{p} - \frac{c_F}{n_F} = .1518 - \frac{18}{182} = .1518 - 0989 = .0529$$

$$\overline{e}_F = \overline{e}\sqrt{\frac{N - n_F}{N - 1}} = .0529\sqrt{\frac{2,000 - 182}{1,999}}$$

$$= .0529 \times .9537 = .0505$$

Inasmuch as $\overline{e}_F$ is very close to $\hat{e}$, it is not considered necessary to calculate $n_F'$, so that $n_F = 182$ is indicated as the Appropriate Sample Size.

### Note

If only $\overline{p}$ is desired, the FPC procedure is the same, except that in Instructions 1 and 3 n and $\overline{p}$ are based on $\gamma_1$ instead of $\gamma_2$.

## EXAMPLE 4-3 _____

### Problem

N = 500
$\hat{p}$ = .10
$\hat{e}$ = .05 at $\gamma_2$ = 95%

Both $\overline{p}$ and p are desired.
Obtain the Appropriate Sample Size $n_F$ for an estimate of a proportion.

### Instructions

1. Based on $\hat{p}$, $\hat{e}$, and $\gamma_2$, obtain n as instructed in Chapter 2.

2. Calculate $n_F = \dfrac{n \times N}{n + N - 1}$.

3. Test $n_F$:
   Calculate $c_F = n_F \hat{p}$.

   Based on $c_F$, $n_F$, and $\gamma_2$, obtain $\overline{p}$ as instructed in Chapter 1.

Calculate $\overline{e} = \overline{p} - \dfrac{c_F}{n_F}$ .

Calculate $\overline{e}_F = \overline{e} \sqrt{\dfrac{N - n_F}{N - 1}}$ .

If $\overline{e}_F$ differs appreciably from $\hat{e}$, calculate

$$n_F' = \frac{n_F \times N}{n_F + (N - n_F) (\hat{e}/\overline{e}_F)^2}$$

## Solution

1. Based on Example 2-1 in Chapter 2, $n = 200$.

2. $n_F = \dfrac{n \times N}{n + N - 1} = \dfrac{200 \times 500}{200 + 500 - 1} = \dfrac{100,000}{699} = 143$

3. Test:

$c_F = n_F \hat{p} = 143 \times .10 = 14$

Based on $c_F = 14$, $n_F = 143$, and $\gamma_2 = 95\%$, and on Example

1-3 in Chapter 1, $\overline{p} = \dfrac{23.490}{147.89} = .1588$ .

$\overline{e} = \overline{p} - \dfrac{c_F}{n_F} = .1588 - \dfrac{14}{143} = .1588 - .0979 = .0609$

$\overline{e}_F = \overline{e} \sqrt{\dfrac{N - n_F}{N - 1}} = .0609 \sqrt{\dfrac{500 - 143}{499}}$

$= .0609 \times .8458 = .0515$

$n_F' = \dfrac{n_F \times N}{n_F + (N - n_F) (\hat{e}/\overline{e}_F)^2}$

$= \dfrac{143 \times 500}{143 + (500 - 143) (.05/.0515)^2}$

$= \dfrac{71,500}{143 + (357 \times .9426)} = \dfrac{71,500}{480} = \boxed{149}$

### Notes

Repeating the above test procedure for $n_F = 149$ (and $c_F = 15$), we obtain $\overline{e}_F = .0502$, which is very close to $\hat{e} = .05$. Trial and error reveals that a slightly better sample size is $n_F = 150$ (and $c_F = 15$). Then we obtain $\overline{e}_F = .0499$.

If only $\overline{p}$ is desired, the FPC procedure is the same, except that in Instructions 1 and 3 n and $\overline{p}$ are based on $\gamma_1$ instead of $\gamma_2$.

### EXAMPLE 4-4 _____

### Problem

$N = 500$
$\hat{p} = .10$
$\hat{e} = .05$ at $\gamma_1 = 95\%$

Only p is desired.
Obtain the Appropriate Sample Size $n_F$ for an estimate of a proportion.

### Instructions

1. Based on $\hat{p}$, $\hat{e}$, and $\gamma_1$, obtain n as instructed in Chapter 2.

2. Calculate $n_F = \dfrac{n \times N}{n + N - 1}$.

3. Test $n_F$:

   Calculate $c_F = n_F \hat{p}$.

   Based on $c_F$, $n_F$, and $\gamma_1$, obtain p as instructed in Chapter 1.

   Calculate $\underline{e} = \dfrac{c_F}{n_F} - \underline{p}$.

   Calculate $\underline{e}_F = \underline{e} \sqrt{\dfrac{N - n_F}{N - 1}}$.

   If $\underline{e}_F$ differs appreciably from $\hat{e}$, calculate

   $$n'_F = \dfrac{n_F \times N}{n_F + (N - n_F)(\hat{e}/\underline{e}_F)^2}$$

*Solution*

1. Based on Example 2-4 in Chapter 2, $n = 80$.

2. $n_F = \dfrac{n \times N}{n + N - 1} = \dfrac{80 \times 500}{80 + 500 - 1} = \dfrac{40,000}{579} = 69$

3. Test:

   $c_F = n_F \hat{p} = 69 \times .10 = 7$

   Based on $c_F = 7$, $n_F = 69$, and $\gamma_1 = 95\%$, and on Example 1-8

   in Chapter 1, $\underline{p} = \dfrac{3.2853}{67.643} = .04857$.

   $\underline{e} = \dfrac{c_F}{n_F} - \underline{p} = \dfrac{7}{69} - .04857 = .10145 - .04857 = .05288$

   $\underline{e}_F = \underline{e} \sqrt{\dfrac{N - n_F}{N - 1}} = .05288 \sqrt{\dfrac{500 - 69}{499}} = .05288$

   $= .05288 \times .9294 = .04915$

   $n_F' = \dfrac{n_F \times N}{n_F + (N - n_F)(\hat{e}/\underline{e}_F)^2}$

   $= \dfrac{69 \times 500}{69 + (500 - 69)(.05/.04915)^2}$

   $= \dfrac{34,500}{69 + (431 \times 1.0349)} = \dfrac{34,500}{515} = \boxed{67}$

*Note*

Repeating the above test procedure for $n_F = 67$ (and $c_F = 7$), we obtain $\underline{e}_F = .05071$, which is quite close to $\hat{e} = .05$. Trial and error reveals that a slightly better sample size is $n_F = 68$ (and $c_F = 7$). Then we obtain $\underline{e}_F = .04991$.

*EXAMPLE 4-5* _____

### Problem

$N$  =  1,000
$p_1$  =    .01
$p_2$  =    .05
$\alpha$  =   5%
$\beta$  =   1%

Obtain the Appropriate Sample Size n and acceptance number $\dot{c}$ for acceptance sampling.

### Instructions

1. Based on $p_1, p_2, \alpha$, and $\beta$, obtain n and $\dot{c}$ as instructed in Chapter 3.

2. Calculate  $n_F = \dfrac{n \times N}{n + N - 1}$ .

3. Test $n_F$ :

   Calculate  $\dot{c}_F = \dfrac{n_F \times \dot{c}}{n}$ .

   Based on $\dot{c}_F$, $n_F$, and $\gamma_1 = 1 - \beta$, obtain $\overline{p}$ as instructed in Chapter 1.

   Calculate  $\overline{p}_F = \dfrac{\dot{c}_F}{n_F} + \left( \overline{p} - \dfrac{\dot{c}_F}{n_F} \right) \sqrt{\dfrac{N - n_F}{N - 1}}$ .

   If $\overline{p}_F$ differs appreciably from $p_2$, use trial and error to find $n_F'$ such that $\overline{p}_F = p_2$ (as nearly as integers for $\dot{c}_F$ and $n_F'$ permit). The procedure may be shortened by using linear interpolation after finding trial values of $n_F$ that bound the desired $n_F'$. The interpolation formula is

   $$n_F' = n_a + \frac{(n_b - n_a)(\overline{p}_a - p_2)}{\overline{p}_a - \overline{p}_b}$$

   where $n_a$ is a bounding value smaller than $n_F'$, $n_b$ is a bounding value larger than $n_F'$, $\overline{p}_a$ is the value of $\overline{p}_F$ corresponding to $n_a$, and $\overline{p}_b$ is the value of $\overline{p}_F$ corresponding to $n_b$.

*Solution*

1. Based on Example 3-1 in Chapter 3, n = 258 and ċ (acceptance number) = 5.

2. $n_F = \dfrac{n \times N}{n + N - 1} = \dfrac{258 \times 1,000}{258 + 999} = \dfrac{258,000}{1,257} = 205$

3. Test:

$$\dot{c}_F = \frac{n_F \times \dot{c}}{n} = \frac{205 \times 5}{258} = \boxed{4}$$

Based on $\dot{c}_F = 4$, $n_F = 205$, and $\gamma_1 = 1 - \beta = 99\%$, and on

Example 1-7 in Chapter 1, $\overline{p} = \dfrac{11.605}{208.975} = .05553.$

$$\overline{p}_F = \frac{\dot{c}_F}{n_F} + \left(\overline{p} - \frac{\dot{c}_F}{n_F}\right) \sqrt{\frac{N - n_F}{N - 1}}$$

$$= .01951 + (.05553 - .01951) \sqrt{\frac{1,000 - 205}{1,000 - 1}}$$

$$= .01951 + (.03602 \times .8921) = .01951 + .03213$$

$$= .05164$$

Inasmuch as $\overline{p}_F$ appreciably exceeds $p_2$, $n_F = 205$ is too small. As a guess, try $n_F = 215$. Then

$$\dot{c}_F = \frac{215 \times 5}{258} = 4$$

Based on $\dot{c}_F = 4$, $n_F = 215$, and $\gamma_1 = 1 - \beta = 99\%$, and on

Example 1-7 in Chapter 1, $\overline{p} = \dfrac{11.605}{218.975} = .05300.$

$$\overline{p}_F = .01860 + (.05300 - .01860) \sqrt{\frac{1,000 - 215}{1,000 - 1}}$$

$$= .01860 + (.03440 \times .8864) = .01860 + .03049$$

$$= .04909$$

Thus we have $n_F = 205$ (with $\overline{p}_F = .05164$) and $n_F = 215$ (with $\overline{p}_F = .04909$) as bounding values of the desired sample size (with $\overline{p}_F = p_2 = .05000$). That is,

$$
\begin{array}{ll}
n_a = 205 & \overline{p}_a = .05164 \\
n_b = 215 & \overline{p}_b = .04909 \\
& p_2 = .05000
\end{array}
$$

Using the interpolation formula in Instruction 3,

$$
n_F' = 205 + \frac{(215 - 205)(.05164 - .05000)}{.05164 - .04909}
$$

$$
= 205 + \frac{10 \times .00164}{.00255} = \boxed{211}
$$

### Notes

If desired, we can test the sample size of 211. Based on $N = 1,000$, $\dot{c}_F = 4$, $n_F = 211$, and $\gamma_1 = 99\%$, we obtain $\overline{p}_F = .05008$, which is very nearly equal to $p_2$.

When acceptance sampling is based only on $p_2$ and $\beta$, as in Examples 3-2 and 3-3 of Chapter 3, the procedure for applying the FPC to sample size is the same as in the present example.

### *EXAMPLE 4-6* _____

### Problem

$$
\begin{array}{ll}
N &= 1,000 \\
p_1 &= \phantom{0}.06 \\
\hat{p} &= \phantom{0}.10 \\
\alpha &= \phantom{0}1\%
\end{array}
$$

Obtain the Appropriate Sample Size n and rejection (decision) number $\dot{c}_1$ for acceptance sampling.

### Instructions

1. Based on $p_1$, $\hat{p}$, and $\alpha$, obtain n and $\dot{c}_1$ as instructed in Chapter 3.

2. Calculate $n_F = \dfrac{n \times N}{n + N - 1}$.

3. Test $n_F$:

Calculate $\dot{c}_{1F} = \dfrac{n_F \times \dot{c}_1}{n}$.

Based on $\dot{c}_{1F}$, $n_F$, and $\gamma_1 = 1 - \alpha$, obtain $\underline{p}$ as instructed in Chapter 1.

Calculate $\underline{p}_F = \dfrac{\dot{c}_{1F}}{n_F} - \left( \dfrac{\dot{c}_{1F}}{n_F} - \underline{p} \right) \sqrt{\dfrac{N - n_F}{N - 1}}$.

If $\underline{p}_F$ differs appreciably from $p_1$, use trial and error to find $n_F'$ such that $\underline{p}_F = p_1$ (as nearly as integers for $\dot{c}_{1F}$ and $n_F'$ permit). The procedure may be shortened by using linear interpolation after finding trial values of $n_F$ that bound the desired $n_F'$. The interpolation formula is

$$n_F' = n_a + \frac{(n_b - n_a)(p_1 - \underline{p}_a)}{\underline{p}_b - \underline{p}_a}$$

where $n_a$ is a bounding value smaller than $n_F'$, $n_b$ is a bounding value larger than $n_F'$, $\underline{p}_a$ is the value of $\underline{p}_F$ corresponding to $n_a$, and $\underline{p}_b$ is the value of $\underline{p}_F$ corresponding to $n_b$.

*Solution*

1. Based on Example 3-4 in Chapter 3, $n = 239$ and $\dot{c}_1$ (rejection number) $= 24$.

2. $n_F = \dfrac{n \times N}{n + N - 1} = \dfrac{239 \times 1{,}000}{239 + 999} = \dfrac{239{,}000}{1{,}238} = 193$

3. Test:

$\dot{c}_{1F} = \dfrac{n_F \times \dot{c}_1}{n} = \dfrac{193 \times 24}{239} = \boxed{19}$

Based on $\dot{c}_{1F} = 19$, $n_F = 193$, and $\gamma_1 = 1 - \alpha = 99\%$, and on

Example 8 in Chapter 2, $\underline{p} = \dfrac{10.346}{189.17} = .05469$.

$$\underline{p}_F = \frac{\dot{c}_{1F}}{n_F} - \left(\frac{\dot{c}_{1F}}{n_F} - \underline{p}\right) \sqrt{\frac{N - n_F}{N - 1}}$$

$$= .09845 - (.09845 - .05469) \sqrt{\frac{1,000 - 193}{1,000 - 1}}$$

$$= .09845 - (.04376 \times .8988) = .09845 - .03933$$

$$= .05912$$

Inasmuch as $\underline{p}_F$ is somewhat less than $p_1$, $n_F = 193$ is not quite the appropriate sample size. If we keep $\dot{c}_{1F} = 19$, we may slightly decrease $n_F$—within the constraint that $\dot{c}_{1F}/n_F$ remains about .10 $(= \hat{p})$—resulting in an increase in $\underline{p}_F$. As a guess, try $n_F = 188$. Then, based on $\dot{c}_{1F} = 19$, $n_F = 188$, and $\gamma_1 = 1 - \alpha = 99\%$ and on Example 1-8 in Chapter 1, $\underline{p} = \dfrac{10.346}{184.17} = .05618$.

$$\underline{p}_F = .10106 - (.10106 - .05618) \sqrt{\frac{1,000 - 188}{1,000 - 1}}$$

$$= .10106 - (.04488 \times .9016) = .10106 - .04046$$

$$= .06060$$

Thus we have $n_F = 193$ (with $\underline{p}_F = .05912$) and $n_F = 188$ (with $\underline{p}_F = .06060$) as bounding values of the appropriate sample size (with $\underline{p}_F = p_1 = .06$). That is,

$$
\begin{array}{ll}
n_a = 188 & \underline{p}_a = .06060 \\
n_b = 193 & \underline{p}_b = .05912 \\
& p_1 = .06000
\end{array}
$$

Using the interpolation formula in Instruction 3,

$$n_F' = 188 + \frac{(193 - 188)(.06000 - .06060)}{.05912 - .06060}$$

$$= 188 + \frac{5 \times .00060}{.00148} = \boxed{190}$$

*Note*

If desired, we can test the sample size of 190. Based on $N = 1,000$, $\dot{c}_{1F} = 19$, $n_F = 190$, and $\gamma_1 = 99\%$, we obtain $\underline{p}_F = .06000$, which is equal to $p_1$.

# PART THREE

# Table 1

**CONSTANTS FOR $\bar{p}'$ (LONG FORMULA), AN APPROXIMATION OF THE UPPER BINOMIAL CONFIDENCE LIMIT $\bar{p}$; $c/n \leqslant .5$, $n \geqslant 20$, $c \leqslant 1,000$**

Approximation Formula: $\quad \bar{p}' = \dfrac{\bar{m}}{n} \times \dfrac{n - \bar{a}}{n - \bar{b}}$

n      sample size.

c      number of events (items having a designated characteristic) in the sample.

$\bar{p}'$      approximation of the upper binomial confidence limit $\bar{p}$; $\bar{p}'$ has relative accuracy of at least .999.

$\bar{m}$      upper confidence limit for the parameter m of a Poisson distribution; each $\bar{m}$ is for a given c and a given $\gamma_1$ or $\gamma_2$.

$\bar{a}, \bar{b}$      approximation constants for a given c and a given $\gamma_1$ or $\gamma_2$.

$\gamma_1$      confidence level for a one-sided (upper or lower) confidence limit.

$\gamma_2$      confidence level for two-sided (upper and lower) confidence limits. ($\gamma_2 = 2\gamma_1 - 100\%$.)

For $c > 1,000$, calculate $\bar{a}$ and $\bar{b}$ from the extension formulas in Table 7–A; and calculate $\bar{m}$ from the extension formula in Table 7–E.

(**Note.** For the purpose of acceptance sampling in conjunction with Tables 8 and 13, c in Table 1 may be read as acceptance number $\dot{c}$.)

| | | |
|---|---|---|
| **Section 1** | $\gamma_1 = 99.5\%$ | $\gamma_2 = 99\%$ |
| **Section 2** | $\gamma_1 = 99\%$ | $\gamma_2 = 98\%$ |
| **Section 3** | $\gamma_1 = 97.5\%$ | $\gamma_2 = 95\%$ |
| **Section 4** | $\gamma_1 = 95\%$ | $\gamma_2 = 90\%$ |
| **Section 5** | $\gamma_1 = 90\%$ | $\gamma_2 = 80\%$ |
| **Section 6** | $\gamma_1 = 80\%$ | $\gamma_2 = 60\%$ |

**Table 1.   CONSTANTS FOR $\bar{p}'$ (LONG FORMULA), AN APPROXIMATION OF THE UPPER BINOMIAL CONFIDENCE LIMIT $\bar{p}$; $c/n \leqslant .5$, $n \geqslant 20$, $c \leqslant 1,000$**

| c | $\bar{m}$ | $\bar{a}$ | $\bar{b}$ | c | $\bar{m}$ | $\bar{a}$ | $\bar{b}$ |
|---|---|---|---|---|---|---|---|
| 0 | 5.2983 | 0.826 | −1.826 | 45 | 65.341 | 16.908 | 6.851 |
| 1 | 7.4302 | 1.234 | −1.983 | 46 | 66.530 | 17.256 | 7.107 |
| 2 | 9.2738 | 1.593 | −2.048 | 47 | 67.717 | 17.641 | 7.406 |
| 3 | 10.978 | 1.941 | −2.054 | 48 | 68.902 | 17.986 | 7.660 |
| 4 | 12.594 | 2.297 | −2.005 | 49 | 70.085 | 18.316 | 7.898 |
| 5 | 14.150 | 2.657 | −1.920 | 50 | 71.266 | 18.610 | 8.098 |
| 6 | 15.660 | 3.026 | −1.802 | 51 | 72.446 | 19.010 | 8.417 |
| 7 | 17.134 | 3.400 | −1.659 | 52 | 73.624 | 19.333 | 8.650 |
| 8 | 18.578 | 3.793 | −1.479 | 53 | 74.800 | 19.702 | 8.935 |
| 9 | 19.998 | 4.193 | −1.279 | 54 | 75.974 | 20.005 | 9.148 |
| 10 | 21.398 | 4.615 | −1.042 | 55 | 77.147 | 20.363 | 9.42 |
| 11 | 22.779 | 4.992 | −0.851 | 56 | 78.319 | 20.692 | 9.67 |
| 12 | 24.145 | 5.360 | −0.662 | 57 | 79.489 | 20.997 | 9.88 |
| 13 | 25.497 | 5.735 | −0.458 | 58 | 80.657 | 21.311 | 10.11 |
| 14 | 26.836 | 6.101 | −0.258 | 59 | 81.824 | 21.693 | 10.42 |
| 15 | 28.164 | 6.464 | −0.057 | 60 | 82.990 | 21.995 | 10.63 |
| 16 | 29.482 | 6.825 | 0.148 | 61 | 84.154 | 22.317 | 10.87 |
| 17 | 30.791 | 7.189 | 0.362 | 62 | 85.317 | 22.65 | 11.12 |
| 18 | 32.091 | 7.542 | 0.566 | 63 | 86.479 | 23.03 | 11.43 |
| 19 | 33.383 | 7.906 | 0.788 | 64 | 87.639 | 23.34 | 11.66 |
| 20 | 34.668 | 8.254 | 0.994 | 65 | 88.799 | 23.67 | 11.90 |
| 21 | 35.946 | 8.615 | 1.220 | 66 | 89.957 | 24.10 | 12.27 |
| 22 | 37.218 | 8.974 | 1.446 | 67 | 91.113 | 24.35 | 12.43 |
| 23 | 38.484 | 9.329 | 1.670 | 68 | 92.269 | 24.79 | 12.81 |
| 24 | 39.745 | 9.678 | 1.890 | 69 | 93.424 | 25.06 | 12.99 |
| 25 | 41.000 | 10.037 | 2.125 | 70 | 94.577 | 25.41 | 13.27 |
| 26 | 42.251 | 10.381 | 2.344 | 71 | 95.729 | 25.73 | 13.51 |
| 27 | 43.497 | 10.720 | 2.560 | 72 | 96.881 | 26.03 | 13.73 |
| 28 | 44.738 | 11.072 | 2.794 | 73 | 98.031 | 26.47 | 14.11 |
| 29 | 45.976 | 11.422 | 3.026 | 74 | 99.180 | 26.70 | 14.26 |
| 30 | 47.209 | 11.764 | 3.253 | 75 | 100.33 | 27.12 | 14.61 |
| 31 | 48.439 | 12.140 | 3.521 | 76 | 101.48 | 27.36 | 14.77 |
| 32 | 49.665 | 12.479 | 3.747 | 77 | 102.62 | 27.73 | 15.07 |
| 33 | 50.888 | 12.834 | 3.994 | 78 | 103.77 | 28.16 | 15.44 |
| 34 | 52.108 | 13.170 | 4.220 | 79 | 104.91 | 28.36 | 15.55 |
| 35 | 53.324 | 13.501 | 4.441 | 80 | 106.06 | 28.80 | 15.94 |
| 36 | 54.537 | 13.865 | 4.703 | 81 | 107.20 | 29.11 | 16.17 |
| 37 | 55.748 | 14.176 | 4.906 | 82 | 108.34 | 29.39 | 16.37 |
| 38 | 56.956 | 14.558 | 5.191 | 83 | 109.48 | 29.71 | 16.63 |
| 39 | 58.161 | 14.904 | 5.436 | 84 | 110.62 | 30.08 | 16.93 |
| 40 | 59.363 | 15.216 | 5.643 | 85 | 111.76 | 30.49 | 17.28 |
| 41 | 60.563 | 15.562 | 5.892 | 86 | 112.90 | 30.76 | 17.47 |
| 42 | 61.761 | 15.909 | 6.142 | 87 | 114.04 | 31.08 | 17.72 |
| 43 | 62.957 | 16.281 | 6.422 | 88 | 115.17 | 31.47 | 18.05 |
| 44 | 64.150 | 16.567 | 6.604 | 89 | 116.31 | 31.69 | 18.20 |

Table 1. CONSTANTS FOR $\overline{p}'$ (LONG FORMULA), AN APPROXIMATION OF THE UPPER BINOMIAL CONFIDENCE LIMIT $\overline{p}$; $c/n \leqslant .5$, $n \geqslant 20$, $c \leqslant 1,000$ (*Continued*)

| c | $\overline{m}$ | $\overline{a}$ | $\overline{b}$ | c | $\overline{m}$ | $\overline{a}$ | $\overline{b}$ |
|---|---|---|---|---|---|---|---|
| 90 | 117.45 | 31.99 | 18.42 | 135 | 167.91 | 41.98 | 25.52 |
| 91 | 118.58 | 32.37 | 18.74 | 136 | 169.02 | 42.26 | 25.74 |
| 92 | 119.71 | 32.81 | 19.13 | 137 | 170.13 | 42.53 | 25.97 |
| 93 | 120.85 | 33.18 | 19.43 | 138 | 171.24 | 42.81 | 26.19 |
| 94 | 121.98 | 33.33 | 19.50 | 139 | 172.35 | 43.09 | 26.41 |
| 95 | 123.11 | 33.62 | 19.72 | 140 | 173.46 | 43.37 | 26.63 |
| 96 | 124.24 | 34.01 | 20.05 | 141 | 174.57 | 43.64 | 26.86 |
| 97 | 125.37 | 34.49 | 20.48 | 142 | 175.68 | 43.92 | 27.08 |
| 98 | 126.50 | 34.63 | 20.54 | 143 | 176.78 | 44.20 | 27.30 |
| 99 | 127.63 | 35.08 | 20.94 | 144 | 177.89 | 44.47 | 27.53 |
| 100 | 128.76 | 35.44 | 21.23 | 145 | 179.00 | 44.75 | 27.75 |
| 101 | 129.89 | 32.47 | 18.03 | 146 | 180.11 | 45.03 | 27.97 |
| 102 | 131.02 | 32.75 | 18.25 | 147 | 181.21 | 45.30 | 28.20 |
| 103 | 132.14 | 33.04 | 18.46 | 148 | 182.32 | 45.58 | 28.42 |
| 104 | 133.27 | 33.32 | 18.68 | 149 | 183.42 | 45.86 | 28.64 |
| 105 | 134.39 | 33.60 | 18.90 | 150 | 184.53 | 46.13 | 28.87 |
| 106 | 135.52 | 33.88 | 19.12 | 151 | 185.63 | 46.41 | 29.09 |
| 107 | 136.64 | 34.16 | 19.34 | 152 | 186.74 | 46.68 | 29.32 |
| 108 | 137.77 | 34.44 | 19.56 | 153 | 187.84 | 46.96 | 29.54 |
| 109 | 138.89 | 34.72 | 19.78 | 154 | 188.94 | 47.24 | 29.76 |
| 110 | 140.01 | 35.00 | 20.00 | 155 | 190.05 | 47.51 | 29.99 |
| 111 | 141.13 | 35.28 | 20.22 | 156 | 191.15 | 47.79 | 30.21 |
| 112 | 142.26 | 35.56 | 20.44 | 157 | 192.25 | 48.06 | 30.44 |
| 113 | 143.38 | 35.85 | 20.65 | 158 | 193.36 | 48.34 | 30.66 |
| 114 | 144.50 | 36.13 | 20.87 | 159 | 194.46 | 48.61 | 30.89 |
| 115 | 145.62 | 36.41 | 21.09 | 160 | 195.56 | 48.89 | 31.11 |
| 116 | 146.74 | 36.68 | 21.32 | 161 | 196.66 | 49.17 | 31.33 |
| 117 | 147.86 | 36.96 | 21.54 | 162 | 197.76 | 49.44 | 31.56 |
| 118 | 148.97 | 37.24 | 21.76 | 163 | 198.86 | 49.72 | 31.78 |
| 119 | 150.09 | 37.52 | 21.98 | 164 | 199.97 | 49.99 | 32.01 |
| 120 | 151.21 | 37.80 | 22.20 | 165 | 201.07 | 50.27 | 32.23 |
| 121 | 152.33 | 38.08 | 22.42 | 166 | 202.16 | 50.54 | 32.46 |
| 122 | 153.44 | 38.36 | 22.64 | 167 | 203.26 | 50.82 | 32.68 |
| 123 | 154.56 | 38.64 | 22.86 | 168 | 204.36 | 51.09 | 32.91 |
| 124 | 155.67 | 38.92 | 23.08 | 169 | 205.46 | 51.37 | 33.13 |
| 125 | 156.79 | 39.20 | 23.30 | 170 | 206.56 | 51.64 | 33.36 |
| 126 | 157.90 | 39.48 | 23.52 | 171 | 207.66 | 51.91 | 33.59 |
| 127 | 159.02 | 39.76 | 23.74 | 172 | 208.76 | 52.19 | 33.81 |
| 128 | 160.13 | 40.03 | 23.97 | 173 | 209.86 | 52.46 | 34.04 |
| 129 | 161.24 | 40.31 | 24.19 | 174 | 210.95 | 52.74 | 34.26 |
| 130 | 162.36 | 40.59 | 24.41 | 175 | 212.05 | 53.01 | 34.49 |
| 131 | 163.47 | 40.87 | 24.63 | 176 | 213.15 | 53.29 | 34.71 |
| 132 | 164.58 | 41.15 | 24.85 | 177 | 214.24 | 53.56 | 34.94 |
| 133 | 165.69 | 41.42 | 25.08 | 178 | 215.34 | 53.84 | 35.16 |
| 134 | 166.80 | 41.70 | 25.30 | 179 | 216.44 | 54.11 | 35.39 |

**Table 1. CONSTANTS FOR $\overline{p}'$ (LONG FORMULA), AN APPROXIMATION OF THE UPPER BINOMIAL CONFIDENCE LIMIT $\overline{p}$; $c/n \leqslant .5$, $n \geqslant 20$, $c \leqslant 1{,}000$ (Continued)**

| c | $\overline{m}$ | $\overline{a}$ | $\overline{b}$ | c | $\overline{m}$ | $\overline{a}$ | $\overline{b}$ |
|---|---|---|---|---|---|---|---|
| 180 | 217.53 | 54.38 | 35.62 | 225 | 266.60 | 66.65 | 45.85 |
| 181 | 218.63 | 54.66 | 35.84 | 226 | 267.69 | 66.92 | 46.08 |
| 182 | 219.72 | 54.93 | 36.07 | 227 | 268.77 | 67.19 | 46.31 |
| 183 | 220.82 | 55.20 | 36.30 | 228 | 269.86 | 67.46 | 46.54 |
| 184 | 221.91 | 55.48 | 36.52 | 229 | 270.94 | 67.74 | 46.76 |
| 185 | 223.01 | 55.75 | 36.75 | 230 | 272.03 | 68.01 | 46.99 |
| 186 | 224.10 | 56.03 | 36.97 | 231 | 273.11 | 68.28 | 47.22 |
| 187 | 225.20 | 56.30 | 37.20 | 232 | 274.20 | 68.55 | 47.45 |
| 188 | 226.29 | 56.57 | 37.43 | 233 | 275.28 | 68.82 | 47.68 |
| 189 | 227.38 | 56.85 | 37.65 | 234 | 276.36 | 69.09 | 47.91 |
| 190 | 228.48 | 57.12 | 37.88 | 235 | 277.45 | 69.36 | 48.14 |
| 191 | 229.57 | 57.39 | 38.11 | 236 | 278.53 | 69.63 | 48.37 |
| 192 | 230.66 | 57.67 | 38.33 | 237 | 279.62 | 69.90 | 48.60 |
| 193 | 231.76 | 57.94 | 38.56 | 238 | 280.70 | 70.17 | 48.83 |
| 194 | 232.85 | 58.21 | 38.79 | 239 | 281.78 | 70.45 | 49.05 |
| 195 | 233.94 | 58.48 | 39.02 | 240 | 282.87 | 70.72 | 49.28 |
| 196 | 235.03 | 58.76 | 39.24 | 241 | 283.95 | 70.99 | 49.51 |
| 197 | 236.12 | 59.03 | 39.47 | 242 | 285.03 | 71.26 | 49.74 |
| 198 | 237.21 | 59.30 | 39.70 | 243 | 286.11 | 71.53 | 49.97 |
| 199 | 238.31 | 59.58 | 39.92 | 244 | 287.20 | 71.80 | 50.20 |
| 200 | 239.40 | 59.85 | 40.15 | 245 | 288.28 | 72.07 | 50.43 |
| 201 | 240.49 | 60.12 | 40.38 | 246 | 289.36 | 72.34 | 50.66 |
| 202 | 241.58 | 60.39 | 40.61 | 247 | 290.44 | 72.61 | 50.89 |
| 203 | 242.67 | 60.67 | 40.83 | 248 | 291.52 | 72.88 | 51.12 |
| 204 | 243.76 | 60.94 | 41.06 | 249 | 292.61 | 73.15 | 51.35 |
| 205 | 244.85 | 61.21 | 41.29 | 250 | 293.69 | 73.42 | 51.58 |
| 206 | 245.94 | 61.48 | 41.52 | 251 | 294.77 | 73.69 | 51.81 |
| 207 | 247.03 | 61.76 | 41.74 | 252 | 295.85 | 73.96 | 52.04 |
| 308 | 248.12 | 62.03 | 41.97 | 253 | 296.93 | 74.23 | 52.27 |
| 209 | 249.21 | 62.30 | 42.20 | 254 | 298.01 | 74.50 | 52.50 |
| 210 | 250.29 | 62.57 | 42.43 | 255 | 299.09 | 74.77 | 52.73 |
| 211 | 251.38 | 62.85 | 42.65 | 256 | 300.17 | 75.04 | 52.96 |
| 212 | 252.47 | 63.12 | 42.88 | 257 | 301.25 | 75.31 | 53.19 |
| 213 | 253.56 | 63.39 | 43.11 | 258 | 302.33 | 75.58 | 53.42 |
| 214 | 254.65 | 63.66 | 43.34 | 259 | 303.41 | 75.85 | 53.65 |
| 215 | 255.73 | 63.93 | 43.57 | 260 | 304.49 | 76.12 | 53.88 |
| 216 | 256.82 | 64.21 | 43.79 | 261 | 305.57 | 76.39 | 54.11 |
| 217 | 257.91 | 64.48 | 44.02 | 262 | 306.65 | 76.66 | 54.34 |
| 218 | 259.00 | 64.75 | 44.25 | 263 | 307.73 | 76.93 | 54.57 |
| 219 | 260.08 | 65.02 | 44.48 | 264 | 308.81 | 77.20 | 54.80 |
| 220 | 261.17 | 65.29 | 44.71 | 265 | 309.89 | 77.47 | 55.03 |
| 221 | 262.26 | 65.56 | 44.94 | 266 | 310.97 | 77.74 | 55.26 |
| 222 | 263.34 | 65.84 | 45.16 | 267 | 312.05 | 78.01 | 55.49 |
| 223 | 264.43 | 66.11 | 45.39 | 268 | 313.12 | 78.28 | 55.72 |
| 224 | 265.52 | 66.38 | 45.62 | 269 | 314.20 | 78.55 | 55.95 |

Table 1. CONSTANTS FOR $\overline{p}'$ (LONG FORMULA), AN APPROXIMATION OF THE UPPER BINOMIAL CONFIDENCE LIMIT $\overline{p}$; $c/n \leqslant .5$, $n \geqslant 20$, $c \leqslant 1,000$ (*Continued*)

| c | $\overline{m}$ | $\overline{a}$ | $\overline{b}$ | c | $\overline{m}$ | $\overline{a}$ | $\overline{b}$ |
|---|---|---|---|---|---|---|---|
| 270 | 315.28 | 78.82 | 56.18 | 315 | 363.67 | 90.92 | 66.58 |
| 271 | 316.36 | 79.09 | 56.41 | 316 | 364.74 | 91.18 | 66.82 |
| 272 | 317.44 | 79.36 | 56.64 | 317 | 365.81 | 91.45 | 67.05 |
| 273 | 318.52 | 79.63 | 56.87 | 318 | 366.88 | 91.72 | 67.28 |
| 274 | 319.59 | 79.90 | 57.10 | 319 | 367.96 | 91.99 | 67.51 |
| 275 | 320.67 | 80.17 | 57.33 | 320 | 369.03 | 92.26 | 67.74 |
| 276 | 321.75 | 80.44 | 57.56 | 321 | 370.10 | 92.52 | 67.98 |
| 277 | 322.83 | 80.71 | 57.79 | 322 | 371.17 | 92.79 | 68.21 |
| 278 | 323.90 | 80.98 | 58.02 | 323 | 372.24 | 93.06 | 68.44 |
| 279 | 324.98 | 81.24 | 58.26 | 324 | 373.31 | 93.33 | 68.67 |
| 280 | 326.06 | 81.51 | 58.49 | 325 | 374.39 | 93.60 | 68.90 |
| 281 | 327.13 | 81.78 | 58.72 | 326 | 375.46 | 93.86 | 69.14 |
| 282 | 328.21 | 82.05 | 58.95 | 327 | 376.53 | 94.13 | 69.37 |
| 283 | 329.29 | 82.32 | 59.18 | 328 | 377.60 | 94.40 | 69.60 |
| 284 | 330.36 | 82.59 | 59.41 | 329 | 378.67 | 94.67 | 69.83 |
| 285 | 331.44 | 82.86 | 59.64 | 330 | 379.74 | 94.94 | 70.06 |
| 286 | 332.52 | 83.13 | 59.87 | 331 | 380.81 | 95.20 | 70.30 |
| 287 | 333.59 | 83.40 | 60.10 | 332 | 381.88 | 95.47 | 70.53 |
| 288 | 334.67 | 83.67 | 60.33 | 333 | 382.95 | 95.74 | 70.76 |
| 289 | 335.74 | 83.94 | 60.56 | 334 | 384.02 | 96.01 | 70.99 |
| 290 | 336.82 | 84.20 | 60.80 | 335 | 385.09 | 96.27 | 71.23 |
| 291 | 337.89 | 84.47 | 61.03 | 336 | 386.16 | 96.54 | 71.46 |
| 292 | 338.97 | 84.74 | 61.26 | 337 | 387.23 | 96.81 | 71.69 |
| 293 | 340.04 | 85.01 | 61.49 | 338 | 388.30 | 97.08 | 71.92 |
| 294 | 341.12 | 85.28 | 61.72 | 339 | 389.37 | 97.34 | 72.16 |
| 295 | 342.19 | 85.55 | 61.95 | 340 | 390.44 | 97.61 | 72.39 |
| 296 | 343.27 | 85.82 | 62.18 | 341 | 391.51 | 97.88 | 72.62 |
| 297 | 344.34 | 86.09 | 62.41 | 342 | 392.58 | 98.15 | 72.85 |
| 298 | 345.42 | 86.35 | 62.65 | 343 | 393.65 | 98.41 | 73.09 |
| 299 | 346.49 | 86.62 | 62.88 | 344 | 394.72 | 98.68 | 73.32 |
| 300 | 347.57 | 86.89 | 63.11 | 345 | 395.79 | 98.95 | 73.55 |
| 301 | 348.64 | 87.16 | 63.34 | 346 | 396.86 | 99.22 | 73.78 |
| 302 | 349.71 | 87.43 | 63.57 | 347 | 397.93 | 99.48 | 74.02 |
| 303 | 350.79 | 87.70 | 63.80 | 348 | 399.00 | 99.75 | 74.25 |
| 304 | 351.86 | 87.97 | 64.03 | 349 | 400.07 | 100.02 | 74.48 |
| 305 | 352.94 | 88.23 | 64.27 | 350 | 401.14 | 100.28 | 74.72 |
| 306 | 354.01 | 88.50 | 64.50 | 351 | 402.20 | 100.55 | 74.95 |
| 307 | 355.08 | 88.77 | 64.73 | 352 | 403.27 | 100.82 | 75.18 |
| 308 | 356.16 | 89.04 | 64.96 | 353 | 404.34 | 101.09 | 75.41 |
| 309 | 357.23 | 89.31 | 65.19 | 354 | 405.41 | 101.35 | 75.65 |
| 310 | 358.30 | 89.58 | 65.42 | 355 | 406.48 | 101.62 | 75.88 |
| 311 | 359.38 | 89.84 | 65.66 | 356 | 407.55 | 101.89 | 76.11 |
| 312 | 360.45 | 90.11 | 65.89 | 357 | 408.61 | 102.15 | 76.35 |
| 313 | 361.52 | 90.38 | 66.12 | 358 | 409.68 | 102.42 | 76.58 |
| 314 | 362.59 | 90.65 | 66.35 | 359 | 410.75 | 102.69 | 76.81 |

Table 1.   CONSTANTS FOR $\overline{p}'$ (LONG FORMULA), AN APPROXIMATION OF
THE UPPER BINOMIAL CONFIDENCE LIMIT $\overline{p}$; $c/n \leqslant .5$, $n \geqslant 20$,
$c \leqslant 1{,}000$ (*Continued*)

| c | $\overline{m}$ | $\overline{a}$ | $\overline{b}$ | c | $\overline{m}$ | $\overline{a}$ | $\overline{b}$ |
|---|---|---|---|---|---|---|---|
| 360 | 411.82 | 102.95 | 77.05 | 405 | 459.78 | 114.94 | 87.56 |
| 361 | 412.89 | 103.22 | 77.28 | 406 | 460.84 | 115.21 | 87.79 |
| 362 | 413.95 | 103.49 | 77.51 | 407 | 461.91 | 115.48 | 88.02 |
| 363 | 415.02 | 103.76 | 77.74 | 408 | 462.97 | 115.74 | 88.26 |
| 364 | 416.09 | 104.02 | 77.98 | 409 | 464.03 | 116.01 | 88.49 |
| 365 | 417.16 | 104.29 | 78.21 | 410 | 465.10 | 116.27 | 88.73 |
| 366 | 418.22 | 104.56 | 78.44 | 411 | 466.16 | 116.54 | 88.96 |
| 367 | 419.29 | 104.82 | 78.68 | 412 | 467.22 | 116.81 | 89.19 |
| 368 | 420.36 | 105.09 | 78.91 | 413 | 468.29 | 117.07 | 89.43 |
| 369 | 421.42 | 105.36 | 79.14 | 414 | 469.35 | 117.34 | 89.66 |
| 370 | 422.49 | 105.62 | 79.38 | 415 | 470.41 | 117.60 | 89.90 |
| 371 | 423.56 | 105.89 | 79.61 | 416 | 471.48 | 117.87 | 90.13 |
| 372 | 424.63 | 106.16 | 79.84 | 417 | 472.54 | 118.14 | 90.36 |
| 373 | 425.69 | 106.42 | 80.08 | 418 | 473.60 | 118.40 | 90.60 |
| 374 | 426.76 | 106.69 | 80.31 | 419 | 474.67 | 118.67 | 90.83 |
| 375 | 427.82 | 106.96 | 80.54 | 420 | 475.73 | 118.93 | 91.07 |
| 376 | 428.89 | 107.22 | 80.78 | 421 | 476.79 | 119.20 | 91.30 |
| 377 | 429.96 | 107.49 | 81.01 | 422 | 477.85 | 119.46 | 91.54 |
| 378 | 431.02 | 107.76 | 81.24 | 423 | 478.92 | 119.73 | 91.77 |
| 379 | 432.09 | 108.02 | 81.48 | 424 | 479.98 | 119.99 | 92.01 |
| 380 | 433.16 | 108.29 | 81.71 | 425 | 481.04 | 120.26 | 92.24 |
| 381 | 434.22 | 108.56 | 81.94 | 426 | 482.10 | 120.53 | 92.47 |
| 382 | 435.29 | 108.82 | 82.18 | 427 | 483.17 | 120.79 | 92.71 |
| 383 | 436.35 | 109.09 | 82.41 | 428 | 484.23 | 121.06 | 92.94 |
| 384 | 437.42 | 109.35 | 82.65 | 429 | 485.29 | 121.32 | 93.18 |
| 385 | 438.48 | 109.62 | 82.88 | 430 | 486.35 | 121.59 | 93.41 |
| 386 | 439.55 | 109.89 | 83.11 | 431 | 487.42 | 121.85 | 93.65 |
| 387 | 440.62 | 110.15 | 83.35 | 432 | 488.48 | 122.12 | 93.88 |
| 388 | 441.68 | 110.42 | 83.58 | 433 | 489.54 | 122.38 | 94.12 |
| 389 | 442.75 | 110.69 | 83.81 | 434 | 490.60 | 122.65 | 94.35 |
| 390 | 443.81 | 110.95 | 84.05 | 435 | 491.66 | 122.92 | 94.58 |
| 391 | 444.88 | 111.22 | 84.28 | 436 | 492.72 | 123.18 | 94.82 |
| 392 | 445.94 | 111.49 | 84.51 | 437 | 493.79 | 123.45 | 95.05 |
| 393 | 447.01 | 111.75 | 84.75 | 438 | 494.85 | 123.71 | 95.29 |
| 394 | 448.07 | 112.02 | 84.98 | 439 | 495.91 | 123.98 | 95.52 |
| 395 | 449.14 | 112.28 | 85.22 | 440 | 496.97 | 124.24 | 95.76 |
| 396 | 450.20 | 112.55 | 85.45 | 441 | 498.03 | 124.51 | 95.99 |
| 397 | 451.27 | 112.82 | 85.68 | 442 | 499.09 | 124.77 | 96.23 |
| 398 | 452.33 | 113.08 | 85.92 | 443 | 500.15 | 125.04 | 96.46 |
| 399 | 453.39 | 113.35 | 86.15 | 444 | 501.21 | 125.30 | 96.70 |
| 400 | 454.46 | 113.61 | 86.39 | 445 | 502.28 | 125.57 | 96.93 |
| 401 | 455.52 | 113.88 | 86.62 | 446 | 503.34 | 125.83 | 97.17 |
| 402 | 456.59 | 114.15 | 86.85 | 447 | 504.40 | 126.10 | 97.40 |
| 403 | 457.65 | 114.41 | 87.09 | 448 | 505.46 | 126.36 | 97.64 |
| 404 | 458.72 | 114.68 | 87.32 | 449 | 506.52 | 126.63 | 97.87 |

Table 1.  CONSTANTS FOR $\overline{p}'$ (LONG FORMULA), AN APPROXIMATION OF THE UPPER BINOMIAL CONFIDENCE LIMIT $\overline{p}$; c/n ⩽ .5, n ⩾ 20, c ⩽ 1,000 (*Continued*)

| c | $\overline{m}$ | $\overline{a}$ | $\overline{b}$ | c | $\overline{m}$ | $\overline{a}$ | $\overline{b}$ |
|---|---|---|---|---|---|---|---|
| 450 | 507.58 | 126.89 | 98.11 | 495 | 555.24 | 138.81 | 108.69 |
| 451 | 508.64 | 127.16 | 98.34 | 496 | 556.30 | 139.08 | 108.92 |
| 452 | 509.70 | 127.43 | 98.57 | 497 | 557.36 | 139.34 | 109.16 |
| 453 | 510.76 | 127.69 | 98.81 | 498 | 558.42 | 139.60 | 109.40 |
| 454 | 511.82 | 127.96 | 99.04 | 499 | 559.47 | 139.87 | 109.63 |
| 455 | 512.88 | 128.22 | 99.28 | 500 | 560.53 | 140.13 | 109.87 |
| 456 | 513.94 | 128.49 | 99.51 | 501 | 561.59 | 140.40 | 110.10 |
| 457 | 515.00 | 128.75 | 99.75 | 502 | 562.65 | 140.66 | 110.34 |
| 458 | 516.06 | 129.02 | 99.98 | 503 | 563.70 | 140.93 | 110.57 |
| 459 | 517.12 | 129.28 | 100.22 | 504 | 564.76 | 141.19 | 110.81 |
| 460 | 518.18 | 129.55 | 100.45 | 505 | 565.82 | 141.45 | 111.05 |
| 461 | 519.24 | 129.81 | 100.69 | 506 | 566.88 | 141.72 | 111.28 |
| 462 | 520.30 | 130.08 | 100.92 | 507 | 567.93 | 141.98 | 111.52 |
| 463 | 521.36 | 130.34 | 101.16 | 508 | 568.99 | 142.25 | 111.75 |
| 464 | 522.42 | 130.61 | 101.39 | 509 | 570.05 | 142.51 | 111.99 |
| 465 | 523.48 | 130.87 | 101.63 | 510 | 571.10 | 142.78 | 112.22 |
| 466 | 524.54 | 131.14 | 101.86 | 511 | 572.16 | 143.04 | 112.46 |
| 467 | 525.60 | 131.40 | 102.10 | 512 | 573.22 | 143.30 | 112.70 |
| 468 | 526.66 | 131.67 | 102.33 | 513 | 574.28 | 143.57 | 112.93 |
| 469 | 527.72 | 131.93 | 102.57 | 514 | 575.33 | 143.83 | 113.17 |
| 470 | 528.78 | 132.19 | 102.81 | 515 | 576.39 | 144.10 | 113.40 |
| 471 | 529.84 | 132.46 | 103.04 | 516 | 577.45 | 144.36 | 113.64 |
| 472 | 530.90 | 132.72 | 103.28 | 517 | 578.50 | 144.63 | 113.87 |
| 473 | 531.96 | 132.99 | 103.51 | 518 | 579.56 | 144.89 | 114.11 |
| 474 | 533.02 | 133.25 | 103.75 | 519 | 580.62 | 145.15 | 114.35 |
| 475 | 534.08 | 133.52 | 103.98 | 520 | 581.67 | 145.42 | 114.58 |
| 476 | 535.13 | 133.78 | 104.22 | 521 | 582.73 | 145.68 | 114.82 |
| 477 | 536.19 | 134.05 | 104.45 | 522 | 583.78 | 145.95 | 115.05 |
| 478 | 537.25 | 134.31 | 104.69 | 523 | 584.84 | 146.21 | 115.29 |
| 479 | 538.31 | 134.58 | 104.92 | 524 | 585.90 | 146.47 | 115.53 |
| 480 | 539.37 | 134.84 | 105.16 | 525 | 586.95 | 146.74 | 115.76 |
| 481 | 540.43 | 135.11 | 105.39 | 526 | 588.01 | 147.00 | 116.00 |
| 482 | 541.49 | 135.37 | 105.63 | 527 | 589.07 | 147.27 | 116.23 |
| 483 | 542.55 | 135.64 | 105.86 | 528 | 590.12 | 147.53 | 116.47 |
| 484 | 543.60 | 135.90 | 106.10 | 529 | 591.18 | 147.79 | 116.71 |
| 485 | 544.66 | 136.17 | 106.33 | 530 | 592.23 | 148.06 | 116.94 |
| 486 | 545.72 | 136.43 | 106.57 | 531 | 593.29 | 148.32 | 117.18 |
| 487 | 546.78 | 136.69 | 106.81 | 532 | 594.35 | 148.59 | 117.41 |
| 488 | 547.84 | 136.96 | 107.04 | 533 | 595.40 | 148.85 | 117.65 |
| 489 | 548.90 | 137.22 | 107.28 | 534 | 596.46 | 149.11 | 117.89 |
| 490 | 549.95 | 137.49 | 107.51 | 535 | 597.51 | 149.38 | 118.12 |
| 491 | 551.01 | 137.75 | 107.75 | 536 | 598.57 | 149.64 | 118.36 |
| 492 | 552.07 | 138.02 | 107.98 | 537 | 599.62 | 149.91 | 118.59 |
| 493 | 553.13 | 138.28 | 108.22 | 538 | 600.68 | 150.17 | 118.83 |
| 494 | 554.19 | 138.55 | 108.45 | 539 | 601.73 | 150.43 | 119.07 |

Table 1. CONSTANTS FOR $\bar{p}'$ (LONG FORMULA), AN APPROXIMATION OF THE UPPER BINOMIAL CONFIDENCE LIMIT $\bar{p}$; $c/n \leqslant .5$, $n \geqslant 20$, $c \leqslant 1,000$ (*Continued*)

Section 1

$\gamma_1 = 99.5\%$
$\gamma_2 = 99\%$

| c | $\bar{m}$ | $\bar{a}$ | $\bar{b}$ | c | $\bar{m}$ | $\bar{a}$ | $\bar{b}$ |
|---|---|---|---|---|---|---|---|
| 540 | 602.79 | 150.70 | 119.30 | 585 | 650.23 | 162.56 | 129.94 |
| 541 | 603.85 | 150.96 | 119.54 | 586 | 651.28 | 162.82 | 130.18 |
| 542 | 604.90 | 151.23 | 119.77 | 587 | 652.34 | 163.08 | 130.42 |
| 543 | 605.96 | 151.49 | 120.01 | 588 | 653.39 | 163.35 | 130.65 |
| 544 | 607.01 | 151.75 | 120.25 | 589 | 654.44 | 163.61 | 130.89 |
| 545 | 608.07 | 152.02 | 120.48 | 590 | 655.50 | 163.87 | 131.13 |
| 546 | 609.12 | 152.28 | 120.72 | 591 | 656.55 | 164.14 | 131.36 |
| 547 | 610.18 | 152.54 | 120.96 | 592 | 657.60 | 164.40 | 131.60 |
| 548 | 611.23 | 152.81 | 121.19 | 593 | 658.66 | 164.66 | 131.84 |
| 549 | 612.29 | 153.07 | 121.43 | 594 | 659.71 | 164.93 | 132.07 |
| 550 | 613.34 | 153.34 | 121.66 | 595 | 660.76 | 165.19 | 132.31 |
| 551 | 614.40 | 153.60 | 121.90 | 596 | 661.81 | 165.45 | 132.55 |
| 552 | 615.45 | 153.86 | 122.14 | 597 | 662.87 | 165.72 | 132.78 |
| 553 | 616.51 | 154.13 | 122.37 | 598 | 663.92 | 165.98 | 133.02 |
| 554 | 617.56 | 154.39 | 122.61 | 599 | 664.97 | 166.24 | 133.26 |
| 555 | 618.61 | 154.65 | 122.85 | 600 | 666.02 | 166.51 | 133.49 |
| 556 | 619.67 | 154.92 | 123.08 | 601 | 667.08 | 166.77 | 133.73 |
| 557 | 620.72 | 155.18 | 123.32 | 602 | 668.13 | 167.03 | 133.97 |
| 558 | 621.78 | 155.44 | 123.56 | 603 | 669.18 | 167.30 | 134.20 |
| 559 | 622.83 | 155.71 | 123.79 | 604 | 670.23 | 167.56 | 134.44 |
| 560 | 623.89 | 155.97 | 124.03 | 605 | 671.29 | 167.82 | 134.68 |
| 561 | 624.94 | 156.24 | 124.26 | 606 | 672.34 | 168.08 | 134.92 |
| 562 | 626.00 | 156.50 | 124.50 | 607 | 673.39 | 168.35 | 135.15 |
| 563 | 627.05 | 156.76 | 124.74 | 608 | 674.44 | 168.61 | 135.39 |
| 564 | 628.10 | 157.03 | 124.97 | 609 | 675.50 | 168.87 | 135.63 |
| 565 | 629.16 | 157.29 | 125.21 | 610 | 676.55 | 169.14 | 135.86 |
| 566 | 630.21 | 157.55 | 125.45 | 611 | 677.60 | 169.40 | 136.10 |
| 567 | 631.27 | 157.82 | 125.68 | 612 | 678.65 | 169.66 | 136.34 |
| 568 | 632.32 | 158.08 | 125.92 | 613 | 679.70 | 169.93 | 136.57 |
| 569 | 633.37 | 158.34 | 126.16 | 614 | 680.76 | 170.19 | 136.81 |
| 570 | 634.43 | 158.61 | 126.39 | 615 | 681.81 | 170.45 | 137.05 |
| 571 | 635.48 | 158.87 | 126.63 | 616 | 682.86 | 170.71 | 137.29 |
| 572 | 636.54 | 159.13 | 126.87 | 617 | 683.91 | 170.98 | 137.52 |
| 573 | 637.59 | 159.40 | 127.10 | 618 | 684.96 | 171.24 | 137.76 |
| 574 | 638.64 | 159.66 | 127.34 | 619 | 686.02 | 171.50 | 138.00 |
| 575 | 639.70 | 159.92 | 127.58 | 620 | 687.07 | 171.77 | 138.23 |
| 576 | 640.75 | 160.19 | 127.81 | 621 | 688.12 | 172.03 | 138.47 |
| 577 | 641.80 | 160.45 | 128.05 | 622 | 689.17 | 172.29 | 138.71 |
| 578 | 642.86 | 160.71 | 128.29 | 623 | 690.22 | 172.56 | 138.94 |
| 579 | 643.91 | 160.98 | 128.52 | 624 | 691.27 | 172.82 | 139.18 |
| 580 | 644.97 | 161.24 | 128.76 | 625 | 692.32 | 173.08 | 139.42 |
| 581 | 646.02 | 161.50 | 129.00 | 626 | 693.38 | 173.34 | 139.66 |
| 582 | 647.07 | 161.77 | 129.23 | 627 | 694.43 | 173.61 | 139.89 |
| 583 | 648.13 | 162.03 | 129.47 | 628 | 695.48 | 173.87 | 140.13 |
| 584 | 649.18 | 162.29 | 129.71 | 629 | 696.53 | 174.13 | 140.37 |

Table 1. **CONSTANTS FOR $\overline{p}'$ (LONG FORMULA), AN APPROXIMATION OF THE UPPER BINOMIAL CONFIDENCE LIMIT $\overline{p}$; $c/n \leqslant .5$, $n \geqslant 20$, $c \leqslant 1,000$ (*Continued*)**

| c | $\overline{m}$ | $\overline{a}$ | $\overline{b}$ | c | $\overline{m}$ | $\overline{a}$ | $\overline{b}$ |
|---|---|---|---|---|---|---|---|
| 630 | 697.58 | 174.40 | 140.60 | 675 | 744.85 | 186.21 | 151.29 |
| 631 | 698.63 | 174.66 | 140.84 | 676 | 745.90 | 186.47 | 151.53 |
| 632 | 699.68 | 174.92 | 141.08 | 677 | 746.95 | 186.74 | 151.76 |
| 633 | 700.74 | 175.18 | 141.32 | 678 | 748.00 | 187.00 | 152.00 |
| 634 | 701.79 | 175.45 | 141.55 | 679 | 749.05 | 187.26 | 152.24 |
| 635 | 702.84 | 175.71 | 141.79 | 680 | 750.10 | 187.52 | 152.48 |
| 636 | 703.89 | 175.97 | 142.03 | 681 | 751.15 | 187.79 | 152.71 |
| 637 | 704.94 | 176.23 | 142.27 | 682 | 752.19 | 188.05 | 152.95 |
| 638 | 705.99 | 176.50 | 142.50 | 683 | 753.24 | 188.31 | 153.19 |
| 639 | 707.04 | 176.76 | 142.74 | 684 | 754.29 | 188.57 | 153.43 |
| 640 | 708.09 | 177.02 | 142.98 | 685 | 755.34 | 188.84 | 153.66 |
| 641 | 709.14 | 177.29 | 143.21 | 686 | 756.39 | 189.10 | 153.90 |
| 642 | 710.19 | 177.55 | 143.45 | 687 | 757.44 | 189.36 | 154.14 |
| 643 | 711.24 | 177.81 | 143.69 | 688 | 758.49 | 189.62 | 154.38 |
| 644 | 712.30 | 178.07 | 143.93 | 689 | 759.54 | 189.88 | 154.62 |
| 645 | 713.35 | 178.34 | 144.16 | 690 | 760.59 | 190.15 | 154.85 |
| 646 | 714.40 | 178.60 | 144.40 | 691 | 761.64 | 190.41 | 155.09 |
| 647 | 715.45 | 178.86 | 144.64 | 692 | 762.69 | 190.67 | 155.33 |
| 648 | 716.50 | 179.12 | 144.88 | 693 | 763.73 | 190.93 | 155.57 |
| 649 | 717.55 | 179.39 | 145.11 | 694 | 764.78 | 191.20 | 155.80 |
| 650 | 718.60 | 179.65 | 145.35 | 695 | 765.83 | 191.46 | 156.04 |
| 651 | 719.65 | 179.91 | 145.59 | 696 | 766.88 | 191.72 | 156.28 |
| 652 | 720.70 | 180.17 | 145.83 | 697 | 767.93 | 191.98 | 156.52 |
| 653 | 721.75 | 180.44 | 146.06 | 698 | 768.98 | 192.24 | 156.76 |
| 654 | 722.80 | 180.70 | 146.30 | 699 | 770.03 | 192.51 | 156.99 |
| 655 | 723.85 | 180.96 | 146.54 | 700 | 771.08 | 192.77 | 157.23 |
| 656 | 724.90 | 181.23 | 146.77 | 701 | 772.12 | 193.03 | 157.47 |
| 657 | 725.95 | 181.49 | 147.01 | 702 | 773.17 | 193.29 | 157.71 |
| 658 | 727.00 | 181.75 | 147.25 | 703 | 774.22 | 193.56 | 157.94 |
| 659 | 728.05 | 182.01 | 147.49 | 704 | 775.27 | 193.82 | 158.18 |
| 660 | 729.10 | 182.28 | 147.72 | 705 | 776.32 | 194.08 | 158.42 |
| 661 | 730.15 | 182.54 | 147.96 | 706 | 777.37 | 194.34 | 158.66 |
| 662 | 731.20 | 182.80 | 148.20 | 707 | 778.42 | 194.60 | 158.90 |
| 663 | 732.25 | 183.06 | 148.44 | 708 | 779.46 | 194.87 | 159.13 |
| 664 | 733.30 | 183.33 | 148.67 | 709 | 780.51 | 195.13 | 159.37 |
| 665 | 734.35 | 183.59 | 148.91 | 710 | 781.56 | 195.39 | 159.61 |
| 666 | 735.40 | 183.85 | 149.15 | 711 | 782.61 | 195.65 | 159.85 |
| 667 | 736.45 | 184.11 | 149.39 | 712 | 783.66 | 195.91 | 160.09 |
| 668 | 737.50 | 184.38 | 149.62 | 713 | 784.71 | 196.18 | 160.32 |
| 669 | 738.55 | 184.64 | 149.86 | 714 | 785.75 | 196.44 | 160.56 |
| 670 | 739.60 | 184.90 | 150.10 | 715 | 786.80 | 196.70 | 160.80 |
| 671 | 740.65 | 185.16 | 150.34 | 716 | 787.85 | 196.96 | 161.04 |
| 672 | 741.70 | 185.43 | 150.57 | 717 | 788.90 | 197.22 | 161.28 |
| 673 | 742.75 | 185.69 | 150.81 | 718 | 789.95 | 197.49 | 161.51 |
| 674 | 743.80 | 185.95 | 151.05 | 719 | 790.99 | 197.75 | 161.75 |

# Table 1. CONSTANTS FOR $\overline{p}'$ (LONG FORMULA), AN APPROXIMATION OF THE UPPER BINOMIAL CONFIDENCE LIMIT $\overline{p}$; c/n $\leqslant$ .5, n $\geqslant$ 20, c $\leqslant$ 1,000 (*Continued*)

| c | $\overline{m}$ | $\overline{a}$ | $\overline{b}$ | c | $\overline{m}$ | $\overline{a}$ | $\overline{b}$ |
|---|---|---|---|---|---|---|---|
| 720 | 792.04 | 198.01 | 161.99 | 765 | 839.17 | 209.79 | 172.71 |
| 721 | 793.09 | 198.27 | 162.23 | 766 | 840.21 | 210.05 | 172.95 |
| 722 | 794.14 | 198.53 | 162.47 | 767 | 841.26 | 210.32 | 173.18 |
| 723 | 795.19 | 198.80 | 162.70 | 768 | 842.31 | 210.58 | 173.42 |
| 724 | 796.23 | 199.06 | 162.94 | 769 | 843.35 | 210.84 | 173.66 |
| 725 | 797.28 | 199.32 | 163.18 | 770 | 844.40 | 211.10 | 173.90 |
| 726 | 798.33 | 199.58 | 163.42 | 771 | 845.45 | 211.36 | 174.14 |
| 727 | 799.38 | 199.84 | 163.66 | 772 | 846.49 | 211.62 | 174.38 |
| 728 | 800.42 | 200.11 | 163.89 | 773 | 847.54 | 211.88 | 174.62 |
| 729 | 801.47 | 200.37 | 164.13 | 774 | 848.59 | 212.15 | 174.85 |
| 730 | 802.52 | 200.63 | 164.37 | 775 | 849.63 | 212.41 | 175.09 |
| 731 | 803.57 | 200.89 | 164.61 | 776 | 850.68 | 212.67 | 175.33 |
| 732 | 804.62 | 201.15 | 164.85 | 777 | 851.72 | 212.93 | 175.57 |
| 733 | 805.66 | 201.42 | 165.08 | 778 | 852.77 | 213.19 | 175.81 |
| 734 | 806.71 | 201.68 | 165.32 | 779 | 853.82 | 213.45 | 176.05 |
| 735 | 807.76 | 201.94 | 165.56 | 780 | 854.86 | 213.72 | 176.28 |
| 736 | 808.81 | 202.20 | 165.80 | 781 | 855.91 | 213.98 | 176.52 |
| 737 | 809.85 | 202.46 | 166.04 | 782 | 856.95 | 214.24 | 176.76 |
| 738 | 810.90 | 202.73 | 166.27 | 783 | 858.00 | 214.50 | 177.00 |
| 739 | 811.95 | 202.99 | 166.51 | 784 | 859.05 | 214.76 | 177.24 |
| 740 | 812.99 | 203.25 | 166.75 | 785 | 860.09 | 215.02 | 177.48 |
| 741 | 814.04 | 203.51 | 166.99 | 786 | 861.14 | 215.28 | 177.72 |
| 742 | 815.09 | 203.77 | 167.23 | 787 | 862.18 | 215.55 | 177.95 |
| 743 | 816.14 | 204.03 | 167.47 | 788 | 863.23 | 215.81 | 178.19 |
| 744 | 817.18 | 204.30 | 167.70 | 789 | 864.28 | 216.07 | 178.43 |
| 745 | 818.23 | 204.56 | 167.94 | 790 | 865.32 | 216.33 | 178.67 |
| 746 | 819.28 | 204.82 | 168.18 | 791 | 866.37 | 216.59 | 178.91 |
| 747 | 820.33 | 205.08 | 168.42 | 792 | 867.41 | 216.85 | 179.15 |
| 748 | 821.37 | 205.34 | 168.66 | 793 | 868.46 | 217.11 | 179.39 |
| 749 | 822.42 | 205.60 | 168.90 | 794 | 869.50 | 217.38 | 179.62 |
| 750 | 823.47 | 205.87 | 169.13 | 795 | 870.55 | 217.64 | 179.86 |
| 751 | 824.51 | 206.13 | 169.37 | 796 | 871.60 | 217.90 | 180.10 |
| 752 | 825.56 | 206.39 | 169.61 | 797 | 872.64 | 218.16 | 180.34 |
| 753 | 826.61 | 206.65 | 169.85 | 798 | 873.69 | 218.42 | 180.58 |
| 754 | 827.65 | 206.91 | 170.09 | 799 | 874.73 | 218.68 | 180.82 |
| 755 | 828.70 | 207.18 | 170.32 | 800 | 875.78 | 218.94 | 181.06 |
| 756 | 829.75 | 207.44 | 170.56 | 801 | 876.82 | 219.21 | 181.29 |
| 757 | 830.79 | 207.70 | 170.80 | 802 | 877.87 | 219.47 | 181.53 |
| 758 | 831.84 | 207.96 | 171.04 | 803 | 878.91 | 219.73 | 181.77 |
| 759 | 832.89 | 208.22 | 171.28 | 804 | 879.96 | 219.99 | 182.01 |
| 760 | 833.93 | 208.48 | 171.52 | 805 | 881.01 | 220.25 | 182.25 |
| 761 | 834.98 | 208.75 | 171.75 | 806 | 882.05 | 220.51 | 182.49 |
| 762 | 836.03 | 209.01 | 171.99 | 807 | 883.10 | 220.77 | 182.73 |
| 763 | 837.07 | 209.27 | 172.23 | 808 | 884.14 | 221.04 | 182.96 |
| 764 | 838.12 | 209.53 | 172.47 | 809 | 885.19 | 221.30 | 183.20 |

Table 1. CONSTANTS FOR $\bar{p}'$ (LONG FORMULA), AN APPROXIMATION OF THE UPPER BINOMIAL CONFIDENCE LIMIT $\bar{p}$; $c/n \leqslant .5$, $n \geqslant 20$, $c \leqslant 1,000$ (*Continued*)

| c | $\bar{m}$ | $\bar{a}$ | $\bar{b}$ | c | $\bar{m}$ | $\bar{a}$ | $\bar{b}$ |
|---|---|---|---|---|---|---|---|
| 810 | 886.23 | 221.56 | 183.44 | 855 | 933.24 | 233.31 | 194.19 |
| 811 | 887.28 | 221.82 | 183.68 | 856 | 934.28 | 233.57 | 194.43 |
| 812 | 888.32 | 222.08 | 183.92 | 857 | 935.33 | 233.83 | 194.67 |
| 813 | 889.37 | 222.34 | 184.16 | 858 | 936.37 | 234.09 | 194.91 |
| 814 | 890.41 | 222.60 | 184.40 | 859 | 937.42 | 234.35 | 195.15 |
| 815 | 891.46 | 222.86 | 184.64 | 860 | 938.46 | 234.61 | 195.39 |
| 816 | 892.50 | 223.13 | 184.87 | 861 | 939.50 | 234.88 | 195.62 |
| 817 | 893.55 | 223.39 | 185.11 | 862 | 940.55 | 235.14 | 195.86 |
| 818 | 894.59 | 223.65 | 185.35 | 863 | 941.59 | 235.40 | 196.10 |
| 819 | 895.64 | 223.91 | 185.59 | 864 | 942.63 | 235.66 | 196.34 |
| 820 | 896.68 | 224.17 | 185.83 | 865 | 943.68 | 235.92 | 196.58 |
| 821 | 897.73 | 224.43 | 186.07 | 866 | 944.72 | 236.18 | 196.82 |
| 822 | 898.77 | 224.69 | 186.31 | 867 | 945.77 | 236.44 | 197.06 |
| 823 | 899.82 | 224.95 | 186.55 | 868 | 946.81 | 236.70 | 197.30 |
| 824 | 900.86 | 225.22 | 186.78 | 869 | 947.85 | 236.96 | 197.54 |
| 825 | 901.91 | 225.48 | 187.02 | 870 | 948.90 | 237.22 | 197.78 |
| 826 | 902.95 | 225.74 | 187.26 | 871 | 949.94 | 237.49 | 198.01 |
| 827 | 904.00 | 226.00 | 187.50 | 872 | 950.98 | 237.75 | 198.25 |
| 828 | 905.04 | 226.26 | 187.74 | 873 | 952.03 | 238.01 | 198.49 |
| 829 | 906.09 | 226.52 | 187.98 | 874 | 953.07 | 238.27 | 198.73 |
| 830 | 907.13 | 226.78 | 188.22 | 875 | 954.11 | 238.53 | 198.97 |
| 831 | 908.18 | 227.04 | 188.46 | 876 | 955.16 | 238.79 | 199.21 |
| 832 | 909.22 | 227.31 | 188.69 | 877 | 956.20 | 239.05 | 199.45 |
| 833 | 910.26 | 227.57 | 188.93 | 878 | 957.25 | 239.31 | 199.69 |
| 834 | 911.31 | 227.83 | 189.17 | 879 | 958.29 | 239.57 | 199.93 |
| 835 | 912.35 | 228.09 | 189.41 | 880 | 959.33 | 239.83 | 200.17 |
| 836 | 913.40 | 228.35 | 189.65 | 881 | 960.38 | 240.09 | 200.41 |
| 837 | 914.44 | 228.61 | 189.89 | 882 | 961.42 | 240.35 | 200.65 |
| 838 | 915.49 | 228.87 | 190.13 | 883 | 962.46 | 240.62 | 200.88 |
| 839 | 916.53 | 229.13 | 190.37 | 884 | 963.51 | 240.88 | 201.12 |
| 840 | 917.58 | 229.39 | 190.61 | 885 | 964.55 | 241.14 | 201.36 |
| 841 | 918.62 | 229.66 | 190.84 | 886 | 965.59 | 241.40 | 201.60 |
| 842 | 919.67 | 229.92 | 191.08 | 887 | 966.64 | 241.66 | 201.84 |
| 843 | 920.71 | 230.18 | 191.32 | 888 | 967.68 | 241.92 | 202.08 |
| 844 | 921.75 | 230.44 | 191.56 | 889 | 968.72 | 242.18 | 202.32 |
| 845 | 922.80 | 230.70 | 191.80 | 890 | 969.76 | 242.44 | 202.56 |
| 846 | 923.84 | 230.96 | 192.04 | 891 | 970.81 | 242.70 | 202.80 |
| 847 | 924.89 | 231.22 | 192.28 | 892 | 971.85 | 242.96 | 203.04 |
| 848 | 925.93 | 231.48 | 192.52 | 893 | 972.89 | 243.22 | 203.28 |
| 849 | 926.98 | 231.74 | 192.76 | 894 | 973.94 | 243.48 | 203.52 |
| 850 | 928.02 | 232.00 | 193.00 | 895 | 974.98 | 243.75 | 203.75 |
| 851 | 929.06 | 232.27 | 193.23 | 896 | 976.02 | 244.01 | 203.99 |
| 852 | 930.11 | 232.53 | 193.47 | 897 | 977.07 | 244.27 | 204.23 |
| 853 | 931.15 | 232.79 | 193.71 | 898 | 978.11 | 244.53 | 204.47 |
| 854 | 932.20 | 233.05 | 193.95 | 899 | 979.15 | 244.79 | 204.71 |

Table 1. CONSTANTS FOR $\bar{p}'$ (LONG FORMULA), AN APPROXIMATION OF
THE UPPER BINOMIAL CONFIDENCE LIMIT $\bar{p}$; c/n ≤ .5, n ≥ 20,
c ≤ 1,000 (*Continued*)

Section 1

$\gamma_1$ = 99.5%
$\gamma_2$ = 99%

| c | $\bar{m}$ | $\bar{a}$ | $\bar{b}$ | c | $\bar{m}$ | $\bar{a}$ | $\bar{b}$ |
|---|---|---|---|---|---|---|---|
| 900 | 980.20 | 245.05 | 204.95 | 945 | 1027.10 | 256.78 | 215.72 |
| 901 | 981.24 | 245.31 | 205.19 | 946 | 1028.14 | 257.04 | 215.96 |
| 902 | 982.28 | 245.57 | 205.43 | 947 | 1029.19 | 257.30 | 216.20 |
| 903 | 983.32 | 245.83 | 205.67 | 948 | 1030.23 | 257.56 | 216.44 |
| 904 | 984.37 | 246.09 | 205.91 | 949 | 1031.27 | 257.82 | 216.68 |
| 905 | 985.41 | 246.35 | 206.15 | 950 | 1032.31 | 258.08 | 216.92 |
| 906 | 986.45 | 246.61 | 206.39 | 951 | 1033.35 | 258.34 | 217.16 |
| 907 | 987.49 | 246.87 | 206.63 | 952 | 1034.40 | 258.60 | 217.40 |
| 908 | 988.54 | 247.13 | 206.87 | 953 | 1035.44 | 258.86 | 217.64 |
| 909 | 989.58 | 247.40 | 207.10 | 954 | 1036.48 | 259.12 | 217.88 |
| 910 | 990.62 | 247.66 | 207.34 | 955 | 1037.52 | 259.38 | 218.12 |
| 911 | 991.67 | 247.92 | 207.58 | 956 | 1038.56 | 259.64 | 218.36 |
| 912 | 992.71 | 248.18 | 207.82 | 957 | 1039.60 | 259.90 | 218.60 |
| 913 | 993.75 | 248.44 | 208.06 | 958 | 1040.64 | 260.16 | 218.84 |
| 914 | 994.79 | 248.70 | 208.30 | 959 | 1041.69 | 260.42 | 219.08 |
| 915 | 995.84 | 248.96 | 208.54 | 960 | 1042.73 | 260.68 | 219.32 |
| 916 | 996.88 | 249.22 | 208.78 | 961 | 1043.77 | 260.94 | 219.56 |
| 917 | 997.92 | 249.48 | 209.02 | 962 | 1044.81 | 261.20 | 219.80 |
| 918 | 998.96 | 249.74 | 209.26 | 963 | 1045.85 | 261.46 | 220.04 |
| 919 | 1000.01 | 250.00 | 209.50 | 964 | 1046.89 | 261.72 | 220.28 |
| 920 | 1001.05 | 250.26 | 209.74 | 965 | 1047.94 | 261.98 | 220.52 |
| 921 | 1002.09 | 250.52 | 209.98 | 966 | 1048.98 | 262.24 | 220.76 |
| 922 | 1003.13 | 250.78 | 210.22 | 967 | 1050.02 | 262.50 | 221.00 |
| 923 | 1004.18 | 251.04 | 210.46 | 968 | 1051.06 | 262.76 | 221.24 |
| 924 | 1005.22 | 251.30 | 210.70 | 969 | 1052.10 | 263.03 | 221.47 |
| 925 | 1006.26 | 251.57 | 210.93 | 970 | 1053.14 | 263.29 | 221.71 |
| 926 | 1007.30 | 251.83 | 211.17 | 971 | 1054.18 | 263.55 | 221.95 |
| 927 | 1008.35 | 252.09 | 211.41 | 972 | 1055.22 | 263.81 | 222.19 |
| 928 | 1009.39 | 252.35 | 211.65 | 973 | 1056.27 | 264.07 | 222.43 |
| 929 | 1010.43 | 252.61 | 211.89 | 974 | 1057.31 | 264.33 | 222.67 |
| 930 | 1011.47 | 252.87 | 212.13 | 975 | 1058.35 | 264.59 | 222.91 |
| 931 | 1012.51 | 253.13 | 212.37 | 976 | 1059.39 | 264.85 | 223.15 |
| 932 | 1013.56 | 253.39 | 212.61 | 977 | 1060.43 | 265.11 | 223.39 |
| 933 | 1014.60 | 253.65 | 212.85 | 978 | 1061.47 | 265.37 | 223.63 |
| 934 | 1015.64 | 253.91 | 213.09 | 979 | 1062.51 | 265.63 | 223.87 |
| 935 | 1016.68 | 254.17 | 213.33 | 980 | 1063.55 | 265.89 | 224.11 |
| 936 | 1017.72 | 254.43 | 213.57 | 981 | 1064.60 | 266.15 | 224.35 |
| 937 | 1018.77 | 254.69 | 213.81 | 982 | 1065.64 | 266.41 | 224.59 |
| 938 | 1019.81 | 254.95 | 214.05 | 983 | 1066.68 | 266.67 | 224.83 |
| 939 | 1020.85 | 255.21 | 214.29 | 984 | 1067.72 | 266.93 | 225.07 |
| 940 | 1021.89 | 255.47 | 214.53 | 985 | 1068.76 | 267.19 | 225.31 |
| 941 | 1022.93 | 255.73 | 214.77 | 986 | 1069.80 | 267.45 | 225.55 |
| 942 | 1023.98 | 255.99 | 215.01 | 987 | 1070.84 | 267.71 | 225.79 |
| 943 | 1025.02 | 256.25 | 215.25 | 988 | 1071.88 | 267.97 | 226.03 |
| 944 | 1026.06 | 256.52 | 215.48 | 989 | 1072.92 | 268.23 | 226.27 |

Table 1.  CONSTANTS FOR $\overline{p}'$ (LONG FORMULA), AN APPROXIMATION OF
THE UPPER BINOMIAL CONFIDENCE LIMIT $\overline{p}$; $c/n \leqslant .5$, $n \geqslant 20$,
$c \leqslant 1,000$ (*Continued*)

| c | $\overline{m}$ | $\overline{a}$ | $\overline{b}$ | c | $\overline{m}$ | $\overline{a}$ | $\overline{b}$ |
|---|---|---|---|---|---|---|---|
| 990 | 1073.96 | 268.49 | 226.51 | 995 | 1079.17 | 269.79 | 227.71 |
| 991 | 1075.01 | 268.75 | 226.75 | 996 | 1080.21 | 270.05 | 227.95 |
| 992 | 1076.05 | 269.01 | 226.99 | 997 | 1081.25 | 270.31 | 228.19 |
| 993 | 1077.09 | 269.27 | 227.23 | 998 | 1082.29 | 270.57 | 228.43 |
| 994 | 1078.13 | 269.53 | 227.47 | 999 | 1083.33 | 270.83 | 228.67 |
| | | | | 1000 | 1084.37 | 271.09 | 228.91 |

**Table 1.** **CONSTANTS FOR $\bar{p}'$ (LONG FORMULA), AN APPROXIMATION OF**      **Section 2**
**THE UPPER BINOMIAL CONFIDENCE LIMIT $\bar{p}$; $c/n \leqslant .5$, $n \geqslant 20$,**      $\gamma_1 = 99\%$
$c \leqslant 1,000$ *(Continued)*      $\gamma_2 = 98\%$

| c | $\bar{m}$ | $\bar{a}$ | $\bar{b}$ | c | $\bar{m}$ | $\bar{a}$ | $\bar{b}$ |
|---|---|---|---|---|---|---|---|
| 0 | 4.6052 | 0.734 | -1.570 | 45 | 63.231 | 16.603 | 7.595 |
| 1 | 6.6384 | 1.094 | -1.730 | 46 | 64.402 | 16.908 | 7.813 |
| 2 | 8.4060 | 1.456 | -1.750 | 47 | 65.571 | 17.269 | 8.092 |
| 3 | 10.045 | 1.813 | -1.711 | 48 | 66.738 | 17.595 | 8.336 |
| 4 | 11.605 | 2.156 | -1.647 | 49 | 67.904 | 17.951 | 8.612 |
| 5 | 13.109 | 2.512 | -1.541 | 50 | 69.067 | 18.328 | 8.913 |
| 6 | 14.571 | 2.873 | -1.407 | 51 | 70.230 | 18.619 | 9.119 |
| 7 | 16.000 | 3.246 | -1.244 | 52 | 71.390 | 18.973 | 9.396 |
| 8 | 17.403 | 3.632 | -1.050 | 53 | 72.550 | 19.280 | 9.620 |
| 9 | 18.783 | 4.036 | -0.826 | 54 | 73.707 | 19.647 | 9.914 |
| 10 | 20.145 | 4.456 | -0.573 | 55 | 74.864 | 20.011 | 10.202 |
| 11 | 21.490 | 4.826 | -0.370 | 56 | 76.019 | 20.253 | 10.36 |
| 12 | 22.821 | 5.184 | -0.176 | 57 | 77.172 | 20.599 | 10.63 |
| 13 | 24.139 | 5.550 | 0.035 | 58 | 78.324 | 20.950 | 10.91 |
| 14 | 25.446 | 5.904 | 0.238 | 59 | 79.475 | 21.330 | 11.22 |
| 15 | 26.743 | 6.269 | 0.458 | 60 | 80.625 | 21.625 | 11.43 |
| 16 | 28.030 | 6.622 | 0.669 | 61 | 81.773 | 21.96 | 11.70 |
| 17 | 29.310 | 6.973 | 0.882 | 62 | 82.921 | 22.33 | 12.00 |
| 18 | 30.581 | 7.327 | 1.103 | 63 | 84.067 | 22.68 | 12.28 |
| 19 | 31.845 | 7.686 | 1.334 | 64 | 85.212 | 22.93 | 12.44 |
| 20 | 33.103 | 8.033 | 1.554 | 65 | 86.356 | 23.25 | 12.69 |
| 21 | 34.355 | 8.390 | 1.787 | 66 | 87.498 | 23.67 | 13.05 |
| 22 | 35.601 | 8.742 | 2.019 | 67 | 88.640 | 24.02 | 13.34 |
| 23 | 36.841 | 9.081 | 2.238 | 68 | 89.781 | 24.23 | 13.46 |
| 24 | 38.077 | 9.422 | 2.461 | 69 | 90.920 | 24.69 | 13.87 |
| 25 | 39.308 | 9.760 | 2.685 | 70 | 92.059 | 24.99 | 14.09 |
| 26 | 40.534 | 10.133 | 2.950 | 71 | 93.197 | 25.26 | 14.29 |
| 27 | 41.757 | 10.458 | 3.163 | 72 | 94.333 | 25.66 | 14.63 |
| 28 | 42.975 | 10.800 | 3.396 | 73 | 95.469 | 25.89 | 14.79 |
| 29 | 44.190 | 11.157 | 3.650 | 74 | 96.604 | 26.32 | 15.15 |
| 30 | 45.401 | 11.503 | 3.892 | 75 | 97.738 | 26.73 | 15.50 |
| 31 | 46.608 | 11.853 | 4.141 | 76 | 98.871 | 26.98 | 15.69 |
| 32 | 47.813 | 12.199 | 4.386 | 77 | 100.00 | 27.22 | 15.85 |
| 33 | 49.014 | 12.520 | 4.605 | 78 | 101.13 | 27.72 | 16.30 |
| 34 | 50.213 | 12.846 | 4.831 | 79 | 102.27 | 28.03 | 16.54 |
| 35 | 51.408 | 13.210 | 5.102 | 80 | 103.40 | 28.37 | 16.82 |
| 36 | 52.601 | 13.528 | 5.322 | 81 | 104.52 | 28.57 | 16.94 |
| 37 | 53.791 | 13.906 | 5.612 | 82 | 105.65 | 29.00 | 17.32 |
| 38 | 54.979 | 14.224 | 5.833 | 83 | 106.78 | 29.26 | 17.51 |
| 39 | 56.165 | 14.575 | 6.094 | 84 | 107.91 | 29.60 | 17.79 |
| 40 | 57.348 | 14.918 | 6.348 | 85 | 109.03 | 30.03 | 18.16 |
| 41 | 58.529 | 15.211 | 6.546 | 86 | 110.16 | 30.28 | 18.35 |
| 42 | 59.707 | 15.541 | 6.786 | 87 | 111.28 | 30.62 | 18.62 |
| 43 | 60.884 | 15.933 | 7.099 | 88 | 112.41 | 31.04 | 18.99 |
| 44 | 62.058 | 16.263 | 7.342 | 89 | 113.53 | 31.31 | 19.19 |

Table 1. **CONSTANTS FOR $\overline{p}'$ (LONG FORMULA), AN APPROXIMATION OF THE UPPER BINOMIAL CONFIDENCE LIMIT $\overline{p}$; $c/n \leqslant .5$, $n \geqslant 20$, $c \leqslant 1,000$** (*Continued*)

| c | $\overline{m}$ | $\overline{a}$ | $\overline{b}$ | c | $\overline{m}$ | $\overline{a}$ | $\overline{b}$ |
|---|---|---|---|---|---|---|---|
| 90 | 114.65 | 31.65 | 19.48 | 135 | 164.59 | 41.15 | 26.35 |
| 91 | 115.77 | 31.91 | 19.67 | 136 | 165.69 | 41.42 | 26.58 |
| 92 | 116.89 | 32.20 | 19.90 | 137 | 166.79 | 41.70 | 26.80 |
| 93 | 118.01 | 32.65 | 20.29 | 138 | 167.89 | 41.97 | 27.03 |
| 94 | 119.13 | 32.95 | 20.54 | 139 | 168.99 | 42.25 | 27.25 |
| 95 | 120.25 | 33.36 | 20.90 | 140 | 170.09 | 42.52 | 27.48 |
| 96 | 121.37 | 33.66 | 21.14 | 141 | 171.18 | 42.80 | 27.70 |
| 97 | 122.49 | 33.84 | 21.24 | 142 | 172.28 | 43.07 | 27.93 |
| 98 | 123.61 | 34.11 | 21.45 | 143 | 173.38 | 43.35 | 28.15 |
| 99 | 124.72 | 34.59 | 21.89 | 144 | 174.47 | 43.62 | 28.38 |
| 100 | 125.84 | 34.92 | 22.15 | 145 | 175.57 | 43.89 | 28.61 |
| 101 | 126.95 | 31.74 | 18.76 | 146 | 176.67 | 44.17 | 28.83 |
| 102 | 128.07 | 32.02 | 18.98 | 147 | 177.76 | 44.44 | 29.06 |
| 103 | 129.18 | 32.30 | 19.20 | 148 | 178.86 | 44.72 | 29.28 |
| 104 | 130.30 | 32.58 | 19.42 | 149 | 179.95 | 44.99 | 29.51 |
| 105 | 131.41 | 32.86 | 19.64 | 150 | 181.05 | 45.26 | 29.74 |
| 106 | 132.52 | 33.13 | 19.87 | 151 | 182.14 | 45.54 | 29.96 |
| 107 | 133.64 | 33.41 | 20.09 | 152 | 183.24 | 45.81 | 30.19 |
| 108 | 134.75 | 33.69 | 20.31 | 153 | 184.33 | 46.08 | 30.42 |
| 109 | 135.86 | 33.97 | 20.53 | 154 | 185.42 | 46.36 | 30.64 |
| 110 | 136.97 | 34.24 | 20.76 | 155 | 186.52 | 46.63 | 30.87 |
| 111 | 138.08 | 34.52 | 20.98 | 156 | 187.61 | 46.90 | 31.10 |
| 112 | 139.19 | 34.80 | 21.20 | 157 | 188.70 | 47.18 | 31.32 |
| 113 | 140.30 | 35.08 | 21.42 | 158 | 189.80 | 47.45 | 31.55 |
| 114 | 141.41 | 35.35 | 21.65 | 159 | 190.89 | 47.72 | 31.78 |
| 115 | 142.52 | 35.63 | 21.87 | 160 | 191.98 | 48.00 | 32.00 |
| 116 | 143.62 | 35.91 | 22.09 | 161 | 193.08 | 48.27 | 32.23 |
| 117 | 144.73 | 36.19 | 22.31 | 162 | 194.17 | 48.54 | 32.46 |
| 118 | 145.84 | 36.46 | 22.54 | 163 | 195.26 | 48.82 | 32.68 |
| 119 | 146.94 | 36.74 | 22.76 | 164 | 196.35 | 49.09 | 32.91 |
| 120 | 148.05 | 37.01 | 22.99 | 165 | 197.44 | 49.36 | 33.14 |
| 121 | 149.16 | 37.29 | 23.21 | 166 | 198.53 | 49.63 | 33.37 |
| 122 | 150.26 | 37.57 | 23.43 | 167 | 199.62 | 49.91 | 33.59 |
| 123 | 151.37 | 37.84 | 23.66 | 168 | 200.71 | 50.18 | 33.82 |
| 124 | 152.47 | 38.12 | 23.88 | 169 | 201.80 | 50.45 | 34.05 |
| 125 | 153.57 | 38.40 | 24.10 | 170 | 202.89 | 50.72 | 34.28 |
| 126 | 154.68 | 38.67 | 24.33 | 171 | 203.98 | 50.99 | 34.51 |
| 127 | i55.78 | 38.95 | 24.55 | 172 | 205.07 | 51.27 | 34.73 |
| 128 | 156.88 | 39.22 | 24.78 | 173 | 206.16 | 51.54 | 34.96 |
| 129 | 157.99 | 39.50 | 25.00 | 174 | 207.24 | 51.81 | 35.19 |
| 130 | 159.09 | 39.77 | 25.23 | 175 | 208.33 | 52.08 | 35.42 |
| 131 | 160.19 | 40.05 | 25.45 | 176 | 209.42 | 52.36 | 35.64 |
| 132 | 161.29 | 40.32 | 25.68 | 177 | 210.51 | 52.63 | 35.87 |
| 133 | 162.39 | 40.60 | 25.90 | 178 | 211.59 | 52.90 | 36.10 |
| 134 | 163.49 | 40.87 | 26.13 | 179 | 212.68 | 53.17 | 36.33 |

Table 1. **CONSTANTS FOR** $\overline{p}'$ **(LONG FORMULA), AN APPROXIMATION OF THE UPPER BINOMIAL CONFIDENCE LIMIT** $\overline{p}$; $c/n \leqslant .5$, $n \geqslant 20$, $c \leqslant 1,000$ (*Continued*)

Section 2

$\gamma_1 = 99\%$

$\gamma_2 = 98\%$

| c | $\overline{m}$ | $\overline{a}$ | $\overline{b}$ | c | $\overline{m}$ | $\overline{a}$ | $\overline{b}$ |
|---|---|---|---|---|---|---|---|
| 180 | 213.77 | 53.44 | 36.56 | 225 | 262.44 | 65.61 | 46.89 |
| 181 | 214.85 | 53.71 | 36.79 | 226 | 263.52 | 65.88 | 47.12 |
| 182 | 215.94 | 53.99 | 37.01 | 227 | 264.60 | 66.15 | 47.35 |
| 183 | 217.03 | 54.26 | 37.24 | 228 | 265.67 | 66.42 | 47.58 |
| 184 | 218.11 | 54.53 | 37.47 | 229 | 266.75 | 66.69 | 47.81 |
| 185 | 219.20 | 54.80 | 37.70 | 230 | 267.83 | 66.96 | 48.04 |
| 186 | 220.28 | 55.07 | 37.93 | 231 | 268.90 | 67.23 | 48.27 |
| 187 | 221.37 | 55.34 | 38.16 | 232 | 269.98 | 67.50 | 48.50 |
| 188 | 222.45 | 55.61 | 38.39 | 233 | 271.06 | 67.76 | 48.74 |
| 189 | 223.54 | 55.88 | 38.62 | 234 | 272.13 | 68.03 | 48.97 |
| 190 | 224.62 | 56.16 | 38.84 | 235 | 273.21 | 68.30 | 49.20 |
| 191 | 225.70 | 56.43 | 39.07 | 236 | 274.28 | 68.57 | 49.43 |
| 192 | 226.79 | 56.70 | 39.30 | 237 | 275.36 | 68.84 | 49.66 |
| 193 | 227.87 | 56.97 | 39.53 | 238 | 276.43 | 69.11 | 49.89 |
| 194 | 228.96 | 57.24 | 39.76 | 239 | 277.51 | 69.38 | 50.12 |
| 195 | 230.04 | 57.51 | 39.99 | 240 | 278.58 | 69.65 | 50.35 |
| 196 | 231.12 | 57.78 | 40.22 | 241 | 279.66 | 69.91 | 50.59 |
| 197 | 232.20 | 58.05 | 40.45 | 242 | 280.73 | 70.18 | 50.82 |
| 198 | 233.29 | 58.32 | 40.68 | 243 | 281.81 | 70.45 | 51.05 |
| 199 | 234.37 | 58.59 | 40.91 | 244 | 282.88 | 70.72 | 51.28 |
| 200 | 235.45 | 58.86 | 41.14 | 245 | 283.96 | 70.99 | 51.51 |
| 201 | 236.53 | 59.13 | 41.37 | 246 | 285.03 | 71.26 | 51.74 |
| 202 | 237.62 | 59.40 | 41.60 | 247 | 286.11 | 71.53 | 51.97 |
| 203 | 238.70 | 59.67 | 41.83 | 248 | 287.18 | 71.79 | 52.21 |
| 204 | 239.78 | 59.94 | 42.06 | 249 | 288.25 | 72.06 | 52.44 |
| 205 | 240.86 | 60.21 | 42.29 | 250 | 289.33 | 72.33 | 52.67 |
| 206 | 241.94 | 60.49 | 42.51 | 251 | 290.40 | 72.60 | 52.90 |
| 207 | 243.02 | 60.76 | 42.74 | 252 | 291.47 | 72.87 | 53.13 |
| 208 | 244.10 | 61.03 | 42.97 | 253 | 292.55 | 73.14 | 53.36 |
| 209 | 245.18 | 61.30 | 43.20 | 254 | 293.62 | 73.40 | 53.60 |
| 210 | 246.26 | 61.57 | 43.43 | 255 | 294.69 | 73.67 | 53.83 |
| 211 | 247.34 | 61.84 | 43.66 | 256 | 295.76 | 73.94 | 54.06 |
| 212 | 248.42 | 62.11 | 43.89 | 257 | 296.84 | 74.21 | 54.29 |
| 213 | 249.50 | 62.38 | 44.12 | 258 | 297.91 | 74.48 | 54.52 |
| 214 | 250.58 | 62.65 | 44.35 | 259 | 298.98 | 74.75 | 54.75 |
| 215 | 251.66 | 62.92 | 44.58 | 260 | 300.05 | 75.01 | 54.99 |
| 216 | 252.74 | 63.18 | 44.82 | 261 | 301.13 | 75.28 | 55.22 |
| 217 | 253.82 | 63.45 | 45.05 | 262 | 302.20 | 75.55 | 55.45 |
| 218 | 254.90 | 63.72 | 45.28 | 263 | 303.27 | 75.82 | 55.68 |
| 219 | 255.98 | 63.99 | 45.51 | 264 | 304.34 | 76.08 | 55.92 |
| 220 | 257.05 | 64.26 | 45.74 | 265 | 305.41 | 76.35 | 56.15 |
| 221 | 258.13 | 64.53 | 45.97 | 266 | 306.48 | 76.62 | 56.38 |
| 222 | 259.21 | 64.80 | 46.20 | 267 | 307.55 | 76.89 | 56.61 |
| 223 | 260.29 | 65.07 | 46.43 | 268 | 308.62 | 77.16 | 56.84 |
| 224 | 261.37 | 65.34 | 46.66 | 269 | 309.70 | 77.42 | 57.08 |

Table 1.   CONSTANTS FOR $\overline{p}'$ (LONG FORMULA), AN APPROXIMATION OF
THE UPPER BINOMIAL CONFIDENCE LIMIT $\overline{p}$; $c/n \leqslant .5$, $n \geqslant 20$,
$c \leqslant 1,000$ (*Continued*)

| c | $\overline{m}$ | $\overline{a}$ | $\overline{b}$ | c | $\overline{m}$ | $\overline{a}$ | $\overline{b}$ |
|---|---|---|---|---|---|---|---|
| 270 | 310.77 | 77.69 | 57.31 | 315 | 358.82 | 89.71 | 67.79 |
| 271 | 311.84 | 77.96 | 57.54 | 316 | 359.89 | 89.97 | 68.03 |
| 272 | 312.91 | 78.23 | 57.77 | 317 | 360.95 | 90.24 | 68.26 |
| 273 | 313.98 | 78.49 | 58.01 | 318 | 362.02 | 90.50 | 68.50 |
| 274 | 315.05 | 78.76 | 58.24 | 319 | 363.08 | 90.77 | 68.73 |
| 275 | 316.12 | 79.03 | 58.47 | 320 | 364.15 | 91.04 | 68.96 |
| 276 | 317.19 | 79.30 | 58.70 | 321 | 365.21 | 91.30 | 69.20 |
| 277 | 318.26 | 79.56 | 58.94 | 322 | 366.28 | 91.57 | 69.43 |
| 278 | 319.33 | 79.83 | 59.17 | 323 | 367.34 | 91.84 | 69.66 |
| 279 | 320.40 | 80.10 | 59.40 | 324 | 368.41 | 92.10 | 69.90 |
| 280 | 321.47 | 80.37 | 59.63 | 325 | 369.47 | 92.37 | 70.13 |
| 281 | 322.54 | 80.63 | 59.87 | 326 | 370.54 | 92.63 | 70.37 |
| 282 | 323.60 | 80.90 | 60.10 | 327 | 371.60 | 92.90 | 70.60 |
| 283 | 324.67 | 81.17 | 60.33 | 328 | 372.67 | 93.17 | 70.83 |
| 284 | 325.74 | 81.44 | 60.56 | 329 | 373.73 | 93.43 | 71.07 |
| 285 | 326.81 | 81.70 | 60.80 | 330 | 374.79 | 93.70 | 71.30 |
| 286 | 327.88 | 81.97 | 61.03 | 331 | 375.86 | 93.96 | 71.54 |
| 287 | 328.95 | 82.24 | 61.26 | 332 | 376.92 | 94.23 | 71.77 |
| 288 | 330.02 | 82.50 | 61.50 | 333 | 377.99 | 94.50 | 72.00 |
| 289 | 331.09 | 82.77 | 61.73 | 334 | 379.05 | 94.76 | 72.24 |
| 290 | 332.15 | 83.04 | 61.96 | 335 | 380.11 | 95.03 | 72.47 |
| 291 | 333.22 | 83.31 | 62.19 | 336 | 381.18 | 95.29 | 72.71 |
| 292 | 334.29 | 83.57 | 62.43 | 337 | 382.24 | 95.56 | 72.94 |
| 293 | 335.36 | 83.84 | 62.66 | 338 | 383.30 | 95.83 | 73.17 |
| 294 | 336.43 | 84.11 | 62.89 | 339 | 384.37 | 96.09 | 73.41 |
| 295 | 337.49 | 84.37 | 63.13 | 340 | 385.43 | 96.36 | 73.64 |
| 296 | 338.56 | 84.64 | 63.36 | 341 | 386.49 | 96.62 | 73.88 |
| 297 | 339.63 | 84.91 | 63.59 | 342 | 387.55 | 96.89 | 74.11 |
| 298 | 340.70 | 85.17 | 63.83 | 343 | 388.62 | 97.15 | 74.35 |
| 299 | 341.76 | 85.44 | 64.06 | 344 | 389.68 | 97.42 | 74.58 |
| 300 | 342.83 | 85.71 | 64.29 | 345 | 390.74 | 97.69 | 74.81 |
| 301 | 343.90 | 85.97 | 64.53 | 346 | 391.80 | 97.95 | 75.05 |
| 302 | 344.96 | 86.24 | 64.76 | 347 | 392.87 | 98.22 | 75.28 |
| 303 | 346.03 | 86.51 | 64.99 | 348 | 393.93 | 98.48 | 75.52 |
| 304 | 347.10 | 86.77 | 65.23 | 349 | 394.99 | 98.75 | 75.75 |
| 305 | 348.16 | 87.04 | 65.46 | 350 | 396.05 | 99.01 | 75.99 |
| 306 | 349.23 | 87.31 | 65.69 | 351 | 397.12 | 99.28 | 76.22 |
| 307 | 350.30 | 87.57 | 65.93 | 352 | 398.18 | 99.54 | 76.46 |
| 308 | 351.36 | 87.84 | 66.16 | 353 | 399.24 | 99.81 | 76.69 |
| 309 | 352.43 | 88.11 | 66.39 | 354 | 400.30 | 100.08 | 76.92 |
| 310 | 353.50 | 88.37 | 66.63 | 355 | 401.36 | 100.34 | 77.16 |
| 311 | 354.56 | 88.64 | 66.86 | 356 | 402.42 | 100.61 | 77.39 |
| 312 | 355.63 | 88.91 | 67.09 | 357 | 403.49 | 100.87 | 77.63 |
| 313 | 356.69 | 89.17 | 67.33 | 358 | 404.55 | 101.14 | 77.86 |
| 314 | 357.76 | 89.44 | 67.56 | 359 | 405.61 | 101.40 | 78.10 |

Table 1.  CONSTANTS FOR $\overline{p}'$ (LONG FORMULA), AN APPROXIMATION OF
THE UPPER BINOMIAL CONFIDENCE LIMIT $\overline{p}$; $c/n \leqslant .5$, $n \geqslant 20$,
$c \leqslant 1,000$ (*Continued*)

Section 2

$\gamma_1 = 99\%$

$\gamma_2 = 98\%$

| c | $\overline{m}$ | $\overline{a}$ | $\overline{b}$ | c | $\overline{m}$ | $\overline{a}$ | $\overline{b}$ |
|---|---|---|---|---|---|---|---|
| 360 | 406.67 | 101.67 | 78.33 | 405 | 454.34 | 113.59 | 88.91 |
| 361 | 407.73 | 101.93 | 78.57 | 406 | 455.40 | 113.85 | 89.15 |
| 362 | 408.79 | 102.20 | 78.80 | 407 | 456.46 | 114.11 | 89.39 |
| 363 | 409.85 | 102.46 | 79.04 | 408 | 457.52 | 114.38 | 89.62 |
| 364 | 410.91 | 102.73 | 79.27 | 409 | 458.57 | 114.64 | 89.86 |
| 365 | 411.98 | 102.99 | 79.51 | 410 | 459.63 | 114.91 | 90.09 |
| 366 | 413.04 | 103.26 | 79.74 | 411 | 460.69 | 115.17 | 90.33 |
| 367 | 414.10 | 103.52 | 79.98 | 412 | 461.75 | 115.44 | 90.56 |
| 368 | 415.16 | 103.79 | 80.21 | 413 | 462.80 | 115.70 | 90.80 |
| 369 | 416.22 | 104.05 | 80.45 | 414 | 463.86 | 115.97 | 91.03 |
| 370 | 417.28 | 104.32 | 80.68 | 415 | 464.92 | 116.23 | 91.27 |
| 371 | 418.34 | 104.58 | 80.92 | 416 | 465.97 | 116.49 | 91.51 |
| 372 | 419.40 | 104.85 | 81.15 | 417 | 467.03 | 116.76 | 91.74 |
| 373 | 420.46 | 105.11 | 81.39 | 418 | 468.09 | 117.02 | 91.98 |
| 374 | 421.52 | 105.38 | 81.62 | 419 | 469.15 | 117.29 | 92.21 |
| 375 | 422.58 | 105.64 | 81.86 | 420 | 470.20 | 117.55 | 92.45 |
| 376 | 423.64 | 105.91 | 82.09 | 421 | 471.26 | 117.81 | 92.69 |
| 377 | 424.70 | 106.17 | 82.33 | 422 | 472.32 | 118.08 | 92.92 |
| 378 | 425.76 | 106.44 | 82.56 | 423 | 473.37 | 118.34 | 93.16 |
| 379 | 426.82 | 106.70 | 82.80 | 424 | 474.43 | 118.61 | 93.39 |
| 380 | 427.88 | 106.97 | 83.03 | 425 | 475.48 | 118.87 | 93.63 |
| 381 | 428.94 | 107.23 | 83.27 | 426 | 476.54 | 119.14 | 93.86 |
| 382 | 430.00 | 107.50 | 83.50 | 427 | 477.60 | 119.40 | 94.10 |
| 383 | 431.06 | 107.76 | 83.74 | 428 | 478.65 | 119.66 | 94.34 |
| 384 | 432.12 | 108.03 | 83.97 | 429 | 479.71 | 119.93 | 94.57 |
| 385 | 433.18 | 108.29 | 84.21 | 430 | 480.77 | 120.19 | 94.81 |
| 386 | 434.23 | 108.56 | 84.44 | 431 | 481.82 | 120.46 | 95.04 |
| 387 | 435.29 | 108.82 | 84.68 | 432 | 482.88 | 120.72 | 95.28 |
| 388 | 436.35 | 109.09 | 84.91 | 433 | 483.93 | 120.98 | 95.52 |
| 389 | 437.41 | 109.35 | 85.15 | 434 | 484.99 | 121.25 | 95.75 |
| 390 | 438.47 | 109.62 | 85.38 | 435 | 486.05 | 121.51 | 95.99 |
| 391 | 439.53 | 109.88 | 85.62 | 436 | 487.10 | 121.78 | 96.22 |
| 392 | 440.59 | 110.15 | 85.85 | 437 | 488.16 | 122.04 | 96.46 |
| 393 | 441.65 | 110.41 | 86.09 | 438 | 489.21 | 122.30 | 96.70 |
| 394 | 442.70 | 110.68 | 86.32 | 439 | 490.27 | 122.57 | 96.93 |
| 395 | 443.76 | 110.94 | 86.56 | 440 | 491.32 | 122.83 | 97.17 |
| 396 | 444.82 | 111.21 | 86.79 | 441 | 492.38 | 123.09 | 97.41 |
| 397 | 445.88 | 111.47 | 87.03 | 442 | 493.43 | 123.36 | 97.64 |
| 398 | 446.94 | 111.73 | 87.27 | 443 | 494.49 | 123.62 | 97.88 |
| 399 | 448.00 | 112.00 | 87.50 | 444 | 495.54 | 123.89 | 98.11 |
| 400 | 449.05 | 112.26 | 87.74 | 445 | 496.60 | 124.15 | 98.35 |
| 401 | 450.11 | 112.53 | 87.97 | 446 | 497.65 | 124.41 | 98.59 |
| 402 | 451.17 | 112.79 | 88.21 | 447 | 498.71 | 124.68 | 98.82 |
| 403 | 452.23 | 113.06 | 88.44 | 448 | 499.76 | 124.94 | 99.06 |
| 404 | 453.29 | 113.32 | 88.68 | 449 | 500.82 | 125.20 | 99.30 |

Table 1.   CONSTANTS FOR $\overline{p}'$ (LONG FORMULA), AN APPROXIMATION OF
THE UPPER BINOMIAL CONFIDENCE LIMIT $\overline{p}$; $c/n \leqslant .5$, $n \geqslant 20$,
$c \leqslant 1,000$ (*Continued*)

| c | $\overline{m}$ | $\overline{a}$ | $\overline{b}$ | c | $\overline{m}$ | $\overline{a}$ | $\overline{b}$ |
|---|---|---|---|---|---|---|---|
| 450 | 501.87 | 125.47 | 99.53 | 495 | 549.28 | 137.32 | 110.18 |
| 451 | 502.93 | 125.73 | 99.77 | 496 | 550.33 | 137.58 | 110.42 |
| 452 | 503.98 | 126.00 | 100.00 | 497 | 551.38 | 137.85 | 110.65 |
| 453 | 505.04 | 126.26 | 100.24 | 498 | 552.44 | 138.11 | 110.89 |
| 454 | 506.09 | 126.52 | 100.48 | 499 | 553.49 | 138.37 | 111.13 |
| 455 | 507.15 | 126.79 | 100.71 | 500 | 554.54 | 138.64 | 111.36 |
| 456 | 508.20 | 127.05 | 100.95 | 501 | 555.59 | 138.90 | 111.60 |
| 457 | 509.26 | 127.31 | 101.19 | 502 | 556.64 | 139.16 | 111.84 |
| 458 | 510.31 | 127.58 | 101.42 | 503 | 557.70 | 139.42 | 112.08 |
| 459 | 511.36 | 127.84 | 101.66 | 504 | 558.75 | 139.69 | 112.31 |
| 460 | 512.42 | 128.10 | 101.90 | 505 | 559.80 | 139.95 | 112.55 |
| 461 | 513.47 | 128.37 | 102.13 | 506 | 560.85 | 140.21 | 112.79 |
| 462 | 514.53 | 128.63 | 102.37 | 507 | 561.90 | 140.48 | 113.02 |
| 463 | 515.58 | 128.90 | 102.60 | 508 | 562.95 | 140.74 | 113.26 |
| 464 | 516.63 | 129.16 | 102.84 | 509 | 564.01 | 141.00 | 113.50 |
| 465 | 517.69 | 129.42 | 103.08 | 510 | 565.06 | 141.26 | 113.74 |
| 466 | 518.74 | 129.69 | 103.31 | 511 | 566.11 | 141.53 | 113.97 |
| 467 | 519.80 | 129.95 | 103.55 | 512 | 567.16 | 141.79 | 114.21 |
| 468 | 520.85 | 130.21 | 103.79 | 513 | 568.21 | 142.05 | 114.45 |
| 469 | 521.90 | 130.48 | 104.02 | 514 | 569.26 | 142.32 | 114.68 |
| 470 | 522.96 | 130.74 | 104.26 | 515 | 570.31 | 142.58 | 114.92 |
| 471 | 524.01 | 131.00 | 104.50 | 516 | 571.37 | 142.84 | 115.16 |
| 472 | 525.06 | 131.27 | 104.73 | 517 | 572.42 | 143.10 | 115.40 |
| 473 | 526.12 | 131.53 | 104.97 | 518 | 573.47 | 143.37 | 115.63 |
| 474 | 527.17 | 131.79 | 105.21 | 519 | 574.52 | 143.63 | 115.87 |
| 475 | 528.22 | 132.06 | 105.44 | 520 | 575.57 | 143.89 | 116.11 |
| 476 | 529.28 | 132.32 | 105.68 | 521 | 576.62 | 144.16 | 116.34 |
| 477 | 530.33 | 132.58 | 105.92 | 522 | 577.67 | 144.42 | 116.58 |
| 478 | 531.38 | 132.85 | 106.15 | 523 | 578.72 | 144.68 | 116.82 |
| 479 | 532.44 | 133.11 | 106.39 | 524 | 579.77 | 144.94 | 117.06 |
| 480 | 533.49 | 133.37 | 106.63 | 525 | 580.82 | 145.21 | 117.29 |
| 481 | 534.54 | 133.64 | 106.86 | 526 | 581.87 | 145.47 | 117.53 |
| 482 | 535.60 | 133.90 | 107.10 | 527 | 582.92 | 145.73 | 117.77 |
| 483 | 536.65 | 134.16 | 107.34 | 528 | 583.98 | 145.99 | 118.01 |
| 484 | 537.70 | 134.43 | 107.57 | 529 | 585.03 | 146.26 | 118.24 |
| 485 | 538.75 | 134.69 | 107.81 | 530 | 586.08 | 146.52 | 118.48 |
| 486 | 539.81 | 134.95 | 108.05 | 531 | 587.13 | 146.78 | 118.72 |
| 487 | 540.86 | 135.22 | 108.28 | 532 | 588.18 | 147.04 | 118.96 |
| 488 | 541.91 | 135.48 | 108.52 | 533 | 589.23 | 147.31 | 119.19 |
| 489 | 542.97 | 135.74 | 108.76 | 534 | 590.28 | 147.57 | 119.43 |
| 490 | 544.02 | 136.00 | 109.00 | 535 | 591.33 | 147.83 | 119.67 |
| 491 | 545.07 | 136.27 | 109.23 | 536 | 592.38 | 148.09 | 119.91 |
| 492 | 546.12 | 136.53 | 109.47 | 537 | 593.43 | 148.36 | 120.14 |
| 493 | 547.18 | 136.79 | 109.71 | 538 | 594.48 | 148.62 | 120.38 |
| 494 | 548.23 | 137.06 | 109.94 | 539 | 595.53 | 148.88 | 120.62 |

**Table 1.** CONSTANTS FOR $\overline{p}'$ (LONG FORMULA), AN APPROXIMATION OF THE UPPER BINOMIAL CONFIDENCE LIMIT $\overline{p}$; $c/n \leqslant .5$, $n \geqslant 20$, $c \leqslant 1,000$ (*Continued*)

Section 2

$\gamma_1 = 99\%$
$\gamma_2 = 98\%$

| c | $\overline{m}$ | $\overline{a}$ | $\overline{b}$ | c | $\overline{m}$ | $\overline{a}$ | $\overline{b}$ |
|---|---|---|---|---|---|---|---|
| 540 | 596.58 | 149.14 | 120.86 | 585 | 643.78 | 160.95 | 131.55 |
| 541 | 597.63 | 149.41 | 121.09 | 586 | 644.83 | 161.21 | 131.79 |
| 542 | 598.68 | 149.67 | 121.33 | 587 | 645.88 | 161.47 | 132.03 |
| 543 | 599.73 | 149.93 | 121.57 | 588 | 646.93 | 161.73 | 132.27 |
| 544 | 600.78 | 150.19 | 121.81 | 589 | 647.98 | 161.99 | 132.51 |
| 545 | 601.83 | 150.46 | 122.04 | 590 | 649.02 | 162.26 | 132.74 |
| 546 | 602.88 | 150.72 | 122.28 | 591 | 650.07 | 162.52 | 132.98 |
| 547 | 603.93 | 150.98 | 122.52 | 592 | 651.12 | 162.78 | 133.22 |
| 548 | 604.98 | 151.24 | 122.76 | 593 | 652.17 | 163.04 | 133.46 |
| 549 | 606.03 | 151.51 | 122.99 | 594 | 653.22 | 163.30 | 133.70 |
| 550 | 607.08 | 151.77 | 123.23 | 595 | 654.26 | 163.57 | 133.93 |
| 551 | 608.13 | 152.03 | 123.47 | 596 | 655.31 | 163.83 | 134.17 |
| 552 | 609.18 | 152.29 | 123.71 | 597 | 656.36 | 164.09 | 134.41 |
| 553 | 610.23 | 152.56 | 123.94 | 598 | 657.41 | 164.35 | 134.65 |
| 554 | 611.27 | 152.82 | 124.18 | 599 | 658.45 | 164.61 | 134.89 |
| 555 | 612.32 | 153.08 | 124.42 | 600 | 659.50 | 164.88 | 135.12 |
| 556 | 613.37 | 153.34 | 124.66 | 601 | 660.55 | 165.14 | 135.36 |
| 557 | 614.42 | 153.61 | 124.89 | 602 | 661.60 | 165.40 | 135.60 |
| 558 | 615.47 | 153.87 | 125.13 | 603 | 662.64 | 165.66 | 135.84 |
| 559 | 616.52 | 154.13 | 125.37 | 604 | 663.69 | 165.92 | 136.08 |
| 560 | 617.57 | 154.39 | 125.61 | 605 | 664.74 | 166.18 | 136.32 |
| 561 | 618.62 | 154.65 | 125.85 | 606 | 665.78 | 166.45 | 136.55 |
| 562 | 619.67 | 154.92 | 126.08 | 607 | 666.83 | 166.71 | 136.79 |
| 563 | 620.72 | 155.18 | 126.32 | 608 | 667.88 | 166.97 | 137.03 |
| 564 | 621.77 | 155.44 | 126.56 | 609 | 668.93 | 167.23 | 137.27 |
| 565 | 622.81 | 155.70 | 126.80 | 610 | 669.97 | 167.49 | 137.51 |
| 566 | 623.86 | 155.97 | 127.03 | 611 | 671.02 | 167.76 | 137.75 |
| 567 | 624.91 | 156.23 | 127.27 | 612 | 672.07 | 168.02 | 137.98 |
| 568 | 625.96 | 156.49 | 127.51 | 613 | 673.11 | 168.28 | 138.22 |
| 569 | 627.01 | 156.75 | 127.75 | 614 | 674.16 | 168.54 | 138.46 |
| 570 | 628.06 | 157.01 | 127.99 | 615 | 675.21 | 168.80 | 138.70 |
| 571 | 629.11 | 157.28 | 128.22 | 616 | 676.25 | 169.06 | 138.94 |
| 572 | 630.16 | 157.54 | 128.46 | 617 | 677.30 | 169.33 | 139.17 |
| 573 | 631.20 | 157.80 | 128.70 | 618 | 678.35 | 169.59 | 139.41 |
| 574 | 632.25 | 158.06 | 128.94 | 619 | 679.39 | 169.85 | 139.65 |
| 575 | 633.30 | 158.33 | 129.17 | 620 | 680.44 | 170.11 | 139.89 |
| 576 | 634.35 | 158.59 | 129.41 | 621 | 681.49 | 170.37 | 140.13 |
| 577 | 635.40 | 158.85 | 129.65 | 622 | 682.53 | 170.63 | 140.37 |
| 578 | 636.45 | 159.11 | 129.89 | 623 | 683.58 | 170.90 | 140.60 |
| 579 | 637.50 | 159.37 | 130.13 | 624 | 684.63 | 171.16 | 140.84 |
| 580 | 638.54 | 159.64 | 130.36 | 625 | 685.67 | 171.42 | 141.08 |
| 581 | 639.59 | 159.90 | 130.60 | 626 | 686.72 | 171.68 | 141.32 |
| 582 | 640.64 | 160.16 | 130.84 | 627 | 687.77 | 171.94 | 141.56 |
| 583 | 641.69 | 160.42 | 131.08 | 628 | 688.81 | 172.20 | 141.80 |
| 584 | 642.74 | 160.68 | 131.32 | 629 | 689.86 | 172.47 | 142.03 |

Table 1.  CONSTANTS FOR $\overline{p}'$ (LONG FORMULA), AN APPROXIMATION OF THE UPPER BINOMIAL CONFIDENCE LIMIT $\overline{p}$; $c/n \leqslant .5$, $n \geqslant 20$, $c \leqslant 1,000$ (*Continued*)

| c | $\overline{m}$ | $\overline{a}$ | $\overline{b}$ | c | $\overline{m}$ | $\overline{a}$ | $\overline{b}$ |
|---|---|---|---|---|---|---|---|
| 630 | 690.91 | 172.73 | 142.27 | 675 | 737.95 | 184.49 | 153.01 |
| 631 | 691.95 | 172.99 | 142.51 | 676 | 739.00 | 184.75 | 153.25 |
| 632 | 693.00 | 173.25 | 142.75 | 677 | 740.04 | 185.01 | 153.49 |
| 633 | 694.05 | 173.51 | 142.99 | 678 | 741.09 | 185.27 | 153.73 |
| 634 | 695.09 | 173.77 | 143.23 | 679 | 742.13 | 185.53 | 153.97 |
| 635 | 696.14 | 174.03 | 143.47 | 680 | 743.18 | 185.79 | 154.21 |
| 636 | 697.18 | 174.30 | 143.70 | 681 | 744.22 | 186.06 | 154.44 |
| 637 | 698.23 | 174.56 | 143.94 | 682 | 745.27 | 186.32 | 154.68 |
| 638 | 699.28 | 174.82 | 144.18 | 683 | 746.31 | 186.58 | 154.92 |
| 639 | 700.32 | 175.08 | 144.42 | 684 | 747.36 | 186.84 | 155.16 |
| 640 | 701.37 | 175.34 | 144.66 | 685 | 748.40 | 187.10 | 155.40 |
| 641 | 702.41 | 175.60 | 144.90 | 686 | 749.44 | 187.36 | 155.64 |
| 642 | 703.46 | 175.86 | 145.14 | 687 | 750.49 | 187.62 | 155.88 |
| 643 | 704.51 | 176.13 | 145.37 | 688 | 751.53 | 187.88 | 156.12 |
| 644 | 705.55 | 176.39 | 145.61 | 689 | 752.58 | 188.14 | 156.36 |
| 645 | 706.60 | 176.65 | 145.85 | 690 | 753.62 | 188.41 | 156.59 |
| 646 | 707.64 | 176.91 | 146.09 | 691 | 754.67 | 188.67 | 156.83 |
| 647 | 708.69 | 177.17 | 146.33 | 692 | 755.71 | 188.93 | 157.07 |
| 648 | 709.73 | 177.43 | 146.57 | 693 | 756.75 | 189.19 | 157.31 |
| 649 | 710.78 | 177.69 | 146.81 | 694 | 757.80 | 189.45 | 157.55 |
| 650 | 711.83 | 177.96 | 147.04 | 695 | 758.84 | 189.71 | 157.79 |
| 651 | 712.87 | 178.22 | 147.28 | 696 | 759.89 | 189.97 | 158.03 |
| 652 | 713.92 | 178.48 | 147.52 | 697 | 760.93 | 190.23 | 158.27 |
| 653 | 714.96 | 178.74 | 147.76 | 698 | 761.97 | 190.49 | 158.51 |
| 654 | 716.01 | 179.00 | 148.00 | 699 | 763.02 | 190.75 | 158.75 |
| 655 | 717.05 | 179.26 | 148.24 | 700 | 764.06 | 191.02 | 158.98 |
| 656 | 718.10 | 179.52 | 148.48 | 701 | 765.11 | 191.28 | 159.22 |
| 657 | 719.14 | 179.79 | 148.71 | 702 | 766.15 | 191.54 | 159.46 |
| 658 | 720.19 | 180.05 | 148.95 | 703 | 767.19 | 191.80 | 159.70 |
| 659 | 721.23 | 180.31 | 149.19 | 704 | 768.24 | 192.06 | 159.94 |
| 660 | 722.28 | 180.57 | 149.43 | 705 | 769.28 | 192.32 | 160.18 |
| 661 | 723.32 | 180.83 | 149.67 | 706 | 770.33 | 192.58 | 160.42 |
| 662 | 724.37 | 181.09 | 149.91 | 707 | 771.37 | 192.84 | 160.66 |
| 663 | 725.42 | 181.35 | 150.15 | 708 | 772.41 | 193.10 | 160.90 |
| 664 | 726.46 | 181.62 | 150.38 | 709 | 773.46 | 193.36 | 161.14 |
| 665 | 727.51 | 181.88 | 150.62 | 710 | 774.50 | 193.63 | 161.37 |
| 666 | 728.55 | 182.14 | 150.86 | 711 | 775.54 | 193.89 | 161.61 |
| 667 | 729.60 | 182.40 | 151.10 | 712 | 776.59 | 194.15 | 161.85 |
| 668 | 730.64 | 182.66 | 151.34 | 713 | 777.63 | 194.41 | 162.09 |
| 669 | 731.69 | 182.92 | 151.58 | 714 | 778.67 | 194.67 | 162.33 |
| 670 | 732.73 | 183.18 | 151.82 | 715 | 779.72 | 194.93 | 162.57 |
| 671 | 733.78 | 183.44 | 152.06 | 716 | 780.76 | 195.19 | 162.81 |
| 672 | 734.82 | 183.70 | 152.30 | 717 | 781.80 | 195.45 | 163.05 |
| 673 | 735.86 | 183.97 | 152.53 | 718 | 782.85 | 195.71 | 163.29 |
| 674 | 736.91 | 184.23 | 152.77 | 719 | 783.89 | 195.97 | 163.53 |

**Table 1.** **CONSTANTS FOR $\overline{p}'$ (LONG FORMULA), AN APPROXIMATION OF THE UPPER BINOMIAL CONFIDENCE LIMIT $\overline{p}$; c/n ≤ .5, n ≥ 20, c ≤ 1,000 (*Continued*)**

| c | $\overline{m}$ | $\overline{a}$ | $\overline{b}$ | c | $\overline{m}$ | $\overline{a}$ | $\overline{b}$ |
|---|---|---|---|---|---|---|---|
| 720 | 784.94 | 196.23 | 163.77 | 765 | 831.85 | 207.96 | 174.54 |
| 721 | 785.98 | 196.49 | 164.01 | 766 | 832.90 | 208.22 | 174.78 |
| 722 | 787.02 | 196.76 | 164.24 | 767 | 833.94 | 208.48 | 175.02 |
| 723 | 788.06 | 197.02 | 164.48 | 768 | 834.98 | 208.75 | 175.25 |
| 724 | 789.11 | 197.28 | 164.72 | 769 | 836.02 | 209.01 | 175.49 |
| 725 | 790.15 | 197.54 | 164.96 | 770 | 837.06 | 209.27 | 175.73 |
| 726 | 791.19 | 197.80 | 165.20 | 771 | 838.11 | 209.53 | 175.97 |
| 727 | 792.24 | 198.06 | 165.44 | 772 | 839.15 | 209.79 | 176.21 |
| 728 | 793.28 | 198.32 | 165.68 | 773 | 840.19 | 210.05 | 176.45 |
| 729 | 794.32 | 198.58 | 165.92 | 774 | 841.23 | 210.31 | 176.69 |
| 730 | 795.37 | 198.84 | 166.16 | 775 | 842.27 | 210.57 | 176.93 |
| 731 | 796.41 | 199.10 | 166.40 | 776 | 843.32 | 210.83 | 177.17 |
| 732 | 797.45 | 199.36 | 166.64 | 777 | 844.36 | 211.09 | 177.41 |
| 733 | 798.50 | 199.62 | 166.88 | 778 | 845.40 | 211.35 | 177.65 |
| 734 | 799.54 | 199.88 | 167.12 | 779 | 846.44 | 211.61 | 177.89 |
| 735 | 800.58 | 200.15 | 167.35 | 780 | 847.48 | 211.87 | 178.13 |
| 736 | 801.62 | 200.41 | 167.59 | 781 | 848.52 | 212.13 | 178.37 |
| 737 | 802.67 | 200.67 | 167.83 | 782 | 849.57 | 212.39 | 178.61 |
| 738 | 803.71 | 200.93 | 168.07 | 783 | 850.61 | 212.65 | 178.85 |
| 739 | 804.75 | 201.19 | 168.31 | 784 | 851.65 | 212.91 | 179.09 |
| 740 | 805.80 | 201.45 | 168.55 | 785 | 852.69 | 213.17 | 179.33 |
| 741 | 806.84 | 201.71 | 168.79 | 786 | 853.73 | 213.43 | 179.57 |
| 742 | 807.88 | 201.97 | 169.03 | 787 | 854.77 | 213.69 | 179.81 |
| 743 | 808.92 | 202.23 | 169.27 | 788 | 855.81 | 213.95 | 180.05 |
| 744 | 809.97 | 202.49 | 169.51 | 789 | 856.86 | 214.21 | 180.29 |
| 745 | 811.01 | 202.75 | 169.75 | 790 | 857.90 | 214.47 | 180.53 |
| 746 | 812.05 | 203.01 | 169.99 | 791 | 858.94 | 214.73 | 180.77 |
| 747 | 813.09 | 203.27 | 170.23 | 792 | 859.98 | 214.99 | 181.01 |
| 748 | 814.14 | 203.53 | 170.47 | 793 | 861.02 | 215.26 | 181.24 |
| 749 | 815.18 | 203.79 | 170.71 | 794 | 862.06 | 215.52 | 181.48 |
| 750 | 816.22 | 204.06 | 170.94 | 795 | 863.10 | 215.78 | 181.72 |
| 751 | 817.26 | 204.32 | 171.18 | 796 | 864.14 | 216.04 | 181.96 |
| 752 | 818.31 | 204.58 | 171.42 | 797 | 865.19 | 216.30 | 182.20 |
| 753 | 819.35 | 204.84 | 171.66 | 798 | 866.23 | 216.56 | 182.44 |
| 754 | 820.39 | 205.10 | 171.90 | 799 | 867.27 | 216.82 | 182.68 |
| 755 | 821.43 | 205.36 | 172.14 | 800 | 868.31 | 217.08 | 182.92 |
| 756 | 822.48 | 205.62 | 172.38 | 801 | 869.35 | 217.34 | 183.16 |
| 757 | 823.52 | 205.88 | 172.62 | 802 | 870.39 | 217.60 | 183.40 |
| 758 | 824.56 | 206.14 | 172.86 | 803 | 871.43 | 217.86 | 183.64 |
| 759 | 825.60 | 206.40 | 173.10 | 804 | 872.47 | 218.12 | 183.88 |
| 760 | 826.64 | 206.66 | 173.34 | 805 | 873.51 | 218.38 | 184.12 |
| 761 | 827.69 | 206.92 | 173.58 | 806 | 874.56 | 218.64 | 184.36 |
| 762 | 828.73 | 207.18 | 173.82 | 807 | 875.60 | 218.90 | 184.60 |
| 763 | 829.77 | 207.44 | 174.06 | 808 | 876.64 | 219.16 | 184.84 |
| 764 | 830.81 | 207.70 | 174.30 | 809 | 877.68 | 219.42 | 185.08 |

Table 1. CONSTANTS FOR $\overline{p}'$ (LONG FORMULA), AN APPROXIMATION OF THE UPPER BINOMIAL CONFIDENCE LIMIT $\overline{p}$; $c/n \leqslant .5$, $n \geqslant 20$, $c \leqslant 1,000$ (*Continued*)

| c | $\overline{m}$ | $\overline{a}$ | $\overline{b}$ | c | $\overline{m}$ | $\overline{a}$ | $\overline{b}$ |
|---|---|---|---|---|---|---|---|
| 810 | 878.72 | 219.68 | 185.32 | 855 | 925.53 | 231.38 | 196.12 |
| 811 | 879.76 | 219.94 | 185.56 | 856 | 926.57 | 231.64 | 196.36 |
| 812 | 880.80 | 220.20 | 185.80 | 857 | 927.61 | 231.90 | 196.60 |
| 813 | 881.84 | 220.46 | 186.04 | 858 | 928.65 | 232.16 | 196.84 |
| 814 | 882.88 | 220.72 | 186.28 | 859 | 929.69 | 232.42 | 197.08 |
| 815 | 883.92 | 220.98 | 186.52 | 860 | 930.73 | 232.68 | 197.32 |
| 816 | 884.96 | 221.24 | 186.76 | 861 | 931.77 | 232.94 | 197.56 |
| 817 | 886.00 | 221.50 | 187.00 | 862 | 932.81 | 233.20 | 197.80 |
| 818 | 887.05 | 221.76 | 187.24 | 863 | 933.85 | 233.46 | 198.04 |
| 819 | 888.09 | 222.02 | 187.48 | 864 | 934.89 | 233.72 | 198.28 |
| 820 | 889.13 | 222.28 | 187.72 | 865 | 935.93 | 233.98 | 198.52 |
| 821 | 890.17 | 222.54 | 187.96 | 866 | 936.97 | 234.24 | 198.76 |
| 822 | 891.21 | 222.80 | 188.20 | 867 | 938.01 | 234.50 | 199.00 |
| 823 | 892.25 | 223.06 | 188.44 | 868 | 939.05 | 234.76 | 199.24 |
| 824 | 893.29 | 223.32 | 188.68 | 869 | 940.09 | 235.02 | 199.48 |
| 825 | 894.33 | 223.58 | 188.92 | 870 | 941.13 | 235.28 | 199.72 |
| 826 | 895.37 | 223.84 | 189.16 | 871 | 942.17 | 235.54 | 199.96 |
| 827 | 896.41 | 224.10 | 189.40 | 872 | 943.20 | 235.80 | 200.20 |
| 828 | 897.45 | 224.36 | 189.64 | 873 | 944.24 | 236.06 | 200.44 |
| 829 | 898.49 | 224.62 | 189.88 | 874 | 945.28 | 236.32 | 200.68 |
| 830 | 899.53 | 224.88 | 190.12 | 875 | 946.32 | 236.58 | 200.92 |
| 831 | 900.57 | 225.14 | 190.36 | 876 | 947.36 | 236.84 | 201.16 |
| 832 | 901.61 | 225.40 | 190.60 | 877 | 948.40 | 237.10 | 201.40 |
| 833 | 902.65 | 225.66 | 190.84 | 878 | 949.44 | 237.36 | 201.64 |
| 834 | 903.69 | 225.92 | 191.08 | 879 | 950.48 | 237.62 | 201.88 |
| 835 | 904.73 | 226.18 | 191.32 | 880 | 951.52 | 237.88 | 202.12 |
| 836 | 905.77 | 226.44 | 191.56 | 881 | 952.56 | 238.14 | 202.36 |
| 837 | 906.81 | 226.70 | 191.80 | 882 | 953.60 | 238.40 | 202.60 |
| 838 | 907.85 | 226.96 | 192.04 | 883 | 954.64 | 238.66 | 202.84 |
| 839 | 908.89 | 227.22 | 192.28 | 884 | 955.68 | 238.92 | 203.08 |
| 840 | 909.93 | 227.48 | 192.52 | 885 | 956.71 | 239.18 | 203.32 |
| 841 | 910.97 | 227.74 | 192.76 | 886 | 957.75 | 239.44 | 203.56 |
| 842 | 912.01 | 228.00 | 193.00 | 887 | 958.79 | 239.70 | 203.80 |
| 843 | 913.05 | 228.26 | 193.24 | 888 | 959.83 | 239.96 | 204.04 |
| 844 | 914.09 | 228.52 | 193.48 | 889 | 960.87 | 240.22 | 204.28 |
| 845 | 915.13 | 228.78 | 193.72 | 890 | 961.91 | 240.48 | 204.52 |
| 846 | 916.17 | 229.04 | 193.96 | 891 | 962.95 | 240.74 | 204.76 |
| 847 | 917.21 | 229.30 | 194.20 | 892 | 963.99 | 241.00 | 205.00 |
| 848 | 918.25 | 229.56 | 194.44 | 893 | 965.03 | 241.26 | 205.24 |
| 849 | 919.29 | 229.82 | 194.68 | 894 | 966.07 | 241.52 | 205.48 |
| 850 | 920.33 | 230.08 | 194.92 | 895 | 967.10 | 241.78 | 205.72 |
| 851 | 921.37 | 230.34 | 195.16 | 896 | 968.14 | 242.04 | 205.96 |
| 852 | 922.41 | 230.60 | 195.40 | 897 | 969.18 | 242.30 | 206.20 |
| 853 | 923.45 | 230.86 | 195.64 | 898 | 970.22 | 242.56 | 206.44 |
| 854 | 924.49 | 231.12 | 195.88 | 899 | 971.26 | 242.81 | 206.69 |

# Table 1.  CONSTANTS FOR $\overline{p}'$ (LONG FORMULA), AN APPROXIMATION OF THE UPPER BINOMIAL CONFIDENCE LIMIT $\overline{p}$; $c/n \leqslant .5$, $n \geqslant 20$, $c \leqslant 1,000$ (*Continued*)

| c | $\overline{m}$ | $\overline{a}$ | $\overline{b}$ | c | $\overline{m}$ | $\overline{a}$ | $\overline{b}$ |
|---|---|---|---|---|---|---|---|
| 900 | 972.30 | 243.07 | 206.93 | 945 | 1019.02 | 254.76 | 217.74 |
| 901 | 973.34 | 243.33 | 207.17 | 946 | 1020.06 | 255.01 | 217.99 |
| 902 | 974.38 | 243.59 | 207.41 | 947 | 1021.10 | 255.27 | 218.23 |
| 903 | 975.41 | 243.85 | 207.65 | 948 | 1022.13 | 255.53 | 218.47 |
| 904 | 976.45 | 244.11 | 207.89 | 949 | 1023.17 | 255.79 | 218.71 |
| 905 | 977.49 | 244.37 | 208.13 | 950 | 1024.21 | 256.05 | 218.95 |
| 906 | 978.53 | 244.63 | 208.37 | 951 | 1025.25 | 256.31 | 219.19 |
| 907 | 979.57 | 244.89 | 208.61 | 952 | 1026.29 | 256.57 | 219.43 |
| 908 | 980.61 | 245.15 | 208.85 | 953 | 1027.32 | 256.83 | 219.67 |
| 909 | 981.65 | 245.41 | 209.09 | 954 | 1028.36 | 257.09 | 219.91 |
| 910 | 982.68 | 245.67 | 209.33 | 955 | 1029.40 | 257.35 | 220.15 |
| 911 | 983.72 | 245.93 | 209.57 | 956 | 1030.44 | 257.61 | 220.39 |
| 912 | 984.76 | 246.19 | 209.81 | 957 | 1031.47 | 257.87 | 220.63 |
| 913 | 985.80 | 246.45 | 210.05 | 958 | 1032.51 | 258.13 | 220.87 |
| 914 | 986.84 | 246.71 | 210.29 | 959 | 1033.55 | 258.39 | 221.11 |
| 915 | 987.88 | 246.97 | 210.53 | 960 | 1034.59 | 258.65 | 221.35 |
| 916 | 988.92 | 247.23 | 210.77 | 961 | 1035.62 | 258.91 | 221.59 |
| 917 | 989.95 | 247.49 | 211.01 | 962 | 1036.66 | 259.17 | 221.83 |
| 918 | 990.99 | 247.75 | 211.25 | 963 | 1037.70 | 259.42 | 222.08 |
| 919 | 992.03 | 248.01 | 211.49 | 964 | 1038.74 | 259.68 | 222.32 |
| 920 | 993.07 | 248.27 | 211.73 | 965 | 1039.77 | 259.94 | 222.56 |
| 921 | 994.11 | 248.53 | 211.97 | 966 | 1040.81 | 260.20 | 222.80 |
| 922 | 995.15 | 248.79 | 212.21 | 967 | 1041.85 | 260.46 | 223.04 |
| 923 | 996.18 | 249.05 | 212.45 | 968 | 1042.89 | 260.72 | 223.28 |
| 924 | 997.22 | 249.31 | 212.69 | 969 | 1043.92 | 260.98 | 223.52 |
| 925 | 998.26 | 249.57 | 212.93 | 970 | 1044.96 | 261.24 | 223.76 |
| 926 | 999.30 | 249.82 | 213.18 | 971 | 1046.00 | 261.50 | 224.00 |
| 927 | 1000.34 | 250.08 | 213.42 | 972 | 1047.03 | 261.76 | 224.24 |
| 928 | 1001.38 | 250.34 | 213.66 | 973 | 1048.07 | 262.02 | 224.48 |
| 929 | 1002.41 | 250.60 | 213.90 | 974 | 1049.11 | 262.28 | 224.72 |
| 930 | 1003.45 | 250.86 | 214.14 | 975 | 1050.15 | 262.54 | 224.96 |
| 931 | 1004.49 | 251.12 | 214.38 | 976 | 1051.18 | 262.80 | 225.20 |
| 932 | 1005.53 | 251.38 | 214.62 | 977 | 1052.22 | 263.06 | 225.44 |
| 933 | 1006.57 | 251.64 | 214.86 | 978 | 1053.26 | 263.31 | 225.69 |
| 934 | 1007.60 | 251.90 | 215.10 | 979 | 1054.30 | 263.57 | 225.93 |
| 935 | 1008.64 | 252.16 | 215.34 | 980 | 1055.33 | 263.83 | 226.17 |
| 936 | 1009.68 | 252.42 | 215.58 | 981 | 1056.37 | 264.09 | 226.41 |
| 937 | 1010.72 | 252.68 | 215.82 | 982 | 1057.41 | 264.35 | 226.65 |
| 938 | 1011.76 | 252.94 | 216.06 | 983 | 1058.44 | 264.61 | 226.89 |
| 939 | 1012.79 | 253.20 | 216.30 | 984 | 1059.48 | 264.87 | 227.13 |
| 940 | 1013.83 | 253.46 | 216.54 | 985 | 1060.52 | 265.13 | 227.37 |
| 941 | 1014.87 | 253.72 | 216.78 | 986 | 1061.55 | 265.39 | 227.61 |
| 942 | 1015.91 | 253.98 | 217.02 | 987 | 1062.59 | 265.65 | 227.85 |
| 943 | 1016.95 | 254.24 | 217.26 | 988 | 1063.63 | 265.91 | 228.09 |
| 944 | 1017.98 | 254.50 | 217.50 | 989 | 1064.67 | 266.17 | 228.33 |

Table 1. **CONSTANTS FOR $\overline{p}'$ (LONG FORMULA), AN APPROXIMATION OF THE UPPER BINOMIAL CONFIDENCE LIMIT $\overline{p}$; $c/n \leqslant .5$, $n \geqslant 20$, $c \leqslant 1,000$** (*Continued*)

| c | $\overline{m}$ | $\overline{a}$ | $\overline{b}$ | c | $\overline{m}$ | $\overline{a}$ | $\overline{b}$ |
|---|---|---|---|---|---|---|---|
| 990 | 1065.70 | 266.43 | 228.57 | 995 | 1070.89 | 267.72 | 229.78 |
| 991 | 1066.74 | 266.68 | 228.82 | 996 | 1071.92 | 267.98 | 230.02 |
| 992 | 1067.78 | 266.94 | 229.06 | 997 | 1072.96 | 268.24 | 230.26 |
| 993 | 1068.81 | 267.20 | 229.30 | 998 | 1074.00 | 268.50 | 230.50 |
| 994 | 1069.85 | 267.46 | 229.54 | 999 | 1075.03 | 268.76 | 230.74 |
| | | | | 1000 | 1076.07 | 269.02 | 230.98 |

# Table 1. CONSTANTS FOR $\overline{p}'$ (LONG FORMULA), AN APPROXIMATION OF THE UPPER BINOMIAL CONFIDENCE LIMIT $\overline{p}$; $c/n \leqslant .5$, $n \geqslant 20$, $c \leqslant 1,000$ (*Continued*)

| c | $\overline{m}$ | $\overline{a}$ | $\overline{b}$ | c | $\overline{m}$ | $\overline{a}$ | $\overline{b}$ |
|---|---|---|---|---|---|---|---|
| 0 | 3.6889 | 0.622 | -1.220 | 45 | 60.214 | 16.120 | 8.609 |
| 1 | 5.5717 | 0.966 | -1.318 | 46 | 61.358 | 16.377 | 8.787 |
| 2 | 7.2247 | 1.291 | -1.321 | 47 | 62.500 | 16.747 | 9.091 |
| 3 | 8.7673 | 1.599 | -1.287 | 48 | 63.641 | 17.041 | 9.312 |
| 4 | 10.242 | 1.957 | -1.162 | 49 | 64.781 | 17.402 | 9.607 |
| 5 | 11.668 | 2.288 | -1.043 | 50 | 65.919 | 17.703 | 9.838 |
| 6 | 13.059 | 2.646 | -0.876 | 51 | 67.056 | 18.113 | 10.187 |
| 7 | 14.423 | 3.016 | -0.682 | 52 | 68.191 | 18.363 | 10.362 |
| 8 | 15.763 | 3.398 | -0.463 | 53 | 69.326 | 18.715 | 10.650 |
| 9 | 17.085 | 3.794 | -0.218 | 54 | 70.459 | 18.976 | 10.839 |
| 10 | 18.390 | 4.211 | 0.059 | 55 | 71.590 | 19.416 | 11.224 |
| 11 | 19.682 | 4.565 | 0.270 | 56 | 72.721 | 19.698 | 11.44 |
| 12 | 20.962 | 4.918 | 0.485 | 57 | 73.850 | 20.066 | 11.74 |
| 13 | 22.230 | 5.262 | 0.697 | 58 | 74.979 | 20.315 | 11.92 |
| 14 | 23.490 | 5.615 | 0.923 | 59 | 76.106 | 20.649 | 12.20 |
| 15 | 24.740 | 5.968 | 1.153 | 60 | 77.232 | 21.050 | 12.54 |
| 16 | 25.983 | 6.310 | 1.374 | 61 | 78.357 | 21.40 | 12.84 |
| 17 | 27.219 | 6.649 | 1.597 | 62 | 79.481 | 21.70 | 13.06 |
| 18 | 28.448 | 6.999 | 1.836 | 63 | 80.605 | 22.09 | 13.40 |
| 19 | 29.671 | 7.349 | 2.076 | 64 | 81.727 | 22.31 | 13.55 |
| 20 | 30.888 | 7.675 | 2.292 | 65 | 82.848 | 22.60 | 13.78 |
| 21 | 32.101 | 8.035 | 2.551 | 66 | 83.968 | 22.96 | 14.08 |
| 22 | 33.308 | 8.351 | 2.761 | 67 | 85.088 | 23.41 | 14.48 |
| 23 | 34.511 | 8.696 | 3.007 | 68 | 86.206 | 23.68 | 14.69 |
| 24 | 35.710 | 9.036 | 3.248 | 69 | 87.324 | 23.96 | 14.91 |
| 25 | 36.905 | 9.370 | 3.486 | 70 | 88.441 | 24.22 | 15.10 |
| 26 | 38.096 | 9.717 | 3.740 | 71 | 89.557 | 24.71 | 15.55 |
| 27 | 39.284 | 10.049 | 3.979 | 72 | 90.672 | 24.99 | 15.77 |
| 28 | 40.468 | 10.418 | 4.260 | 73 | 91.787 | 25.35 | 16.08 |
| 29 | 41.649 | 10.727 | 4.477 | 74 | 92.900 | 25.71 | 16.39 |
| 30 | 42.827 | 11.068 | 4.731 | 75 | 94.013 | 25.89 | 16.50 |
| 31 | 44.002 | 11.384 | 4.959 | 76 | 95.126 | 26.34 | 16.90 |
| 32 | 45.174 | 11.711 | 5.199 | 77 | 96.237 | 26.57 | 17.07 |
| 33 | 46.344 | 12.089 | 5.498 | 78 | 97.348 | 26.93 | 17.38 |
| 34 | 47.512 | 12.413 | 5.739 | 79 | 98.458 | 27.30 | 17.70 |
| 35 | 48.677 | 12.746 | 5.990 | 80 | 99.567 | 27.55 | 17.89 |
| 36 | 49.839 | 13.092 | 6.256 | 81 | 100.68 | 27.90 | 18.18 |
| 37 | 51.000 | 13.423 | 6.507 | 82 | 101.78 | 28.26 | 18.49 |
| 38 | 52.158 | 13.743 | 6.748 | 83 | 102.89 | 28.50 | 18.67 |
| 39 | 53.315 | 14.095 | 7.025 | 84 | 104.00 | 28.86 | 18.98 |
| 40 | 54.469 | 14.435 | 7.290 | 85 | 105.10 | 29.30 | 19.37 |
| 41 | 55.621 | 14.751 | 7.529 | 86 | 106.21 | 29.58 | 19.60 |
| 42 | 56.772 | 15.100 | 7.805 | 87 | 107.31 | 30.06 | 20.04 |
| 43 | 57.921 | 15.418 | 8.049 | 88 | 108.42 | 30.28 | 20.20 |
| 44 | 59.068 | 15.767 | 8.326 | 89 | 109.52 | 30.40 | 20.26 |

Table 1.   CONSTANTS FOR $\bar{p}'$ (LONG FORMULA), AN APPROXIMATION OF
THE UPPER BINOMIAL CONFIDENCE LIMIT $\bar{p}$; $c/n \leqslant .5$, $n \geqslant 20$,
$c \leqslant 1,000$ (*Continued*)

| c | $\bar{m}$ | $\bar{a}$ | $\bar{b}$ | c | $\bar{m}$ | $\bar{a}$ | $\bar{b}$ |
|---|---|---|---|---|---|---|---|
| 90 | 110.63 | 30.94 | 20.75 | 135 | 159.79 | 39.95 | 27.55 |
| 91 | 111.73 | 31.08 | 20.84 | 136 | 160.87 | 40.22 | 27.78 |
| 92 | 112.83 | 31.60 | 21.32 | 137 | 161.96 | 40.49 | 28.01 |
| 93 | 113.93 | 31.73 | 21.38 | 138 | 163.04 | 40.76 | 28.24 |
| 94 | 115.03 | 32.06 | 21.66 | 139 | 164.12 | 41.03 | 28.47 |
| 95 | 116.13 | 32.46 | 22.02 | 140 | 165.21 | 41.31 | 28.69 |
| 96 | 117.23 | 32.77 | 22.28 | 141 | 166.29 | 41.58 | 28.92 |
| 97 | 118.33 | 33.00 | 22.46 | 142 | 167.37 | 41.85 | 29.15 |
| 98 | 119.43 | 33.39 | 22.80 | 143 | 168.45 | 42.12 | 29.38 |
| 99 | 120.53 | 33.95 | 23.33 | 144 | 169.53 | 42.39 | 29.61 |
| 100 | 121.63 | 34.07 | 23.39 | 145 | 170.62 | 42.66 | 29.84 |
| 101 | 122.72 | 30.69 | 19.81 | 146 | 171.70 | 42.93 | 30.07 |
| 102 | 123.82 | 30.96 | 20.04 | 147 | 172.78 | 43.20 | 30.30 |
| 103 | 124.92 | 31.23 | 20.27 | 148 | 173.86 | 43.47 | 30.53 |
| 104 | 126.01 | 31.51 | 20.49 | 149 | 174.94 | 43.74 | 30.76 |
| 105 | 127.11 | 31.78 | 20.72 | 150 | 176.02 | 44.01 | 30.99 |
| 106 | 128.20 | 32.06 | 20.94 | 151 | 177.10 | 44.28 | 31.22 |
| 107 | 129.30 | 32.33 | 21.17 | 152 | 178.18 | 44.55 | 31.45 |
| 108 | 130.39 | 32.60 | 21.40 | 153 | 179.26 | 44.82 | 31.68 |
| 109 | 131.49 | 32.88 | 21.62 | 154 | 180.33 | 45.09 | 31.91 |
| 110 | 132.58 | 33.15 | 21.85 | 155 | 181.41 | 45.36 | 32.14 |
| 111 | 133.67 | 33.42 | 22.08 | 156 | 182.49 | 45.63 | 32.37 |
| 112 | 134.77 | 33.70 | 22.30 | 157 | 183.57 | 45.90 | 32.60 |
| 113 | 135.86 | 33.97 | 22.53 | 158 | 184.65 | 46.17 | 32.83 |
| 114 | 136.95 | 34.24 | 22.76 | 159 | 185.73 | 46.43 | 33.07 |
| 115 | 138.04 | 34.51 | 22.99 | 160 | 186.80 | 46.70 | 33.30 |
| 116 | 139.13 | 34.79 | 23.21 | 161 | 187.89 | 46.97 | 33.53 |
| 117 | 140.22 | 35.06 | 23.44 | 162 | 188.97 | 47.24 | 33.76 |
| 118 | 141.31 | 35.33 | 23.67 | 163 | 190.05 | 47.51 | 33.99 |
| 119 | 142.40 | 35.60 | 23.90 | 164 | 191.12 | 47.78 | 34.22 |
| 120 | 143.49 | 35.88 | 24.12 | 165 | 192.20 | 48.05 | 34.45 |
| 121 | 144.58 | 36.15 | 24.35 | 166 | 193.28 | 48.32 | 34.68 |
| 122 | 145.67 | 36.42 | 24.58 | 167 | 194.35 | 48.59 | 34.91 |
| 123 | 146.76 | 36.69 | 24.81 | 168 | 195.43 | 48.86 | 35.14 |
| 124 | 147.84 | 36.97 | 25.03 | 169 | 196.50 | 49.13 | 35.37 |
| 125 | 148.93 | 37.24 | 25.26 | 170 | 197.58 | 49.39 | 35.61 |
| 126 | 150.02 | 37.51 | 25.49 | 171 | 198.65 | 49.66 | 35.84 |
| 127 | 151.11 | 37.78 | 25.72 | 172 | 199.73 | 49.93 | 36.07 |
| 128 | 152.19 | 38.05 | 25.95 | 173 | 200.80 | 50.20 | 36.30 |
| 129 | 153.28 | 38.32 | 26.18 | 174 | 201.88 | 50.47 | 36.53 |
| 130 | 154.36 | 38.60 | 26.40 | 175 | 202.95 | 50.74 | 36.76 |
| 131 | 155.45 | 38.87 | 26.63 | 176 | 204.02 | 51.01 | 36.99 |
| 132 | 156.54 | 39.14 | 26.86 | 177 | 205.10 | 51.27 | 37.23 |
| 133 | 157.62 | 39.41 | 27.09 | 178 | 206.17 | 51.54 | 37.46 |
| 134 | 158.70 | 39.68 | 27.32 | 179 | 207.24 | 51.81 | 37.69 |

Table 1. CONSTANTS FOR $\bar{p}'$ (LONG FORMULA), AN APPROXIMATION OF THE UPPER BINOMIAL CONFIDENCE LIMIT $\bar{p}$; $c/n \leqslant .5$, $n \geqslant 20$, $c \leqslant 1,000$ (*Continued*)

Section 3

$\gamma_1 = 97.5\%$
$\gamma_2 = 95\%$

| c | $\bar{m}$ | $\bar{a}$ | $\bar{b}$ | c | $\bar{m}$ | $\bar{a}$ | $\bar{b}$ |
|---|---|---|---|---|---|---|---|
| 180 | 208.32 | 52.08 | 37.92 | 225 | 256.41 | 64.10 | 48.40 |
| 181 | 209.39 | 52.35 | 38.15 | 226 | 257.48 | 64.37 | 48.63 |
| 182 | 210.46 | 52.62 | 38.38 | 227 | 258.54 | 64.64 | 48.86 |
| 183 | 211.53 | 52.88 | 38.62 | 228 | 259.61 | 64.90 | 49.10 |
| 184 | 212.61 | 53.15 | 38.85 | 229 | 260.67 | 65.17 | 49.33 |
| 185 | 213.68 | 53.42 | 39.08 | 230 | 261.74 | 65.43 | 49.57 |
| 186 | 214.75 | 53.69 | 39.31 | 231 | 262.80 | 65.70 | 49.80 |
| 187 | 215.82 | 53.96 | 39.54 | 232 | 263.87 | 65.97 | 50.03 |
| 188 | 216.89 | 54.22 | 39.78 | 233 | 264.93 | 66.23 | 50.27 |
| 189 | 217.96 | 54.49 | 40.01 | 234 | 265.99 | 66.50 | 50.50 |
| 190 | 219.03 | 54.76 | 40.24 | 235 | 267.06 | 66.76 | 50.74 |
| 191 | 220.11 | 55.03 | 40.47 | 236 | 268.12 | 67.03 | 50.97 |
| 192 | 221.18 | 55.29 | 40.71 | 237 | 269.18 | 67.30 | 51.20 |
| 193 | 222.25 | 55.56 | 40.94 | 238 | 270.25 | 67.56 | 51.44 |
| 194 | 223.32 | 55.83 | 41.17 | 239 | 271.31 | 67.83 | 51.67 |
| 195 | 224.39 | 56.10 | 41.40 | 240 | 272.37 | 68.09 | 51.91 |
| 196 | 225.46 | 56.36 | 41.64 | 241 | 273.44 | 68.36 | 52.14 |
| 197 | 226.53 | 56.63 | 41.87 | 242 | 274.50 | 68.63 | 52.37 |
| 198 | 227.60 | 56.90 | 42.10 | 243 | 275.56 | 68.89 | 52.61 |
| 199 | 228.67 | 57.17 | 42.33 | 244 | 276.63 | 69.16 | 52.84 |
| 200 | 229.73 | 57.43 | 42.57 | 245 | 277.69 | 69.42 | 53.08 |
| 201 | 230.80 | 57.70 | 42.80 | 246 | 278.75 | 69.69 | 53.31 |
| 202 | 231.87 | 57.97 | 43.03 | 247 | 279.81 | 69.95 | 53.55 |
| 203 | 232.94 | 58.24 | 43.26 | 248 | 280.88 | 70.22 | 53.78 |
| 204 | 234.01 | 58.50 | 43.50 | 249 | 281.94 | 70.48 | 54.02 |
| 205 | 235.08 | 58.77 | 43.73 | 250 | 283.00 | 70.75 | 54.25 |
| 206 | 236.15 | 59.04 | 43.96 | 251 | 284.06 | 71.02 | 54.48 |
| 207 | 237.21 | 59.30 | 44.20 | 252 | 285.12 | 71.28 | 54.72 |
| 208 | 238.28 | 59.57 | 44.43 | 253 | 286.18 | 71.55 | 54.95 |
| 209 | 239.35 | 59.84 | 44.66 | 254 | 287.25 | 71.81 | 55.19 |
| 210 | 240.42 | 60.10 | 44.90 | 255 | 288.31 | 72.08 | 55.42 |
| 211 | 241.49 | 60.37 | 45.13 | 256 | 289.37 | 72.34 | 55.66 |
| 212 | 242.55 | 60.64 | 45.36 | 257 | 290.43 | 72.61 | 55.89 |
| 213 | 243.62 | 60.90 | 45.60 | 258 | 291.49 | 72.87 | 56.13 |
| 214 | 244.69 | 61.17 | 45.83 | 259 | 292.55 | 73.14 | 56.36 |
| 215 | 245.75 | 61.44 | 46.06 | 260 | 293.61 | 73.40 | 56.60 |
| 216 | 246.82 | 61.70 | 46.30 | 261 | 294.67 | 73.67 | 56.83 |
| 217 | 247.89 | 61.97 | 46.53 | 262 | 295.73 | 73.93 | 57.07 |
| 218 | 248.95 | 62.24 | 46.76 | 263 | 296.79 | 74.20 | 57.30 |
| 219 | 250.02 | 62.50 | 47.00 | 264 | 297.85 | 74.46 | 57.54 |
| 220 | 251.08 | 62.77 | 47.23 | 265 | 298.91 | 74.73 | 57.77 |
| 221 | 252.15 | 63.04 | 47.46 | 266 | 299.97 | 74.99 | 58.01 |
| 222 | 253.22 | 63.30 | 47.70 | 267 | 301.03 | 75.26 | 58.24 |
| 223 | 254.28 | 63.57 | 47.93 | 268 | 302.09 | 75.52 | 58.48 |
| 224 | 255.35 | 63.84 | 48.16 | 269 | 303.15 | 75.79 | 58.71 |

**Table 1. CONSTANTS FOR $\bar{p}'$ (LONG FORMULA), AN APPROXIMATION OF THE UPPER BINOMIAL CONFIDENCE LIMIT $\bar{p}$; $c/n \leqslant .5$, $n \geqslant 20$. $c \leqslant 1,000$ (Continued)**

| c | $\overline{m}$ | $\overline{a}$ | $\overline{b}$ | c | $\overline{m}$ | $\overline{a}$ | $\overline{b}$ |
|---|---|---|---|---|---|---|---|
| 270 | 304.21 | 76.05 | 58.95 | 315 | 351.79 | 87.95 | 69.: |
| 271 | 305.27 | 76.32 | 59.18 | 316 | 352.84 | 88.21 | 69.7 |
| 272 | 306.33 | 76.58 | 59.42 | 317 | 353.90 | 88.47 | 70.0 |
| 273 | 307.39 | 76.85 | 59.65 | 318 | 354.95 | 88.74 | 70.2 |
| 274 | 308.45 | 77.11 | 59.89 | 319 | 356.01 | 89.00 | 70.5 |
| 275 | 309.51 | 77.38 | 60.12 | 320 | 357.06 | 89.27 | 70.7 |
| 276 | 310.57 | 77.64 | 60.36 | 321 | 358.12 | 89.53 | 70.9 |
| 277 | 311.63 | 77.91 | 60.59 | 322 | 359.17 | 89.79 | 71.2 |
| 278 | 312.69 | 78.17 | 60.83 | 323 | 360.23 | 90.06 | 71.4 |
| 279 | 313.74 | 78.44 | 61.06 | 324 | 361.28 | 90.32 | 71.6 |
| 280 | 314.80 | 78.70 | 61.30 | 325 | 362.34 | 90.58 | 71.9 |
| 281 | 315.86 | 78.97 | 61.53 | 326 | 363.39 | 90.85 | 72.1 |
| 282 | 316.92 | 79.23 | 61.77 | 327 | 364.44 | 91.11 | 72.3 |
| 283 | 317.98 | 79.49 | 62.01 | 328 | 365.50 | 91.37 | 72.6 |
| 284 | 319.04 | 79.76 | 62.24 | 329 | 366.55 | 91.64 | 72.8 |
| 285 | 320.09 | 80.02 | 62.48 | 330 | 367.61 | 91.90 | 73.1 |
| 286 | 321.15 | 80.29 | 62.71 | 331 | 368.66 | 92.17 | 73.3 |
| 287 | 322.21 | 80.55 | 62.95 | 332 | 369.71 | 92.43 | 73.5 |
| 288 | 323.27 | 80.82 | 63.18 | 333 | 370.77 | 92.69 | 73.8 |
| 289 | 324.32 | 81.08 | 63.42 | 334 | 371.82 | 92.96 | 74.0 |
| 290 | 325.38 | 81.35 | 63.65 | 335 | 372.87 | 93.22 | 74.2 |
| 291 | 326.44 | 81.61 | 63.89 | 336 | 373.93 | 93.48 | 74.5 |
| 292 | 327.50 | 81.87 | 64.13 | 337 | 374.98 | 93.75 | 74.7 |
| 293 | 328.55 | 82.14 | 64.36 | 338 | 376.03 | 94.01 | 74.9 |
| 294 | 329.61 | 82.40 | 64.60 | 339 | 377.09 | 94.27 | 75.2 |
| 295 | 330.67 | 82.67 | 64.83 | 340 | 378.14 | 94.54 | 75.4 |
| 296 | 331.73 | 82.93 | 65.07 | 341 | 379.19 | 94.80 | 75.7 |
| 297 | 332.78 | 83.20 | 65.30 | 342 | 380.25 | 95.06 | 75.9 |
| 298 | 333.84 | 83.46 | 65.54 | 343 | 381.30 | 95.32 | 76.1 |
| 299 | 334.90 | 83.72 | 65.78 | 344 | 382.35 | 95.59 | 76.4 |
| 300 | 335.95 | 83.99 | 66.01 | 345 | 383.41 | 95.85 | 76.0 |
| 301 | 337.01 | 84.25 | 66.25 | 346 | 384.46 | 96.11 | 76.8 |
| 302 | 338.06 | 84.52 | 66.48 | 347 | 385.51 | 96.38 | 77.1 |
| 303 | 339.12 | 84.78 | 66.72 | 348 | 386.56 | 96.64 | 77.3 |
| 304 | 340.18 | 85.04 | 66.96 | 349 | 387.62 | 96.90 | 77.6 |
| 305 | 341.23 | 85.31 | 67.19 | 350 | 388.67 | 97.17 | 77.8 |
| 306 | 342.29 | 85.57 | 67.43 | 351 | 389.72 | 97.43 | 78.0 |
| 307 | 343.35 | 85.84 | 67.66 | 352 | 390.77 | 97.69 | 78.3 |
| 308 | 344.40 | 86.10 | 67.90 | 353 | 391.82 | 97.96 | 78.5 |
| 309 | 345.46 | 86.36 | 68.14 | 354 | 392.88 | 98.22 | 78.7 |
| 310 | 346.51 | 86.63 | 68.37 | 355 | 393.93 | 98.48 | 79.0 |
| 311 | 347.57 | 86.89 | 68.61 | 356 | 394.98 | 98.75 | 79.2 |
| 312 | 348.62 | 87.16 | 68.84 | 357 | 396.03 | 99.01 | 79.4 |
| 313 | 349.68 | 87.42 | 69.08 | 358 | 397.08 | 99.27 | 79.7 |
| 314 | 350.73 | 87.68 | 69.32 | 359 | 398.14 | 99.53 | 79.9 |

Table 1. CONSTANTS FOR $\overline{p}'$ (LONG FORMULA), AN APPROXIMATION OF
THE UPPER BINOMIAL CONFIDENCE LIMIT $\overline{p}$; $c/n \leqslant .5$, $n \geqslant 20$,
$c \leqslant 1,000$ (*Continued*)

Section 3

$\gamma_1 = 97.5\%$
$\gamma_2 = 95\%$

| c | $\overline{m}$ | $\overline{a}$ | $\overline{b}$ | c | $\overline{m}$ | $\overline{a}$ | $\overline{b}$ |
|---|---|---|---|---|---|---|---|
| 360 | 399.19 | 99.80 | 80.20 | 405 | 446.44 | 111.61 | 90.89 |
| 361 | 400.24 | 100.06 | 80.44 | 406 | 447.49 | 111.87 | 91.13 |
| 362 | 401.29 | 100.32 | 80.68 | 407 | 448.54 | 112.13 | 91.37 |
| 363 | 402.34 | 100.59 | 80.91 | 408 | 449.59 | 112.40 | 91.60 |
| 364 | 403.39 | 100.85 | 81.15 | 409 | 450.63 | 112.66 | 91.84 |
| 365 | 404.44 | 101.11 | 81.39 | 410 | 451.68 | 112.92 | 92.08 |
| 366 | 405.50 | 101.37 | 81.63 | 411 | 452.73 | 113.18 | 92.32 |
| 367 | 406.55 | 101.64 | 81.86 | 412 | 453.78 | 113.44 | 92.56 |
| 368 | 407.60 | 101.90 | 82.10 | 413 | 454.83 | 113.71 | 92.79 |
| 369 | 408.65 | 102.16 | 82.34 | 414 | 455.88 | 113.97 | 93.03 |
| 370 | 409.70 | 102.42 | 82.58 | 415 | 456.92 | 114.23 | 93.27 |
| 371 | 410.75 | 102.69 | 82.81 | 416 | 457.97 | 114.49 | 93.51 |
| 372 | 411.80 | 102.95 | 83.05 | 417 | 459.02 | 114.75 | 93.75 |
| 373 | 412.85 | 103.21 | 83.29 | 418 | 460.07 | 115.02 | 93.98 |
| 374 | 413.90 | 103.48 | 83.52 | 419 | 461.12 | 115.28 | 94.22 |
| 375 | 414.95 | 103.74 | 83.76 | 420 | 462.16 | 115.54 | 94.46 |
| 376 | 416.00 | 104.00 | 84.00 | 421 | 463.21 | 115.80 | 94.70 |
| 377 | 417.05 | 104.26 | 84.24 | 422 | 464.26 | 116.06 | 94.94 |
| 378 | 418.10 | 104.53 | 84.47 | 423 | 465.31 | 116.33 | 95.17 |
| 379 | 419.15 | 104.79 | 84.71 | 424 | 466.35 | 116.59 | 95.41 |
| 380 | 420.20 | 105.05 | 84.95 | 425 | 467.40 | 116.85 | 95.65 |
| 381 | 421.26 | 105.31 | 85.19 | 426 | 468.45 | 117.11 | 95.89 |
| 382 | 422.31 | 105.58 | 85.42 | 427 | 469.50 | 117.37 | 96.13 |
| 383 | 423.36 | 105.84 | 85.66 | 428 | 470.54 | 117.64 | 96.36 |
| 384 | 424.41 | 106.10 | 85.90 | 429 | 471.59 | 117.90 | 96.60 |
| 385 | 425.46 | 106.36 | 86.14 | 430 | 472.64 | 118.16 | 96.84 |
| 386 | 426.50 | 106.63 | 86.37 | 431 | 473.69 | 118.42 | 97.08 |
| 387 | 427.55 | 106.89 | 86.61 | 432 | 474.73 | 118.68 | 97.32 |
| 388 | 428.60 | 107.15 | 86.85 | 433 | 475.78 | 118.94 | 97.56 |
| 389 | 429.65 | 107.41 | 87.09 | 434 | 476.83 | 119.21 | 97.79 |
| 390 | 430.70 | 107.68 | 87.32 | 435 | 477.87 | 119.47 | 98.03 |
| 391 | 431.75 | 107.94 | 87.56 | 436 | 478.92 | 119.73 | 98.27 |
| 392 | 432.80 | 108.20 | 87.80 | 437 | 479.97 | 119.99 | 98.51 |
| 393 | 433.85 | 108.46 | 88.04 | 438 | 481.01 | 120.25 | 98.75 |
| 394 | 434.90 | 108.73 | 88.27 | 439 | 482.06 | 120.52 | 98.98 |
| 395 | 435.95 | 108.99 | 88.51 | 440 | 483.11 | 120.78 | 99.22 |
| 396 | 437.00 | 109.25 | 88.75 | 441 | 484.15 | 121.04 | 99.46 |
| 397 | 438.05 | 109.51 | 88.99 | 442 | 485.20 | 121.30 | 99.70 |
| 398 | 439.10 | 109.77 | 89.23 | 443 | 486.25 | 121.56 | 99.94 |
| 399 | 440.15 | 110.04 | 89.46 | 444 | 487.29 | 121.82 | 100.18 |
| 400 | 441.20 | 110.30 | 89.70 | 445 | 488.34 | 122.08 | 100.42 |
| 401 | 442.25 | 110.56 | 89.94 | 446 | 489.39 | 122.35 | 100.65 |
| 402 | 443.29 | 110.82 | 90.18 | 447 | 490.43 | 122.61 | 100.89 |
| 403 | 444.34 | 111.09 | 90.41 | 448 | 491.48 | 122.87 | 101.13 |
| 404 | 445.39 | 111.35 | 90.65 | 449 | 492.53 | 123.13 | 101.37 |

Table 1.  CONSTANTS FOR $\bar{p}'$ (LONG FORMULA), AN APPROXIMATION OF THE UPPER BINOMIAL CONFIDENCE LIMIT $\bar{p}$; $c/n \leqslant .5$, $n \geqslant 20$, $c \leqslant 1,000$ (*Continued*)

| c | $\overline{m}$ | $\overline{a}$ | $\overline{b}$ | c | $\overline{m}$ | $\overline{a}$ | $\overline{b}$ |
|---|---|---|---|---|---|---|---|
| 450 | 493.57 | 123.39 | 101.61 | 495 | 540.60 | 135.15 | 112.35 |
| 451 | 494.62 | 123.65 | 101.85 | 496 | 541.64 | 135.41 | 112.59 |
| 452 | 495.66 | 123.92 | 102.08 | 497 | 542.69 | 135.67 | 112.83 |
| 453 | 496.71 | 124.18 | 102.32 | 498 | 543.73 | 135.93 | 113.07 |
| 454 | 497.76 | 124.44 | 102.56 | 499 | 544.77 | 136.19 | 113.31 |
| 455 | 498.80 | 124.70 | 102.80 | 500 | 545.82 | 136.45 | 113.55 |
| 456 | 499.85 | 124.96 | 103.04 | 501 | 546.86 | 136.72 | 113.78 |
| 457 | 500.89 | 125.22 | 103.28 | 502 | 547.91 | 136.98 | 114.02 |
| 458 | 501.94 | 125.48 | 103.52 | 503 | 548.95 | 137.24 | 114.26 |
| 459 | 502.98 | 125.75 | 103.75 | 504 | 549.99 | 137.50 | 114.50 |
| 460 | 504.03 | 126.01 | 103.99 | 505 | 551.04 | 137.76 | 114.74 |
| 461 | 505.08 | 126.27 | 104.23 | 506 | 552.08 | 138.02 | 114.98 |
| 462 | 506.12 | 126.53 | 104.47 | 507 | 553.12 | 138.28 | 115.22 |
| 463 | 507.17 | 126.79 | 104.71 | 508 | 554.17 | 138.54 | 115.46 |
| 464 | 508.21 | 127.05 | 104.95 | 509 | 555.21 | 138.80 | 115.70 |
| 465 | 509.26 | 127.31 | 105.19 | 510 | 556.25 | 139.06 | 115.94 |
| 466 | 510.30 | 127.58 | 105.42 | 511 | 557.30 | 139.32 | 116.18 |
| 467 | 511.35 | 127.84 | 105.66 | 512 | 558.34 | 139.59 | 116.41 |
| 468 | 512.39 | 128.10 | 105.90 | 513 | 559.38 | 139.85 | 116.65 |
| 469 | 513.44 | 128.36 | 106.14 | 514 | 560.43 | 140.11 | 116.89 |
| 470 | 514.48 | 128.62 | 106.38 | 515 | 561.47 | 140.37 | 117.13 |
| 471 | 515.53 | 128.88 | 106.62 | 516 | 562.51 | 140.63 | 117.37 |
| 472 | 516.57 | 129.14 | 106.86 | 517 | 563.56 | 140.89 | 117.61 |
| 473 | 517.62 | 129.40 | 107.10 | 518 | 564.60 | 141.15 | 117.85 |
| 474 | 518.66 | 129.67 | 107.33 | 519 | 565.64 | 141.41 | 118.09 |
| 475 | 519.71 | 129.93 | 107.57 | 520 | 566.68 | 141.67 | 118.33 |
| 476 | 520.75 | 130.19 | 107.81 | 521 | 567.73 | 141.93 | 118.57 |
| 477 | 521.80 | 130.45 | 108.05 | 522 | 568.77 | 142.19 | 118.81 |
| 478 | 522.84 | 130.71 | 108.29 | 523 | 569.81 | 142.45 | 119.05 |
| 479 | 523.89 | 130.97 | 108.53 | 524 | 570.86 | 142.71 | 119.29 |
| 480 | 524.93 | 131.23 | 108.77 | 525 | 571.90 | 142.97 | 119.53 |
| 481 | 525.98 | 131.49 | 109.01 | 526 | 572.94 | 143.24 | 119.76 |
| 482 | 527.02 | 131.76 | 109.24 | 527 | 573.98 | 143.50 | 120.00 |
| 483 | 528.07 | 132.02 | 109.48 | 528 | 575.03 | 143.76 | 120.24 |
| 484 | 529.11 | 132.28 | 109.72 | 529 | 576.07 | 144.02 | 120.48 |
| 485 | 530.16 | 132.54 | 109.96 | 530 | 577.11 | 144.28 | 120.72 |
| 486 | 531.20 | 132.80 | 110.20 | 531 | 578.15 | 144.54 | 120.96 |
| 487 | 532.25 | 133.06 | 110.44 | 532 | 579.20 | 144.80 | 121.20 |
| 488 | 533.29 | 133.32 | 110.68 | 533 | 580.24 | 145.06 | 121.44 |
| 489 | 534.33 | 133.58 | 110.92 | 534 | 581.28 | 145.32 | 121.68 |
| 490 | 535.38 | 133.84 | 111.16 | 535 | 582.32 | 145.58 | 121.92 |
| 491 | 536.42 | 134.11 | 111.39 | 536 | 583.37 | 145.84 | 122.16 |
| 492 | 537.47 | 134.37 | 111.63 | 537 | 584.41 | 146.10 | 122.40 |
| 493 | 538.51 | 134.63 | 111.87 | 538 | 585.45 | 146.36 | 122.64 |
| 494 | 539.55 | 134.89 | 112.11 | 539 | 586.49 | 146.62 | 122.88 |

Table 1. CONSTANTS FOR $\bar{p}'$ (LONG FORMULA), AN APPROXIMATION OF THE UPPER BINOMIAL CONFIDENCE LIMIT $\bar{p}$; $c/n \leqslant .5$, $n \geqslant 20$, $c \leqslant 1{,}000$ (*Continued*)

| c | $\bar{m}$ | $\bar{a}$ | $\bar{b}$ | c | $\bar{m}$ | $\bar{a}$ | $\bar{b}$ |
|---|---|---|---|---|---|---|---|
| 540 | 587.54 | 146.88 | 123.12 | 585 | 634.39 | 158.60 | 133.90 |
| 541 | 588.58 | 147.14 | 123.36 | 586 | 635.43 | 158.86 | 134.14 |
| 542 | 589.62 | 147.40 | 123.60 | 587 | 636.47 | 159.12 | 134.38 |
| 543 | 590.66 | 147.67 | 123.83 | 588 | 637.52 | 159.38 | 134.62 |
| 544 | 591.70 | 147.93 | 124.07 | 589 | 638.56 | 159.64 | 134.86 |
| 545 | 592.75 | 148.19 | 124.31 | 590 | 639.60 | 159.90 | 135.10 |
| 546 | 593.79 | 148.45 | 124.55 | 591 | 640.64 | 160.16 | 135.34 |
| 547 | 594.83 | 148.71 | 124.79 | 592 | 641.68 | 160.42 | 135.58 |
| 548 | 595.87 | 148.97 | 125.03 | 593 | 642.72 | 160.68 | 135.82 |
| 549 | 596.91 | 149.23 | 125.27 | 594 | 643.76 | 160.94 | 136.06 |
| 550 | 597.95 | 149.49 | 125.51 | 595 | 644.80 | 161.20 | 136.30 |
| 551 | 599.00 | 149.75 | 125.75 | 596 | 645.84 | 161.46 | 136.54 |
| 552 | 600.04 | 150.01 | 125.99 | 597 | 646.88 | 161.72 | 136.78 |
| 553 | 601.08 | 150.27 | 126.23 | 598 | 647.92 | 161.98 | 137.02 |
| 554 | 602.12 | 150.53 | 126.47 | 599 | 648.96 | 162.24 | 137.26 |
| 555 | 603.16 | 150.79 | 126.71 | 600 | 650.00 | 162.50 | 137.50 |
| 556 | 604.20 | 151.05 | 126.95 | 601 | 651.04 | 162.76 | 137.74 |
| 557 | 605.25 | 151.31 | 127.19 | 602 | 652.08 | 163.02 | 137.98 |
| 558 | 606.29 | 151.57 | 127.43 | 603 | 653.12 | 163.28 | 138.22 |
| 559 | 607.33 | 151.83 | 127.67 | 604 | 654.16 | 163.54 | 138.46 |
| 560 | 608.37 | 152.09 | 127.91 | 605 | 655.20 | 163.80 | 138.70 |
| 561 | 609.41 | 152.35 | 128.15 | 606 | 656.24 | 164.06 | 138.94 |
| 562 | 610.45 | 152.61 | 128.39 | 607 | 657.28 | 164.32 | 139.18 |
| 563 | 611.49 | 152.87 | 128.63 | 608 | 658.32 | 164.58 | 139.42 |
| 564 | 612.54 | 153.13 | 128.87 | 609 | 659.36 | 164.84 | 139.66 |
| 565 | 613.58 | 153.39 | 129.11 | 610 | 660.40 | 165.10 | 139.90 |
| 566 | 614.62 | 153.65 | 129.35 | 611 | 661.43 | 165.36 | 140.14 |
| 567 | 615.66 | 153.91 | 129.59 | 612 | 662.47 | 165.62 | 140.38 |
| 568 | 616.70 | 154.18 | 129.82 | 613 | 663.51 | 165.88 | 140.62 |
| 569 | 617.74 | 154.44 | 130.06 | 614 | 664.55 | 166.14 | 140.86 |
| 570 | 618.78 | 154.70 | 130.30 | 615 | 665.59 | 166.40 | 141.10 |
| 571 | 619.82 | 154.96 | 130.54 | 616 | 666.63 | 166.66 | 141.34 |
| 572 | 620.86 | 155.22 | 130.78 | 617 | 667.67 | 166.92 | 141.58 |
| 573 | 621.91 | 155.48 | 131.02 | 618 | 668.71 | 167.18 | 141.82 |
| 574 | 622.95 | 155.74 | 131.26 | 619 | 669.75 | 167.44 | 142.06 |
| 575 | 623.99 | 156.00 | 131.50 | 620 | 670.79 | 167.70 | 142.30 |
| 576 | 625.03 | 156.26 | 131.74 | 621 | 671.83 | 167.96 | 142.54 |
| 577 | 626.07 | 156.52 | 131.98 | 622 | 672.87 | 168.22 | 142.78 |
| 578 | 627.11 | 156.78 | 132.22 | 623 | 673.91 | 168.48 | 143.02 |
| 579 | 628.15 | 157.04 | 132.46 | 624 | 674.95 | 168.74 | 143.26 |
| 580 | 629.19 | 157.30 | 132.70 | 625 | 675.99 | 169.00 | 143.50 |
| 581 | 630.23 | 157.56 | 132.94 | 626 | 677.03 | 169.26 | 143.74 |
| 582 | 631.27 | 157.82 | 133.18 | 627 | 678.06 | 169.52 | 143.98 |
| 583 | 632.31 | 158.08 | 133.42 | 628 | 679.10 | 169.78 | 144.22 |
| 584 | 633.35 | 158.34 | 133.66 | 629 | 680.14 | 170.04 | 144.46 |

Table 1. CONSTANTS FOR $\bar{p}'$ (LONG FORMULA), AN APPROXIMATION OF THE UPPER BINOMIAL CONFIDENCE LIMIT $\bar{p}$; c/n ≤ .5, n ≥ 20, c ≤ 1,000 (*Continued*)

| c | $\overline{m}$ | $\overline{a}$ | $\overline{b}$ | c | $\overline{m}$ | $\overline{a}$ | $\overline{b}$ |
|---|---|---|---|---|---|---|---|
| 630 | 681.18 | 170.30 | 144.70 | 675 | 727.91 | 181.98 | 155.52 |
| 631 | 682.22 | 170.56 | 144.94 | 676 | 728.94 | 182.24 | 155.76 |
| 632 | 683.26 | 170.81 | 145.19 | 677 | 729.98 | 182.50 | 156.00 |
| 633 | 684.30 | 171.07 | 145.43 | 678 | 731.02 | 182.76 | 156.24 |
| 634 | 685.34 | 171.33 | 145.67 | 679 | 732.06 | 183.01 | 156.49 |
| 635 | 686.38 | 171.59 | 145.91 | 680 | 733.10 | 183.27 | 156.73 |
| 636 | 687.42 | 171.85 | 146.15 | 681 | 734.13 | 183.53 | 156.97 |
| 637 | 688.45 | 172.11 | 146.39 | 682 | 735.17 | 183.79 | 157.21 |
| 638 | 689.49 | 172.37 | 146.63 | 683 | 736.21 | 184.05 | 157.45 |
| 639 | 690.53 | 172.63 | 146.87 | 684 | 737.25 | 184.31 | 157.69 |
| 640 | 691.57 | 172.89 | 147.11 | 685 | 738.28 | 184.57 | 157.93 |
| 641 | 692.61 | 173.15 | 147.35 | 686 | 739.32 | 184.83 | 158.17 |
| 642 | 693.65 | 173.41 | 147.59 | 687 | 740.36 | 185.09 | 158.41 |
| 643 | 694.69 | 173.67 | 147.83 | 688 | 741.39 | 185.35 | 158.65 |
| 644 | 695.72 | 173.93 | 148.07 | 689 | 742.43 | 185.61 | 158.89 |
| 645 | 696.76 | 174.19 | 148.31 | 690 | 743.47 | 185.87 | 159.13 |
| 646 | 697.80 | 174.45 | 148.55 | 691 | 744.51 | 186.13 | 159.37 |
| 647 | 698.84 | 174.71 | 148.79 | 692 | 745.54 | 186.39 | 159.61 |
| 648 | 699.88 | 174.97 | 149.03 | 693 | 746.58 | 186.65 | 159.85 |
| 649 | 700.92 | 175.23 | 149.27 | 694 | 747.62 | 186.90 | 160.10 |
| 650 | 701.96 | 175.49 | 149.51 | 695 | 748.66 | 187.16 | 160.34 |
| 651 | 702.99 | 175.75 | 149.75 | 696 | 749.69 | 187.42 | 160.58 |
| 652 | 704.03 | 176.01 | 149.99 | 697 | 750.73 | 187.68 | 160.82 |
| 653 | 705.07 | 176.27 | 150.23 | 698 | 751.77 | 187.94 | 161.06 |
| 654 | 706.11 | 176.53 | 150.47 | 699 | 752.80 | 188.20 | 161.30 |
| 655 | 707.15 | 176.79 | 150.71 | 700 | 753.84 | 188.46 | 161.54 |
| 656 | 708.19 | 177.05 | 150.95 | 701 | 754.88 | 188.72 | 161.78 |
| 657 | 709.22 | 177.31 | 151.19 | 702 | 755.91 | 188.98 | 162.02 |
| 658 | 710.26 | 177.57 | 151.43 | 703 | 756.95 | 189.24 | 162.26 |
| 659 | 711.30 | 177.83 | 151.67 | 704 | 757.99 | 189.50 | 162.50 |
| 660 | 712.34 | 178.08 | 151.92 | 705 | 759.03 | 189.76 | 162.74 |
| 661 | 713.38 | 178.34 | 152.16 | 706 | 760.06 | 190.02 | 162.98 |
| 662 | 714.41 | 178.60 | 152.40 | 707 | 761.10 | 190.27 | 163.23 |
| 663 | 715.45 | 178.86 | 152.64 | 708 | 762.14 | 190.53 | 163.47 |
| 664 | 716.49 | 179.12 | 152.88 | 709 | 763.17 | 190.79 | 163.71 |
| 665 | 717.53 | 179.38 | 153.12 | 710 | 764.21 | 191.05 | 163.95 |
| 666 | 718.57 | 179.64 | 153.36 | 711 | 765.25 | 191.31 | 164.19 |
| 667 | 719.60 | 179.90 | 153.60 | 712 | 766.28 | 191.57 | 164.43 |
| 668 | 720.64 | 180.16 | 153.84 | 713 | 767.32 | 191.83 | 164.67 |
| 669 | 721.68 | 180.42 | 154.08 | 714 | 768.36 | 192.09 | 164.91 |
| 670 | 722.72 | 180.68 | 154.32 | 715 | 769.39 | 192.35 | 165.15 |
| 671 | 723.76 | 180.94 | 154.56 | 716 | 770.43 | 192.61 | 165.39 |
| 672 | 724.79 | 181.20 | 154.80 | 717 | 771.47 | 192.87 | 165.63 |
| 673 | 725.83 | 181.46 | 155.04 | 718 | 772.50 | 193.13 | 165.87 |
| 674 | 726.87 | 181.72 | 155.28 | 719 | 773.54 | 193.38 | 166.12 |

Table 1. CONSTANTS FOR p̄′ (LONG FORMULA), AN APPROXIMATION OF
THE UPPER BINOMIAL CONFIDENCE LIMIT p̄; c/n ⩽ .5, n ⩾ 20,
c ⩽ 1,000 (*Continued*)

Section 3

$\gamma_1 = 97.5\%$

$\gamma_2 = 95\%$

| c | m̄ | ā | b̄ | c | m̄ | ā | b̄ |
|---|---|---|---|---|---|---|---|
| 720 | 774.58 | 193.64 | 166.36 | 765 | 821.19 | 205.30 | 177.20 |
| 721 | 775.61 | 193.90 | 166.60 | 766 | 822.23 | 205.56 | 177.44 |
| 722 | 776.65 | 194.16 | 166.84 | 767 | 823.26 | 205.82 | 177.68 |
| 723 | 777.69 | 194.42 | 167.08 | 768 | 824.30 | 206.07 | 177.93 |
| 724 | 778.72 | 194.68 | 167.32 | 769 | 825.33 | 206.33 | 178.17 |
| 725 | 779.76 | 194.94 | 167.56 | 770 | 826.37 | 206.59 | 178.41 |
| 726 | 780.79 | 195.20 | 167.80 | 771 | 827.41 | 206.85 | 178.65 |
| 727 | 781.83 | 195.46 | 168.04 | 772 | 828.44 | 207.11 | 178.89 |
| 728 | 782.87 | 195.72 | 168.28 | 773 | 829.48 | 207.37 | 179.13 |
| 729 | 783.90 | 195.98 | 168.52 | 774 | 830.51 | 207.63 | 179.37 |
| 730 | 784.94 | 196.23 | 168.77 | 775 | 831.55 | 207.89 | 179.61 |
| 731 | 785.98 | 196.49 | 169.01 | 776 | 832.58 | 208.15 | 179.85 |
| 732 | 787.01 | 196.75 | 169.25 | 777 | 833.62 | 208.40 | 180.10 |
| 733 | 788.05 | 197.01 | 169.49 | 778 | 834.65 | 208.66 | 180.34 |
| 734 | 789.08 | 197.27 | 169.73 | 779 | 835.69 | 208.92 | 180.58 |
| 735 | 790.12 | 197.53 | 169.97 | 780 | 836.72 | 209.18 | 180.82 |
| 736 | 791.16 | 197.79 | 170.21 | 781 | 837.76 | 209.44 | 181.06 |
| 737 | 792.19 | 198.05 | 170.45 | 782 | 838.79 | 209.70 | 181.30 |
| 738 | 793.23 | 198.31 | 170.69 | 783 | 839.83 | 209.96 | 181.54 |
| 739 | 794.26 | 198.57 | 170.93 | 784 | 840.86 | 210.22 | 181.78 |
| 740 | 795.30 | 198.83 | 171.17 | 785 | 841.90 | 210.47 | 182.03 |
| 741 | 796.34 | 199.08 | 171.42 | 786 | 842.93 | 210.73 | 182.27 |
| 742 | 797.37 | 199.34 | 171.66 | 787 | 843.97 | 210.99 | 182.51 |
| 743 | 798.41 | 199.60 | 171.90 | 788 | 845.00 | 211.25 | 182.75 |
| 744 | 799.44 | 199.86 | 172.14 | 789 | 846.04 | 211.51 | 182.99 |
| 745 | 800.48 | 200.12 | 172.38 | 790 | 847.07 | 211.77 | 183.23 |
| 746 | 801.52 | 200.38 | 172.62 | 791 | 848.11 | 212.03 | 183.47 |
| 747 | 802.55 | 200.64 | 172.86 | 792 | 849.14 | 212.29 | 183.71 |
| 748 | 803.59 | 200.90 | 173.10 | 793 | 850.18 | 212.54 | 183.96 |
| 749 | 804.62 | 201.16 | 173.34 | 794 | 851.21 | 212.80 | 184.20 |
| 750 | 805.66 | 201.41 | 173.59 | 795 | 852.25 | 213.06 | 184.44 |
| 751 | 806.70 | 201.67 | 173.83 | 796 | 853.28 | 213.32 | 184.68 |
| 752 | 807.73 | 201.93 | 174.07 | 797 | 854.31 | 213.58 | 184.92 |
| 753 | 808.77 | 202.19 | 174.31 | 798 | 855.35 | 213.84 | 185.16 |
| 754 | 809.80 | 202.45 | 174.55 | 799 | 856.38 | 214.10 | 185.40 |
| 755 | 810.84 | 202.71 | 174.79 | 800 | 857.42 | 214.35 | 185.65 |
| 756 | 811.87 | 202.97 | 175.03 | 801 | 858.45 | 214.61 | 185.89 |
| 757 | 812.91 | 203.23 | 175.27 | 802 | 859.49 | 214.87 | 186.13 |
| 758 | 813.95 | 203.49 | 175.51 | 803 | 860.52 | 215.13 | 186.37 |
| 759 | 814.98 | 203.75 | 175.75 | 804 | 861.56 | 215.39 | 186.61 |
| 760 | 816.02 | 204.00 | 176.00 | 805 | 862.59 | 215.65 | 186.85 |
| 761 | 817.05 | 204.26 | 176.24 | 806 | 863.63 | 215.91 | 187.09 |
| 762 | 818.09 | 204.52 | 176.48 | 807 | 864.66 | 216.17 | 187.33 |
| 763 | 819.12 | 204.78 | 176.72 | 808 | 865.70 | 216.42 | 187.58 |
| 764 | 820.16 | 205.04 | 176.96 | 809 | 866.73 | 216.68 | 187.82 |

Table 1. CONSTANTS FOR $\overline{p}'$ (LONG FORMULA), AN APPROXIMATION OF THE UPPER BINOMIAL CONFIDENCE LIMIT $\overline{p}$; $c/n \leqslant .5$, $n \geqslant 20$, $c \leqslant 1,000$ (*Continued*)

| c | $\overline{m}$ | $\overline{a}$ | $\overline{b}$ | c | $\overline{m}$ | $\overline{a}$ | $\overline{b}$ |
|---|---|---|---|---|---|---|---|
| 810 | 867.76 | 216.94 | 188.06 | 855 | 914.29 | 228.57 | 198.93 |
| 811 | 868.80 | 217.20 | 188.30 | 856 | 915.33 | 228.83 | 199.17 |
| 812 | 869.83 | 217.46 | 188.54 | 857 | 916.36 | 229.09 | 199.41 |
| 813 | 870.87 | 217.72 | 188.78 | 858 | 917.39 | 229.35 | 199.65 |
| 814 | 871.90 | 217.98 | 189.02 | 859 | 918.43 | 229.61 | 199.89 |
| 815 | 872.94 | 218.23 | 189.27 | 860 | 919.46 | 229.86 | 200.14 |
| 816 | 873.97 | 218.49 | 189.51 | 861 | 920.49 | 230.12 | 200.38 |
| 817 | 875.00 | 218.75 | 189.75 | 862 | 921.53 | 230.38 | 200.62 |
| 818 | 876.04 | 219.01 | 189.99 | 863 | 922.56 | 230.64 | 200.86 |
| 819 | 877.07 | 219.27 | 190.23 | 864 | 923.59 | 230.90 | 201.10 |
| 820 | 878.11 | 219.53 | 190.47 | 865 | 924.63 | 231.16 | 201.34 |
| 821 | 879.14 | 219.79 | 190.71 | 866 | 925.66 | 231.41 | 201.59 |
| 822 | 880.18 | 220.04 | 190.96 | 867 | 926.69 | 231.67 | 201.83 |
| 823 | 881.21 | 220.30 | 191.20 | 868 | 927.73 | 231.93 | 202.07 |
| 824 | 882.24 | 220.56 | 191.44 | 869 | 928.76 | 232.19 | 202.31 |
| 825 | 883.28 | 220.82 | 191.68 | 870 | 929.79 | 232.45 | 202.55 |
| 826 | 884.31 | 221.08 | 191.92 | 871 | 930.83 | 232.71 | 202.79 |
| 827 | 885.35 | 221.34 | 192.16 | 872 | 931.86 | 232.96 | 203.04 |
| 828 | 886.38 | 221.60 | 192.40 | 873 | 932.89 | 233.22 | 203.28 |
| 829 | 887.41 | 221.85 | 192.65 | 874 | 933.92 | 233.48 | 203.52 |
| 830 | 888.45 | 222.11 | 192.89 | 875 | 934.96 | 233.74 | 203.76 |
| 831 | 889.48 | 222.37 | 193.13 | 876 | 935.99 | 234.00 | 204.00 |
| 832 | 890.52 | 222.63 | 193.37 | 877 | 937.02 | 234.26 | 204.24 |
| 833 | 891.55 | 222.89 | 193.61 | 878 | 938.06 | 234.51 | 204.49 |
| 834 | 892.58 | 223.15 | 193.85 | 879 | 939.09 | 234.77 | 204.73 |
| 835 | 893.62 | 223.40 | 194.10 | 880 | 940.12 | 235.03 | 204.97 |
| 836 | 894.65 | 223.66 | 194.34 | 881 | 941.16 | 235.29 | 205.21 |
| 837 | 895.69 | 223.92 | 194.58 | 882 | 942.19 | 235.55 | 205.45 |
| 838 | 896.72 | 224.18 | 194.82 | 883 | 943.22 | 235.81 | 205.69 |
| 839 | 897.75 | 224.44 | 195.06 | 884 | 944.26 | 236.06 | 205.94 |
| 840 | 898.79 | 224.70 | 195.30 | 885 | 945.29 | 236.32 | 206.18 |
| 841 | 899.82 | 224.96 | 195.54 | 886 | 946.32 | 236.58 | 206.42 |
| 842 | 900.85 | 225.21 | 195.79 | 887 | 947.35 | 236.84 | 206.66 |
| 843 | 901.89 | 225.47 | 196.03 | 888 | 948.39 | 237.10 | 206.90 |
| 844 | 902.92 | 225.73 | 196.27 | 889 | 949.42 | 237.35 | 207.15 |
| 845 | 903.96 | 225.99 | 196.51 | 890 | 950.45 | 237.61 | 207.39 |
| 846 | 904.99 | 226.25 | 196.75 | 891 | 951.49 | 237.87 | 207.63 |
| 847 | 906.02 | 226.51 | 196.99 | 892 | 952.52 | 238.13 | 207.87 |
| 848 | 907.06 | 226.76 | 197.24 | 893 | 953.55 | 238.39 | 208.11 |
| 849 | 908.09 | 227.02 | 197.48 | 894 | 954.58 | 238.65 | 208.35 |
| 850 | 909.12 | 227.28 | 197.72 | 895 | 955.62 | 238.90 | 208.60 |
| 851 | 910.16 | 227.54 | 197.96 | 896 | 956.65 | 239.16 | 208.84 |
| 852 | 911.19 | 227.80 | 198.20 | 897 | 957.68 | 239.42 | 209.08 |
| 853 | 912.22 | 228.06 | 198.44 | 898 | 958.71 | 239.68 | 209.32 |
| 854 | 913.26 | 228.31 | 198.69 | 899 | 959.75 | 239.94 | 209.56 |

# Table 1. CONSTANTS FOR $\overline{p}'$ (LONG FORMULA), AN APPROXIMATION OF THE UPPER BINOMIAL CONFIDENCE LIMIT $\overline{p}$; c/n ⩽ .5, n ⩾ 20, c ⩽ 1,000 (*Continued*)

| c | $\overline{m}$ | $\overline{a}$ | $\overline{b}$ | c | $\overline{m}$ | $\overline{a}$ | $\overline{b}$ |
|---|---|---|---|---|---|---|---|
| 900 | 960.78 | 240.19 | 209.81 | 945 | 1007.23 | 251.81 | 220.69 |
| 901 | 961.81 | 240.45 | 210.05 | 946 | 1008.26 | 252.07 | 220.93 |
| 902 | 962.85 | 240.71 | 210.29 | 947 | 1009.29 | 252.32 | 221.18 |
| 903 | 963.88 | 240.97 | 210.53 | 948 | 1010.33 | 252.58 | 221.42 |
| 904 | 964.91 | 241.23 | 210.77 | 949 | 1011.36 | 252.84 | 221.66 |
| 905 | 965.94 | 241.49 | 211.01 | 950 | 1012.39 | 253.10 | 221.90 |
| 906 | 966.98 | 241.74 | 211.26 | 951 | 1013.42 | 253.36 | 222.14 |
| 907 | 968.01 | 242.00 | 211.50 | 952 | 1014.45 | 253.61 | 222.39 |
| 908 | 969.04 | 242.26 | 211.74 | 953 | 1015.49 | 253.87 | 222.63 |
| 909 | 970.07 | 242.52 | 211.98 | 954 | 1016.52 | 254.13 | 222.87 |
| 910 | 971.11 | 242.78 | 212.22 | 955 | 1017.55 | 254.39 | 223.11 |
| 911 | 972.14 | 243.03 | 212.47 | 956 | 1018.58 | 254.65 | 223.35 |
| 912 | 973.17 | 243.29 | 212.71 | 957 | 1019.61 | 254.90 | 223.60 |
| 913 | 974.20 | 243.55 | 212.95 | 958 | 1020.64 | 255.16 | 223.84 |
| 914 | 975.24 | 243.81 | 213.19 | 959 | 1021.68 | 255.42 | 224.08 |
| 915 | 976.27 | 244.07 | 213.43 | 960 | 1022.71 | 255.68 | 224.32 |
| 916 | 977.30 | 244.32 | 213.68 | 961 | 1023.74 | 255.93 | 224.57 |
| 917 | 978.33 | 244.58 | 213.92 | 962 | 1024.77 | 256.19 | 224.81 |
| 918 | 979.36 | 244.84 | 214.16 | 963 | 1025.80 | 256.45 | 225.05 |
| 919 | 980.40 | 245.10 | 214.40 | 964 | 1026.83 | 256.71 | 225.29 |
| 920 | 981.43 | 245.36 | 214.64 | 965 | 1027.87 | 256.97 | 225.53 |
| 921 | 982.46 | 245.62 | 214.88 | 966 | 1028.90 | 257.22 | 225.78 |
| 922 | 983.49 | 245.87 | 215.13 | 967 | 1029.93 | 257.48 | 226.02 |
| 923 | 984.53 | 246.13 | 215.37 | 968 | 1030.96 | 257.74 | 226.26 |
| 924 | 985.56 | 246.39 | 215.61 | 969 | 1031.99 | 258.00 | 226.50 |
| 925 | 986.59 | 246.65 | 215.85 | 970 | 1033.02 | 258.26 | 226.74 |
| 926 | 987.62 | 246.91 | 216.09 | 971 | 1034.05 | 258.51 | 226.99 |
| 927 | 988.65 | 247.16 | 216.34 | 972 | 1035.09 | 258.77 | 227.23 |
| 928 | 989.69 | 247.42 | 216.58 | 973 | 1036.12 | 259.03 | 227.47 |
| 929 | 990.72 | 247.68 | 216.82 | 974 | 1037.15 | 259.29 | 227.71 |
| 930 | 991.75 | 247.94 | 217.06 | 975 | 1038.18 | 259.54 | 227.96 |
| 931 | 992.78 | 248.20 | 217.30 | 976 | 1039.21 | 259.80 | 228.20 |
| 932 | 993.82 | 248.45 | 217.55 | 977 | 1040.24 | 260.06 | 228.44 |
| 933 | 994.85 | 248.71 | 217.79 | 978 | 1041.27 | 260.32 | 228.68 |
| 934 | 995.88 | 248.97 | 218.03 | 979 | 1042.30 | 260.58 | 228.92 |
| 935 | 996.91 | 249.23 | 218.27 | 980 | 1043.34 | 260.83 | 229.17 |
| 936 | 997.94 | 249.49 | 218.51 | 981 | 1044.37 | 261.09 | 229.41 |
| 937 | 998.98 | 249.74 | 218.76 | 982 | 1045.40 | 261.35 | 229.65 |
| 938 | 1000.01 | 250.00 | 219.00 | 983 | 1046.43 | 261.61 | 229.89 |
| 939 | 1001.04 | 250.26 | 219.24 | 984 | 1047.46 | 261.87 | 230.13 |
| 940 | 1002.07 | 250.52 | 219.48 | 985 | 1048.49 | 262.12 | 230.38 |
| 941 | 1003.10 | 250.78 | 219.72 | 986 | 1049.52 | 262.38 | 230.62 |
| 942 | 1004.14 | 251.03 | 219.97 | 987 | 1050.55 | 262.64 | 230.86 |
| 943 | 1005.17 | 251.29 | 220.21 | 988 | 1051.59 | 262.90 | 231.10 |
| 944 | 1006.20 | 251.55 | 220.45 | 989 | 1052.62 | 263.15 | 231.35 |

Table 1.  CONSTANTS FOR $\bar{p}'$ (LONG FORMULA), AN APPROXIMATION OF THE UPPER BINOMIAL CONFIDENCE LIMIT $\bar{p}$; $c/n \leqslant .5$, $n \geqslant 20$, $c \leqslant 1,000$ (*Continued*)

| c | $\bar{m}$ | $\bar{a}$ | $\bar{b}$ | c | $\bar{m}$ | $\bar{a}$ | $\bar{b}$ |
|---|---|---|---|---|---|---|---|
| 990 | 1053.65 | 263.41 | 231.59 | 995 | 1058.80 | 264.70 | 232.80 |
| 991 | 1054.68 | 263.67 | 231.83 | 996 | 1059.83 | 264.96 | 233.04 |
| 992 | 1055.71 | 263.93 | 232.07 | 997 | 1060.87 | 265.22 | 233.28 |
| 993 | 1056.74 | 264.19 | 232.31 | 998 | 1061.90 | 265.47 | 233.53 |
| 994 | 1057.77 | 264.44 | 232.56 | 999 | 1062.93 | 265.73 | 233.77 |
|  |  |  |  | 1000 | 1063.96 | 265.99 | 234.01 |

# Table 1. CONSTANTS FOR $\overline{p}'$ (LONG FORMULA), AN APPROXIMATION OF THE UPPER BINOMIAL CONFIDENCE LIMIT $\overline{p}$; $c/n \leqslant .5$, $n \geqslant 20$, $c \leqslant 1,000$ (*Continued*)

| c | $\overline{m}$ | $\overline{a}$ | $\overline{b}$ | c | $\overline{m}$ | $\overline{a}$ | $\overline{b}$ |
|---|---|---|---|---|---|---|---|
| 0 | 2.9957 | 0.511 | -0.985 | 45 | 57.695 | 15.572 | 9.300 |
| 1 | 4.7439 | 0.815 | -1.056 | 46 | 58.816 | 15.937 | 9.607 |
| 2 | 6.2958 | 1.116 | -1.033 | 47 | 59.936 | 16.244 | 9.854 |
| 3 | 7.7537 | 1.462 | -0.913 | 48 | 61.054 | 16.615 | 10.171 |
| 4 | 9.1535 | 1.778 | -0.796 | 49 | 62.171 | 16.879 | 10.372 |
| 5 | 10.513 | 2.107 | -0.645 | 50 | 63.287 | 17.268 | 10.708 |
| 6 | 11.842 | 2.444 | -0.470 | 51 | 64.402 | 17.512 | 10.889 |
| 7 | 13.148 | 2.819 | -0.241 | 52 | 65.516 | 17.968 | 11.300 |
| 8 | 14.435 | 3.186 | -0.011 | 53 | 66.629 | 18.276 | 11.550 |
| 9 | 15.705 | 3.574 | 0.250 | 54 | 67.740 | 18.511 | 11.723 |
| 10 | 16.962 | 3.989 | 0.547 | 55 | 68.851 | 18.781 | 11.935 |
| 11 | 18.208 | 4.336 | 0.774 | 56 | 69.961 | 19.164 | 12.267 |
| 12 | 19.443 | 4.677 | 0.999 | 57 | 71.069 | 19.456 | 12.50 |
| 13 | 20.669 | 5.015 | 1.226 | 58 | 72.177 | 19.888 | 12.89 |
| 14 | 21.887 | 5.355 | 1.459 | 59 | 73.284 | 20.243 | 13.19 |
| 15 | 23.097 | 5.687 | 1.686 | 60 | 74.390 | 20.573 | 13.47 |
| 16 | 24.301 | 6.029 | 1.928 | 61 | 75.495 | 20.86 | 13.70 |
| 17 | 25.499 | 6.358 | 2.158 | 62 | 76.599 | 21.08 | 13.86 |
| 18 | 26.692 | 6.703 | 2.411 | 63 | 77.703 | 21.52 | 14.26 |
| 19 | 27.879 | 7.041 | 2.655 | 64 | 78.805 | 21.73 | 14.41 |
| 20 | 29.062 | 7.348 | 2.869 | 65 | 79.907 | 22.16 | 14.80 |
| 21 | 30.240 | 7.684 | 3.118 | 66 | 81.008 | 22.36 | 14.94 |
| 22 | 31.415 | 8.041 | 3.392 | 67 | 82.108 | 22.83 | 15.37 |
| 23 | 32.585 | 8.355 | 3.619 | 68 | 83.208 | 23.10 | 15.59 |
| 24 | 33.752 | 8.688 | 3.870 | 69 | 84.307 | 23.35 | 15.78 |
| 25 | 34.916 | 9.025 | 4.126 | 70 | 85.405 | 23.66 | 16.05 |
| 26 | 36.077 | 9.370 | 4.394 | 71 | 86.502 | 24.03 | 16.37 |
| 27 | 37.234 | 9.693 | 4.638 | 72 | 87.599 | 24.45 | 16.76 |
| 28 | 38.389 | 10.037 | 4.907 | 73 | 88.695 | 24.71 | 16.96 |
| 29 | 39.541 | 10.341 | 5.134 | 74 | 89.791 | 25.04 | 17.24 |
| 30 | 40.691 | 10.657 | 5.374 | 75 | 90.885 | 25.47 | 17.63 |
| 31 | 41.838 | 11.017 | 5.665 | 76 | 91.980 | 25.61 | 17.72 |
| 32 | 42.982 | 11.340 | 5.916 | 77 | 93.073 | 25.94 | 18.00 |
| 33 | 44.125 | 11.642 | 6.145 | 78 | 94.166 | 26.36 | 18.38 |
| 34 | 45.266 | 11.983 | 6.417 | 79 | 95.258 | 26.62 | 18.59 |
| 35 | 46.404 | 12.310 | 6.676 | 80 | 96.350 | 26.97 | 18.90 |
| 36 | 47.541 | 12.625 | 6.922 | 81 | 97.442 | 27.20 | 19.07 |
| 37 | 48.676 | 13.008 | 7.244 | 82 | 98.532 | 27.50 | 19.33 |
| 38 | 49.808 | 13.286 | 7.451 | 83 | 99.622 | 28.02 | 19.81 |
| 39 | 50.940 | 13.656 | 7.761 | 84 | 100.71 | 28.34 | 20.10 |
| 40 | 52.070 | 13.949 | 7.987 | 85 | 101.80 | 28.45 | 20.15 |
| 41 | 53.198 | 14.333 | 8.312 | 86 | 102.89 | 28.82 | 20.47 |
| 42 | 54.324 | 14.579 | 8.488 | 87 | 103.98 | 29.26 | 20.88 |
| 43 | 55.449 | 14.908 | 8.756 | 88 | 105.07 | 29.69 | 21.27 |
| 44 | 56.573 | 15.288 | 9.080 | 89 | 106.15 | 29.85 | 21.37 |

Table 1. CONSTANTS FOR $\overline{p}'$ (LONG FORMULA), AN APPROXIMATION OF THE UPPER BINOMIAL CONFIDENCE LIMIT $\overline{p}$; $c/n \leqslant .5$, $n \geqslant 20$, $c \leqslant 1,000$ (*Continued*)

| c | $\overline{m}$ | $\overline{a}$ | $\overline{b}$ | c | $\overline{m}$ | $\overline{a}$ | $\overline{b}$ |
|---|---|---|---|---|---|---|---|
| 90 | 107.24 | 30.26 | 21.75 | 135 | 155.73 | 38.94 | 28.56 |
| 91 | 108.32 | 30.49 | 21.94 | 136 | 156.80 | 39.21 | 28.79 |
| 92 | 109.41 | 31.01 | 22.42 | 137 | 157.87 | 39.47 | 29.03 |
| 93 | 110.50 | 31.31 | 22.67 | 138 | 158.94 | 39.74 | 29.26 |
| 94 | 111.58 | 31.44 | 22.76 | 139 | 160.01 | 40.01 | 29.49 |
| 95 | 112.66 | 31.85 | 23.13 | 140 | 161.08 | 40.28 | 29.72 |
| 96 | 113.75 | 32.21 | 23.44 | 141 | 162.15 | 40.54 | 29.96 |
| 97 | 114.83 | 32.34 | 23.52 | 142 | 163.22 | 40.81 | 30.19 |
| 98 | 115.91 | 32.70 | 23.84 | 143 | 164.29 | 41.08 | 30.42 |
| 99 | 117.00 | 32.97 | 24.07 | 144 | 165.36 | 41.34 | 30.66 |
| 100 | 118.08 | 33.46 | 24.53 | 145 | 166.43 | 41.61 | 30.89 |
| 101 | 119.16 | 29.80 | 20.70 | 146 | 167.50 | 41.88 | 31.12 |
| 102 | 120.24 | 30.07 | 20.93 | 147 | 168.56 | 42.14 | 31.36 |
| 103 | 121.32 | 30.34 | 21.16 | 148 | 169.63 | 42.41 | 31.59 |
| 104 | 122.40 | 30.61 | 21.39 | 149 | 170.70 | 42.68 | 31.82 |
| 105 | 123.48 | 30.88 | 21.62 | 150 | 171.76 | 42.95 | 32.05 |
| 106 | 124.56 | 31.15 | 21.85 | 151 | 172.83 | 43.21 | 32.29 |
| 107 | 125.64 | 31.42 | 22.08 | 152 | 173.90 | 43.48 | 32.52 |
| 108 | 126.72 | 31.69 | 22.31 | 153 | 174.96 | 43.75 | 32.75 |
| 109 | 127.80 | 31.96 | 22.54 | 154 | 176.03 | 44.01 | 32.99 |
| 110 | 128.88 | 32.22 | 22.78 | 155 | 177.10 | 44.28 | 33.22 |
| 111 | 129.96 | 32.49 | 23.01 | 156 | 178.16 | 44.54 | 33.46 |
| 112 | 131.04 | 32.76 | 23.24 | 157 | 179.23 | 44.81 | 33.69 |
| 113 | 132.11 | 33.03 | 23.47 | 158 | 180.29 | 45.08 | 33.92 |
| 114 | 133.19 | 33.30 | 23.70 | 159 | 181.36 | 45.34 | 34.16 |
| 115 | 134.27 | 33.57 | 23.93 | 160 | 182.42 | 45.61 | 34.39 |
| 116 | 135.34 | 33.84 | 24.16 | 161 | 183.50 | 45.88 | 34.62 |
| 117 | 136.42 | 34.11 | 24.39 | 162 | 184.57 | 46.14 | 34.86 |
| 118 | 137.49 | 34.38 | 24.62 | 163 | 185.63 | 46.41 | 35.09 |
| 119 | 138.57 | 34.65 | 24.85 | 164 | 186.70 | 46.67 | 35.33 |
| 120 | 139.64 | 34.92 | 25.08 | 165 | 187.76 | 46.94 | 35.56 |
| 121 | 140.72 | 35.18 | 25.32 | 166 | 188.83 | 47.21 | 35.79 |
| 122 | 141.79 | 35.45 | 25.55 | 167 | 189.89 | 47.47 | 36.03 |
| 123 | 142.87 | 35.72 | 25.78 | 168 | 190.95 | 47.74 | 36.26 |
| 124 | 143.94 | 35.99 | 26.01 | 169 | 192.02 | 48.00 | 36.50 |
| 125 | 145.01 | 36.26 | 26.24 | 170 | 193.08 | 48.27 | 36.73 |
| 126 | 146.09 | 36.53 | 26.47 | 171 | 194.14 | 48.54 | 36.96 |
| 127 | 147.16 | 36.79 | 26.71 | 172 | 195.20 | 48.80 | 37.20 |
| 128 | 148.23 | 37.06 | 26.94 | 173 | 196.27 | 49.07 | 37.43 |
| 129 | 149.31 | 37.33 | 27.17 | 174 | 197.33 | 49.33 | 37.67 |
| 130 | 150.38 | 37.60 | 27.40 | 175 | 198.39 | 49.60 | 37.90 |
| 131 | 151.45 | 37.87 | 27.63 | 176 | 199.45 | 49.86 | 38.14 |
| 132 | 152.52 | 38.13 | 27.87 | 177 | 200.51 | 50.13 | 38.37 |
| 133 | 153.59 | 38.40 | 28.10 | 178 | 201.58 | 50.39 | 38.61 |
| 134 | 154.66 | 38.67 | 28.33 | 179 | 202.64 | 50.66 | 38.84 |

**Table 1. CONSTANTS FOR $\bar{p}'$ (LONG FORMULA), AN APPROXIMATION OF THE UPPER BINOMIAL CONFIDENCE LIMIT $\bar{p}$; $c/n \leqslant .5$, $n \geqslant 20$, $c \leqslant 1,000$ (*Continued*)**

| c | $\overline{m}$ | $\overline{a}$ | $\overline{b}$ | c | $\overline{m}$ | $\overline{a}$ | $\overline{b}$ |
|---|---|---|---|---|---|---|---|
| 180 | 203.70 | 50.92 | 39.08 | 225 | 251.30 | 62.82 | 49.68 |
| 181 | 204.76 | 51.19 | 39.31 | 226 | 252.35 | 63.09 | 49.91 |
| 182 | 205.82 | 51.46 | 39.54 | 227 | 253.41 | 63.35 | 50.15 |
| 183 | 206.88 | 51.72 | 39.78 | 228 | 254.46 | 63.62 | 50.38 |
| 184 | 207.94 | 51.99 | 40.01 | 229 | 255.51 | 63.88 | 50.62 |
| 185 | 209.00 | 52.25 | 40.25 | 230 | 256.57 | 64.14 | 50.86 |
| 186 | 210.06 | 52.52 | 40.48 | 231 | 257.62 | 64.41 | 51.09 |
| 187 | 211.12 | 52.78 | 40.72 | 232 | 258.68 | 64.67 | 51.33 |
| 188 | 212.18 | 53.05 | 40.95 | 233 | 259.73 | 64.93 | 51.57 |
| 189 | 213.24 | 53.31 | 41.19 | 234 | 260.78 | 65.20 | 51.80 |
| 190 | 214.30 | 53.58 | 41.42 | 235 | 261.84 | 65.46 | 52.04 |
| 191 | 215.36 | 53.84 | 41.66 | 236 | 262.89 | 65.72 | 52.28 |
| 192 | 216.42 | 54.11 | 41.89 | 237 | 263.94 | 65.99 | 52.51 |
| 193 | 217.48 | 54.37 | 42.13 | 238 | 265.00 | 66.25 | 52.75 |
| 194 | 218.54 | 54.63 | 42.37 | 239 | 266.05 | 66.51 | 52.99 |
| 195 | 219.60 | 54.90 | 42.60 | 240 | 267.10 | 66.78 | 53.22 |
| 196 | 220.66 | 55.16 | 42.84 | 241 | 268.16 | 67.04 | 53.46 |
| 197 | 221.71 | 55.43 | 43.07 | 242 | 269.21 | 67.30 | 53.70 |
| 198 | 222.77 | 55.69 | 43.31 | 243 | 270.26 | 67.57 | 53.93 |
| 199 | 223.83 | 55.96 | 43.54 | 244 | 271.32 | 67.83 | 54.17 |
| 200 | 224.89 | 56.22 | 43.78 | 245 | 272.37 | 68.09 | 54.41 |
| 201 | 225.95 | 56.49 | 44.01 | 246 | 273.42 | 68.36 | 54.64 |
| 202 | 227.00 | 56.75 | 44.25 | 247 | 274.47 | 68.62 | 54.88 |
| 203 | 228.06 | 57.02 | 44.48 | 248 | 275.52 | 68.88 | 55.12 |
| 204 | 229.12 | 57.28 | 44.72 | 249 | 276.58 | 69.14 | 55.36 |
| 205 | 230.18 | 57.54 | 44.96 | 250 | 277.63 | 69.41 | 55.59 |
| 206 | 231.23 | 57.81 | 45.19 | 251 | 278.68 | 69.67 | 55.83 |
| 207 | 232.29 | 58.07 | 45.43 | 252 | 279.73 | 69.93 | 56.07 |
| 208 | 233.35 | 58.34 | 45.66 | 253 | 280.78 | 70.20 | 56.30 |
| 209 | 234.41 | 58.60 | 45.90 | 254 | 281.84 | 70.46 | 56.54 |
| 210 | 235.46 | 58.87 | 46.13 | 255 | 282.89 | 70.72 | 56.78 |
| 211 | 236.52 | 59.13 | 46.37 | 256 | 283.94 | 70.98 | 57.02 |
| 212 | 237.58 | 59.39 | 46.61 | 257 | 284.99 | 71.25 | 57.25 |
| 213 | 238.63 | 59.66 | 46.84 | 258 | 286.04 | 71.51 | 57.49 |
| 214 | 239.69 | 59.92 | 47.08 | 259 | 287.09 | 71.77 | 57.73 |
| 215 | 240.74 | 60.19 | 47.31 | 260 | 288.14 | 72.04 | 57.96 |
| 216 | 241.80 | 60.45 | 47.55 | 261 | 289.19 | 72.30 | 58.20 |
| 217 | 242.86 | 60.71 | 47.79 | 262 | 290.24 | 72.56 | 58.44 |
| 218 | 243.91 | 60.98 | 48.02 | 263 | 291.29 | 72.82 | 58.68 |
| 219 | 244.97 | 61.24 | 48.26 | 264 | 292.35 | 73.09 | 58.91 |
| 220 | 246.02 | 61.51 | 48.49 | 265 | 293.40 | 73.35 | 59.15 |
| 221 | 247.08 | 61.77 | 48.73 | 266 | 294.45 | 73.61 | 59.39 |
| 222 | 248.13 | 62.03 | 48.97 | 267 | 295.50 | 73.87 | 59.63 |
| 223 | 249.19 | 62.30 | 49.20 | 268 | 296.55 | 74.14 | 59.86 |
| 224 | 250.24 | 62.56 | 49.44 | 269 | 297.60 | 74.40 | 60.10 |

Table 1.  CONSTANTS FOR $\bar{p}'$ (LONG FORMULA), AN APPROXIMATION OF THE UPPER BINOMIAL CONFIDENCE LIMIT $\bar{p}$; $c/n \leqslant .5$, $n \geqslant 20$, $c \leqslant 1,000$ (*Continued*)

| c | $\bar{m}$ | $\bar{a}$ | $\bar{b}$ | c | $\bar{m}$ | $\bar{a}$ | $\bar{b}$ |
|---|---|---|---|---|---|---|---|
| 270 | 298.65 | 74.66 | 60.34 | 315 | 345.81 | 86.45 | 71.05 |
| 271 | 299.70 | 74.92 | 60.58 | 316 | 346.86 | 86.71 | 71.29 |
| 272 | 300.75 | 75.19 | 60.81 | 317 | 347.90 | 86.98 | 71.52 |
| 273 | 301.80 | 75.45 | 61.05 | 318 | 348.95 | 87.24 | 71.76 |
| 274 | 302.85 | 75.71 | 61.29 | 319 | 349.99 | 87.50 | 72.00 |
| 275 | 303.90 | 75.97 | 61.53 | 320 | 351.04 | 87.76 | 72.24 |
| 276 | 304.95 | 76.24 | 61.76 | 321 | 352.09 | 88.02 | 72.48 |
| 277 | 305.99 | 76.50 | 62.00 | 322 | 353.13 | 88.28 | 72.72 |
| 278 | 307.04 | 76.76 | 62.24 | 323 | 354.18 | 88.54 | 72.96 |
| 279 | 308.09 | 77.02 | 62.48 | 324 | 355.22 | 88.81 | 73.19 |
| 280 | 309.14 | 77.29 | 62.71 | 325 | 356.27 | 89.07 | 73.43 |
| 281 | 310.19 | 77.55 | 62.95 | 326 | 357.31 | 89.33 | 73.67 |
| 282 | 311.24 | 77.81 | 63.19 | 327 | 358.36 | 89.59 | 73.91 |
| 283 | 312.29 | 78.07 | 63.43 | 328 | 359.40 | 89.85 | 74.15 |
| 284 | 313.34 | 78.33 | 63.67 | 329 | 360.45 | 90.11 | 74.39 |
| 285 | 314.39 | 78.60 | 63.90 | 330 | 361.49 | 90.37 | 74.63 |
| 286 | 315.43 | 78.86 | 64.14 | 331 | 362.54 | 90.63 | 74.87 |
| 287 | 316.48 | 79.12 | 64.38 | 332 | 363.59 | 90.90 | 75.10 |
| 288 | 317.53 | 79.38 | 64.62 | 333 | 364.63 | 91.16 | 75.34 |
| 289 | 318.58 | 79.65 | 64.85 | 334 | 365.68 | 91.42 | 75.58 |
| 290 | 319.63 | 79.91 | 65.09 | 335 | 366.72 | 91.68 | 75.82 |
| 291 | 320.68 | 80.17 | 65.33 | 336 | 367.76 | 91.94 | 76.06 |
| 292 | 321.72 | 80.43 | 65.57 | 337 | 368.81 | 92.20 | 76.30 |
| 293 | 322.77 | 80.69 | 65.81 | 338 | 369.85 | 92.46 | 76.54 |
| 294 | 323.82 | 80.96 | 66.04 | 339 | 370.90 | 92.72 | 76.78 |
| 295 | 324.87 | 81.22 | 66.28 | 340 | 371.94 | 92.99 | 77.01 |
| 296 | 325.92 | 81.48 | 66.52 | 341 | 372.99 | 93.25 | 77.25 |
| 297 | 326.96 | 81.74 | 66.76 | 342 | 374.03 | 93.51 | 77.49 |
| 298 | 328.01 | 82.00 | 67.00 | 343 | 375.08 | 93.77 | 77.73 |
| 299 | 329.06 | 82.26 | 67.24 | 344 | 376.12 | 94.03 | 77.97 |
| 300 | 330.11 | 82.53 | 67.47 | 345 | 377.17 | 94.29 | 78.21 |
| 301 | 331.15 | 82.79 | 67.71 | 346 | 378.21 | 94.55 | 78.45 |
| 302 | 332.20 | 83.05 | 67.95 | 347 | 379.25 | 94.81 | 78.69 |
| 303 | 333.25 | 83.31 | 68.19 | 348 | 380.30 | 95.07 | 78.93 |
| 304 | 334.30 | 83.57 | 68.43 | 349 | 381.34 | 95.34 | 79.16 |
| 305 | 335.34 | 83.84 | 68.66 | 350 | 382.39 | 95.60 | 79.40 |
| 306 | 336.39 | 84.10 | 68.90 | 351 | 383.43 | 95.86 | 79.64 |
| 307 | 337.44 | 84.36 | 69.14 | 352 | 384.47 | 96.12 | 79.88 |
| 308 | 338.48 | 84.62 | 69.38 | 353 | 385.52 | 96.38 | 80.12 |
| 309 | 339.53 | 84.88 | 69.62 | 354 | 386.56 | 96.64 | 80.36 |
| 310 | 340.58 | 85.14 | 69.86 | 355 | 387.60 | 96.90 | 80.60 |
| 311 | 341.62 | 85.41 | 70.09 | 356 | 388.65 | 97.16 | 80.84 |
| 312 | 342.67 | 85.67 | 70.33 | 357 | 389.69 | 97.42 | 81.08 |
| 313 | 343.72 | 85.93 | 70.57 | 358 | 390.73 | 97.68 | 81.32 |
| 314 | 344.76 | 86.19 | 70.81 | 359 | 391.78 | 97.94 | 81.56 |

**Table 1. CONSTANTS FOR $\overline{p}'$ (LONG FORMULA), AN APPROXIMATION OF THE UPPER BINOMIAL CONFIDENCE LIMIT $\overline{p}$; $c/n \leqslant .5$, $n \geqslant 20$, $c \leqslant 1{,}000$ (Continued)**

| c | $\overline{m}$ | $\overline{a}$ | $\overline{b}$ | c | $\overline{m}$ | $\overline{a}$ | $\overline{b}$ |
|---|---|---|---|---|---|---|---|
| 360 | 392.82 | 98.21 | 81.79 | 405 | 439.71 | 109.93 | 92.57 |
| 361 | 393.86 | 98.47 | 82.03 | 406 | 440.75 | 110.19 | 92.81 |
| 362 | 394.91 | 98.73 | 82.27 | 407 | 441.79 | 110.45 | 93.05 |
| 363 | 395.95 | 98.99 | 82.51 | 408 | 442.83 | 110.71 | 93.29 |
| 364 | 396.99 | 99.25 | 82.75 | 409 | 443.88 | 110.97 | 93.53 |
| 365 | 398.04 | 99.51 | 82.99 | 410 | 444.92 | 111.23 | 93.77 |
| 366 | 399.08 | 99.77 | 83.23 | 411 | 445.96 | 111.49 | 94.01 |
| 367 | 400.12 | 100.03 | 83.47 | 412 | 447.00 | 111.75 | 94.25 |
| 368 | 401.17 | 100.29 | 83.71 | 413 | 448.04 | 112.01 | 94.49 |
| 369 | 402.21 | 100.55 | 83.95 | 414 | 449.08 | 112.27 | 94.73 |
| 370 | 403.25 | 100.81 | 84.19 | 415 | 450.12 | 112.53 | 94.97 |
| 371 | 404.29 | 101.07 | 84.43 | 416 | 451.16 | 112.79 | 95.21 |
| 372 | 405.34 | 101.33 | 84.67 | 417 | 452.20 | 113.05 | 95.45 |
| 373 | 406.38 | 101.59 | 84.91 | 418 | 453.24 | 113.31 | 95.69 |
| 374 | 407.42 | 101.86 | 85.14 | 419 | 454.28 | 113.57 | 95.93 |
| 375 | 408.46 | 102.12 | 85.38 | 420 | 455.32 | 113.83 | 96.17 |
| 376 | 409.51 | 102.38 | 85.62 | 421 | 456.36 | 114.09 | 96.41 |
| 377 | 410.55 | 102.64 | 85.86 | 422 | 457.40 | 114.35 | 96.65 |
| 378 | 411.59 | 102.90 | 86.10 | 423 | 458.44 | 114.61 | 96.89 |
| 379 | 412.63 | 103.16 | 86.34 | 424 | 459.48 | 114.87 | 97.13 |
| 380 | 413.68 | 103.42 | 86.58 | 425 | 460.52 | 115.13 | 97.37 |
| 381 | 414.72 | 103.68 | 86.82 | 426 | 461.56 | 115.39 | 97.61 |
| 382 | 415.76 | 103.94 | 87.06 | 427 | 462.60 | 115.65 | 97.85 |
| 383 | 416.80 | 104.20 | 87.30 | 428 | 463.64 | 115.91 | 98.09 |
| 384 | 417.84 | 104.46 | 87.54 | 429 | 464.68 | 116.17 | 98.33 |
| 385 | 418.89 | 104.72 | 87.78 | 430 | 465.72 | 116.43 | 98.57 |
| 386 | 419.93 | 104.98 | 88.02 | 431 | 466.76 | 116.69 | 98.81 |
| 387 | 420.97 | 105.24 | 88.26 | 432 | 467.80 | 116.95 | 99.05 |
| 388 | 422.01 | 105.50 | 88.50 | 433 | 468.84 | 117.21 | 99.29 |
| 389 | 423.05 | 105.76 | 88.74 | 434 | 469.88 | 117.47 | 99.53 |
| 390 | 424.09 | 106.02 | 88.98 | 435 | 470.91 | 117.73 | 99.77 |
| 391 | 425.14 | 106.28 | 89.22 | 436 | 471.95 | 117.99 | 100.01 |
| 392 | 426.18 | 106.54 | 89.46 | 437 | 472.99 | 118.25 | 100.25 |
| 393 | 427.22 | 106.80 | 89.70 | 438 | 474.03 | 118.51 | 100.49 |
| 394 | 428.26 | 107.07 | 89.93 | 439 | 475.07 | 118.77 | 100.73 |
| 395 | 429.30 | 107.33 | 90.17 | 440 | 476.11 | 119.03 | 100.97 |
| 396 | 430.34 | 107.59 | 90.41 | 441 | 477.15 | 119.29 | 101.21 |
| 397 | 431.38 | 107.85 | 90.65 | 442 | 478.19 | 119.55 | 101.45 |
| 398 | 432.43 | 108.11 | 90.89 | 443 | 479.23 | 119.81 | 101.69 |
| 399 | 433.47 | 108.37 | 91.13 | 444 | 480.27 | 120.07 | 101.93 |
| 400 | 434.51 | 108.63 | 91.37 | 445 | 481.31 | 120.33 | 102.17 |
| 401 | 435.55 | 108.89 | 91.61 | 446 | 482.35 | 120.59 | 102.41 |
| 402 | 436.59 | 109.15 | 91.85 | 447 | 483.38 | 120.85 | 102.65 |
| 403 | 437.63 | 109.41 | 92.09 | 448 | 484.42 | 121.11 | 102.89 |
| 404 | 438.67 | 109.67 | 92.33 | 449 | 485.46 | 121.37 | 103.13 |

Table 1. CONSTANTS FOR $\bar{p}'$ (LONG FORMULA), AN APPROXIMATION OF THE UPPER BINOMIAL CONFIDENCE LIMIT $\bar{p}$; $c/n \leqslant .5$, $n \geqslant 20$, $c \leqslant 1,000$ (*Continued*)

| c | $\overline{m}$ | $\overline{a}$ | $\overline{b}$ | c | $\overline{m}$ | $\overline{a}$ | $\overline{b}$ |
|---|---|---|---|---|---|---|---|
| 450 | 486.50 | 121.63 | 103.37 | 495 | 533.20 | 133.30 | 114.20 |
| 451 | 487.54 | 121.88 | 103.62 | 496 | 534.24 | 133.56 | 114.44 |
| 452 | 488.58 | 122.14 | 103.86 | 497 | 535.28 | 133.82 | 114.68 |
| 453 | 489.62 | 122.40 | 104.10 | 498 | 536.31 | 134.08 | 114.92 |
| 454 | 490.66 | 122.66 | 104.34 | 499 | 537.35 | 134.34 | 115.16 |
| 455 | 491.69 | 122.92 | 104.58 | 500 | 538.39 | 134.60 | 115.40 |
| 456 | 492.73 | 123.18 | 104.82 | 501 | 539.42 | 134.86 | 115.64 |
| 457 | 493.77 | 123.44 | 105.06 | 502 | 540.46 | 135.11 | 115.89 |
| 458 | 494.81 | 123.70 | 105.30 | 503 | 541.50 | 135.37 | 116.13 |
| 459 | 495.85 | 123.96 | 105.54 | 504 | 542.53 | 135.63 | 116.37 |
| 460 | 496.89 | 124.22 | 105.78 | 505 | 543.57 | 135.89 | 116.61 |
| 461 | 497.92 | 124.48 | 106.02 | 506 | 544.61 | 136.15 | 116.85 |
| 462 | 498.96 | 124.74 | 106.26 | 507 | 545.64 | 136.41 | 117.09 |
| 463 | 500.00 | 125.00 | 106.50 | 508 | 546.68 | 136.67 | 117.33 |
| 464 | 501.04 | 125.26 | 106.74 | 509 | 547.72 | 136.93 | 117.57 |
| 465 | 502.08 | 125.52 | 106.98 | 510 | 548.75 | 137.19 | 117.81 |
| 466 | 503.12 | 125.78 | 107.22 | 511 | 549.79 | 137.45 | 118.05 |
| 467 | 504.15 | 126.04 | 107.46 | 512 | 550.82 | 137.71 | 118.29 |
| 468 | 505.19 | 126.30 | 107.70 | 513 | 551.86 | 137.97 | 118.53 |
| 469 | 506.23 | 126.56 | 107.94 | 514 | 552.90 | 138.22 | 118.78 |
| 470 | 507.27 | 126.82 | 108.18 | 515 | 553.93 | 138.48 | 119.02 |
| 471 | 508.30 | 127.08 | 108.42 | 516 | 554.97 | 138.74 | 119.26 |
| 472 | 509.34 | 127.34 | 108.66 | 517 | 556.01 | 139.00 | 119.50 |
| 473 | 510.38 | 127.60 | 108.90 | 518 | 557.04 | 139.26 | 119.74 |
| 474 | 511.42 | 127.85 | 109.15 | 519 | 558.08 | 139.52 | 119.98 |
| 475 | 512.46 | 128.11 | 109.39 | 520 | 559.11 | 139.78 | 120.22 |
| 476 | 513.49 | 128.37 | 109.63 | 521 | 560.15 | 140.04 | 120.46 |
| 477 | 514.53 | 128.63 | 109.87 | 522 | 561.19 | 140.30 | 120.70 |
| 478 | 515.57 | 128.89 | 110.11 | 523 | 562.22 | 140.56 | 120.94 |
| 479 | 516.61 | 129.15 | 110.35 | 524 | 563.26 | 140.81 | 121.19 |
| 480 | 517.64 | 129.41 | 110.59 | 525 | 564.29 | 141.07 | 121.43 |
| 481 | 518.68 | 129.67 | 110.83 | 526 | 565.33 | 141.33 | 121.67 |
| 482 | 519.72 | 129.93 | 111.07 | 527 | 566.37 | 141.59 | 121.91 |
| 483 | 520.76 | 130.19 | 111.31 | 528 | 567.40 | 141.85 | 122.15 |
| 484 | 521.79 | 130.45 | 111.55 | 529 | 568.44 | 142.11 | 122.39 |
| 485 | 522.83 | 130.71 | 111.79 | 530 | 569.47 | 142.37 | 122.63 |
| 486 | 523.87 | 130.97 | 112.03 | 531 | 570.51 | 142.63 | 122.87 |
| 487 | 524.91 | 131.23 | 112.27 | 532 | 571.54 | 142.89 | 123.11 |
| 488 | 525.94 | 131.49 | 112.51 | 533 | 572.58 | 143.14 | 123.36 |
| 489 | 526.98 | 131.74 | 112.76 | 534 | 573.62 | 143.40 | 123.60 |
| 490 | 528.02 | 132.00 | 113.00 | 535 | 574.65 | 143.66 | 123.84 |
| 491 | 529.05 | 132.26 | 113.24 | 536 | 575.69 | 143.92 | 124.08 |
| 492 | 530.09 | 132.52 | 113.48 | 537 | 576.72 | 144.18 | 124.32 |
| 493 | 531.13 | 132.78 | 113.72 | 538 | 577.76 | 144.44 | 124.56 |
| 494 | 532.17 | 133.04 | 113.96 | 539 | 578.79 | 144.70 | 124.80 |

Table 1. CONSTANTS FOR $\bar{p}'$ (LONG FORMULA), AN APPROXIMATION OF THE UPPER BINOMIAL CONFIDENCE LIMIT $\bar{p}$; $c/n \leqslant .5$, $n \geqslant 20$, $c \leqslant 1,000$ (*Continued*)

Section 4
$\gamma_1 = 95\%$
$\gamma_2 = 90\%$

| c | $\bar{m}$ | $\bar{a}$ | $\bar{b}$ | c | $\bar{m}$ | $\bar{a}$ | $\bar{b}$ |
|---|---|---|---|---|---|---|---|
| 540 | 579.83 | 144.96 | 125.04 | 585 | 626.39 | 156.60 | 135.90 |
| 541 | 580.86 | 145.22 | 125.28 | 586 | 627.42 | 156.86 | 136.14 |
| 542 | 581.90 | 145.47 | 125.53 | 587 | 628.46 | 157.11 | 136.39 |
| 543 | 582.93 | 145.73 | 125.77 | 588 | 629.49 | 157.37 | 136.63 |
| 544 | 583.97 | 145.99 | 126.01 | 589 | 630.52 | 157.63 | 136.87 |
| 545 | 585.00 | 146.25 | 126.25 | 590 | 631.56 | 157.89 | 137.11 |
| 546 | 586.04 | 146.51 | 126.49 | 591 | 632.59 | 158.15 | 137.35 |
| 547 | 587.07 | 146.77 | 126.73 | 592 | 633.62 | 158.41 | 137.59 |
| 548 | 588.11 | 147.03 | 126.97 | 593 | 634.66 | 158.66 | 137.84 |
| 549 | 589.14 | 147.29 | 127.21 | 594 | 635.69 | 158.92 | 138.08 |
| 550 | 590.18 | 147.54 | 127.46 | 595 | 636.73 | 159.18 | 138.22 |
| 551 | 591.21 | 147.80 | 127.70 | 596 | 637.76 | 159.44 | 138.56 |
| 552 | 592.25 | 148.06 | 127.94 | 597 | 638.79 | 159.70 | 138.80 |
| 553 | 593.28 | 148.32 | 128.18 | 598 | 639.83 | 159.96 | 139.04 |
| 554 | 594.32 | 148.58 | 128.42 | 599 | 640.86 | 160.22 | 139.28 |
| 555 | 595.35 | 148.84 | 128.66 | 600 | 641.89 | 160.47 | 139.53 |
| 556 | 596.39 | 149.10 | 128.90 | 601 | 642.93 | 160.73 | 139.77 |
| 557 | 597.42 | 149.36 | 129.14 | 602 | 643.96 | 160.99 | 140.01 |
| 558 | 598.46 | 149.61 | 129.39 | 603 | 644.99 | 161.25 | 140.25 |
| 559 | 599.49 | 149.87 | 129.63 | 604 | 646.03 | 161.51 | 140.49 |
| 560 | 600.53 | 150.13 | 129.87 | 605 | 647.06 | 161.77 | 140.73 |
| 561 | 601.56 | 150.39 | 130.11 | 606 | 648.09 | 162.02 | 140.98 |
| 562 | 602.60 | 150.65 | 130.35 | 607 | 649.13 | 162.28 | 141.22 |
| 563 | 603.63 | 150.91 | 130.59 | 608 | 650.16 | 162.54 | 141.46 |
| 564 | 604.67 | 151.17 | 130.83 | 609 | 651.19 | 162.80 | 141.70 |
| 565 | 605.70 | 151.43 | 131.07 | 610 | 652.23 | 163.06 | 141.94 |
| 566 | 606.74 | 151.68 | 131.32 | 611 | 653.26 | 163.32 | 142.18 |
| 567 | 607.77 | 151.94 | 131.56 | 612 | 654.29 | 163.57 | 142.43 |
| 568 | 608.81 | 152.20 | 131.80 | 613 | 655.33 | 163.83 | 142.67 |
| 569 | 609.84 | 152.46 | 132.04 | 614 | 656.36 | 164.09 | 142.91 |
| 570 | 610.87 | 152.72 | 132.28 | 615 | 657.39 | 164.35 | 143.15 |
| 571 | 611.91 | 152.98 | 132.52 | 616 | 658.43 | 164.61 | 143.39 |
| 572 | 612.94 | 153.24 | 132.76 | 617 | 659.46 | 164.86 | 143.64 |
| 573 | 613.98 | 153.49 | 133.01 | 618 | 660.49 | 165.12 | 143.88 |
| 574 | 615.01 | 153.75 | 133.25 | 619 | 661.53 | 165.38 | 144.12 |
| 575 | 616.05 | 154.01 | 133.49 | 620 | 662.56 | 165.64 | 144.36 |
| 576 | 617.08 | 154.27 | 133.73 | 621 | 663.59 | 165.90 | 144.60 |
| 577 | 618.11 | 154.53 | 133.97 | 622 | 664.63 | 166.16 | 144.84 |
| 578 | 619.15 | 154.79 | 134.21 | 623 | 665.66 | 166.41 | 145.09 |
| 579 | 620.18 | 155.05 | 134.45 | 624 | 666.69 | 166.67 | 145.33 |
| 580 | 621.22 | 155.30 | 134.70 | 625 | 667.72 | 166.93 | 145.57 |
| 581 | 622.25 | 155.56 | 134.94 | 626 | 668.76 | 167.19 | 145.81 |
| 582 | 623.29 | 155.82 | 135.18 | 627 | 669.79 | 167.45 | 146.05 |
| 583 | 624.32 | 156.08 | 135.42 | 628 | 670.82 | 167.71 | 146.29 |
| 584 | 625.35 | 156.34 | 135.66 | 629 | 671.86 | 167.96 | 146.54 |

Table 1.   CONSTANTS FOR $\overline{p}'$ (LONG FORMULA), AN APPROXIMATION O
THE UPPER BINOMIAL CONFIDENCE LIMIT $\overline{p}$; $c/n \leqslant .5$, $n \geqslant 2$
$c \leqslant 1,000$ (*Continued*)

| c | $\overline{m}$ | $\overline{a}$ | $\overline{b}$ | c | $\overline{m}$ | $\overline{a}$ | |
|---|---|---|---|---|---|---|---|
| 630 | 672.89 | 168.22 | 146.78 | 675 | 719.34 | 179.83 | 157 |
| 631 | 673.92 | 168.48 | 147.02 | 676 | 720.37 | 180.09 | 157 |
| 632 | 674.95 | 168.74 | 147.26 | 677 | 721.40 | 180.35 | 158 |
| 633 | 675.99 | 169.00 | 147.50 | 678 | 722.43 | 180.61 | 158 |
| 634 | 677.02 | 169.25 | 147.75 | 679 | 723.46 | 180.87 | 158 |
| 635 | 678.05 | 169.51 | 147.99 | 680 | 724.49 | 181.12 | 158 |
| 636 | 679.08 | 169.77 | 148.23 | 681 | 725.53 | 181.38 | 159 |
| 637 | 680.12 | 170.03 | 148.47 | 682 | 726.56 | 181.64 | 159 |
| 638 | 681.15 | 170.29 | 148.71 | 683 | 727.59 | 181.90 | 159 |
| 639 | 682.18 | 170.55 | 148.95 | 684 | 728.62 | 182.15 | 159 |
| 640 | 683.21 | 170.80 | 149.20 | 685 | 729.65 | 182.41 | 160 |
| 641 | 684.25 | 171.06 | 149.44 | 686 | 730.68 | 182.67 | 160 |
| 642 | 685.28 | 171.32 | 149.68 | 687 | 731.71 | 182.93 | 160 |
| 643 | 686.31 | 171.58 | 149.92 | 688 | 732.75 | 183.19 | 160 |
| 644 | 687.34 | 171.84 | 150.16 | 689 | 733.78 | 183.44 | 161 |
| 645 | 688.38 | 172.09 | 150.41 | 690 | 734.81 | 183.70 | 161 |
| 646 | 689.41 | 172.35 | 150.65 | 691 | 735.84 | 183.96 | 161 |
| 647 | 690.44 | 172.61 | 150.89 | 692 | 736.87 | 184.22 | 161 |
| 648 | 691.47 | 172.87 | 151.13 | 693 | 737.90 | 184.48 | 162 |
| 649 | 692.51 | 173.13 | 151.37 | 694 | 738.93 | 184.73 | 162 |
| 650 | 693.54 | 173.38 | 151.62 | 695 | 739.96 | 184.99 | 162 |
| 651 | 694.57 | 173.64 | 151.86 | 696 | 741.00 | 185.25 | 162 |
| 652 | 695.60 | 173.90 | 152.10 | 697 | 742.03 | 185.51 | 162 |
| 653 | 696.63 | 174.16 | 152.34 | 698 | 743.06 | 185.76 | 163 |
| 654 | 697.67 | 174.42 | 152.58 | 699 | 744.09 | 186.02 | 163 |
| 655 | 698.70 | 174.67 | 152.83 | 700 | 745.12 | 186.28 | 163 |
| 656 | 699.73 | 174.93 | 153.07 | 701 | 746.15 | 186.54 | 163 |
| 657 | 700.76 | 175.19 | 153.31 | 702 | 747.18 | 186.80 | 164 |
| 658 | 701.79 | 175.45 | 153.55 | 703 | 748.21 | 187.05 | 164 |
| 659 | 702.83 | 175.71 | 153.79 | 704 | 749.24 | 187.31 | 164 |
| 660 | 703.86 | 175.96 | 154.04 | 705 | 750.27 | 187.57 | 164 |
| 661 | 704.89 | 176.22 | 154.28 | 706 | 751.31 | 187.83 | 165 |
| 662 | 705.92 | 176.48 | 154.52 | 707 | 752.34 | 188.08 | 165 |
| 663 | 706.95 | 176.74 | 154.76 | 708 | 753.37 | 188.34 | 165 |
| 664 | 707.99 | 177.00 | 155.00 | 709 | 754.40 | 188.60 | 165 |
| 665 | 709.02 | 177.25 | 155.25 | 710 | 755.43 | 188.86 | 166 |
| 666 | 710.05 | 177.51 | 155.49 | 711 | 756.46 | 189.11 | 166 |
| 667 | 711.08 | 177.77 | 155.73 | 712 | 757.49 | 189.37 | 166 |
| 668 | 712.11 | 178.03 | 155.97 | 713 | 758.52 | 189.63 | 166 |
| 669 | 713.15 | 178.29 | 156.21 | 714 | 759.55 | 189.89 | 167 |
| 670 | 714.18 | 178.54 | 156.46 | 715 | 760.58 | 190.15 | 167 |
| 671 | 715.21 | 178.80 | 156.70 | 716 | 761.61 | 190.40 | 167 |
| 672 | 716.24 | 179.06 | 156.94 | 717 | 762.64 | 190.66 | 167 |
| 673 | 717.27 | 179.32 | 157.18 | 718 | 763.68 | 190.92 | 168 |
| 674 | 718.30 | 179.58 | 157.42 | 719 | 764.71 | 191.18 | 168 |

Table 1.  CONSTANTS FOR $\overline{p}'$ (LONG FORMULA), AN APPROXIMATION OF
THE UPPER BINOMIAL CONFIDENCE LIMIT $\overline{p}$; $c/n \leqslant .5$, $n \geqslant 20$,
$c \leqslant 1,000$ (*Continued*)

Section 4

$\gamma_1 = 95\%$
$\gamma_2 = 90\%$

| c | $\overline{m}$ | $\overline{a}$ | $\overline{b}$ | c | $\overline{m}$ | $\overline{a}$ | $\overline{b}$ |
|---|---|---|---|---|---|---|---|
| 720 | 765.74 | 191.43 | 168.57 | 765 | 812.09 | 203.02 | 179.48 |
| 721 | 766.77 | 191.69 | 168.81 | 766 | 813.12 | 203.28 | 179.72 |
| 722 | 767.80 | 191.95 | 169.05 | 767 | 814.15 | 203.54 | 179.96 |
| 723 | 768.83 | 192.21 | 169.29 | 768 | 815.18 | 203.80 | 180.20 |
| 724 | 769.86 | 192.46 | 169.54 | 769 | 816.21 | 204.05 | 180.45 |
| 725 | 770.89 | 192.72 | 169.78 | 770 | 817.24 | 204.31 | 180.69 |
| 726 | 771.92 | 192.98 | 170.02 | 771 | 818.27 | 204.57 | 180.93 |
| 727 | 772.95 | 193.24 | 170.26 | 772 | 819.30 | 204.83 | 181.17 |
| 728 | 773.98 | 193.50 | 170.50 | 773 | 820.33 | 205.08 | 181.42 |
| 729 | 775.01 | 193.75 | 170.75 | 774 | 821.36 | 205.34 | 181.66 |
| 730 | 776.04 | 194.01 | 170.99 | 775 | 822.39 | 205.60 | 181.90 |
| 731 | 777.07 | 194.27 | 171.23 | 776 | 823.42 | 205.85 | 182.15 |
| 732 | 778.10 | 194.53 | 171.47 | 777 | 824.45 | 206.11 | 182.39 |
| 733 | 779.13 | 194.78 | 171.72 | 778 | 825.48 | 206.37 | 182.63 |
| 734 | 780.16 | 195.04 | 171.96 | 779 | 826.51 | 206.63 | 182.87 |
| 735 | 781.19 | 195.30 | 172.20 | 780 | 827.54 | 206.88 | 183.12 |
| 736 | 782.22 | 195.56 | 172.44 | 781 | 828.57 | 207.14 | 183.36 |
| 737 | 783.25 | 195.81 | 172.69 | 782 | 829.60 | 207.40 | 183.60 |
| 738 | 784.28 | 196.07 | 172.93 | 783 | 830.63 | 207.66 | 183.84 |
| 739 | 785.31 | 196.33 | 173.17 | 784 | 831.66 | 207.91 | 184.09 |
| 740 | 786.34 | 196.59 | 173.41 | 785 | 832.68 | 208.17 | 184.33 |
| 741 | 787.38 | 196.84 | 173.66 | 786 | 833.71 | 208.43 | 184.57 |
| 742 | 788.41 | 197.10 | 173.90 | 787 | 834.74 | 208.69 | 184.81 |
| 743 | 789.44 | 197.36 | 174.14 | 788 | 835.77 | 208.94 | 185.06 |
| 744 | 790.47 | 197.62 | 174.38 | 789 | 836.80 | 209.20 | 185.30 |
| 745 | 791.50 | 197.87 | 174.63 | 790 | 837.83 | 209.46 | 185.54 |
| 746 | 792.53 | 198.13 | 174.87 | 791 | 838.86 | 209.72 | 185.78 |
| 747 | 793.56 | 198.39 | 175.11 | 792 | 839.89 | 209.97 | 186.03 |
| 748 | 794.59 | 198.65 | 175.35 | 793 | 840.92 | 210.23 | 186.27 |
| 749 | 795.62 | 198.90 | 175.60 | 794 | 841.95 | 210.49 | 186.51 |
| 750 | 796.65 | 199.16 | 175.84 | 795 | 842.98 | 210.74 | 186.76 |
| 751 | 797.68 | 199.42 | 176.08 | 796 | 844.01 | 211.00 | 187.00 |
| 752 | 798.71 | 199.68 | 176.32 | 797 | 845.04 | 211.26 | 187.24 |
| 753 | 799.74 | 199.93 | 176.57 | 798 | 846.06 | 211.52 | 187.48 |
| 754 | 800.77 | 200.19 | 176.81 | 799 | 847.09 | 211.77 | 187.73 |
| 755 | 801.80 | 200.45 | 177.05 | 800 | 848.12 | 212.03 | 187.97 |
| 756 | 802.83 | 200.71 | 177.29 | 801 | 849.15 | 212.29 | 188.21 |
| 757 | 803.86 | 200.96 | 177.54 | 802 | 850.18 | 212.55 | 188.45 |
| 758 | 804.89 | 201.22 | 177.78 | 803 | 851.21 | 212.80 | 188.70 |
| 759 | 805.92 | 201.48 | 178.02 | 804 | 852.24 | 213.06 | 188.94 |
| 760 | 806.95 | 201.74 | 178.26 | 805 | 853.27 | 213.32 | 189.18 |
| 761 | 807.97 | 201.99 | 178.51 | 806 | 854.30 | 213.57 | 189.43 |
| 762 | 809.00 | 202.25 | 178.75 | 807 | 855.33 | 213.83 | 189.67 |
| 763 | 810.03 | 202.51 | 178.99 | 808 | 856.35 | 214.09 | 189.91 |
| 764 | 811.06 | 202.77 | 179.23 | 809 | 857.38 | 214.35 | 190.15 |

Table 1. CONSTANTS FOR $\overline{p}'$ (LONG FORMULA), AN APPROXIMATION OF THE UPPER BINOMIAL CONFIDENCE LIMIT $\overline{p}$; $c/n \leqslant .5$, $n \geqslant 20$ $c \leqslant 1,000$ (*Continued*)

| c | $\overline{m}$ | $\overline{a}$ | $\overline{b}$ | c | $\overline{m}$ | $\overline{a}$ | $\overline{b}$ |
|---|---|---|---|---|---|---|---|
| 810 | 858.41 | 214.60 | 190.40 | 855 | 904.69 | 226.17 | 201.. |
| 811 | 859.44 | 214.86 | 190.64 | 856 | 905.72 | 226.43 | 201.. |
| 812 | 860.47 | 215.12 | 190.88 | 857 | 906.75 | 226.69 | 201.. |
| 813 | 861.50 | 215.37 | 191.13 | 858 | 907.78 | 226.94 | 202.. |
| 814 | 862.53 | 215.63 | 191.37 | 859 | 908.81 | 227.20 | 202.. |
| 815 | 863.56 | 215.89 | 191.61 | 860 | 909.83 | 227.46 | 202. |
| 816 | 864.58 | 216.15 | 191.85 | 861 | 910.86 | 227.72 | 202. |
| 817 | 865.61 | 216.40 | 192.10 | 862 | 911.89 | 227.97 | 203.. |
| 818 | 866.64 | 216.66 | 192.34 | 863 | 912.92 | 228.23 | 203. |
| 819 | 867.67 | 216.92 | 192.58 | 864 | 913.95 | 228.49 | 203. |
| 820 | 868.70 | 217.17 | 192.83 | 865 | 914.97 | 228.74 | 203.. |
| 821 | 869.73 | 217.43 | 193.07 | 866 | 916.00 | 229.00 | 204.. |
| 822 | 870.76 | 217.69 | 193.31 | 867 | 917.03 | 229.26 | 204. |
| 823 | 871.79 | 217.95 | 193.55 | 868 | 918.06 | 229.51 | 204.. |
| 824 | 872.81 | 218.20 | 193.80 | 869 | 919.09 | 229.77 | 204.. |
| 825 | 873.84 | 218.46 | 194.04 | 870 | 920.11 | 230.03 | 204. |
| 826 | 874.87 | 218.72 | 194.28 | 871 | 921.14 | 230.29 | 205.. |
| 827 | 875.90 | 218.98 | 194.52 | 872 | 922.17 | 230.54 | 205.. |
| 828 | 876.93 | 219.23 | 194.77 | 873 | 923.20 | 230.80 | 205. |
| 829 | 877.96 | 219.49 | 195.01 | 874 | 924.23 | 231.06 | 205. |
| 830 | 878.99 | 219.75 | 195.25 | 875 | 925.25 | 231.31 | 206. |
| 831 | 880.01 | 220.00 | 195.50 | 876 | 926.28 | 231.57 | 206.. |
| 832 | 881.04 | 220.26 | 195.74 | 877 | 927.31 | 231.83 | 206.. |
| 833 | 882.07 | 220.52 | 195.98 | 878 | 928.34 | 232.08 | 206. |
| 834 | 883.10 | 220.78 | 196.22 | 879 | 929.36 | 232.34 | 207. |
| 835 | 884.13 | 221.03 | 196.47 | 880 | 930.39 | 232.60 | 207.. |
| 836 | 885.16 | 221.29 | 196.71 | 881 | 931.42 | 232.85 | 207.. |
| 837 | 886.19 | 221.55 | 196.95 | 882 | 932.45 | 233.11 | 207. |
| 838 | 887.21 | 221.80 | 197.20 | 883 | 933.47 | 233.37 | 208. |
| 839 | 888.24 | 222.06 | 197.44 | 884 | 934.50 | 233.63 | 208. |
| 840 | 889.27 | 222.32 | 197.68 | 885 | 935.53 | 233.88 | 208. |
| 841 | 890.30 | 222.57 | 197.93 | 886 | 936.56 | 234.14 | 208. |
| 842 | 891.33 | 222.83 | 198.17 | 887 | 937.59 | 234.40 | 209. |
| 843 | 892.36 | 223.09 | 198.41 | 888 | 938.61 | 234.65 | 209.. |
| 844 | 893.38 | 223.35 | 198.65 | 889 | 939.64 | 234.91 | 209. |
| 845 | 894.41 | 223.60 | 198.90 | 890 | 940.67 | 235.17 | 209. |
| 846 | 895.44 | 223.86 | 199.14 | 891 | 941.70 | 235.42 | 210.. |
| 847 | 896.47 | 224.12 | 199.38 | 892 | 942.72 | 235.68 | 210. |
| 848 | 897.50 | 224.37 | 199.63 | 893 | 943.75 | 235.94 | 210. |
| 849 | 898.53 | 224.63 | 199.87 | 894 | 944.78 | 236.19 | 210. |
| 850 | 899.55 | 224.89 | 200.11 | 895 | 945.81 | 236.45 | 211.. |
| 851 | 900.58 | 225.15 | 200.35 | 896 | 946.83 | 236.71 | 211. |
| 852 | 901.61 | 225.40 | 200.60 | 897 | 947.86 | 236.97 | 211. |
| 853 | 902.64 | 225.66 | 200.84 | 898 | 948.89 | 237.22 | 211. |
| 854 | 903.67 | 225.92 | 201.08 | 899 | 949.92 | 237.48 | 212.. |

Table 1. CONSTANTS FOR $\bar{p}'$ (LONG FORMULA), AN APPROXIMATION OF Section 4
THE UPPER BINOMIAL CONFIDENCE LIMIT $\bar{p}$; $c/n \leqslant .5$, $n \geqslant 20$, $\gamma_1 = 95\%$
$c \leqslant 1{,}000$ (*Continued*) $\gamma_2 = 90\%$

| c | $\bar{m}$ | $\bar{a}$ | $\bar{b}$ | c | $\bar{m}$ | $\bar{a}$ | $\bar{b}$ |
|---|---|---|---|---|---|---|---|
| 900 | 950.94 | 237.74 | 212.26 | 945 | 997.16 | 249.29 | 223.21 |
| 901 | 951.97 | 237.99 | 212.51 | 946 | 998.19 | 249.55 | 223.45 |
| 902 | 953.00 | 238.25 | 212.75 | 947 | 999.21 | 249.80 | 223.70 |
| 903 | 954.03 | 238.51 | 212.99 | 948 | 1000.24 | 250.06 | 223.94 |
| 904 | 955.05 | 238.76 | 213.24 | 949 | 1001.27 | 250.32 | 224.18 |
| 905 | 956.08 | 239.02 | 213.48 | 950 | 1002.29 | 250.57 | 224.43 |
| 906 | 957.11 | 239.28 | 213.72 | 951 | 1003.32 | 250.83 | 224.67 |
| 907 | 958.13 | 239.53 | 213.97 | 952 | 1004.35 | 251.09 | 224.91 |
| 908 | 959.16 | 239.79 | 214.21 | 953 | 1005.37 | 251.34 | 225.16 |
| 909 | 960.19 | 240.05 | 214.45 | 954 | 1006.40 | 251.60 | 225.40 |
| 910 | 961.22 | 240.30 | 214.70 | 955 | 1007.43 | 251.86 | 225.64 |
| 911 | 962.24 | 240.56 | 214.94 | 956 | 1008.45 | 252.11 | 225.89 |
| 912 | 963.27 | 240.82 | 215.18 | 957 | 1009.48 | 252.37 | 226.13 |
| 913 | 964.30 | 241.07 | 215.43 | 958 | 1010.51 | 252.63 | 226.37 |
| 914 | 965.32 | 241.33 | 215.67 | 959 | 1011.53 | 252.88 | 226.62 |
| 915 | 966.35 | 241.59 | 215.91 | 960 | 1012.56 | 253.14 | 226.86 |
| 916 | 967.38 | 241.84 | 216.16 | 961 | 1013.59 | 253.40 | 227.10 |
| 917 | 968.41 | 242.10 | 216.40 | 962 | 1014.61 | 253.65 | 227.35 |
| 918 | 969.43 | 242.36 | 216.64 | 963 | 1015.64 | 253.91 | 227.59 |
| 919 | 970.46 | 242.62 | 216.88 | 964 | 1016.67 | 254.17 | 227.83 |
| 920 | 971.49 | 242.87 | 217.13 | 965 | 1017.69 | 254.42 | 228.08 |
| 921 | 972.51 | 243.13 | 217.37 | 966 | 1018.72 | 254.68 | 228.32 |
| 922 | 973.54 | 243.39 | 217.61 | 967 | 1019.75 | 254.94 | 228.56 |
| 923 | 974.57 | 243.64 | 217.86 | 968 | 1020.77 | 255.19 | 228.81 |
| 924 | 975.60 | 243.90 | 218.10 | 969 | 1021.80 | 255.45 | 229.05 |
| 925 | 976.62 | 244.16 | 218.34 | 970 | 1022.82 | 255.71 | 229.29 |
| 926 | 977.65 | 244.41 | 218.59 | 971 | 1023.85 | 255.96 | 229.54 |
| 927 | 978.68 | 244.67 | 218.83 | 972 | 1024.88 | 256.22 | 229.78 |
| 928 | 979.70 | 244.93 | 219.07 | 973 | 1025.90 | 256.48 | 230.02 |
| 929 | 980.73 | 245.18 | 219.32 | 974 | 1026.93 | 256.73 | 230.27 |
| 930 | 981.76 | 245.44 | 219.56 | 975 | 1027.96 | 256.99 | 230.51 |
| 931 | 982.79 | 245.70 | 219.80 | 976 | 1028.98 | 257.25 | 230.75 |
| 932 | 983.81 | 245.95 | 220.05 | 977 | 1030.01 | 257.50 | 231.00 |
| 933 | 984.84 | 246.21 | 220.29 | 978 | 1031.04 | 257.76 | 231.24 |
| 934 | 985.87 | 246.47 | 220.53 | 979 | 1032.06 | 258.02 | 231.48 |
| 935 | 986.89 | 246.72 | 220.78 | 980 | 1033.09 | 258.27 | 231.73 |
| 936 | 987.92 | 246.98 | 221.02 | 981 | 1034.11 | 258.53 | 231.97 |
| 937 | 988.95 | 247.24 | 221.26 | 982 | 1035.14 | 258.79 | 232.21 |
| 938 | 989.97 | 247.49 | 221.51 | 983 | 1036.17 | 259.04 | 232.46 |
| 939 | 991.00 | 247.75 | 221.75 | 984 | 1037.19 | 259.30 | 232.70 |
| 940 | 992.03 | 248.01 | 221.99 | 985 | 1038.22 | 259.55 | 232.95 |
| 941 | 993.05 | 248.26 | 222.24 | 986 | 1039.25 | 259.81 | 233.19 |
| 942 | 994.08 | 248.52 | 222.48 | 987 | 1040.27 | 260.07 | 233.43 |
| 943 | 995.11 | 248.78 | 222.72 | 988 | 1041.30 | 260.32 | 233.68 |
| 944 | 996.13 | 249.03 | 222.97 | 989 | 1042.32 | 260.58 | 233.92 |

Table 1. CONSTANTS FOR $\overline{p}'$ (LONG FORMULA), AN APPROXIMATION OF THE UPPER BINOMIAL CONFIDENCE LIMIT $\overline{p}$; c/n ⩽ .5, n ⩾ 20, c ⩽ 1,000 (*Continued*)

| c | $\overline{m}$ | $\overline{a}$ | $\overline{b}$ | c | $\overline{m}$ | $\overline{a}$ | $\overline{b}$ |
|---|---|---|---|---|---|---|---|
| 990 | 1043.35 | 260.84 | 234.16 | 995 | 1048.48 | 262.12 | 235.38 |
| 991 | 1044.38 | 261.09 | 234.41 | 996 | 1049.51 | 262.38 | 235.62 |
| 992 | 1045.40 | 261.35 | 234.65 | 997 | 1050.53 | 262.63 | 235.87 |
| 993 | 1046.43 | 261.61 | 234.89 | 998 | 1051.56 | 262.89 | 236.11 |
| 994 | 1047.45 | 261.86 | 235.14 | 999 | 1052.58 | 263.15 | 236.35 |
| | | | | 1000 | 1053.61 | 263.40 | 236.60 |

Table 1. CONSTANTS FOR $\bar{p}'$ (LONG FORMULA), AN APPROXIMATION OF THE UPPER BINOMIAL CONFIDENCE LIMIT $\bar{p}$; c/n $\leqslant$ .5, n $\geqslant$ 20, c $\leqslant$ 1,000 (*Continued*)

Section 5

$\gamma_1$ = 90%

$\gamma_2$ = 80%

| c | $\bar{m}$ | $\bar{a}$ | $\bar{b}$ | c | $\bar{m}$ | $\bar{a}$ | $\bar{b}$ |
|---|-----------|-----------|-----------|---|-----------|-----------|-----------|
| 0 | 2.3026 | 0.387 | -0.764 | 45 | 54.878 | 14.942 | 10.061 |
| 1 | 3.8897 | 0.702 | -0.741 | 46 | 55.972 | 15.363 | 10.442 |
| 2 | 5.3223 | 0.964 | -0.696 | 47 | 57.066 | 15.603 | 10.631 |
| 3 | 6.6808 | 1.255 | -0.585 | 48 | 58.158 | 15.929 | 10.911 |
| 4 | 7.9936 | 1.560 | -0.435 | 49 | 59.249 | 16.269 | 11.206 |
| 5 | 9.2747 | 1.883 | -0.249 | 50 | 60.340 | 16.707 | 11.607 |
| 6 | 10.532 | 2.210 | -0.049 | 51 | 61.429 | 16.822 | 11.665 |
| 7 | 11.771 | 2.568 | 0.195 | 52 | 62.518 | 17.256 | 12.063 |
| 8 | 12.995 | 2.928 | 0.449 | 53 | 63.606 | 17.653 | 12.421 |
| 9 | 14.206 | 3.310 | 0.731 | 54 | 64.693 | 17.943 | 12.667 |
| 10 | 15.407 | 3.705 | 1.033 | 55 | 65.779 | 18.153 | 12.828 |
| 11 | 16.598 | 4.041 | 1.278 | 56 | 66.865 | 18.600 | 13.239 |
| 12 | 17.782 | 4.372 | 1.517 | 57 | 67.949 | 18.780 | 13.37 |
| 13 | 18.958 | 4.696 | 1.754 | 58 | 69.033 | 19.201 | 13.75 |
| 14 | 20.128 | 5.023 | 1.997 | 59 | 70.117 | 19.580 | 14.09 |
| 15 | 21.292 | 5.343 | 2.235 | 60 | 71.199 | 19.69 | 14.15 |
| 16 | 22.452 | 5.677 | 2.491 | 61 | 72.281 | 20.09 | 14.51 |
| 17 | 23.606 | 6.003 | 2.741 | 62 | 73.362 | 20.56 | 14.95 |
| 18 | 24.756 | 6.338 | 3.003 | 63 | 74.443 | 20.73 | 15.08 |
| 19 | 25.903 | 6.656 | 3.248 | 64 | 75.523 | 20.96 | 15.27 |
| 20 | 27.045 | 6.962 | 3.483 | 65 | 76.602 | 21.25 | 15.52 |
| 21 | 28.184 | 7.273 | 3.723 | 66 | 77.681 | 21.65 | 15.88 |
| 22 | 29.320 | 7.631 | 4.017 | 67 | 78.759 | 22.10 | 16.29 |
| 23 | 30.453 | 7.937 | 4.257 | 68 | 79.837 | 22.29 | 16.44 |
| 24 | 31.584 | 8.264 | 4.520 | 69 | 80.914 | 22.65 | 16.76 |
| 25 | 32.711 | 8.574 | 4.765 | 70 | 81.990 | 23.04 | 17.12 |
| 26 | 33.836 | 8.924 | 5.056 | 71 | 83.066 | 23.30 | 17.34 |
| 27 | 34.959 | 9.231 | 5.301 | 72 | 84.142 | 23.68 | 17.69 |
| 28 | 36.080 | 9.575 | 5.588 | 73 | 85.216 | 23.88 | 17.84 |
| 29 | 37.199 | 9.845 | 5.795 | 74 | 86.291 | 24.18 | 18.10 |
| 30 | 38.315 | 10.165 | 6.056 | 75 | 87.365 | 24.63 | 18.53 |
| 31 | 39.430 | 10.506 | 6.343 | 76 | 88.438 | 24.87 | 18.73 |
| 32 | 40.543 | 10.814 | 6.594 | 77 | 89.511 | 25.38 | 19.21 |
| 33 | 41.654 | 11.132 | 6.857 | 78 | 90.584 | 25.67 | 19.46 |
| 34 | 42.764 | 11.511 | 7.186 | 79 | 91.656 | 25.74 | 19.49 |
| 35 | 43.872 | 11.769 | 7.386 | 80 | 92.727 | 26:00 | 19.71 |
| 36 | 44.978 | 12.138 | 7.706 | 81 | 93.798 | 26.57 | 20.25 |
| 37 | 46.083 | 12.449 | 7.964 | 82 | 94.869 | 26.92 | 20.57 |
| 38 | 47.187 | 12.713 | 8.174 | 83 | 95.939 | 27.13 | 20.74 |
| 39 | 48.289 | 13.101 | 8.515 | 84 | 97.009 | 27.60 | 21.18 |
| 40 | 49.390 | 13.349 | 8.708 | 85 | 98.078 | 27.82 | 21.37 |
| 41 | 50.490 | 13.799 | 9.117 | 86 | 99.147 | 27.94 | 21.44 |
| 42 | 51.589 | 13.999 | 9.261 | 87 | 100.22 | 28.26 | 21.73 |
| 43 | 52.686 | 14.327 | 9.541 | 88 | 101.28 | 28.89 | 22.34 |
| 44 | 53.783 | 14.741 | 9.913 | 89 | 102.35 | 28.90 | 22.30 |

Table 1.  CONSTANTS FOR $\overline{p}'$ (LONG FORMULA), AN APPROXIMATION OF
THE UPPER BINOMIAL CONFIDENCE LIMIT $\overline{p}$; $c/n \leqslant .5$, $n \geqslant 20$,
$c \leqslant 1,000$ (*Continued*)

| c | $\overline{m}$ | $\overline{a}$ | $\overline{b}$ | c | $\overline{m}$ | $\overline{a}$ | $\overline{b}$ |
|---|---|---|---|---|---|---|---|
| 90 | 103.42 | 29.30 | 22.67 | 135 | 151.14 | 37.79 | 29.7? |
| 91 | 104.49 | 29.48 | 22.82 | 136 | 152.20 | 38.05 | 29.9? |
| 92 | 105.55 | 29.96 | 23.27 | 137 | 153.25 | 38.32 | 30.1? |
| 93 | 106.62 | 30.44 | 23.72 | 138 | 154.31 | 38.58 | 30.4? |
| 94 | 107.69 | 30.57 | 23.81 | 139 | 155.36 | 38.84 | 30.6? |
| 95 | 108.75 | 31.16 | 24.38 | 140 | 156.42 | 39.11 | 30.8? |
| 96 | 109.82 | 31.07 | 24.24 | 141 | 157.47 | 39.37 | 31.1? |
| 97 | 110.88 | 31.31 | 24.44 | 142 | 158.52 | 39.63 | 31.3? |
| 98 | 111.95 | 32.03 | 25.15 | 143 | 159.58 | 39.90 | 31.6? |
| 99 | 113.01 | 32.06 | 25.13 | 144 | 160.63 | 40.16 | 31.8? |
| 100 | 114.07 | 32.45 | 25.49 | 145 | 161.68 | 40.42 | 32.0? |
| 101 | 115.14 | 28.79 | 21.71 | 146 | 162.74 | 40.69 | 32.3? |
| 102 | 116.20 | 29.06 | 21.94 | 147 | 163.79 | 40.95 | 32.5? |
| 103 | 117.26 | 29.32 | 22.18 | 148 | 164.84 | 41.21 | 32.7? |
| 104 | 118.33 | 29.59 | 22.41 | 149 | 165.89 | 41.48 | 33.0? |
| 105 | 119.39 | 29.85 | 22.65 | 150 | 166.95 | 41.74 | 33.2? |
| 106 | 120.45 | 30.12 | 22.88 | 151 | 168.00 | 42.00 | 33.5? |
| 107 | 121.51 | 30.38 | 23.12 | 152 | 169.05 | 42.27 | 33.7? |
| 108 | 122.58 | 30.65 | 23.35 | 153 | 170.10 | 42.53 | 33.9? |
| 109 | 123.64 | 30.91 | 23.59 | 154 | 171.15 | 42.79 | 34.2? |
| 110 | 124.70 | 31.18 | 23.82 | 155 | 172.21 | 43.06 | 34.4? |
| 111 | 125.76 | 31.44 | 24.06 | 156 | 173.26 | 43.32 | 34.6? |
| 112 | 126.82 | 31.71 | 24.29 | 157 | 174.31 | 43.58 | 34.9? |
| 113 | 127.88 | 31.97 | 24.53 | 158 | 175.36 | 43.84 | 35.1? |
| 114 | 128.94 | 32.24 | 24.76 | 159 | 176.41 | 44.11 | 35.3? |
| 115 | 130.00 | 32.50 | 25.00 | 160 | 177.46 | 44.37 | 35.6? |
| 116 | 131.06 | 32.77 | 25.23 | 161 | 178.53 | 44.63 | 35.8? |
| 117 | 132.12 | 33.03 | 25.47 | 162 | 179.58 | 44.89 | 36.1? |
| 118 | 133.18 | 33.30 | 25.70 | 163 | 180.63 | 45.16 | 36.3? |
| 119 | 134.24 | 33.56 | 25.94 | 164 | 181.68 | 45.42 | 36.5? |
| 120 | 135.29 | 33.83 | 26.17 | 165 | 182.73 | 45.68 | 36.8? |
| 121 | 136.35 | 34.09 | 26.41 | 166 | 183.78 | 45.94 | 37.0? |
| 122 | 137.41 | 34.36 | 26.64 | 167 | 184.83 | 46.21 | 37.2? |
| 123 | 138.47 | 34.62 | 26.88 | 168 | 185.87 | 46.47 | 37.5? |
| 124 | 139.53 | 34.89 | 27.11 | 169 | 186.92 | 46.73 | 37.7? |
| 125 | 140.58 | 35.15 | 27.35 | 170 | 187.97 | 46.99 | 38.0? |
| 126 | 141.64 | 35.41 | 27.59 | 171 | 189.02 | 47.26 | 38.2? |
| 127 | 142.70 | 35.68 | 27.82 | 172 | 190.07 | 47.52 | 38.4? |
| 128 | 143.75 | 35.94 | 28.06 | 173 | 191.12 | 47.78 | 38.7? |
| 129 | 144.81 | 36.21 | 28.29 | 174 | 192.17 | 48.04 | 38.9? |
| 130 | 145.87 | 36.47 | 28.53 | 175 | 193.22 | 48.30 | 39.2? |
| 131 | 146.92 | 36.73 | 28.77 | 176 | 194.26 | 48.57 | 39.4? |
| 132 | 147.98 | 37.00 | 29.00 | 177 | 195.31 | 48.83 | 39.6? |
| 133 | 149.03 | 37.26 | 29.24 | 178 | 196.36 | 49.09 | 39.9? |
| 134 | 150.09 | 37.53 | 29.47 | 179 | 197.41 | 49.35 | 40.1? |

**Table 1. CONSTANTS FOR $\overline{p}'$ (LONG FORMULA), AN APPROXIMATION OF THE UPPER BINOMIAL CONFIDENCE LIMIT $\overline{p}$; $c/n \leqslant .5$, $n \geqslant 20$, $c \leqslant 1{,}000$ (Continued)**

| c | $\overline{m}$ | $\overline{a}$ | $\overline{b}$ | c | $\overline{m}$ | $\overline{a}$ | $\overline{b}$ |
|---|---|---|---|---|---|---|---|
| 180 | 198.46 | 49.61 | 40.39 | 225 | 245.48 | 61.37 | 51.13 |
| 181 | 199.50 | 49.88 | 40.62 | 226 | 246.52 | 61.63 | 51.37 |
| 182 | 200.55 | 50.14 | 40.86 | 227 | 247.57 | 61.89 | 51.61 |
| 183 | 201.60 | 50.40 | 41.10 | 228 | 248.61 | 62.15 | 51.85 |
| 184 | 202.65 | 50.66 | 41.34 | 229 | 249.65 | 62.41 | 52.09 |
| 185 | 203.69 | 50.92 | 41.58 | 230 | 250.69 | 62.67 | 52.33 |
| 186 | 204.74 | 51.18 | 41.82 | 231 | 251.73 | 62.93 | 52.57 |
| 187 | 205.79 | 51.45 | 42.05 | 232 | 252.78 | 63.19 | 52.81 |
| 188 | 206.83 | 51.71 | 42.29 | 233 | 253.82 | 63.45 | 53.05 |
| 189 | 207.88 | 51.97 | 42.53 | 234 | 254.86 | 63.72 | 53.28 |
| 190 | 208.93 | 52.23 | 42.77 | 235 | 255.90 | 63.98 | 53.52 |
| 191 | 209.97 | 52.49 | 43.01 | 236 | 256.94 | 64.24 | 53.76 |
| 192 | 211.02 | 52.75 | 43.25 | 237 | 257.99 | 64.50 | 54.00 |
| 193 | 212.06 | 53.02 | 43.48 | 238 | 259.03 | 64.76 | 54.24 |
| 194 | 213.11 | 53.28 | 43.72 | 239 | 260.07 | 65.02 | 54.48 |
| 195 | 214.16 | 53.54 | 43.96 | 240 | 261.11 | 65.28 | 54.72 |
| 196 | 215.20 | 53.80 | 44.20 | 241 | 262.15 | 65.54 | 54.96 |
| 197 | 216.25 | 54.06 | 44.44 | 242 | 263.19 | 65.80 | 55.20 |
| 198 | 217.29 | 54.32 | 44.68 | 243 | 264.23 | 66.06 | 55.44 |
| 199 | 218.34 | 54.58 | 44.92 | 244 | 265.27 | 66.32 | 55.68 |
| 200 | 219.38 | 54.85 | 45.15 | 245 | 266.32 | 66.58 | 55.92 |
| 201 | 220.43 | 55.11 | 45.39 | 246 | 267.36 | 66.84 | 56.16 |
| 202 | 221.47 | 55.37 | 45.63 | 247 | 268.40 | 67.10 | 56.40 |
| 203 | 222.52 | 55.63 | 45.87 | 248 | 269.44 | 67.36 | 56.64 |
| 204 | 223.56 | 55.89 | 46.11 | 249 | 270.48 | 67.62 | 56.88 |
| 205 | 224.61 | 56.15 | 46.35 | 250 | 271.52 | 67.88 | 57.12 |
| 206 | 225.65 | 56.41 | 46.59 | 251 | 272.56 | 68.14 | 57.36 |
| 207 | 226.70 | 56.67 | 46.83 | 252 | 273.60 | 68.40 | 57.60 |
| 208 | 227.74 | 56.94 | 47.06 | 253 | 274.64 | 68.66 | 57.84 |
| 209 | 228.79 | 57.20 | 47.30 | 254 | 275.68 | 68.92 | 58.08 |
| 210 | 229.83 | 57.46 | 47.54 | 255 | 276.72 | 69.18 | 58.32 |
| 211 | 230.87 | 57.72 | 47.78 | 256 | 277.76 | 69.44 | 58.56 |
| 212 | 231.92 | 57.98 | 48.02 | 257 | 278.80 | 69.70 | 58.80 |
| 213 | 232.96 | 58.24 | 48.26 | 258 | 279.84 | 69.96 | 59.04 |
| 214 | 234.01 | 58.50 | 48.50 | 259 | 280.88 | 70.22 | 59.28 |
| 215 | 235.05 | 58.76 | 48.74 | 260 | 281.92 | 70.48 | 59.52 |
| 216 | 236.09 | 59.02 | 48.98 | 261 | 282.96 | 70.74 | 59.76 |
| 217 | 237.14 | 59.28 | 49.22 | 262 | 284.00 | 71.00 | 60.00 |
| 218 | 238.18 | 59.55 | 49.45 | 263 | 285.04 | 71.26 | 60.24 |
| 219 | 239.22 | 59.81 | 49.69 | 264 | 286.08 | 71.52 | 60.48 |
| 220 | 240.27 | 60.07 | 49.93 | 265 | 287.12 | 71.78 | 60.72 |
| 221 | 241.31 | 60.33 | 50.17 | 266 | 288.16 | 72.04 | 60.96 |
| 222 | 242.35 | 60.59 | 50.41 | 267 | 289.19 | 72.30 | 61.20 |
| 223 | 243.40 | 60.85 | 50.65 | 268 | 290.23 | 72.56 | 61.44 |
| 224 | 244.44 | 61.11 | 50.89 | 269 | 291.27 | 72.82 | 61.68 |

Table 1. CONSTANTS FOR $\overline{p}'$ (LONG FORMULA), AN APPROXIMATION OF THE UPPER BINOMIAL CONFIDENCE LIMIT $\overline{p}$; $c/n \leqslant .5$, $n \geqslant 20$, $c \leqslant 1,000$ (*Continued*)

| c | $\overline{m}$ | $\overline{a}$ | $\overline{b}$ | c | $\overline{m}$ | $\overline{a}$ | $\overline{b}$ |
|---|---|---|---|---|---|---|---|
| 270 | 292.31 | 73.08 | 61.92 | 315 | 339.00 | 84.75 | 72.75 |
| 271 | 293.35 | 73.34 | 62.16 | 316 | 340.03 | 85.01 | 72.99 |
| 272 | 294.39 | 73.60 | 62.40 | 317 | 341.07 | 85.27 | 73.23 |
| 273 | 295.43 | 73.86 | 62.64 | 318 | 342.10 | 85.53 | 73.47 |
| 274 | 296.47 | 74.12 | 62.88 | 319 | 343.14 | 85.79 | 73.71 |
| 275 | 297.51 | 74.38 | 63.12 | 320 | 344.18 | 86.04 | 73.96 |
| 276 | 298.54 | 74.64 | 63.36 | 321 | 345.21 | 86.30 | 74.20 |
| 277 | 299.58 | 74.90 | 63.60 | 322 | 346.25 | 86.56 | 74.44 |
| 278 | 300.62 | 75.16 | 63.84 | 323 | 347.28 | 86.82 | 74.68 |
| 279 | 301.66 | 75.41 | 64.09 | 324 | 348.32 | 87.08 | 74.92 |
| 280 | 302.70 | 75.67 | 64.33 | 325 | 349.35 | 87.34 | 75.16 |
| 281 | 303.74 | 75.93 | 64.57 | 326 | 350.39 | 87.60 | 75.40 |
| 282 | 304.77 | 76.19 | 64.81 | 327 | 351.42 | 87.86 | 75.64 |
| 283 | 305.81 | 76.45 | 65.05 | 328 | 352.46 | 88.12 | 75.88 |
| 284 | 306.85 | 76.71 | 65.29 | 329 | 353.50 | 88.37 | 76.13 |
| 285 | 307.89 | 76.97 | 65.53 | 330 | 354.53 | 88.63 | 76.37 |
| 286 | 308.93 | 77.23 | 65.77 | 331 | 355.57 | 88.89 | 76.61 |
| 287 | 309.96 | 77.49 | 66.01 | 332 | 356.60 | 89.15 | 76.85 |
| 288 | 311.00 | 77.75 | 66.25 | 333 | 357.64 | 89.41 | 77.09 |
| 289 | 312.04 | 78.01 | 66.49 | 334 | 358.67 | 89.67 | 77.33 |
| 290 | 313.08 | 78.27 | 66.73 | 335 | 359.71 | 89.93 | 77.57 |
| 291 | 314.11 | 78.53 | 66.97 | 336 | 360.74 | 90.19 | 77.81 |
| 292 | 315.15 | 78.79 | 67.21 | 337 | 361.78 | 90.44 | 78.06 |
| 293 | 316.19 | 79.05 | 67.45 | 338 | 362.81 | 90.70 | 78.30 |
| 294 | 317.23 | 79.31 | 67.69 | 339 | 363.85 | 90.96 | 78.54 |
| 295 | 318.26 | 79.57 | 67.93 | 340 | 364.88 | 91.22 | 78.78 |
| 296 | 319.30 | 79.83 | 68.17 | 341 | 365.91 | 91.48 | 79.02 |
| 297 | 320.34 | 80.08 | 68.42 | 342 | 366.95 | 91.74 | 79.26 |
| 298 | 321.38 | 80.34 | 68.66 | 343 | 367.98 | 92.00 | 79.50 |
| 299 | 322.41 | 80.60 | 68.90 | 344 | 369.02 | 92.25 | 79.75 |
| 300 | 323.45 | 80.86 | 69.14 | 345 | 370.05 | 92.51 | 79.99 |
| 301 | 324.49 | 81.12 | 69.38 | 346 | 371.09 | 92.77 | 80.23 |
| 302 | 325.52 | 81.38 | 69.62 | 347 | 372.12 | 93.03 | 80.47 |
| 303 | 326.56 | 81.64 | 69.86 | 348 | 373.16 | 93.29 | 80.71 |
| 304 | 327.60 | 81.90 | 70.10 | 349 | 374.19 | 93.55 | 80.95 |
| 305 | 328.63 | 82.16 | 70.34 | 350 | 375.22 | 93.81 | 81.19 |
| 306 | 329.67 | 82.42 | 70.58 | 351 | 376.26 | 94.06 | 81.44 |
| 307 | 330.71 | 82.68 | 70.82 | 352 | 377.29 | 94.32 | 81.68 |
| 308 | 331.74 | 82.94 | 71.06 | 353 | 378.33 | 94.58 | 81.92 |
| 309 | 332.78 | 83.19 | 71.31 | 354 | 379.36 | 94.84 | 82.16 |
| 310 | 333.82 | 83.45 | 71.55 | 355 | 380.40 | 95.10 | 82.40 |
| 311 | 334.85 | 83.71 | 71.79 | 356 | 381.43 | 95.36 | 82.64 |
| 312 | 335.89 | 83.97 | 72.03 | 357 | 382.46 | 95.62 | 82.88 |
| 313 | 336.92 | 84.23 | 72.27 | 358 | 383.50 | 95.87 | 83.13 |
| 314 | 337.96 | 84.49 | 72.51 | 359 | 384.53 | 96.13 | 83.37 |

# Table 1. CONSTANTS FOR $\overline{p}'$ (LONG FORMULA), AN APPROXIMATION OF THE UPPER BINOMIAL CONFIDENCE LIMIT $\overline{p}$; $c/n \leqslant .5$, $n \geqslant 20$, $c \leqslant 1,000$ (*Continued*)

| c | $\overline{m}$ | $\overline{a}$ | $\overline{b}$ | c | $\overline{m}$ | $\overline{a}$ | $\overline{b}$ |
|---|---|---|---|---|---|---|---|
| 360 | 385.56 | 96.39 | 83.61 | 405 | 432.04 | 108.01 | 94.49 |
| 361 | 386.60 | 96.65 | 83.85 | 406 | 433.07 | 108.27 | 94.73 |
| 362 | 387.63 | 96.91 | 84.09 | 407 | 434.10 | 108.53 | 94.97 |
| 363 | 388.67 | 97.17 | 84.33 | 408 | 435.13 | 108.78 | 95.22 |
| 364 | 389.70 | 97.42 | 84.58 | 409 | 436.16 | 109.04 | 95.46 |
| 365 | 390.73 | 97.68 | 84.82 | 410 | 437.20 | 109.30 | 95.70 |
| 366 | 391.77 | 97.94 | 85.06 | 411 | 438.23 | 109.56 | 95.94 |
| 367 | 392.80 | 98.20 | 85.30 | 412 | 439.26 | 109.81 | 96.19 |
| 368 | 393.83 | 98.46 | 85.54 | 413 | 440.29 | 110.07 | 96.43 |
| 369 | 394.87 | 98.72 | 85.78 | 414 | 441.32 | 110.33 | 96.67 |
| 370 | 395.90 | 98.97 | 86.03 | 415 | 442.35 | 110.59 | 96.91 |
| 371 | 396.93 | 99.23 | 86.27 | 416 | 443.39 | 110.85 | 97.15 |
| 372 | 397.97 | 99.49 | 86.51 | 417 | 444.42 | 111.10 | 97.40 |
| 373 | 399.00 | 99.75 | 86.75 | 418 | 445.45 | 111.36 | 97.64 |
| 374 | 400.03 | 100.01 | 86.99 | 419 | 446.48 | 111.62 | 97.88 |
| 375 | 401.07 | 100.27 | 87.23 | 420 | 447.51 | 111.88 | 98.12 |
| 376 | 402.10 | 100.52 | 87.48 | 421 | 448.54 | 112.14 | 98.36 |
| 377 | 403.13 | 100.78 | 87.72 | 422 | 449.57 | 112.39 | 98.61 |
| 378 | 404.16 | 101.04 | 87.96 | 423 | 450.60 | 112.65 | 98.85 |
| 379 | 405.20 | 101.30 | 88.20 | 424 | 451.63 | 112.91 | 99.09 |
| 380 | 406.23 | 101.56 | 88.44 | 425 | 452.67 | 113.17 | 99.33 |
| 381 | 407.26 | 101.82 | 88.68 | 426 | 453.70 | 113.42 | 99.58 |
| 382 | 408.30 | 102.07 | 88.93 | 427 | 454.73 | 113.68 | 99.82 |
| 383 | 409.33 | 102.33 | 89.17 | 428 | 455.76 | 113.94 | 100.06 |
| 384 | 410.36 | 102.59 | 89.41 | 429 | 456.79 | 114.20 | 100.30 |
| 385 | 411.39 | 102.85 | 89.65 | 430 | 457.82 | 114.46 | 100.54 |
| 386 | 412.43 | 103.11 | 89.89 | 431 | 458.85 | 114.71 | 100.79 |
| 387 | 413.46 | 103.36 | 90.14 | 432 | 459.88 | 114.97 | 101.03 |
| 388 | 414.49 | 103.62 | 90.38 | 433 | 460.91 | 115.23 | 101.27 |
| 389 | 415.52 | 103.88 | 90.62 | 434 | 461.94 | 115.49 | 101.51 |
| 390 | 416.56 | 104.14 | 90.86 | 435 | 462.97 | 115.74 | 101.76 |
| 391 | 417.59 | 104.40 | 91.10 | 436 | 464.01 | 116.00 | 102.00 |
| 392 | 418.62 | 104.66 | 91.34 | 437 | 465.04 | 116.26 | 102.24 |
| 393 | 419.65 | 104.91 | 91.59 | 438 | 466.07 | 116.52 | 102.48 |
| 394 | 420.69 | 105.17 | 91.83 | 439 | 467.10 | 116.77 | 102.73 |
| 395 | 421.72 | 105.43 | 92.07 | 440 | 468.13 | 117.03 | 102.97 |
| 396 | 422.75 | 105.69 | 92.31 | 441 | 469.16 | 117.29 | 103.21 |
| 397 | 423.78 | 105.95 | 92.55 | 442 | 470.19 | 117.55 | 103.45 |
| 398 | 424.81 | 106.20 | 92.80 | 443 | 471.22 | 117.80 | 103.70 |
| 399 | 425.85 | 106.46 | 93.04 | 444 | 472.25 | 118.06 | 103.94 |
| 400 | 426.88 | 106.72 | 93.28 | 445 | 473.28 | 118.32 | 104.18 |
| 401 | 427.91 | 106.98 | 93.52 | 446 | 474.31 | 118.58 | 104.42 |
| 402 | 428.94 | 107.24 | 93.76 | 447 | 475.34 | 118.84 | 104.66 |
| 403 | 429.97 | 107.49 | 94.01 | 448 | 476.37 | 119.09 | 104.91 |
| 404 | 431.01 | 107.75 | 94.25 | 449 | 477.40 | 119.35 | 105.15 |

Table 1.   CONSTANTS FOR $\overline{p}'$ (LONG FORMULA), AN APPROXIMATION OF THE UPPER BINOMIAL CONFIDENCE LIMIT $\overline{p}$; $c/n \leqslant .5$, $n \geqslant 20$, $c \leqslant 1{,}000$ (*Continued*)

| c | $\overline{m}$ | $\overline{a}$ | $\overline{b}$ | c | $\overline{m}$ | $\overline{a}$ | $\overline{b}$ |
|---|---|---|---|---|---|---|---|
| 450 | 478.43 | 119.61 | 105.39 | 495 | 524.76 | 131.19 | 116.31 |
| 451 | 479.46 | 119.87 | 105.63 | 496 | 525.79 | 131.45 | 116.55 |
| 452 | 480.49 | 120.12 | 105.88 | 497 | 526.81 | 131.70 | 116.80 |
| 453 | 481.52 | 120.38 | 106.12 | 498 | 527.84 | 131.96 | 117.04 |
| 454 | 482.55 | 120.64 | 106.36 | 499 | 528.87 | 132.22 | 117.28 |
| 455 | 483.58 | 120.90 | 106.60 | 500 | 529.90 | 132.48 | 117.52 |
| 456 | 484.61 | 121.15 | 106.85 | 501 | 530.93 | 132.73 | 117.77 |
| 457 | 485.64 | 121.41 | 107.09 | 502 | 531.96 | 132.99 | 118.01 |
| 458 | 486.67 | 121.67 | 107.33 | 503 | 532.99 | 133.25 | 118.25 |
| 459 | 487.70 | 121.93 | 107.57 | 504 | 534.01 | 133.50 | 118.50 |
| 460 | 488.73 | 122.18 | 107.82 | 505 | 535.04 | 133.76 | 118.74 |
| 461 | 489.76 | 122.44 | 108.06 | 506 | 536.07 | 134.02 | 118.98 |
| 462 | 490.79 | 122.70 | 108.30 | 507 | 537.10 | 134.27 | 119.23 |
| 463 | 491.82 | 122.96 | 108.54 | 508 | 538.13 | 134.53 | 119.47 |
| 464 | 492.85 | 123.21 | 108.79 | 509 | 539.16 | 134.79 | 119.71 |
| 465 | 493.88 | 123.47 | 109.03 | 510 | 540.19 | 135.05 | 119.95 |
| 466 | 494.91 | 123.73 | 109.27 | 511 | 541.21 | 135.30 | 120.20 |
| 467 | 495.94 | 123.98 | 109.52 | 512 | 542.24 | 135.56 | 120.44 |
| 468 | 496.97 | 124.24 | 109.76 | 513 | 543.27 | 135.82 | 120.68 |
| 469 | 498.00 | 124.50 | 110.00 | 514 | 544.30 | 136.07 | 120.93 |
| 470 | 499.03 | 124.76 | 110.24 | 515 | 545.33 | 136.33 | 121.17 |
| 471 | 500.06 | 125.01 | 110.49 | 516 | 546.35 | 136.59 | 121.41 |
| 472 | 501.09 | 125.27 | 110.73 | 517 | 547.38 | 136.85 | 121.65 |
| 473 | 502.12 | 125.53 | 110.97 | 518 | 548.41 | 137.10 | 121.90 |
| 474 | 503.15 | 125.79 | 111.21 | 519 | 549.44 | 137.36 | 122.14 |
| 475 | 504.18 | 126.04 | 111.46 | 520 | 550.47 | 137.62 | 122.38 |
| 476 | 505.20 | 126.30 | 111.70 | 521 | 551.50 | 137.87 | 122.63 |
| 477 | 506.23 | 126.56 | 111.94 | 522 | 552.52 | 138.13 | 122.87 |
| 478 | 507.26 | 126.82 | 112.18 | 523 | 553.55 | 138.39 | 123.11 |
| 479 | 508.29 | 127.07 | 112.43 | 524 | 554.58 | 138.64 | 123.36 |
| 480 | 509.32 | 127.33 | 112.67 | 525 | 555.61 | 138.90 | 123.60 |
| 481 | 510.35 | 127.59 | 112.91 | 526 | 556.64 | 139.16 | 123.84 |
| 482 | 511.38 | 127.85 | 113.15 | 527 | 557.66 | 139.42 | 124.08 |
| 483 | 512.41 | 128.10 | 113.40 | 528 | 558.69 | 139.67 | 124.33 |
| 484 | 513.44 | 128.36 | 113.64 | 529 | 559.72 | 139.93 | 124.57 |
| 485 | 514.47 | 128.62 | 113.88 | 530 | 560.75 | 140.19 | 124.81 |
| 486 | 515.50 | 128.87 | 114.13 | 531 | 561.77 | 140.44 | 125.06 |
| 487 | 516.53 | 129.13 | 114.37 | 532 | 562.80 | 140.70 | 125.30 |
| 488 | 517.55 | 129.39 | 114.61 | 533 | 563.83 | 140.96 | 125.54 |
| 489 | 518.58 | 129.65 | 114.85 | 534 | 564.86 | 141.21 | 125.79 |
| 490 | 519.61 | 129.90 | 115.10 | 535 | 565.89 | 141.47 | 126.03 |
| 491 | 520.64 | 130.16 | 115.34 | 536 | 566.91 | 141.73 | 126.27 |
| 492 | 521.67 | 130.42 | 115.58 | 537 | 567.94 | 141.99 | 126.51 |
| 493 | 522.70 | 130.67 | 115.83 | 538 | 568.97 | 142.24 | 126.76 |
| 494 | 523.73 | 130.93 | 116.07 | 539 | 570.00 | 142.50 | 127.00 |

Table 1.   CONSTANTS FOR p̄′ (LONG FORMULA), AN APPROXIMATION OF    Section 5
THE UPPER BINOMIAL CONFIDENCE LIMIT p̄; c/n ≤ .5, n ≥ 20,    γ₁ = 90%
c ≤ 1,000 (*Continued*)    γ₂ = 80%

| c | m̄ | ā | b̄ | c | m̄ | ā | b̄ |
|---|---|---|---|---|---|---|---|
| 540 | 571.02 | 142.76 | 127.24 | 585 | 617.24 | 154.31 | 138.19 |
| 541 | 572.05 | 143.01 | 127.49 | 586 | 618.26 | 154.57 | 138.43 |
| 542 | 573.08 | 143.27 | 127.73 | 587 | 619.29 | 154.82 | 138.68 |
| 543 | 574.11 | 143.53 | 127.97 | 588 | 620.32 | 155.08 | 138.92 |
| 544 | 575.13 | 143.78 | 128.22 | 589 | 621.34 | 155.34 | 139.16 |
| 545 | 576.16 | 144.04 | 128.46 | 590 | 622.37 | 155.59 | 139.41 |
| 546 | 577.19 | 144.30 | 128.70 | 591 | 623.40 | 155.85 | 139.65 |
| 547 | 578.22 | 144.55 | 128.95 | 592 | 624.42 | 156.11 | 139.89 |
| 548 | 579.24 | 144.81 | 129.19 | 593 | 625.45 | 156.36 | 140.14 |
| 549 | 580.27 | 145.07 | 129.43 | 594 | 626.48 | 156.62 | 140.38 |
| 550 | 581.30 | 145.32 | 129.68 | 595 | 627.50 | 156.88 | 140.62 |
| 551 | 582.32 | 145.58 | 129.92 | 596 | 628.53 | 157.13 | 140.87 |
| 552 | 583.35 | 145.84 | 130.16 | 597 | 629.55 | 157.39 | 141.11 |
| 553 | 584.38 | 146.09 | 130.41 | 598 | 630.58 | 157.65 | 141.35 |
| 554 | 585.41 | 146.35 | 130.65 | 599 | 631.61 | 157.90 | 141.60 |
| 555 | 586.43 | 146.61 | 130.89 | 600 | 632.63 | 158.16 | 141.84 |
| 556 | 587.46 | 146.87 | 131.13 | 601 | 633.66 | 158.41 | 142.09 |
| 557 | 588.49 | 147.12 | 131.38 | 602 | 634.69 | 158.67 | 142.33 |
| 558 | 589.52 | 147.38 | 131.62 | 603 | 635.71 | 158.93 | 142.57 |
| 559 | 590.54 | 147.64 | 131.86 | 604 | 636.74 | 159.18 | 142.82 |
| 560 | 591.57 | 147.89 | 132.11 | 605 | 637.76 | 159.44 | 143.06 |
| 561 | 592.60 | 148.15 | 132.35 | 606 | 638.79 | 159.70 | 143.30 |
| 562 | 593.62 | 148.41 | 132.59 | 607 | 639.82 | 159.95 | 143.55 |
| 563 | 594.65 | 148.66 | 132.84 | 608 | 640.84 | 160.21 | 143.79 |
| 564 | 595.68 | 148.92 | 133.08 | 609 | 641.87 | 160.47 | 144.03 |
| 565 | 596.70 | 149.18 | 133.32 | 610 | 642.89 | 160.72 | 144.28 |
| 566 | 597.73 | 149.43 | 133.57 | 611 | 643.92 | 160.98 | 144.52 |
| 567 | 598.76 | 149.69 | 133.81 | 612 | 644.94 | 161.24 | 144.76 |
| 568 | 599.78 | 149.95 | 134.05 | 613 | 645.97 | 161.49 | 145.01 |
| 569 | 600.81 | 150.20 | 134.30 | 614 | 647.00 | 161.75 | 145.25 |
| 570 | 601.84 | 150.46 | 134.54 | 615 | 648.02 | 162.01 | 145.49 |
| 571 | 602.87 | 150.72 | 134.78 | 616 | 649.05 | 162.26 | 145.74 |
| 572 | 603.89 | 150.97 | 135.03 | 617 | 650.07 | 162.52 | 145.98 |
| 573 | 604.92 | 151.23 | 135.27 | 618 | 651.10 | 162.77 | 146.23 |
| 574 | 605.95 | 151.49 | 135.51 | 619 | 652.13 | 163.03 | 146.47 |
| 575 | 606.97 | 151.74 | 135.76 | 620 | 653.15 | 163.29 | 146.71 |
| 576 | 608.00 | 152.00 | 136.00 | 621 | 654.18 | 163.54 | 146.96 |
| 577 | 609.03 | 152.26 | 136.24 | 622 | 655.20 | 163.80 | 147.20 |
| 578 | 610.05 | 152.51 | 136.49 | 623 | 656.23 | 164.06 | 147.44 |
| 579 | 611.08 | 152.77 | 136.73 | 624 | 657.25 | 164.31 | 147.69 |
| 580 | 612.11 | 153.03 | 136.97 | 625 | 658.28 | 164.57 | 147.93 |
| 581 | 613.13 | 153.28 | 137.22 | 626 | 659.31 | 164.83 | 148.17 |
| 582 | 614.16 | 153.54 | 137.46 | 627 | 660.33 | 165.08 | 148.42 |
| 583 | 615.19 | 153.80 | 137.70 | 628 | 661.36 | 165.34 | 148.66 |
| 584 | 616.21 | 154.05 | 137.95 | 629 | 662.38 | 165.60 | 148.90 |

Table 1.   CONSTANTS FOR $\overline{p}'$ (LONG FORMULA), AN APPROXIMATION OF THE UPPER BINOMIAL CONFIDENCE LIMIT $\overline{p}$; $c/n \leqslant .5, n \geqslant 20$, $c \leqslant 1{,}000$ (*Continued*)

| c | $\overline{m}$ | $\overline{a}$ | $\overline{b}$ | c | $\overline{m}$ | $\overline{a}$ | $\overline{b}$ |
|---|---|---|---|---|---|---|---|
| 630 | 663.41 | 165.85 | 149.15 | 675 | 709.54 | 177.38 | 160.12 |
| 631 | 664.43 | 166.11 | 149.39 | 676 | 710.56 | 177.64 | 160.36 |
| 632 | 665.46 | 166.36 | 149.64 | 677 | 711.58 | 177.90 | 160.60 |
| 633 | 666.48 | 166.62 | 149.88 | 678 | 712.61 | 178.15 | 160.85 |
| 634 | 667.51 | 166.88 | 150.12 | 679 | 713.63 | 178.41 | 161.09 |
| 635 | 668.53 | 167.13 | 150.37 | 680 | 714.66 | 178.66 | 161.34 |
| 636 | 669.56 | 167.39 | 150.61 | 681 | 715.68 | 178.92 | 161.58 |
| 637 | 670.59 | 167.65 | 150.85 | 682 | 716.71 | 179.18 | 161.82 |
| 638 | 671.61 | 167.90 | 151.10 | 683 | 717.73 | 179.43 | 162.07 |
| 639 | 672.64 | 168.16 | 151.34 | 684 | 718.76 | 179.69 | 162.31 |
| 640 | 673.66 | 168.42 | 151.58 | 685 | 719.78 | 179.95 | 162.55 |
| 641 | 674.69 | 168.67 | 151.83 | 686 | 720.81 | 180.20 | 162.80 |
| 642 | 675.71 | 168.93 | 152.07 | 687 | 721.83 | 180.46 | 163.04 |
| 643 | 676.74 | 169.18 | 152.32 | 688 | 722.85 | 180.71 | 163.29 |
| 644 | 677.76 | 169.44 | 152.56 | 689 | 723.88 | 180.97 | 163.53 |
| 645 | 678.79 | 169.70 | 152.80 | 690 | 724.90 | 181.23 | 163.77 |
| 646 | 679.81 | 169.95 | 153.05 | 691 | 725.93 | 181.48 | 164.02 |
| 647 | 680.84 | 170.21 | 153.29 | 692 | 726.95 | 181.74 | 164.26 |
| 648 | 681.86 | 170.47 | 153.53 | 693 | 727.98 | 181.99 | 164.51 |
| 649 | 682.89 | 170.72 | 153.78 | 694 | 729.00 | 182.25 | 164.75 |
| 650 | 683.91 | 170.98 | 154.02 | 695 | 730.03 | 182.51 | 164.99 |
| 651 | 684.94 | 171.23 | 154.27 | 696 | 731.05 | 182.76 | 165.24 |
| 652 | 685.96 | 171.49 | 154.51 | 697 | 732.07 | 183.02 | 165.48 |
| 653 | 686.99 | 171.75 | 154.75 | 698 | 733.10 | 183.27 | 165.73 |
| 654 | 688.01 | 172.00 | 155.00 | 699 | 734.12 | 183.53 | 165.97 |
| 655 | 689.04 | 172.26 | 155.24 | 700 | 735.15 | 183.79 | 166.21 |
| 656 | 690.06 | 172.52 | 155.48 | 701 | 736.17 | 184.04 | 166.46 |
| 657 | 691.09 | 172.77 | 155.73 | 702 | 737.19 | 184.30 | 166.70 |
| 658 | 692.11 | 173.03 | 155.97 | 703 | 738.22 | 184.55 | 166.95 |
| 659 | 693.14 | 173.28 | 156.22 | 704 | 739.24 | 184.81 | 167.19 |
| 660 | 694.16 | 173.54 | 156.46 | 705 | 740.27 | 185.07 | 167.43 |
| 661 | 695.19 | 173.80 | 156.70 | 706 | 741.29 | 185.32 | 167.68 |
| 662 | 696.21 | 174.05 | 156.95 | 707 | 742.32 | 185.58 | 167.92 |
| 663 | 697.24 | 174.31 | 157.19 | 708 | 743.34 | 185.83 | 168.17 |
| 664 | 698.26 | 174.57 | 157.43 | 709 | 744.36 | 186.09 | 168.41 |
| 665 | 699.29 | 174.82 | 157.68 | 710 | 745.39 | 186.35 | 168.65 |
| 666 | 700.31 | 175.08 | 157.92 | 711 | 746.41 | 186.60 | 168.90 |
| 667 | 701.34 | 175.33 | 158.17 | 712 | 747.44 | 186.86 | 169.14 |
| 668 | 702.36 | 175.59 | 158.41 | 713 | 748.46 | 187.11 | 169.39 |
| 669 | 703.39 | 175.85 | 158.65 | 714 | 749.48 | 187.37 | 169.63 |
| 670 | 704.41 | 176.10 | 158.90 | 715 | 750.51 | 187.63 | 169.87 |
| 671 | 705.44 | 176.36 | 159.14 | 716 | 751.53 | 187.88 | 170.12 |
| 672 | 706.46 | 176.62 | 159.38 | 717 | 752.56 | 188.14 | 170.36 |
| 673 | 707.49 | 176.87 | 159.63 | 718 | 753.58 | 188.39 | 170.61 |
| 674 | 708.51 | 177.13 | 159.87 | 719 | 754.60 | 188.65 | 170.85 |

Table 1. **CONSTANTS FOR $\overline{p}'$ (LONG FORMULA), AN APPROXIMATION OF THE UPPER BINOMIAL CONFIDENCE LIMIT $\overline{p}$; $c/n \leqslant .5$, $n \geqslant 20$, $c \leqslant 1,000$** (*Continued*)

Section 5
$\gamma_1 = 90\%$
$\gamma_2 = 80\%$

| c | $\overline{m}$ | $\overline{a}$ | $\overline{b}$ | c | $\overline{m}$ | $\overline{a}$ | $\overline{b}$ |
|---|---|---|---|---|---|---|---|
| 720 | 755.63 | 188.91 | 171.09 | 765 | 801.68 | 200.42 | 182.08 |
| 721 | 756.65 | 189.16 | 171.34 | 766 | 802.71 | 200.68 | 182.32 |
| 722 | 757.67 | 189.42 | 171.58 | 767 | 803.73 | 200.93 | 182.57 |
| 723 | 758.70 | 189.67 | 171.83 | 768 | 804.75 | 201.19 | 182.81 |
| 724 | 759.72 | 189.93 | 172.07 | 769 | 805.78 | 201.44 | 183.06 |
| 725 | 760.75 | 190.19 | 172.31 | 770 | 806.80 | 201.70 | 183.30 |
| 726 | 761.77 | 190.44 | 172.56 | 771 | 807.82 | 201.96 | 183.54 |
| 727 | 762.79 | 190.70 | 172.80 | 772 | 808.85 | 202.21 | 183.79 |
| 728 | 763.82 | 190.95 | 173.05 | 773 | 809.87 | 202.47 | 184.03 |
| 729 | 764.84 | 191.21 | 173.29 | 774 | 810.89 | 202.72 | 184.28 |
| 730 | 765.86 | 191.47 | 173.53 | 775 | 811.92 | 202.98 | 184.52 |
| 731 | 766.89 | 191.72 | 173.78 | 776 | 812.94 | 203.23 | 184.77 |
| 732 | 767.91 | 191.98 | 174.02 | 777 | 813.96 | 203.49 | 185.01 |
| 733 | 768.94 | 192.23 | 174.27 | 778 | 814.98 | 203.75 | 185.25 |
| 734 | 769.96 | 192.49 | 174.51 | 779 | 816.01 | 204.00 | 185.50 |
| 735 | 770.98 | 192.75 | 174.75 | 780 | 817.03 | 204.26 | 185.74 |
| 736 | 772.01 | 193.00 | 175.00 | 781 | 818.05 | 204.51 | 185.99 |
| 737 | 773.03 | 193.26 | 175.24 | 782 | 819.08 | 204.77 | 186.23 |
| 738 | 774.05 | 193.51 | 175.49 | 783 | 820.10 | 205.02 | 186.48 |
| 739 | 775.08 | 193.77 | 175.73 | 784 | 821.12 | 205.28 | 186.72 |
| 740 | 776.10 | 194.03 | 175.97 | 785 | 822.14 | 205.54 | 186.96 |
| 741 | 777.12 | 194.28 | 176.22 | 786 | 823.17 | 205.79 | 187.21 |
| 742 | 778.15 | 194.54 | 176.46 | 787 | 824.19 | 206.05 | 187.45 |
| 743 | 779.17 | 194.79 | 176.71 | 788 | 825.21 | 206.30 | 187.70 |
| 744 | 780.19 | 195.05 | 176.95 | 789 | 826.24 | 206.56 | 187.94 |
| 745 | 781.22 | 195.30 | 177.20 | 790 | 827.26 | 206.81 | 188.19 |
| 746 | 782.24 | 195.56 | 177.44 | 791 | 828.28 | 207.07 | 188.43 |
| 747 | 783.27 | 195.82 | 177.68 | 792 | 829.30 | 207.33 | 188.67 |
| 748 | 784.29 | 196.07 | 177.93 | 793 | 830.33 | 207.58 | 188.92 |
| 749 | 785.31 | 196.33 | 178.17 | 794 | 831.35 | 207.84 | 189.16 |
| 750 | 786.34 | 196.58 | 178.42 | 795 | 832.37 | 208.09 | 189.41 |
| 751 | 787.36 | 196.84 | 178.66 | 796 | 833.40 | 208.35 | 189.65 |
| 752 | 788.38 | 197.10 | 178.90 | 797 | 834.42 | 208.60 | 189.90 |
| 753 | 789.41 | 197.35 | 179.15 | 798 | 835.44 | 208.86 | 190.14 |
| 754 | 790.43 | 197.61 | 179.39 | 799 | 836.46 | 209.12 | 190.38 |
| 755 | 791.45 | 197.86 | 179.64 | 800 | 837.49 | 209.37 | 190.63 |
| 756 | 792.48 | 198.12 | 179.88 | 801 | 838.51 | 209.63 | 190.87 |
| 757 | 793.50 | 198.37 | 180.13 | 802 | 839.53 | 209.88 | 191.12 |
| 758 | 794.52 | 198.63 | 180.37 | 803 | 840.55 | 210.14 | 191.36 |
| 759 | 795.55 | 198.89 | 180.61 | 804 | 841.58 | 210.39 | 191.61 |
| 760 | 796.57 | 199.14 | 180.86 | 805 | 842.60 | 210.65 | 191.85 |
| 761 | 797.59 | 199.40 | 181.10 | 806 | 843.62 | 210.91 | 192.09 |
| 762 | 798.61 | 199.65 | 181.35 | 807 | 844.64 | 211.16 | 192.34 |
| 763 | 799.64 | 199.91 | 181.59 | 808 | 845.67 | 211.42 | 192.58 |
| 764 | 800.66 | 200.17 | 181.83 | 809 | 846.69 | 211.67 | 192.83 |

Table 1. CONSTANTS FOR $\overline{p}'$ (LONG FORMULA), AN APPROXIMATION OF THE UPPER BINOMIAL CONFIDENCE LIMIT $\overline{p}$; $c/n \leqslant .5$, $n \geqslant 20$, $c \leqslant 1,000$ (*Continued*)

| c | $\overline{m}$ | $\overline{a}$ | $\overline{b}$ | c | $\overline{m}$ | $\overline{a}$ | $\overline{b}$ |
|---|---|---|---|---|---|---|---|
| 810 | 847.71 | 211.93 | 193.07 | 855 | 893.71 | 223.43 | 204.07 |
| 811 | 848.73 | 212.18 | 193.32 | 856 | 894.73 | 223.68 | 204.32 |
| 812 | 849.76 | 212.44 | 193.56 | 857 | 895.75 | 223.94 | 204.56 |
| 813 | 850.78 | 212.69 | 193.81 | 858 | 896.78 | 224.19 | 204.81 |
| 814 | 851.80 | 212.95 | 194.05 | 859 | 897.80 | 224.45 | 205.05 |
| 815 | 852.82 | 213.21 | 194.29 | 860 | 898.82 | 224.70 | 205.30 |
| 816 | 853.85 | 213.46 | 194.54 | 861 | 899.84 | 224.96 | 205.54 |
| 817 | 854.87 | 213.72 | 194.78 | 862 | 900.86 | 225.22 | 205.78 |
| 818 | 855.89 | 213.97 | 195.03 | 863 | 901.89 | 225.47 | 206.03 |
| 819 | 856.91 | 214.23 | 195.27 | 864 | 902.91 | 225.73 | 206.27 |
| 820 | 857.94 | 214.48 | 195.52 | 865 | 903.93 | 225.98 | 206.52 |
| 821 | 858.96 | 214.74 | 195.76 | 866 | 904.95 | 226.24 | 206.76 |
| 822 | 859.98 | 215.00 | 196.00 | 867 | 905.97 | 226.49 | 207.01 |
| 823 | 861.00 | 215.25 | 196.25 | 868 | 906.99 | 226.75 | 207.25 |
| 824 | 862.03 | 215.51 | 196.49 | 869 | 908.02 | 227.00 | 207.50 |
| 825 | 863.05 | 215.76 | 196.74 | 870 | 909.04 | 227.26 | 207.74 |
| 826 | 864.07 | 216.02 | 196.98 | 871 | 910.06 | 227.51 | 207.99 |
| 827 | 865.09 | 216.27 | 197.23 | 872 | 911.08 | 227.77 | 208.23 |
| 828 | 866.11 | 216.53 | 197.47 | 873 | 912.10 | 228.03 | 208.47 |
| 829 | 867.14 | 216.78 | 197.72 | 874 | 913.12 | 228.28 | 208.72 |
| 830 | 868.16 | 217.04 | 197.96 | 875 | 914.15 | 228.54 | 208.96 |
| 831 | 869.18 | 217.30 | 198.20 | 876 | 915.17 | 228.79 | 209.21 |
| 832 | 870.20 | 217.55 | 198.45 | 877 | 916.19 | 229.05 | 209.45 |
| 833 | 871.23 | 217.81 | 198.69 | 878 | 917.21 | 229.30 | 209.70 |
| 834 | 872.25 | 218.06 | 198.94 | 879 | 918.23 | 229.56 | 209.94 |
| 835 | 873.27 | 218.32 | 199.18 | 880 | 919.25 | 229.81 | 210.19 |
| 836 | 874.29 | 218.57 | 199.43 | 881 | 920.28 | 230.07 | 210.43 |
| 837 | 875.31 | 218.83 | 199.67 | 882 | 921.30 | 230.32 | 210.68 |
| 838 | 876.34 | 219.08 | 199.92 | 883 | 922.32 | 230.58 | 210.92 |
| 839 | 877.36 | 219.34 | 200.16 | 884 | 923.34 | 230.84 | 211.16 |
| 840 | 878.38 | 219.60 | 200.40 | 885 | 924.36 | 231.09 | 211.41 |
| 841 | 879.40 | 219.85 | 200.65 | 886 | 925.38 | 231.35 | 211.65 |
| 842 | 880.42 | 220.11 | 200.89 | 887 | 926.40 | 231.60 | 211.90 |
| 843 | 881.45 | 220.36 | 201.14 | 888 | 927.43 | 231.86 | 212.14 |
| 844 | 882.47 | 220.62 | 201.38 | 889 | 928.45 | 232.11 | 212.39 |
| 845 | 883.49 | 220.87 | 201.63 | 890 | 929.47 | 232.37 | 212.63 |
| 846 | 884.51 | 221.13 | 201.87 | 891 | 930.49 | 232.62 | 212.88 |
| 847 | 885.53 | 221.38 | 202.12 | 892 | 931.51 | 232.88 | 213.12 |
| 848 | 886.56 | 221.64 | 202.36 | 893 | 932.53 | 233.13 | 213.37 |
| 849 | 887.58 | 221.89 | 202.61 | 894 | 933.56 | 233.39 | 213.61 |
| 850 | 888.60 | 222.15 | 202.85 | 895 | 934.58 | 233.64 | 213.86 |
| 851 | 889.62 | 222.41 | 203.09 | 896 | 935.60 | 233.90 | 214.10 |
| 852 | 890.64 | 222.66 | 203.34 | 897 | 936.62 | 234.15 | 214.35 |
| 853 | 891.67 | 222.92 | 203.58 | 898 | 937.64 | 234.41 | 214.59 |
| 854 | 892.69 | 223.17 | 203.83 | 899 | 938.66 | 234.67 | 214.83 |

Table 1. CONSTANTS FOR $\overline{p}'$ (LONG FORMULA), AN APPROXIMATION OF THE UPPER BINOMIAL CONFIDENCE LIMIT $\overline{p}$; $c/n \leqslant .5$, $n \geqslant 20$, $c \leqslant 1,000$ (*Continued*)

Section 5
$\gamma_1 = 90\%$
$\gamma_2 = 80\%$

| c | $\overline{m}$ | $\overline{a}$ | $\overline{b}$ | c | $\overline{m}$ | $\overline{a}$ | $\overline{b}$ |
|---|---|---|---|---|---|---|---|
| 900 | 939.68 | 234.92 | 215.08 | 945 | 985.63 | 246.41 | 226.09 |
| 901 | 940.70 | 235.18 | 215.32 | 946 | 986.65 | 246.66 | 226.34 |
| 902 | 941.73 | 235.43 | 215.57 | 947 | 987.67 | 246.92 | 226.58 |
| 903 | 942.75 | 235.69 | 215.81 | 948 | 988.69 | 247.17 | 226.83 |
| 904 | 943.77 | 235.94 | 216.06 | 949 | 989.72 | 247.43 | 227.07 |
| 905 | 944.79 | 236.20 | 216.30 | 950 | 990.74 | 247.68 | 227.32 |
| 906 | 945.81 | 236.45 | 216.55 | 951 | 991.76 | 247.94 | 227.56 |
| 907 | 946.83 | 236.71 | 216.79 | 952 | 992.78 | 248.19 | 227.81 |
| 908 | 947.85 | 236.96 | 217.04 | 953 | 993.80 | 248.45 | 228.05 |
| 909 | 948.88 | 237.22 | 217.28 | 954 | 994.82 | 248.70 | 228.30 |
| 910 | 949.90 | 237.47 | 217.53 | 955 | 995.84 | 248.96 | 228.54 |
| 911 | 950.92 | 237.73 | 217.77 | 956 | 996.86 | 249.22 | 228.78 |
| 912 | 951.94 | 237.98 | 218.02 | 957 | 997.88 | 249.47 | 229.03 |
| 913 | 952.96 | 238.24 | 218.26 | 958 | 998.90 | 249.73 | 229.27 |
| 914 | 953.98 | 238.50 | 218.50 | 959 | 999.92 | 249.98 | 229.52 |
| 915 | 955.00 | 238.75 | 218.75 | 960 | 1000.94 | 250.24 | 229.76 |
| 916 | 956.02 | 239.01 | 218.99 | 961 | 1001.96 | 250.49 | 230.01 |
| 917 | 957.04 | 239.26 | 219.24 | 962 | 1002.98 | 250.75 | 230.25 |
| 918 | 958.07 | 239.52 | 219.48 | 963 | 1004.01 | 251.00 | 230.50 |
| 919 | 959.09 | 239.77 | 219.73 | 964 | 1005.03 | 251.26 | 230.74 |
| 920 | 960.11 | 240.03 | 219.97 | 965 | 1006.05 | 251.51 | 230.99 |
| 921 | 961.13 | 240.28 | 220.22 | 966 | 1007.07 | 251.77 | 231.23 |
| 922 | 962.15 | 240.54 | 220.46 | 967 | 1008.09 | 252.02 | 231.48 |
| 923 | 963.17 | 240.79 | 220.71 | 968 | 1009.11 | 252.28 | 231.72 |
| 924 | 964.19 | 241.05 | 220.95 | 969 | 1010.13 | 252.53 | 231.97 |
| 925 | 965.21 | 241.30 | 221.20 | 970 | 1011.15 | 252.79 | 232.21 |
| 926 | 966.23 | 241.56 | 221.44 | 971 | 1012.17 | 253.04 | 232.46 |
| 927 | 967.26 | 241.81 | 221.69 | 972 | 1013.19 | 253.30 | 232.70 |
| 928 | 968.28 | 242.07 | 221.93 | 973 | 1014.21 | 253.55 | 232.95 |
| 929 | 969.30 | 242.32 | 222.18 | 974 | 1015.23 | 253.81 | 233.19 |
| 930 | 970.32 | 242.58 | 222.42 | 975 | 1016.25 | 254.06 | 233.44 |
| 931 | 971.34 | 242.83 | 222.67 | 976 | 1017.27 | 254.32 | 233.68 |
| 932 | 972.36 | 243.09 | 222.91 | 977 | 1018.29 | 254.57 | 233.93 |
| 933 | 973.38 | 243.35 | 223.15 | 978 | 1019.31 | 254.83 | 234.17 |
| 934 | 974.40 | 243.60 | 223.40 | 979 | 1020.33 | 255.08 | 234.42 |
| 935 | 975.42 | 243.86 | 223.64 | 980 | 1021.35 | 255.34 | 234.66 |
| 936 | 976.44 | 244.11 | 223.89 | 981 | 1022.38 | 255.59 | 234.91 |
| 937 | 977.47 | 244.37 | 224.13 | 982 | 1023.40 | 255.85 | 235.15 |
| 938 | 978.49 | 244.62 | 224.38 | 983 | 1024.42 | 256.10 | 235.40 |
| 939 | 979.51 | 244.88 | 224.62 | 984 | 1025.44 | 256.36 | 235.64 |
| 940 | 980.53 | 245.13 | 224.87 | 985 | 1026.46 | 256.61 | 235.89 |
| 941 | 981.55 | 245.39 | 225.11 | 986 | 1027.48 | 256.87 | 236.13 |
| 942 | 982.57 | 245.64 | 225.36 | 987 | 1028.50 | 257.12 | 236.38 |
| 943 | 983.59 | 245.90 | 225.60 | 988 | 1029.52 | 257.38 | 236.62 |
| 944 | 984.61 | 246.15 | 225.85 | 989 | 1030.54 | 257.63 | 236.87 |

Table 1.  CONSTANTS FOR $\overline{p}'$ (LONG FORMULA), AN APPROXIMATION OF THE UPPER BINOMIAL CONFIDENCE LIMIT $\overline{p}$; $c/n \leqslant .5$, $n \geqslant 20$, $c \leqslant 1,000$ (*Continued*)

| c | $\overline{m}$ | $\overline{a}$ | $\overline{b}$ | c | $\overline{m}$ | $\overline{a}$ | $\overline{b}$ |
|---|---|---|---|---|---|---|---|
| 990 | 1031.56 | 257.89 | 237.11 | 995 | 1036.66 | 259.17 | 238.3? |
| 991 | 1032.58 | 258.14 | 237.36 | 996 | 1037.68 | 259.42 | 238.5? |
| 992 | 1033.60 | 258.40 | 237.60 | 997 | 1038.70 | 259.68 | 238.8? |
| 993 | 1034.62 | 258.65 | 237.85 | 998 | 1039.72 | 259.93 | 239.0? |
| 994 | 1035.64 | 258.91 | 238.09 | 999 | 1040.74 | 260.19 | 239.3? |
|  |  |  |  | 1000 | 1041.76 | 260.44 | 239.5? |

**Table 1. CONSTANTS FOR $\overline{p}'$ (LONG FORMULA), AN APPROXIMATION OF THE UPPER BINOMIAL CONFIDENCE LIMIT $\overline{p}$; $c/n \leqslant .5$, $n \geqslant 20$, $c \leqslant 1,000$ (*Continued*)**

| c | $\overline{m}$ | $\overline{a}$ | $\overline{b}$ | c | $\overline{m}$ | $\overline{a}$ | $\overline{b}$ |
|---|---|---|---|---|---|---|---|
| 0 | 1.6094 | 0.389 | -0.410 | 45 | 51.590 | 14.244 | 10.996 |
| 1 | 2.9943 | 0.414 | -0.588 | 46 | 52.652 | 14.377 | 11.090 |
| 2 | 4.2790 | 0.762 | -0.377 | 47 | 53.713 | 14.647 | 11.330 |
| 3 | 5.5151 | 0.993 | -0.265 | 48 | 54.774 | 15.143 | 11.803 |
| 4 | 6.7210 | 1.305 | -0.053 | 49 | 55.834 | 15.320 | 11.945 |
| 5 | 7.9060 | 1.562 | 0.111 | 50 | 56.893 | 15.731 | 12.330 |
| 6 | 9.0754 | 1.901 | 0.370 | 51 | 57.952 | 15.818 | 12.381 |
| 7 | 10.233 | 2.197 | 0.588 | 52 | 59.010 | 16.213 | 12.750 |
| 8 | 11.380 | 2.571 | 0.895 | 53 | 60.068 | 16.704 | 13.219 |
| 9 | 12.519 | 2.911 | 1.168 | 54 | 61.125 | 16.796 | 13.274 |
| 10 | 13.651 | 3.303 | 1.501 | 55 | 62.182 | 17.351 | 13.811 |
| 11 | 14.777 | 3.604 | 1.739 | 56 | 63.238 | 17.641 | 14.071 |
| 12 | 15.897 | 3.913 | 1.988 | 57 | 64.294 | 17.896 | 14.297 |
| 13 | 17.013 | 4.226 | 2.245 | 58 | 65.349 | 17.916 | 14.28 |
| 14 | 18.125 | 4.527 | 2.489 | 59 | 66.403 | 18.449 | 14.79 |
| 15 | 19.233 | 4.843 | 2.753 | 60 | 67.458 | 18.68 | 14.99 |
| 16 | 20.338 | 5.146 | 3.004 | 61 | 68.511 | 19.10 | 15.39 |
| 17 | 21.439 | 5.459 | 3.267 | 62 | 69.565 | 19.21 | 15.47 |
| 18 | 22.538 | 5.775 | 3.535 | 63 | 70.618 | 19.49 | 15.72 |
| 19 | 23.634 | 6.052 | 3.762 | 64 | 71.670 | 20.06 | 16.28 |
| 20 | 24.728 | 6.384 | 4.049 | 65 | 72.722 | 20.27 | 16.46 |
| 21 | 25.819 | 6.712 | 4.334 | 66 | 73.774 | 20.76 | 16.93 |
| 22 | 26.909 | 7.006 | 4.582 | 67 | 74.826 | 20.88 | 17.02 |
| 23 | 27.996 | 7.348 | 4.884 | 68 | 75.877 | 21.35 | 17.46 |
| 24 | 29.082 | 7.583 | 5.072 | 69 | 76.927 | 21.68 | 17.77 |
| 25 | 30.166 | 7.939 | 5.390 | 70 | 77.977 | 21.63 | 17.69 |
| 26 | 31.248 | 8.258 | 5.668 | 71 | 79.027 | 21.92 | 17.95 |
| 27 | 32.329 | 8.576 | 5.947 | 72 | 80.077 | 22.49 | 18.51 |
| 28 | 33.408 | 8.849 | 6.178 | 73 | 81.126 | 22.95 | 18.95 |
| 29 | 34.486 | 9.107 | 6.396 | 74 | 82.175 | 23.10 | 19.07 |
| 30 | 35.563 | 9.456 | 6.709 | 75 | 83.223 | 23.58 | 19.52 |
| 31 | 36.638 | 9.799 | 7.016 | 76 | 84.272 | 23.88 | 19.81 |
| 32 | 37.712 | 10.024 | 7.200 | 77 | 85.320 | 23.93 | 19.82 |
| 33 | 38.785 | 10.332 | 7.472 | 78 | 86.367 | 24.48 | 20.35 |
| 34 | 39.857 | 10.769 | 7.880 | 79 | 87.414 | 24.77 | 20.62 |
| 35 | 40.928 | 11.037 | 8.112 | 80 | 88.461 | 24.90 | 20.72 |
| 36 | 41.998 | 11.401 | 8.443 | 81 | 89.508 | 25.37 | 21.17 |
| 37 | 43.067 | 11.616 | 8.620 | 82 | 90.555 | 25.74 | 21.52 |
| 38 | 44.135 | 11.977 | 8.950 | 83 | 91.601 | 25.85 | 21.61 |
| 39 | 45.203 | 12.322 | 9.263 | 84 | 92.647 | 25.80 | 21.52 |
| 40 | 46.269 | 12.669 | 9.578 | 85 | 93.692 | 26.35 | 22.06 |
| 41 | 47.335 | 12.814 | 9.683 | 86 | 94.738 | 26.59 | 22.27 |
| 42 | 48.400 | 13.205 | 10.046 | 87 | 95.783 | 27.48 | 23.16 |
| 43 | 49.464 | 13.419 | 10.224 | 88 | 96.827 | 27.37 | 23.00 |
| 44 | 50.527 | 13.746 | 10.521 | 89 | 97.872 | 27.86 | 23.48 |

Table 1. CONSTANTS FOR $\overline{p}'$ (LONG FORMULA), AN APPROXIMATION OF THE UPPER BINOMIAL CONFIDENCE LIMIT $\overline{p}$; $c/n \leqslant .5$, $n \geqslant 20$, $c \leqslant 1{,}000$ (*Continued*)

| c | $\overline{m}$ | $\overline{a}$ | $\overline{b}$ | c | $\overline{m}$ | $\overline{a}$ | $\overline{b}$ |
|---|---|---|---|---|---|---|---|
| 90 | 98.916 | 28.03 | 23.62 | 135 | 145.71 | 36.43 | 31.07 |
| 91 | 99.960 | 28.29 | 23.87 | 136 | 146.74 | 36.69 | 31.31 |
| 92 | 101.00 | 28.85 | 24.41 | 137 | 147.78 | 36.95 | 31.55 |
| 93 | 102.05 | 28.70 | 24.22 | 138 | 148.81 | 37.21 | 31.79 |
| 94 | 103.09 | 29.04 | 24.55 | 139 | 149.85 | 37.47 | 32.03 |
| 95 | 104.13 | 29.31 | 24.79 | 140 | 150.88 | 37.72 | 32.28 |
| 96 | 105.18 | 30.10 | 25.58 | 141 | 151.92 | 37.98 | 32.52 |
| 97 | 106.22 | 30.16 | 25.61 | 142 | 152.96 | 38.24 | 32.76 |
| 98 | 107.26 | 30.77 | 26.21 | 143 | 153.99 | 38.50 | 33.00 |
| 99 | 108.30 | 31.05 | 26.46 | 144 | 155.03 | 38.76 | 33.24 |
| 100 | 109.35 | 31.47 | 26.86 | 145 | 156.06 | 39.02 | 33.48 |
| 101 | 110.39 | 27.60 | 22.90 | 146 | 157.10 | 39.28 | 33.72 |
| 102 | 111.43 | 27.86 | 23.14 | 147 | 158.13 | 39.54 | 33.96 |
| 103 | 112.47 | 28.12 | 23.38 | 148 | 159.16 | 39.79 | 34.21 |
| 104 | 113.51 | 28.38 | 23.62 | 149 | 160.20 | 40.05 | 34.45 |
| 105 | 114.55 | 28.64 | 23.86 | 150 | 161.23 | 40.31 | 34.69 |
| 106 | 115.59 | 28.90 | 24.10 | 151 | 162.27 | 40.57 | 34.93 |
| 107 | 116.64 | 29.16 | 24.34 | 152 | 163.30 | 40.83 | 35.17 |
| 108 | 117.68 | 29.42 | 24.58 | 153 | 164.34 | 41.09 | 35.41 |
| 109 | 118.72 | 29.68 | 24.82 | 154 | 165.37 | 41.35 | 35.65 |
| 110 | 119.76 | 29.94 | 25.06 | 155 | 166.40 | 41.60 | 35.90 |
| 111 | 120.80 | 30.20 | 25.30 | 156 | 167.44 | 41.86 | 36.14 |
| 112 | 121.84 | 30.46 | 25.54 | 157 | 168.47 | 42.12 | 36.38 |
| 113 | 122.88 | 30.72 | 25.78 | 158 | 169.50 | 42.38 | 36.62 |
| 114 | 123.91 | 30.98 | 26.02 | 159 | 170.54 | 42.64 | 36.86 |
| 115 | 124.95 | 31.24 | 26.26 | 160 | 171.57 | 42.90 | 37.10 |
| 116 | 125.99 | 31.50 | 26.50 | 161 | 172.61 | 43.15 | 37.35 |
| 117 | 127.03 | 31.76 | 26.74 | 162 | 173.65 | 43.41 | 37.59 |
| 118 | 128.07 | 32.02 | 26.98 | 163 | 174.68 | 43.67 | 37.83 |
| 119 | 129.11 | 32.28 | 27.22 | 164 | 175.71 | 43.93 | 38.07 |
| 120 | 130.15 | 32.54 | 27.46 | 165 | 176.75 | 44.19 | 38.31 |
| 121 | 131.19 | 32.80 | 27.70 | 166 | 177.78 | 44.44 | 38.56 |
| 122 | 132.22 | 33.06 | 27.94 | 167 | 178.81 | 44.70 | 38.80 |
| 123 | 133.26 | 33.32 | 28.18 | 168 | 179.84 | 44.96 | 39.04 |
| 124 | 134.30 | 33.58 | 28.42 | 169 | 180.88 | 45.22 | 39.28 |
| 125 | 135.34 | 33.84 | 28.66 | 170 | 181.91 | 45.48 | 39.52 |
| 126 | 136.37 | 34.10 | 28.90 | 171 | 182.94 | 45.74 | 39.76 |
| 127 | 137.41 | 34.36 | 29.14 | 172 | 183.97 | 45.99 | 40.01 |
| 128 | 138.45 | 34.62 | 29.38 | 173 | 185.00 | 46.25 | 40.25 |
| 129 | 139.49 | 34.87 | 29.63 | 174 | 186.04 | 46.51 | 40.49 |
| 130 | 140.52 | 35.13 | 29.87 | 175 | 187.07 | 46.77 | 40.73 |
| 131 | 141.56 | 35.39 | 30.11 | 176 | 188.10 | 47.02 | 40.98 |
| 132 | 142.60 | 35.65 | 30.35 | 177 | 189.13 | 47.28 | 41.22 |
| 133 | 143.63 | 35.91 | 30.59 | 178 | 190.16 | 47.54 | 41.46 |
| 134 | 144.67 | 36.17 | 30.83 | 179 | 191.19 | 47.80 | 41.70 |

Table 1. **CONSTANTS FOR $\bar{p}'$ (LONG FORMULA), AN APPROXIMATION OF THE UPPER BINOMIAL CONFIDENCE LIMIT** $\bar{p}$; c/n ≤ .5, n ≥ 20, c ≤ 1,000 (*Continued*)

Section 6

$\gamma_1 = 80\%$
$\gamma_2 = 60\%$

| c | $\overline{m}$ | $\overline{a}$ | $\overline{b}$ | c | $\overline{m}$ | $\overline{a}$ | $\overline{b}$ |
|---|---|---|---|---|---|---|---|
| 180 | 192.23 | 48.06 | 41.94 | 225 | 238.56 | 59.64 | 52.86 |
| 181 | 193.26 | 48.31 | 42.19 | 226 | 239.58 | 59.90 | 53.10 |
| 182 | 194.29 | 48.57 | 42.43 | 227 | 240.61 | 60.15 | 53.35 |
| 183 | 195.32 | 48.83 | 42.67 | 228 | 241.64 | 60.41 | 53.59 |
| 184 | 196.35 | 49.09 | 42.91 | 229 | 242.67 | 60.67 | 53.83 |
| 185 | 197.38 | 49.35 | 43.15 | 230 | 243.69 | 60.92 | 54.08 |
| 186 | 198.41 | 49.60 | 43.40 | 231 | 244.72 | 61.18 | 54.32 |
| 187 | 199.44 | 49.86 | 43.64 | 232 | 245.75 | 61.44 | 54.56 |
| 188 | 200.47 | 50.12 | 43.88 | 233 | 246.78 | 61.69 | 54.81 |
| 189 | 201.50 | 50.38 | 44.12 | 234 | 247.80 | 61.95 | 55.05 |
| 190 | 202.53 | 50.63 | 44.37 | 235 | 248.83 | 62.21 | 55.29 |
| 191 | 203.56 | 50.89 | 44.61 | 236 | 249.86 | 62.46 | 55.54 |
| 192 | 204.59 | 51.15 | 44.85 | 237 | 250.89 | 62.72 | 55.78 |
| 193 | 205.63 | 51.41 | 45.09 | 238 | 251.91 | 62.98 | 56.02 |
| 194 | 206.66 | 51.66 | 45.34 | 239 | 252.94 | 63.24 | 56.26 |
| 195 | 207.69 | 51.92 | 45.58 | 240 | 253.97 | 63.49 | 56.51 |
| 196 | 208.72 | 52.18 | 45.82 | 241 | 255.00 | 63.75 | 56.75 |
| 197 | 209.75 | 52.44 | 46.06 | 242 | 256.02 | 64.01 | 56.99 |
| 198 | 210.78 | 52.69 | 46.31 | 243 | 257.05 | 64.26 | 57.24 |
| 199 | 211.81 | 52.95 | 46.55 | 244 | 258.08 | 64.52 | 57.48 |
| 200 | 212.83 | 53.21 | 46.79 | 245 | 259.10 | 64.78 | 57.72 |
| 201 | 213.86 | 53.47 | 47.03 | 246 | 260.13 | 65.03 | 57.97 |
| 202 | 214.89 | 53.72 | 47.28 | 247 | 261.16 | 65.29 | 58.21 |
| 203 | 215.92 | 53.98 | 47.52 | 248 | 262.18 | 65.55 | 58.45 |
| 204 | 216.95 | 54.24 | 47.76 | 249 | 263.21 | 65.80 | 58.70 |
| 205 | 217.98 | 54.50 | 48.00 | 250 | 264.24 | 66.06 | 58.94 |
| 206 | 219.01 | 54.75 | 48.25 | 251 | 265.26 | 66.32 | 59.18 |
| 207 | 220.04 | 55.01 | 48.49 | 252 | 266.29 | 66.57 | 59.43 |
| 208 | 221.07 | 55.27 | 48.73 | 253 | 267.32 | 66.83 | 59.67 |
| 209 | 222.10 | 55.52 | 48.98 | 254 | 268.34 | 67.09 | 59.91 |
| 210 | 223.13 | 55.78 | 49.22 | 255 | 269.37 | 67.34 | 60.16 |
| 211 | 224.16 | 56.04 | 49.46 | 256 | 270.39 | 67.60 | 60.40 |
| 212 | 225.19 | 56.30 | 49.70 | 257 | 271.42 | 67.86 | 60.64 |
| 213 | 226.21 | 56.55 | 49.95 | 258 | 272.45 | 68.11 | 60.89 |
| 214 | 227.24 | 56.81 | 50.19 | 259 | 273.47 | 68.37 | 61.13 |
| 215 | 228.27 | 57.07 | 50.43 | 260 | 274.50 | 68.62 | 61.38 |
| 216 | 229.30 | 57.33 | 50.67 | 261 | 275.53 | 68.88 | 61.62 |
| 217 | 230.33 | 57.58 | 50.92 | 262 | 276.55 | 69.14 | 61.86 |
| 218 | 231.36 | 57.84 | 51.16 | 263 | 277.58 | 69.39 | 62.11 |
| 219 | 232.39 | 58.10 | 51.40 | 264 | 278.60 | 69.65 | 62.35 |
| 220 | 233.41 | 58.35 | 51.65 | 265 | 279.63 | 69.91 | 62.59 |
| 221 | 234.44 | 58.61 | 51.89 | 266 | 280.65 | 70.16 | 62.84 |
| 222 | 235.47 | 58.87 | 52.13 | 267 | 281.68 | 70.42 | 63.08 |
| 223 | 236.50 | 59.12 | 52.38 | 268 | 282.71 | 70.68 | 63.32 |
| 224 | 237.53 | 59.38 | 52.62 | 269 | 283.73 | 70.93 | 63.57 |

Table 1.  **CONSTANTS FOR $\overline{p}'$ (LONG FORMULA), AN APPROXIMATION OF THE UPPER BINOMIAL CONFIDENCE LIMIT** $\overline{p}$; c/n ≤ .5, n ≥ 20, c ≤ 1,000 (*Continued*)

| c | $\overline{m}$ | $\overline{a}$ | $\overline{b}$ | c | $\overline{m}$ | $\overline{a}$ | $\overline{b}$ |
|---|---|---|---|---|---|---|---|
| 270 | 284.76 | 71.19 | 63.81 | 315 | 330.86 | 82.72 | 74.78 |
| 271 | 285.78 | 71.45 | 64.05 | 316 | 331.89 | 82.97 | 75.03 |
| 272 | 286.81 | 71.70 | 64.30 | 317 | 332.91 | 83.23 | 75.27 |
| 273 | 287.83 | 71.96 | 64.54 | 318 | 333.93 | 83.48 | 75.52 |
| 274 | 288.86 | 72.21 | 64.79 | 319 | 334.96 | 83.74 | 75.76 |
| 275 | 289.88 | 72.47 | 65.03 | 320 | 335.98 | 84.00 | 76.00 |
| 276 | 290.91 | 72.73 | 65.27 | 321 | 337.01 | 84.25 | 76.25 |
| 277 | 291.94 | 72.98 | 65.52 | 322 | 338.03 | 84.51 | 76.49 |
| 278 | 292.96 | 73.24 | 65.76 | 323 | 339.05 | 84.76 | 76.74 |
| 279 | 293.99 | 73.50 | 66.00 | 324 | 340.08 | 85.02 | 76.98 |
| 280 | 295.01 | 73.75 | 66.25 | 325 | 341.10 | 85.27 | 77.23 |
| 281 | 296.04 | 74.01 | 66.49 | 326 | 342.12 | 85.53 | 77.47 |
| 282 | 297.06 | 74.27 | 66.73 | 327 | 343.15 | 85.79 | 77.71 |
| 283 | 298.09 | 74.52 | 66.98 | 328 | 344.17 | 86.04 | 77.96 |
| 284 | 299.11 | 74.78 | 67.22 | 329 | 345.19 | 86.30 | 78.20 |
| 285 | 300.14 | 75.03 | 67.47 | 330 | 346.21 | 86.55 | 78.45 |
| 286 | 301.16 | 75.29 | 67.71 | 331 | 347.24 | 86.81 | 78.69 |
| 287 | 302.19 | 75.55 | 67.95 | 332 | 348.26 | 87.07 | 78.93 |
| 288 | 303.21 | 75.80 | 68.20 | 333 | 349.28 | 87.32 | 79.18 |
| 289 | 304.24 | 76.06 | 68.44 | 334 | 350.31 | 87.58 | 79.42 |
| 290 | 305.26 | 76.31 | 68.69 | 335 | 351.33 | 87.83 | 79.67 |
| 291 | 306.28 | 76.57 | 68.93 | 336 | 352.35 | 88.09 | 79.91 |
| 292 | 307.31 | 76.83 | 69.17 | 337 | 353.38 | 88.34 | 80.16 |
| 293 | 308.33 | 77.08 | 69.42 | 338 | 354.40 | 88.60 | 80.40 |
| 294 | 309.36 | 77.34 | 69.66 | 339 | 355.42 | 88.86 | 80.64 |
| 295 | 310.38 | 77.60 | 69.90 | 340 | 356.44 | 89.11 | 80.89 |
| 296 | 311.41 | 77.85 | 70.15 | 341 | 357.47 | 89.37 | 81.13 |
| 297 | 312.43 | 78.11 | 70.39 | 342 | 358.49 | 89.62 | 81.38 |
| 298 | 313.46 | 78.36 | 70.64 | 343 | 359.51 | 89.88 | 81.62 |
| 299 | 314.48 | 78.62 | 70.88 | 344 | 360.54 | 90.13 | 81.87 |
| 300 | 315.50 | 78.88 | 71.12 | 345 | 361.56 | 90.39 | 82.11 |
| 301 | 316.53 | 79.13 | 71.37 | 346 | 362.58 | 90.65 | 82.35 |
| 302 | 317.55 | 79.39 | 71.61 | 347 | 363.60 | 90.90 | 82.60 |
| 303 | 318.58 | 79.64 | 71.86 | 348 | 364.63 | 91.16 | 82.84 |
| 304 | 319.60 | 79.90 | 72.10 | 349 | 365.65 | 91.41 | 83.09 |
| 305 | 320.63 | 80.16 | 72.34 | 350 | 366.67 | 91.67 | 83.33 |
| 306 | 321.65 | 80.41 | 72.59 | 351 | 367.69 | 91.92 | 83.58 |
| 307 | 322.67 | 80.67 | 72.83 | 352 | 368.72 | 92.18 | 83.82 |
| 308 | 323.70 | 80.92 | 73.08 | 353 | 369.74 | 92.43 | 84.07 |
| 309 | 324.72 | 81.18 | 73.32 | 354 | 370.76 | 92.69 | 84.31 |
| 310 | 325.74 | 81.44 | 73.56 | 355 | 371.78 | 92.95 | 84.55 |
| 311 | 326.77 | 81.69 | 73.81 | 356 | 372.80 | 93.20 | 84.80 |
| 312 | 327.79 | 81.95 | 74.05 | 357 | 373.83 | 93.46 | 85.04 |
| 313 | 328.82 | 82.20 | 74.30 | 358 | 374.85 | 93.71 | 85.29 |
| 314 | 329.84 | 82.46 | 74.54 | 359 | 375.87 | 93.97 | 85.53 |

**Table 1.** **CONSTANTS FOR $\bar{p}'$ (LONG FORMULA), AN APPROXIMATION OF THE UPPER BINOMIAL CONFIDENCE LIMIT $\bar{p}$; $c/n \leqslant .5$, $n \geqslant 20$, $c \leqslant 1,000$ (Continued)**

| c | $\bar{m}$ | $\bar{a}$ | $\bar{b}$ | c | $\bar{m}$ | $\bar{a}$ | $\bar{b}$ |
|---|---|---|---|---|---|---|---|
| 360 | 376.89 | 94.22 | 85.78 | 405 | 422.86 | 105.72 | 96.78 |
| 361 | 377.92 | 94.48 | 86.02 | 406 | 423.88 | 105.97 | 97.03 |
| 362 | 378.94 | 94.73 | 86.27 | 407 | 424.90 | 106.23 | 97.27 |
| 363 | 379.96 | 94.99 | 86.51 | 408 | 425.92 | 106.48 | 97.52 |
| 364 | 380.98 | 95.25 | 86.75 | 409 | 426.94 | 106.74 | 97.76 |
| 365 | 382.00 | 95.50 | 87.00 | 410 | 427.97 | 106.99 | 98.01 |
| 366 | 383.03 | 95.76 | 87.24 | 411 | 428.99 | 107.25 | 98.25 |
| 367 | 384.05 | 96.01 | 87.49 | 412 | 430.01 | 107.50 | 98.50 |
| 368 | 385.07 | 96.27 | 87.73 | 413 | 431.03 | 107.76 | 98.74 |
| 369 | 386.09 | 96.52 | 87.98 | 414 | 432.05 | 108.01 | 98.99 |
| 370 | 387.11 | 96.78 | 88.22 | 415 | 433.07 | 108.27 | 99.23 |
| 371 | 388.14 | 97.03 | 88.47 | 416 | 434.09 | 108.52 | 99.48 |
| 372 | 389.16 | 97.29 | 88.71 | 417 | 435.11 | 108.78 | 99.72 |
| 373 | 390.18 | 97.54 | 88.96 | 418 | 436.13 | 109.03 | 99.97 |
| 374 | 391.20 | 97.80 | 89.20 | 419 | 437.15 | 109.29 | 100.21 |
| 375 | 392.22 | 98.06 | 89.44 | 420 | 438.17 | 109.54 | 100.46 |
| 376 | 393.24 | 98.31 | 89.69 | 421 | 439.19 | 109.80 | 100.70 |
| 377 | 394.27 | 98.57 | 89.93 | 422 | 440.21 | 110.05 | 100.95 |
| 378 | 395.29 | 98.82 | 90.18 | 423 | 441.23 | 110.31 | 101.19 |
| 379 | 396.31 | 99.08 | 90.42 | 424 | 442.25 | 110.56 | 101.44 |
| 380 | 397.33 | 99.33 | 90.67 | 425 | 443.27 | 110.82 | 101.68 |
| 381 | 398.35 | 99.59 | 90.91 | 426 | 444.29 | 111.07 | 101.93 |
| 382 | 399.37 | 99.84 | 91.16 | 427 | 445.31 | 111.33 | 102.17 |
| 383 | 400.40 | 100.10 | 91.40 | 428 | 446.33 | 111.58 | 102.42 |
| 384 | 401.42 | 100.35 | 91.65 | 429 | 447.35 | 111.84 | 102.66 |
| 385 | 402.44 | 100.61 | 91.89 | 430 | 448.38 | 112.09 | 102.91 |
| 386 | 403.46 | 100.86 | 92.14 | 431 | 449.40 | 112.35 | 103.15 |
| 387 | 404.48 | 101.12 | 92.38 | 432 | 450.42 | 112.60 | 103.40 |
| 388 | 405.50 | 101.38 | 92.62 | 433 | 451.44 | 112.86 | 103.64 |
| 389 | 406.52 | 101.63 | 92.87 | 434 | 452.46 | 113.11 | 103.89 |
| 390 | 407.54 | 101.89 | 93.11 | 435 | 453.48 | 113.37 | 104.13 |
| 391 | 408.57 | 102.14 | 93.36 | 436 | 454.50 | 113.62 | 104.38 |
| 392 | 409.59 | 102.40 | 93.60 | 437 | 455.52 | 113.88 | 104.62 |
| 393 | 410.61 | 102.65 | 93.85 | 438 | 456.54 | 114.13 | 104.87 |
| 394 | 411.63 | 102.91 | 94.09 | 439 | 457.56 | 114.39 | 105.11 |
| 395 | 412.65 | 103.16 | 94.34 | 440 | 458.58 | 114.64 | 105.36 |
| 396 | 413.67 | 103.42 | 94.58 | 441 | 459.60 | 114.90 | 105.60 |
| 397 | 414.69 | 103.67 | 94.83 | 442 | 460.62 | 115.15 | 105.85 |
| 398 | 415.71 | 103.93 | 95.07 | 443 | 461.64 | 115.41 | 106.09 |
| 399 | 416.74 | 104.18 | 95.32 | 444 | 462.66 | 115.66 | 106.34 |
| 400 | 417.76 | 104.44 | 95.56 | 445 | 463.68 | 115.92 | 106.58 |
| 401 | 418.78 | 104.69 | 95.81 | 446 | 464.70 | 116.17 | 106.83 |
| 402 | 419.80 | 104.95 | 96.05 | 447 | 465.72 | 116.43 | 107.07 |
| 403 | 420.82 | 105.20 | 96.30 | 448 | 466.74 | 116.68 | 107.32 |
| 404 | 421.84 | 105.46 | 96.54 | 449 | 467.76 | 116.94 | 107.56 |

Table 1. **CONSTANTS FOR $\overline{p}'$ (LONG FORMULA), AN APPROXIMATION OF THE UPPER BINOMIAL CONFIDENCE LIMIT $\overline{p}$; $c/n \leqslant .5$, $n \geqslant 20$, $c \leqslant 1,000$ (*Continued*)**

| c | $\overline{m}$ | $\overline{a}$ | $\overline{b}$ | c | $\overline{m}$ | $\overline{a}$ | $\overline{b}$ |
|---|---|---|---|---|---|---|---|
| 450 | 468.78 | 117.19 | 107.81 | 495 | 514.65 | 128.66 | 118.84 |
| 451 | 469.80 | 117.45 | 108.05 | 496 | 515.67 | 128.92 | 119.08 |
| 452 | 470.82 | 117.70 | 108.30 | 497 | 516.68 | 129.17 | 119.33 |
| 453 | 471.84 | 117.96 | 108.54 | 498 | 517.70 | 129.43 | 119.57 |
| 454 | 472.86 | 118.21 | 108.79 | 499 | 518.72 | 129.68 | 119.82 |
| 455 | 473.87 | 118.47 | 109.03 | 500 | 519.74 | 129.94 | 120.06 |
| 456 | 474.89 | 118.72 | 109.28 | 501 | 520.76 | 130.19 | 120.31 |
| 457 | 475.91 | 118.98 | 109.52 | 502 | 521.78 | 130.44 | 120.56 |
| 458 | 476.93 | 119.23 | 109.77 | 503 | 522.80 | 130.70 | 120.80 |
| 459 | 477.95 | 119.49 | 110.01 | 504 | 523.82 | 130.95 | 121.05 |
| 460 | 478.97 | 119.74 | 110.26 | 505 | 524.83 | 131.21 | 121.29 |
| 461 | 479.99 | 120.00 | 110.50 | 506 | 525.85 | 131.46 | 121.54 |
| 462 | 481.01 | 120.25 | 110.75 | 507 | 526.87 | 131.72 | 121.78 |
| 463 | 482.03 | 120.51 | 110.99 | 508 | 527.89 | 131.97 | 122.03 |
| 464 | 483.05 | 120.76 | 111.24 | 509 | 528.91 | 132.23 | 122.27 |
| 465 | 484.07 | 121.02 | 111.48 | 510 | 529.93 | 132.48 | 122.52 |
| 466 | 485.09 | 121.27 | 111.73 | 511 | 530.95 | 132.74 | 122.76 |
| 467 | 486.11 | 121.53 | 111.97 | 512 | 531.96 | 132.99 | 123.01 |
| 468 | 487.13 | 121.78 | 112.22 | 513 | 532.98 | 133.25 | 123.25 |
| 469 | 488.15 | 122.04 | 112.46 | 514 | 534.00 | 133.50 | 123.50 |
| 470 | 489.17 | 122.29 | 112.71 | 515 | 535.02 | 133.76 | 123.74 |
| 471 | 490.19 | 122.55 | 112.95 | 516 | 536.04 | 134.01 | 123.99 |
| 472 | 491.21 | 122.80 | 113.20 | 517 | 537.06 | 134.26 | 124.24 |
| 473 | 492.23 | 123.06 | 113.44 | 518 | 538.08 | 134.52 | 124.48 |
| 474 | 493.25 | 123.31 | 113.69 | 519 | 539.09 | 134.77 | 124.73 |
| 475 | 494.26 | 123.57 | 113.93 | 520 | 540.11 | 135.03 | 124.97 |
| 476 | 495.28 | 123.82 | 114.18 | 521 | 541.13 | 135.28 | 125.22 |
| 477 | 496.30 | 124.08 | 114.42 | 522 | 542.15 | 135.54 | 125.46 |
| 478 | 497.32 | 124.33 | 114.67 | 523 | 543.17 | 135.79 | 125.71 |
| 479 | 498.34 | 124.59 | 114.91 | 524 | 544.19 | 136:05 | 125.95 |
| 480 | 499.36 | 124.84 | 115.16 | 525 | 545.20 | 136.30 | 126.20 |
| 481 | 500.38 | 125.10 | 115.40 | 526 | 546.22 | 136.56 | 126.44 |
| 482 | 501.40 | 125.35 | 115.65 | 527 | 547.24 | 136.81 | 126.69 |
| 483 | 502.42 | 125.60 | 115.90 | 528 | 548.26 | 137.07 | 126.93 |
| 484 | 503.44 | 125.86 | 116.14 | 529 | 549.28 | 137.32 | 127.18 |
| 485 | 504.46 | 126.11 | 116.39 | 530 | 550.30 | 137.57 | 127.43 |
| 486 | 505.48 | 126.37 | 116.63 | 531 | 551.31 | 137.83 | 127.67 |
| 487 | 506.49 | 126.62 | 116.88 | 532 | 552.33 | 138.08 | 127.92 |
| 488 | 507.51 | 126.88 | 117.12 | 533 | 553.35 | 138.34 | 128.16 |
| 489 | 508.53 | 127.13 | 117.37 | 534 | 554.37 | 138.59 | 128.41 |
| 490 | 509.55 | 127.39 | 117.61 | 535 | 555.39 | 138.85 | 128.65 |
| 491 | 510.57 | 127.64 | 117.86 | 536 | 556.41 | 139.10 | 128.90 |
| 492 | 511.59 | 127.90 | 118.10 | 537 | 557.42 | 139.36 | 129.14 |
| 493 | 512.61 | 128.15 | 118.35 | 538 | 558.44 | 139.61 | 129.39 |
| 494 | 513.63 | 128.41 | 118.59 | 539 | 559.46 | 139.87 | 129.63 |

Table 1. **CONSTANTS FOR $\overline{p}'$ (LONG FORMULA), AN APPROXIMATION OF** **Section 6**
**THE UPPER BINOMIAL CONFIDENCE LIMIT $\overline{p}$; c/n ≤ .5, n ≥ 20,** $\gamma_1 = 80\%$
c ≤ 1,000 (*Continued*) $\gamma_2 = 60\%$

| c | $\overline{m}$ | $\overline{a}$ | $\overline{b}$ | c | $\overline{m}$ | $\overline{a}$ | $\overline{b}$ |
|---|---|---|---|---|---|---|---|
| 540 | 560.48 | 140.12 | 129.88 | 585 | 606.28 | 151.57 | 140.93 |
| 541 | 561.50 | 140.37 | 130.13 | 586 | 607.29 | 151.82 | 141.18 |
| 542 | 562.51 | 140.63 | 130.37 | 587 | 608.31 | 152.08 | 141.42 |
| 543 | 563.53 | 140.88 | 130.62 | 588 | 609.33 | 152.33 | 141.67 |
| 544 | 564.55 | 141.14 | 130.86 | 589 | 610.35 | 152.59 | 141.91 |
| 545 | 565.57 | 141.39 | 131.11 | 590 | 611.36 | 152.84 | 142.16 |
| 546 | 566.59 | 141.65 | 131.35 | 591 | 612.38 | 153.10 | 142.40 |
| 547 | 567.60 | 141.90 | 131.60 | 592 | 613.40 | 153.35 | 142.65 |
| 548 | 568.62 | 142.16 | 131.84 | 593 | 614.41 | 153.60 | 142.90 |
| 549 | 569.64 | 142.41 | 132.09 | 594 | 615.43 | 153.86 | 143.14 |
| 550 | 570.66 | 142.66 | 132.34 | 595 | 616.45 | 154.11 | 143.39 |
| 551 | 571.68 | 142.92 | 132.58 | 596 | 617.47 | 154.37 | 143.63 |
| 552 | 572.69 | 143.17 | 132.83 | 597 | 618.48 | 154.62 | 143.88 |
| 553 | 573.71 | 143.43 | 133.07 | 598 | 619.50 | 154.88 | 144.12 |
| 554 | 574.73 | 143.68 | 133.32 | 599 | 620.52 | 155.13 | 144.37 |
| 555 | 575.75 | 143.94 | 133.56 | 600 | 621.54 | 155.38 | 144.62 |
| 556 | 576.77 | 144.19 | 133.81 | 601 | 622.55 | 155.64 | 144.86 |
| 557 | 577.78 | 144.45 | 134.05 | 602 | 623.57 | 155.89 | 145.11 |
| 558 | 578.80 | 144.70 | 134.30 | 603 | 624.59 | 156.15 | 145.35 |
| 559 | 579.82 | 144.95 | 134.55 | 604 | 625.60 | 156.40 | 145.60 |
| 560 | 580.84 | 145.21 | 134.79 | 605 | 626.62 | 156.66 | 145.84 |
| 561 | 581.85 | 145.46 | 135.04 | 606 | 627.64 | 156.91 | 146.09 |
| 562 | 582.87 | 145.72 | 135.28 | 607 | 628.66 | 157.16 | 146.34 |
| 563 | 583.89 | 145.97 | 135.53 | 608 | 629.67 | 157.42 | 146.58 |
| 564 | 584.91 | 146.23 | 135.77 | 609 | 630.69 | 157.67 | 146.83 |
| 565 | 585.93 | 146.48 | 136.02 | 610 | 631.71 | 157.93 | 147.07 |
| 566 | 586.94 | 146.74 | 136.26 | 611 | 632.72 | 158.18 | 147.32 |
| 567 | 587.96 | 146.99 | 136.51 | 612 | 633.74 | 158.44 | 147.56 |
| 568 | 588.98 | 147.24 | 136.76 | 613 | 634.76 | 158.69 | 147.81 |
| 569 | 590.00 | 147.50 | 137.00 | 614 | 635.77 | 158.94 | 148.06 |
| 570 | 591.01 | 147.75 | 137.25 | 615 | 636.79 | 159.20 | 148.30 |
| 571 | 592.03 | 148.01 | 137.49 | 616 | 637.81 | 159.45 | 148.55 |
| 572 | 593.05 | 148.26 | 137.74 | 617 | 638.83 | 159.71 | 148.79 |
| 573 | 594.07 | 148.52 | 137.98 | 618 | 639.84 | 159.96 | 149.04 |
| 574 | 595.08 | 148.77 | 138.23 | 619 | 640.86 | 160.21 | 149.29 |
| 575 | 596.10 | 149.03 | 138.47 | 620 | 641.88 | 160.47 | 149.53 |
| 576 | 597.12 | 149.28 | 138.72 | 621 | 642.89 | 160.72 | 149.78 |
| 577 | 598.14 | 149.53 | 138.97 | 622 | 643.91 | 160.98 | 150.02 |
| 578 | 599.15 | 149.79 | 139.21 | 623 | 644.93 | 161.23 | 150.27 |
| 579 | 600.17 | 150.04 | 139.46 | 624 | 645.94 | 161.49 | 150.51 |
| 580 | 601.19 | 150.30 | 139.70 | 625 | 646.96 | 161.74 | 150.76 |
| 581 | 602.21 | 150.55 | 139.95 | 626 | 647.98 | 161.99 | 151.01 |
| 582 | 603.22 | 150.81 | 140.19 | 627 | 648.99 | 162.25 | 151.25 |
| 583 | 604.24 | 151.06 | 140.44 | 628 | 650.01 | 162.50 | 151.50 |
| 584 | 605.26 | 151.31 | 140.69 | 629 | 651.03 | 162.76 | 151.74 |

Table 1.  CONSTANTS FOR $\overline{p}'$ (LONG FORMULA), AN APPROXIMATION OF THE UPPER BINOMIAL CONFIDENCE LIMIT $\overline{p}$; $c/n \leqslant .5$, $n \geqslant 20$, $c \leqslant 1,000$ (*Continued*)

| c | $\overline{m}$ | $\overline{a}$ | $\overline{b}$ | c | $\overline{m}$ | $\overline{a}$ | $\overline{b}$ |
|---|---|---|---|---|---|---|---|
| 630 | 652.04 | 163.01 | 151.99 | 675 | 697.78 | 174.45 | 163.05 |
| 631 | 653.06 | 163.27 | 152.23 | 676 | 698.80 | 174.70 | 163.30 |
| 632 | 654.08 | 163.52 | 152.48 | 677 | 699.82 | 174.95 | 163.55 |
| 633 | 655.09 | 163.77 | 152.73 | 678 | 700.83 | 175.21 | 163.79 |
| 634 | 656.11 | 164.03 | 152.97 | 679 | 701.85 | 175.46 | 164.04 |
| 635 | 657.13 | 164.28 | 153.22 | 680 | 702.87 | 175.72 | 164.28 |
| 636 | 658.14 | 164.54 | 153.46 | 681 | 703.88 | 175.97 | 164.53 |
| 637 | 659.16 | 164.79 | 153.71 | 682 | 704.90 | 176.22 | 164.78 |
| 638 | 660.18 | 165.04 | 153.96 | 683 | 705.91 | 176.48 | 165.02 |
| 639 | 661.19 | 165.30 | 154.20 | 684 | 706.93 | 176.73 | 165.27 |
| 640 | 662.21 | 165.55 | 154.45 | 685 | 707.95 | 176.99 | 165.51 |
| 641 | 663.23 | 165.81 | 154.69 | 686 | 708.96 | 177.24 | 165.76 |
| 642 | 664.24 | 166.06 | 154.94 | 687 | 709.98 | 177.49 | 166.01 |
| 643 | 665.26 | 166.32 | 155.18 | 688 | 710.99 | 177.75 | 166.25 |
| 644 | 666.28 | 166.57 | 155.43 | 689 | 712.01 | 178.00 | 166.50 |
| 645 | 667.29 | 166.82 | 155.68 | 690 | 713.03 | 178.26 | 166.74 |
| 646 | 668.31 | 167.08 | 155.92 | 691 | 714.04 | 178.51 | 166.99 |
| 647 | 669.33 | 167.33 | 156.17 | 692 | 715.06 | 178.76 | 167.24 |
| 648 | 670.34 | 167.59 | 156.41 | 693 | 716.07 | 179.02 | 167.48 |
| 649 | 671.36 | 167.84 | 156.66 | 694 | 717.09 | 179.27 | 167.73 |
| 650 | 672.38 | 168.09 | 156.91 | 695 | 718.11 | 179.53 | 167.97 |
| 651 | 673.39 | 168.35 | 157.15 | 696 | 719.12 | 179.78 | 168.22 |
| 652 | 674.41 | 168.60 | 157.40 | 697 | 720.14 | 180.03 | 168.47 |
| 653 | 675.43 | 168.86 | 157.64 | 698 | 721.15 | 180.29 | 168.71 |
| 654 | 676.44 | 169.11 | 157.89 | 699 | 722.17 | 180.54 | 168.96 |
| 655 | 677.46 | 169.36 | 158.14 | 700 | 723.19 | 180.80 | 169.20 |
| 656 | 678.48 | 169.62 | 158.38 | 701 | 724.20 | 181.05 | 169.45 |
| 657 | 679.49 | 169.87 | 158.63 | 702 | 725.22 | 181.30 | 169.70 |
| 658 | 680.51 | 170.13 | 158.87 | 703 | 726.23 | 181.56 | 169.94 |
| 659 | 681.52 | 170.38 | 159.12 | 704 | 727.25 | 181.81 | 170.19 |
| 660 | 682.54 | 170.64 | 159.36 | 705 | 728.27 | 182.07 | 170.43 |
| 661 | 683.56 | 170.89 | 159.61 | 706 | 729.28 | 182.32 | 170.68 |
| 662 | 684.57 | 171.14 | 159.86 | 707 | 730.30 | 182.57 | 170.93 |
| 663 | 685.59 | 171.40 | 160.10 | 708 | 731.31 | 182.83 | 171.17 |
| 664 | 686.61 | 171.65 | 160.35 | 709 | 732.33 | 183.08 | 171.42 |
| 665 | 687.62 | 171.91 | 160.59 | 710 | 733.34 | 183.34 | 171.66 |
| 666 | 688.64 | 172.16 | 160.84 | 711 | 734.36 | 183.59 | 171.91 |
| 667 | 689.66 | 172.41 | 161.09 | 712 | 735.38 | 183.84 | 172.16 |
| 668 | 690.67 | 172.67 | 161.33 | 713 | 736.39 | 184.10 | 172.40 |
| 669 | 691.69 | 172.92 | 161.58 | 714 | 737.41 | 184.35 | 172.65 |
| 670 | 692.70 | 173.18 | 161.82 | 715 | 738.42 | 184.61 | 172.89 |
| 671 | 693.72 | 173.43 | 162.07 | 716 | 739.44 | 184.86 | 173.14 |
| 672 | 694.74 | 173.68 | 162.32 | 717 | 740.45 | 185.11 | 173.39 |
| 673 | 695.75 | 173.94 | 162.56 | 718 | 741.47 | 185.37 | 173.63 |
| 674 | 696.77 | 174.19 | 162.81 | 719 | 742.49 | 185.62 | 173.88 |

Table 1.  CONSTANTS FOR $\overline{p}'$ (LONG FORMULA), AN APPROXIMATION OF
THE UPPER BINOMIAL CONFIDENCE LIMIT $\overline{p}$; $c/n \leqslant .5$, $n \geqslant 20$,
$c \leqslant 1,000$ (*Continued*)

| c | $\overline{m}$ | $\overline{a}$ | $\overline{b}$ | c | $\overline{m}$ | $\overline{a}$ | $\overline{b}$ |
|---|---|---|---|---|---|---|---|
| 720 | 743.50 | 185.88 | 174.12 | 765 | 789.20 | 197.30 | 185.20 |
| 721 | 744.52 | 186.13 | 174.37 | 766 | 790.21 | 197.55 | 185.45 |
| 722 | 745.53 | 186.38 | 174.62 | 767 | 791.23 | 197.81 | 185.69 |
| 723 | 746.55 | 186.64 | 174.86 | 768 | 792.24 | 198.06 | 185.94 |
| 724 | 747.56 | 186.89 | 175.11 | 769 | 793.26 | 198.31 | 186.19 |
| 725 | 748.58 | 187.14 | 175.36 | 770 | 794.27 | 198.57 | 186.43 |
| 726 | 749.60 | 187.40 | 175.60 | 771 | 795.29 | 198.82 | 186.68 |
| 727 | 750.61 | 187.65 | 175.85 | 772 | 796.30 | 199.08 | 186.92 |
| 728 | 751.63 | 187.91 | 176.09 | 773 | 797.32 | 199.33 | 187.17 |
| 729 | 752.64 | 188.16 | 176.34 | 774 | 798.33 | 199.58 | 187.42 |
| 730 | 753.66 | 188.41 | 176.59 | 775 | 799.35 | 199.84 | 187.66 |
| 731 | 754.67 | 188.67 | 176.83 | 776 | 800.36 | 200.09 | 187.91 |
| 732 | 755.69 | 188.92 | 177.08 | 777 | 801.38 | 200.34 | 188.16 |
| 733 | 756.70 | 189.18 | 177.32 | 778 | 802.39 | 200.60 | 188.40 |
| 734 | 757.72 | 189.43 | 177.57 | 779 | 803.41 | 200.85 | 188.65 |
| 735 | 758.74 | 189.68 | 177.82 | 780 | 804.42 | 201.11 | 188.89 |
| 736 | 759.75 | 189.94 | 178.06 | 781 | 805.44 | 201.36 | 189.14 |
| 737 | 760.77 | 190.19 | 178.31 | 782 | 806.45 | 201.61 | 189.39 |
| 738 | 761.78 | 190.45 | 178.55 | 783 | 807.47 | 201.87 | 189.63 |
| 739 | 762.80 | 190.70 | 178.80 | 784 | 808.48 | 202.12 | 189.88 |
| 740 | 763.81 | 190.95 | 179.05 | 785 | 809.50 | 202.37 | 190.13 |
| 741 | 764.83 | 191.21 | 179.29 | 786 | 810.51 | 202.63 | 190.37 |
| 742 | 765.84 | 191.46 | 179.54 | 787 | 811.53 | 202.88 | 190.62 |
| 743 | 766.86 | 191.71 | 179.79 | 788 | 812.54 | 203.14 | 190.86 |
| 744 | 767.87 | 191.97 | 180.03 | 789 | 813.56 | 203.39 | 191.11 |
| 745 | 768.89 | 192.22 | 180.28 | 790 | 814.57 | 203.64 | 191.36 |
| 746 | 769.91 | 192.48 | 180.52 | 791 | 815.59 | 203.90 | 191.60 |
| 747 | 770.92 | 192.73 | 180.77 | 792 | 816.60 | 204.15 | 191.85 |
| 748 | 771.94 | 192.98 | 181.02 | 793 | 817.62 | 204.40 | 192.10 |
| 749 | 772.95 | 193.24 | 181.26 | 794 | 818.63 | 204.66 | 192.34 |
| 750 | 773.97 | 193.49 | 181.51 | 795 | 819.65 | 204.91 | 192.59 |
| 751 | 774.98 | 193.75 | 181.75 | 796 | 820.66 | 205.17 | 192.83 |
| 752 | 776.00 | 194.00 | 182.00 | 797 | 821.68 | 205.42 | 193.08 |
| 753 | 777.01 | 194.25 | 182.25 | 798 | 822.69 | 205.67 | 193.33 |
| 754 | 778.03 | 194.51 | 182.49 | 799 | 823.71 | 205.93 | 193.57 |
| 755 | 779.04 | 194.76 | 182.74 | 800 | 824.72 | 206.18 | 193.82 |
| 756 | 780.06 | 195.01 | 182.99 | 801 | 825.74 | 206.43 | 194.07 |
| 757 | 781.07 | 195.27 | 183.23 | 802 | 826.75 | 206.69 | 194.31 |
| 758 | 782.09 | 195.52 | 183.48 | 803 | 827.77 | 206.94 | 194.56 |
| 759 | 783.10 | 195.78 | 183.72 | 804 | 828.78 | 207.20 | 194.80 |
| 760 | 784.12 | 196.03 | 183.97 | 805 | 829.80 | 207.45 | 195.05 |
| 761 | 785.14 | 196.28 | 184.22 | 806 | 830.81 | 207.70 | 195.30 |
| 762 | 786.15 | 196.54 | 184.46 | 807 | 831.83 | 207.96 | 195.54 |
| 763 | 787.17 | 196.79 | 184.71 | 808 | 832.84 | 208.21 | 195.79 |
| 764 | 788.18 | 197.05 | 184.95 | 809 | 833.86 | 208.46 | 196.04 |

Table 1.   CONSTANTS FOR $\overline{p}'$ (LONG FORMULA), AN APPROXIMATION OF THE UPPER BINOMIAL CONFIDENCE LIMIT $\overline{p}$; $c/n \leqslant .5$, $n \geqslant 20$, $c \leqslant 1,000$ (*Continued*)

| c | $\overline{m}$ | $\overline{a}$ | $\overline{b}$ | c | $\overline{m}$ | $\overline{a}$ | $\overline{b}$ |
|---|---|---|---|---|---|---|---|
| 810 | 834.87 | 208.72 | 196.28 | 855 | 880.53 | 220.13 | 207.37 |
| 811 | 835.89 | 208.97 | 196.53 | 856 | 881.54 | 220.39 | 207.61 |
| 812 | 836.90 | 209.22 | 196.78 | 857 | 882.56 | 220.64 | 207.86 |
| 813 | 837.91 | 209.48 | 197.02 | 858 | 883.57 | 220.89 | 208.11 |
| 814 | 838.93 | 209.73 | 197.27 | 859 | 884.58 | 221.15 | 208.35 |
| 815 | 839.94 | 209.99 | 197.51 | 860 | 885.60 | 221.40 | 208.60 |
| 816 | 840.96 | 210.24 | 197.76 | 861 | 886.61 | 221.65 | 208.85 |
| 817 | 841.97 | 210.49 | 198.01 | 862 | 887.63 | 221.91 | 209.09 |
| 818 | 842.99 | 210.75 | 198.25 | 863 | 888.64 | 222.16 | 209.34 |
| 819 | 844.00 | 211.00 | 198.50 | 864 | 889.66 | 222.41 | 209.59 |
| 820 | 845.02 | 211.25 | 198.75 | 865 | 890.67 | 222.67 | 209.83 |
| 821 | 846.03 | 211.51 | 198.99 | 866 | 891.68 | 222.92 | 210.08 |
| 822 | 847.05 | 211.76 | 199.24 | 867 | 892.70 | 223.17 | 210.33 |
| 823 | 848.06 | 212.02 | 199.48 | 868 | 893.71 | 223.43 | 210.57 |
| 824 | 849.08 | 212.27 | 199.73 | 869 | 894.73 | 223.68 | 210.82 |
| 825 | 850.09 | 212.52 | 199.98 | 870 | 895.74 | 223.94 | 211.06 |
| 826 | 851.11 | 212.78 | 200.22 | 871 | 896.76 | 224.19 | 211.31 |
| 827 | 852.12 | 213.03 | 200.47 | 872 | 897.77 | 224.44 | 211.56 |
| 828 | 853.13 | 213.28 | 200.72 | 873 | 898.78 | 224.70 | 211.80 |
| 829 | 854.15 | 213.54 | 200.96 | 874 | 899.80 | 224.95 | 212.05 |
| 830 | 855.16 | 213.79 | 201.21 | 875 | 900.81 | 225.20 | 212.30 |
| 831 | 856.18 | 214.04 | 201.46 | 876 | 901.83 | 225.46 | 212.54 |
| 832 | 857.19 | 214.30 | 201.70 | 877 | 902.84 | 225.71 | 212.79 |
| 833 | 858.21 | 214.55 | 201.95 | 878 | 903.85 | 225.96 | 213.04 |
| 834 | 859.22 | 214.81 | 202.19 | 879 | 904.87 | 226.22 | 213.28 |
| 835 | 860.24 | 215.06 | 202.44 | 880 | 905.88 | 226.47 | 213.53 |
| 836 | 861.25 | 215.31 | 202.69 | 881 | 906.90 | 226.72 | 213.78 |
| 837 | 862.27 | 215.57 | 202.93 | 882 | 907.91 | 226.98 | 214.02 |
| 838 | 863.28 | 215.82 | 203.18 | 883 | 908.93 | 227.23 | 214.27 |
| 839 | 864.30 | 216.07 | 203.43 | 884 | 909.94 | 227.49 | 214.51 |
| 840 | 865.31 | 216.33 | 203.67 | 885 | 910.95 | 227.74 | 214.76 |
| 841 | 866.32 | 216.58 | 203.92 | 886 | 911.97 | 227.99 | 215.01 |
| 842 | 867.34 | 216.83 | 204.17 | 887 | 912.98 | 228.25 | 215.25 |
| 843 | 868.35 | 217.09 | 204.41 | 888 | 914.00 | 228.50 | 215.50 |
| 844 | 869.37 | 217.34 | 204.66 | 889 | 915.01 | 228.75 | 215.75 |
| 845 | 870.38 | 217.60 | 204.90 | 890 | 916.02 | 229.01 | 215.99 |
| 846 | 871.40 | 217.85 | 205.15 | 891 | 917.04 | 229.26 | 216.24 |
| 847 | 872.41 | 218.10 | 205.40 | 892 | 918.05 | 229.51 | 216.49 |
| 848 | 873.43 | 218.36 | 205.64 | 893 | 919.07 | 229.77 | 216.73 |
| 849 | 874.44 | 218.61 | 205.89 | 894 | 920.08 | 230.02 | 216.98 |
| 850 | 875.45 | 218.86 | 206.14 | 895 | 921.10 | 230.27 | 217.23 |
| 851 | 876.47 | 219.12 | 206.38 | 896 | 922.11 | 230.53 | 217.47 |
| 852 | 877.48 | 219.37 | 206.63 | 897 | 923.12 | 230.78 | 217.72 |
| 853 | 878.50 | 219.62 | 206.88 | 898 | 924.14 | 231.03 | 217.97 |
| 854 | 879.51 | 219.88 | 207.12 | 899 | 925.15 | 231.29 | 218.21 |

Table 1. CONSTANTS FOR $\overline{p}'$ (LONG FORMULA), AN APPROXIMATION OF
THE UPPER BINOMIAL CONFIDENCE LIMIT $\overline{p}$; $c/n \leqslant .5$, $n \geqslant 20$,
$c \leqslant 1,000$ (*Continued*)

Section 6
$\gamma_1 = 80\%$
$\gamma_2 = 60\%$

| c | $\overline{m}$ | $\overline{a}$ | $\overline{b}$ | c | $\overline{m}$ | $\overline{a}$ | $\overline{b}$ |
|---|------|--------|--------|-----|---------|--------|--------|
| 900 | 926.17 | 231.54 | 218.46 | 945 | 971.79 | 242.95 | 229.55 |
| 901 | 927.18 | 231.79 | 218.71 | 946 | 972.80 | 243.20 | 229.80 |
| 902 | 928.19 | 232.05 | 218.95 | 947 | 973.82 | 243.45 | 230.05 |
| 903 | 929.21 | 232.30 | 219.20 | 948 | 974.83 | 243.71 | 230.29 |
| 904 | 930.22 | 232.56 | 219.44 | 949 | 975.84 | 243.96 | 230.54 |
| 905 | 931.24 | 232.81 | 219.69 | 950 | 976.86 | 244.21 | 230.79 |
| 906 | 932.25 | 233.06 | 219.94 | 951 | 977.87 | 244.47 | 231.03 |
| 907 | 933.26 | 233.32 | 220.18 | 952 | 978.88 | 244.72 | 231.28 |
| 908 | 934.28 | 233.57 | 220.43 | 953 | 979.90 | 244.97 | 231.53 |
| 909 | 935.29 | 233.82 | 220.68 | 954 | 980.91 | 245.23 | 231.77 |
| 910 | 936.31 | 234.08 | 220.92 | 955 | 981.92 | 245.48 | 232.02 |
| 911 | 937.32 | 234.33 | 221.17 | 956 | 982.94 | 245.73 | 232.27 |
| 912 | 938.33 | 234.58 | 221.42 | 957 | 983.95 | 245.99 | 232.51 |
| 913 | 939.35 | 234.84 | 221.66 | 958 | 984.97 | 246.24 | 232.76 |
| 914 | 940.36 | 235.09 | 221.91 | 959 | 985.98 | 246.49 | 233.01 |
| 915 | 941.37 | 235.34 | 222.16 | 960 | 986.99 | 246.75 | 233.25 |
| 916 | 942.39 | 235.60 | 222.40 | 961 | 988.01 | 247.00 | 233.50 |
| 917 | 943.40 | 235.85 | 222.65 | 962 | 989.02 | 247.26 | 233.74 |
| 918 | 944.42 | 236.10 | 222.90 | 963 | 990.03 | 247.51 | 233.99 |
| 919 | 945.43 | 236.36 | 223.14 | 964 | 991.05 | 247.76 | 234.24 |
| 920 | 946.44 | 236.61 | 223.39 | 965 | 992.06 | 248.02 | 234.48 |
| 921 | 947.46 | 236.86 | 223.64 | 966 | 993.07 | 248.27 | 234.73 |
| 922 | 948.47 | 237.12 | 223.88 | 967 | 994.09 | 248.52 | 234.98 |
| 923 | 949.49 | 237.37 | 224.13 | 968 | 995.10 | 248.78 | 235.22 |
| 924 | 950.50 | 237.62 | 224.38 | 969 | 996.11 | 249.03 | 235.47 |
| 925 | 951.51 | 237.88 | 224.62 | 970 | 997.13 | 249.28 | 235.72 |
| 926 | 952.53 | 238.13 | 224.87 | 971 | 998.14 | 249.54 | 235.96 |
| 927 | 953.54 | 238.39 | 225.11 | 972 | 999.16 | 249.79 | 236.21 |
| 928 | 954.55 | 238.64 | 225.36 | 973 | 1000.17 | 250.04 | 236.46 |
| 929 | 955.57 | 238.89 | 225.61 | 974 | 1001.18 | 250.30 | 236.70 |
| 930 | 956.58 | 239.15 | 225.85 | 975 | 1002.20 | 250.55 | 236.95 |
| 931 | 957.60 | 239.40 | 226.10 | 976 | 1003.21 | 250.80 | 237.20 |
| 932 | 958.61 | 239.65 | 226.35 | 977 | 1004.22 | 251.06 | 237.44 |
| 933 | 959.62 | 239.91 | 226.59 | 978 | 1005.24 | 251.31 | 237.69 |
| 934 | 960.64 | 240.16 | 226.84 | 979 | 1006.25 | 251.56 | 237.94 |
| 935 | 961.65 | 240.41 | 227.09 | 980 | 1007.26 | 251.82 | 238.18 |
| 936 | 962.67 | 240.67 | 227.33 | 981 | 1008.28 | 252.07 | 238.43 |
| 937 | 963.68 | 240.92 | 227.58 | 982 | 1009.29 | 252.32 | 238.68 |
| 938 | 964.69 | 241.17 | 227.83 | 983 | 1010.30 | 252.58 | 238.92 |
| 939 | 965.71 | 241.43 | 228.07 | 984 | 1011.32 | 252.83 | 239.17 |
| 940 | 966.72 | 241.68 | 228.32 | 985 | 1012.33 | 253.08 | 239.42 |
| 941 | 967.73 | 241.93 | 228.57 | 986 | 1013.34 | 253.34 | 239.66 |
| 942 | 968.75 | 242.19 | 228.81 | 987 | 1014.36 | 253.59 | 239.91 |
| 943 | 969.76 | 242.44 | 229.06 | 988 | 1015.37 | 253.84 | 240.16 |
| 944 | 970.77 | 242.69 | 229.31 | 989 | 1016.38 | 254.10 | 240.40 |

Table 1. **CONSTANTS FOR $\bar{p}'$ (LONG FORMULA), AN APPROXIMATION OF THE UPPER BINOMIAL CONFIDENCE LIMIT $\bar{p}$; c/n $\leqslant$ .5, n $\geqslant$ 20,** c $\leqslant$ 1,000 (*Continued*)

| c | $\bar{m}$ | $\bar{a}$ | $\bar{b}$ | c | $\bar{m}$ | $\bar{a}$ | $\bar{b}$ |
|---|---|---|---|---|---|---|---|
| 990 | 1017.40 | 254.35 | 240.65 | 995 | 1022.46 | 255.62 | 241.88 |
| 991 | 1018.41 | 254.60 | 240.90 | 996 | 1023.48 | 255.87 | 242.13 |
| 992 | 1019.42 | 254.86 | 241.14 | 997 | 1024.49 | 256.12 | 242.38 |
| 993 | 1020.44 | 255.11 | 241.39 | 998 | 1025.50 | 256.38 | 242.62 |
| 994 | 1021.45 | 255.36 | 241.64 | 999 | 1026.52 | 256.63 | 242.87 |
| | | | | 1000 | 1027.53 | 256.88 | 243.12 |

# Table 2

Approximation Formula: $\underline{p}' = \dfrac{m}{n} \times \dfrac{n - \underline{a}}{n - \underline{b}}$

| | |
|---|---|
| n | sample size. |
| c | number of events (items having a designated characteristic) in the sample. |
| $\underline{p}'$ | approximation of the lower binomial confidence limit $\underline{p}$; $\underline{p}'$ has relative accuracy of at least .999. |
| $\underline{m}$ | lower confidence limit for the parameter m of a Poisson distribution; each $\underline{m}$ is for a given c and a given $\gamma_1$ or $\gamma_2$. |
| $\underline{a}$, $\underline{b}$ | approximation constants for a given c and a given $\gamma_1$ or $\gamma_2$. |
| $\gamma_1$ | confidence level for a one-sided (upper or lower) confidence limit. |
| $\gamma_2$ | confidence level for two-sided (upper and lower) confidence limits. ($\gamma_2 = 2\gamma_1 - 100\%$.) |

For $c > 1,000$, calculate $\underline{a}$ and $\underline{b}$ from the extension formulas in Table 7–B; and calculate $\underline{m}$ from the extension formula in Table 7–E.

(Note. For the purpose of acceptance sampling in conjunction with Table 9, c and $\underline{m}$ in Table 2 may respectively be read as rejection number $\dot{c}_1$ and $\underline{m}_1$.)

| | | |
|---|---|---|
| Section 1 | $\gamma_1 = 99.5\%$ | $\gamma_2 = 99\%$ |
| Section 2 | $\gamma_1 = 99\%$ | $\gamma_2 = 98\%$ |
| Section 3 | $\gamma_1 = 97.5\%$ | $\gamma_2 = 95\%$ |
| Section 4 | $\gamma_1 = 95\%$ | $\gamma_2 = 90\%$ |
| Section 5 | $\gamma_1 = 90\%$ | $\gamma_2 = 80\%$ |
| Section 6 | $\gamma_1 = 80\%$ | $\gamma_2 = 60\%$ |

Table 2. **CONSTANTS FOR $p'$ (LONG FORMULA), AN APPROXIMATION OF THE LOWER BINOMIAL CONFIDENCE LIMIT $\underline{p}$; $c/n \leqslant .5$, $n \geqslant 20$, $c \leqslant 1,000$**

| c | m | a | b | c | m | a | b |
|---|---|---|---|---|---|---|---|
| 1 | 0.0050126 | 0.0025 | 0.0 | 46 | 30.407 | 11.149 | 18.446 |
| 2 | 0.10350 | 0.237 | 0.685 | 47 | 31.219 | 11.428 | 18.818 |
| 3 | 0.33786 | 0.450 | 1.281 | 48 | 32.032 | 11.696 | 19.180 |
| 4 | 0.67221 | 0.603 | 1.767 | 49 | 32.847 | 11.969 | 19.546 |
| 5 | 1.0779 | 0.794 | 2.255 | 50 | 33.664 | 12.220 | 19.888 |
| 6 | 1.5369 | 0.998 | 2.729 | 51 | 34.483 | 12.492 | 20.251 |
| 7 | 2.0373 | 1.218 | 3.199 | 52 | 35.303 | 12.756 | 20.604 |
| 8 | 2.5711 | 1.448 | 3.662 | 53 | 36.125 | 13.035 | 20.972 |
| 9 | 3.1324 | 1.703 | 4.137 | 54 | 36.949 | 13.294 | 21.320 |
| 10 | 3.7169 | 1.971 | 4.613 | 55 | 37.775 | 13.561 | 21.673 |
| 11 | 4.3214 | 2.206 | 5.045 | 56 | 38.602 | 13.827 | 22.026 |
| 12 | 4.9431 | 2.439 | 5.467 | 57 | 39.431 | 14.092 | 22.377 |
| 13 | 5.5801 | 2.682 | 5.892 | 58 | 40.261 | 14.38 | 22.752 |
| 14 | 6.2307 | 2.921 | 6.305 | 59 | 41.093 | 14.63 | 23.086 |
| 15 | 6.8934 | 3.166 | 6.720 | 60 | 41.926 | 14.90 | 23.436 |
| 16 | 7.5670 | 3.409 | 7.126 | 61 | 42.760 | 15.18 | 23.801 |
| 17 | 8.2507 | 3.657 | 7.532 | 62 | 43.596 | 15.44 | 24.145 |
| 18 | 8.9434 | 3.903 | 7.932 | 63 | 44.433 | 15.71 | 24.50 |
| 19 | 9.6445 | 4.153 | 8.331 | 64 | 45.272 | 15.99 | 24.86 |
| 20 | 10.353 | 4.400 | 8.724 | 65 | 46.111 | 16.24 | 25.18 |
| 21 | 11.069 | 4.658 | 9.123 | 66 | 46.952 | 16.52 | 25.55 |
| 22 | 11.792 | 4.904 | 9.509 | 67 | 47.794 | 16.80 | 25.91 |
| 23 | 12.521 | 5.166 | 9.906 | 68 | 48.637 | 17.08 | 26.26 |
| 24 | 13.255 | 5.415 | 10.287 | 69 | 49.482 | 17.35 | 26.61 |
| 25 | 13.995 | 5.670 | 10.672 | 70 | 50.328 | 17.61 | 26.95 |
| 26 | 14.741 | 5.928 | 11.058 | 71 | 51.174 | 17.87 | 27.28 |
| 27 | 15.491 | 6.183 | 11.438 | 72 | 52.022 | 18.13 | 27.62 |
| 28 | 16.245 | 6.439 | 11.817 | 73 | 52.871 | 18.44 | 28.00 |
| 29 | 17.004 | 6.703 | 12.201 | 74 | 53.721 | 18.67 | 28.31 |
| 30 | 17.767 | 6.959 | 12.576 | 75 | 54.571 | 18.97 | 28.69 |
| 31 | 18.534 | 7.220 | 12.952 | 76 | 55.423 | 19.24 | 29.03 |
| 32 | 19.305 | 7.483 | 13.331 | 77 | 56.276 | 19.50 | 29.36 |
| 33 | 20.079 | 7.739 | 13.699 | 78 | 57.130 | 19.80 | 29.73 |
| 34 | 20.857 | 8.001 | 14.073 | 79 | 57.984 | 20.03 | 30.04 |
| 35 | 21.638 | 8.262 | 14.444 | 80 | 58.840 | 20.32 | 30.40 |
| 36 | 22.422 | 8.527 | 14.816 | 81 | 59.696 | 20.62 | 30.77 |
| 37 | 23.209 | 8.783 | 15.179 | 82 | 60.554 | 20.87 | 31.09 |
| 38 | 23.998 | 9.041 | 15.542 | 83 | 61.412 | 21.14 | 31.44 |
| 39 | 24.791 | 9.307 | 15.912 | 84 | 62.271 | 21.41 | 31.78 |
| 40 | 25.586 | 9.576 | 16.283 | 85 | 63.131 | 21.68 | 32.12 |
| 41 | 26.384 | 9.838 | 16.646 | 86 | 63.992 | 21.95 | 32.46 |
| 42 | 27.184 | 10.104 | 17.012 | 87 | 64.853 | 22.23 | 32.81 |
| 43 | 27.986 | 10.370 | 17.376 | 88 | 65.715 | 22.52 | 33.16 |
| 44 | 28.791 | 10.626 | 17.730 | 89 | 66.579 | 22.78 | 33.49 |
| 45 | 29.598 | 10.888 | 18.089 | 90 | 67.442 | 23.06 | 33.84 |

**Table 2. CONSTANTS FOR $\underline{p}'$ (LONG FORMULA), AN APPROXIMATION OF THE LOWER BINOMIAL CONFIDENCE LIMIT $\underline{p}$; $c/n \leqslant .5$, $n \geqslant 20$ $c \leqslant 1,000$ (Continued)**

| c | m | a | b | c | m | a | b |
|---|---|---|---|---|---|---|---|
| 91 | 68.307 | 23.32 | 34.17 | 136 | 107.84 | 35.95 | 49.53 |
| 92 | 69.172 | 23.59 | 34.50 | 137 | 108.73 | 36.24 | 49.88 |
| 93 | 70.039 | 23.89 | 34.87 | 138 | 109.62 | 36.54 | 50.23 |
| 94 | 70.905 | 24.14 | 35.19 | 139 | 110.51 | 36.84 | 50.58 |
| 95 | 71.773 | 24.43 | 35.54 | 140 | 111.40 | 37.13 | 50.93 |
| 96 | 72.641 | 24.70 | 35.88 | 141 | 112.29 | 37.43 | 51.28 |
| 97 | 73.510 | 24.96 | 36.20 | 142 | 113.18 | 37.73 | 51.64 |
| 98 | 74.380 | 25.27 | 36.58 | 143 | 114.08 | 38.03 | 51.99 |
| 99 | 75.250 | 25.51 | 36.88 | 144 | 114.97 | 38.32 | 52.34 |
| 100 | 76.121 | 25.78 | 37.22 | 145 | 115.86 | 38.62 | 52.69 |
| 101 | 76.992 | 25.66 | 37.17 | 146 | 116.76 | 38.92 | 53.04 |
| 102 | 77.864 | 25.95 | 37.52 | 147 | 117.65 | 39.22 | 53.39 |
| 103 | 78.737 | 26.25 | 37.88 | 148 | 118.54 | 39.51 | 53.74 |
| 104 | 79.611 | 26.54 | 38.23 | 149 | 119.44 | 39.81 | 54.09 |
| 105 | 80.485 | 26.83 | 38.59 | 150 | 120.33 | 40.11 | 54.44 |
| 106 | 81.359 | 27.12 | 38.94 | 151 | 121.23 | 40.41 | 54.80 |
| 107 | 82.235 | 27.41 | 39.29 | 152 | 122.12 | 40.71 | 55.15 |
| 108 | 83.110 | 27.70 | 39.65 | 153 | 123.02 | 41.01 | 55.50 |
| 109 | 83.987 | 28.00 | 40.00 | 154 | 123.91 | 41.30 | 55.85 |
| 110 | 84.864 | 28.29 | 40.36 | 155 | 124.81 | 41.60 | 56.20 |
| 111 | 85.741 | 28.58 | 40.71 | 156 | 125.71 | 41.90 | 56.55 |
| 112 | 86.619 | 28.87 | 41.06 | 157 | 126.60 | 42.20 | 56.90 |
| 113 | 87.498 | 29.17 | 41.42 | 158 | 127.50 | 42.50 | 57.25 |
| 114 | 88.377 | 29.46 | 41.77 | 159 | 128.40 | 42.80 | 57.60 |
| 115 | 89.256 | 29.75 | 42.12 | 160 | 129.30 | 43.10 | 57.95 |
| 116 | 90.137 | 30.05 | 42.48 | 161 | 130.20 | 43.40 | 58.30 |
| 117 | 91.017 | 30.34 | 42.83 | 162 | 131.09 | 43.70 | 58.65 |
| 118 | 91.898 | 30.63 | 43.18 | 163 | 131.99 | 44.00 | 59.00 |
| 119 | 92.780 | 30.93 | 43.54 | 164 | 132.89 | 44.30 | 59.35 |
| 120 | 93.662 | 31.22 | 43.89 | 165 | 133.79 | 44.60 | 59.70 |
| 121 | 94.545 | 31.51 | 44.24 | 166 | 134.69 | 44.90 | 60.05 |
| 122 | 95.428 | 31.81 | 44.60 | 167 | 135.59 | 45.20 | 60.40 |
| 123 | 96.312 | 32.10 | 44.95 | 168 | 136.49 | 45.50 | 60.75 |
| 124 | 97.196 | 32.40 | 45.30 | 169 | 137.39 | 45.80 | 61.10 |
| 125 | 98.081 | 32.69 | 45.65 | 170 | 138.29 | 46.10 | 61.45 |
| 126 | 98.966 | 32.99 | 46.01 | 171 | 139.20 | 46.40 | 61.80 |
| 127 | 99.851 | 33.28 | 46.36 | 172 | 140.10 | 46.70 | 62.15 |
| 128 | 100.74 | 33.58 | 46.71 | 173 | 141.00 | 47.00 | 62.50 |
| 129 | 101.62 | 33.87 | 47.06 | 174 | 141.90 | 47.30 | 62.85 |
| 130 | 102.51 | 34.17 | 47.42 | 175 | 142.80 | 47.60 | 63.20 |
| 131 | 103.40 | 34.47 | 47.77 | 176 | 143.71 | 47.90 | 63.55 |
| 132 | 104.29 | 34.76 | 48.12 | 177 | 144.61 | 48.20 | 63.90 |
| 133 | 105.17 | 35.06 | 48.47 | 178 | 145.51 | 48.50 | 64.25 |
| 134 | 106.06 | 35.35 | 48.82 | 179 | 146.42 | 48.81 | 64.60 |
| 135 | 106.95 | 35.65 | 49.17 | 180 | 147.32 | 49.11 | 64.95 |

**Table 2. CONSTANTS FOR $p'$ (LONG FORMULA), AN APPROXIMATION OF THE LOWER BINOMIAL CONFIDENCE LIMIT $\underline{p}$; c/n ⩽ .5, n ⩾ 20, c ⩽ 1,000 (*Continued*)**

| c | m | a | b | c | m | a | b |
|---|---|---|---|---|---|---|---|
| 181 | 148.22 | 49.41 | 65.30 | 226 | 189.16 | 63.05 | 80.97 |
| 182 | 149.13 | 49.71 | 65.65 | 227 | 190.07 | 63.36 | 81.32 |
| 183 | 150.03 | 50.01 | 65.99 | 228 | 190.98 | 63.66 | 81.67 |
| 184 | 150.94 | 50.31 | 66.34 | 229 | 191.90 | 63.97 | 82.02 |
| 185 | 151.84 | 50.61 | 66.69 | 230 | 192.81 | 64.27 | 82.36 |
| 186 | 152.75 | 50.92 | 67.04 | 231 | 193.73 | 64.58 | 82.71 |
| 187 | 153.65 | 51.22 | 67.39 | 232 | 194.64 | 64.88 | 83.06 |
| 188 | 154.56 | 51.52 | 67.74 | 233 | 195.56 | 65.19 | 83.41 |
| 189 | 155.47 | 51.82 | 68.09 | 234 | 196.48 | 65.49 | 83.75 |
| 190 | 156.37 | 52.12 | 68.44 | 235 | 197.39 | 65.80 | 84.10 |
| 191 | 157.28 | 52.43 | 68.79 | 236 | 198.31 | 66.10 | 84.45 |
| 192 | 158.19 | 52.73 | 69.14 | 237 | 199.22 | 66.41 | 84.80 |
| 193 | 159.09 | 53.03 | 69.48 | 238 | 200.14 | 66.71 | 85.14 |
| 194 | 160.00 | 53.33 | 69.83 | 239 | 201.06 | 67.02 | 85.49 |
| 195 | 160.91 | 53.64 | 70.18 | 240 | 201.97 | 67.32 | 85.84 |
| 196 | 161.82 | 53.94 | 70.53 | 241 | 202.89 | 67.63 | 86.18 |
| 197 | 162.73 | 54.24 | 70.88 | 242 | 203.81 | 67.94 | 86.53 |
| 198 | 163.63 | 54.54 | 71.23 | 243 | 204.73 | 68.24 | 86.88 |
| 199 | 164.54 | 54.85 | 71.58 | 244 | 205.64 | 68.55 | 87.23 |
| 200 | 165.45 | 55.15 | 71.92 | 245 | 206.56 | 68.85 | 87.57 |
| 201 | 166.36 | 55.45 | 72.27 | 246 | 207.48 | 69.16 | 87.92 |
| 202 | 167.27 | 55.76 | 72.62 | 247 | 208.40 | 69.47 | 88.27 |
| 203 | 168.18 | 56.06 | 72.97 | 248 | 209.31 | 69.77 | 88.61 |
| 204 | 169.09 | 56.36 | 73.32 | 249 | 210.23 | 70.08 | 88.96 |
| 205 | 170.00 | 56.67 | 73.67 | 250 | 211.15 | 70.38 | 89.31 |
| 206 | 170.91 | 56.97 | 74.02 | 251 | 212.07 | 70.69 | 89.66 |
| 207 | 171.82 | 57.27 | 74.36 | 252 | 212.99 | 71.00 | 90.00 |
| 208 | 172.73 | 57.58 | 74.71 | 253 | 213.91 | 71.30 | 90.35 |
| 209 | 173.64 | 57.88 | 75.06 | 254 | 214.83 | 71.61 | 90.70 |
| 210 | 174.55 | 58.18 | 75.41 | 255 | 215.75 | 71.92 | 91.04 |
| 211 | 175.46 | 58.49 | 75.76 | 256 | 216.67 | 72.22 | 91.39 |
| 212 | 176.37 | 58.79 | 76.10 | 257 | 217.59 | 72.53 | 91.74 |
| 213 | 177.29 | 59.10 | 76.45 | 258 | 218.50 | 72.83 | 92.08 |
| 214 | 178.20 | 59.40 | 76.80 | 259 | 219.42 | 73.14 | 92.43 |
| 215 | 179.11 | 59.70 | 77.15 | 260 | 220.34 | 73.45 | 92.78 |
| 216 | 180.02 | 60.01 | 77.50 | 261 | 221.26 | 73.75 | 93.12 |
| 217 | 180.93 | 60.31 | 77.84 | 262 | 222.19 | 74.06 | 93.47 |
| 218 | 181.85 | 60.62 | 78.19 | 263 | 223.11 | 74.37 | 93.82 |
| 219 | 182.76 | 60.92 | 78.54 | 264 | 224.03 | 74.68 | 94.16 |
| 220 | 183.67 | 61.22 | 78.89 | 265 | 224.95 | 74.98 | 94.51 |
| 221 | 184.59 | 61.53 | 79.24 | 266 | 225.87 | 75.29 | 94.86 |
| 222 | 185.50 | 61.83 | 79.58 | 267 | 226.79 | 75.60 | 95.20 |
| 223 | 186.41 | 62.14 | 79.93 | 268 | 227.71 | 75.90 | 95.55 |
| 224 | 187.33 | 62.44 | 80.28 | 269 | 228.63 | 76.21 | 95.89 |
| 225 | 188.24 | 62.75 | 80.63 | 270 | 229.55 | 76.52 | 96.24 |

Table 2. CONSTANTS FOR p' (LONG FORMULA), AN APPROXIMATION OF THE LOWER BINOMIAL CONFIDENCE LIMIT $\underline{p}$; c/n ⩽ .5, n ⩾ 20, c ⩽ 1,000 (*Continued*)

| c | $\underline{m}$ | $\underline{a}$ | $\underline{b}$ | c | $\underline{m}$ | $\underline{a}$ | $\underline{b}$ |
|---|---|---|---|---|---|---|---|
| 271 | 230.48 | 76.83 | 96.59 | 316 | 272.09 | 90.70 | 112.15 |
| 272 | 231.40 | 77.13 | 96.93 | 317 | 273.02 | 91.01 | 112.50 |
| 273 | 232.32 | 77.44 | 97.28 | 318 | 273.95 | 91.32 | 112.84 |
| 274 | 233.24 | 77.75 | 97.63 | 319 | 274.87 | 91.62 | 113.19 |
| 275 | 234.16 | 78.05 | 97.97 | 320 | 275.80 | 91.93 | 113.53 |
| 276 | 235.09 | 78.36 | 98.32 | 321 | 276.73 | 92.24 | 113.88 |
| 277 | 236.01 | 78.67 | 98.67 | 322 | 277.66 | 92.55 | 114.22 |
| 278 | 236.93 | 78.98 | 99.01 | 323 | 278.59 | 92.86 | 114.57 |
| 279 | 237.85 | 79.28 | 99.36 | 324 | 279.51 | 93.17 | 114.91 |
| 280 | 238.78 | 79.59 | 99.70 | 325 | 280.44 | 93.48 | 115.26 |
| 281 | 239.70 | 79.90 | 100.05 | 326 | 281.37 | 93.79 | 115.60 |
| 282 | 240.62 | 80.21 | 100.40 | 327 | 282.30 | 94.10 | 115.95 |
| 283 | 241.55 | 80.52 | 100.74 | 328 | 283.23 | 94.41 | 116.30 |
| 284 | 242.47 | 80.82 | 101.09 | 329 | 284.16 | 94.72 | 116.64 |
| 285 | 243.39 | 81.13 | 101.43 | 330 | 285.09 | 95.03 | 116.99 |
| 286 | 244.32 | 81.44 | 101.78 | 331 | 286.02 | 95.34 | 117.33 |
| 287 | 245.24 | 81.75 | 102.13 | 332 | 286.94 | 95.65 | 117.68 |
| 288 | 246.17 | 82.06 | 102.47 | 333 | 287.87 | 95.96 | 118.02 |
| 289 | 247.09 | 82.36 | 102.82 | 334 | 288.80 | 96.27 | 118.37 |
| 290 | 248.01 | 82.67 | 103.16 | 335 | 289.73 | 96.58 | 118.71 |
| 291 | 248.94 | 82.98 | 103.51 | 336 | 290.66 | 96.89 | 119.06 |
| 292 | 249.86 | 83.29 | 103.86 | 337 | 291.59 | 97.20 | 119.40 |
| 293 | 250.79 | 83.60 | 104.20 | 338 | 292.52 | 97.51 | 119.75 |
| 294 | 251.71 | 83.90 | 104.55 | 339 | 293.45 | 97.82 | 120.09 |
| 295 | 252.64 | 84.21 | 104.89 | 340 | 294.38 | 98.13 | 120.44 |
| 296 | 253.56 | 84.52 | 105.24 | 341 | 295.31 | 98.44 | 120.78 |
| 297 | 254.49 | 84.83 | 105.59 | 342 | 296.24 | 98.75 | 121.13 |
| 298 | 255.41 | 85.14 | 105.93 | 343 | 297.17 | 99.06 | 121.47 |
| 299 | 256.34 | 85.45 | 106.28 | 344 | 298.10 | 99.37 | 121.82 |
| 300 | 257.26 | 85.75 | 106.62 | 345 | 299.03 | 99.68 | 122.16 |
| 301 | 258.19 | 86.06 | 106.97 | 346 | 299.97 | 99.99 | 122.51 |
| 302 | 259.12 | 86.37 | 107.31 | 347 | 300.90 | 100.30 | 122.85 |
| 303 | 260.04 | 86.68 | 107.66 | 348 | 301.83 | 100.61 | 123.20 |
| 304 | 260.97 | 86.99 | 108.01 | 349 | 302.76 | 100.92 | 123.54 |
| 305 | 261.89 | 87.30 | 108.35 | 350 | 303.69 | 101.23 | 123.89 |
| 306 | 262.82 | 87.61 | 108.70 | 351 | 304.62 | 101.54 | 124.23 |
| 307 | 263.75 | 87.92 | 109.04 | 352 | 305.55 | 101.85 | 124.57 |
| 308 | 264.67 | 88.22 | 109.39 | 353 | 306.48 | 102.16 | 124.92 |
| 309 | 265.60 | 88.53 | 109.73 | 354 | 307.41 | 102.47 | 125.26 |
| 310 | 266.53 | 88.84 | 110.08 | 355 | 308.35 | 102.78 | 125.61 |
| 311 | 267.45 | 89.15 | 110.42 | 356 | 309.28 | 103.09 | 125.95 |
| 312 | 268.38 | 89.46 | 110.77 | 357 | 310.21 | 103.40 | 126.30 |
| 313 | 269.31 | 89.77 | 111.12 | 358 | 311.14 | 103.71 | 126.64 |
| 314 | 270.23 | 90.08 | 111.46 | 359 | 312.07 | 104.02 | 126.99 |
| 315 | 271.16 | 90.39 | 111.81 | 360 | 313.01 | 104.34 | 127.33 |

Table 2. CONSTANTS FOR $p'$ (LONG FORMULA), AN APPROXIMATION OF THE LOWER BINOMIAL CONFIDENCE LIMIT $\underline{p}$; $c/n \leqslant .5$, $n \geqslant 20$, $c \leqslant 1,000$ (*Continued*)

Section 1

$\gamma_1 = 99.5\%$

$\gamma_2 = 99\%$

| c | m | a | b | c | m | a | b |
|---|---|---|---|---|---|---|---|
| 361 | 313.94 | 104.65 | 127.68 | 406 | 355.98 | 118.66 | 143.17 |
| 362 | 314.87 | 104.96 | 128.02 | 407 | 356.91 | 118.97 | 143.51 |
| 363 | 315.80 | 105.27 | 128.37 | 408 | 357.85 | 119.28 | 143.86 |
| 364 | 316.74 | 105.58 | 128.71 | 409 | 358.79 | 119.60 | 144.20 |
| 365 | 317.67 | 105.89 | 129.06 | 410 | 359.72 | 119.91 | 144.55 |
| 366 | 318.60 | 106.20 | 129.40 | 411 | 360.66 | 120.22 | 144.89 |
| 367 | 319.53 | 106.51 | 129.74 | 412 | 361.60 | 120.53 | 145.23 |
| 368 | 320.47 | 106.82 | 130.09 | 413 | 362.53 | 120.84 | 145.58 |
| 369 | 321.40 | 107.13 | 130.43 | 414 | 363.47 | 121.16 | 145.92 |
| 370 | 322.33 | 107.44 | 130.78 | 415 | 364.41 | 121.47 | 146.27 |
| 371 | 323.26 | 107.75 | 131.12 | 416 | 365.34 | 121.78 | 146.61 |
| 372 | 324.20 | 108.07 | 131.47 | 417 | 366.28 | 122.09 | 146.95 |
| 373 | 325.13 | 108.38 | 131.81 | 418 | 367.22 | 122.41 | 147.30 |
| 374 | 326.06 | 108.69 | 132.16 | 419 | 368.15 | 122.72 | 147.64 |
| 375 | 327.00 | 109.00 | 132.50 | 420 | 369.09 | 123.03 | 147.98 |
| 376 | 327.93 | 109.31 | 132.84 | 421 | 370.03 | 123.34 | 148.33 |
| 377 | 328.87 | 109.62 | 133.19 | 422 | 370.96 | 123.65 | 148.67 |
| 378 | 329.80 | 109.93 | 133.53 | 423 | 371.90 | 123.97 | 149.02 |
| 379 | 330.73 | 110.24 | 133.88 | 424 | 372.84 | 124.28 | 149.36 |
| 380 | 331.67 | 110.56 | 134.22 | 425 | 373.78 | 124.59 | 149.70 |
| 381 | 332.60 | 110.87 | 134.57 | 426 | 374.71 | 124.90 | 150.05 |
| 382 | 333.53 | 111.18 | 134.91 | 427 | 375.65 | 125.22 | 150.39 |
| 383 | 334.47 | 111.49 | 135.26 | 428 | 376.59 | 125.53 | 150.74 |
| 384 | 335.40 | 111.80 | 135.60 | 429 | 377.53 | 125.84 | 151.08 |
| 385 | 336.34 | 112.11 | 135.94 | 430 | 378.47 | 126.16 | 151.42 |
| 386 | 337.27 | 112.42 | 136.29 | 431 | 379.40 | 126.47 | 151.77 |
| 387 | 338.21 | 112.74 | 136.63 | 432 | 380.34 | 126.78 | 152.11 |
| 388 | 339.14 | 113.05 | 136.98 | 433 | 381.28 | 127.09 | 152.45 |
| 389 | 340.08 | 113.36 | 137.32 | 434 | 382.22 | 127.41 | 152.80 |
| 390 | 341.01 | 113.67 | 137.66 | 435 | 383.16 | 127.72 | 153.14 |
| 391 | 341.95 | 113.98 | 138.01 | 436 | 384.09 | 128.03 | 153.48 |
| 392 | 342.88 | 114.29 | 138.35 | 437 | 385.03 | 128.34 | 153.83 |
| 393 | 343.82 | 114.61 | 138.70 | 438 | 385.97 | 128.66 | 154.17 |
| 394 | 344.75 | 114.92 | 139.04 | 439 | 386.91 | 128.97 | 154.52 |
| 395 | 345.69 | 115.23 | 139.39 | 440 | 387.85. | 129.28 | 154.86 |
| 396 | 346.62 | 115.54 | 139.73 | 441 | 388.79 | 129.60 | 155.20 |
| 397 | 347.56 | 115.85 | 140.07 | 442 | 389.73 | 129.91 | 155.55 |
| 398 | 348.49 | 116.16 | 140.42 | 443 | 390.66 | 130.22 | 155.89 |
| 399 | 349.43 | 116.48 | 140.76 | 444 | 391.60 | 130.53 | 156.23 |
| 400 | 350.36 | 116.79 | 141.11 | 445 | 392.54 | 130.85 | 156.58 |
| 401 | 351.30 | 117.10 | 141.45 | 446 | 393.48 | 131.16 | 156.92 |
| 402 | 352.23 | 117.41 | 141.79 | 447 | 394.42 | 131.47 | 157.26 |
| 403 | 353.17 | 117.72 | 142.14 | 448 | 395.36 | 131.79 | 157.61 |
| 404 | 354.11 | 118.04 | 142.48 | 449 | 396.30 | 132.10 | 157.95 |
| 405 | 355.04 | 118.35 | 142.83 | 450 | 397.24 | 132.41 | 158.29 |

Table 2.  CONSTANTS FOR p' (LONG FORMULA), AN APPROXIMATION OF THE LOWER BINOMIAL CONFIDENCE LIMIT p̲; c/n ⩽ .5, n ⩾ 20, c ⩽ 1,000 (*Continued*)

| c | m | a | b | c | m | a | b |
|---|---|---|---|---|---|---|---|
| 451 | 398.18 | 132.73 | 158.64 | 496 | 440.51 | 146.84 | 174.08 |
| 452 | 399.12 | 133.04 | 158.98 | 497 | 441.45 | 147.15 | 174.42 |
| 453 | 400.06 | 133.35 | 159.32 | 498 | 442.40 | 147.47 | 174.77 |
| 454 | 401.00 | 133.66 | 159.67 | 499 | 443.34 | 147.78 | 175.11 |
| 455 | 401.93 | 133.98 | 160.01 | 500 | 444.28 | 148.09 | 175.45 |
| 456 | 402.87 | 134.29 | 160.35 | 501 | 445.22 | 148.41 | 175.80 |
| 457 | 403.81 | 134.60 | 160.70 | 502 | 446.17 | 148.72 | 176.14 |
| 458 | 404.75 | 134.92 | 161.04 | 503 | 447.11 | 149.04 | 176.48 |
| 459 | 405.69 | 135.23 | 161.38 | 504 | 448.05 | 149.35 | 176.82 |
| 460 | 406.63 | 135.54 | 161.73 | 505 | 448.99 | 149.66 | 177.17 |
| 461 | 407.57 | 135.86 | 162.07 | 506 | 449.94 | 149.98 | 177.51 |
| 462 | 408.51 | 136.17 | 162.41 | 507 | 450.88 | 150.29 | 177.85 |
| 463 | 409.45 | 136.48 | 162.76 | 508 | 451.82 | 150.61 | 178.20 |
| 464 | 410.39 | 136.80 | 163.10 | 509 | 452.77 | 150.92 | 178.54 |
| 465 | 411.33 | 137.11 | 163.44 | 510 | 453.71 | 151.24 | 178.88 |
| 466 | 412.27 | 137.42 | 163.79 | 511 | 454.65 | 151.55 | 179.22 |
| 467 | 413.21 | 137.74 | 164.13 | 512 | 455.59 | 151.86 | 179.57 |
| 468 | 414.16 | 138.05 | 164.47 | 513 | 456.54 | 152.18 | 179.91 |
| 469 | 415.10 | 138.37 | 164.82 | 514 | 457.48 | 152.49 | 180.25 |
| 470 | 416.04 | 138.68 | 165.16 | 515 | 458.42 | 152.81 | 180.60 |
| 471 | 416.98 | 138.99 | 165.50 | 516 | 459.37 | 153.12 | 180.94 |
| 472 | 417.92 | 139.31 | 165.85 | 517 | 460.31 | 153.44 | 181.28 |
| 473 | 418.86 | 139.62 | 166.19 | 518 | 461.25 | 153.75 | 181.62 |
| 474 | 419.80 | 139.93 | 166.53 | 519 | 462.20 | 154.07 | 181.97 |
| 475 | 420.74 | 140.25 | 166.88 | 520 | 463.14 | 154.38 | 182.31 |
| 476 | 421.68 | 140.56 | 167.22 | 521 | 464.08 | 154.69 | 182.65 |
| 477 | 422.62 | 140.87 | 167.56 | 522 | 465.03 | 155.01 | 183.00 |
| 478 | 423.56 | 141.19 | 167.91 | 523 | 465.97 | 155.32 | 183.34 |
| 479 | 424.50 | 141.50 | 168.25 | 524 | 466.92 | 155.64 | 183.68 |
| 480 | 425.45 | 141.82 | 168.59 | 525 | 467.86 | 155.95 | 184.02 |
| 481 | 426.39 | 142.13 | 168.94 | 526 | 468.80 | 156.27 | 184.37 |
| 482 | 427.33 | 142.44 | 169.28 | 527 | 469.75 | 156.58 | 184.71 |
| 483 | 428.27 | 142.76 | 169.62 | 528 | 470.69 | 156.90 | 185.05 |
| 484 | 429.21 | 143.07 | 169.96 | 529 | 471.63 | 157.21 | 185.39 |
| 485 | 430.15 | 143.38 | 170.31 | 530 | 472.58 | 157.53 | 185.74 |
| 486 | 431.09 | 143.70 | 170.65 | 531 | 473.52 | 157.84 | 186.08 |
| 487 | 432.04 | 144.01 | 170.99 | 532 | 474.47 | 158.16 | 186.42 |
| 488 | 432.98 | 144.33 | 171.34 | 533 | 475.41 | 158.47 | 186.76 |
| 489 | 433.92 | 144.64 | 171.68 | 534 | 476.36 | 158.79 | 187.11 |
| 490 | 434.86 | 144.95 | 172.02 | 535 | 477.30 | 159.10 | 187.45 |
| 491 | 435.80 | 145.27 | 172.37 | 536 | 478.24 | 159.41 | 187.79 |
| 492 | 436.74 | 145.58 | 172.71 | 537 | 479.19 | 159.73 | 188.14 |
| 493 | 437.69 | 145.90 | 173.05 | 538 | 480.13 | 160.04 | 188.48 |
| 494 | 438.63 | 146.21 | 173.40 | 539 | 481.08 | 160.36 | 188.82 |
| 495 | 439.57 | 146.52 | 173.74 | 540 | 482.02 | 160.67 | 189.16 |

Table 2. CONSTANTS FOR p′ (LONG FORMULA), AN APPROXIMATION OF
THE LOWER BINOMIAL CONFIDENCE LIMIT p; c/n ≤ .5, n ≥ 20,
c ≤ 1,000 (*Continued*)

| c | m | a | b | c | m | a | b |
|---|---|---|---|---|---|---|---|
| 541 | 482.97 | 160.99 | 189.51 | 586 | 525.52 | 175.17 | 204.91 |
| 542 | 483.91 | 161.30 | 189.85 | 587 | 526.47 | 175.49 | 205.25 |
| 543 | 484.86 | 161.62 | 190.19 | 588 | 527.42 | 175.81 | 205.60 |
| 544 | 485.80 | 161.93 | 190.53 | 589 | 528.37 | 176.12 | 205.94 |
| 545 | 486.75 | 162.25 | 190.88 | 590 | 529.31 | 176.44 | 206.28 |
| 546 | 487.69 | 162.56 | 191.22 | 591 | 530.26 | 176.75 | 206.62 |
| 547 | 488.64 | 162.88 | 191.56 | 592 | 531.21 | 177.07 | 206.97 |
| 548 | 489.58 | 163.19 | 191.90 | 593 | 532.15 | 177.38 | 207.31 |
| 549 | 490.53 | 163.51 | 192.25 | 594 | 533.10 | 177.70 | 207.65 |
| 550 | 491.47 | 163.82 | 192.59 | 595 | 534.05 | 178.02 | 207.99 |
| 551 | 492.42 | 164.14 | 192.93 | 596 | 535.00 | 178.33 | 208.33 |
| 552 | 493.36 | 164.45 | 193.27 | 597 | 535.94 | 178.65 | 208.68 |
| 553 | 494.31 | 164.77 | 193.62 | 598 | 536.89 | 178.96 | 209.02 |
| 554 | 495.25 | 165.08 | 193.96 | 599 | 537.84 | 179.28 | 209.36 |
| 555 | 496.20 | 165.40 | 194.30 | 600 | 538.78 | 179.59 | 209.70 |
| 556 | 497.14 | 165.71 | 194.64 | 601 | 539.73 | 179.91 | 210.04 |
| 557 | 498.09 | 166.03 | 194.99 | 602 | 540.68 | 180.23 | 210.39 |
| 558 | 499.03 | 166.34 | 195.33 | 603 | 541.63 | 180.54 | 210.73 |
| 559 | 499.98 | 166.66 | 195.67 | 604 | 542.57 | 180.86 | 211.07 |
| 560 | 500.92 | 166.97 | 196.01 | 605 | 543.52 | 181.17 | 211.41 |
| 561 | 501.87 | 167.29 | 196.36 | 606 | 544.47 | 181.49 | 211.76 |
| 562 | 502.81 | 167.60 | 196.70 | 607 | 545.42 | 181.81 | 212.10 |
| 563 | 503.76 | 167.92 | 197.04 | 608 | 546.37 | 182.12 | 212.44 |
| 564 | 504.71 | 168.24 | 197.38 | 609 | 547.31 | 182.44 | 212.78 |
| 565 | 505.65 | 168.55 | 197.72 | 610 | 548.26 | 182.75 | 213.12 |
| 566 | 506.60 | 168.87 | 198.07 | 611 | 549.21 | 183.07 | 213.47 |
| 567 | 507.54 | 169.18 | 198.41 | 612 | 550.16 | 183.39 | 213.81 |
| 568 | 508.49 | 169.50 | 198.75 | 613 | 551.10 | 183.70 | 214.15 |
| 569 | 509.44 | 169.81 | 199.09 | 614 | 552.05 | 184.02 | 214.49 |
| 570 | 510.38 | 170.13 | 199.44 | 615 | 553.00 | 184.33 | 214.83 |
| 571 | 511.33 | 170.44 | 199.78 | 616 | 553.95 | 184.65 | 215.18 |
| 572 | 512.27 | 170.76 | 200.12 | 617 | 554.90 | 184.97 | 215.52 |
| 573 | 513.22 | 171.07 | 200.46 | 618 | 555.84 | 185.28 | 215.86 |
| 574 | 514.17 | 171.39 | 200.81 | 619 | 556.79 | 185.60 | 216.20 |
| 575 | 515.11 | 171.70 | 201.15 | 620 | 557.74 | 185.91 | 216.54 |
| 576 | 516.06 | 172.02 | 201.49 | 621 | 558.69 | 186.23 | 216.89 |
| 577 | 517.01 | 172.34 | 201.83 | 622 | 559.64 | 186.55 | 217.23 |
| 578 | 517.95 | 172.65 | 202.17 | 623 | 560.59 | 186.86 | 217.57 |
| 579 | 518.90 | 172.97 | 202.52 | 624 | 561.53 | 187.18 | 217.91 |
| 580 | 519.84 | 173.28 | 202.86 | 625 | 562.48 | 187.49 | 218.25 |
| 581 | 520.79 | 173.60 | 203.20 | 626 | 563.43 | 187.81 | 218.59 |
| 582 | 521.74 | 173.91 | 203.54 | 627 | 564.38 | 188.13 | 218.94 |
| 583 | 522.68 | 174.23 | 203.89 | 628 | 565.33 | 188.44 | 219.28 |
| 584 | 523.63 | 174.54 | 204.23 | 629 | 566.28 | 188.76 | 219.62 |
| 585 | 524.58 | 174.86 | 204.57 | 630 | 567.23 | 189.08 | 219.96 |

Table 2.  CONSTANTS FOR $p'$ (LONG FORMULA), AN APPROXIMATION OF THE LOWER BINOMIAL CONFIDENCE LIMIT $\underline{p}$; $c/n \leqslant .5$, $n \geqslant 20$, $c \leqslant 1,000$ (*Continued*)

| c | m | a | b | c | m | a | b |
|---|---|---|---|---|---|---|---|
| 631 | 568.17 | 189.39 | 220.30 | 676 | 610.91 | 203.64 | 235.68 |
| 632 | 569.12 | 189.71 | 220.65 | 677 | 611.86 | 203.95 | 236.02 |
| 633 | 570.07 | 190.02 | 220.99 | 678 | 612.81 | 204.27 | 236.37 |
| 634 | 571.02 | 190.34 | 221.33 | 679 | 613.76 | 204.59 | 236.71 |
| 635 | 571.97 | 190.66 | 221.67 | 680 | 614.71 | 204.90 | 237.05 |
| 636 | 572.92 | 190.97 | 222.01 | 681 | 615.66 | 205.22 | 237.39 |
| 637 | 573.87 | 191.29 | 222.36 | 682 | 616.61 | 205.54 | 237.73 |
| 638 | 574.82 | 191.61 | 222.70 | 683 | 617.56 | 205.85 | 238.07 |
| 639 | 575.77 | 191.92 | 223.04 | 684 | 618.51 | 206.17 | 238.41 |
| 640 | 576.72 | 192.24 | 223.38 | 685 | 619.46 | 206.49 | 238.76 |
| 641 | 577.66 | 192.55 | 223.72 | 686 | 620.41 | 206.80 | 239.10 |
| 642 | 578.61 | 192.87 | 224.06 | 687 | 621.36 | 207.12 | 239.44 |
| 643 | 579.56 | 193.19 | 224.41 | 688 | 622.32 | 207.44 | 239.78 |
| 644 | 580.51 | 193.50 | 224.75 | 689 | 623.27 | 207.76 | 240.12 |
| 645 | 581.46 | 193.82 | 225.09 | 690 | 624.22 | 208.07 | 240.46 |
| 646 | 582.41 | 194.14 | 225.43 | 691 | 625.17 | 208.39 | 240.81 |
| 647 | 583.36 | 194.45 | 225.77 | 692 | 626.12 | 208.71 | 241.15 |
| 648 | 584.31 | 194.77 | 226.12 | 693 | 627.07 | 209.02 | 241.49 |
| 649 | 585.26 | 195.09 | 226.46 | 694 | 628.02 | 209.34 | 241.83 |
| 650 | 586.21 | 195.40 | 226.80 | 695 | 628.97 | 209.66 | 242.17 |
| 651 | 587.16 | 195.72 | 227.14 | 696 | 629.92 | 209.97 | 242.51 |
| 652 | 588.11 | 196.04 | 227.48 | 697 | 630.88 | 210.29 | 242.85 |
| 653 | 589.06 | 196.35 | 227.82 | 698 | 631.83 | 210.61 | 243.20 |
| 654 | 590.01 | 196.67 | 228.17 | 699 | 632.78 | 210.93 | 243.54 |
| 655 | 590.96 | 196.99 | 228.51 | 700 | 633.73 | 211.24 | 243.88 |
| 656 | 591.91 | 197.30 | 228.85 | 701 | 634.68 | 211.56 | 244.22 |
| 657 | 592.86 | 197.62 | 229.19 | 702 | 635.63 | 211.88 | 244.56 |
| 658 | 593.81 | 197.94 | 229.53 | 703 | 636.58 | 212.19 | 244.90 |
| 659 | 594.75 | 198.25 | 229.87 | 704 | 637.53 | 212.51 | 245.24 |
| 660 | 595.70 | 198.57 | 230.22 | 705 | 638.49 | 212.83 | 245.59 |
| 661 | 596.65 | 198.88 | 230.56 | 706 | 639.44 | 213.15 | 245.93 |
| 662 | 597.60 | 199.20 | 230.90 | 707 | 640.39 | 213.46 | 246.27 |
| 663 | 598.55 | 199.52 | 231.24 | 708 | 641.34 | 213.78 | 246.61 |
| 664 | 599.50 | 199.83 | 231.58 | 709 | 642.29 | 214.10 | 246.95 |
| 665 | 600.45 | 200.15 | 231.92 | 710 | 643.24 | 214.41 | 247.29 |
| 666 | 601.40 | 200.47 | 232.27 | 711 | 644.20 | 214.73 | 247.63 |
| 667 | 602.35 | 200.78 | 232.61 | 712 | 645.15 | 215.05 | 247.98 |
| 668 | 603.30 | 201.10 | 232.95 | 713 | 646.10 | 215.37 | 248.32 |
| 669 | 604.26 | 201.42 | 233.29 | 714 | 647.05 | 215.68 | 248.66 |
| 670 | 605.21 | 201.74 | 233.63 | 715 | 648.00 | 216.00 | 249.00 |
| 671 | 606.16 | 202.05 | 233.97 | 716 | 648.95 | 216.32 | 249.34 |
| 672 | 607.11 | 202.37 | 234.32 | 717 | 649.91 | 216.64 | 249.68 |
| 673 | 608.06 | 202.69 | 234.66 | 718 | 650.86 | 216.95 | 250.02 |
| 674 | 609.01 | 203.00 | 235.00 | 719 | 651.81 | 217.27 | 250.36 |
| 675 | 609.96 | 203.32 | 235.34 | 720 | 652.76 | 217.59 | 250.71 |

Table 2. CONSTANTS FOR p′ (LONG FORMULA), AN APPROXIMATION OF
THE LOWER BINOMIAL CONFIDENCE LIMIT p; c/n ≤ .5, n ≥ 20,
c ≤ 1,000 (*Continued*)

Section 1

$\gamma_1 = 99.5\%$
$\gamma_2 = 99\%$

| c | m | a | b | c | m | a | b |
|---|---|---|---|---|---|---|---|
| 721 | 653.71 | 217.90 | 251.05 | 766 | 696.59 | 232.20 | 266.40 |
| 722 | 654.67 | 218.22 | 251.39 | 767 | 697.54 | 232.51 | 266.74 |
| 723 | 655.62 | 218.54 | 251.73 | 768 | 698.50 | 232.83 | 267.08 |
| 724 | 656.57 | 218.86 | 252.07 | 769 | 699.45 | 233.15 | 267.43 |
| 725 | 657.52 | 219.17 | 252.41 | 770 | 700.40 | 233.47 | 267.77 |
| 726 | 658.47 | 219.49 | 252.75 | 771 | 701.36 | 233.79 | 268.11 |
| 727 | 659.43 | 219.81 | 253.10 | 772 | 702.31 | 234.10 | 268.45 |
| 728 | 660.38 | 220.13 | 253.44 | 773 | 703.26 | 234.42 | 268.79 |
| 729 | 661.33 | 220.44 | 253.78 | 774 | 704.22 | 234.74 | 269.13 |
| 730 | 662.28 | 220.76 | 254.12 | 775 | 705.17 | 235.06 | 269.47 |
| 731 | 663.24 | 221.08 | 254.46 | 776 | 706.12 | 235.37 | 269.81 |
| 732 | 664.19 | 221.40 | 254.80 | 777 | 707.08 | 235.69 | 270.15 |
| 733 | 665.14 | 221.71 | 255.14 | 778 | 708.03 | 236.01 | 270.49 |
| 734 | 666.09 | 222.03 | 255.48 | 779 | 708.99 | 236.33 | 270.84 |
| 735 | 667.05 | 222.35 | 255.83 | 780 | 709.94 | 236.65 | 271.18 |
| 736 | 668.00 | 222.67 | 256.17 | 781 | 710.89 | 236.96 | 271.52 |
| 737 | 668.95 | 222.98 | 256.51 | 782 | 711.85 | 237.28 | 271.86 |
| 738 | 669.90 | 223.30 | 256.85 | 783 | 712.80 | 237.60 | 272.20 |
| 739 | 670.86 | 223.62 | 257.19 | 784 | 713.76 | 237.92 | 272.54 |
| 740 | 671.81 | 223.94 | 257.53 | 785 | 714.71 | 238.24 | 272.88 |
| 741 | 672.76 | 224.25 | 257.87 | 786 | 715.66 | 238.55 | 273.22 |
| 742 | 673.71 | 224.57 | 258.21 | 787 | 716.62 | 238.87 | 273.56 |
| 743 | 674.67 | 224.89 | 258.56 | 788 | 717.57 | 239.19 | 273.90 |
| 744 | 675.62 | 225.21 | 258.90 | 789 | 718.53 | 239.51 | 274.25 |
| 745 | 676.57 | 225.52 | 259.24 | 790 | 719.48 | 239.83 | 274.59 |
| 746 | 677.53 | 225.84 | 259.58 | 791 | 720.43 | 240.14 | 274.93 |
| 747 | 678.48 | 226.16 | 259.92 | 792 | 721.39 | 240.46 | 275.27 |
| 748 | 679.43 | 226.48 | 260.26 | 793 | 722.34 | 240.78 | 275.61 |
| 749 | 680.38 | 226.79 | 260.60 | 794 | 723.30 | 241.10 | 275.95 |
| 750 | 681.34 | 227.11 | 260.94 | 795 | 724.25 | 241.42 | 276.29 |
| 751 | 682.29 | 227.43 | 261.29 | 796 | 725.21 | 241.74 | 276.63 |
| 752 | 683.24 | 227.75 | 261.63 | 797 | 726.16 | 242.05 | 276.97 |
| 753 | 684.20 | 228.07 | 261.97 | 798 | 727.11 | 242.37 | 277.31 |
| 754 | 685.15 | 228.38 | 262.31 | 799 | 728.07 | 242.69 | 277.66 |
| 755 | 686.10 | 228.70 | 262.65 | 800 | 729.02 | 243.01 | 278.00 |
| 756 | 687.06 | 229.02 | 262.99 | 801 | 729.98 | 243.33 | 278.34 |
| 757 | 688.01 | 229.34 | 263.33 | 802 | 730.93 | 243.64 | 278.68 |
| 758 | 688.96 | 229.65 | 263.67 | 803 | 731.89 | 243.96 | 279.02 |
| 759 | 689.92 | 229.97 | 264.01 | 804 | 732.84 | 244.28 | 279.36 |
| 760 | 690.87 | 230.29 | 264.36 | 805 | 733.80 | 244.60 | 279.70 |
| 761 | 691.82 | 230.61 | 264.70 | 806 | 734.75 | 244.92 | 280.04 |
| 762 | 692.77 | 230.92 | 265.04 | 807 | 735.71 | 245.24 | 280.38 |
| 763 | 693.73 | 231.24 | 265.38 | 808 | 736.66 | 245.55 | 280.72 |
| 764 | 694.68 | 231.56 | 265.72 | 809 | 737.61 | 245.87 | 281.06 |
| 765 | 695.64 | 231.88 | 266.06 | 810 | 738.57 | 246.19 | 281.41 |

Table 2.  CONSTANTS FOR p′ (LONG FORMULA), AN APPROXIMATION OF THE LOWER BINOMIAL CONFIDENCE LIMIT p̲; c/n ⩽ .5, n ⩾ 20, c ⩽ 1,000 (*Continued*)

| c | m | a | b | c | m | a | b |
|---|---|---|---|---|---|---|---|
| 811 | 739.52 | 246.51 | 281.75 | 856 | 782.52 | 260.84 | 297.08 |
| 812 | 740.48 | 246.83 | 282.09 | 857 | 783.47 | 261.16 | 297.42 |
| 813 | 741.43 | 247.14 | 282.43 | 858 | 784.43 | 261.48 | 297.76 |
| 814 | 742.39 | 247.46 | 282.77 | 859 | 785.38 | 261.79 | 298.10 |
| 815 | 743.34 | 247.78 | 283.11 | 860 | 786.34 | 262.11 | 298.44 |
| 816 | 744.30 | 248.10 | 283.45 | 861 | 787.30 | 262.43 | 298.78 |
| 817 | 745.25 | 248.42 | 283.79 | 862 | 788.25 | 262.75 | 299.12 |
| 818 | 746.21 | 248.74 | 284.13 | 863 | 789.21 | 263.07 | 299.47 |
| 819 | 747.16 | 249.05 | 284.47 | 864 | 790.17 | 263.39 | 299.81 |
| 820 | 748.12 | 249.37 | 284.81 | 865 | 791.12 | 263.71 | 300.15 |
| 821 | 749.07 | 249.69 | 285.15 | 866 | 792.08 | 264.03 | 300.49 |
| 822 | 750.03 | 250.01 | 285.50 | 867 | 793.03 | 264.34 | 300.83 |
| 823 | 750.98 | 250.33 | 285.84 | 868 | 793.99 | 264.66 | 301.17 |
| 824 | 751.94 | 250.65 | 286.18 | 869 | 794.95 | 264.98 | 301.51 |
| 825 | 752.89 | 250.96 | 286.52 | 870 | 795.90 | 265.30 | 301.85 |
| 826 | 753.85 | 251.28 | 286.86 | 871 | 796.86 | 265.62 | 302.19 |
| 827 | 754.80 | 251.60 | 287.20 | 872 | 797.82 | 265.94 | 302.53 |
| 828 | 755.76 | 251.92 | 287.54 | 873 | 798.77 | 266.26 | 302.87 |
| 829 | 756.71 | 252.24 | 287.88 | 874 | 799.73 | 266.58 | 303.21 |
| 830 | 757.67 | 252.56 | 288.22 | 875 | 800.69 | 266.90 | 303.55 |
| 831 | 758.63 | 252.88 | 288.56 | 876 | 801.64 | 267.21 | 303.89 |
| 832 | 759.58 | 253.19 | 288.90 | 877 | 802.60 | 267.53 | 304.23 |
| 833 | 760.54 | 253.51 | 289.24 | 878 | 803.55 | 267.85 | 304.57 |
| 834 | 761.49 | 253.83 | 289.58 | 879 | 804.51 | 268.17 | 304.91 |
| 835 | 762.45 | 254.15 | 289.93 | 880 | 805.47 | 268.49 | 305.26 |
| 836 | 763.40 | 254.47 | 290.27 | 881 | 806.42 | 268.81 | 305.60 |
| 837 | 764.36 | 254.79 | 290.61 | 882 | 807.38 | 269.13 | 305.94 |
| 838 | 765.31 | 255.10 | 290.95 | 883 | 808.34 | 269.45 | 306.28 |
| 839 | 766.27 | 255.42 | 291.29 | 884 | 809.29 | 269.76 | 306.62 |
| 840 | 767.22 | 255.74 | 291.63 | 885 | 810.25 | 270.08 | 306.96 |
| 841 | 768.18 | 256.06 | 291.97 | 886 | 811.21 | 270.40 | 307.30 |
| 842 | 769.14 | 256.38 | 292.31 | 887 | 812.16 | 270.72 | 307.64 |
| 843 | 770.09 | 256.70 | 292.65 | 888 | 813.12 | 271.04 | 307.98 |
| 844 | 771.05 | 257.02 | 292.99 | 889 | 814.08 | 271.36 | 308.32 |
| 845 | 772.00 | 257.33 | 293.33 | 890 | 815.03 | 271.68 | 308.66 |
| 846 | 772.96 | 257.65 | 293.67 | 891 | 815.99 | 272.00 | 309.00 |
| 847 | 773.91 | 257.97 | 294.01 | 892 | 816.95 | 272.32 | 309.34 |
| 848 | 774.87 | 258.29 | 294.36 | 893 | 817.91 | 272.64 | 309.68 |
| 849 | 775.83 | 258.61 | 294.70 | 894 | 818.86 | 272.95 | 310.02 |
| 850 | 776.78 | 258.93 | 295.04 | 895 | 819.82 | 273.27 | 310.36 |
| 851 | 777.74 | 259.25 | 295.38 | 896 | 820.78 | 273.59 | 310.70 |
| 852 | 778.69 | 259.56 | 295.72 | 897 | 821.73 | 273.91 | 311.04 |
| 853 | 779.65 | 259.88 | 296.06 | 898 | 822.69 | 274.23 | 311.39 |
| 854 | 780.60 | 260.20 | 296.40 | 899 | 823.65 | 274.55 | 311.73 |
| 855 | 781.56 | 260.52 | 296.74 | 900 | 824.60 | 274.87 | 312.07 |

Table 2. CONSTANTS FOR p′ (LONG FORMULA), AN APPROXIMATION OF THE LOWER BINOMIAL CONFIDENCE LIMIT p; c/n ≤ .5, n ≥ 20, c ≤ 1,000 (*Continued*)

Section 1

$\gamma_1$ = 99.5%
$\gamma_2$ = 99%

| c | m | a | b | c | m | a | b |
|---|---|---|---|---|---|---|---|
| 901 | 825.56 | 275.19 | 312.41 | 946 | 868.65 | 289.55 | 327.72 |
| 902 | 826.52 | 275.51 | 312.75 | 947 | 869.61 | 289.87 | 328.06 |
| 903 | 827.48 | 275.83 | 313.09 | 948 | 870.57 | 290.19 | 328.40 |
| 904 | 828.43 | 276.14 | 313.43 | 949 | 871.53 | 290.51 | 328.75 |
| 905 | 829.39 | 276.46 | 313.77 | 950 | 872.49 | 290.83 | 329.09 |
| 906 | 830.35 | 276.78 | 314.11 | 951 | 873.44 | 291.15 | 329.43 |
| 907 | 831.30 | 277.10 | 314.45 | 952 | 874.40 | 291.47 | 329.77 |
| 908 | 832.26 | 277.42 | 314.79 | 953 | 875.36 | 291.79 | 330.11 |
| 909 | 833.22 | 277.74 | 315.13 | 954 | 876.32 | 292.11 | 330.45 |
| 910 | 834.18 | 278.06 | 315.47 | 955 | 877.28 | 292.43 | 330.79 |
| 911 | 835.13 | 278.38 | 315.81 | 956 | 878.24 | 292.75 | 331.13 |
| 912 | 836.09 | 278.70 | 316.15 | 957 | 879.19 | 293.06 | 331.47 |
| 913 | 837.05 | 279.02 | 316.49 | 958 | 880.15 | 293.38 | 331.81 |
| 914 | 838.01 | 279.34 | 316.83 | 959 | 881.11 | 293.70 | 332.15 |
| 915 | 838.96 | 279.65 | 317.17 | 960 | 882.07 | 294.02 | 332.49 |
| 916 | 839.92 | 279.97 | 317.51 | 961 | 883.03 | 294.34 | 332.83 |
| 917 | 840.88 | 280.29 | 317.85 | 962 | 883.99 | 294.66 | 333.17 |
| 918 | 841.84 | 280.61 | 318.19 | 963 | 884.95 | 294.98 | 333.51 |
| 919 | 842.79 | 280.93 | 318.53 | 964 | 885.90 | 295.30 | 333.85 |
| 920 | 843.75 | 281.25 | 318.87 | 965 | 886.86 | 295.62 | 334.19 |
| 921 | 844.71 | 281.57 | 319.22 | 966 | 887.82 | 295.94 | 334.53 |
| 922 | 845.67 | 281.89 | 319.56 | 967 | 888.78 | 296.26 | 334.87 |
| 923 | 846.62 | 282.21 | 319.90 | 968 | 889.74 | 296.58 | 335.21 |
| 924 | 847.58 | 282.53 | 320.24 | 969 | 890.70 | 296.90 | 335.55 |
| 925 | 848.54 | 282.85 | 320.58 | 970 | 891.66 | 297.22 | 335.89 |
| 926 | 849.50 | 283.17 | 320.92 | 971 | 892.61 | 297.54 | 336.23 |
| 927 | 850.45 | 283.48 | 321.26 | 972 | 893.57 | 297.86 | 336.57 |
| 928 | 851.41 | 283.80 | 321.60 | 973 | 894.53 | 298.18 | 336.91 |
| 929 | 852.37 | 284.12 | 321.94 | 974 | 895.49 | 298.50 | 337.25 |
| 930 | 853.33 | 284.44 | 322.28 | 975 | 896.45 | 298.82 | 337.59 |
| 931 | 854.28 | 284.76 | 322.62 | 976 | 897.41 | 299.14 | 337.93 |
| 932 | 855.24 | 285.08 | 322.96 | 977 | 898.37 | 299.46 | 338.27 |
| 933 | 856.20 | 285.40 | 323.30 | 978 | 899.33 | 299.78 | 338.61 |
| 934 | 857.16 | 285.72 | 323.64 | 979 | 900.28 | 300.09 | 338.95 |
| 935 | 858.12 | 286.04 | 323.98 | 980 | 901.24 | 300.41 | 339.29 |
| 936 | 859.07 | 286.36 | 324.32 | 981 | 902.20 | 300.73 | 339.63 |
| 937 | 860.03 | 286.68 | 324.66 | 982 | 903.16 | 301.05 | 339.97 |
| 938 | 860.99 | 287.00 | 325.00 | 983 | 904.12 | 301.37 | 340.31 |
| 939 | 861.95 | 287.32 | 325.34 | 984 | 905.08 | 301.69 | 340.65 |
| 940 | 862.91 | 287.64 | 325.68 | 985 | 906.04 | 302.01 | 340.99 |
| 941 | 863.86 | 287.95 | 326.02 | 986 | 907.00 | 302.33 | 341.33 |
| 942 | 864.82 | 288.27 | 326.36 | 987 | 907.96 | 302.65 | 341.67 |
| 943 | 865.78 | 288.59 | 326.70 | 988 | 908.91 | 302.97 | 342.01 |
| 944 | 866.74 | 288.91 | 327.04 | 989 | 909.87 | 303.29 | 342.35 |
| 945 | 867.70 | 289.23 | 327.38 | 990 | 910.83 | 303.61 | 342.69 |

Table 2.   **CONSTANTS FOR $\underline{p}'$ (LONG FORMULA), AN APPROXIMATION OF THE LOWER BINOMIAL CONFIDENCE LIMIT $\underline{p}$; $c/n \leqslant .5$, $n \geqslant 20$, $c \leqslant 1,000$ (*Continued*)**

| c | $\underline{m}$ | $\underline{a}$ | $\underline{b}$ | c | $\underline{m}$ | $\underline{a}$ | $\underline{b}$ |
|---|---|---|---|---|---|---|---|
| 991 | 911.79 | 303.93 | 343.03 | 996 | 916.59 | 305.53 | 344.74 |
| 992 | 912.75 | 304.25 | 343.37 | 997 | 917.55 | 305.85 | 345.08 |
| 993 | 913.71 | 304.57 | 343.72 | 998 | 918.51 | 306.17 | 345.42 |
| 994 | 914.67 | 304.89 | 344.06 | 999 | 919.47 | 306.49 | 345.76 |
| 995 | 915.63 | 305.21 | 344.40 | 1000 | 920.42 | 306.81 | 346.10 |

**Table 2. CONSTANTS FOR p′ (LONG FORMULA), AN APPROXIMATION OF THE LOWER BINOMIAL CONFIDENCE LIMIT p; c/n ≤ .5, n ≥ 20, c ≤ 1,000 (Continued)**

| c | m | a | b | c | m | a | b |
|---|---|---|---|---|---|---|---|
| 1 | 0.010051 | 0.005 | 0.0 | 46 | 31.705 | 11.442 | 18.090 |
| 2 | 0.14856 | 0.281 | 0.706 | 47 | 32.534 | 11.717 | 18.450 |
| 3 | 0.43605 | 0.452 | 1.234 | 48 | 33.365 | 11.973 | 18.790 |
| 4 | 0.82325 | 0.644 | 1.733 | 49 | 34.198 | 12.252 | 19.153 |
| 5 | 1.2791 | 0.845 | 2.205 | 50 | 35.032 | 12.524 | 19.508 |
| 6 | 1.7853 | 1.059 | 2.666 | 51 | 35.869 | 12.786 | 19.852 |
| 7 | 2.3302 | 1.291 | 3.126 | 52 | 36.707 | 13.060 | 20.207 |
| 8 | 2.9061 | 1.533 | 3.580 | 53 | 37.546 | 13.340 | 20.567 |
| 9 | 3.5075 | 1.790 | 4.036 | 54 | 38.387 | 13.594 | 20.901 |
| 10 | 4.1302 | 2.070 | 4.504 | 55 | 39.229 | 13.877 | 21.262 |
| 11 | 4.7713 | 2.310 | 4.924 | 56 | 40.073 | 14.155 | 21.618 |
| 12 | 5.4282 | 2.553 | 5.339 | 57 | 40.918 | 14.424 | 21.965 |
| 13 | 6.0991 | 2.801 | 5.751 | 58 | 41.765 | 14.68 | 22.297 |
| 14 | 6.7824 | 3.047 | 6.156 | 59 | 42.613 | 14.95 | 22.644 |
| 15 | 7.4768 | 3.296 | 6.557 | 60 | 43.462 | 15.23 | 23.003 |
| 16 | 8.1811 | 3.551 | 6.960 | 61 | 44.312 | 15.49 | 23.337 |
| 17 | 8.8946 | 3.806 | 7.358 | 62 | 45.164 | 15.76 | 23.68 |
| 18 | 9.6164 | 4.059 | 7.751 | 63 | 46.016 | 16.06 | 24.06 |
| 19 | 10.346 | 4.313 | 8.140 | 64 | 46.870 | 16.33 | 24.40 |
| 20 | 11.082 | 4.565 | 8.524 | 65 | 47.726 | 16.60 | 24.74 |
| 21 | 11.825 | 4.825 | 8.912 | 66 | 48.582 | 16.87 | 25.08 |
| 22 | 12.574 | 5.088 | 9.301 | 67 | 49.439 | 17.12 | 25.40 |
| 23 | 13.329 | 5.343 | 9.679 | 68 | 50.298 | 17.40 | 25.75 |
| 24 | 14.089 | 5.603 | 10.059 | 69 | 51.157 | 17.67 | 26.09 |
| 25 | 14.853 | 5.856 | 10.429 | 70 | 52.017 | 17.98 | 26.47 |
| 26 | 15.623 | 6.124 | 10.813 | 71 | 52.879 | 18.25 | 26.81 |
| 27 | 16.397 | 6.389 | 11.190 | 72 | 53.741 | 18.51 | 27.14 |
| 28 | 17.175 | 6.640 | 11.553 | 73 | 54.605 | 18.78 | 27.48 |
| 29 | 17.957 | 6.915 | 11.936 | 74 | 55.469 | 19.08 | 27.85 |
| 30 | 18.742 | 7.170 | 12.298 | 75 | 56.334 | 19.34 | 28.17 |
| 31 | 19.532 | 7.432 | 12.666 | 76 | 57.200 | 19.62 | 28.52 |
| 32 | 20.324 | 7.701 | 13.038 | 77 | 58.067 | 19.87 | 28.84 |
| 33 | 21.120 | 7.966 | 13.406 | 78 | 58.935 | 20.18 | 29.21 |
| 34 | 21.919 | 8.229 | 13.769 | 79 | 59.804 | 20.44 | 29.54 |
| 35 | 22.721 | 8.495 | 14.134 | 80 | 60.673 | 20.71 | 29.87 |
| 36 | 23.526 | 8.754 | 14.491 | 81 | 61.543 | 21.00 | 30.23 |
| 37 | 24.333 | 9.031 | 14.865 | 82 | 62.414 | 21.28 | 30.58 |
| 38 | 25.143 | 9.299 | 15.228 | 83 | 63.286 | 21.51 | 30.87 |
| 39 | 25.955 | 9.569 | 15.591 | 84 | 64.159 | 21.79 | 31.21 |
| 40 | 26.770 | 9.829 | 15.944 | 85 | 65.032 | 22.09 | 31.57 |
| 41 | 27.587 | 10.094 | 16.300 | 86 | 65.907 | 22.36 | 31.91 |
| 42 | 28.407 | 10.355 | 16.651 | 87 | 66.781 | 22.66 | 32.26 |
| 43 | 29.228 | 10.627 | 17.013 | 88 | 67.657 | 22.92 | 32.59 |
| 44 | 30.052 | 10.902 | 17.376 | 89 | 68.533 | 23.20 | 32.93 |
| 45 | 30.877 | 11.176 | 17.737 | 90 | 69.410 | 23.46 | 33.25 |

Table 2. CONSTANTS FOR p′ (LONG FORMULA), AN APPROXIMATION OF THE LOWER BINOMIAL CONFIDENCE LIMIT p̲; c/n ≤ .5, n ≥ 20, c ≤ 1,000 (*Continued*)

| c | m | a | b | c | m | a | b |
|---|---|---|---|---|---|---|---|
| 91 | 70.288 | 23.76 | 33.62 | 136 | 110.35 | 36.78 | 49.11 |
| 92 | 71.166 | 24.01 | 33.93 | 137 | 111.25 | 37.08 | 49.46 |
| 93 | 72.045 | 24.30 | 34.27 | 138 | 112.15 | 37.38 | 49.81 |
| 94 | 72.925 | 24.56 | 34.60 | 139 | 113.05 | 37.68 | 50.16 |
| 95 | 73.805 | 24.82 | 34.92 | 140 | 113.95 | 37.98 | 50.51 |
| 96 | 74.686 | 25.12 | 35.28 | 141 | 114.85 | 38.28 | 50.86 |
| 97 | 75.568 | 25.41 | 35.62 | 142 | 115.76 | 38.58 | 51.21 |
| 98 | 76.450 | 25.66 | 35.94 | 143 | 116.66 | 38.88 | 51.56 |
| 99 | 77.333 | 25.98 | 36.31 | 144 | 117.56 | 39.19 | 51.91 |
| 100 | 78.216 | 26.26 | 36.65 | 145 | 118.47 | 39.49 | 52.26 |
| 101 | 79.100 | 26.36 | 36.82 | 146 | 119.37 | 39.79 | 52.61 |
| 102 | 79.985 | 26.66 | 37.17 | 147 | 120.27 | 40.09 | 52.96 |
| 103 | 80.870 | 26.95 | 37.52 | 148 | 121.18 | 40.39 | 53.31 |
| 104 | 81.755 | 27.25 | 37.88 | 149 | 122.08 | 40.69 | 53.65 |
| 105 | 82.642 | 27.54 | 38.23 | 150 | 122.99 | 40.99 | 54.00 |
| 106 | 83.528 | 27.84 | 38.58 | 151 | 123.89 | 41.29 | 54.35 |
| 107 | 84.416 | 28.14 | 38.93 | 152 | 124.80 | 41.60 | 54.70 |
| 108 | 85.303 | 28.43 | 39.28 | 153 | 125.70 | 41.90 | 55.05 |
| 109 | 86.192 | 28.73 | 39.64 | 154 | 126.61 | 42.20 | 55.40 |
| 110 | 87.080 | 29.02 | 39.99 | 155 | 127.52 | 42.50 | 55.75 |
| 111 | 87.970 | 29.32 | 40.34 | 156 | 128.42 | 42.81 | 56.10 |
| 112 | 88.860 | 29.62 | 40.69 | 157 | 129.33 | 43.11 | 56.45 |
| 113 | 89.750 | 29.91 | 41.04 | 158 | 130.24 | 43.41 | 56.80 |
| 114 | 90.641 | 30.21 | 41.39 | 159 | 131.14 | 43.71 | 57.14 |
| 115 | 91.532 | 30.51 | 41.75 | 160 | 132.05 | 44.01 | 57.49 |
| 116 | 92.424 | 30.81 | 42.10 | 161 | 132.95 | 44.32 | 57.84 |
| 117 | 93.316 | 31.10 | 42.45 | 162 | 133.86 | 44.62 | 58.19 |
| 118 | 94.209 | 31.40 | 42.80 | 163 | 134.77 | 44.92 | 58.54 |
| 119 | 95.102 | 31.70 | 43.15 | 164 | 135.68 | 45.23 | 58.89 |
| 120 | 95.995 | 32.00 | 43.50 | 165 | 136.59 | 45.53 | 59.24 |
| 121 | 96.889 | 32.29 | 43.85 | 166 | 137.50 | 45.83 | 59.58 |
| 122 | 97.784 | 32.59 | 44.20 | 167 | 138.41 | 46.14 | 59.93 |
| 123 | 98.679 | 32.89 | 44.55 | 168 | 139.32 | 46.44 | 60.28 |
| 124 | 99.574 | 33.19 | 44.91 | 169 | 140.23 | 46.74 | 60.63 |
| 125 | 100.47 | 33.49 | 45.26 | 170 | 141.14 | 47.05 | 60.98 |
| 126 | 101.37 | 33.79 | 45.61 | 171 | 142.05 | 47.35 | 61.32 |
| 127 | 102.26 | 34.08 | 45.96 | 172 | 142.96 | 47.65 | 61.67 |
| 128 | 103.16 | 34.38 | 46.31 | 173 | 143.87 | 47.96 | 62.02 |
| 129 | 104.06 | 34.68 | 46.66 | 174 | 144.78 | 48.26 | 62.37 |
| 130 | 104.95 | 34.98 | 47.01 | 175 | 145.70 | 48.57 | 62.72 |
| 131 | 105.85 | 35.28 | 47.36 | 176 | 146.61 | 48.87 | 63.07 |
| 132 | 106.75 | 35.58 | 47.71 | 177 | 147.52 | 49.17 | 63.41 |
| 133 | 107.65 | 35.88 | 48.06 | 178 | 148.43 | 49.48 | 63.76 |
| 134 | 108.55 | 36.18 | 48.41 | 179 | 149.35 | 49.78 | 64.11 |
| 135 | 109.45 | 36.48 | 48.76 | 180 | 150.26 | 50.09 | 64.46 |

Table 2. **CONSTANTS FOR p′ (LONG FORMULA), AN APPROXIMATION OF THE LOWER BINOMIAL CONFIDENCE LIMIT** p; c/n ≤ .5, n ≥ 20, c ≤ 1,000 (*Continued*)

Section 2
$\gamma_1 = 99\%$
$\gamma_2 = 98\%$

| c | m | a | b | c | m | a | b |
|---|---|---|---|---|---|---|---|
| 181 | 151.17 | 50.39 | 64.80 | 226 | 192.50 | 64.17 | 80.42 |
| 182 | 152.09 | 50.70 | 65.15 | 227 | 193.42 | 64.47 | 80.76 |
| 183 | 153.00 | 51.00 | 65.50 | 228 | 194.34 | 64.78 | 81.11 |
| 184 | 153.92 | 51.31 | 65.85 | 229 | 195.27 | 65.09 | 81.46 |
| 185 | 154.83 | 51.61 | 66.20 | 230 | 196.19 | 65.40 | 81.80 |
| 186 | 155.74 | 51.91 | 66.54 | 231 | 197.11 | 65.70 | 82.15 |
| 187 | 156.66 | 52.22 | 66.89 | 232 | 198.04 | 66.01 | 82.49 |
| 188 | 157.57 | 52.52 | 67.24 | 233 | 198.96 | 66.32 | 82.84 |
| 189 | 158.49 | 52.83 | 67.59 | 234 | 199.89 | 66.63 | 83.19 |
| 190 | 159.40 | 53.13 | 67.93 | 235 | 200.81 | 66.94 | 83.53 |
| 191 | 160.32 | 53.44 | 68.28 | 236 | 201.73 | 67.24 | 83.88 |
| 192 | 161.24 | 53.75 | 68.63 | 237 | 202.66 | 67.55 | 84.22 |
| 193 | 162.15 | 54.05 | 68.97 | 238 | 203.58 | 67.86 | 84.57 |
| 194 | 163.07 | 54.36 | 69.32 | 239 | 204.51 | 68.17 | 84.92 |
| 195 | 163.99 | 54.66 | 69.67 | 240 | 205.43 | 68.48 | 85.26 |
| 196 | 164.90 | 54.97 | 70.02 | 241 | 206.36 | 68.79 | 85.61 |
| 197 | 165.82 | 55.27 | 70.36 | 242 | 207.28 | 69.09 | 85.95 |
| 198 | 166.74 | 55.58 | 70.71 | 243 | 208.21 | 69.40 | 86.30 |
| 199 | 167.65 | 55.88 | 71.06 | 244 | 209.13 | 69.71 | 86.64 |
| 200 | 168.57 | 56.19 | 71.40 | 245 | 210.06 | 70.02 | 86.99 |
| 201 | 169.49 | 56.50 | 71.75 | 246 | 210.98 | 70.33 | 87.34 |
| 202 | 170.41 | 56.80 | 72.10 | 247 | 211.91 | 70.64 | 87.68 |
| 203 | 171.33 | 57.11 | 72.45 | 248 | 212.84 | 70.95 | 88.03 |
| 204 | 172.24 | 57.41 | 72.79 | 249 | 213.76 | 71.25 | 88.37 |
| 205 | 173.16 | 57.72 | 73.14 | 250 | 214.69 | 71.56 | 88.72 |
| 206 | 174.08 | 58.03 | 73.49 | 251 | 215.62 | 71.87 | 89.06 |
| 207 | 175.00 | 58.33 | 73.83 | 252 | 216.54 | 72.18 | 89.41 |
| 208 | 175.92 | 58.64 | 74.18 | 253 | 217.47 | 72.49 | 89.76 |
| 209 | 176.84 | 58.95 | 74.53 | 254 | 218.40 | 72.80 | 90.10 |
| 210 | 177.76 | 59.25 | 74.87 | 255 | 219.32 | 73.11 | 90.45 |
| 211 | 178.68 | 59.56 | 75.22 | 256 | 220.25 | 73.42 | 90.79 |
| 212 | 179.60 | 59.87 | 75.57 | 257 | 221.18 | 73.73 | 91.14 |
| 213 | 180.52 | 60.17 | 75.91 | 258 | 222.10 | 74.03 | 91.48 |
| 214 | 181.44 | 60.48 | 76.26 | 259 | 223.03 | 74.34 | 91.83 |
| 215 | 182.36 | 60.79 | 76.61 | 260 | 223.96 | 74.65 | 92.17 |
| 216 | 183.28 | 61.09 | 76.95 | 261 | 224.89 | 74.96 | 92.52 |
| 217 | 184.20 | 61.40 | 77.30 | 262 | 225.82 | 75.27 | 92.86 |
| 218 | 185.12 | 61.71 | 77.65 | 263 | 226.74 | 75.58 | 93.21 |
| 219 | 186.04 | 62.01 | 77.99 | 264 | 227.67 | 75.89 | 93.55 |
| 220 | 186.97 | 62.32 | 78.34 | 265 | 228.60 | 76.20 | 93.90 |
| 221 | 187.89 | 62.63 | 78.69 | 266 | 229.53 | 76.51 | 94.25 |
| 222 | 188.81 | 62.94 | 79.03 | 267 | 230.46 | 76.82 | 94.59 |
| 223 | 189.73 | 63.24 | 79.38 | 268 | 231.39 | 77.13 | 94.94 |
| 224 | 190.65 | 63.55 | 79.72 | 269 | 232.32 | 77.44 | 95.28 |
| 225 | 191.58 | 63.86 | 80.07 | 270 | 233.25 | 77.75 | 95.63 |

Table 2. CONSTANTS FOR $p'$ (LONG FORMULA), AN APPROXIMATION OF THE LOWER BINOMIAL CONFIDENCE LIMIT $\underline{p}$; $c/n \leqslant .5$, $n \geqslant 20$, $c \leqslant 1,000$ (*Continued*)

| c | m | a | b | c | m | a | b |
|---|---|---|---|---|---|---|---|
| 271 | 234.17 | 78.06 | 95.97 | 316 | 276.12 | 92.04 | 111.48 |
| 272 | 235.10 | 78.37 | 96.32 | 317 | 277.05 | 92.35 | 111.82 |
| 273 | 236.03 | 78.68 | 96.66 | 318 | 277.99 | 92.66 | 112.17 |
| 274 | 236.96 | 78.99 | 97.01 | 319 | 278.92 | 92.97 | 112.51 |
| 275 | 237.89 | 79.30 | 97.35 | 320 | 279.86 | 93.29 | 112.86 |
| 276 | 238.82 | 79.61 | 97.70 | 321 | 280.79 | 93.60 | 113.20 |
| 277 | 239.75 | 79.92 | 98.04 | 322 | 281.73 | 93.91 | 113.55 |
| 278 | 240.68 | 80.23 | 98.39 | 323 | 282.66 | 94.22 | 113.89 |
| 279 | 241.61 | 80.54 | 98.73 | 324 | 283.60 | 94.53 | 114.23 |
| 280 | 242.54 | 80.85 | 99.08 | 325 | 284.53 | 94.84 | 114.58 |
| 281 | 243.47 | 81.16 | 99.42 | 326 | 285.47 | 95.16 | 114.92 |
| 282 | 244.41 | 81.47 | 99.77 | 327 | 286.40 | 95.47 | 115.27 |
| 283 | 245.34 | 81.78 | 100.11 | 328 | 287.34 | 95.78 | 115.61 |
| 284 | 246.27 | 82.09 | 100.46 | 329 | 288.28 | 96.09 | 115.95 |
| 285 | 247.20 | 82.40 | 100.80 | 330 | 289.21 | 96.40 | 116.30 |
| 286 | 248.13 | 82.71 | 101.15 | 331 | 290.15 | 96.72 | 116.64 |
| 287 | 249.06 | 83.02 | 101.49 | 332 | 291.08 | 97.03 | 116.99 |
| 288 | 249.99 | 83.33 | 101.83 | 333 | 292.02 | 97.34 | 117.33 |
| 289 | 250.92 | 83.64 | 102.18 | 334 | 292.96 | 97.65 | 117.67 |
| 290 | 251.86 | 83.95 | 102.52 | 335 | 293.89 | 97.96 | 118.02 |
| 291 | 252.79 | 84.26 | 102.87 | 336 | 294.83 | 98.28 | 118.36 |
| 292 | 253.72 | 84.57 | 103.21 | 337 | 295.77 | 98.59 | 118.71 |
| 293 | 254.65 | 84.88 | 103.56 | 338 | 296.70 | 98.90 | 119.05 |
| 294 | 255.58 | 85.19 | 103.90 | 339 | 297.64 | 99.21 | 119.39 |
| 295 | 256.52 | 85.51 | 104.25 | 340 | 298.58 | 99.53 | 119.74 |
| 296 | 257.45 | 85.82 | 104.59 | 341 | 299.51 | 99.84 | 120.08 |
| 297 | 258.38 | 86.13 | 104.94 | 342 | 300.45 | 100.15 | 120.43 |
| 298 | 259.31 | 86.44 | 105.28 | 343 | 301.39 | 100.46 | 120.77 |
| 299 | 260.25 | 86.75 | 105.63 | 344 | 302.32 | 100.77 | 121.11 |
| 300 | 261.18 | 87.06 | 105.97 | 345 | 303.26 | 101.09 | 121.46 |
| 301 | 262.11 | 87.37 | 106.31 | 346 | 304.20 | 101.40 | 121.80 |
| 302 | 263.04 | 87.68 | 106.66 | 347 | 305.14 | 101.71 | 122.14 |
| 303 | 263.98 | 87.99 | 107.00 | 348 | 306.07 | 102.02 | 122.49 |
| 304 | 264.91 | 88.30 | 107.35 | 349 | 307.01 | 102.34 | 122.83 |
| 305 | 265.84 | 88.61 | 107.69 | 350 | 307.95 | 102.65 | 123.18 |
| 306 | 266.78 | 88.93 | 108.04 | 351 | 308.89 | 102.96 | 123.52 |
| 307 | 267.71 | 89.24 | 108.38 | 352 | 309.83 | 103.28 | 123.86 |
| 308 | 268.64 | 89.55 | 108.73 | 353 | 310.76 | 103.59 | 124.21 |
| 309 | 269.58 | 89.86 | 109.07 | 354 | 311.70 | 103.90 | 124.55 |
| 310 | 270.51 | 90.17 | 109.41 | 355 | 312.64 | 104.21 | 124.89 |
| 311 | 271.45 | 90.48 | 109.76 | 356 | 313.58 | 104.53 | 125.24 |
| 312 | 272.38 | 90.79 | 110.10 | 357 | 314.52 | 104.84 | 125.58 |
| 313 | 273.31 | 91.10 | 110.45 | 358 | 315.45 | 105.15 | 125.92 |
| 314 | 274.25 | 91.42 | 110.79 | 359 | 316.39 | 105.46 | 126.27 |
| 315 | 275.18 | 91.73 | 111.14 | 360 | 317.33 | 105.78 | 126.61 |

Table 2. CONSTANTS FOR p′ (LONG FORMULA), AN APPROXIMATION OF
THE LOWER BINOMIAL CONFIDENCE LIMIT p; c/n ⩽ .5, n ⩾ 20,
c ⩽ 1,000 (*Continued*)

Section 2

$\gamma_1 = 99\%$
$\gamma_2 = 98\%$

| c | m | a | b | c | m | a | b |
|---|---|---|---|---|---|---|---|
| 361 | 318.27 | 106.09 | 126.95 | 406 | 360.00 | 120.20 | 142.40 |
| 362 | 319.21 | 106.40 | 127.30 | 407 | 361.54 | 120.51 | 142.74 |
| 363 | 320.15 | 106.72 | 127.64 | 408 | 362.48 | 120.83 | 143.09 |
| 364 | 321.09 | 107.03 | 127.99 | 409 | 363.42 | 121.14 | 143.43 |
| 365 | 322.03 | 107.34 | 128.33 | 410 | 364.37 | 121.46 | 143.77 |
| 366 | 322.97 | 107.66 | 128.67 | 411 | 365.31 | 121.77 | 144.12 |
| 367 | 323.91 | 107.97 | 129.02 | 412 | 366.25 | 122.08 | 144.46 |
| 368 | 324.84 | 108.28 | 129.36 | 413 | 367.19 | 122.40 | 144.80 |
| 369 | 325.78 | 108.59 | 129.70 | 414 | 368.14 | 122.71 | 145.14 |
| 370 | 326.72 | 108.91 | 130.05 | 415 | 369.08 | 123.03 | 145.49 |
| 371 | 327.66 | 109.22 | 130.39 | 416 | 370.02 | 123.34 | 145.83 |
| 372 | 328.60 | 109.53 | 130.73 | 417 | 370.97 | 123.66 | 146.17 |
| 373 | 329.54 | 109.85 | 131.08 | 418 | 371.91 | 123.97 | 146.52 |
| 374 | 330.48 | 110.16 | 131.42 | 419 | 372.85 | 124.28 | 146.86 |
| 375 | 331.42 | 110.47 | 131.76 | 420 | 373.80 | 124.60 | 147.20 |
| 376 | 332.36 | 110.79 | 132.11 | 421 | 374.74 | 124.91 | 147.54 |
| 377 | 333.30 | 111.10 | 132.45 | 422 | 375.68 | 125.23 | 147.89 |
| 378 | 334.24 | 111.41 | 132.79 | 423 | 376.63 | 125.54 | 148.23 |
| 379 | 335.18 | 111.73 | 133.14 | 424 | 377.57 | 125.86 | 148.57 |
| 380 | 336.12 | 112.04 | 133.48 | 425 | 378.51 | 126.17 | 148.91 |
| 381 | 337.06 | 112.35 | 133.82 | 426 | 379.46 | 126.49 | 149.26 |
| 382 | 338.00 | 112.67 | 134.17 | 427 | 380.40 | 126.80 | 149.60 |
| 383 | 338.94 | 112.98 | 134.51 | 428 | 381.34 | 127.11 | 149.94 |
| 384 | 339.88 | 113.29 | 134.85 | 429 | 382.29 | 127.43 | 150.29 |
| 385 | 340.83 | 113.61 | 135.20 | 430 | 383.23 | 127.74 | 150.63 |
| 386 | 341.77 | 113.92 | 135.54 | 431 | 384.18 | 128.06 | 150.97 |
| 387 | 342.71 | 114.24 | 135.88 | 432 | 385.12 | 128.37 | 151.31 |
| 388 | 343.65 | 114.55 | 136.23 | 433 | 386.06 | 128.69 | 151.66 |
| 389 | 344.59 | 114.86 | 136.57 | 434 | 387.01 | 129.00 | 152.00 |
| 390 | 345.53 | 115.18 | 136.91 | 435 | 387.95 | 129.32 | 152.34 |
| 391 | 346.47 | 115.49 | 137.25 | 436 | 388.90 | 129.63 | 152.68 |
| 392 | 347.41 | 115.80 | 137.60 | 437 | 389.84 | 129.95 | 153.03 |
| 393 | 348.35 | 116.12 | 137.94 | 438 | 390.78 | 130.26 | 153.37 |
| 394 | 349.29 | 116.43 | 138.28 | 439 | 391.73 | 130.58 | 153.71 |
| 395 | 350.24 | 116.75 | 138.63 | 440 | 392.67 | 130.89 | 154.05 |
| 396 | 351.18 | 117.06 | 138.97 | 441 | 393.62 | 131.21 | 154.40 |
| 397 | 352.12 | 117.37 | 139.31 | 442 | 394.56 | 131.52 | 154.74 |
| 398 | 353.06 | 117.69 | 139.66 | 443 | 395.51 | 131.84 | 155.08 |
| 399 | 354.00 | 118.00 | 140.00 | 444 | 396.45 | 132.15 | 155.42 |
| 400 | 354.94 | 118.31 | 140.34 | 445 | 397.40 | 132.47 | 155.77 |
| 401 | 355.89 | 118.63 | 140.69 | 446 | 398.34 | 132.78 | 156.11 |
| 402 | 356.83 | 118.94 | 141.03 | 447 | 399.29 | 133.10 | 156.45 |
| 403 | 357.77 | 119.26 | 141.37 | 448 | 400.23 | 133.41 | 156.79 |
| 404 | 358.71 | 119.57 | 141.71 | 449 | 401.18 | 133.73 | 157.14 |
| 405 | 359.65 | 119.88 | 142.06 | 450 | 402.12 | 134.04 | 157.48 |

Table 2.  **CONSTANTS FOR p′ (LONG FORMULA), AN APPROXIMATION OF THE LOWER BINOMIAL CONFIDENCE LIMIT** p; c/n ≤ .5, n ≥ 20, c ≤ 1,000 (*Continued*)

| c | m | a | b | c | m | a | b |
|---|---|---|---|---|---|---|---|
| 451 | 403.07 | 134.36 | 157.82 | 496 | 445.66 | 148.55 | 173.22 |
| 452 | 404.01 | 134.67 | 158.16 | 497 | 446.61 | 148.87 | 173.57 |
| 453 | 404.96 | 134.99 | 158.51 | 498 | 447.56 | 149.19 | 173.91 |
| 454 | 405.90 | 135.30 | 158.85 | 499 | 448.51 | 149.50 | 174.25 |
| 455 | 406.85 | 135.62 | 159.19 | 500 | 449.45 | 149.82 | 174.59 |
| 456 | 407.79 | 135.93 | 159.53 | 501 | 450.40 | 150.13 | 174.93 |
| 457 | 408.74 | 136.25 | 159.88 | 502 | 451.35 | 150.45 | 175.28 |
| 458 | 409.69 | 136.56 | 160.22 | 503 | 452.30 | 150.77 | 175.62 |
| 459 | 410.63 | 136.88 | 160.56 | 504 | 453.25 | 151.08 | 175.96 |
| 460 | 411.58 | 137.19 | 160.90 | 505 | 454.19 | 151.40 | 176.30 |
| 461 | 412.52 | 137.51 | 161.25 | 506 | 455.14 | 151.71 | 176.64 |
| 462 | 413.47 | 137.82 | 161.59 | 507 | 456.09 | 152.03 | 176.98 |
| 463 | 414.41 | 138.14 | 161.93 | 508 | 457.04 | 152.35 | 177.33 |
| 464 | 415.36 | 138.45 | 162.27 | 509 | 457.99 | 152.66 | 177.67 |
| 465 | 416.31 | 138.77 | 162.62 | 510 | 458.94 | 152.98 | 178.01 |
| 466 | 417.25 | 139.08 | 162.96 | 511 | 459.88 | 153.29 | 178.35 |
| 467 | 418.20 | 139.40 | 163.30 | 512 | 460.83 | 153.61 | 178.69 |
| 468 | 419.15 | 139.72 | 163.64 | 513 | 461.78 | 153.93 | 179.04 |
| 469 | 420.09 | 140.03 | 163.98 | 514 | 462.73 | 154.24 | 179.38 |
| 470 | 421.04 | 140.35 | 164.33 | 515 | 463.68 | 154.56 | 179.72 |
| 471 | 421.98 | 140.66 | 164.67 | 516 | 464.63 | 154.88 | 180.06 |
| 472 | 422.93 | 140.98 | 165.01 | 517 | 465.58 | 155.19 | 180.40 |
| 473 | 423.88 | 141.29 | 165.35 | 518 | 466.52 | 155.51 | 180.75 |
| 474 | 424.82 | 141.61 | 165.70 | 519 | 467.47 | 155.82 | 181.09 |
| 475 | 425.77 | 141.92 | 166.04 | 520 | 468.42 | 156.14 | 181.43 |
| 476 | 426.72 | 142.24 | 166.38 | 521 | 469.37 | 156.46 | 181.77 |
| 477 | 427.66 | 142.55 | 166.72 | 522 | 470.32 | 156.77 | 182.11 |
| 478 | 428.61 | 142.87 | 167.06 | 523 | 471.27 | 157.09 | 182.46 |
| 479 | 429.56 | 143.19 | 167.41 | 524 | 472.22 | 157.41 | 182.80 |
| 480 | 430.50 | 143.50 | 167.75 | 525 | 473.17 | 157.72 | 183.14 |
| 481 | 431.45 | 143.82 | 168.09 | 526 | 474.12 | 158.04 | 183.48 |
| 482 | 432.40 | 144.13 | 168.43 | 527 | 475.07 | 158.36 | 183.82 |
| 483 | 433.34 | 144.45 | 168.78 | 528 | 476.02 | 158.67 | 184.16 |
| 484 | 434.29 | 144.76 | 169.12 | 529 | 476.97 | 158.99 | 184.51 |
| 485 | 435.24 | 145.08 | 169.46 | 530 | 477.92 | 159.31 | 184.85 |
| 486 | 436.19 | 145.40 | 169.80 | 531 | 478.86 | 159.62 | 185.19 |
| 487 | 437.13 | 145.71 | 170.14 | 532 | 479.81 | 159.94 | 185.53 |
| 488 | 438.08 | 146.03 | 170.49 | 533 | 480.76 | 160.25 | 185.87 |
| 489 | 439.03 | 146.34 | 170.83 | 534 | 481.71 | 160.57 | 186.21 |
| 490 | 439.98 | 146.66 | 171.17 | 535 | 482.66 | 160.89 | 186.56 |
| 491 | 440.92 | 146.97 | 171.51 | 536 | 483.61 | 161.20 | 186.90 |
| 492 | 441.87 | 147.29 | 171.85 | 537 | 484.56 | 161.52 | 187.24 |
| 493 | 442.82 | 147.61 | 172.20 | 538 | 485.51 | 161.84 | 187.58 |
| 494 | 443.77 | 147.92 | 172.54 | 539 | 486.46 | 162.15 | 187.92 |
| 495 | 444.71 | 148.24 | 172.88 | 540 | 487.41 | 162.47 | 188.26 |

**Table 2. CONSTANTS FOR p′ (LONG FORMULA), AN APPROXIMATION OF THE LOWER BINOMIAL CONFIDENCE LIMIT p; c/n ≤ .5, n ≥ 20, c ≤ 1,000 (Continued)**

| c | m | a | b | c | m | a | b |
|---|---|---|---|---|---|---|---|
| 541 | 488.36 | 162.79 | 188.61 | 586 | 531.16 | 177.05 | 203.97 |
| 542 | 489.31 | 163.10 | 188.95 | 587 | 532.11 | 177.37 | 204.32 |
| 543 | 490.26 | 163.42 | 189.29 | 588 | 533.06 | 177.69 | 204.66 |
| 544 | 491.21 | 163.74 | 189.63 | 589 | 534.01 | 178.00 | 205.00 |
| 545 | 492.16 | 164.05 | 189.97 | 590 | 534.96 | 178.32 | 205.34 |
| 546 | 493.11 | 164.37 | 190.31 | 591 | 535.92 | 178.64 | 205.68 |
| 547 | 494.06 | 164.69 | 190.66 | 592 | 536.87 | 178.96 | 206.02 |
| 548 | 495.01 | 165.00 | 191.00 | 593 | 537.82 | 179.27 | 206.36 |
| 549 | 495.96 | 165.32 | 191.34 | 594 | 538.77 | 179.59 | 206.70 |
| 550 | 496.91 | 165.64 | 191.68 | 595 | 539.73 | 179.91 | 207.05 |
| 551 | 497.86 | 165.95 | 192.02 | 596 | 540.68 | 180.23 | 207.39 |
| 552 | 498.81 | 166.27 | 192.36 | 597 | 541.63 | 180.54 | 207.73 |
| 553 | 499.77 | 166.59 | 192.71 | 598 | 542.58 | 180.86 | 208.07 |
| 554 | 500.72 | 166.91 | 193.05 | 599 | 543.54 | 181.18 | 208.41 |
| 555 | 501.67 | 167.22 | 193.39 | 600 | 544.49 | 181.50 | 208.75 |
| 556 | 502.62 | 167.54 | 193.73 | 601 | 545.44 | 181.81 | 209.09 |
| 557 | 503.57 | 167.86 | 194.07 | 602 | 546.39 | 182.13 | 209.43 |
| 558 | 504.52 | 168.17 | 194.41 | 603 | 547.35 | 182.45 | 209.78 |
| 559 | 505.47 | 168.49 | 194.76 | 604 | 548.30 | 182.77 | 210.12 |
| 560 | 506.42 | 168.81 | 195.10 | 605 | 549.25 | 183.08 | 210.46 |
| 561 | 507.37 | 169.12 | 195.44 | 606 | 550.20 | 183.40 | 210.80 |
| 562 | 508.32 | 169.44 | 195.78 | 607 | 551.16 | 183.72 | 211.14 |
| 563 | 509.27 | 169.76 | 196.12 | 608 | 552.11 | 184.04 | 211.48 |
| 564 | 510.22 | 170.07 | 196.46 | 609 | 553.06 | 184.35 | 211.82 |
| 565 | 511.18 | 170.39 | 196.80 | 610 | 554.02 | 184.67 | 212.16 |
| 566 | 512.13 | 170.71 | 197.15 | 611 | 554.97 | 184.99 | 212.51 |
| 567 | 513.08 | 171.03 | 197.49 | 612 | 555.92 | 185.31 | 212.85 |
| 568 | 514.03 | 171.34 | 197.83 | 613 | 556.87 | 185.62 | 213.19 |
| 569 | 514.98 | 171.66 | 198.17 | 614 | 557.83 | 185.94 | 213.53 |
| 570 | 515.93 | 171.98 | 198.51 | 615 | 558.78 | 186.26 | 213.87 |
| 571 | 516.88 | 172.29 | 198.85 | 616 | 559.73 | 186.58 | 214.21 |
| 572 | 517.83 | 172.61 | 199.19 | 617 | 560.69 | 186.90 | 214.55 |
| 573 | 518.78 | 172.93 | 199.54 | 618 | 561.64 | 187.21 | 214.89 |
| 574 | 519.74 | 173.25 | 199.88 | 619 | 562.59 | 187.53 | 215.23 |
| 575 | 520.69 | 173.56 | 200.22 | 620 | 563.55 | 187.85 | 215.58 |
| 576 | 521.64 | 173.88 | 200.56 | 621 | 564.50 | 188.17 | 215.92 |
| 577 | 522.59 | 174.20 | 200.90 | 622 | 565.45 | 188.48 | 216.26 |
| 578 | 523.54 | 174.51 | 201.24 | 623 | 566.41 | 188.80 | 216.60 |
| 579 | 524.49 | 174.83 | 201.58 | 624 | 567.36 | 189.12 | 216.94 |
| 580 | 525.45 | 175.15 | 201.93 | 625 | 568.31 | 189.44 | 217.28 |
| 581 | 526.40 | 175.47 | 202.27 | 626 | 569.27 | 189.76 | 217.62 |
| 582 | 527.35 | 175.78 | 202.61 | 627 | 570.22 | 190.07 | 217.96 |
| 583 | 528.30 | 176.10 | 202.95 | 628 | 571.17 | 190.39 | 218.30 |
| 584 | 529.25 | 176.42 | 203.29 | 629 | 572.13 | 190.71 | 218.65 |
| 585 | 530.20 | 176.73 | 203.63 | 630 | 573.08 | 191.03 | 218.99 |

Table 2. CONSTANTS FOR p′ (LONG FORMULA), AN APPROXIMATION OF THE LOWER BINOMIAL CONFIDENCE LIMIT p; c/n ⩽ .5, n ⩾ 20, c ⩽ 1,000 (*Continued*)

| c | m | a | b | c | m | a | b |
|---|---|---|---|---|---|---|---|
| 631 | 574.03 | 191.34 | 219.33 | 676 | 616.99 | 205.66 | 234.67 |
| 632 | 574.99 | 191.66 | 219.67 | 677 | 617.94 | 205.98 | 235.01 |
| 633 | 575.94 | 191.98 | 220.01 | 678 | 618.90 | 206.30 | 235.35 |
| 634 | 576.90 | 192.30 | 220.35 | 679 | 619.85 | 206.62 | 235.69 |
| 635 | 577.85 | 192.62 | 220.69 | 680 | 620.81 | 206.94 | 236.03 |
| 636 | 578.80 | 192.93 | 221.03 | 681 | 621.76 | 207.25 | 236.37 |
| 637 | 579.76 | 193.25 | 221.37 | 682 | 622.72 | 207.57 | 236.71 |
| 638 | 580.71 | 193.57 | 221.71 | 683 | 623.67 | 207.89 | 237.05 |
| 639 | 581.67 | 193.89 | 222.06 | 684 | 624.63 | 208.21 | 237.40 |
| 640 | 582.62 | 194.21 | 222.40 | 685 | 625.59 | 208.53 | 237.74 |
| 641 | 583.57 | 194.52 | 222.74 | 686 | 626.54 | 208.85 | 238.08 |
| 642 | 584.53 | 194.84 | 223.08 | 687 | 627.50 | 209.17 | 238.42 |
| 643 | 585.48 | 195.16 | 223.42 | 688 | 628.45 | 209.48 | 238.76 |
| 644 | 586.44 | 195.48 | 223.76 | 689 | 629.41 | 209.80 | 239.10 |
| 645 | 587.39 | 195.80 | 224.10 | 690 | 630.36 | 210.12 | 239.44 |
| 646 | 588.34 | 196.11 | 224.44 | 691 | 631.32 | 210.44 | 239.78 |
| 647 | 589.30 | 196.43 | 224.78 | 692 | 632.28 | 210.76 | 240.12 |
| 648 | 590.25 | 196.75 | 225.12 | 693 | 633.23 | 211.08 | 240.46 |
| 649 | 591.21 | 197.07 | 225.47 | 694 | 634.19 | 211.40 | 240.80 |
| 650 | 592.16 | 197.39 | 225.81 | 695 | 635.14 | 211.71 | 241.14 |
| 651 | 593.12 | 197.71 | 226.15 | 696 | 636.10 | 212.03 | 241.48 |
| 652 | 594.07 | 198.02 | 226.49 | 697 | 637.05 | 212.35 | 241.82 |
| 653 | 595.02 | 198.34 | 226.83 | 698 | 638.01 | 212.67 | 242.16 |
| 654 | 595.98 | 198.66 | 227.17 | 699 | 638.97 | 212.99 | 242.51 |
| 655 | 596.93 | 198.98 | 227.51 | 700 | 639.92 | 213.31 | 242.85 |
| 656 | 597.89 | 199.30 | 227.85 | 701 | 640.88 | 213.63 | 243.19 |
| 657 | 598.84 | 199.61 | 228.19 | 702 | 641.83 | 213.94 | 243.53 |
| 658 | 599.80 | 199.93 | 228.53 | 703 | 642.79 | 214.26 | 243.87 |
| 659 | 600.75 | 200.25 | 228.87 | 704 | 643.75 | 214.58 | 244.21 |
| 660 | 601.71 | 200.57 | 229.22 | 705 | 644.70 | 214.90 | 244.55 |
| 661 | 602.66 | 200.89 | 229.56 | 706 | 645.66 | 215.22 | 244.89 |
| 662 | 603.62 | 201.21 | 229.90 | 707 | 646.62 | 215.54 | 245.23 |
| 663 | 604.57 | 201.52 | 230.24 | 708 | 647.57 | 215.86 | 245.57 |
| 664 | 605.53 | 201.84 | 230.58 | 709 | 648.53 | 216.18 | 245.91 |
| 665 | 606.48 | 202.16 | 230.92 | 710 | 649.48 | 216.49 | 246.25 |
| 666 | 607.44 | 202.48 | 231.26 | 711 | 650.44 | 216.81 | 246.59 |
| 667 | 608.39 | 202.80 | 231.60 | 712 | 651.40 | 217.13 | 246.93 |
| 668 | 609.35 | 203.12 | 231.94 | 713 | 652.35 | 217.45 | 247.27 |
| 669 | 610.30 | 203.43 | 232.28 | 714 | 653.31 | 217.77 | 247.61 |
| 670 | 611.26 | 203.75 | 232.62 | 715 | 654.27 | 218.09 | 247.96 |
| 671 | 612.21 | 204.07 | 232.96 | 716 | 655.22 | 218.41 | 248.30 |
| 672 | 613.17 | 204.39 | 233.31 | 717 | 656.18 | 218.73 | 248.64 |
| 673 | 614.12 | 204.71 | 233.65 | 718 | 657.14 | 219.05 | 248.98 |
| 674 | 615.08 | 205.03 | 233.99 | 719 | 658.09 | 219.36 | 249.32 |
| 675 | 616.03 | 205.34 | 234.33 | 720 | 659.05 | 219.68 | 249.66 |

Table 2. CONSTANTS FOR p′ (LONG FORMULA), AN APPROXIMATION OF
THE LOWER BINOMIAL CONFIDENCE LIMIT p; c/n ≤ .5, n ≥ 20,
c ≤ 1,000 (*Continued*)

Section 2

$\gamma_1 = 99\%$
$\gamma_2 = 98\%$

| c | m | a | b | c | m | a | b |
|---|---|---|---|---|---|---|---|
| 721 | 660.01 | 220.00 | 250.00 | 766 | 703.09 | 234.36 | 265.32 |
| 722 | 660.96 | 220.32 | 250.34 | 767 | 704.04 | 234.68 | 265.66 |
| 723 | 661.92 | 220.64 | 250.68 | 768 | 705.00 | 235.00 | 266.00 |
| 724 | 662.88 | 220.96 | 251.02 | 769 | 705.96 | 235.32 | 266.34 |
| 725 | 663.83 | 221.28 | 251.36 | 770 | 706.92 | 235.64 | 266.68 |
| 726 | 664.79 | 221.60 | 251.70 | 771 | 707.88 | 235.96 | 267.02 |
| 727 | 665.75 | 221.92 | 252.04 | 772 | 708.83 | 236.28 | 267.36 |
| 728 | 666.70 | 222.23 | 252.38 | 773 | 709.79 | 236.60 | 267.70 |
| 729 | 667.66 | 222.55 | 252.72 | 774 | 710.75 | 236.92 | 268.04 |
| 730 | 668.62 | 222.87 | 253.06 | 775 | 711.71 | 237.24 | 268.38 |
| 731 | 669.57 | 223.19 | 253.40 | 776 | 712.67 | 237.56 | 268.72 |
| 732 | 670.53 | 223.51 | 253.74 | 777 | 713.63 | 237.88 | 269.06 |
| 733 | 671.49 | 223.83 | 254.09 | 778 | 714.58 | 238.19 | 269.40 |
| 734 | 672.45 | 224.15 | 254.43 | 779 | 715.54 | 238.51 | 269.74 |
| 735 | 673.40 | 224.47 | 254.77 | 780 | 716.50 | 238.83 | 270.08 |
| 736 | 674.36 | 224.79 | 255.11 | 781 | 717.46 | 239.15 | 270.42 |
| 737 | 675.32 | 225.11 | 255.45 | 782 | 718.42 | 239.47 | 270.76 |
| 738 | 676.27 | 225.42 | 255.79 | 783 | 719.38 | 239.79 | 271.10 |
| 739 | 677.23 | 225.74 | 256.13 | 784 | 720.33 | 240.11 | 271.44 |
| 740 | 678.19 | 226.06 | 256.47 | 785 | 721.29 | 240.43 | 271.78 |
| 741 | 679.15 | 226.38 | 256.81 | 786 | 722.25 | 240.75 | 272.12 |
| 742 | 680.10 | 226.70 | 257.15 | 787 | 723.21 | 241.07 | 272.47 |
| 743 | 681.06 | 227.02 | 257.49 | 788 | 724.17 | 241.39 | 272.81 |
| 744 | 682.02 | 227.34 | 257.83 | 789 | 725.13 | 241.71 | 273.15 |
| 745 | 682.97 | 227.66 | 258.17 | 790 | 726.09 | 242.03 | 273.49 |
| 746 | 683.93 | 227.98 | 258.51 | 791 | 727.04 | 242.35 | 273.83 |
| 747 | 684.89 | 228.30 | 258.85 | 792 | 728.00 | 242.67 | 274.17 |
| 748 | 685.85 | 228.62 | 259.19 | 793 | 728.96 | 242.99 | 274.51 |
| 749 | 686.80 | 228.93 | 259.53 | 794 | 729.92 | 243.31 | 274.85 |
| 750 | 687.76 | 229.25 | 259.87 | 795 | 730.88 | 243.63 | 275.19 |
| 751 | 688.72 | 229.57 | 260.21 | 796 | 731.84 | 243.95 | 275.53 |
| 752 | 689.68 | 229.89 | 260.55 | 797 | 732.80 | 244.27 | 275.87 |
| 753 | 690.63 | 230.21 | 260.89 | 798 | 733.76 | 244.59 | 276.21 |
| 754 | 691.59 | 230.53 | 261.23 | 799 | 734.71 | 244.90 | 276.55 |
| 755 | 692.55 | 230.85 | 261.58 | 800 | 735.67 | 245.22 | 276.89 |
| 756 | 693.51 | 231.17 | 261.92 | 801 | 736.63 | 245.54 | 277.23 |
| 757 | 694.47 | 231.49 | 262.26 | 802 | 737.59 | 245.86 | 277.57 |
| 758 | 695.42 | 231.81 | 262.60 | 803 | 738.55 | 246.18 | 277.91 |
| 759 | 696.38 | 232.13 | 262.94 | 804 | 739.51 | 246.50 | 278.25 |
| 760 | 697.34 | 232.45 | 263.28 | 805 | 740.47 | 246.82 | 278.59 |
| 761 | 698.30 | 232.77 | 263.62 | 806 | 741.43 | 247.14 | 278.93 |
| 762 | 699.25 | 233.08 | 263.96 | 807 | 742.39 | 247.46 | 279.27 |
| 763 | 700.21 | 233.40 | 264.30 | 808 | 743.34 | 247.78 | 279.61 |
| 764 | 701.17 | 233.72 | 264.64 | 809 | 744.30 | 248.10 | 279.95 |
| 765 | 702.13 | 234.04 | 264.98 | 810 | 745.26 | 248.42 | 280.29 |

Table 2. CONSTANTS FOR $p'$ (LONG FORMULA), AN APPROXIMATION OF THE LOWER BINOMIAL CONFIDENCE LIMIT $\underline{p}$; $c/n \leq .5$, $n \geq 20$, $c \leq 1,000$ (*Continued*)

| c | m | a | b | c | m | a | b |
|---|---|---|---|---|---|---|---|
| 811 | 746.22 | 248.74 | 280.63 | 856 | 789.41 | 263.14 | 295.93 |
| 812 | 747.18 | 249.06 | 280.97 | 857 | 790.37 | 263.46 | 296.27 |
| 813 | 748.14 | 249.38 | 281.31 | 858 | 791.33 | 263.78 | 296.61 |
| 814 | 749.10 | 249.70 | 281.65 | 859 | 792.29 | 264.10 | 296.95 |
| 815 | 750.06 | 250.02 | 281.99 | 860 | 793.25 | 264.42 | 297.29 |
| 816 | 751.02 | 250.34 | 282.33 | 861 | 794.21 | 264.74 | 297.63 |
| 817 | 751.98 | 250.66 | 282.67 | 862 | 795.17 | 265.06 | 297.97 |
| 818 | 752.94 | 250.98 | 283.01 | 863 | 796.13 | 265.38 | 298.31 |
| 819 | 753.90 | 251.30 | 283.35 | 864 | 797.09 | 265.70 | 298.65 |
| 820 | 754.86 | 251.62 | 283.69 | 865 | 798.05 | 266.02 | 298.99 |
| 821 | 755.81 | 251.94 | 284.03 | 866 | 799.01 | 266.34 | 299.33 |
| 822 | 756.77 | 252.26 | 284.37 | 867 | 799.97 | 266.66 | 299.67 |
| 823 | 757.73 | 252.58 | 284.71 | 868 | 800.93 | 266.98 | 300.01 |
| 824 | 758.69 | 252.90 | 285.05 | 869 | 801.89 | 267.30 | 300.35 |
| 825 | 759.65 | 253.22 | 285.39 | 870 | 802.85 | 267.62 | 300.69 |
| 826 | 760.61 | 253.54 | 285.73 | 871 | 803.82 | 267.94 | 301.03 |
| 827 | 761.57 | 253.86 | 286.07 | 872 | 804.78 | 268.26 | 301.37 |
| 828 | 762.53 | 254.18 | 286.41 | 873 | 805.74 | 268.58 | 301.71 |
| 829 | 763.49 | 254.50 | 286.75 | 874 | 806.70 | 268.90 | 302.05 |
| 830 | 764.45 | 254.82 | 287.09 | 875 | 807.66 | 269.22 | 302.39 |
| 831 | 765.41 | 255.14 | 287.43 | 876 | 808.62 | 269.54 | 302.73 |
| 832 | 766.37 | 255.46 | 287.77 | 877 | 809.58 | 269.86 | 303.07 |
| 833 | 767.33 | 255.78 | 288.11 | 878 | 810.54 | 270.18 | 303.41 |
| 834 | 768.29 | 256.10 | 288.45 | 879 | 811.50 | 270.50 | 303.75 |
| 835 | 769.25 | 256.42 | 288.79 | 880 | 812.46 | 270.82 | 304.09 |
| 836 | 770.21 | 256.74 | 289.13 | 881 | 813.42 | 271.14 | 304.43 |
| 837 | 771.17 | 257.06 | 289.47 | 882 | 814.38 | 271.46 | 304.77 |
| 838 | 772.13 | 257.38 | 289.81 | 883 | 815.34 | 271.78 | 305.11 |
| 839 | 773.09 | 257.70 | 290.15 | 884 | 816.30 | 272.10 | 305.45 |
| 840 | 774.05 | 258.02 | 290.49 | 885 | 817.27 | 272.42 | 305.79 |
| 841 | 775.01 | 258.34 | 290.83 | 886 | 818.23 | 272.74 | 306.13 |
| 842 | 775.97 | 258.66 | 291.17 | 887 | 819.19 | 273.06 | 306.47 |
| 843 | 776.93 | 258.98 | 291.51 | 888 | 820.15 | 273.38 | 306.81 |
| 844 | 777.89 | 259.30 | 291.85 | 889 | 821.11 | 273.70 | 307.15 |
| 845 | 778.85 | 259.62 | 292.19 | 890 | 822.07 | 274.02 | 307.49 |
| 846 | 779.81 | 259.94 | 292.53 | 891 | 823.03 | 274.34 | 307.83 |
| 847 | 780.77 | 260.26 | 292.87 | 892 | 823.99 | 274.66 | 308.17 |
| 848 | 781.73 | 260.58 | 293.21 | 893 | 824.95 | 274.98 | 308.51 |
| 849 | 782.69 | 260.90 | 293.55 | 894 | 825.91 | 275.30 | 308.85 |
| 850 | 783.65 | 261.22 | 293.89 | 895 | 826.88 | 275.63 | 309.19 |
| 851 | 784.61 | 261.54 | 294.23 | 896 | 827.84 | 275.95 | 309.53 |
| 852 | 785.57 | 261.86 | 294.57 | 897 | 828.80 | 276.27 | 309.87 |
| 853 | 786.53 | 262.18 | 294.91 | 898 | 829.76 | 276.59 | 310.21 |
| 854 | 787.49 | 262.50 | 295.25 | 899 | 830.72 | 276.91 | 310.55 |
| 855 | 788.45 | 262.82 | 295.59 | 900 | 831.68 | 277.23 | 310.89 |

**Table 2. CONSTANTS FOR** $p'$ **(LONG FORMULA), AN APPROXIMATION OF THE LOWER BINOMIAL CONFIDENCE LIMIT** $\underline{p}$; $c/n \leqslant .5$, $n \geqslant 20$, $c \leqslant 1{,}000$ **(Continued)**

| c | m | a | b | c | m | a | b |
|---|---|---|---|---|---|---|---|
| 901 | 832.64 | 277.55 | 311.23 | 946 | 875.92 | 291.97 | 326.51 |
| 902 | 833.60 | 277.87 | 311.57 | 947 | 876.88 | 292.29 | 326.85 |
| 903 | 834.57 | 278.19 | 311.91 | 948 | 877.84 | 292.61 | 327.19 |
| 904 | 835.53 | 278.51 | 312.25 | 949 | 878.81 | 292.94 | 327.53 |
| 905 | 836.49 | 278.83 | 312.59 | 950 | 879.77 | 293.26 | 327.87 |
| 906 | 837.45 | 279.15 | 312.93 | 951 | 880.73 | 293.58 | 328.21 |
| 907 | 838.41 | 279.47 | 313.26 | 952 | 881.69 | 293.90 | 328.55 |
| 908 | 839.37 | 279.79 | 313.60 | 953 | 882.66 | 294.22 | 328.89 |
| 909 | 840.33 | 280.11 | 313.94 | 954 | 883.62 | 294.54 | 329.23 |
| 910 | 841.29 | 280.43 | 314.28 | 955 | 884.58 | 294.86 | 329.57 |
| 911 | 842.26 | 280.75 | 314.62 | 956 | 885.54 | 295.18 | 329.91 |
| 912 | 843.22 | 281.07 | 314.96 | 957 | 886.51 | 295.50 | 330.25 |
| 913 | 844.18 | 281.39 | 315.30 | 958 | 887.47 | 295.82 | 330.59 |
| 914 | 845.14 | 281.71 | 315.64 | 959 | 888.43 | 296.14 | 330.93 |
| 915 | 846.10 | 282.03 | 315.98 | 960 | 889.39 | 296.46 | 331.27 |
| 916 | 847.06 | 282.35 | 316.32 | 961 | 890.36 | 296.79 | 331.61 |
| 917 | 848.03 | 282.68 | 316.66 | 962 | 891.32 | 297.11 | 331.95 |
| 918 | 848.99 | 283.00 | 317.00 | 963 | 892.28 | 297.43 | 332.29 |
| 919 | 849.95 | 283.32 | 317.34 | 964 | 893.24 | 297.75 | 332.63 |
| 920 | 850.91 | 283.64 | 317.68 | 965 | 894.21 | 298.07 | 332.97 |
| 921 | 851.87 | 283.96 | 318.02 | 966 | 895.17 | 298.39 | 333.31 |
| 922 | 852.83 | 284.28 | 318.36 | 967 | 896.13 | 298.71 | 333.64 |
| 923 | 853.80 | 284.60 | 318.70 | 968 | 897.09 | 299.03 | 333.98 |
| 924 | 854.76 | 284.92 | 319.04 | 969 | 898.06 | 299.35 | 334.32 |
| 925 | 855.72 | 285.24 | 319.38 | 970 | 899.02 | 299.67 | 334.66 |
| 926 | 856.68 | 285.56 | 319.72 | 971 | 899.98 | 299.99 | 335.00 |
| 927 | 857.64 | 285.88 | 320.06 | 972 | 900.94 | 300.31 | 335.34 |
| 928 | 858.60 | 286.20 | 320.40 | 973 | 901.91 | 300.64 | 335.68 |
| 929 | 859.57 | 286.52 | 320.74 | 974 | 902.87 | 300.96 | 336.02 |
| 930 | 860.53 | 286.84 | 321.08 | 975 | 903.83 | 301.28 | 336.36 |
| 931 | 861.49 | 287.16 | 321.42 | 976 | 904.79 | 301.60 | 336.70 |
| 932 | 862.45 | 287.48 | 321.76 | 977 | 905.76 | 301.92 | 337.04 |
| 933 | 863.41 | 287.80 | 322.10 | 978 | 906.72 | 302.24 | 337.38 |
| 934 | 864.38 | 288.13 | 322.44 | 979 | 907.68 | 302.56 | 337.72 |
| 935 | 865.34 | 288.45 | 322.78 | 980 | 908.65 | 302.88 | 338.06 |
| 936 | 866.30 | 288.77 | 323.12 | 981 | 909.61 | 303.20 | 338.40 |
| 937 | 867.26 | 289.09 | 323.46 | 982 | 910.57 | 303.52 | 338.74 |
| 938 | 868.22 | 289.41 | 323.80 | 983 | 911.53 | 303.84 | 339.08 |
| 939 | 869.19 | 289.73 | 324.14 | 984 | 912.50 | 304.17 | 339.42 |
| 940 | 870.15 | 290.05 | 324.48 | 985 | 913.46 | 304.49 | 339.76 |
| 941 | 871.11 | 290.37 | 324.82 | 986 | 914.42 | 304.81 | 340.10 |
| 942 | 872.07 | 290.69 | 325.15 | 987 | 915.39 | 305.13 | 340.44 |
| 943 | 873.03 | 291.01 | 325.49 | 988 | 916.35 | 305.45 | 340.78 |
| 944 | 874.00 | 291.33 | 325.83 | 989 | 917.31 | 305.77 | 341.11 |
| 945 | 874.96 | 291.65 | 326.17 | 990 | 918.28 | 306.09 | 341.45 |

Table 2.  CONSTANTS FOR $\underline{p}'$ (LONG FORMULA), AN APPROXIMATION OF
THE LOWER BINOMIAL CONFIDENCE LIMIT $\underline{p}$; $c/n \leqslant .5$, $n \geqslant 20$,
$c \leqslant 1,000$ (*Continued*)

| c | $\underline{m}$ | $\underline{a}$ | $\underline{b}$ | c | $\underline{m}$ | $\underline{a}$ | $\underline{b}$ |
|---|---|---|---|---|---|---|---|
| 991 | 919.24 | 306.41 | 341.79 | 996 | 924.05 | 308.02 | 343.49 |
| 992 | 920.20 | 306.73 | 342.13 | 997 | 925.02 | 308.34 | 343.83 |
| 993 | 921.16 | 307.05 | 342.47 | 998 | 925.98 | 308.66 | 344.17 |
| 994 | 922.13 | 307.38 | 342.81 | 999 | 926.94 | 308.98 | 344.51 |
| 995 | 923.09 | 307.70 | 343.15 | 1000 | 927.91 | 309.30 | 344.85 |

Table 2.  **CONSTANTS FOR** p′ **(LONG FORMULA), AN APPROXIMATION OF THE LOWER BINOMIAL CONFIDENCE LIMIT** $\underline{p}$; $c/n \leqslant .5$, $n \geqslant 20$, $c \leqslant 1,000$ (*Continued*)

| c | m | a | b | c | m | a | b |
|---|---|---|---|---|---|---|---|
| 1 | 0.025318 | 0.013 | 0.0 | 46 | 33.678 | 11.894 | 17.555 |
| 2 | 0.24221 | 0.382 | 0.761 | 47 | 34.534 | 12.171 | 17.904 |
| 3 | 0.61867 | 0.527 | 1.218 | 48 | 35.391 | 12.440 | 18.244 |
| 4 | 1.0899 | 0.723 | 1.678 | 49 | 36.250 | 12.721 | 18.596 |
| 5 | 1.6235 | 0.938 | 2.126 | 50 | 37.111 | 12.981 | 18.926 |
| 6 | 2.2019 | 1.167 | 2.566 | 51 | 37.973 | 13.275 | 19.289 |
| 7 | 2.8144 | 1.416 | 3.009 | 52 | 38.836 | 13.539 | 19.621 |
| 8 | 3.4538 | 1.674 | 3.447 | 53 | 39.701 | 13.797 | 19.947 |
| 9 | 4.1154 | 1.949 | 3.892 | 54 | 40.566 | 14.078 | 20.295 |
| 10 | 4.7954 | 2.239 | 4.341 | 55 | 41.434 | 14.378 | 20.662 |
| 11 | 5.4912 | 2.491 | 4.746 | 56 | 42.302 | 14.628 | 20.977 |
| 12 | 6.2006 | 2.745 | 5.145 | 57 | 43.171 | 14.925 | 21.339 |
| 13 | 6.9220 | 3.003 | 5.542 | 58 | 44.042 | 15.19 | 21.674 |
| 14 | 7.6540 | 3.262 | 5.935 | 59 | 44.914 | 15.47 | 22.015 |
| 15 | 8.3954 | 3.522 | 6.324 | 60 | 45.786 | 15.75 | 22.358 |
| 16 | 9.1454 | 3.787 | 6.714 | 61 | 46.660 | 16.03 | 22.700 |
| 17 | 9.9032 | 4.046 | 7.094 | 62 | 47.535 | 16.31 | 23.04 |
| 18 | 10.668 | 4.307 | 7.473 | 63 | 48.411 | 16.57 | 23.37 |
| 19 | 11.439 | 4.573 | 7.854 | 64 | 49.288 | 16.87 | 23.73 |
| 20 | 12.217 | 4.840 | 8.232 | 65 | 50.166 | 17.12 | 24.04 |
| 21 | 12.999 | 5.105 | 8.605 | 66 | 51.045 | 17.43 | 24.41 |
| 22 | 13.787 | 5.367 | 8.973 | 67 | 51.924 | 17.69 | 24.73 |
| 23 | 14.580 | 5.637 | 9.347 | 68 | 52.805 | 17.97 | 25.06 |
| 24 | 15.377 | 5.901 | 9.713 | 69 | 53.686 | 18.26 | 25.42 |
| 25 | 16.179 | 6.167 | 10.077 | 70 | 54.569 | 18.52 | 25.74 |
| 26 | 16.984 | 6.439 | 10.447 | 71 | 55.452 | 18.81 | 26.09 |
| 27 | 17.793 | 6.709 | 10.812 | 72 | 56.336 | 19.06 | 26.40 |
| 28 | 18.606 | 6.984 | 11.181 | 73 | 57.221 | 19.37 | 26.76 |
| 29 | 19.422 | 7.249 | 11.538 | 74 | 58.106 | 19.65 | 27.09 |
| 30 | 20.241 | 7.515 | 11.895 | 75 | 58.993 | 19.93 | 27.43 |
| 31 | 21.063 | 7.782 | 12.251 | 76 | 59.880 | 20.22 | 27.78 |
| 32 | 21.888 | 8.054 | 12.610 | 77 | 60.767 | 20.47 | 28.09 |
| 33 | 22.716 | 8.335 | 12.977 | 78 | 61.656 | 20.76 | 28.43 |
| 34 | 23.546 | 8.595 | 13.322 | 79 | 62.545 | 21.01 | 28.74 |
| 35 | 24.379 | 8.877 | 13.687 | 80 | 63.435 | 21.32 | 29.10 |
| 36 | 25.214 | 9.152 | 14.045 | 81 | 64.326 | 21.62 | 29.46 |
| 37 | 26.051 | 9.429 | 14.404 | 82 | 65.217 | 21.86 | 29.75 |
| 38 | 26.891 | 9.699 | 14.753 | 83 | 66.109 | 22.15 | 30.10 |
| 39 | 27.733 | 9.973 | 15.107 | 84 | 67.002 | 22.44 | 30.44 |
| 40 | 28.577 | 10.243 | 15.454 | 85 | 67.895 | 22.71 | 30.76 |
| 41 | 29.422 | 10.501 | 15.790 | 86 | 68.789 | 22.98 | 31.08 |
| 42 | 30.270 | 10.785 | 16.150 | 87 | 69.684 | 23.29 | 31.45 |
| 43 | 31.119 | 11.066 | 16.506 | 88 | 70.579 | 23.54 | 31.75 |
| 44 | 31.970 | 11.336 | 16.851 | 89 | 71.475 | 23.82 | 32.08 |
| 45 | 32.823 | 11.614 | 17.202 | 90 | 72.371 | 24.10 | 32.42 |

Table 2. CONSTANTS FOR p′ (LONG FORMULA), AN APPROXIMATION OF THE LOWER BINOMIAL CONFIDENCE LIMIT p; c/n ⩽ .5, n ⩾ 20, c ⩽ 1,000 (*Continued*)

| c | m | a | b | c | m | a | b |
|---|---|---|---|---|---|---|---|
| 91 | 73.268 | 24.42 | 32.78 | 136 | 114.10 | 38.03 | 48.49 |
| 92 | 74.165 | 24.65 | 33.07 | 137 | 115.02 | 38.34 | 48.83 |
| 93 | 75.063 | 24.93 | 33.40 | 138 | 115.94 | 38.64 | 49.18 |
| 94 | 75.962 | 25.26 | 33.78 | 139 | 116.85 | 38.95 | 49.53 |
| 95 | 76.861 | 25.49 | 34.06 | 140 | 117.77 | 39.25 | 49.87 |
| 96 | 77.761 | 25.75 | 34.37 | 141 | 118.69 | 39.56 | 50.22 |
| 97 | 78.661 | 26.08 | 34.75 | 142 | 119.61 | 39.86 | 50.57 |
| 98 | 79.561 | 26.37 | 35.09 | 143 | 120.52 | 40.17 | 50.92 |
| 99 | 80.463 | 26.62 | 35.39 | 144 | 121.44 | 40.48 | 51.26 |
| 100 | 81.364 | 26.93 | 35.74 | 145 | 122.36 | 40.78 | 51.61 |
| 101 | 82.266 | 27.42 | 36.29 | 146 | 123.28 | 41.09 | 51.96 |
| 102 | 83.169 | 27.72 | 36.64 | 147 | 124.20 | 41.39 | 52.30 |
| 103 | 84.072 | 28.02 | 36.99 | 148 | 125.12 | 41.70 | 52.65 |
| 104 | 84.976 | 28.32 | 37.34 | 149 | 126.04 | 42.01 | 53.00 |
| 105 | 85.880 | 28.62 | 37.69 | 150 | 126.96 | 42.31 | 53.34 |
| 106 | 86.784 | 28.92 | 38.04 | 151 | 127.88 | 42.62 | 53.69 |
| 107 | 87.689 | 29.22 | 38.39 | 152 | 128.80 | 42.93 | 54.04 |
| 108 | 88.595 | 29.53 | 38.74 | 153 | 129.72 | 43.23 | 54.38 |
| 109 | 89.501 | 29.83 | 39.09 | 154 | 130.64 | 43.54 | 54.73 |
| 110 | 90.407 | 30.13 | 39.43 | 155 | 131.56 | 43.85 | 55.08 |
| 111 | 91.314 | 30.43 | 39.78 | 156 | 132.48 | 44.16 | 55.42 |
| 112 | 92.221 | 30.73 | 40.13 | 157 | 133.40 | 44.46 | 55.77 |
| 113 | 93.128 | 31.04 | 40.48 | 158 | 134.32 | 44.77 | 56.11 |
| 114 | 94.036 | 31.34 | 40.83 | 159 | 135.25 | 45.08 | 56.46 |
| 115 | 94.945 | 31.64 | 41.18 | 160 | 136.17 | 45.38 | 56.81 |
| 116 | 95.853 | 31.95 | 41.53 | 161 | 137.08 | 45.69 | 57.15 |
| 117 | 96.762 | 32.25 | 41.88 | 162 | 138.00 | 46.00 | 57.50 |
| 118 | 97.672 | 32.55 | 42.22 | 163 | 138.92 | 46.31 | 57.85 |
| 119 | 98.582 | 32.86 | 42.57 | 164 | 139.85 | 46.62 | 58.19 |
| 120 | 99.492 | 33.16 | 42.92 | 165 | 140.77 | 46.92 | 58.54 |
| 121 | 100.40 | 33.46 | 43.27 | 166 | 141.69 | 47.23 | 58.88 |
| 122 | 101.31 | 33.77 | 43.62 | 167 | 142.62 | 47.54 | 59.23 |
| 123 | 102.23 | 34.07 | 43.97 | 168 | 143.54 | 47.85 | 59.58 |
| 124 | 103.14 | 34.37 | 44.31 | 169 | 144.47 | 48.16 | 59.92 |
| 125 | 104.05 | 34.68 | 44.66 | 170 | 145.39 | 48.46 | 60.27 |
| 126 | 104.96 | 34.98 | 45.01 | 171 | 146.32 | 48.77 | 60.61 |
| 127 | 105.87 | 35.29 | 45.36 | 172 | 147.24 | 49.08 | 60.96 |
| 128 | 106.79 | 35.59 | 45.70 | 173 | 148.17 | 49.39 | 61.31 |
| 129 | 107.70 | 35.90 | 46.05 | 174 | 149.09 | 49.70 | 61.65 |
| 130 | 108.61 | 36.20 | 46.40 | 175 | 150.02 | 50.01 | 62.00 |
| 131 | 109.53 | 36.50 | 46.75 | 176 | 150.94 | 50.31 | 62.34 |
| 132 | 110.44 | 36.81 | 47.10 | 177 | 151.87 | 50.62 | 62.69 |
| 133 | 111.36 | 37.11 | 47.44 | 178 | 152.80 | 50.93 | 63.03 |
| 134 | 112.27 | 37.42 | 47.79 | 179 | 153.72 | 51.24 | 63.38 |
| 135 | 113.19 | 37.72 | 48.14 | 180 | 154.65 | 51.55 | 63.72 |

Table 2.  CONSTANTS FOR $\underline{p}'$ (LONG FORMULA), AN APPROXIMATION OF
THE LOWER BINOMIAL CONFIDENCE LIMIT $\underline{p}$; $c/n \leqslant .5$, $n \geqslant 20$,
$c \leqslant 1,000$ (*Continued*)

| c | m | a | b | c | m | a | b |
|---|---|---|---|---|---|---|---|
| 181 | 155.58 | 51.86 | 64.07 | 226 | 197.48 | 65.83 | 79.59 |
| 182 | 156.51 | 52.17 | 64.42 | 227 | 198.42 | 66.14 | 79.93 |
| 183 | 157.43 | 52.48 | 64.76 | 228 | 199.35 | 66.45 | 80.27 |
| 184 | 158.36 | 52.79 | 65.11 | 229 | 200.29 | 66.76 | 80.62 |
| 185 | 159.29 | 53.10 | 65.45 | 230 | 201.22 | 67.07 | 80.96 |
| 186 | 160.22 | 53.41 | 65.80 | 231 | 202.16 | 67.39 | 81.31 |
| 187 | 161.14 | 53.71 | 66.14 | 232 | 203.09 | 67.70 | 81.65 |
| 188 | 162.07 | 54.02 | 66.49 | 233 | 204.03 | 68.01 | 82.00 |
| 189 | 163.00 | 54.33 | 66.83 | 234 | 204.96 | 68.32 | 82.34 |
| 190 | 163.93 | 54.64 | 67.18 | 235 | 205.90 | 68.63 | 82.68 |
| 191 | 164.86 | 54.95 | 67.52 | 236 | 206.84 | 68.95 | 83.03 |
| 192 | 165.79 | 55.26 | 67.87 | 237 | 207.77 | 69.26 | 83.37 |
| 193 | 166.72 | 55.57 | 68.21 | 238 | 208.71 | 69.57 | 83.72 |
| 194 | 167.65 | 55.88 | 68.56 | 239 | 209.65 | 69.88 | 84.06 |
| 195 | 168.58 | 56.19 | 68.90 | 240 | 210.58 | 70.19 | 84.40 |
| 196 | 169.51 | 56.50 | 69.25 | 241 | 211.52 | 70.51 | 84.75 |
| 197 | 170.44 | 56.81 | 69.59 | 242 | 212.46 | 70.82 | 85.09 |
| 198 | 171.37 | 57.12 | 69.94 | 243 | 213.39 | 71.13 | 85.43 |
| 199 | 172.30 | 57.43 | 70.28 | 244 | 214.33 | 71.44 | 85.78 |
| 200 | 173.23 | 57.74 | 70.63 | 245 | 215.27 | 71.76 | 86.12 |
| 201 | 174.16 | 58.05 | 70.97 | 246 | 216.21 | 72.07 | 86.47 |
| 202 | 175.09 | 58.36 | 71.32 | 247 | 217.14 | 72.38 | 86.81 |
| 203 | 176.02 | 58.67 | 71.66 | 248 | 218.08 | 72.69 | 87.15 |
| 204 | 176.95 | 58.98 | 72.01 | 249 | 219.02 | 73.01 | 87.50 |
| 205 | 177.88 | 59.29 | 72.35 | 250 | 219.96 | 73.32 | 87.84 |
| 206 | 178.82 | 59.61 | 72.70 | 251 | 220.89 | 73.63 | 88.18 |
| 207 | 179.75 | 59.92 | 73.04 | 252 | 221.83 | 73.94 | 88.53 |
| 208 | 180.68 | 60.23 | 73.39 | 253 | 222.77 | 74.26 | 88.87 |
| 209 | 181.61 | 60.54 | 73.73 | 254 | 223.71 | 74.57 | 89.22 |
| 210 | 182.54 | 60.85 | 74.08 | 255 | 224.65 | 74.88 | 89.56 |
| 211 | 183.48 | 61.16 | 74.42 | 256 | 225.59 | 75.20 | 89.90 |
| 212 | 184.41 | 61.47 | 74.77 | 257 | 226.53 | 75.51 | 90.25 |
| 213 | 185.34 | 61.78 | 75.11 | 258 | 227.46 | 75.82 | 90.59 |
| 214 | 186.27 | 62.09 | 75.45 | 259 | 228.40 | 76.13 | 90.93 |
| 215 | 187.21 | 62.40 | 75.80 | 260 | 229.34 | 76.45 | 91.28 |
| 216 | 188.14 | 62.71 | 76.14 | 261 | 230.28 | 76.76 | 91.62 |
| 217 | 189.07 | 63.02 | 76.49 | 262 | 231.22 | 77.07 | 91.96 |
| 218 | 190.01 | 63.34 | 76.83 | 263 | 232.16 | 77.39 | 92.31 |
| 219 | 190.94 | 63.65 | 77.18 | 264 | 233.10 | 77.70 | 92.65 |
| 220 | 191.88 | 63.96 | 77.52 | 265 | 234.04 | 78.01 | 92.99 |
| 221 | 192.81 | 64.27 | 77.87 | 266 | 234.98 | 78.33 | 93.34 |
| 222 | 193.74 | 64.58 | 78.21 | 267 | 235.92 | 78.64 | 93.68 |
| 223 | 194.68 | 64.89 | 78.55 | 268 | 236.86 | 78.95 | 94.02 |
| 224 | 195.61 | 65.20 | 78.90 | 269 | 237.80 | 79.27 | 94.37 |
| 225 | 196.55 | 65.52 | 79.24 | 270 | 238.74 | 79.58 | 94.71 |

Table 2.  CONSTANTS FOR p′ (LONG FORMULA), AN APPROXIMATION OF THE LOWER BINOMIAL CONFIDENCE LIMIT p; c/n ≤ .5, n ≥ 20, c ≤ 1,000 (*Continued*)

| c | m | a | b | c | m | a | b |
|---|---|---|---|---|---|---|---|
| 271 | 239.68 | 79.89 | 95.05 | 316 | 282.11 | 94.04 | 110.48 |
| 272 | 240.62 | 80.21 | 95.40 | 317 | 283.05 | 94.35 | 110.82 |
| 273 | 241.56 | 80.52 | 95.74 | 318 | 284.00 | 94.67 | 111.17 |
| 274 | 242.50 | 80.83 | 96.08 | 319 | 284.94 | 94.98 | 111.51 |
| 275 | 243.44 | 81.15 | 96.43 | 320 | 285.89 | 95.30 | 111.85 |
| 276 | 244.39 | 81.46 | 96.77 | 321 | 286.83 | 95.61 | 112.19 |
| 277 | 245.33 | 81.78 | 97.11 | 322 | 287.78 | 95.93 | 112.54 |
| 278 | 246.27 | 82.09 | 97.46 | 323 | 288.72 | 96.24 | 112.88 |
| 279 | 247.21 | 82.40 | 97.80 | 324 | 289.67 | 96.56 | 113.22 |
| 280 | 248.15 | 82.72 | 98.14 | 325 | 290.61 | 96.87 | 113.56 |
| 281 | 249.09 | 83.03 | 98.48 | 326 | 291.56 | 97.19 | 113.91 |
| 282 | 250.03 | 83.34 | 98.83 | 327 | 292.50 | 97.50 | 114.25 |
| 283 | 250.97 | 83.66 | 99.17 | 328 | 293.45 | 97.82 | 114.59 |
| 284 | 251.92 | 83.97 | 99.51 | 329 | 294.40 | 98.13 | 114.93 |
| 285 | 252.86 | 84.29 | 99.86 | 330 | 295.34 | 98.45 | 115.28 |
| 286 | 253.80 | 84.60 | 100.20 | 331 | 296.29 | 98.76 | 115.62 |
| 287 | 254.74 | 84.91 | 100.54 | 332 | 297.23 | 99.08 | 115.96 |
| 288 | 255.68 | 85.23 | 100.89 | 333 | 298.18 | 99.39 | 116.30 |
| 289 | 256.63 | 85.54 | 101.23 | 334 | 299.13 | 99.71 | 116.65 |
| 290 | 257.57 | 85.86 | 101.57 | 335 | 300.07 | 100.02 | 116.99 |
| 291 | 258.51 | 86.17 | 101.91 | 336 | 301.02 | 100.34 | 117.33 |
| 292 | 259.45 | 86.48 | 102.26 | 337 | 301.97 | 100.66 | 117.67 |
| 293 | 260.40 | 86.80 | 102.60 | 338 | 302.91 | 100.97 | 118.01 |
| 294 | 261.34 | 87.11 | 102.94 | 339 | 303.86 | 101.29 | 118.36 |
| 295 | 262.28 | 87.43 | 103.29 | 340 | 304.81 | 101.60 | 118.70 |
| 296 | 263.23 | 87.74 | 103.63 | 341 | 305.75 | 101.92 | 119.04 |
| 297 | 264.17 | 88.06 | 103.97 | 342 | 306.70 | 102.23 | 119.38 |
| 298 | 265.11 | 88.37 | 104.31 | 343 | 307.65 | 102.55 | 119.73 |
| 299 | 266.06 | 88.69 | 104.66 | 344 | 308.59 | 102.86 | 120.07 |
| 300 | 267.00 | 89.00 | 105.00 | 345 | 309.54 | 103.18 | 120.41 |
| 301 | 267.94 | 89.31 | 105.34 | 346 | 310.49 | 103.50 | 120.75 |
| 302 | 268.89 | 89.63 | 105.69 | 347 | 311.44 | 103.81 | 121.09 |
| 303 | 269.83 | 89.94 | 106.03 | 348 | 312.38 | 104.13 | 121.44 |
| 304 | 270.77 | 90.26 | 106.37 | 349 | 313.33 | 104.44 | 121.78 |
| 305 | 271.72 | 90.57 | 106.71 | 350 | 314.28 | 104.76 | 122.12 |
| 306 | 272.66 | 90.89 | 107.06 | 351 | 315.23 | 105.08 | 122.46 |
| 307 | 273.61 | 91.20 | 107.40 | 352 | 316.17 | 105.39 | 122.80 |
| 308 | 274.55 | 91.52 | 107.74 | 353 | 317.12 | 105.71 | 123.15 |
| 309 | 275.49 | 91.83 | 108.08 | 354 | 318.07 | 106.02 | 123.49 |
| 310 | 276.44 | 92.15 | 108.43 | 355 | 319.02 | 106.34 | 123.83 |
| 311 | 277.38 | 92.46 | 108.77 | 356 | 319.97 | 106.66 | 124.17 |
| 312 | 278.33 | 92.78 | 109.11 | 357 | 320.91 | 106.97 | 124.51 |
| 313 | 279.27 | 93.09 | 109.45 | 358 | 321.86 | 107.29 | 124.86 |
| 314 | 280.22 | 93.41 | 109.80 | 359 | 322.81 | 107.60 | 125.20 |
| 315 | 281.16 | 93.72 | 110.14 | 360 | 323.76 | 107.92 | 125.54 |

Table 2. CONSTANTS FOR p′ (LONG FORMULA), AN APPROXIMATION OF
THE LOWER BINOMIAL CONFIDENCE LIMIT p; c/n ≤ .5, n ≥ 20,
c ≤ 1,000 (*Continued*)

Section 3
$\gamma_1 = 97.5\%$
$\gamma_2 = 95\%$

| c | m | a | b | c | m | a | b |
|---|---|---|---|---|---|---|---|
| 361 | 324.71 | 108.24 | 125.88 | 406 | 367.45 | 122.48 | 141.26 |
| 362 | 325.66 | 108.55 | 126.22 | 407 | 368.41 | 122.80 | 141.60 |
| 363 | 326.60 | 108.87 | 126.57 | 408 | 369.36 | 123.12 | 141.94 |
| 364 | 327.55 | 109.18 | 126.91 | 409 | 370.31 | 123.44 | 142.28 |
| 365 | 328.50 | 109.50 | 127.25 | 410 | 371.26 | 123.75 | 142.62 |
| 366 | 329.45 | 109.82 | 127.59 | 411 | 372.21 | 124.07 | 142.96 |
| 367 | 330.40 | 110.13 | 127.93 | 412 | 373.16 | 124.39 | 143.31 |
| 368 | 331.35 | 110.45 | 128.28 | 413 | 374.12 | 124.71 | 143.65 |
| 369 | 332.30 | 110.77 | 128.62 | 414 | 375.07 | 125.02 | 143.99 |
| 370 | 333.25 | 111.08 | 128.96 | 415 | 376.02 | 125.34 | 144.33 |
| 371 | 334.19 | 111.40 | 129.30 | 416 | 376.97 | 125.66 | 144.67 |
| 372 | 335.14 | 111.71 | 129.64 | 417 | 377.92 | 125.97 | 145.01 |
| 373 | 336.09 | 112.03 | 129.98 | 418 | 378.87 | 126.29 | 145.35 |
| 374 | 337.04 | 112.35 | 130.33 | 419 | 379.83 | 126.61 | 145.70 |
| 375 | 337.99 | 112.66 | 130.67 | 420 | 380.78 | 126.93 | 146.04 |
| 376 | 338.94 | 112.98 | 131.01 | 421 | 381.73 | 127.24 | 146.38 |
| 377 | 339.89 | 113.30 | 131.35 | 422 | 382.68 | 127.56 | 146.72 |
| 378 | 340.84 | 113.61 | 131.69 | 423 | 383.64 | 127.88 | 147.06 |
| 379 | 341.79 | 113.93 | 132.04 | 424 | 384.59 | 128.20 | 147.40 |
| 380 | 342.74 | 114.25 | 132.38 | 425 | 385.54 | 128.51 | 147.74 |
| 381 | 343.69 | 114.56 | 132.72 | 426 | 386.49 | 128.83 | 148.08 |
| 382 | 344.64 | 114.88 | 133.06 | 427 | 387.45 | 129.15 | 148.43 |
| 383 | 345.59 | 115.20 | 133.40 | 428 | 388.40 | 129.47 | 148.77 |
| 384 | 346.54 | 115.51 | 133.74 | 429 | 389.35 | 129.78 | 149.11 |
| 385 | 347.49 | 115.83 | 134.09 | 430 | 390.30 | 130.10 | 149.45 |
| 386 | 348.44 | 116.15 | 134.43 | 431 | 391.26 | 130.42 | 149.79 |
| 387 | 349.39 | 116.46 | 134.77 | 432 | 392.21 | 130.74 | 150.13 |
| 388 | 350.34 | 116.78 | 135.11 | 433 | 393.16 | 131.05 | 150.47 |
| 389 | 351.29 | 117.10 | 135.45 | 434 | 394.12 | 131.37 | 150.81 |
| 390 | 352.24 | 117.41 | 135.79 | 435 | 395.07 | 131.69 | 151.16 |
| 391 | 353.19 | 117.73 | 136.13 | 436 | 396.02 | 132.01 | 151.50 |
| 392 | 354.14 | 118.05 | 136.48 | 437 | 396.97 | 132.32 | 151.84 |
| 393 | 355.09 | 118.36 | 136.82 | 438 | 397.93 | 132.64 | 152.18 |
| 394 | 356.04 | 118.68 | 137.16 | 439 | 398.88 | 132.96 | 152.52 |
| 395 | 356.99 | 119.00 | 137.50 | 440 | 399.83 | 133.28 | 152.86 |
| 396 | 357.94 | 119.31 | 137.84 | 441 | 400.79 | 133.60 | 153.20 |
| 397 | 358.89 | 119.63 | 138.18 | 442 | 401.74 | 133.91 | 153.54 |
| 398 | 359.85 | 119.95 | 138.53 | 443 | 402.69 | 134.23 | 153.88 |
| 399 | 360.80 | 120.27 | 138.87 | 444 | 403.65 | 134.55 | 154.23 |
| 400 | 361.75 | 120.58 | 139.21 | 445 | 404.60 | 134.87 | 154.57 |
| 401 | 362.70 | 120.90 | 139.55 | 446 | 405.55 | 135.18 | 154.91 |
| 402 | 363.65 | 121.22 | 139.89 | 447 | 406.51 | 135.50 | 155.25 |
| 403 | 364.60 | 121.53 | 140.23 | 448 | 407.46 | 135.82 | 155.59 |
| 404 | 365.55 | 121.85 | 140.57 | 449 | 408.42 | 136.14 | 155.93 |
| 405 | 366.50 | 122.17 | 140.92 | 450 | 409.37 | 136.46 | 156.27 |

Table 2.  CONSTANTS FOR $\underline{p}'$ (LONG FORMULA), AN APPROXIMATION OF THE LOWER BINOMIAL CONFIDENCE LIMIT $\underline{p}$; $c/n \leqslant .5$, $n \geqslant 20$, $c \leqslant 1,000$ (*Continued*)

| c | m | a | b | c | m | a | b |
|---|---|---|---|---|---|---|---|
| 451 | 410.32 | 136.77 | 156.61 | 496 | 453.30 | 151.10 | 171.95 |
| 452 | 411.28 | 137.09 | 156.95 | 497 | 454.25 | 151.42 | 172.29 |
| 453 | 412.23 | 137.41 | 157.29 | 498 | 455.21 | 151.74 | 172.63 |
| 454 | 413.18 | 137.73 | 157.64 | 499 | 456.16 | 152.05 | 172.97 |
| 455 | 414.14 | 138.05 | 157.98 | 500 | 457.12 | 152.37 | 173.31 |
| 456 | 415.09 | 138.36 | 158.32 | 501 | 458.08 | 152.69 | 173.65 |
| 457 | 416.05 | 138.68 | 158.66 | 502 | 459.03 | 153.01 | 173.99 |
| 458 | 417.00 | 139.00 | 159.00 | 503 | 459.99 | 153.33 | 174.34 |
| 459 | 417.96 | 139.32 | 159.34 | 504 | 460.95 | 153.65 | 174.68 |
| 460 | 418.91 | 139.64 | 159.68 | 505 | 461.90 | 153.97 | 175.02 |
| 461 | 419.86 | 139.95 | 160.02 | 506 | 462.86 | 154.29 | 175.36 |
| 462 | 420.82 | 140.27 | 160.36 | 507 | 463.81 | 154.60 | 175.70 |
| 463 | 421.77 | 140.59 | 160.70 | 508 | 464.77 | 154.92 | 176.04 |
| 464 | 422.73 | 140.91 | 161.05 | 509 | 465.73 | 155.24 | 176.38 |
| 465 | 423.68 | 141.23 | 161.39 | 510 | 466.68 | 155.56 | 176.72 |
| 466 | 424.64 | 141.55 | 161.73 | 511 | 467.64 | 155.88 | 177.06 |
| 467 | 425.59 | 141.86 | 162.07 | 512 | 468.60 | 156.20 | 177.40 |
| 468 | 426.55 | 142.18 | 162.41 | 513 | 469.55 | 156.52 | 177.74 |
| 469 | 427.50 | 142.50 | 162.75 | 514 | 470.51 | 156.84 | 178.08 |
| 470 | 428.46 | 142.82 | 163.09 | 515 | 471.47 | 157.16 | 178.42 |
| 471 | 429.41 | 143.14 | 163.43 | 516 | 472.42 | 157.47 | 178.76 |
| 472 | 430.37 | 143.46 | 163.77 | 517 | 473.38 | 157.79 | 179.10 |
| 473 | 431.32 | 143.77 | 164.11 | 518 | 474.34 | 158.11 | 179.44 |
| 474 | 432.27 | 144.09 | 164.45 | 519 | 475.30 | 158.43 | 179.78 |
| 475 | 433.23 | 144.41 | 164.80 | 520 | 476.25 | 158.75 | 180.12 |
| 476 | 434.19 | 144.73 | 165.14 | 521 | 477.21 | 159.07 | 180.47 |
| 477 | 435.14 | 145.05 | 165.48 | 522 | 478.17 | 159.39 | 180.81 |
| 478 | 436.10 | 145.37 | 165.82 | 523 | 479.12 | 159.71 | 181.15 |
| 479 | 437.05 | 145.68 | 166.16 | 524 | 480.08 | 160.03 | 181.49 |
| 480 | 438.01 | 146.00 | 166.50 | 525 | 481.04 | 160.35 | 181.83 |
| 481 | 438.96 | 146.32 | 166.84 | 526 | 482.00 | 160.67 | 182.17 |
| 482 | 439.92 | 146.64 | 167.18 | 527 | 482.95 | 160.98 | 182.51 |
| 483 | 440.87 | 146.96 | 167.52 | 528 | 483.91 | 161.30 | 182.85 |
| 484 | 441.83 | 147.28 | 167.86 | 529 | 484.87 | 161.62 | 183.19 |
| 485 | 442.78 | 147.59 | 168.20 | 530 | 485.82 | 161.94 | 183.53 |
| 486 | 443.74 | 147.91 | 168.54 | 531 | 486.78 | 162.26 | 183.87 |
| 487 | 444.69 | 148.23 | 168.88 | 532 | 487.74 | 162.58 | 184.21 |
| 488 | 445.65 | 148.55 | 169.23 | 533 | 488.70 | 162.90 | 184.55 |
| 489 | 446.61 | 148.87 | 169.57 | 534 | 489.65 | 163.22 | 184.89 |
| 490 | 447.56 | 149.19 | 169.91 | 535 | 490.61 | 163.54 | 185.23 |
| 491 | 448.52 | 149.51 | 170.25 | 536 | 491.57 | 163.86 | 185.57 |
| 492 | 449.47 | 149.82 | 170.59 | 537 | 492.53 | 164.18 | 185.91 |
| 493 | 450.43 | 150.14 | 170.93 | 538 | 493.49 | 164.50 | 186.25 |
| 494 | 451.38 | 150.46 | 171.27 | 539 | 494.44 | 164.81 | 186.59 |
| 495 | 452.34 | 150.78 | 171.61 | 540 | 495.40 | 165.13 | 186.93 |

Table 2.  CONSTANTS FOR $p'$ (LONG FORMULA), AN APPROXIMATION OF THE LOWER BINOMIAL CONFIDENCE LIMIT $\underline{p}$; $c/n \leqslant .5$, $n \geqslant 20$, $c \leqslant 1,000$ (*Continued*)

Section 3

$\gamma_1 = 97.5\%$
$\gamma_2 = 95\%$

| c | m | a | b | c | m | a | b |
|---|---|---|---|---|---|---|---|
| 541 | 496.36 | 165.45 | 187.27 | 586 | 539.50 | 179.83 | 202.58 |
| 542 | 497.32 | 165.77 | 187.61 | 587 | 540.46 | 180.15 | 202.92 |
| 543 | 498.27 | 166.09 | 187.95 | 588 | 541.42 | 180.47 | 203.26 |
| 544 | 499.23 | 166.41 | 188.29 | 589 | 542.38 | 180.79 | 203.60 |
| 545 | 500.19 | 166.73 | 188.63 | 590 | 543.34 | 181.11 | 203.94 |
| 546 | 501.15 | 167.05 | 188.98 | 591 | 544.30 | 181.43 | 204.28 |
| 547 | 502.11 | 167.37 | 189.32 | 592 | 545.26 | 181.75 | 204.62 |
| 548 | 503.06 | 167.69 | 189.66 | 593 | 546.22 | 182.07 | 204.96 |
| 549 | 504.02 | 168.01 | 190.00 | 594 | 547.18 | 182.39 | 205.30 |
| 550 | 504.98 | 168.33 | 190.34 | 595 | 548.14 | 182.71 | 205.64 |
| 551 | 505.94 | 168.65 | 190.68 | 596 | 549.10 | 183.03 | 205.98 |
| 552 | 506.90 | 168.97 | 191.02 | 597 | 550.06 | 183.35 | 206.32 |
| 553 | 507.86 | 169.29 | 191.36 | 598 | 551.02 | 183.67 | 206.66 |
| 554 | 508.81 | 169.60 | 191.70 | 599 | 551.98 | 183.99 | 207.00 |
| 555 | 509.77 | 169.92 | 192.04 | 600 | 552.94 | 184.31 | 207.34 |
| 556 | 510.73 | 170.24 | 192.38 | 601 | 553.90 | 184.63 | 207.68 |
| 557 | 511.69 | 170.56 | 192.72 | 602 | 554.86 | 184.95 | 208.02 |
| 558 | 512.65 | 170.88 | 193.06 | 603 | 555.82 | 185.27 | 208.36 |
| 559 | 513.61 | 171.20 | 193.40 | 604 | 556.78 | 185.59 | 208.70 |
| 560 | 514.57 | 171.52 | 193.74 | 605 | 557.74 | 185.91 | 209.04 |
| 561 | 515.52 | 171.84 | 194.08 | 606 | 558.70 | 186.23 | 209.38 |
| 562 | 516.48 | 172.16 | 194.42 | 607 | 559.66 | 186.55 | 209.72 |
| 563 | 517.44 | 172.48 | 194.76 | 608 | 560.62 | 186.87 | 210.06 |
| 564 | 518.40 | 172.80 | 195.10 | 609 | 561.58 | 187.19 | 210.40 |
| 565 | 519.36 | 173.12 | 195.44 | 610 | 562.54 | 187.51 | 210.74 |
| 566 | 520.32 | 173.44 | 195.78 | 611 | 563.50 | 187.83 | 211.08 |
| 567 | 521.28 | 173.76 | 196.12 | 612 | 564.46 | 188.15 | 211.42 |
| 568 | 522.23 | 174.08 | 196.46 | 613 | 565.42 | 188.47 | 211.76 |
| 569 | 523.19 | 174.40 | 196.80 | 614 | 566.38 | 188.79 | 212.10 |
| 570 | 524.15 | 174.72 | 197.14 | 615 | 567.34 | 189.11 | 212.44 |
| 571 | 525.11 | 175.04 | 197.48 | 616 | 568.30 | 189.43 | 212.78 |
| 572 | 526.07 | 175.36 | 197.82 | 617 | 569.26 | 189.75 | 213.12 |
| 573 | 527.03 | 175.68 | 198.16 | 618 | 570.22 | 190.07 | 213.46 |
| 574 | 527.99 | 176.00 | 198.50 | 619 | 571.18 | 190.39 | 213.80 |
| 575 | 528.95 | 176.32 | 198.84 | 620 | 572.14 | 190.71 | 214.14 |
| 576 | 529.91 | 176.64 | 199.18 | 621 | 573.10 | 191.03 | 214.48 |
| 577 | 530.87 | 176.96 | 199.52 | 622 | 574.06 | 191.35 | 214.82 |
| 578 | 531.83 | 177.28 | 199.86 | 623 | 575.03 | 191.68 | 215.16 |
| 579 | 532.78 | 177.59 | 200.20 | 624 | 575.99 | 192.00 | 215.50 |
| 580 | 533.74 | 177.91 | 200.54 | 625 | 576.95 | 192.32 | 215.84 |
| 581 | 534.70 | 178.23 | 200.88 | 626 | 577.91 | 192.64 | 216.18 |
| 582 | 535.66 | 178.55 | 201.22 | 627 | 578.87 | 192.96 | 216.52 |
| 583 | 536.62 | 178.87 | 201.56 | 628 | 579.83 | 193.28 | 216.86 |
| 584 | 537.58 | 179.19 | 201.90 | 629 | 580.79 | 193.60 | 217.20 |
| 585 | 538.54 | 179.51 | 202.24 | 630 | 581.75 | 193.92 | 217.54 |

Table 2.  CONSTANTS FOR p′ (LONG FORMULA), AN APPROXIMATION OF
THE LOWER BINOMIAL CONFIDENCE LIMIT p̲; c/n ⩽ .5, n ⩾ 20,
c ⩽ 1,000 (*Continued*)

| c | m | a | b | c | m | a | b |
|---|---|---|---|---|---|---|---|
| 631 | 582.71 | 194.24 | 217.88 | 676 | 625.99 | 208.66 | 233.17 |
| 632 | 583.67 | 194.56 | 218.22 | 677 | 626.95 | 208.98 | 233.51 |
| 633 | 584.63 | 194.88 | 218.56 | 678 | 627.91 | 209.30 | 233.85 |
| 634 | 585.60 | 195.20 | 218.90 | 679 | 628.87 | 209.62 | 234.19 |
| 635 | 586.56 | 195.52 | 219.24 | 680 | 629.84 | 209.95 | 234.53 |
| 636 | 587.52 | 195.84 | 219.58 | 681 | 630.80 | 210.27 | 234.87 |
| 637 | 588.48 | 196.16 | 219.92 | 682 | 631.76 | 210.59 | 235.21 |
| 638 | 589.44 | 196.48 | 220.26 | 683 | 632.72 | 210.91 | 235.55 |
| 639 | 590.40 | 196.80 | 220.60 | 684 | 633.69 | 211.23 | 235.89 |
| 640 | 591.36 | 197.12 | 220.94 | 685 | 634.65 | 211.55 | 236.23 |
| 641 | 592.32 | 197.44 | 221.28 | 686 | 635.61 | 211.87 | 236.56 |
| 642 | 593.29 | 197.76 | 221.62 | 687 | 636.57 | 212.19 | 236.90 |
| 643 | 594.25 | 198.08 | 221.96 | 688 | 637.54 | 212.51 | 237.24 |
| 644 | 595.21 | 198.40 | 222.30 | 689 | 638.50 | 212.83 | 237.58 |
| 645 | 596.17 | 198.72 | 222.64 | 690 | 639.46 | 213.15 | 237.92 |
| 646 | 597.13 | 199.04 | 222.98 | 691 | 640.42 | 213.47 | 238.26 |
| 647 | 598.09 | 199.36 | 223.32 | 692 | 641.39 | 213.80 | 238.60 |
| 648 | 599.05 | 199.68 | 223.66 | 693 | 642.35 | 214.12 | 238.94 |
| 649 | 600.02 | 200.01 | 224.00 | 694 | 643.31 | 214.44 | 239.28 |
| 650 | 600.98 | 200.33 | 224.34 | 695 | 644.28 | 214.76 | 239.62 |
| 651 | 601.94 | 200.65 | 224.68 | 696 | 645.24 | 215.08 | 239.96 |
| 652 | 602.90 | 200.97 | 225.02 | 697 | 646.20 | 215.40 | 240.30 |
| 653 | 603.86 | 201.29 | 225.36 | 698 | 647.16 | 215.72 | 240.64 |
| 654 | 604.82 | 201.61 | 225.70 | 699 | 648.13 | 216.04 | 240.98 |
| 655 | 605.78 | 201.93 | 226.04 | 700 | 649.09 | 216.36 | 241.32 |
| 656 | 606.75 | 202.25 | 226.38 | 701 | 650.05 | 216.68 | 241.66 |
| 657 | 607.71 | 202.57 | 226.72 | 702 | 651.02 | 217.01 | 242.00 |
| 658 | 608.67 | 202.89 | 227.05 | 703 | 651.98 | 217.33 | 242.34 |
| 659 | 609.63 | 203.21 | 227.39 | 704 | 652.94 | 217.65 | 242.68 |
| 660 | 610.59 | 203.53 | 227.73 | 705 | 653.91 | 217.97 | 243.02 |
| 661 | 611.56 | 203.85 | 228.07 | 706 | 654.87 | 218.29 | 243.36 |
| 662 | 612.52 | 204.17 | 228.41 | 707 | 655.83 | 218.61 | 243.69 |
| 663 | 613.48 | 204.49 | 228.75 | 708 | 656.79 | 218.93 | 244.03 |
| 664 | 614.44 | 204.81 | 229.09 | 709 | 657.76 | 219.25 | 244.37 |
| 665 | 615.40 | 205.13 | 229.43 | 710 | 658.72 | 219.57 | 244.71 |
| 666 | 616.37 | 205.46 | 229.77 | 711 | 659.68 | 219.89 | 245.05 |
| 667 | 617.33 | 205.78 | 230.11 | 712 | 660.65 | 220.22 | 245.39 |
| 668 | 618.29 | 206.10 | 230.45 | 713 | 661.61 | 220.54 | 245.73 |
| 669 | 619.25 | 206.42 | 230.79 | 714 | 662.57 | 220.86 | 246.07 |
| 670 | 620.21 | 206.74 | 231.13 | 715 | 663.54 | 221.18 | 246.41 |
| 671 | 621.18 | 207.06 | 231.47 | 716 | 664.50 | 221.50 | 246.75 |
| 672 | 622.14 | 207.38 | 231.81 | 717 | 665.46 | 221.82 | 247.09 |
| 673 | 623.10 | 207.70 | 232.15 | 718 | 666.43 | 222.14 | 247.43 |
| 674 | 624.06 | 208.02 | 232.49 | 719 | 667.39 | 222.46 | 247.77 |
| 675 | 625.02 | 208.34 | 232.83 | 720 | 668.35 | 222.78 | 248.11 |

Table 2.  CONSTANTS FOR $p'$ (LONG FORMULA), AN APPROXIMATION OF
THE LOWER BINOMIAL CONFIDENCE LIMIT $\underline{p}$; $c/n \leqslant .5$, $n \geqslant 20$,
$c \leqslant 1,000$ (*Continued*)

| c | m | a | b | c | m | a | b |
|---|---|---|---|---|---|---|---|
| 721 | 669.32 | 223.11 | 248.45 | 766 | 712.70 | 237.57 | 263.72 |
| 722 | 670.28 | 223.43 | 248.79 | 767 | 713.67 | 237.89 | 264.06 |
| 723 | 671.25 | 223.75 | 249.13 | 768 | 714.63 | 238.21 | 264.40 |
| 724 | 672.21 | 224.07 | 249.47 | 769 | 715.59 | 238.53 | 264.73 |
| 725 | 673.17 | 224.39 | 249.80 | 770 | 716.56 | 238.85 | 265.07 |
| 726 | 674.14 | 224.71 | 250.14 | 771 | 717.52 | 239.17 | 265.41 |
| 727 | 675.10 | 225.03 | 250.48 | 772 | 718.49 | 239.50 | 265.75 |
| 728 | 676.06 | 225.35 | 250.82 | 773 | 719.45 | 239.82 | 266.09 |
| 729 | 677.03 | 225.68 | 251.16 | 774 | 720.42 | 240.14 | 266.43 |
| 730 | 677.99 | 226.00 | 251.50 | 775 | 721.38 | 240.46 | 266.77 |
| 731 | 678.95 | 226.32 | 251.84 | 776 | 722.35 | 240.78 | 267.11 |
| 732 | 679.92 | 226.64 | 252.18 | 777 | 723.31 | 241.10 | 267.45 |
| 733 | 680.88 | 226.96 | 252.52 | 778 | 724.28 | 241.43 | 267.79 |
| 734 | 681.85 | 227.28 | 252.86 | 779 | 725.24 | 241.75 | 268.13 |
| 735 | 682.81 | 227.60 | 253.20 | 780 | 726.21 | 242.07 | 268.47 |
| 736 | 683.77 | 227.92 | 253.54 | 781 | 727.17 | 242.39 | 268.80 |
| 737 | 684.74 | 228.25 | 253.88 | 782 | 728.14 | 242.71 | 269.14 |
| 738 | 685.70 | 228.57 | 254.22 | 783 | 729.10 | 243.03 | 269.48 |
| 739 | 686.67 | 228.89 | 254.56 | 784 | 730.07 | 243.36 | 269.82 |
| 740 | 687.63 | 229.21 | 254.90 | 785 | 731.03 | 243.68 | 270.16 |
| 741 | 688.59 | 229.53 | 255.23 | 786 | 732.00 | 244.00 | 270.50 |
| 742 | 689.56 | 229.85 | 255.57 | 787 | 732.96 | 244.32 | 270.84 |
| 743 | 690.52 | 230.17 | 255.91 | 788 | 733.93 | 244.64 | 271.18 |
| 744 | 691.49 | 230.50 | 256.25 | 789 | 734.89 | 244.96 | 271.52 |
| 745 | 692.45 | 230.82 | 256.59 | 790 | 735.86 | 245.29 | 271.86 |
| 746 | 693.41 | 231.14 | 256.93 | 791 | 736.82 | 245.61 | 272.20 |
| 747 | 694.38 | 231.46 | 257.27 | 792 | 737.79 | 245.93 | 272.54 |
| 748 | 695.34 | 231.78 | 257.61 | 793 | 738.75 | 246.25 | 272.87 |
| 749 | 696.31 | 232.10 | 257.95 | 794 | 739.72 | 246.57 | 273.21 |
| 750 | 697.27 | 232.42 | 258.29 | 795 | 740.68 | 246.89 | 273.55 |
| 751 | 698.23 | 232.74 | 258.63 | 796 | 741.65 | 247.22 | 273.89 |
| 752 | 699.20 | 233.07 | 258.97 | 797 | 742.61 | 247.54 | 274.23 |
| 753 | 700.16 | 233.39 | 259.31 | 798 | 743.58 | 247.86 | 274.57 |
| 754 | 701.13 | 233.71 | 259.65 | 799 | 744.54 | 248.18 | 274.91 |
| 755 | 702.09 | 234.03 | 259.98 | 800 | 745.51 | 248.50 | 275.25 |
| 756 | 703.06 | 234.35 | 260.32 | 801 | 746.48 | 248.83 | 275.59 |
| 757 | 704.02 | 234.67 | 260.66 | 802 | 747.44 | 249.15 | 275.93 |
| 758 | 704.98 | 234.99 | 261.00 | 803 | 748.41 | 249.47 | 276.27 |
| 759 | 705.95 | 235.32 | 261.34 | 804 | 749.37 | 249.79 | 276.60 |
| 760 | 706.91 | 235.64 | 261.68 | 805 | 750.34 | 250.11 | 276.94 |
| 761 | 707.88 | 235.96 | 262.02 | 806 | 751.30 | 250.43 | 277.28 |
| 762 | 708.84 | 236.28 | 262.36 | 807 | 752.27 | 250.76 | 277.62 |
| 763 | 709.81 | 236.60 | 262.70 | 808 | 753.23 | 251.08 | 277.96 |
| 764 | 710.77 | 236.92 | 263.04 | 809 | 754.20 | 251.40 | 278.30 |
| 765 | 711.74 | 237.25 | 263.38 | 810 | 755.16 | 251.72 | 278.64 |

Table 2.  CONSTANTS FOR $\underline{p}'$ (LONG FORMULA), AN APPROXIMATION OF THE LOWER BINOMIAL CONFIDENCE LIMIT $\underline{p}$; $c/n \leqslant .5$, $n \geqslant 20$ $c \leqslant 1,000$ (*Continued*)

| c | m | a | b | c | m | a | b |
|---|---|---|---|---|---|---|---|
| 811 | 756.13 | 252.04 | 278.98 | 856 | 799.60 | 266.53 | 294.23 |
| 812 | 757.10 | 252.37 | 279.32 | 857 | 800.57 | 266.86 | 294.57 |
| 813 | 758.06 | 252.69 | 279.66 | 858 | 801.54 | 267.18 | 294.91 |
| 814 | 759.03 | 253.01 | 280.00 | 859 | 802.50 | 267.50 | 295.25 |
| 815 | 759.99 | 253.33 | 280.33 | 860 | 803.47 | 267.82 | 295.59 |
| 816 | 760.96 | 253.65 | 280.67 | 861 | 804.44 | 268.15 | 295.93 |
| 817 | 761.92 | 253.97 | 281.01 | 862 | 805.40 | 268.47 | 296.27 |
| 818 | 762.89 | 254.30 | 281.35 | 863 | 806.37 | 268.79 | 296.61 |
| 819 | 763.86 | 254.62 | 281.69 | 864 | 807.34 | 269.11 | 296.94 |
| 820 | 764.82 | 254.94 | 282.03 | 865 | 808.30 | 269.43 | 297.28 |
| 821 | 765.79 | 255.26 | 282.37 | 866 | 809.27 | 269.76 | 297.62 |
| 822 | 766.75 | 255.58 | 282.71 | 867 | 810.24 | 270.08 | 297.96 |
| 823 | 767.72 | 255.91 | 283.05 | 868 | 811.20 | 270.40 | 298.30 |
| 824 | 768.68 | 256.23 | 283.39 | 869 | 812.17 | 270.72 | 298.64 |
| 825 | 769.65 | 256.55 | 283.72 | 870 | 813.14 | 271.05 | 298.98 |
| 826 | 770.62 | 256.87 | 284.06 | 871 | 814.10 | 271.37 | 299.32 |
| 827 | 771.58 | 257.19 | 284.40 | 872 | 815.07 | 271.69 | 299.66 |
| 828 | 772.55 | 257.52 | 284.74 | 873 | 816.04 | 272.01 | 299.99 |
| 829 | 773.51 | 257.84 | 285.08 | 874 | 817.00 | 272.33 | 300.33 |
| 830 | 774.48 | 258.16 | 285.42 | 875 | 817.97 | 272.66 | 300.67 |
| 831 | 775.45 | 258.48 | 285.76 | 876 | 818.94 | 272.98 | 301.01 |
| 832 | 776.41 | 258.80 | 286.10 | 877 | 819.90 | 273.30 | 301.35 |
| 833 | 777.38 | 259.13 | 286.44 | 878 | 820.87 | 273.62 | 301.69 |
| 834 | 778.34 | 259.45 | 286.78 | 879 | 821.84 | 273.95 | 302.03 |
| 835 | 779.31 | 259.77 | 287.11 | 880 | 822.80 | 274.27 | 302.37 |
| 836 | 780.28 | 260.09 | 287.45 | 881 | 823.77 | 274.59 | 302.70 |
| 837 | 781.24 | 260.41 | 287.79 | 882 | 824.74 | 274.91 | 303.04 |
| 838 | 782.21 | 260.74 | 288.13 | 883 | 825.71 | 275.23 | 303.38 |
| 839 | 783.17 | 261.06 | 288.47 | 884 | 826.67 | 275.56 | 303.72 |
| 840 | 784.14 | 261.38 | 288.81 | 885 | 827.64 | 275.88 | 304.06 |
| 841 | 785.11 | 261.70 | 289.15 | 886 | 828.61 | 276.20 | 304.40 |
| 842 | 786.07 | 262.02 | 289.49 | 887 | 829.57 | 276.52 | 304.74 |
| 843 | 787.04 | 262.35 | 289.83 | 888 | 830.54 | 276.85 | 305.08 |
| 844 | 788.01 | 262.67 | 290.17 | 889 | 831.51 | 277.17 | 305.42 |
| 845 | 788.97 | 262.99 | 290.50 | 890 | 832.47 | 277.49 | 305.75 |
| 846 | 789.94 | 263.31 | 290.84 | 891 | 833.44 | 277.81 | 306.09 |
| 847 | 790.90 | 263.63 | 291.18 | 892 | 834.41 | 278.14 | 306.43 |
| 848 | 791.87 | 263.96 | 291.52 | 893 | 835.38 | 278.46 | 306.77 |
| 849 | 792.84 | 264.28 | 291.86 | 894 | 836.34 | 278.78 | 307.11 |
| 850 | 793.80 | 264.60 | 292.20 | 895 | 837.31 | 279.10 | 307.45 |
| 851 | 794.77 | 264.92 | 292.54 | 896 | 838.28 | 279.43 | 307.79 |
| 852 | 795.74 | 265.25 | 292.88 | 897 | 839.25 | 279.75 | 308.13 |
| 853 | 796.70 | 265.57 | 293.22 | 898 | 840.21 | 280.07 | 308.46 |
| 854 | 797.67 | 265.89 | 293.56 | 899 | 841.18 | 280.39 | 308.80 |
| 855 | 798.64 | 266.21 | 293.89 | 900 | 842.15 | 280.72 | 309.14 |

Table 2.  CONSTANTS FOR p′ (LONG FORMULA), AN APPROXIMATION OF
THE LOWER BINOMIAL CONFIDENCE LIMIT p̲; c/n ≤ .5, n ≥ 20,
c ≤ 1,000 (*Continued*)

Section 3

$\gamma_1 = 97.5\%$
$\gamma_2 = 95\%$

| c | m | a | b | c | m | a | b |
|---|---|---|---|---|---|---|---|
| 901 | 843.11 | 281.04 | 309.48 | 946 | 886.66 | 295.55 | 324.72 |
| 902 | 844.08 | 281.36 | 309.82 | 947 | 887.63 | 295.88 | 325.06 |
| 903 | 845.05 | 281.68 | 310.16 | 948 | 888.60 | 296.20 | 325.40 |
| 904 | 846.02 | 282.01 | 310.50 | 949 | 889.57 | 296.52 | 325.74 |
| 905 | 846.98 | 282.33 | 310.84 | 950 | 890.54 | 296.85 | 326.08 |
| 906 | 847.95 | 282.65 | 311.17 | 951 | 891.50 | 297.17 | 326.42 |
| 907 | 848.92 | 282.97 | 311.51 | 952 | 892.47 | 297.49 | 326.75 |
| 908 | 849.89 | 283.30 | 311.85 | 953 | 893.44 | 297.81 | 327.09 |
| 909 | 850.85 | 283.62 | 312.19 | 954 | 894.41 | 298.14 | 327.43 |
| 910 | 851.82 | 283.94 | 312.53 | 955 | 895.38 | 298.46 | 327.77 |
| 911 | 852.79 | 284.26 | 312.87 | 956 | 896.35 | 298.78 | 328.11 |
| 912 | 853.76 | 284.59 | 313.21 | 957 | 897.31 | 299.10 | 328.45 |
| 913 | 854.72 | 284.91 | 313.55 | 958 | 898.28 | 299.43 | 328.79 |
| 914 | 855.69 | 285.23 | 313.88 | 959 | 899.25 | 299.75 | 329.12 |
| 915 | 856.66 | 285.55 | 314.22 | 960 | 900.22 | 300.07 | 329.46 |
| 916 | 857.63 | 285.88 | 314.56 | 961 | 901.19 | 300.40 | 329.80 |
| 917 | 858.59 | 286.20 | 314.90 | 962 | 902.16 | 300.72 | 330.14 |
| 918 | 859.56 | 286.52 | 315.24 | 963 | 903.12 | 301.04 | 330.48 |
| 919 | 860.53 | 286.84 | 315.58 | 964 | 904.09 | 301.36 | 330.82 |
| 920 | 861.50 | 287.17 | 315.92 | 965 | 905.06 | 301.69 | 331.16 |
| 921 | 862.47 | 287.49 | 316.26 | 966 | 906.03 | 302.01 | 331.50 |
| 922 | 863.43 | 287.81 | 316.59 | 967 | 907.00 | 302.33 | 331.83 |
| 923 | 864.40 | 288.13 | 316.93 | 968 | 907.97 | 302.66 | 332.17 |
| 924 | 865.37 | 288.46 | 317.27 | 969 | 908.93 | 302.98 | 332.51 |
| 925 | 866.34 | 288.78 | 317.61 | 970 | 909.90 | 303.30 | 332.85 |
| 926 | 867.30 | 289.10 | 317.95 | 971 | 910.87 | 303.62 | 333.19 |
| 927 | 868.27 | 289.42 | 318.29 | 972 | 911.84 | 303.95 | 333.53 |
| 928 | 869.24 | 289.75 | 318.63 | 973 | 912.81 | 304.27 | 333.87 |
| 929 | 870.21 | 290.07 | 318.97 | 974 | 913.78 | 304.59 | 334.20 |
| 930 | 871.18 | 290.39 | 319.30 | 975 | 914.75 | 304.92 | 334.54 |
| 931 | 872.14 | 290.71 | 319.64 | 976 | 915.71 | 305.24 | 334.88 |
| 932 | 873.11 | 291.04 | 319.98 | 977 | 916.68 | 305.56 | 335.22 |
| 933 | 874.08 | 291.36 | 320.32 | 978 | 917.65 | 305.88 | 335.56 |
| 934 | 875.05 | 291.68 | 320.66 | 979 | 918.62 | 306.21 | 335.90 |
| 935 | 876.01 | 292.00 | 321.00 | 980 | 919.59 | 306.53 | 336.24 |
| 936 | 876.98 | 292.33 | 321.34 | 981 | 920.56 | 306.85 | 336.57 |
| 937 | 877.95 | 292.65 | 321.67 | 982 | 921.53 | 307.18 | 336.91 |
| 938 | 878.92 | 292.97 | 322.01 | 983 | 922.50 | 307.50 | 337.25 |
| 939 | 879.89 | 293.30 | 322.35 | 984 | 923.46 | 307.82 | 337.59 |
| 940 | 880.85 | 293.62 | 322.69 | 985 | 924.43 | 308.14 | 337.93 |
| 941 | 881.82 | 293.94 | 323.03 | 986 | 925.40 | 308.47 | 338.27 |
| 942 | 882.79 | 294.26 | 323.37 | 987 | 926.37 | 308.79 | 338.60 |
| 943 | 883.76 | 294.59 | 323.71 | 988 | 927.34 | 309.11 | 338.94 |
| 944 | 884.73 | 294.91 | 324.05 | 989 | 928.31 | 309.44 | 339.28 |
| 945 | 885.70 | 295.23 | 324.38 | 990 | 929.28 | 309.76 | 339.62 |

Table 2.  **CONSTANTS FOR p′ (LONG FORMULA), AN APPROXIMATION OF
THE LOWER BINOMIAL CONFIDENCE LIMIT** $\underline{p}$; $c/n \leqslant .5$, $n \geqslant 20$,
$c \leqslant 1{,}000$ (*Continued*)

| c | m | a | b | c | m | a | b |
|---|---|---|---|---|---|---|---|
| 991 | 930.25 | 310.08 | 339.96 | 996 | 935.09 | 311.70 | 341.65 |
| 992 | 931.21 | 310.40 | 340.30 | 997 | 936.06 | 312.02 | 341.99 |
| 993 | 932.18 | 310.73 | 340.64 | 998 | 937.03 | 312.34 | 342.33 |
| 994 | 933.15 | 311.05 | 340.97 | 999 | 938.00 | 312.67 | 342.67 |
| 995 | 934.12 | 311.37 | 341.31 | 1000 | 938.97 | 312.99 | 343.01 |

# Table 2. CONSTANTS FOR p′ (LONG FORMULA), AN APPROXIMATION OF THE LOWER BINOMIAL CONFIDENCE LIMIT p; c/n ≤ .5, n ≥ 20, c ≤ 1,000 (*Continued*)

| c | m | a | b | c | m | a | b |
|---|---|---|---|---|---|---|---|
| 1 | 0.051294 | 0.026 | 0.0 | 46 | 35.441 | 12.325 | 17.104 |
| 2 | 0.35536 | 0.395 | 0.717 | 47 | 36.320 | 12.604 | 17.444 |
| 3 | 0.81769 | 0.593 | 1.184 | 48 | 37.200 | 12.859 | 17.759 |
| 4 | 1.3663 | 0.815 | 1.632 | 49 | 38.082 | 13.144 | 18.103 |
| 5 | 1.9702 | 1.044 | 2.059 | 50 | 38.965 | 13.426 | 18.444 |
| 6 | 2.6130 | 1.284 | 2.478 | 51 | 39.849 | 13.699 | 18.775 |
| 7 | 3.2853 | 1.549 | 2.907 | 52 | 40.734 | 13.993 | 19.127 |
| 8 | 3.9808 | 1.824 | 3.334 | 53 | 41.620 | 14.269 | 19.459 |
| 9 | 4.6952 | 2.109 | 3.761 | 54 | 42.507 | 14.559 | 19.805 |
| 10 | 5.4254 | 2.410 | 4.197 | 55 | 43.396 | 14.819 | 20.122 |
| 11 | 6.1690 | 2.674 | 4.590 | 56 | 44.285 | 15.124 | 20.481 |
| 12 | 6.9242 | 2.942 | 4.980 | 57 | 45.176 | 15.392 | 20.804 |
| 13 | 7.6896 | 3.207 | 5.363 | 58 | 46.067 | 15.70 | 21.165 |
| 14 | 8.4640 | 3.477 | 5.745 | 59 | 46.959 | 15.96 | 21.480 |
| 15 | 9.2464 | 3.744 | 6.121 | 60 | 47.852 | 16.25 | 21.829 |
| 16 | 10.036 | 4.018 | 6.500 | 61 | 48.746 | 16.54 | 22.164 |
| 17 | 10.832 | 4.290 | 6.873 | 62 | 49.641 | 16.80 | 22.48 |
| 18 | 11.634 | 4.555 | 7.238 | 63 | 50.537 | 17.10 | 22.83 |
| 19 | 12.442 | 4.830 | 7.609 | 64 | 51.434 | 17.38 | 23.16 |
| 20 | 13.255 | 5.102 | 7.975 | 65 | 52.331 | 17.67 | 23.50 |
| 21 | 14.072 | 5.375 | 8.339 | 66 | 53.230 | 17.93 | 23.81 |
| 22 | 14.894 | 5.646 | 8.699 | 67 | 54.129 | 18.21 | 24.14 |
| 23 | 15.720 | 5.922 | 9.062 | 68 | 55.028 | 18.51 | 24.50 |
| 24 | 16.549 | 6.199 | 9.425 | 69 | 55.929 | 18.77 | 24.81 |
| 25 | 17.382 | 6.480 | 9.789 | 70 | 56.830 | 19.05 | 25.14 |
| 26 | 18.219 | 6.755 | 10.146 | 71 | 57.732 | 19.33 | 25.46 |
| 27 | 19.058 | 7.031 | 10.502 | 72 | 58.634 | 19.62 | 25.80 |
| 28 | 19.901 | 7.304 | 10.854 | 73 | 59.538 | 19.90 | 26.13 |
| 29 | 20.746 | 7.572 | 11.199 | 74 | 60.442 | 20.20 | 26.47 |
| 30 | 21.594 | 7.852 | 11.555 | 75 | 61.346 | 20.49 | 26.82 |
| 31 | 22.445 | 8.131 | 11.909 | 76 | 62.251 | 20.74 | 27.11 |
| 32 | 23.297 | 8.411 | 12.262 | 77 | 63.157 | 21.03 | 27.45 |
| 33 | 24.153 | 8.679 | 12.602 | 78 | 64.064 | 21.31 | 27.78 |
| 34 | 25.010 | 8.963 | 12.958 | 79 | 64.971 | 21.59 | 28.11 |
| 35 | 25.870 | 9.250 | 13.315 | 80 | 65.878 | 21.87 | 28.43 |
| 36 | 26.731 | 9.511 | 13.645 | 81 | 66.786 | 22.20 | 28.81 |
| 37 | 27.595 | 9.798 | 14.001 | 82 | 67.695 | 22.45 | 29.10 |
| 38 | 28.460 | 10.081 | 14.352 | 83 | 68.605 | 22.74 | 29.44 |
| 39 | 29.327 | 10.353 | 14.689 | 84 | 69.514 | 23.04 | 29.78 |
| 40 | 30.196 | 10.625 | 15.027 | 85 | 70.425 | 23.30 | 30.09 |
| 41 | 31.066 | 10.918 | 15.385 | 86 | 71.336 | 23.62 | 30.45 |
| 42 | 31.938 | 11.199 | 15.730 | 87 | 72.247 | 23.84 | 30.72 |
| 43 | 32.812 | 11.459 | 16.053 | 88 | 73.159 | 24.19 | 31.11 |
| 44 | 33.687 | 11.747 | 16.403 | 89 | 74.072 | 24.43 | 31.40 |
| 45 | 34.563 | 12.027 | 16.746 | 90 | 74.985 | 24.73 | 31.73 |

Table 2. CONSTANTS FOR $p'$ (LONG FORMULA), AN APPROXIMATION OF THE LOWER BINOMIAL CONFIDENCE LIMIT $\underline{p}$; $c/n \leqslant .5$, $n \geqslant 20$, $c \leqslant 1,000$ (*Continued*)

| c | m | a | b | c | m | a | b |
|---|---|---|---|---|---|---|---|
| 91 | 75.898 | 24.99 | 32.04 | 136 | 117.40 | 39.13 | 47.94 |
| 92 | 76.812 | 25.30 | 32.39 | 137 | 118.33 | 39.44 | 48.28 |
| 93 | 77.726 | 25.59 | 32.73 | 138 | 119.26 | 39.75 | 48.63 |
| 94 | 78.641 | 25.83 | 33.01 | 139 | 120.19 | 40.06 | 48.97 |
| 95 | 79.557 | 26.11 | 33.34 | 140 | 121.12 | 40.37 | 49.32 |
| 96 | 80.472 | 26.46 | 33.72 | 141 | 122.05 | 40.68 | 49.66 |
| 97 | 81.388 | 26.76 | 34.06 | 142 | 122.98 | 40.99 | 50.01 |
| 98 | 82.305 | 27.02 | 34.36 | 143 | 123.92 | 41.30 | 50.35 |
| 99 | 83.222 | 27.30 | 34.69 | 144 | 124.85 | 41.61 | 50.70 |
| 100 | 84.139 | 27.57 | 35.00 | 145 | 125.78 | 41.92 | 51.04 |
| 101 | 85.057 | 28.35 | 35.83 | 146 | 126.71 | 42.23 | 51.38 |
| 102 | 85.976 | 28.65 | 36.17 | 147 | 127.64 | 42.54 | 51.73 |
| 103 | 86.894 | 28.96 | 36.52 | 148 | 128.57 | 42.85 | 52.07 |
| 104 | 87.813 | 29.26 | 36.87 | 149 | 129.51 | 43.16 | 52.42 |
| 105 | 88.733 | 29.57 | 37.21 | 150 | 130.44 | 43.47 | 52.76 |
| 106 | 89.653 | 29.88 | 37.56 | 151 | 131.37 | 43.79 | 53.11 |
| 107 | 90.573 | 30.18 | 37.91 | 152 | 132.31 | 44.10 | 53.45 |
| 108 | 91.493 | 30.49 | 38.25 | 153 | 133.24 | 44.41 | 53.80 |
| 109 | 92.414 | 30.80 | 38.60 | 154 | 134.17 | 44.72 | 54.14 |
| 110 | 93.336 | 31.11 | 38.95 | 155 | 135.11 | 45.03 | 54.49 |
| 111 | 94.257 | 31.41 | 39.29 | 156 | 136.04 | 45.34 | 54.83 |
| 112 | 95.179 | 31.72 | 39.64 | 157 | 136.97 | 45.65 | 55.17 |
| 113 | 96.102 | 32.03 | 39.99 | 158 | 137.91 | 45.96 | 55.52 |
| 114 | 97.025 | 32.34 | 40.33 | 159 | 138.84 | 46.28 | 55.86 |
| 115 | 97.948 | 32.64 | 40.68 | 160 | 139.78 | 46.59 | 56.21 |
| 116 | 98.871 | 32.95 | 41.02 | 161 | 140.70 | 46.90 | 56.55 |
| 117 | 99.795 | 33.26 | 41.37 | 162 | 141.63 | 47.21 | 56.89 |
| 118 | 100.72 | 33.57 | 41.72 | 163 | 142.57 | 47.52 | 57.24 |
| 119 | 101.64 | 33.87 | 42.06 | 164 | 143.50 | 47.83 | 57.58 |
| 120 | 102.57 | 34.18 | 42.41 | 165 | 144.44 | 48.15 | 57.93 |
| 121 | 103.49 | 34.49 | 42.75 | 166 | 145.38 | 48.46 | 58.27 |
| 122 | 104.42 | 34.80 | 43.10 | 167 | 146.31 | 48.77 | 58.61 |
| 123 | 105.34 | 35.11 | 43.45 | 168 | 147.25 | 49.08 | 58.96 |
| 124 | 106.27 | 35.42 | 43.79 | 169 | 148.18 | 49.39 | 59.30 |
| 125 | 107.20 | 35.73 | 44.14 | 170 | 149.12 | 49.71 | 59.65 |
| 126 | 108.12 | 36.03 | 44.48 | 171 | 150.06 | 50.02 | 59.99 |
| 127 | 109.05 | 36.34 | 44.83 | 172 | 151.00 | 50.33 | 60.33 |
| 128 | 109.98 | 36.65 | 45.17 | 173 | 151.93 | 50.64 | 60.68 |
| 129 | 110.90 | 36.96 | 45.52 | 174 | 152.87 | 50.96 | 61.02 |
| 130 | 111.83 | 37.27 | 45.86 | 175 | 153.81 | 51.27 | 61.37 |
| 131 | 112.76 | 37.58 | 46.21 | 176 | 154.75 | 51.58 | 61.71 |
| 132 | 113.69 | 37.89 | 46.55 | 177 | 155.68 | 51.89 | 62.05 |
| 133 | 114.62 | 38.20 | 46.90 | 178 | 156.62 | 52.21 | 62.40 |
| 134 | 115.54 | 38.51 | 47.25 | 179 | 157.56 | 52.52 | 62.74 |
| 135 | 116.47 | 38.82 | 47.59 | 180 | 158.50 | 52.83 | 63.08 |

# Table 2. CONSTANTS FOR $p'$ (LONG FORMULA), AN APPROXIMATION OF THE LOWER BINOMIAL CONFIDENCE LIMIT $\underline{p}$; $c/n \leqslant .5$, $n \geqslant 20$, $c \leqslant 1,000$ (*Continued*)

| c | m | a | b | c | m | a | b |
|---|---|---|---|---|---|---|---|
| 181 | 159.44 | 53.15 | 63.43 | 226 | 201.84 | 67.28 | 78.86 |
| 182 | 160.38 | 53.46 | 63.77 | 227 | 202.79 | 67.60 | 79.20 |
| 183 | 161.32 | 53.77 | 64.11 | 228 | 203.73 | 67.91 | 79.54 |
| 184 | 162.26 | 54.09 | 64.46 | 229 | 204.68 | 68.23 | 79.89 |
| 185 | 163.20 | 54.40 | 64.80 | 230 | 205.62 | 68.54 | 80.23 |
| 186 | 164.14 | 54.71 | 65.14 | 231 | 206.57 | 68.86 | 80.57 |
| 187 | 165.07 | 55.02 | 65.49 | 232 | 207.51 | 69.17 | 80.91 |
| 188 | 166.01 | 55.34 | 65.83 | 233 | 208.46 | 69.49 | 81.26 |
| 189 | 166.95 | 55.65 | 66.17 | 234 | 209.41 | 69.80 | 81.60 |
| 190 | 167.90 | 55.97 | 66.52 | 235 | 210.35 | 70.12 | 81.94 |
| 191 | 168.84 | 56.28 | 66.86 | 236 | 211.30 | 70.43 | 82.28 |
| 192 | 169.78 | 56.59 | 67.20 | 237 | 212.25 | 70.75 | 82.63 |
| 193 | 170.72 | 56.91 | 67.55 | 238 | 213.19 | 71.06 | 82.97 |
| 194 | 171.66 | 57.22 | 67.89 | 239 | 214.14 | 71.38 | 83.31 |
| 195 | 172.60 | 57.53 | 68.23 | 240 | 215.09 | 71.70 | 83.65 |
| 196 | 173.54 | 57.85 | 68.58 | 241 | 216.03 | 72.01 | 83.99 |
| 197 | 174.48 | 58.16 | 68.92 | 242 | 216.98 | 72.33 | 84.34 |
| 198 | 175.42 | 58.47 | 69.26 | 243 | 217.93 | 72.64 | 84.68 |
| 199 | 176.36 | 58.79 | 69.61 | 244 | 218.87 | 72.96 | 85.02 |
| 200 | 177.31 | 59.10 | 69.95 | 245 | 219.82 | 73.27 | 85.36 |
| 201 | 178.25 | 59.42 | 70.29 | 246 | 220.77 | 73.59 | 85.71 |
| 202 | 179.19 | 59.73 | 70.63 | 247 | 221.72 | 73.91 | 86.05 |
| 203 | 180.13 | 60.04 | 70.98 | 248 | 222.66 | 74.22 | 86.39 |
| 204 | 181.07 | 60.36 | 71.32 | 249 | 223.61 | 74.54 | 86.73 |
| 205 | 182.02 | 60.67 | 71.66 | 250 | 224.56 | 74.85 | 87.07 |
| 206 | 182.96 | 60.99 | 72.01 | 251 | 225.51 | 75.17 | 87.42 |
| 207 | 183.90 | 61.30 | 72.35 | 252 | 226.46 | 75.49 | 87.76 |
| 208 | 184.85 | 61.62 | 72.69 | 253 | 227.40 | 75.80 | 88.10 |
| 209 | 185.79 | 61.93 | 73.04 | 254 | 228.35 | 76.12 | 88.44 |
| 210 | 186.73 | 62.24 | 73.38 | 255 | 229.30 | 76.43 | 88.78 |
| 211 | 187.67 | 62.56 | 73.72 | 256 | 230.25 | 76.75 | 89.12 |
| 212 | 188.62 | 62.87 | 74.06 | 257 | 231.20 | 77.07 | 89.47 |
| 213 | 189.56 | 63.19 | 74.41 | 258 | 232.15 | 77.38 | 89.81 |
| 214 | 190.51 | 63.50 | 74.75 | 259 | 233.10 | 77.70 | 90.15 |
| 215 | 191.45 | 63.82 | 75.09 | 260 | 234.05 | 78.02 | 90.49 |
| 216 | 192.39 | 64.13 | 75.43 | 261 | 234.99 | 78.33 | 90.83 |
| 217 | 193.34 | 64.45 | 75.78 | 262 | 235.94 | 78.65 | 91.18 |
| 218 | 194.28 | 64.76 | 76.12 | 263 | 236.89 | 78.96 | 91.52 |
| 219 | 195.23 | 65.08 | 76.46 | 264 | 237.84 | 79.28 | 91.86 |
| 220 | 196.17 | 65.39 | 76.80 | 265 | 238.79 | 79.60 | 92.20 |
| 221 | 197.12 | 65.71 | 77.15 | 266 | 239.74 | 79.91 | 92.54 |
| 222 | 198.06 | 66.02 | 77.49 | 267 | 240.69 | 80.23 | 92.88 |
| 223 | 199.00 | 66.33 | 77.83 | 268 | 241.64 | 80.55 | 93.23 |
| 224 | 199.95 | 66.65 | 78.18 | 269 | 242.59 | 80.86 | 93.57 |
| 225 | 200.89 | 66.97 | 78.52 | 270 | 243.54 | 81.18 | 93.91 |

Table 2. CONSTANTS FOR p′ (LONG FORMULA), AN APPROXIMATION OF THE LOWER BINOMIAL CONFIDENCE LIMIT p̲; c/n ⩽ .5, n ⩾ 20, c ⩽ 1,000 (*Continued*)

| c | m | a | b | c | m | a | b |
|---|---|---|---|---|---|---|---|
| 271 | 244.49 | 81.50 | 94.25 | 316 | 287.33 | 95.78 | 109.61 |
| 272 | 245.44 | 81.81 | 94.59 | 317 | 288.28 | 96.09 | 109.95 |
| 273 | 246.39 | 82.13 | 94.93 | 318 | 289.24 | 96.41 | 110.29 |
| 274 | 247.34 | 82.45 | 95.28 | 319 | 290.19 | 96.73 | 110.64 |
| 275 | 248.29 | 82.76 | 95.62 | 320 | 291.14 | 97.05 | 110.98 |
| 276 | 249.24 | 83.08 | 95.96 | 321 | 292.10 | 97.37 | 111.32 |
| 277 | 250.19 | 83.40 | 96.30 | 322 | 293.05 | 97.68 | 111.66 |
| 278 | 251.14 | 83.71 | 96.64 | 323 | 294.01 | 98.00 | 112.00 |
| 279 | 252.09 | 84.03 | 96.98 | 324 | 294.96 | 98.32 | 112.34 |
| 280 | 253.04 | 84.35 | 97.33 | 325 | 295.91 | 98.64 | 112.68 |
| 281 | 253.99 | 84.66 | 97.67 | 326 | 296.87 | 98.96 | 113.02 |
| 282 | 254.95 | 84.98 | 98.01 | 327 | 297.82 | 99.27 | 113.36 |
| 283 | 255.90 | 85.30 | 98.35 | 328 | 298.78 | 99.59 | 113.70 |
| 284 | 256.85 | 85.62 | 98.69 | 329 | 299.73 | 99.91 | 114.04 |
| 285 | 257.80 | 85.93 | 99.03 | 330 | 300.69 | 100.23 | 114.39 |
| 286 | 258.75 | 86.25 | 99.37 | 331 | 301.64 | 100.55 | 114.73 |
| 287 | 259.70 | 86.57 | 99.72 | 332 | 302.60 | 100.87 | 115.07 |
| 288 | 260.65 | 86.88 | 100.06 | 333 | 303.55 | 101.18 | 115.41 |
| 289 | 261.61 | 87.20 | 100.40 | 334 | 304.51 | 101.50 | 115.75 |
| 290 | 262.56 | 87.52 | 100.74 | 335 | 305.46 | 101.82 | 116.09 |
| 291 | 263.51 | 87.84 | 101.08 | 336 | 306.42 | 102.14 | 116.43 |
| 292 | 264.46 | 88.15 | 101.42 | 337 | 307.37 | 102.46 | 116.77 |
| 293 | 265.41 | 88.47 | 101.76 | 338 | 308.33 | 102.78 | 117.11 |
| 294 | 266.36 | 88.79 | 102.11 | 339 | 309.28 | 103.09 | 117.45 |
| 295 | 267.32 | 89.11 | 102.45 | 340 | 310.24 | 103.41 | 117.79 |
| 296 | 268.27 | 89.42 | 102.79 | 341 | 311.19 | 103.73 | 118.13 |
| 297 | 269.22 | 89.74 | 103.13 | 342 | 312.15 | 104.05 | 118.48 |
| 298 | 270.17 | 90.06 | 103.47 | 343 | 313.10 | 104.37 | 118.82 |
| 299 | 271.13 | 90.38 | 103.81 | 344 | 314.06 | 104.69 | 119.16 |
| 300 | 272.08 | 90.69 | 104.15 | 345 | 315.02 | 105.01 | 119.50 |
| 301 | 273.03 | 91.01 | 104.49 | 346 | 315.97 | 105.32 | 119.84 |
| 302 | 273.98 | 91.33 | 104.84 | 347 | 316.93 | 105.64 | 120.18 |
| 303 | 274.94 | 91.65 | 105.18 | 348 | 317.88 | 105.96 | 120.52 |
| 304 | 275.89 | 91.96 | 105.52 | 349 | 318.84 | 106.28 | 120.86 |
| 305 | 276.84 | 92.28 | 105.86 | 350 | 319.80 | 106.60 | 121.20 |
| 306 | 277.79 | 92.60 | 106.20 | 351 | 320.75 | 106.92 | 121.54 |
| 307 | 278.75 | 92.92 | 106.54 | 352 | 321.71 | 107.24 | 121.88 |
| 308 | 279.70 | 93.23 | 106.88 | 353 | 322.66 | 107.55 | 122.22 |
| 309 | 280.65 | 93.55 | 107.22 | 354 | 323.62 | 107.87 | 122.56 |
| 310 | 281.61 | 93.87 | 107.57 | 355 | 324.58 | 108.19 | 122.90 |
| 311 | 282.56 | 94.19 | 107.91 | 356 | 325.53 | 108.51 | 123.24 |
| 312 | 283.51 | 94.50 | 108.25 | 357 | 326.49 | 108.83 | 123.59 |
| 313 | 284.47 | 94.82 | 108.59 | 358 | 327.45 | 109.15 | 123.93 |
| 314 | 285.42 | 95.14 | 108.93 | 359 | 328.40 | 109.47 | 124.27 |
| 315 | 286.37 | 95.46 | 109.27 | 360 | 329.36 | 109.79 | 124.61 |

**Table 2. CONSTANTS FOR p′ (LONG FORMULA), AN APPROXIMATION OF THE LOWER BINOMIAL CONFIDENCE LIMIT p; c/n ⩽ .5, n ⩾ 20, c ⩽ 1,000 (*Continued*)**

| c | m | a | b | c | m | a | b |
|---|---|---|---|---|---|---|---|
| 361 | 330.32 | 110.11 | 124.95 | 406 | 373.42 | 124.47 | 140.26 |
| 362 | 331.27 | 110.42 | 125.29 | 407 | 374.38 | 124.79 | 140.60 |
| 363 | 332.23 | 110.74 | 125.63 | 408 | 375.34 | 125.11 | 140.94 |
| 364 | 333.19 | 111.06 | 125.97 | 409 | 376.30 | 125.43 | 141.28 |
| 365 | 334.14 | 111.38 | 126.31 | 410 | 377.26 | 125.75 | 141.62 |
| 366 | 335.10 | 111.70 | 126.65 | 411 | 378.22 | 126.07 | 141.96 |
| 367 | 336.06 | 112.02 | 126.99 | 412 | 379.18 | 126.39 | 142.30 |
| 368 | 337.01 | 112.34 | 127.33 | 413 | 380.14 | 126.71 | 142.64 |
| 369 | 337.97 | 112.66 | 127.67 | 414 | 381.10 | 127.03 | 142.98 |
| 370 | 338.93 | 112.98 | 128.01 | 415 | 382.06 | 127.35 | 143.32 |
| 371 | 339.89 | 113.30 | 128.35 | 416 | 383.02 | 127.67 | 143.66 |
| 372 | 340.84 | 113.61 | 128.69 | 417 | 383.98 | 127.99 | 144.00 |
| 373 | 341.80 | 113.93 | 129.03 | 418 | 384.94 | 128.31 | 144.34 |
| 374 | 342.76 | 114.25 | 129.37 | 419 | 385.90 | 128.63 | 144.68 |
| 375 | 343.72 | 114.57 | 129.71 | 420 | 386.86 | 128.95 | 145.02 |
| 376 | 344.67 | 114.89 | 130.05 | 421 | 387.82 | 129.27 | 145.36 |
| 377 | 345.63 | 115.21 | 130.39 | 422 | 388.78 | 129.59 | 145.70 |
| 378 | 346.59 | 115.53 | 130.74 | 423 | 389.74 | 129.91 | 146.04 |
| 379 | 347.55 | 115.85 | 131.08 | 424 | 390.70 | 130.23 | 146.38 |
| 380 | 348.50 | 116.17 | 131.42 | 425 | 391.66 | 130.55 | 146.72 |
| 381 | 349.46 | 116.49 | 131.76 | 426 | 392.62 | 130.87 | 147.06 |
| 382 | 350.42 | 116.81 | 132.10 | 427 | 393.58 | 131.19 | 147.40 |
| 383 | 351.38 | 117.13 | 132.44 | 428 | 394.54 | 131.51 | 147.74 |
| 384 | 352.34 | 117.45 | 132.78 | 429 | 395.50 | 131.83 | 148.08 |
| 385 | 353.29 | 117.76 | 133.12 | 430 | 396.46 | 132.15 | 148.42 |
| 386 | 354.25 | 118.08 | 133.46 | 431 | 397.42 | 132.47 | 148.76 |
| 387 | 355.21 | 118.40 | 133.80 | 432 | 398.38 | 132.79 | 149.10 |
| 388 | 356.17 | 118.72 | 134.14 | 433 | 399.34 | 133.11 | 149.44 |
| 389 | 357.13 | 119.04 | 134.48 | 434 | 400.30 | 133.43 | 149.78 |
| 390 | 358.08 | 119.36 | 134.82 | 435 | 401.26 | 133.75 | 150.12 |
| 391 | 359.04 | 119.68 | 135.16 | 436 | 402.22 | 134.07 | 150.46 |
| 392 | 360.00 | 120.00 | 135.50 | 437 | 403.18 | 134.39 | 150.80 |
| 393 | 360.96 | 120.32 | 135.84 | 438 | 404.14 | 134.71 | 151.14 |
| 394 | 361.92 | 120.64 | 136.18 | 439 | 405.10 | 135.03 | 151.48 |
| 395 | 362.88 | 120.96 | 136.52 | 440 | 406.06 | 135.35 | 151.82 |
| 396 | 363.84 | 121.28 | 136.86 | 441 | 407.03 | 135.68 | 152.16 |
| 397 | 364.79 | 121.60 | 137.20 | 442 | 407.99 | 136.00 | 152.50 |
| 398 | 365.75 | 121.92 | 137.54 | 443 | 408.95 | 136.32 | 152.84 |
| 399 | 366.71 | 122.24 | 137.88 | 444 | 409.91 | 136.64 | 153.18 |
| 400 | 367.67 | 122.56 | 138.22 | 445 | 410.87 | 136.96 | 153.52 |
| 401 | 368.63 | 122.88 | 138.56 | 446 | 411.83 | 137.28 | 153.86 |
| 402 | 369.59 | 123.20 | 138.90 | 447 | 412.79 | 137.60 | 154.20 |
| 403 | 370.55 | 123.52 | 139.24 | 448 | 413.75 | 137.92 | 154.54 |
| 404 | 371.51 | 123.84 | 139.58 | 449 | 414.71 | 138.24 | 154.88 |
| 405 | 372.47 | 124.16 | 139.92 | 450 | 415.67 | 138.56 | 155.22 |

Table 2. CONSTANTS FOR $p'$ (LONG FORMULA), AN APPROXIMATION OF THE LOWER BINOMIAL CONFIDENCE LIMIT $\underline{p}$; $c/n \leqslant .5$, $n \geqslant 20$, $c \leqslant 1{,}000$ (*Continued*)

| c | m | a | b | c | m | a | b |
|---|---|---|---|---|---|---|---|
| 451 | 416.64 | 138.88 | 155.56 | 496 | 459.93 | 153.31 | 170.84 |
| 452 | 417.60 | 139.20 | 155.90 | 497 | 460.90 | 153.63 | 171.18 |
| 453 | 418.56 | 139.52 | 156.24 | 498 | 461.86 | 153.95 | 171.52 |
| 454 | 419.52 | 139.84 | 156.58 | 499 | 462.82 | 154.27 | 171.86 |
| 455 | 420.48 | 140.16 | 156.92 | 500 | 463.79 | 154.60 | 172.20 |
| 456 | 421.44 | 140.48 | 157.26 | 501 | 464.75 | 154.92 | 172.54 |
| 457 | 422.40 | 140.80 | 157.60 | 502 | 465.71 | 155.24 | 172.88 |
| 458 | 423.37 | 141.12 | 157.94 | 503 | 466.68 | 155.56 | 173.22 |
| 459 | 424.33 | 141.44 | 158.28 | 504 | 467.64 | 155.88 | 173.56 |
| 460 | 425.29 | 141.76 | 158.62 | 505 | 468.60 | 156.20 | 173.90 |
| 461 | 426.25 | 142.08 | 158.96 | 506 | 469.57 | 156.52 | 174.24 |
| 462 | 427.21 | 142.40 | 159.30 | 507 | 470.53 | 156.84 | 174.58 |
| 463 | 428.17 | 142.72 | 159.64 | 508 | 471.49 | 157.16 | 174.92 |
| 464 | 429.14 | 143.05 | 159.98 | 509 | 472.46 | 157.49 | 175.26 |
| 465 | 430.10 | 143.37 | 160.32 | 510 | 473.42 | 157.81 | 175.60 |
| 466 | 431.06 | 143.69 | 160.66 | 511 | 474.39 | 158.13 | 175.94 |
| 467 | 432.02 | 144.01 | 161.00 | 512 | 475.35 | 158.45 | 176.28 |
| 468 | 432.98 | 144.33 | 161.34 | 513 | 476.31 | 158.77 | 176.61 |
| 469 | 433.95 | 144.65 | 161.68 | 514 | 477.28 | 159.09 | 176.95 |
| 470 | 434.91 | 144.97 | 162.02 | 515 | 478.24 | 159.41 | 177.29 |
| 471 | 435.87 | 145.29 | 162.36 | 516 | 479.20 | 159.73 | 177.63 |
| 472 | 436.83 | 145.61 | 162.69 | 517 | 480.17 | 160.06 | 177.97 |
| 473 | 437.79 | 145.93 | 163.03 | 518 | 481.13 | 160.38 | 178.31 |
| 474 | 438.76 | 146.25 | 163.37 | 519 | 482.10 | 160.70 | 178.65 |
| 475 | 439.72 | 146.57 | 163.71 | 520 | 483.06 | 161.02 | 178.99 |
| 476 | 440.68 | 146.89 | 164.05 | 521 | 484.02 | 161.34 | 179.33 |
| 477 | 441.64 | 147.21 | 164.39 | 522 | 484.99 | 161.66 | 179.67 |
| 478 | 442.61 | 147.54 | 164.73 | 523 | 485.95 | 161.98 | 180.01 |
| 479 | 443.57 | 147.86 | 165.07 | 524 | 486.91 | 162.30 | 180.35 |
| 480 | 444.53 | 148.18 | 165.41 | 525 | 487.88 | 162.63 | 180.69 |
| 481 | 445.49 | 148.50 | 165.75 | 526 | 488.84 | 162.95 | 181.03 |
| 482 | 446.46 | 148.82 | 166.09 | 527 | 489.81 | 163.27 | 181.37 |
| 483 | 447.42 | 149.14 | 166.43 | 528 | 490.77 | 163.59 | 181.70 |
| 484 | 448.38 | 149.46 | 166.77 | 529 | 491.74 | 163.91 | 182.04 |
| 485 | 449.34 | 149.78 | 167.11 | 530 | 492.70 | 164.23 | 182.38 |
| 486 | 450.31 | 150.10 | 167.45 | 531 | 493.66 | 164.55 | 182.72 |
| 487 | 451.27 | 150.42 | 167.79 | 532 | 494.63 | 164.88 | 183.06 |
| 488 | 452.23 | 150.74 | 168.13 | 533 | 495.59 | 165.20 | 183.40 |
| 489 | 453.19 | 151.06 | 168.47 | 534 | 496.56 | 165.52 | 183.74 |
| 490 | 454.16 | 151.39 | 168.81 | 535 | 497.52 | 165.84 | 184.08 |
| 491 | 455.12 | 151.71 | 169.15 | 536 | 498.49 | 166.16 | 184.42 |
| 492 | 456.08 | 152.03 | 169.49 | 537 | 499.45 | 166.48 | 184.76 |
| 493 | 457.05 | 152.35 | 169.83 | 538 | 500.42 | 166.81 | 185.10 |
| 494 | 458.01 | 152.67 | 170.17 | 539 | 501.38 | 167.13 | 185.44 |
| 495 | 458.97 | 152.99 | 170.50 | 540 | 502.34 | 167.45 | 185.78 |

Table 2.  CONSTANTS FOR $p'$ (LONG FORMULA), AN APPROXIMATION OF
THE LOWER BINOMIAL CONFIDENCE LIMIT $\underline{p}$; $c/n \leqslant .5$, $n \geqslant 20$,
$c \leqslant 1,000$ (*Continued*)

Section 4

$\gamma_1 = 95\%$

$\gamma_2 = 90\%$

| c | $\underline{m}$ | $\underline{a}$ | $\underline{b}$ | c | $\underline{m}$ | $\underline{a}$ | $\underline{b}$ |
|---|---|---|---|---|---|---|---|
| 541 | 503.31 | 167.77 | 186.12 | 586 | 546.75 | 182.25 | 201.38 |
| 542 | 504.27 | 168.09 | 186.45 | 587 | 547.72 | 182.57 | 201.71 |
| 543 | 505.24 | 168.41 | 186.79 | 588 | 548.68 | 182.89 | 202.05 |
| 544 | 506.20 | 168.73 | 187.13 | 589 | 549.65 | 183.22 | 202.39 |
| 545 | 507.17 | 169.06 | 187.47 | 590 | 550.61 | 183.54 | 202.73 |
| 546 | 508.13 | 169.38 | 187.81 | 591 | 551.58 | 183.86 | 203.07 |
| 547 | 509.10 | 169.70 | 188.15 | 592 | 552.55 | 184.18 | 203.41 |
| 548 | 510.06 | 170.02 | 188.49 | 593 | 553.51 | 184.50 | 203.75 |
| 549 | 511.03 | 170.34 | 188.83 | 594 | 554.48 | 184.83 | 204.09 |
| 550 | 511.99 | 170.66 | 189.17 | 595 | 555.45 | 185.15 | 204.43 |
| 551 | 512.96 | 170.99 | 189.51 | 596 | 556.41 | 185.47 | 204.76 |
| 552 | 513.92 | 171.31 | 189.85 | 597 | 557.38· | 185.79 | 205.10 |
| 553 | 514.89 | 171.63 | 190.19 | 598 | 558.34 | 186.11 | 205.44 |
| 554 | 515.85 | 171.95 | 190.52 | 599 | 559.31 | 186.44 | 205.78 |
| 555 | 516.82 | 172.27 | 190.86 | 600 | 560.28 | 186.76 | 206.12 |
| 556 | 517.78 | 172.59 | 191.20 | 601 | 561.24 | 187.08 | 206.46 |
| 557 | 518.75 | 172.92 | 191.54 | 602 | 562.21 | 187.40 | 206.80 |
| 558 | 519.71 | 173.24 | 191.88 | 603 | 563.18 | 187.73 | 207.14 |
| 559 | 520.68 | 173.56 | 192.22 | 604 | 564.14 | 188.05 | 207.48 |
| 560 | 521.64 | 173.88 | 192.56 | 605 | 565.11 | 188.37 | 207.82 |
| 561 | 522.61 | 174.20 | 192.90 | 606 | 566.08 | 188.69 | 208.15 |
| 562 | 523.57 | 174.52 | 193.24 | 607 | 567.04 | 189.01 | 208.49 |
| 563 | 524.54 | 174.85 | 193.58 | 608 | 568.01 | 189.34 | 208.83 |
| 564 | 525.50 | 175.17 | 193.92 | 609 | 568.98 | 189.66 | 209.17 |
| 565 | 526.47 | 175.49 | 194.26 | 610 | 569.94 | 189.98 | 209.51 |
| 566 | 527.44 | 175.81 | 194.59 | 611 | 570.91 | 190.30 | 209.85 |
| 567 | 528.40 | 176.13 | 194.93 | 612 | 571.88 | 190.63 | 210.19 |
| 568 | 529.37 | 176.46 | 195.27 | 613 | 572.84 | 190.95 | 210.53 |
| 569 | 530.33 | 176.78 | 195.61 | 614 | 573.81 | 191.27 | 210.87 |
| 570 | 531.30 | 177.10 | 195.95 | 615 | 574.78 | 191.59 | 211.20 |
| 571 | 532.26 | 177.42 | 196.29 | 616 | 575.74 | 191.91 | 211.54 |
| 572 | 533.23 | 177.74 | 196.63 | 617 | 576.71 | 192.24 | 211.88 |
| 573 | 534.19 | 178.06 | 196.97 | 618 | 577.68 | 192.56 | 212.22 |
| 574 | 535.16 | 178.39 | 197.31 | 619 | 578.64 | 192.88 | 212.56 |
| 575 | 536.13 | 178.71 | 197.65 | 620 | 579.61 | 193.20 | 212.90 |
| 576 | 537.09 | 179.03 | 197.98 | 621 | 580.58 | 193.53 | 213.24 |
| 577 | 538.06 | 179.35 | 198.32 | 622 | 581.54 | 193.85 | 213.58 |
| 578 | 539.02 | 179.67 | 198.66 | 623 | 582.51 | 194.17 | 213.91 |
| 579 | 539.99 | 180.00 | 199.00 | 624 | 583.48 | 194.49 | 214.25 |
| 580 | 540.95 | 180.32 | 199.34 | 625 | 584.45 | 194.82 | 214.59 |
| 581 | 541.92 | 180.64 | 199.68 | 626 | 585.41 | 195.14 | 214.93 |
| 582 | 542.89 | 180.96 | 200.02 | 627 | 586.38 | 195.46 | 215.27 |
| 583 | 543.85 | 181.28 | 200.36 | 628 | 587.35 | 195.78 | 215.61 |
| 584 | 544.82 | 181.61 | 200.70 | 629 | 588.31 | 196.10 | 215.95 |
| 585 | 545.78 | 181.93 | 201.04 | 630 | 589.28 | 196.43 | 216.29 |

Table 2.  CONSTANTS FOR p′ (LONG FORMULA), AN APPROXIMATION OF THE LOWER BINOMIAL CONFIDENCE LIMIT p; c/n ⩽ .5, n ⩾ 20, c ⩽ 1,000 (*Continued*)

| c | m | a | b | c | m | a | b |
|---|---|---|---|---|---|---|---|
| 631 | 590.25 | 196.75 | 216.63 | 676 | 633.80 | 211.27 | 231.87 |
| 632 | 591.22 | 197.07 | 216.96 | 677 | 634.77 | 211.59 | 232.21 |
| 633 | 592.18 | 197.39 | 217.30 | 678 | 635.74 | 211.91 | 232.54 |
| 634 | 593.15 | 197.72 | 217.64 | 679 | 636.71 | 212.24 | 232.88 |
| 635 | 594.12 | 198.04 | 217.98 | 680 | 637.67 | 212.56 | 233.22 |
| 636 | 595.09 | 198.36 | 218.32 | 681 | 638.64 | 212.88 | 233.56 |
| 637 | 596.05 | 198.68 | 218.66 | 682 | 639.61 | 213.20 | 233.90 |
| 638 | 597.02 | 199.01 | 219.00 | 683 | 640.58 | 213.53 | 234.24 |
| 639 | 597.99 | 199.33 | 219.34 | 684 | 641.55 | 213.85 | 234.58 |
| 640 | 598.96 | 199.65 | 219.67 | 685 | 642.52 | 214.17 | 234.91 |
| 641 | 599.92 | 199.97 | 220.01 | 686 | 643.49 | 214.50 | 235.25 |
| 642 | 600.89 | 200.30 | 220.35 | 687 | 644.45 | 214.82 | 235.59 |
| 643 | 601.86 | 200.62 | 220.69 | 688 | 645.42 | 215.14 | 235.93 |
| 644 | 602.83 | 200.94 | 221.03 | 689 | 646.39 | 215.46 | 236.27 |
| 645 | 603.79 | 201.26 | 221.37 | 690 | 647.36 | 215.79 | 236.61 |
| 646 | 604.76 | 201.59 | 221.71 | 691 | 648.33 | 216.11 | 236.95 |
| 647 | 605.73 | 201.91 | 222.05 | 692 | 649.30 | 216.43 | 237.28 |
| 648 | 606.70 | 202.23 | 222.38 | 693 | 650.27 | 216.76 | 237.62 |
| 649 | 607.66 | 202.55 | 222.72 | 694 | 651.24 | 217.08 | 237.96 |
| 650 | 608.63 | 202.88 | 223.06 | 695 | 652.20 | 217.40 | 238.30 |
| 651 | 609.60 | 203.20 | 223.40 | 696 | 653.17 | 217.72 | 238.64 |
| 652 | 610.57 | 203.52 | 223.74 | 697 | 654.14 | 218.05 | 238.98 |
| 653 | 611.53 | 203.84 | 224.08 | 698 | 655.11 | 218.37 | 239.31 |
| 654 | 612.50 | 204.17 | 224.42 | 699 | 656.08 | 218.69 | 239.65 |
| 655 | 613.47 | 204.49 | 224.75 | 700 | 657.05 | 219.02 | 239.99 |
| 656 | 614.44 | 204.81 | 225.09 | 701 | 658.02 | 219.34 | 240.33 |
| 657 | 615.41 | 205.14 | 225.43 | 702 | 658.99 | 219.66 | 240.67 |
| 658 | 616.37 | 205.46 | 225.77 | 703 | 659.96 | 219.99 | 241.01 |
| 659 | 617.34 | 205.78 | 226.11 | 704 | 660.92 | 220.31 | 241.35 |
| 660 | 618.31 | 206.10 | 226.45 | 705 | 661.89 | 220.63 | 241.68 |
| 661 | 619.28 | 206.43 | 226.79 | 706 | 662.86 | 220.95 | 242.02 |
| 662 | 620.25 | 206.75 | 227.13 | 707 | 663.83 | 221.28 | 242.36 |
| 663 | 621.21 | 207.07 | 227.46 | 708 | 664.80 | 221.60 | 242.70 |
| 664 | 622.18 | 207.39 | 227.80 | 709 | 665.77 | 221.92 | 243.04 |
| 665 | 623.15 | 207.72 | 228.14 | 710 | 666.74 | 222.25 | 243.38 |
| 666 | 624.12 | 208.04 | 228.48 | 711 | 667.71 | 222.57 | 243.72 |
| 667 | 625.09 | 208.36 | 228.82 | 712 | 668.68 | 222.89 | 244.05 |
| 668 | 626.05 | 208.68 | 229.16 | 713 | 669.65 | 223.22 | 244.39 |
| 669 | 627.02 | 209.01 | 229.50 | 714 | 670.62 | 223.54 | 244.73 |
| 670 | 627.99 | 209.33 | 229.83 | 715 | 671.58 | 223.86 | 245.07 |
| 671 | 628.96 | 209.65 | 230.17 | 716 | 672.55 | 224.18 | 245.41 |
| 672 | 629.93 | 209.98 | 230.51 | 717 | 673.52 | 224.51 | 245.75 |
| 673 | 630.90 | 210.30 | 230.85 | 718 | 674.49 | 224.83 | 246.08 |
| 674 | 631.86 | 210.62 | 231.19 | 719 | 675.46 | 225.15 | 246.42 |
| 675 | 632.83 | 210.94 | 231.53 | 720 | 676.43 | 225.48 | 246.76 |

Table 2. CONSTANTS FOR $p'$ (LONG FORMULA), AN APPROXIMATION OF THE LOWER BINOMIAL CONFIDENCE LIMIT $\underline{p}$; $c/n \leqslant .5$, $n \geqslant 20$, $c \leqslant 1,000$ (*Continued*)

Section 4

$\gamma_1 = 95\%$
$\gamma_2 = 90\%$

| c | m | a | b | c | m | a | b |
|---|---|---|---|---|---|---|---|
| 721 | 677.40 | 225.80 | 247.10 | 766 | 721.04 | 240.35 | 262.33 |
| 722 | 678.37 | 226.12 | 247.44 | 767 | 722.01 | 240.67 | 262.66 |
| 723 | 679.34 | 226.45 | 247.78 | 768 | 722.98 | 240.99 | 263.00 |
| 724 | 680.31 | 226.77 | 248.12 | 769 | 723.95 | 241.32 | 263.34 |
| 725 | 681.28 | 227.09 | 248.45 | 770 | 724.92 | 241.64 | 263.68 |
| 726 | 682.25 | 227.42 | 248.79 | 771 | 725.89 | 241.96 | 264.02 |
| 727 | 683.22 | 227.74 | 249.13 | 772 | 726.87 | 242.29 | 264.36 |
| 728 | 684.19 | 228.06 | 249.47 | 773 | 727.84 | 242.61 | 264.69 |
| 729 | 685.16 | 228.39 | 249.81 | 774 | 728.81 | 242.94 | 265.03 |
| 730 | 686.13 | 228.71 | 250.15 | 775 | 729.78 | 243.26 | 265.37 |
| 731 | 687.10 | 229.03 | 250.48 | 776 | 730.75 | 243.58 | 265.71 |
| 732 | 688.06 | 229.35 | 250.82 | 777 | 731.72 | 243.91 | 266.05 |
| 733 | 689.03 | 229.68 | 251.16 | 778 | 732.69 | 244.23 | 266.39 |
| 734 | 690.00 | 230.00 | 251.50 | 779 | 733.66 | 244.55 | 266.72 |
| 735 | 690.97 | 230.32 | 251.84 | 780 | 734.63 | 244.88 | 267.06 |
| 736 | 691.94 | 230.65 | 252.18 | 781 | 735.60 | 245.20 | 267.40 |
| 737 | 692.91 | 230.97 | 252.51 | 782 | 736.57 | 245.52 | 267.74 |
| 738 | 693.88 | 231.29 | 252.85 | 783 | 737.54 | 245.85 | 268.08 |
| 739 | 694.85 | 231.62 | 253.19 | 784 | 738.51 | 246.17 | 268.41 |
| 740 | 695.82 | 231.94 | 253.53 | 785 | 739.48 | 246.49 | 268.75 |
| 741 | 696.79 | 232.26 | 253.87 | 786 | 740.45 | 246.82 | 269.09 |
| 742 | 697.76 | 232.59 | 254.21 | 787 | 741.42 | 247.14 | 269.43 |
| 743 | 698.73 | 232.91 | 254.54 | 788 | 742.39 | 247.46 | 269.77 |
| 744 | 699.70 | 233.23 | 254.88 | 789 | 743.36 | 247.79 | 270.11 |
| 745 | 700.67 | 233.56 | 255.22 | 790 | 744.34 | 248.11 | 270.44 |
| 746 | 701.64 | 233.88 | 255.56 | 791 | 745.31 | 248.44 | 270.78 |
| 747 | 702.61 | 234.20 | 255.90 | 792 | 746.28 | 248.76 | 271.12 |
| 748 | 703.58 | 234.53 | 256.24 | 793 | 747.25 | 249.08 | 271.46 |
| 749 | 704.55 | 234.85 | 256.57 | 794 | 748.22 | 249.41 | 271.80 |
| 750 | 705.52 | 235.17 | 256.91 | 795 | 749.19 | 249.73 | 272.14 |
| 751 | 706.49 | 235.50 | 257.25 | 796 | 750.16 | 250.05 | 272.47 |
| 752 | 707.46 | 235.82 | 257.59 | 797 | 751.13 | 250.38 | 272.81 |
| 753 | 708.43 | 236.14 | 257.93 | 798 | 752.10 | 250.70 | 273.15 |
| 754 | 709.40 | 236.47 | 258.27 | 799 | 753.07 | 251.02 | 273.49 |
| 755 | 710.37 | 236.79 | 258.60 | 800 | 754.04 | 251.35 | 273.83 |
| 756 | 711.34 | 237.11 | 258.94 | 801 | 755.01 | 251.67 | 274.16 |
| 757 | 712.31 | 237.44 | 259.28 | 802 | 755.99 | 252.00 | 274.50 |
| 758 | 713.28 | 237.76 | 259.62 | 803 | 756.96 | 252.32 | 274.84 |
| 759 | 714.25 | 238.08 | 259.96 | 804 | 757.93 | 252.64 | 275.18 |
| 760 | 715.22 | 238.41 | 260.30 | 805 | 758.90 | 252.97 | 275.52 |
| 761 | 716.19 | 238.73 | 260.63 | 806 | 759.87 | 253.29 | 275.86 |
| 762 | 717.16 | 239.05 | 260.97 | 807 | 760.84 | 253.61 | 276.19 |
| 763 | 718.13 | 239.38 | 261.31 | 808 | 761.81 | 253.94 | 276.53 |
| 764 | 719.10 | 239.70 | 261.65 | 809 | 762.78 | 254.26 | 276.87 |
| 765 | 720.07 | 240.02 | 261.99 | 810 | 763.75 | 254.58 | 277.21 |

Table 2. **CONSTANTS FOR p′ (LONG FORMULA), AN APPROXIMATION OF THE LOWER BINOMIAL CONFIDENCE LIMIT** p; c/n ⩽ .5, n ⩾ 20, c ⩽ 1,000 (*Continued*)

| c | m | a | b | c | m | a | b |
|---|---|---|---|---|---|---|---|
| 811 | 764.72 | 254.91 | 277.55 | 856 | 808.44 | 269.48 | 292.76 |
| 812 | 765.70 | 255.23 | 277.88 | 857 | 809.41 | 269.80 | 293.10 |
| 813 | 766.67 | 255.56 | 278.22 | 858 | 810.39 | 270.13 | 293.44 |
| 814 | 767.64 | 255.88 | 278.56 | 859 | 811.36 | 270.45 | 293.77 |
| 815 | 768.61 | 256.20 | 278.90 | 860 | 812.33 | 270.78 | 294.11 |
| 816 | 769.58 | 256.53 | 279.24 | 861 | 813.30 | 271.10 | 294.45 |
| 817 | 770.55 | 256.85 | 279.57 | 862 | 814.27 | 271.42 | 294.79 |
| 818 | 771.52 | 257.17 | 279.91 | 863 | 815.25 | 271.75 | 295.13 |
| 819 | 772.49 | 257.50 | 280.25 | 864 | 816.22 | 272.07 | 295.46 |
| 820 | 773.47 | 257.82 | 280.59 | 865 | 817.19 | 272.40 | 295.80 |
| 821 | 774.44 | 258.15 | 280.93 | 866 | 818.16 | 272.72 | 296.14 |
| 822 | 775.41 | 258.47 | 281.27 | 867 | 819.13 | 273.04 | 296.48 |
| 823 | 776.38 | 258.79 | 281.60 | 868 | 820.11 | 273.37 | 296.82 |
| 824 | 777.35 | 259.12 | 281.94 | 869 | 821.08 | 273.69 | 297.15 |
| 825 | 778.32 | 259.44 | 282.28 | 870 | 822.05 | 274.02 | 297.49 |
| 826 | 779.29 | 259.76 | 282.62 | 871 | 823.02 | 274.34 | 297.83 |
| 827 | 780.27 | 260.09 | 282.96 | 872 | 824.00 | 274.67 | 298.17 |
| 828 | 781.24 | 260.41 | 283.29 | 873 | 824.97 | 274.99 | 298.51 |
| 829 | 782.21 | 260.74 | 283.63 | 874 | 825.94 | 275.31 | 298.84 |
| 830 | 783.18 | 261.06 | 283.97 | 875 | 826.91 | 275.64 | 299.18 |
| 831 | 784.15 | 261.38 | 284.31 | 876 | 827.88 | 275.96 | 299.52 |
| 832 | 785.12 | 261.71 | 284.65 | 877 | 828.86 | 276.29 | 299.86 |
| 833 | 786.09 | 262.03 | 284.98 | 878 | 829.83 | 276.61 | 300.20 |
| 834 | 787.07 | 262.36 | 285.32 | 879 | 830.80 | 276.93 | 300.53 |
| 835 | 788.04 | 262.68 | 285.66 | 880 | 831.77 | 277.26 | 300.87 |
| 836 | 789.01 | 263.00 | 286.00 | 881 | 832.75 | 277.58 | 301.21 |
| 837 | 789.98 | 263.33 | 286.34 | 882 | 833.72 | 277.91 | 301.55 |
| 838 | 790.95 | 263.65 | 286.67 | 883 | 834.69 | 278.23 | 301.89 |
| 839 | 791.92 | 263.97 | 287.01 | 884 | 835.66 | 278.55 | 302.22 |
| 840 | 792.89 | 264.30 | 287.35 | 885 | 836.63 | 278.88 | 302.56 |
| 841 | 793.87 | 264.62 | 287.69 | 886 | 837.61 | 279.20 | 302.90 |
| 842 | 794.84 | 264.95 | 288.03 | 887 | 838.58 | 279.53 | 303.24 |
| 843 | 795.81 | 265.27 | 288.37 | 888 | 839.55 | 279.85 | 303.57 |
| 844 | 796.78 | 265.59 | 288.70 | 889 | 840.52 | 280.17 | 303.91 |
| 845 | 797.75 | 265.92 | 289.04 | 890 | 841.50 | 280.50 | 304.25 |
| 846 | 798.72 | 266.24 | 289.38 | 891 | 842.47 | 280.82 | 304.59 |
| 847 | 799.70 | 266.57 | 289.72 | 892 | 843.44 | 281.15 | 304.93 |
| 848 | 800.67 | 266.89 | 290.06 | 893 | 844.41 | 281.47 | 305.26 |
| 849 | 801.64 | 267.21 | 290.39 | 894 | 845.39 | 281.80 | 305.60 |
| 850 | 802.61 | 267.54 | 290.73 | 895 | 846.36 | 282.12 | 305.94 |
| 851 | 803.58 | 267.86 | 291.07 | 896 | 847.33 | 282.44 | 306.28 |
| 852 | 804.56 | 268.19 | 291.41 | 897 | 848.30 | 282.77 | 306.62 |
| 853 | 805.53 | 268.51 | 291.75 | 898 | 849.28 | 283.09 | 306.95 |
| 854 | 806.50 | 268.83 | 292.08 | 899 | 850.25 | 283.42 | 307.29 |
| 855 | 807.47 | 269.16 | 292.42 | 900 | 851.22 | 283.74 | 307.63 |

Table 2.  CONSTANTS FOR p′ (LONG FORMULA), AN APPROXIMATION OF
THE LOWER BINOMIAL CONFIDENCE LIMIT p; c/n ⩽ .5, n ⩾ 20,
c ⩽ 1,000 (*Continued*)

Section 4

$\gamma_1$ = 95%
$\gamma_2$ = 90%

| c | m | a | b | c | m | a | b |
|---|---|---|---|---|---|---|---|
| 901 | 852.19 | 284.06 | 307.97 | 946 | 895.98 | 298.66 | 323.17 |
| 902 | 853.17 | 284.39 | 308.31 | 947 | 896.95 | 298.98 | 323.51 |
| 903 | 854.14 | 284.71 | 308.64 | 948 | 897.92 | 299.31 | 323.85 |
| 904 | 855.11 | 285.04 | 308.98 | 949 | 898.90 | 299.63 | 324.18 |
| 905 | 856.08 | 285.36 | 309.32 | 950 | 899.87 | 299.96 | 324.52 |
| 906 | 857.06 | 285.69 | 309.66 | 951 | 900.84 | 300.28 | 324.86 |
| 907 | 858.03 | 286.01 | 310.00 | 952 | 901.82 | 300.61 | 325.20 |
| 908 | 859.00 | 286.33 | 310.33 | 953 | 902.79 | 300.93 | 325.54 |
| 909 | 859.98 | 286.66 | 310.67 | 954 | 903.76 | 301.25 | 325.87 |
| 910 | 860.95 | 286.98 | 311.01 | 955 | 904.74 | 301.58 | 326.21 |
| 911 | 861.92 | 287.31 | 311.35 | 956 | 905.71 | 301.90 | 326.55 |
| 912 | 862.89 | 287.63 | 311.68 | 957 | 906.68 | 302.23 | 326.89 |
| 913 | 863.87 | 287.96 | 312.02 | 958 | 907.66 | 302.55 | 327.22 |
| 914 | 864.84 | 288.28 | 312.36 | 959 | 908.63 | 302.88 | 327.56 |
| 915 | 865.81 | 288.60 | 312.70 | 960 | 909.60 | 303.20 | 327.90 |
| 916 | 866.78 | 288.93 | 313.04 | 961 | 910.58 | 303.53 | 328.24 |
| 917 | 867.76 | 289.25 | 313.37 | 962 | 911.55 | 303.85 | 328.58 |
| 918 | 868.73 | 289.58 | 313.71 | 963 | 912.52 | 304.17 | 328.91 |
| 919 | 869.70 | 289.90 | 314.05 | 964 | 913.50 | 304.50 | 329.25 |
| 920 | 870.68 | 290.23 | 314.39 | 965 | 914.47 | 304.82 | 329.59 |
| 921 | 871.65 | 290.55 | 314.73 | 966 | 915.44 | 305.15 | 329.93 |
| 922 | 872.62 | 290.87 | 315.06 | 967 | 916.42 | 305.47 | 330.26 |
| 923 | 873.59 | 291.20 | 315.40 | 968 | 917.39 | 305.80 | 330.60 |
| 924 | 874.57 | 291.52 | 315.74 | 969 | 918.36 | 306.12 | 330.94 |
| 925 | 875.54 | 291.85 | 316.08 | 970 | 919.34 | 306.45 | 331.28 |
| 926 | 876.51 | 292.17 | 316.41 | 971 | 920.31 | 306.77 | 331.61 |
| 927 | 877.49 | 292.50 | 316.75 | 972 | 921.29 | 307.10 | 331.95 |
| 928 | 878.46 | 292.82 | 317.09 | 973 | 922.26 | 307.42 | 332.29 |
| 929 | 879.43 | 293.14 | 317.43 | 974 | 923.23 | 307.74 | 332.63 |
| 930 | 880.41 | 293.47 | 317.77 | 975 | 924.21 | 308.07 | 332.97 |
| 931 | 881.38 | 293.79 | 318.10 | 976 | 925.18 | 308.39 | 333.30 |
| 932 | 882.35 | 294.12 | 318.44 | 977 | 926.15 | 308.72 | 333.64 |
| 933 | 883.32 | 294.44 | 318.78 | 978 | 927.13 | 309.04 | 333.98 |
| 934 | 884.30 | 294.77 | 319.12 | 979 | 928.10 | 309.37 | 334.32 |
| 935 | 885.27 | 295.09 | 319.45 | 980 | 929.07 | 309.69 | 334.65 |
| 936 | 886.24 | 295.41 | 319.79 | 981 | 930.05 | 310.02 | 334.99 |
| 937 | 887.22 | 295.74 | 320.13 | 982 | 931.02 | 310.34 | 335.33 |
| 938 | 888.19 | 296.06 | 320.47 | 983 | 932.00 | 310.67 | 335.67 |
| 939 | 889.16 | 296.39 | 320.81 | 984 | 932.97 | 310.99 | 336.01 |
| 940 | 890.14 | 296.71 | 321.14 | 985 | 933.94 | 311.31 | 336.34 |
| 941 | 891.11 | 297.04 | 321.48 | 986 | 934.92 | 311.64 | 336.68 |
| 942 | 892.08 | 297.36 | 321.82 | 987 | 935.89 | 311.96 | 337.02 |
| 943 | 893.06 | 297.69 | 322.16 | 988 | 936.87 | 312.29 | 337.36 |
| 944 | 894.03 | 298.01 | 322.50 | 989 | 937.84 | 312.61 | 337.69 |
| 945 | 895.00 | 298.33 | 322.83 | 990 | 938.81 | 312.94 | 338.03 |

Table 2.  CONSTANTS FOR $\underline{p}'$ (LONG FORMULA), AN APPROXIMATION OF THE LOWER BINOMIAL CONFIDENCE LIMIT $\underline{p}$; $c/n \leqslant .5$, $n \geqslant 20$, $c \leqslant 1,000$ (*Continued*)

| c | m | a | b | c | m | a | b |
|---|---|---|---|---|---|---|---|
| 991 | 939.79 | 313.26 | 338.37 | 996 | 944.66 | 314.89 | 340.06 |
| 992 | 940.76 | 313.59 | 338.71 | 997 | 945.63 | 315.21 | 340.39 |
| 993 | 941.73 | 313.91 | 339.04 | 998 | 946.60 | 315.53 | 340.73 |
| 994 | 942.71 | 314.24 | 339.38 | 999 | 947.58 | 315.86 | 341.07 |
| 995 | 943.68 | 314.56 | 339.72 | 1000 | 948.55 | 316.18 | 341.41 |

**Table 2.** **CONSTANTS FOR** $p'$ **(LONG FORMULA), AN APPROXIMATION OF THE LOWER BINOMIAL CONFIDENCE LIMIT** $\underline{p}$; $c/n \leqslant .5$, $n \geqslant 20$, $c \leqslant 1,000$ (*Continued*)

| c | m | a | b | c | m | a | b |
|---|---|---|---|---|---|---|---|
| 1 | 0.10536 | 0.053 | 0.0 | 46 | 37.550 | 12.881 | 16.606 |
| 2 | 0.53181 | 0.524 | 0.758 | 47 | 38.456 | 13.162 | 16.934 |
| 3 | 1.1021 | 0.731 | 1.180 | 48 | 39.363 | 13.468 | 17.287 |
| 4 | 1.7448 | 0.965 | 1.593 | 49 | 40.270 | 13.754 | 17.619 |
| 5 | 2.4326 | 1.233 | 2.016 | 50 | 41.179 | 14.055 | 17.965 |
| 6 | 3.1519 | 1.478 | 2.402 | 51 | 42.089 | 14.323 | 18.279 |
| 7 | 3.8948 | 1.753 | 2.805 | 52 | 42.999 | 14.597 | 18.598 |
| 8 | 4.6561 | 2.042 | 3.214 | 53 | 43.910 | 14.918 | 18.962 |
| 9 | 5.4325 | 2.351 | 3.635 | 54 | 44.823 | 15.190 | 19.279 |
| 10 | 6.2213 | 2.666 | 4.055 | 55 | 45.736 | 15.455 | 19.587 |
| 11 | 7.0208 | 2.946 | 4.436 | 56 | 46.649 | 15.755 | 19.930 |
| 12 | 7.8294 | 3.223 | 4.808 | 57 | 47.564 | 16.039 | 20.257 |
| 13 | 8.6460 | 3.501 | 5.178 | 58 | 48.479 | 16.354 | 20.614 |
| 14 | 9.4696 | 3.783 | 5.548 | 59 | 49.395 | 16.65 | 20.949 |
| 15 | 10.300 | 4.071 | 5.921 | 60 | 50.312 | 16.91 | 21.249 |
| 16 | 11.135 | 4.346 | 6.278 | 61 | 51.229 | 17.21 | 21.59 |
| 17 | 11.976 | 4.626 | 6.638 | 62 | 52.148 | 17.50 | 21.92 |
| 18 | 12.822 | 4.917 | 7.006 | 63 | 53.066 | 17.77 | 22.24 |
| 19 | 13.671 | 5.198 | 7.363 | 64 | 53.986 | 18.05 | 22.56 |
| 20 | 14.525 | 5.482 | 7.719 | 65 | 54.906 | 18.34 | 22.89 |
| 21 | 15.383 | 5.768 | 8.076 | 66 | 55.826 | 18.65 | 23.23 |
| 22 | 16.244 | 6.045 | 8.423 | 67 | 56.748 | 18.94 | 23.57 |
| 23 | 17.108 | 6.323 | 8.769 | 68 | 57.669 | 19.20 | 23.87 |
| 24 | 17.975 | 6.606 | 9.119 | 69 | 58.592 | 19.51 | 24.22 |
| 25 | 18.844 | 6.897 | 9.475 | 70 | 59.515 | 19.76 | 24.50 |
| 26 | 19.717 | 7.189 | 9.831 | 71 | 60.438 | 20.07 | 24.85 |
| 27 | 20.592 | 7.463 | 10.168 | 72 | 61.362 | 20.38 | 25.20 |
| 28 | 21.469 | 7.751 | 10.517 | 73 | 62.287 | 20.66 | 25.52 |
| 29 | 22.348 | 8.035 | 10.861 | 74 | 63.212 | 20.95 | 25.85 |
| 30 | 23.229 | 8.319 | 11.204 | 75 | 64.138 | 21.23 | 26.16 |
| 31 | 24.113 | 8.609 | 11.553 | 76 | 65.064 | 21.50 | 26.46 |
| 32 | 24.998 | 8.889 | 11.890 | 77 | 65.990 | 21.82 | 26.82 |
| 33 | 25.885 | 9.169 | 12.227 | 78 | 66.918 | 22.10 | 27.14 |
| 34 | 26.774 | 9.462 | 12.575 | 79 | 67.845 | 22.38 | 27.45 |
| 35 | 27.664 | 9.754 | 12.922 | 80 | 68.773 | 22.63 | 27.74 |
| 36 | 28.556 | 10.035 | 13.257 | 81 | 69.702 | 22.93 | 28.08 |
| 37 | 29.450 | 10.320 | 13.595 | 82 | 70.630 | 23.23 | 28.41 |
| 38 | 30.345 | 10.602 | 13.930 | 83 | 71.560 | 23.55 | 28.77 |
| 39 | 31.241 | 10.898 | 14.278 | 84 | 72.490 | 23.84 | 29.10 |
| 40 | 32.139 | 11.172 | 14.602 | 85 | 73.420 | 24.11 | 29.41 |
| 41 | 33.038 | 11.475 | 14.956 | 86 | 74.350 | 24.43 | 29.76 |
| 42 | 33.938 | 11.736 | 15.267 | 87 | 75.281 | 24.71 | 30.07 |
| 43 | 34.839 | 12.048 | 15.628 | 88 | 76.213 | 24.94 | 30.34 |
| 44 | 35.742 | 12.337 | 15.966 | 89 | 77.144 | 25.24 | 30.67 |
| 45 | 36.646 | 12.592 | 16.269 | 90 | 78.077 | 25.56 | 31.02 |

Table 2.  CONSTANTS FOR $p'$ (LONG FORMULA), AN APPROXIMATION OF THE LOWER BINOMIAL CONFIDENCE LIMIT $\underline{p}$; $c/n \leqslant .5$, $n \geqslant 20$, $c \leqslant 1{,}000$ (*Continued*)

| c | m | a | b | c | m | a | b |
|---|---|---|---|---|---|---|---|
| 91 | 79.009 | 25.84 | 31.34 | 136 | 121.29 | 40.42 | 47.29 |
| 92 | 79.942 | 26.07 | 31.60 | 137 | 122.23 | 40.74 | 47.63 |
| 93 | 80.875 | 26.46 | 32.03 | 138 | 123.18 | 41.05 | 47.97 |
| 94 | 81.809 | 26.70 | 32.29 | 139 | 124.12 | 41.37 | 48.32 |
| 95 | 82.743 | 26.97 | 32.60 | 140 | 125.07 | 41.68 | 48.66 |
| 96 | 83.677 | 27.28 | 32.94 | 141 | 126.01 | 42.00 | 49.00 |
| 97 | 84.612 | 27.54 | 33.24 | 142 | 126.96 | 42.31 | 49.34 |
| 98 | 85.547 | 27.84 | 33.57 | 143 | 127.91 | 42.63 | 49.69 |
| 99 | 86.482 | 28.18 | 33.94 | 144 | 128.85 | 42.94 | 50.03 |
| 100 | 87.418 | 28.47 | 34.26 | 145 | 129.80 | 43.26 | 50.37 |
| 101 | 88.354 | 29.44 | 35.28 | 146 | 130.75 | 43.58 | 50.71 |
| 102 | 89.290 | 29.76 | 35.62 | 147 | 131.69 | 43.89 | 51.05 |
| 103 | 90.227 | 30.07 | 35.97 | 148 | 132.64 | 44.21 | 51.40 |
| 104 | 91.164 | 30.38 | 36.31 | 149 | 133.59 | 44.52 | 51.74 |
| 105 | 92.101 | 30.69 | 36.65 | 150 | 134.53 | 44.84 | 52.08 |
| 106 | 93.038 | 31.01 | 37.00 | 151 | 135.48 | 45.16 | 52.42 |
| 107 | 93.976 | 31.32 | 37.34 | 152 | 136.43 | 45.47 | 52.76 |
| 108 | 94.914 | 31.63 | 37.68 | 153 | 137.38 | 45.79 | 53.11 |
| 109 | 95.853 | 31.94 | 38.03 | 154 | 138.33 | 46.10 | 53.45 |
| 110 | 96.792 | 32.26 | 38.37 | 155 | 139.27 | 46.42 | 53.79 |
| 111 | 97.730 | 32.57 | 38.71 | 156 | 140.22 | 46.74 | 54.13 |
| 112 | 98.670 | 32.88 | 39.06 | 157 | 141.17 | 47.05 | 54.47 |
| 113 | 99.609 | 33.20 | 39.40 | 158 | 142.12 | 47.37 | 54.82 |
| 114 | 100.55 | 33.51 | 39.74 | 159 | 143.07 | 47.68 | 55.16 |
| 115 | 101.49 | 33.82 | 40.09 | 160 | 144.02 | 48.00 | 55.50 |
| 116 | 102.43 | 34.14 | 40.43 | 161 | 144.95 | 48.32 | 55.84 |
| 117 | 103.37 | 34.45 | 40.77 | 162 | 145.90 | 48.63 | 56.18 |
| 118 | 104.31 | 34.76 | 41.12 | 163 | 146.85 | 48.95 | 56.52 |
| 119 | 105.25 | 35.08 | 41.46 | 164 | 147.80 | 49.27 | 56.87 |
| 120 | 106.19 | 35.39 | 41.80 | 165 | 148.75 | 49.58 | 57.21 |
| 121 | 107.13 | 35.71 | 42.15 | 166 | 149.70 | 49.90 | 57.55 |
| 122 | 108.08 | 36.02 | 42.49 | 167 | 150.65 | 50.22 | 57.89 |
| 123 | 109.02 | 36.33 | 42.83 | 168 | 151.60 | 50.53 | 58.23 |
| 124 | 109.96 | 36.65 | 43.18 | 169 | 152.55 | 50.85 | 58.57 |
| 125 | 110.90 | 36.96 | 43.52 | 170 | 153.50 | 51.17 | 58.92 |
| 126 | 111.85 | 37.28 | 43.86 | 171 | 154.45 | 51.48 | 59.26 |
| 127 | 112.79 | 37.59 | 44.20 | 172 | 155.41 | 51.80 | 59.60 |
| 128 | 113.73 | 37.90 | 44.55 | 173 | 156.36 | 52.12 | 59.94 |
| 129 | 114.68 | 38.22 | 44.89 | 174 | 157.31 | 52.44 | 60.28 |
| 130 | 115.62 | 38.53 | 45.23 | 175 | 158.26 | 52.75 | 60.62 |
| 131 | 116.56 | 38.85 | 45.58 | 176 | 159.21 | 53.07 | 60.96 |
| 132 | 117.51 | 39.16 | 45.92 | 177 | 160.16 | 53.39 | 61.31 |
| 133 | 118.45 | 39.48 | 46.26 | 178 | 161.12 | 53.71 | 61.65 |
| 134 | 119.40 | 39.79 | 46.60 | 179 | 162.07 | 54.02 | 61.99 |
| 135 | 120.34 | 40.11 | 46.95 | 180 | 163.02 | 54.34 | 62.33 |

Table 2. **CONSTANTS FOR $p'$ (LONG FORMULA), AN APPROXIMATION OF THE LOWER BINOMIAL CONFIDENCE LIMIT $\underline{p}$; $c/n \leqslant .5$, $n \geqslant 20$, $c \leqslant 1{,}000$ (*Continued*)**

| c | m | a | b | c | m | a | b |
|---|---|---|---|---|---|---|---|
| 181 | 163.97 | 54.66 | 62.67 | 226 | 206.95 | 68.98 | 78.01 |
| 182 | 164.92 | 54.97 | 63.01 | 227 | 207.90 | 69.30 | 78.35 |
| 183 | 165.88 | 55.29 | 63.35 | 228 | 208.86 | 69.62 | 78.69 |
| 184 | 166.83 | 55.61 | 63.70 | 229 | 209.82 | 69.94 | 79.03 |
| 185 | 167.78 | 55.93 | 64.04 | 230 | 210.78 | 70.26 | 79.37 |
| 186 | 168.74 | 56.25 | 64.38 | 231 | 211.74 | 70.58 | 79.71 |
| 187 | 169.69 | 56.56 | 64.72 | 232 | 212.69 | 70.90 | 80.05 |
| 188 | 170.64 | 56.88 | 65.06 | 233 | 213.65 | 71.22 | 80.39 |
| 189 | 171.60 | 57.20 | 65.40 | 234 | 214.61 | 71.54 | 80.73 |
| 190 | 172.55 | 57.52 | 65.74 | 235 | 215.57 | 71.86 | 81.07 |
| 191 | 173.50 | 57.83 | 66.08 | 236 | 216.53 | 72.18 | 81.41 |
| 192 | 174.46 | 58.15 | 66.42 | 237 | 217.48 | 72.49 | 81.75 |
| 193 | 175.41 | 58.47 | 66.77 | 238 | 218.44 | 72.81 | 82.09 |
| 194 | 176.36 | 58.79 | 67.11 | 239 | 219.40 | 73.13 | 82.43 |
| 195 | 177.32 | 59.11 | 67.45 | 240 | 220.36 | 73.45 | 82.77 |
| 196 | 178.27 | 59.42 | 67.79 | 241 | 221.32 | 73.77 | 83.11 |
| 197 | 179.23 | 59.74 | 68.13 | 242 | 222.28 | 74.09 | 83.45 |
| 198 | 180.18 | 60.06 | 68.47 | 243 | 223.24 | 74.41 | 83.79 |
| 199 | 181.13 | 60.38 | 68.81 | 244 | 224.19 | 74.73 | 84.13 |
| 200 | 182.09 | 60.70 | 69.15 | 245 | 225.15 | 75.05 | 84.47 |
| 201 | 183.04 | 61.01 | 69.49 | 246 | 226.11 | 75.37 | 84.81 |
| 202 | 184.00 | 61.33 | 69.83 | 247 | 227.07 | 75.69 | 85.15 |
| 203 | 184.95 | 61.65 | 70.17 | 248 | 228.03 | 76.01 | 85.49 |
| 204 | 185.91 | 61.97 | 70.52 | 249 | 228.99 | 76.33 | 85.83 |
| 205 | 186.86 | 62.29 | 70.86 | 250 | 229.95 | 76.65 | 86.17 |
| 206 | 187.82 | 62.61 | 71.20 | 251 | 230.91 | 76.97 | 86.52 |
| 207 | 188.78 | 62.93 | 71.54 | 252 | 231.87 | 77.29 | 86.86 |
| 208 | 189.73 | 63.24 | 71.88 | 253 | 232.83 | 77.61 | 87.20 |
| 209 | 190.69 | 63.56 | 72.22 | 254 | 233.79 | 77.93 | 87.54 |
| 210 | 191.64 | 63.88 | 72.56 | 255 | 234.75 | 78.25 | 87.88 |
| 211 | 192.60 | 64.20 | 72.90 | 256 | 235.71 | 78.57 | 88.22 |
| 212 | 193.55 | 64.52 | 73.24 | 257 | 236.67 | 78.89 | 88.56 |
| 213 | 194.51 | 64.84 | 73.58 | 258 | 237.63 | 79.21 | 88.90 |
| 214 | 195.47 | 65.16 | 73.92 | 259 | 238.59 | 79.53 | 89.24 |
| 215 | 196.42 | 65.47 | 74.26 | 260 | 239.55 | 79.85 | 89.58 |
| 216 | 197.38 | 65.79 | 74.60 | 261 | 240.51 | 80.17 | 89.92 |
| 217 | 198.33 | 66.11 | 74.94 | 262 | 241.47 | 80.49 | 90.26 |
| 218 | 199.29 | 66.43 | 75.28 | 263 | 242.43 | 80.81 | 90.60 |
| 219 | 200.25 | 66.75 | 75.63 | 264 | 243.39 | 81.13 | 90.93 |
| 220 | 201.20 | 67.07 | 75.97 | 265 | 244.35 | 81.45 | 91.27 |
| 221 | 202.16 | 67.39 | 76.31 | 266 | 245.31 | 81.77 | 91.61 |
| 222 | 203.12 | 67.71 | 76.65 | 267 | 246.27 | 82.09 | 91.95 |
| 223 | 204.08 | 68.03 | 76.99 | 268 | 247.23 | 82.41 | 92.29 |
| 224 | 205.03 | 68.34 | 77.33 | 269 | 248.19 | 82.73 | 92.63 |
| 225 | 205.99 | 68.66 | 77.67 | 270 | 249.16 | 83.05 | 92.97 |

Table 2. CONSTANTS FOR $p'$ (LONG FORMULA), AN APPROXIMATION OF THE LOWER BINOMIAL CONFIDENCE LIMIT $\underline{p}$; $c/n \leqslant .5$, $n \geqslant 20$, $c \leqslant 1,000$ (*Continued*)

| c | m | a | b | c | m | a | b |
|---|---|---|---|---|---|---|---|
| 271 | 250.12 | 83.37 | 93.31 | 316 | 293.43 | 97.81 | 108.59 |
| 272 | 251.08 | 83.69 | 93.65 | 317 | 294.40 | 98.13 | 108.93 |
| 273 | 252.04 | 84.01 | 93.99 | 318 | 295.36 | 98.45 | 109.27 |
| 274 | 253.00 | 84.33 | 94.33 | 319 | 296.32 | 98.77 | 109.61 |
| 275 | 253.96 | 84.65 | 94.67 | 320 | 297.29 | 99.10 | 109.95 |
| 276 | 254.92 | 84.97 | 95.01 | 321 | 298.25 | 99.42 | 110.29 |
| 277 | 255.88 | 85.29 | 95.35 | 322 | 299.22 | 99.74 | 110.63 |
| 278 | 256.85 | 85.62 | 95.69 | 323 | 300.18 | 100.06 | 110.97 |
| 279 | 257.81 | 85.94 | 96.03 | 324 | 301.15 | 100.38 | 111.31 |
| 280 | 258.77 | 86.26 | 96.37 | 325 | 302.11 | 100.70 | 111.65 |
| 281 | 259.73 | 86.58 | 96.71 | 326 | 303.07 | 101.02 | 111.99 |
| 282 | 260.69 | 86.90 | 97.05 | 327 | 304.04 | 101.35 | 112.33 |
| 283 | 261.65 | 87.22 | 97.39 | 328 | 305.00 | 101.67 | 112.67 |
| 284 | 262.62 | 87.54 | 97.73 | 329 | 305.97 | 101.99 | 113.01 |
| 285 | 263.58 | 87.86 | 98.07 | 330 | 306.93 | 102.31 | 113.34 |
| 286 | 264.54 | 88.18 | 98.41 | 331 | 307.90 | 102.63 | 113.68 |
| 287 | 265.50 | 88.50 | 98.75 | 332 | 308.86 | 102.95 | 114.02 |
| 288 | 266.46 | 88.82 | 99.09 | 333 | 309.83 | 103.28 | 114.36 |
| 289 | 267.43 | 89.14 | 99.43 | 334 | 310.79 | 103.60 | 114.70 |
| 290 | 268.39 | 89.46 | 99.77 | 335 | 311.76 | 103.92 | 115.04 |
| 291 | 269.35 | 89.78 | 100.11 | 336 | 312.72 | 104.24 | 115.38 |
| 292 | 270.31 | 90.10 | 100.45 | 337 | 313.69 | 104.56 | 115.72 |
| 293 | 271.28 | 90.43 | 100.79 | 338 | 314.65 | 104.88 | 116.06 |
| 294 | 272.24 | 90.75 | 101.13 | 339 | 315.62 | 105.21 | 116.40 |
| 295 | 273.20 | 91.07 | 101.47 | 340 | 316.58 | 105.53 | 116.74 |
| 296 | 274.16 | 91.39 | 101.81 | 341 | 317.55 | 105.85 | 117.08 |
| 297 | 275.13 | 91.71 | 102.15 | 342 | 318.51 | 106.17 | 117.41 |
| 298 | 276.09 | 92.03 | 102.48 | 343 | 319.48 | 106.49 | 117.75 |
| 299 | 277.05 | 92.35 | 102.82 | 344 | 320.44 | 106.81 | 118.09 |
| 300 | 278.02 | 92.67 | 103.16 | 345 | 321.41 | 107.14 | 118.43 |
| 301 | 278.98 | 92.99 | 103.50 | 346 | 322.37 | 107.46 | 118.77 |
| 302 | 279.94 | 93.31 | 103.84 | 347 | 323.34 | 107.78 | 119.11 |
| 303 | 280.91 | 93.64 | 104.18 | 348 | 324.31 | 108.10 | 119.45 |
| 304 | 281.87 | 93.96 | 104.52 | 349 | 325.27 | 108.42 | 119.79 |
| 305 | 282.83 | 94.28 | 104.86 | 350 | 326.24 | 108.75 | 120.13 |
| 306 | 283.80 | 94.60 | 105.20 | 351 | 327.20 | 109.07 | 120.47 |
| 307 | 284.76 | 94.92 | 105.54 | 352 | 328.17 | 109.39 | 120.81 |
| 308 | 285.72 | 95.24 | 105.88 | 353 | 329.14 | 109.71 | 121.14 |
| 309 | 286.69 | 95.56 | 106.22 | 354 | 330.10 | 110.03 | 121.48 |
| 310 | 287.65 | 95.88 | 106.56 | 355 | 331.07 | 110.36 | 121.82 |
| 311 | 288.61 | 96.20 | 106.90 | 356 | 332.03 | 110.68 | 122.16 |
| 312 | 289.58 | 96.53 | 107.24 | 357 | 333.00 | 111.00 | 122.50 |
| 313 | 290.54 | 96.85 | 107.58 | 358 | 333.97 | 111.32 | 122.84 |
| 314 | 291.50 | 97.17 | 107.92 | 359 | 334.93 | 111.64 | 123.18 |
| 315 | 292.47 | 97.49 | 108.26 | 360 | 335.90 | 111.97 | 123.52 |

Table 2. CONSTANTS FOR p′ (LONG FORMULA), AN APPROXIMATION OF
THE LOWER BINOMIAL CONFIDENCE LIMIT p; c/n ≤ .5, n ≥ 20,
c ≤ 1,000 (*Continued*)

| c | m | a | b | c | m | a | b |
|---|---|---|---|---|---|---|---|
| 361 | 336.86 | 112.29 | 123.86 | 406 | 380.39 | 126.80 | 139.10 |
| 362 | 337.83 | 112.61 | 124.20 | 407 | 381.36 | 127.12 | 139.44 |
| 363 | 338.80 | 112.93 | 124.53 | 408 | 382.33 | 127.44 | 139.78 |
| 364 | 339.76 | 113.25 | 124.87 | 409 | 383.30 | 127.77 | 140.12 |
| 365 | 340.73 | 113.58 | 125.21 | 410 | 384.26 | 128.09 | 140.46 |
| 366 | 341.70 | 113.90 | 125.55 | 411 | 385.23 | 128.41 | 140.79 |
| 367 | 342.66 | 114.22 | 125.89 | 412 | 386.20 | 128.73 | 141.13 |
| 368 | 343.63 | 114.54 | 126.23 | 413 | 387.17 | 129.06 | 141.47 |
| 369 | 344.60 | 114.87 | 126.57 | 414 | 388.14 | 129.38 | 141.81 |
| 370 | 345.56 | 115.19 | 126.91 | 415 | 389.11 | 129.70 | 142.15 |
| 371 | 346.53 | 115.51 | 127.25 | 416 | 390.07 | 130.02 | 142.49 |
| 372 | 347.50 | 115.83 | 127.58 | 417 | 391.04 | 130.35 | 142.83 |
| 373 | 348.46 | 116.15 | 127.92 | 418 | 392.01 | 130.67 | 143.16 |
| 374 | 349.43 | 116.48 | 128.26 | 419 | 392.98 | 130.99 | 143.50 |
| 375 | 350.40 | 116.80 | 128.60 | 420 | 393.95 | 131.32 | 143.84 |
| 376 | 351.36 | 117.12 | 128.94 | 421 | 394.92 | 131.64 | 144.18 |
| 377 | 352.33 | 117.44 | 129.28 | 422 | 395.89 | 131.96 | 144.52 |
| 378 | 353.30 | 117.77 | 129.62 | 423 | 396.86 | 132.29 | 144.86 |
| 379 | 354.26 | 118.09 | 129.96 | 424 | 397.82 | 132.61 | 145.20 |
| 380 | 355.23 | 118.41 | 130.29 | 425 | 398.79 | 132.93 | 145.53 |
| 381 | 356.20 | 118.73 | 130.63 | 426 | 399.76 | 133.25 | 145.87 |
| 382 | 357.17 | 119.06 | 130.97 | 427 | 400.73 | 133.58 | 146.21 |
| 383 | 358.13 | 119.38 | 131.31 | 428 | 401.70 | 133.90 | 146.55 |
| 384 | 359.10 | 119.70 | 131.65 | 429 | 402.67 | 134.22 | 146.89 |
| 385 | 360.07 | 120.02 | 131.99 | 430 | 403.64 | 134.55 | 147.23 |
| 386 | 361.03 | 120.34 | 132.33 | 431 | 404.61 | 134.87 | 147.57 |
| 387 | 362.00 | 120.67 | 132.67 | 432 | 405.58 | 135.19 | 147.90 |
| 388 | 362.97 | 120.99 | 133.01 | 433 | 406.55 | 135.52 | 148.24 |
| 389 | 363.94 | 121.31 | 133.34 | 434 | 407.51 | 135.84 | 148.58 |
| 390 | 364.90 | 121.63 | 133.68 | 435 | 408.48 | 136.16 | 148.92 |
| 391 | 365.87 | 121.96 | 134.02 | 436 | 409.45 | 136.48 | 149.26 |
| 392 | 366.84 | 122.28 | 134.36 | 437 | 410.42 | 136.81 | 149.60 |
| 393 | 367.81 | 122.60 | 134.70 | 438 | 411.39 | 137.13 | 149.93 |
| 394 | 368.78 | 122.93 | 135.04 | 439 | 412.36 | 137.45 | 150.27 |
| 395 | 369.74 | 123.25 | 135.38 | 440 | 413.33 | 137.78 | 150.61 |
| 396 | 370.71 | 123.57 | 135.71 | 441 | 414.30 | 138.10 | 150.95 |
| 397 | 371.68 | 123.89 | 136.05 | 442 | 415.27 | 138.42 | 151.29 |
| 398 | 372.65 | 124.22 | 136.39 | 443 | 416.24 | 138.75 | 151.63 |
| 399 | 373.61 | 124.54 | 136.73 | 444 | 417.21 | 139.07 | 151.97 |
| 400 | 374.58 | 124.86 | 137.07 | 445 | 418.18 | 139.39 | 152.30 |
| 401 | 375.55 | 125.18 | 137.41 | 446 | 419.15 | 139.72 | 152.64 |
| 402 | 376.52 | 125.51 | 137.75 | 447 | 420.12 | 140.04 | 152.98 |
| 403 | 377.49 | 125.83 | 138.09 | 448 | 421.09 | 140.36 | 153.32 |
| 404 | 378.45 | 126.15 | 138.42 | 449 | 422.06 | 140.69 | 153.66 |
| 405 | 379.42 | 126.47 | 138.76 | 450 | 423.03 | 141.01 | 154.00 |

Table 2. CONSTANTS FOR $p'$ (LONG FORMULA), AN APPROXIMATION OF THE LOWER BINOMIAL CONFIDENCE LIMIT $\underline{p}$; $c/n \leqslant .5$, $n \geqslant 20$, $c \leqslant 1,000$ (*Continued*)

| c | m | a | b | c | m | a | b |
|---|---|---|---|---|---|---|---|
| 451 | 424.00 | 141.33 | 154.33 | 496 | 467.67 | 155.89 | 169.55 |
| 452 | 424.97 | 141.66 | 154.67 | 497 | 468.64 | 156.21 | 169.89 |
| 453 | 425.94 | 141.98 | 155.01 | 498 | 469.61 | 156.54 | 170.23 |
| 454 | 426.91 | 142.30 | 155.35 | 499 | 470.59 | 156.86 | 170.57 |
| 455 | 427.88 | 142.63 | 155.69 | 500 | 471.56 | 157.19 | 170.91 |
| 456 | 428.85 | 142.95 | 156.03 | 501 | 472.53 | 157.51 | 171.25 |
| 457 | 429.82 | 143.27 | 156.36 | 502 | 473.50 | 157.83 | 171.58 |
| 458 | 430.79 | 143.60 | 156.70 | 503 | 474.47 | 158.16 | 171.92 |
| 459 | 431.76 | 143.92 | 157.04 | 504 | 475.44 | 158.48 | 172.26 |
| 460 | 432.73 | 144.24 | 157.38 | 505 | 476.41 | 158.80 | 172.60 |
| 461 | 433.70 | 144.57 | 157.72 | 506 | 477.39 | 159.13 | 172.94 |
| 462 | 434.67 | 144.89 | 158.06 | 507 | 478.36 | 159.45 | 173.27 |
| 463 | 435.64 | 145.21 | 158.39 | 508 | 479.33 | 159.78 | 173.61 |
| 464 | 436.61 | 145.54 | 158.73 | 509 | 480.30 | 160.10 | 173.95 |
| 465 | 437.58 | 145.86 | 159.07 | 510 | 481.27 | 160.42 | 174.29 |
| 466 | 438.55 | 146.18 | 159.41 | 511 | 482.24 | 160.75 | 174.63 |
| 467 | 439.52 | 146.51 | 159.75 | 512 | 483.21 | 161.07 | 174.96 |
| 468 | 440.49 | 146.83 | 160.09 | 513 | 484.19 | 161.40 | 175.30 |
| 469 | 441.46 | 147.15 | 160.42 | 514 | 485.16 | 161.72 | 175.64 |
| 470 | 442.43 | 147.48 | 160.76 | 515 | 486.13 | 162.04 | 175.98 |
| 471 | 443.40 | 147.80 | 161.10 | 516 | 487.10 | 162.37 | 176.32 |
| 472 | 444.37 | 148.12 | 161.44 | 517 | 488.07 | 162.69 | 176.65 |
| 473 | 445.34 | 148.45 | 161.78 | 518 | 489.05 | 163.02 | 176.99 |
| 474 | 446.31 | 148.77 | 162.11 | 519 | 490.02 | 163.34 | 177.33 |
| 475 | 447.28 | 149.09 | 162.45 | 520 | 490.99 | 163.66 | 177.67 |
| 476 | 448.25 | 149.42 | 162.79 | 521 | 491.96 | 163.99 | 178.01 |
| 477 | 449.22 | 149.74 | 163.13 | 522 | 492.93 | 164.31 | 178.34 |
| 478 | 450.19 | 150.06 | 163.47 | 523 | 493.90 | 164.63 | 178.68 |
| 479 | 451.16 | 150.39 | 163.81 | 524 | 494.88 | 164.96 | 179.02 |
| 480 | 452.14 | 150.71 | 164.14 | 525 | 495.85 | 165.28 | 179.36 |
| 481 | 453.11 | 151.04 | 164.48 | 526 | 496.82 | 165.61 | 179.70 |
| 482 | 454.08 | 151.36 | 164.82 | 527 | 497.79 | 165.93 | 180.03 |
| 483 | 455.05 | 151.68 | 165.16 | 528 | 498.77 | 166.26 | 180.37 |
| 484 | 456.02 | 152.01 | 165.50 | 529 | 499.74 | 166.58 | 180.71 |
| 485 | 456.99 | 152.33 | 165.84 | 530 | 500.71 | 166.90 | 181.05 |
| 486 | 457.96 | 152.65 | 166.17 | 531 | 501.68 | 167.23 | 181.39 |
| 487 | 458.93 | 152.98 | 166.51 | 532 | 502.65 | 167.55 | 181.72 |
| 488 | 459.90 | 153.30 | 166.85 | 533 | 503.63 | 167.88 | 182.06 |
| 489 | 460.87 | 153.62 | 167.19 | 534 | 504.60 | 168.20 | 182.40 |
| 490 | 461.84 | 153.95 | 167.53 | 535 | 505.57 | 168.52 | 182.74 |
| 491 | 462.82 | 154.27 | 167.86 | 536 | 506.54 | 168.85 | 183.08 |
| 492 | 463.79 | 154.60 | 168.20 | 537 | 507.52 | 169.17 | 183.41 |
| 493 | 464.76 | 154.92 | 168.54 | 538 | 508.49 | 169.50 | 183.75 |
| 494 | 465.73 | 155.24 | 168.88 | 539 | 509.46 | 169.82 | 184.09 |
| 495 | 466.70 | 155.57 | 169.22 | 540 | 510.43 | 170.14 | 184.43 |

Table 2.  CONSTANTS FOR $p'$ (LONG FORMULA), AN APPROXIMATION OF
THE LOWER BINOMIAL CONFIDENCE LIMIT $\underline{p}$; $c/n \leqslant .5$, $n \geqslant 20$,
$c \leqslant 1,000$ (*Continued*)

| c | m | a | b | c | m | a | b |
|---|---|---|---|---|---|---|---|
| 541 | 511.40 | 170.47 | 184.77 | 586 | 555.19 | 185.06 | 199.97 |
| 542 | 512.38 | 170.79 | 185.10 | 587 | 556.16 | 185.39 | 200.31 |
| 543 | 513.35 | 171.12 | 185.44 | 588 | 557.14 | 185.71 | 200.64 |
| 544 | 514.32 | 171.44 | 185.78 | 589 | 558.11 | 186.04 | 200.98 |
| 545 | 515.29 | 171.76 | 186.12 | 590 | 559.08 | 186.36 | 201.32 |
| 546 | 516.27 | 172.09 | 186.46 | 591 | 560.06 | 186.69 | 201.66 |
| 547 | 517.24 | 172.41 | 186.79 | 592 | 561.03 | 187.01 | 201.99 |
| 548 | 518.21 | 172.74 | 187.13 | 593 | 562.01 | 187.33 | 202.33 |
| 549 | 519.19 | 173.06 | 187.47 | 594 | 562.98 | 187.66 | 202.67 |
| 550 | 520.16 | 173.39 | 187.81 | 595 | 563.95 | 187.98 | 203.01 |
| 551 | 521.13 | 173.71 | 188.14 | 596 | 564.93 | 188.31 | 203.35 |
| 552 | 522.10 | 174.03 | 188.48 | 597 | 565.90 | 188.63 | 203.68 |
| 553 | 523.08 | 174.36 | 188.82 | 598 | 566.87 | 188.96 | 204.02 |
| 554 | 524.05 | 174.68 | 189.16 | 599 | 567.85 | 189.28 | 204.36 |
| 555 | 525.02 | 175.01 | 189.50 | 600 | 568.82 | 189.61 | 204.70 |
| 556 | 525.99 | 175.33 | 189.83 | 601 | 569.80 | 189.93 | 205.03 |
| 557 | 526.97 | 175.66 | 190.17 | 602 | 570.77 | 190.26 | 205.37 |
| 558 | 527.94 | 175.98 | 190.51 | 603 | 571.74 | 190.58 | 205.71 |
| 559 | 528.91 | 176.30 | 190.85 | 604 | 572.72 | 190.91 | 206.05 |
| 560 | 529.89 | 176.63 | 191.19 | 605 | 573.69 | 191.23 | 206.38 |
| 561 | 530.86 | 176.95 | 191.52 | 606 | 574.66 | 191.55 | 206.72 |
| 562 | 531.83 | 177.28 | 191.86 | 607 | 575.64 | 191.88 | 207.06 |
| 563 | 532.80 | 177.60 | 192.20 | 608 | 576.61 | 192.20 | 207.40 |
| 564 | 533.78 | 177.93 | 192.54 | 609 | 577.59 | 192.53 | 207.74 |
| 565 | 534.75 | 178.25 | 192.87 | 610 | 578.56 | 192.85 | 208.07 |
| 566 | 535.72 | 178.57 | 193.21 | 611 | 579.53 | 193.18 | 208.41 |
| 567 | 536.70 | 178.90 | 193.55 | 612 | 580.51 | 193.50 | 208.75 |
| 568 | 537.67 | 179.22 | 193.89 | 613 | 581.48 | 193.83 | 209.09 |
| 569 | 538.64 | 179.55 | 194.23 | 614 | 582.46 | 194.15 | 209.42 |
| 570 | 539.62 | 179.87 | 194.56 | 615 | 583.43 | 194.48 | 209.76 |
| 571 | 540.59 | 180.20 | 194.90 | 616 | 584.41 | 194.80 | 210.10 |
| 572 | 541.56 | 180.52 | 195.24 | 617 | 585.38 | 195.13 | 210.44 |
| 573 | 542.54 | 180.85 | 195.58 | 618 | 586.35 | 195.45 | 210.77 |
| 574 | 543.51 | 181.17 | 195.92 | 619 | 587.33 | 195.78 | 211.11 |
| 575 | 544.48 | 181.49 | 196.25 | 620 | 588.30 | 196.10 | 211.45 |
| 576 | 545.46 | 181.82 | 196.59 | 621 | 589.28 | 196.43 | 211.79 |
| 577 | 546.43 | 182.14 | 196.93 | 622 | 590.25 | 196.75 | 212.12 |
| 578 | 547.40 | 182.47 | 197.27 | 623 | 591.23 | 197.08 | 212.46 |
| 579 | 548.38 | 182.79 | 197.60 | 624 | 592.20 | 197.40 | 212.80 |
| 580 | 549.35 | 183.12 | 197.94 | 625 | 593.17 | 197.72 | 213.14 |
| 581 | 550.32 | 183.44 | 198.28 | 626 | 594.15 | 198.05 | 213.48 |
| 582 | 551.30 | 183.77 | 198.62 | 627 | 595.12 | 198.37 | 213.81 |
| 583 | 552.27 | 184.09 | 198.96 | 628 | 596.10 | 198.70 | 214.15 |
| 584 | 553.24 | 184.41 | 199.29 | 629 | 597.07 | 199.02 | 214.49 |
| 585 | 554.22 | 184.74 | 199.63 | 630 | 598.05 | 199.35 | 214.83 |

Table 2.  CONSTANTS FOR $\underline{p}'$ (LONG FORMULA), AN APPROXIMATION OF THE LOWER BINOMIAL CONFIDENCE LIMIT $\underline{p}$; $c/n \leqslant .5$, $n \geqslant 20$, $c \leqslant 1,000$ (*Continued*)

| c | m | a | b | c | m | a | b |
|---|---|---|---|---|---|---|---|
| 631 | 599.02 | 199.67 | 215.16 | 676 | 642.89 | 214.30 | 230.35 |
| 632 | 600.00 | 200.00 | 215.50 | 677 | 643.87 | 214.62 | 230.69 |
| 633 | 600.97 | 200.32 | 215.84 | 678 | 644.84 | 214.95 | 231.03 |
| 634 | 601.94 | 200.65 | 216.18 | 679 | 645.82 | 215.27 | 231.36 |
| 635 | 602.92 | 200.97 | 216.51 | 680 | 646.79 | 215.60 | 231.70 |
| 636 | 603.89 | 201.30 | 216.85 | 681 | 647.77 | 215.92 | 232.04 |
| 637 | 604.87 | 201.62 | 217.19 | 682 | 648.74 | 216.25 | 232.38 |
| 638 | 605.84 | 201.95 | 217.53 | 683 | 649.72 | 216.57 | 232.71 |
| 639 | 606.82 | 202.27 | 217.86 | 684 | 650.70 | 216.90 | 233.05 |
| 640 | 607.79 | 202.60 | 218.20 | 685 | 651.67 | 217.22 | 233.39 |
| 641 | 608.77 | 202.92 | 218.54 | 686 | 652.65 | 217.55 | 233.73 |
| 642 | 609.74 | 203.25 | 218.88 | 687 | 653.62 | 217.87 | 234.06 |
| 643 | 610.72 | 203.57 | 219.21 | 688 | 654.60 | 218.20 | 234.40 |
| 644 | 611.69 | 203.90 | 219.55 | 689 | 655.57 | 218.52 | 234.74 |
| 645 | 612.67 | 204.22 | 219.89 | 690 | 656.55 | 218.85 | 235.08 |
| 646 | 613.64 | 204.55 | 220.23 | 691 | 657.52 | 219.17 | 235.41 |
| 647 | 614.61 | 204.87 | 220.56 | 692 | 658.50 | 219.50 | 235.75 |
| 648 | 615.59 | 205.20 | 220.90 | 693 | 659.48 | 219.83 | 236.09 |
| 649 | 616.56 | 205.52 | 221.24 | 694 | 660.45 | 220.15 | 236.42 |
| 650 | 617.54 | 205.85 | 221.58 | 695 | 661.43 | 220.48 | 236.76 |
| 651 | 618.51 | 206.17 | 221.91 | 696 | 662.40 | 220.80 | 237.10 |
| 652 | 619.49 | 206.50 | 222.25 | 697 | 663.38 | 221.13 | 237.44 |
| 653 | 620.46 | 206.82 | 222.59 | 698 | 664.35 | 221.45 | 237.77 |
| 654 | 621.44 | 207.15 | 222.93 | 699 | 665.33 | 221.78 | 238.11 |
| 655 | 622.41 | 207.47 | 223.26 | 700 | 666.31 | 222.10 | 238.45 |
| 656 | 623.39 | 207.80 | 223.60 | 701 | 667.28 | 222.43 | 238.79 |
| 657 | 624.36 | 208.12 | 223.94 | 702 | 668.26 | 222.75 | 239.12 |
| 658 | 625.34 | 208.45 | 224.28 | 703 | 669.23 | 223.08 | 239.46 |
| 659 | 626.31 | 208.77 | 224.61 | 704 | 670.21 | 223.40 | 239.80 |
| 660 | 627.29 | 209.10 | 224.95 | 705 | 671.19 | 223.73 | 240.14 |
| 661 | 628.26 | 209.42 | 225.29 | 706 | 672.16 | 224.05 | 240.47 |
| 662 | 629.24 | 209.75 | 225.63 | 707 | 673.14 | 224.38 | 240.81 |
| 663 | 630.21 | 210.07 | 225.96 | 708 | 674.11 | 224.70 | 241.15 |
| 664 | 631.19 | 210.40 | 226.30 | 709 | 675.09 | 225.03 | 241.49 |
| 665 | 632.16 | 210.72 | 226.64 | 710 | 676.06 | 225.35 | 241.82 |
| 666 | 633.14 | 211.05 | 226.98 | 711 | 677.04 | 225.68 | 242.16 |
| 667 | 634.11 | 211.37 | 227.31 | 712 | 678.02 | 226.01 | 242.50 |
| 668 | 635.09 | 211.70 | 227.65 | 713 | 678.99 | 226.33 | 242.83 |
| 669 | 636.07 | 212.02 | 227.99 | 714 | 679.97 | 226.66 | 243.17 |
| 670 | 637.04 | 212.35 | 228.33 | 715 | 680.94 | 226.98 | 243.51 |
| 671 | 638.02 | 212.67 | 228.66 | 716 | 681.92 | 227.31 | 243.85 |
| 672 | 638.99 | 213.00 | 229.00 | 717 | 682.90 | 227.63 | 244.18 |
| 673 | 639.97 | 213.32 | 229.34 | 718 | 683.87 | 227.96 | 244.52 |
| 674 | 640.94 | 213.65 | 229.68 | 719 | 684.85 | 228.28 | 244.86 |
| 675 | 641.92 | 213.97 | 230.01 | 720 | 685.83 | 228.61 | 245.20 |

**Table 2. CONSTANTS FOR p' (LONG FORMULA), AN APPROXIMATION OF THE LOWER BINOMIAL CONFIDENCE LIMIT p; c/n ≤ .5, n ≥ 20, c ≤ 1,000 (Continued)**

| c | m | a | b | c | m | a | b |
|---|---|---|---|---|---|---|---|
| 721 | 686.80 | 228.93 | 245.53 | 766 | 730.74 | 243.58 | 260.71 |
| 722 | 687.78 | 229.26 | 245.87 | 767 | 731.72 | 243.91 | 261.05 |
| 723 | 688.75 | 229.58 | 246.21 | 768 | 732.70 | 244.23 | 261.38 |
| 724 | 689.73 | 229.91 | 246.55 | 769 | 733.67 | 244.56 | 261.72 |
| 725 | 690.71 | 230.24 | 246.88 | 770 | 734.65 | 244.88 | 262.06 |
| 726 | 691.68 | 230.56 | 247.22 | 771 | 735.63 | 245.21 | 262.40 |
| 727 | 692.66 | 230.89 | 247.56 | 772 | 736.60 | 245.53 | 262.73 |
| 728 | 693.63 | 231.21 | 247.89 | 773 | 737.58 | 245.86 | 263.07 |
| 729 | 694.61 | 231.54 | 248.23 | 774 | 738.56 | 246.19 | 263.41 |
| 730 | 695.59 | 231.86 | 248.57 | 775 | 739.54 | 246.51 | 263.74 |
| 731 | 696.56 | 232.19 | 248.91 | 776 | 740.51 | 246.84 | 264.08 |
| 732 | 697.54 | 232.51 | 249.24 | 777 | 741.49 | 247.16 | 264.42 |
| 733 | 698.52 | 232.84 | 249.58 | 778 | 742.47 | 247.49 | 264.76 |
| 734 | 699.49 | 233.16 | 249.92 | 779 | 743.44 | 247.81 | 265.09 |
| 735 | 700.47 | 233.49 | 250.26 | 780 | 744.42 | 248.14 | 265.43 |
| 736 | 701.45 | 233.82 | 250.59 | 781 | 745.40 | 248.47 | 265.77 |
| 737 | 702.42 | 234.14 | 250.93 | 782 | 746.38 | 248.79 | 266.10 |
| 738 | 703.40 | 234.47 | 251.27 | 783 | 747.35 | 249.12 | 266.44 |
| 739 | 704.37 | 234.79 | 251.60 | 784 | 748.33 | 249.44 | 266.78 |
| 740 | 705.35 | 235.12 | 251.94 | 785 | 749.31 | 249.77 | 267.12 |
| 741 | 706.33 | 235.44 | 252.28 | 786 | 750.28 | 250.09 | 267.45 |
| 742 | 707.30 | 235.77 | 252.62 | 787 | 751.26 | 250.42 | 267.79 |
| 743 | 708.28 | 236.09 | 252.95 | 788 | 752.24 | 250.75 | 268.13 |
| 744 | 709.26 | 236.42 | 253.29 | 789 | 753.21 | 251.07 | 268.46 |
| 745 | 710.23 | 236.74 | 253.63 | 790 | 754.19 | 251.40 | 268.80 |
| 746 | 711.21 | 237.07 | 253.97 | 791 | 755.17 | 251.72 | 269.14 |
| 747 | 712.19 | 237.40 | 254.30 | 792 | 756.15 | 252.05 | 269.48 |
| 748 | 713.16 | 237.72 | 254.64 | 793 | 757.12 | 252.37 | 269.81 |
| 749 | 714.14 | 238.05 | 254.98 | 794 | 758.10 | 252.70 | 270.15 |
| 750 | 715.12 | 238.37 | 255.31 | 795 | 759.08 | 253.03 | 270.49 |
| 751 | 716.09 | 238.70 | 255.65 | 796 | 760.06 | 253.35 | 270.82 |
| 752 | 717.07 | 239.02 | 255.99 | 797 | 761.03 | 253.68 | 271.16 |
| 753 | 718.05 | 239.35 | 256.33 | 798 | 762.01 | 254.00 | 271.50 |
| 754 | 719.02 | 239.67 | 256.66 | 799 | 762.99 | 254.33 | 271.84 |
| 755 | 720.00 | 240.00 | 257.00 | 800 | 763.96 | 254.65 | 272.17 |
| 756 | 720.98 | 240.33 | 257.34 | 801 | 764.94 | 254.98 | 272.51 |
| 757 | 721.95 | 240.65 | 257.67 | 802 | 765.92 | 255.31 | 272.85 |
| 758 | 722.93 | 240.98 | 258.01 | 803 | 766.90 | 255.63 | 273.18 |
| 759 | 723.91 | 241.30 | 258.35 | 804 | 767.87 | 255.96 | 273.52 |
| 760 | 724.88 | 241.63 | 258.69 | 805 | 768.85 | 256.28 | 273.86 |
| 761 | 725.86 | 241.95 | 259.02 | 806 | 769.83 | 256.61 | 274.20 |
| 762 | 726.84 | 242.28 | 259.36 | 807 | 770.81 | 256.94 | 274.53 |
| 763 | 727.81 | 242.60 | 259.70 | 808 | 771.78 | 257.26 | 274.87 |
| 764 | 728.79 | 242.93 | 260.04 | 809 | 772.76 | 257.59 | 275.21 |
| 765 | 729.77 | 243.26 | 260.37 | 810 | 773.74 | 257.91 | 275.54 |

Table 2.  CONSTANTS FOR $p'$ (LONG FORMULA), AN APPROXIMATION OF THE LOWER BINOMIAL CONFIDENCE LIMIT $\underline{p}$; $c/n \leqslant .5$, $n \geqslant 20$, $c \leqslant 1,000$ (*Continued*)

| c | m | a | b | c | m | a | b |
|---|---|---|---|---|---|---|---|
| 811 | 774.72 | 258.24 | 275.88 | 856 | 818.72 | 272.91 | 291.05 |
| 812 | 775.69 | 258.56 | 276.22 | 857 | 819.70 | 273.23 | 291.38 |
| 813 | 776.67 | 258.89 | 276.55 | 858 | 820.67 | 273.56 | 291.72 |
| 814 | 777.65 | 259.22 | 276.89 | 859 | 821.65 | 273.88 | 292.06 |
| 815 | 778.63 | 259.54 | 277.23 | 860 | 822.63 | 274.21 | 292.40 |
| 816 | 779.60 | 259.87 | 277.57 | 861 | 823.61 | 274.54 | 292.73 |
| 817 | 780.58 | 260.19 | 277.90 | 862 | 824.59 | 274.86 | 293.07 |
| 818 | 781.56 | 260.52 | 278.24 | 863 | 825.56 | 275.19 | 293.41 |
| 819 | 782.54 | 260.85 | 278.58 | 864 | 826.54 | 275.51 | 293.74 |
| 820 | 783.51 | 261.17 | 278.91 | 865 | 827.52 | 275.84 | 294.08 |
| 821 | 784.49 | 261.50 | 279.25 | 866 | 828.50 | 276.17 | 294.42 |
| 822 | 785.47 | 261.82 | 279.59 | 867 | 829.48 | 276.49 | 294.75 |
| 823 | 786.45 | 262.15 | 279.93 | 868 | 830.46 | 276.82 | 295.09 |
| 824 | 787.43 | 262.48 | 280.26 | 869 | 831.43 | 277.14 | 295.43 |
| 825 | 788.40 | 262.80 | 280.60 | 870 | 832.41 | 277.47 | 295.76 |
| 826 | 789.38 | 263.13 | 280.94 | 871 | 833.39 | 277.80 | 296.10 |
| 827 | 790.36 | 263.45 | 281.27 | 872 | 834.37 | 278.12 | 296.44 |
| 828 | 791.34 | 263.78 | 281.61 | 873 | 835.35 | 278.45 | 296.78 |
| 829 | 792.31 | 264.10 | 281.95 | 874 | 836.33 | 278.78 | 297.11 |
| 830 | 793.29 | 264.43 | 282.28 | 875 | 837.30 | 279.10 | 297.45 |
| 831 | 794.27 | 264.76 | 282.62 | 876 | 838.28 | 279.43 | 297.79 |
| 832 | 795.25 | 265.08 | 282.96 | 877 | 839.26 | 279.75 | 298.12 |
| 833 | 796.22 | 265.41 | 283.30 | 878 | 840.24 | 280.08 | 298.46 |
| 834 | 797.20 | 265.73 | 283.63 | 879 | 841.22 | 280.41 | 298.80 |
| 835 | 798.18 | 266.06 | 283.97 | 880 | 842.20 | 280.73 | 299.13 |
| 836 | 799.16 | 266.39 | 284.31 | 881 | 843.17 | 281.06 | 299.47 |
| 837 | 800.14 | 266.71 | 284.64 | 882 | 844.15 | 281.38 | 299.81 |
| 838 | 801.11 | 267.04 | 284.98 | 883 | 845.13 | 281.71 | 300.14 |
| 839 | 802.09 | 267.36 | 285.32 | 884 | 846.11 | 282.04 | 300.48 |
| 840 | 803.07 | 267.69 | 285.66 | 885 | 847.09 | 282.36 | 300.82 |
| 841 | 804.05 | 268.02 | 285.99 | 886 | 848.07 | 282.69 | 301.16 |
| 842 | 805.03 | 268.34 | 286.33 | 887 | 849.04 | 283.01 | 301.49 |
| 843 | 806.00 | 268.67 | 286.67 | 888 | 850.02 | 283.34 | 301.83 |
| 844 | 806.98 | 268.99 | 287.00 | 889 | 851.00 | 283.67 | 302.17 |
| 845 | 807.96 | 269.32 | 287.34 | 890 | 851.98 | 283.99 | 302.50 |
| 846 | 808.94 | 269.65 | 287.68 | 891 | 852.96 | 284.32 | 302.84 |
| 847 | 809.92 | 269.97 | 288.01 | 892 | 853.94 | 284.65 | 303.18 |
| 848 | 810.89 | 270.30 | 288.35 | 893 | 854.92 | 284.97 | 303.51 |
| 849 | 811.87 | 270.62 | 288.69 | 894 | 855.89 | 285.30 | 303.85 |
| 850 | 812.85 | 270.95 | 289.03 | 895 | 856.87 | 285.62 | 304.19 |
| 851 | 813.83 | 271.28 | 289.36 | 896 | 857.85 | 285.95 | 304.52 |
| 852 | 814.81 | 271.60 | 289.70 | 897 | 858.83 | 286.28 | 304.86 |
| 853 | 815.78 | 271.93 | 290.04 | 898 | 859.81 | 286.60 | 305.20 |
| 854 | 816.76 | 272.25 | 290.37 | 899 | 860.79 | 286.93 | 305.54 |
| 855 | 817.74 | 272.58 | 290.71 | 900 | 861.77 | 287.26 | 305.87 |

Table 2. CONSTANTS FOR p′ (LONG FORMULA), AN APPROXIMATION OF THE LOWER BINOMIAL CONFIDENCE LIMIT p; c/n ≤ .5, n ≥ 20, c ≤ 1,000 (*Continued*)

Section 5

$\gamma_1 = 90\%$
$\gamma_2 = 80\%$

| c | m | a | b | c | m | a | b |
|---|---|---|---|---|---|---|---|
| 901 | 862.74 | 287.58 | 306.21 | 946 | 906.80 | 302.27 | 321.37 |
| 902 | 863.72 | 287.91 | 306.55 | 947 | 907.77 | 302.59 | 321.70 |
| 903 | 864.70 | 288.23 | 306.88 | 948 | 908.75 | 302.92 | 322.04 |
| 904 | 865.68 | 288.56 | 307.22 | 949 | 909.73 | 303.24 | 322.38 |
| 905 | 866.66 | 288.89 | 307.56 | 950 | 910.71 | 303.57 | 322.71 |
| 906 | 867.64 | 289.21 | 307.89 | 951 | 911.69 | 303.90 | 323.05 |
| 907 | 868.62 | 289.54 | 308.23 | 952 | 912.67 | 304.22 | 323.39 |
| 908 | 869.60 | 289.87 | 308.57 | 953 | 913.65 | 304.55 | 323.72 |
| 909 | 870.57 | 290.19 | 308.90 | 954 | 914.63 | 304.88 | 324.06 |
| 910 | 871.55 | 290.52 | 309.24 | 955 | 915.61 | 305.20 | 324.40 |
| 911 | 872.53 | 290.84 | 309.58 | 956 | 916.59 | 305.53 | 324.74 |
| 912 | 873.51 | 291.17 | 309.91 | 957 | 917.57 | 305.86 | 325.07 |
| 913 | 874.49 | 291.50 | 310.25 | 958 | 918.55 | 306.18 | 325.41 |
| 914 | 875.47 | 291.82 | 310.59 | 959 | 919.53 | 306.51 | 325.75 |
| 915 | 876.45 | 292.15 | 310.93 | 960 | 920.51 | 306.84 | 326.08 |
| 916 | 877.43 | 292.48 | 311.26 | 961 | 921.48 | 307.16 | 326.42 |
| 917 | 878.40 | 292.80 | 311.60 | 962 | 922.46 | 307.49 | 326.76 |
| 918 | 879.38 | 293.13 | 311.94 | 963 | 923.44 | 307.81 | 327.09 |
| 919 | 880.36 | 293.45 | 312.27 | 964 | 924.42 | 308.14 | 327.43 |
| 920 | 881.34 | 293.78 | 312.61 | 965 | 925.40 | 308.47 | 327.77 |
| 921 | 882.32 | 294.11 | 312.95 | 966 | 926.38 | 308.79 | 328.10 |
| 922 | 883.30 | 294.43 | 313.28 | 967 | 927.36 | 309.12 | 328.44 |
| 923 | 884.28 | 294.76 | 313.62 | 968 | 928.34 | 309.45 | 328.78 |
| 924 | 885.26 | 295.09 | 313.96 | 969 | 929.32 | 309.77 | 329.11 |
| 925 | 886.24 | 295.41 | 314.29 | 970 | 930.30 | 310.10 | 329.45 |
| 926 | 887.21 | 295.74 | 314.63 | 971 | 931.28 | 310.43 | 329.79 |
| 927 | 888.19 | 296.06 | 314.97 | 972 | 932.26 | 310.75 | 330.12 |
| 928 | 889.17 | 296.39 | 315.30 | 973 | 933.24 | 311.08 | 330.46 |
| 929 | 890.15 | 296.72 | 315.64 | 974 | 934.22 | 311.41 | 330.80 |
| 930 | 891.13 | 297.04 | 315.98 | 975 | 935.20 | 311.73 | 331.13 |
| 931 | 892.11 | 297.37 | 316.32 | 976 | 936.18 | 312.06 | 331.47 |
| 932 | 893.09 | 297.70 | 316.65 | 977 | 937.16 | 312.39 | 331.81 |
| 933 | 894.07 | 298.02 | 316.99 | 978 | 938.13 | 312.71 | 332.14 |
| 934 | 895.05 | 298.35 | 317.33 | 979 | 939.11 | 313.04 | 332.48 |
| 935 | 896.03 | 298.68 | 317.66 | 980 | 940.09 | 313.36 | 332.82 |
| 936 | 897.00 | 299.00 | 318.00 | 981 | 941.07 | 313.69 | 333.15 |
| 937 | 897.98 | 299.33 | 318.34 | 982 | 942.05 | 314.02 | 333.49 |
| 938 | 898.96 | 299.65 | 318.67 | 983 | 943.03 | 314.34 | 333.83 |
| 939 | 899.94 | 299.98 | 319.01 | 984 | 944.01 | 314.67 | 334.16 |
| 940 | 900.92 | 300.31 | 319.35 | 985 | 944.99 | 315.00 | 334.50 |
| 941 | 901.90 | 300.63 | 319.68 | 986 | 945.97 | 315.32 | 334.84 |
| 942 | 902.88 | 300.96 | 320.02 | 987 | 946.95 | 315.65 | 335.17 |
| 943 | 903.86 | 301.29 | 320.36 | 988 | 947.93 | 315.98 | 335.51 |
| 944 | 904.84 | 301.61 | 320.69 | 989 | 948.91 | 316.30 | 335.85 |
| 945 | 905.82 | 301.94 | 321.03 | 990 | 949.89 | 316.63 | 336.19 |

Table 2. **CONSTANTS FOR $p'$ (LONG FORMULA), AN APPROXIMATION OF THE LOWER BINOMIAL CONFIDENCE LIMIT $\underline{p}$; $c/n \leqslant .5$, $n \geqslant 20$, $c \leqslant 1,000$** (*Continued*)

| c | m | a | b | c | m | a | b |
|---|---|---|---|---|---|---|---|
| 991 | 950.87 | 316.96 | 336.52 | 996 | 955.77 | 318.59 | 338.21 |
| 992 | 951.85 | 317.28 | 336.86 | 997 | 956.75 | 318.92 | 338.54 |
| 993 | 952.83 | 317.61 | 337.20 | 998 | 957.73 | 319.24 | 338.88 |
| 994 | 953.81 | 317.94 | 337.53 | 999 | 958.71 | 319.57 | 339.22 |
| 995 | 954.79 | 318.26 | 337.87 | 1000 | 959.69 | 319.90 | 339.55 |

# Table 2. CONSTANTS FOR p′ (LONG FORMULA), AN APPROXIMATION OF THE LOWER BINOMIAL CONFIDENCE LIMIT p; c/n ≤ .5, n ≥ 20, c ≤ 1,000 (*Continued*)

| c | m | a | b | c | m | a | b |
|---|---|---|---|---|---|---|---|
| 1 | 0.22314 | 0.112 | 0.0 | 46 | 40.217 | 13.919 | 16.311 |
| 2 | 0.82439 | 1.111 | 1.199 | 47 | 41.155 | 14.181 | 16.604 |
| 3 | 1.5351 | 1.133 | 1.365 | 48 | 42.093 | 14.494 | 16.947 |
| 4 | 2.2968 | 1.382 | 1.733 | 49 | 43.033 | 14.796 | 17.280 |
| 5 | 3.0895 | 1.619 | 2.074 | 50 | 43.973 | 15.082 | 17.596 |
| 6 | 3.9037 | 1.902 | 2.450 | 51 | 44.913 | 15.341 | 17.885 |
| 7 | 4.7337 | 2.194 | 2.827 | 52 | 45.854 | 15.635 | 18.208 |
| 8 | 5.5761 | 2.505 | 3.217 | 53 | 46.796 | 15.961 | 18.563 |
| 9 | 6.4285 | 2.832 | 3.618 | 54 | 47.738 | 16.245 | 18.876 |
| 10 | 7.2892 | 3.165 | 4.021 | 55 | 48.681 | 16.553 | 19.213 |
| 11 | 8.1570 | 3.464 | 4.386 | 56 | 49.625 | 16.880 | 19.568 |
| 12 | 9.0309 | 3.763 | 4.747 | 57 | 50.569 | 17.146 | 19.861 |
| 13 | 9.9101 | 4.061 | 5.106 | 58 | 51.513 | 17.443 | 20.187 |
| 14 | 10.794 | 4.364 | 5.467 | 59 | 52.458 | 17.72 | 20.495 |
| 15 | 11.682 | 4.666 | 5.825 | 60 | 53.403 | 18.01 | 20.804 |
| 16 | 12.574 | 4.960 | 6.173 | 61 | 54.349 | 18.31 | 21.14 |
| 17 | 13.469 | 5.264 | 6.529 | 62 | 55.295 | 18.61 | 21.47 |
| 18 | 14.368 | 5.553 | 6.869 | 63 | 56.242 | 18.88 | 21.76 |
| 19 | 15.269 | 5.864 | 7.230 | 64 | 57.189 | 19.24 | 22.15 |
| 20 | 16.173 | 6.160 | 7.574 | 65 | 58.136 | 19.51 | 22.44 |
| 21 | 17.079 | 6.452 | 7.912 | 66 | 59.084 | 19.86 | 22.82 |
| 22 | 17.987 | 6.753 | 8.259 | 67 | 60.032 | 20.08 | 23.06 |
| 23 | 18.898 | 7.051 | 8.602 | 68 | 60.981 | 20.38 | 23.39 |
| 24 | 19.810 | 7.356 | 8.951 | 69 | 61.930 | 20.75 | 23.78 |
| 25 | 20.725 | 7.637 | 9.274 | 70 | 62.879 | 20.98 | 24.04 |
| 26 | 21.641 | 7.937 | 9.617 | 71 | 63.829 | 21.28 | 24.36 |
| 27 | 22.558 | 8.238 | 9.959 | 72 | 64.779 | 21.56 | 24.67 |
| 28 | 23.478 | 8.539 | 10.301 | 73 | 65.730 | 21.89 | 25.03 |
| 29 | 24.398 | 8.824 | 10.625 | 74 | 66.680 | 22.21 | 25.37 |
| 30 | 25.320 | 9.128 | 10.968 | 75 | 67.632 | 22.44 | 25.62 |
| 31 | 26.244 | 9.437 | 11.316 | 76 | 68.583 | 22.78 | 25.99 |
| 32 | 27.168 | 9.735 | 11.651 | 77 | 69.535 | 23.03 | 26.27 |
| 33 | 28.094 | 10.039 | 11.992 | 78 | 70.487 | 23.36 | 26.62 |
| 34 | 29.021 | 10.337 | 12.327 | 79 | 71.438 | 23.68 | 26.96 |
| 35 | 29.949 | 10.615 | 12.641 | 80 | 72.393 | 23.91 | 27.21 |
| 36 | 30.878 | 10.935 | 12.996 | 81 | 73.345 | 24.23 | 27.55 |
| 37 | 31.808 | 11.207 | 13.303 | 82 | 74.298 | 24.54 | 27.89 |
| 38 | 32.739 | 11.501 | 13.631 | 83 | 75.252 | 24.79 | 28.17 |
| 39 | 33.671 | 11.807 | 13.971 | 84 | 76.206 | 25.09 | 28.49 |
| 40 | 34.603 | 12.121 | 14.319 | 85 | 77.160 | 25.46 | 28.88 |
| 41 | 35.537 | 12.430 | 14.661 | 86 | 78.114 | 25.74 | 29.19 |
| 42 | 36.471 | 12.726 | 14.991 | 87 | 79.069 | 26.09 | 29.56 |
| 43 | 37.407 | 13.002 | 15.299 | 88 | 80.024 | 26.35 | 29.84 |
| 44 | 38.343 | 13.296 | 15.624 | 89 | 80.979 | 26.67 | 30.18 |
| 45 | 39.279 | 13.604 | 15.965 | 90 | 81.934 | 26.88 | 30.42 |

Table 2. CONSTANTS FOR $\underline{p}'$ (LONG FORMULA), AN APPROXIMATION OF THE LOWER BINOMIAL CONFIDENCE LIMIT $\underline{p}$; $c/n \leqslant .5$, $n \geqslant 20$, $c \leqslant 1,000$ (*Continued*)

| c | m | a | b | c | m | a | b |
|---|---|---|---|---|---|---|---|
| 91 | 82.890 | 27.28 | 30.83 | 136 | 126.10 | 42.03 | 46.49 |
| 92 | 83.846 | 27.48 | 31.05 | 137 | 127.06 | 42.35 | 46.82 |
| 93 | 84.802 | 27.82 | 31.42 | 138 | 128.03 | 42.67 | 47.16 |
| 94 | 85.759 | 28.09 | 31.71 | 139 | 128.99 | 42.99 | 47.50 |
| 95 | 86.715 | 28.41 | 32.05 | 140 | 129.96 | 43.31 | 47.84 |
| 96 | 87.672 | 28.62 | 32.28 | 141 | 130.92 | 43.64 | 48.18 |
| 97 | 88.629 | 29.03 | 32.72 | 142 | 131.89 | 43.96 | 48.52 |
| 98 | 89.586 | 29.35 | 33.06 | 143 | 132.85 | 44.28 | 48.86 |
| 99 | 90.544 | 29.54 | 33.27 | 144 | 133.82 | 44.60 | 49.20 |
| 100 | 91.502 | 29.92 | 33.67 | 145 | 134.78 | 44.92 | 49.54 |
| 101 | 92.460 | 30.81 | 34.59 | 146 | 135.75 | 45.24 | 49.88 |
| 102 | 93.418 | 31.13 | 34.93 | 147 | 136.71 | 45.57 | 50.22 |
| 103 | 94.376 | 31.45 | 35.27 | 148 | 137.68 | 45.89 | 50.56 |
| 104 | 95.335 | 31.77 | 35.61 | 149 | 138.64 | 46.21 | 50.90 |
| 105 | 96.293 | 32.09 | 35.95 | 150 | 139.61 | 46.53 | 51.23 |
| 106 | 97.252 | 32.41 | 36.29 | 151 | 140.57 | 46.85 | 51.57 |
| 107 | 98.212 | 32.73 | 36.63 | 152 | 141.54 | 47.18 | 51.91 |
| 108 | 99.171 | 33.05 | 36.97 | 153 | 142.50 | 47.50 | 52.25 |
| 109 | 100.13 | 33.37 | 37.31 | 154 | 143.47 | 47.82 | 52.59 |
| 110 | 101.09 | 33.69 | 37.65 | 155 | 144.44 | 48.14 | 52.93 |
| 111 | 102.05 | 34.01 | 37.99 | 156 | 145.40 | 48.46 | 53.27 |
| 112 | 103.01 | 34.33 | 38.33 | 157 | 146.37 | 48.79 | 53.61 |
| 113 | 103.97 | 34.65 | 38.67 | 158 | 147.34 | 49.11 | 53.95 |
| 114 | 104.93 | 34.97 | 39.01 | 159 | 148.30 | 49.43 | 54.28 |
| 115 | 105.89 | 35.29 | 39.35 | 160 | 149.27 | 49.75 | 54.62 |
| 116 | 106.85 | 35.61 | 39.69 | 161 | 150.22 | 50.07 | 54.96 |
| 117 | 107.81 | 35.93 | 40.03 | 162 | 151.19 | 50.40 | 55.30 |
| 118 | 108.77 | 36.25 | 40.37 | 163 | 152.16 | 50.72 | 55.64 |
| 119 | 109.74 | 36.57 | 40.71 | 164 | 153.12 | 51.04 | 55.98 |
| 120 | 110.70 | 36.89 | 41.05 | 165 | 154.09 | 51.36 | 56.32 |
| 121 | 111.66 | 37.21 | 41.39 | 166 | 155.06 | 51.69 | 56.66 |
| 122 | 112.62 | 37.54 | 41.73 | 167 | 156.03 | 52.01 | 57.00 |
| 123 | 113.58 | 37.86 | 42.07 | 168 | 156.99 | 52.33 | 57.33 |
| 124 | 114.54 | 38.18 | 42.41 | 169 | 157.96 | 52.65 | 57.67 |
| 125 | 115.51 | 38.50 | 42.75 | 170 | 158.93 | 52.98 | 58.01 |
| 126 | 116.47 | 38.82 | 43.09 | 171 | 159.90 | 53.30 | 58.35 |
| 127 | 117.43 | 39.14 | 43.43 | 172 | 160.87 | 53.62 | 58.69 |
| 128 | 118.39 | 39.46 | 43.77 | 173 | 161.83 | 53.94 | 59.03 |
| 129 | 119.36 | 39.78 | 44.11 | 174 | 162.80 | 54.27 | 59.37 |
| 130 | 120.32 | 40.10 | 44.45 | 175 | 163.77 | 54.59 | 59.71 |
| 131 | 121.28 | 40.42 | 44.79 | 176 | 164.74 | 54.91 | 60.04 |
| 132 | 122.25 | 40.74 | 45.13 | 177 | 165.71 | 55.24 | 60.38 |
| 133 | 123.21 | 41.07 | 45.47 | 178 | 166.67 | 55.56 | 60.72 |
| 134 | 124.17 | 41.39 | 45.81 | 179 | 167.64 | 55.88 | 61.06 |
| 135 | 125.14 | 41.71 | 46.15 | 180 | 168.61 | 56.20 | 61.40 |

**Table 2. CONSTANTS FOR p′ (LONG FORMULA), AN APPROXIMATION OF THE LOWER BINOMIAL CONFIDENCE LIMIT p; c/n ⩽ .5, n ⩾ 20, c ⩽ 1,000 (*Continued*)**

| c | m | a | b | c | m | a | b |
|---|---|---|---|---|---|---|---|
| 181 | 169.58 | 56.53 | 61.74 | 226 | 213.25 | 71.08 | 76.96 |
| 182 | 170.55 | 56.85 | 62.08 | 227 | 214.22 | 71.41 | 77.30 |
| 183 | 171.52 | 57.17 | 62.41 | 228 | 215.19 | 71.73 | 77.63 |
| 184 | 172.49 | 57.50 | 62.75 | 229 | 216.17 | 72.06 | 77.97 |
| 185 | 173.46 | 57.82 | 63.09 | 230 | 217.14 | 72.38 | 78.31 |
| 186 | 174.42 | 58.14 | 63.43 | 231 | 218.11 | 72.70 | 78.65 |
| 187 | 175.39 | 58.46 | 63.77 | 232 | 219.08 | 73.03 | 78.99 |
| 188 | 176.36 | 58.79 | 64.11 | 233 | 220.06 | 73.35 | 79.32 |
| 189 | 177.33 | 59.11 | 64.44 | 234 | 221.03 | 73.68 | 79.66 |
| 190 | 178.30 | 59.43 | 64.78 | 235 | 222.00 | 74.00 | 80.00 |
| 191 | 179.27 | 59.76 | 65.12 | 236 | 222.97 | 74.32 | 80.34 |
| 192 | 180.24 | 60.08 | 65.46 | 237 | 223.95 | 74.65 | 80.68 |
| 193 | 181.21 | 60.40 | 65.80 | 238 | 224.92 | 74.97 | 81.01 |
| 194 | 182.18 | 60.73 | 66.14 | 239 | 225.89 | 75.30 | 81.35 |
| 195 | 183.15 | 61.05 | 66.47 | 240 | 226.86 | 75.62 | 81.69 |
| 196 | 184.12 | 61.37 | 66.81 | 241 | 227.84 | 75.95 | 82.03 |
| 197 | 185.09 | 61.70 | 67.15 | 242 | 228.81 | 76.27 | 82.36 |
| 198 | 186.06 | 62.02 | 67.49 | 243 | 229.78 | 76.59 | 82.70 |
| 199 | 187.03 | 62.34 | 67.83 | 244 | 230.76 | 76.92 | 83.04 |
| 200 | 188.00 | 62.67 | 68.17 | 245 | 231.73 | 77.24 | 83.38 |
| 201 | 188.97 | 62.99 | 68.50 | 246 | 232.70 | 77.57 | 83.72 |
| 202 | 189.94 | 63.31 | 68.84 | 247 | 233.68 | 77.89 | 84.05 |
| 203 | 190.91 | 63.64 | 69.18 | 248 | 234.65 | 78.22 | 84.39 |
| 204 | 191.88 | 63.96 | 69.52 | 249 | 235.62 | 78.54 | 84.73 |
| 205 | 192.85 | 64.28 | 69.86 | 250 | 236.60 | 78.87 | 85.07 |
| 206 | 193.82 | 64.61 | 70.20 | 251 | 237.57 | 79.19 | 85.41 |
| 207 | 194.79 | 64.93 | 70.53 | 252 | 238.54 | 79.51 | 85.74 |
| 208 | 195.76 | 65.25 | 70.87 | 253 | 239.52 | 79.84 | 86.08 |
| 209 | 196.74 | 65.58 | 71.21 | 254 | 240.49 | 80.16 | 86.42 |
| 210 | 197.71 | 65.90 | 71.55 | 255 | 241.46 | 80.49 | 86.76 |
| 211 | 198.68 | 66.23 | 71.89 | 256 | 242.44 | 80.81 | 87.09 |
| 212 | 199.65 | 66.55 | 72.23 | 257 | 243.41 | 81.14 | 87.43 |
| 213 | 200.62 | 66.87 | 72.56 | 258 | 244.38 | 81.46 | 87.77 |
| 214 | 201.59 | 67.20 | 72.90 | 259 | 245.36 | 81.79 | 88.11 |
| 215 | 202.56 | 67.52 | 73.24 | 260 | 246.33 | 82.11 | 88.44 |
| 216 | 203.53 | 67.84 | 73.58 | 261 | 247.31 | 82.44 | 88.78 |
| 217 | 204.50 | 68.17 | 73.92 | 262 | 248.28 | 82.76 | 89.12 |
| 218 | 205.48 | 68.49 | 74.25 | 263 | 249.25 | 83.08 | 89.46 |
| 219 | 206.45 | 68.82 | 74.59 | 264 | 250.23 | 83.41 | 89.80 |
| 220 | 207.42 | 69.14 | 74.93 | 265 | 251.20 | 83.73 | 90.13 |
| 221 | 208.39 | 69.46 | 75.27 | 266 | 252.18 | 84.06 | 90.47 |
| 222 | 209.36 | 69.79 | 75.61 | 267 | 253.15 | 84.38 | 90.81 |
| 223 | 210.33 | 70.11 | 75.94 | 268 | 254.12 | 84.71 | 91.15 |
| 224 | 211.31 | 70.44 | 76.28 | 269 | 255.10 | 85.03 | 91.48 |
| 225 | 212.28 | 70.76 | 76.62 | 270 | 256.07 | 85.36 | 91.82 |

Table 2.  CONSTANTS FOR p′ (LONG FORMULA), AN APPROXIMATION OF THE LOWER BINOMIAL CONFIDENCE LIMIT $\underline{p}$; $c/n \leqslant .5$, $n \geqslant 20$, $c \leqslant 1,000$ (*Continued*)

| c | m | a | b | c | m | a | b |
|---|---|---|---|---|---|---|---|
| 271 | 257.05 | 85.68 | 92.16 | 316 | 300.94 | 100.31 | 107.34 |
| 272 | 258.02 | 86.01 | 92.50 | 317 | 301.92 | 100.64 | 107.68 |
| 273 | 259.00 | 86.33 | 92.83 | 318 | 302.89 | 100.96 | 108.02 |
| 274 | 259.97 | 86.66 | 93.17 | 319 | 303.87 | 101.29 | 108.35 |
| 275 | 260.95 | 86.98 | 93.51 | 320 | 304.85 | 101.62 | 108.69 |
| 276 | 261.92 | 87.31 | 93.85 | 321 | 305.82 | 101.94 | 109.03 |
| 277 | 262.90 | 87.63 | 94.18 | 322 | 306.80 | 102.27 | 109.37 |
| 278 | 263.87 | 87.96 | 94.52 | 323 | 307.78 | 102.59 | 109.70 |
| 279 | 264.84 | 88.28 | 94.86 | 324 | 308.75 | 102.92 | 110.04 |
| 280 | 265.82 | 88.61 | 95.20 | 325 | 309.73 | 103.24 | 110.38 |
| 281 | 266.79 | 88.93 | 95.53 | 326 | 310.71 | 103.57 | 110.72 |
| 282 | 267.77 | 89.26 | 95.87 | 327 | 311.68 | 103.89 | 111.05 |
| 283 | 268.74 | 89.58 | 96.21 | 328 | 312.66 | 104.22 | 111.39 |
| 284 | 269.72 | 89.91 | 96.55 | 329 | 313.64 | 104.55 | 111.73 |
| 285 | 270.69 | 90.23 | 96.88 | 330 | 314.61 | 104.87 | 112.06 |
| 286 | 271.67 | 90.56 | 97.22 | 331 | 315.59 | 105.20 | 112.40 |
| 287 | 272.64 | 90.88 | 97.56 | 332 | 316.57 | 105.52 | 112.74 |
| 288 | 273.62 | 91.21 | 97.90 | 333 | 317.54 | 105.85 | 113.08 |
| 289 | 274.60 | 91.53 | 98.23 | 334 | 318.52 | 106.17 | 113.41 |
| 290 | 275.57 | 91.86 | 98.57 | 335 | 319.50 | 106.50 | 113.75 |
| 291 | 276.55 | 92.18 | 98.91 | 336 | 320.48 | 106.83 | 114.09 |
| 292 | 277.52 | 92.51 | 99.25 | 337 | 321.45 | 107.15 | 114.42 |
| 293 | 278.50 | 92.83 | 99.58 | 338 | 322.43 | 107.48 | 114.76 |
| 294 | 279.47 | 93.16 | 99.92 | 339 | 323.41 | 107.80 | 115.10 |
| 295 | 280.45 | 93.48 | 100.26 | 340 | 324.38 | 108.13 | 115.44 |
| 296 | 281.42 | 93.81 | 100.60 | 341 | 325.36 | 108.45 | 115.77 |
| 297 | 282.40 | 94.13 | 100.93 | 342 | 326.34 | 108.78 | 116.11 |
| 298 | 283.37 | 94.46 | 101.27 | 343 | 327.32 | 109.11 | 116.45 |
| 299 | 284.35 | 94.78 | 101.61 | 344 | 328.29 | 109.43 | 116.78 |
| 300 | 285.33 | 95.11 | 101.95 | 345 | 329.27 | 109.76 | 117.12 |
| 301 | 286.30 | 95.43 | 102.28 | 346 | 330.25 | 110.08 | 117.46 |
| 302 | 287.28 | 95.76 | 102.62 | 347 | 331.23 | 110.41 | 117.80 |
| 303 | 288.25 | 96.08 | 102.96 | 348 | 332.20 | 110.73 | 118.13 |
| 304 | 289.23 | 96.41 | 103.30 | 349 | 333.18 | 111.06 | 118.47 |
| 305 | 290.20 | 96.73 | 103.63 | 350 | 334.16 | 111.39 | 118.81 |
| 306 | 291.18 | 97.06 | 103.97 | 351 | 335.14 | 111.71 | 119.14 |
| 307 | 292.16 | 97.39 | 104.31 | 352 | 336.11 | 112.04 | 119.48 |
| 308 | 293.13 | 97.71 | 104.64 | 353 | 337.09 | 112.36 | 119.82 |
| 309 | 294.11 | 98.04 | 104.98 | 354 | 338.07 | 112.69 | 120.16 |
| 310 | 295.08 | 98.36 | 105.32 | 355 | 339.05 | 113.02 | 120.49 |
| 311 | 296.06 | 98.69 | 105.66 | 356 | 340.02 | 113.34 | 120.83 |
| 312 | 297.04 | 99.01 | 105.99 | 357 | 341.00 | 113.67 | 121.17 |
| 313 | 298.01 | 99.34 | 106.33 | 358 | 341.98 | 113.99 | 121.50 |
| 314 | 298.99 | 99.66 | 106.67 | 359 | 342.96 | 114.32 | 121.84 |
| 315 | 299.97 | 99.99 | 107.01 | 360 | 343.93 | 114.64 | 122.18 |

Table 2. CONSTANTS FOR p′ (LONG FORMULA), AN APPROXIMATION OF THE LOWER BINOMIAL CONFIDENCE LIMIT p; c/n ⩽ .5, n ⩾ 20, c ⩽ 1,000 (*Continued*)

Section 6
$\gamma_1 = 80\%$
$\gamma_2 = 60\%$

| c | m | a | b | c | m | a | b |
|---|---|---|---|---|---|---|---|
| 361 | 344.91 | 114.97 | 122.51 | 406 | 388.94 | 129.65 | 137.68 |
| 362 | 345.89 | 115.30 | 122.85 | 407 | 389.92 | 129.97 | 138.01 |
| 363 | 346.87 | 115.62 | 123.19 | 408 | 390.90 | 130.30 | 138.35 |
| 364 | 347.85 | 115.95 | 123.53 | 409 | 391.88 | 130.63 | 138.69 |
| 365 | 348.82 | 116.27 | 123.86 | 410 | 392.86 | 130.95 | 139.02 |
| 366 | 349.80 | 116.60 | 124.20 | 411 | 393.84 | 131.28 | 139.36 |
| 367 | 350.78 | 116.93 | 124.54 | 412 | 394.82 | 131.61 | 139.70 |
| 368 | 351.76 | 117.25 | 124.87 | 413 | 395.80 | 131.93 | 140.03 |
| 369 | 352.74 | 117.58 | 125.21 | 414 | 396.78 | 132.26 | 140.37 |
| 370 | 353.71 | 117.90 | 125.55 | 415 | 397.76 | 132.59 | 140.71 |
| 371 | 354.69 | 118.23 | 125.88 | 416 | 398.74 | 132.91 | 141.04 |
| 372 | 355.67 | 118.56 | 126.22 | 417 | 399.72 | 133.24 | 141.38 |
| 373 | 356.65 | 118.88 | 126.56 | 418 | 400.70 | 133.57 | 141.72 |
| 374 | 357.63 | 119.21 | 126.90 | 419 | 401.68 | 133.89 | 142.05 |
| 375 | 358.60 | 119.53 | 127.23 | 420 | 402.65 | 134.22 | 142.39 |
| 376 | 359.58 | 119.86 | 127.57 | 421 | 403.63 | 134.54 | 142.73 |
| 377 | 360.56 | 120.19 | 127.91 | 422 | 404.61 | 134.87 | 143.06 |
| 378 | 361.54 | 120.51 | 128.24 | 423 | 405.59 | 135.20 | 143.40 |
| 379 | 362.52 | 120.84 | 128.58 | 424 | 406.57 | 135.52 | 143.74 |
| 380 | 363.50 | 121.17 | 128.92 | 425 | 407.55 | 135.85 | 144.07 |
| 381 | 364.47 | 121.49 | 129.25 | 426 | 408.53 | 136.18 | 144.41 |
| 382 | 365.45 | 121.82 | 129.59 | 427 | 409.51 | 136.50 | 144.75 |
| 383 | 366.43 | 122.14 | 129.93 | 428 | 410.49 | 136.83 | 145.08 |
| 384 | 367.41 | 122.47 | 130.26 | 429 | 411.47 | 137.16 | 145.42 |
| 385 | 368.39 | 122.80 | 130.60 | 430 | 412.45 | 137.48 | 145.76 |
| 386 | 369.37 | 123.12 | 130.94 | 431 | 413.43 | 137.81 | 146.09 |
| 387 | 370.35 | 123.45 | 131.28 | 432 | 414.41 | 138.14 | 146.43 |
| 388 | 371.32 | 123.77 | 131.61 | 433 | 415.39 | 138.46 | 146.77 |
| 389 | 372.30 | 124.10 | 131.95 | 434 | 416.37 | 138.79 | 147.11 |
| 390 | 373.28 | 124.43 | 132.29 | 435 | 417.35 | 139.12 | 147.44 |
| 391 | 374.26 | 124.75 | 132.62 | 436 | 418.33 | 139.44 | 147.78 |
| 392 | 375.24 | 125.08 | 132.96 | 437 | 419.31 | 139.77 | 148.12 |
| 393 | 376.22 | 125.41 | 133.30 | 438 | 420.29 | 140.10 | 148.45 |
| 394 | 377.20 | 125.73 | 133.63 | 439 | 421.27 | 140.42 | 148.79 |
| 395 | 378.18 | 126.06 | 133.97 | 440 | 422.25 | 140.75 | 149.13 |
| 396 | 379.15 | 126.38 | 134.31 | 441 | 423.23 | 141.08 | 149.46 |
| 397 | 380.13 | 126.71 | 134.64 | 442 | 424.21 | 141.40 | 149.80 |
| 398 | 381.11 | 127.04 | 134.98 | 443 | 425.19 | 141.73 | 150.14 |
| 399 | 382.09 | 127.36 | 135.32 | 444 | 426.17 | 142.06 | 150.47 |
| 400 | 383.07 | 127.69 | 135.65 | 445 | 427.15 | 142.38 | 150.81 |
| 401 | 384.05 | 128.02 | 135.99 | 446 | 428.13 | 142.71 | 151.15 |
| 402 | 385.03 | 128.34 | 136.33 | 447 | 429.11 | 143.04 | 151.48 |
| 403 | 386.01 | 128.67 | 136.67 | 448 | 430.09 | 143.36 | 151.82 |
| 404 | 386.99 | 129.00 | 137.00 | 449 | 431.07 | 143.69 | 152.16 |
| 405 | 387.97 | 129.32 | 137.34 | 450 | 432.05 | 144.02 | 152.49 |

Table 2. **CONSTANTS FOR** $p'$ **(LONG FORMULA), AN APPROXIMATION OF THE LOWER BINOMIAL CONFIDENCE LIMIT** $\underline{p}$; $c/n \leqslant .5$, $n \geqslant 20$, $c \leqslant 1,000$ (*Continued*)

| c | m | a | b | c | m | a | b |
|---|---|---|---|---|---|---|---|
| 451 | 433.03 | 144.34 | 152.83 | 496 | 477.16 | 159.05 | 167.97 |
| 452 | 434.01 | 144.67 | 153.17 | 497 | 478.14 | 159.38 | 168.31 |
| 453 | 434.99 | 145.00 | 153.50 | 498 | 479.12 | 159.71 | 168.65 |
| 454 | 435.97 | 145.32 | 153.84 | 499 | 480.10 | 160.03 | 168.98 |
| 455 | 436.95 | 145.65 | 154.17 | 500 | 481.08 | 160.36 | 169.32 |
| 456 | 437.93 | 145.98 | 154.51 | 501 | 482.06 | 160.69 | 169.66 |
| 457 | 438.91 | 146.30 | 154.85 | 502 | 483.05 | 161.02 | 169.99 |
| 458 | 439.89 | 146.63 | 155.18 | 503 | 484.03 | 161.34 | 170.33 |
| 459 | 440.87 | 146.96 | 155.52 | 504 | 485.01 | 161.67 | 170.67 |
| 460 | 441.85 | 147.28 | 155.86 | 505 | 485.99 | 162.00 | 171.00 |
| 461 | 442.83 | 147.61 | 156.19 | 506 | 486.97 | 162.32 | 171.34 |
| 462 | 443.81 | 147.94 | 156.53 | 507 | 487.95 | 162.65 | 171.67 |
| 463 | 444.79 | 148.26 | 156.87 | 508 | 488.93 | 162.98 | 172.01 |
| 464 | 445.77 | 148.59 | 157.20 | 509 | 489.91 | 163.30 | 172.35 |
| 465 | 446.75 | 148.92 | 157.54 | 510 | 490.90 | 163.63 | 172.68 |
| 466 | 447.73 | 149.24 | 157.88 | 511 | 491.88 | 163.96 | 173.02 |
| 467 | 448.72 | 149.57 | 158.21 | 512 | 492.86 | 164.29 | 173.36 |
| 468 | 449.70 | 149.90 | 158.55 | 513 | 493.84 | 164.61 | 173.69 |
| 469 | 450.68 | 150.23 | 158.89 | 514 | 494.82 | 164.94 | 174.03 |
| 470 | 451.66 | 150.55 | 159.22 | 515 | 495.80 | 165.27 | 174.37 |
| 471 | 452.64 | 150.88 | 159.56 | 516 | 496.78 | 165.59 | 174.70 |
| 472 | 453.62 | 151.21 | 159.90 | 517 | 497.77 | 165.92 | 175.04 |
| 473 | 454.60 | 151.53 | 160.23 | 518 | 498.75 | 166.25 | 175.38 |
| 474 | 455.58 | 151.86 | 160.57 | 519 | 499.73 | 166.58 | 175.71 |
| 475 | 456.56 | 152.19 | 160.91 | 520 | 500.71 | 166.90 | 176.05 |
| 476 | 457.54 | 152.51 | 161.24 | 521 | 501.69 | 167.23 | 176.38 |
| 477 | 458.52 | 152.84 | 161.58 | 522 | 502.67 | 167.56 | 176.72 |
| 478 | 459.50 | 153.17 | 161.92 | 523 | 503.66 | 167.89 | 177.06 |
| 479 | 460.48 | 153.49 | 162.25 | 524 | 504.64 | 168.21 | 177.39 |
| 480 | 461.46 | 153.82 | 162.59 | 525 | 505.62 | 168.54 | 177.73 |
| 481 | 462.44 | 154.15 | 162.93 | 526 | 506.60 | 168.87 | 178.07 |
| 482 | 463.43 | 154.48 | 163.26 | 527 | 507.58 | 169.19 | 178.40 |
| 483 | 464.41 | 154.80 | 163.60 | 528 | 508.56 | 169.52 | 178.74 |
| 484 | 465.39 | 155.13 | 163.94 | 529 | 509.55 | 169.85 | 179.08 |
| 485 | 466.37 | 155.46 | 164.27 | 530 | 510.53 | 170.18 | 179.41 |
| 486 | 467.35 | 155.78 | 164.61 | 531 | 511.51 | 170.50 | 179.75 |
| 487 | 468.33 | 156.11 | 164.95 | 532 | 512.49 | 170.83 | 180.08 |
| 488 | 469.31 | 156.44 | 165.28 | 533 | 513.47 | 171.16 | 180.42 |
| 489 | 470.29 | 156.76 | 165.62 | 534 | 514.45 | 171.48 | 180.76 |
| 490 | 471.27 | 157.09 | 165.95 | 535 | 515.44 | 171.81 | 181.09 |
| 491 | 472.25 | 157.42 | 166.29 | 536 | 516.42 | 172.14 | 181.43 |
| 492 | 473.23 | 157.74 | 166.63 | 537 | 517.40 | 172.47 | 181.77 |
| 493 | 474.22 | 158.07 | 166.96 | 538 | 518.38 | 172.79 | 182.10 |
| 494 | 475.20 | 158.40 | 167.30 | 539 | 519.36 | 173.12 | 182.44 |
| 495 | 476.18 | 158.73 | 167.64 | 540 | 520.35 | 173.45 | 182.78 |

Table 2. CONSTANTS FOR p′ (LONG FORMULA), AN APPROXIMATION OF
THE LOWER BINOMIAL CONFIDENCE LIMIT p; c/n ≤ .5, n ≥ 20,
c ≤ 1,000 (*Continued*)

Section 6
$\gamma_1$ = 80%
$\gamma_2$ = 60%

| c | m | a | b | c | m | a | b |
|---|---|---|---|---|---|---|---|
| 541 | 521.33 | 173.78 | 183.11 | 586 | 565.53 | 188.51 | 198.25 |
| 542 | 522.31 | 174.10 | 183.45 | 587 | 566.51 | 188.84 | 198.58 |
| 543 | 523.29 | 174.43 | 183.78 | 588 | 567.49 | 189.16 | 198.92 |
| 544 | 524.27 | 174.76 | 184.12 | 589 | 568.48 | 189.49 | 199.25 |
| 545 | 525.25 | 175.08 | 184.46 | 590 | 569.46 | 189.82 | 199.59 |
| 546 | 526.24 | 175.41 | 184.79 | 591 | 570.44 | 190.15 | 199.93 |
| 547 | 527.22 | 175.74 | 185.13 | 592 | 571.43 | 190.48 | 200.26 |
| 548 | 528.20 | 176.07 | 185.47 | 593 | 572.41 | 190.80 | 200.60 |
| 549 | 529.18 | 176.39 | 185.80 | 594 | 573.39 | 191.13 | 200.93 |
| 550 | 530.17 | 176.72 | 186.14 | 595 | 574.37 | 191.46 | 201.27 |
| 551 | 531.15 | 177.05 | 186.48 | 596 | 575.36 | 191.79 | 201.61 |
| 552 | 532.13 | 177.38 | 186.81 | 597 | 576.34 | 192.11 | 201.94 |
| 553 | 533.11 | 177.70 | 187.15 | 598 | 577.32 | 192.44 | 202.28 |
| 554 | 534.09 | 178.03 | 187.48 | 599 | 578.30 | 192.77 | 202.62 |
| 555 | 535.08 | 178.36 | 187.82 | 600 | 579.29 | 193.10 | 202.95 |
| 556 | 536.06 | 178.69 | 188.16 | 601 | 580.27 | 193.42 | 203.29 |
| 557 | 537.04 | 179.01 | 188.49 | 602 | 581.25 | 193.75 | 203.62 |
| 558 | 538.02 | 179.34 | 188.83 | 603 | 582.24 | 194.08 | 203.96 |
| 559 | 539.00 | 179.67 | 189.17 | 604 | 583.22 | 194.41 | 204.30 |
| 560 | 539.99 | 180.00 | 189.50 | 605 | 584.20 | 194.73 | 204.63 |
| 561 | 540.97 | 180.32 | 189.84 | 606 | 585.18 | 195.06 | 204.97 |
| 562 | 541.95 | 180.65 | 190.17 | 607 | 586.17 | 195.39 | 205.31 |
| 563 | 542.93 | 180.98 | 190.51 | 608 | 587.15 | 195.72 | 205.64 |
| 564 | 543.92 | 181.31 | 190.85 | 609 | 588.13 | 196.04 | 205.98 |
| 565 | 544.90 | 181.63 | 191.18 | 610 | 589.12 | 196.37 | 206.31 |
| 566 | 545.88 | 181.96 | 191.52 | 611 | 590.10 | 196.70 | 206.65 |
| 567 | 546.86 | 182.29 | 191.86 | 612 | 591.08 | 197.03 | 206.99 |
| 568 | 547.84 | 182.61 | 192.19 | 613 | 592.07 | 197.36 | 207.32 |
| 569 | 548.83 | 182.94 | 192.53 | 614 | 593.05 | 197.68 | 207.66 |
| 570 | 549.81 | 183.27 | 192.87 | 615 | 594.03 | 198.01 | 207.99 |
| 571 | 550.79 | 183.60 | 193.20 | 616 | 595.01 | 198.34 | 208.33 |
| 572 | 551.77 | 183.92 | 193.54 | 617 | 596.00 | 198.67 | 208.67 |
| 573 | 552.76 | 184.25 | 193.87 | 618 | 596.98 | 198.99 | 209.00 |
| 574 | 553.74 | 184.58 | 194.21 | 619 | 597.96 | 199.32 | 209.34 |
| 575 | 554.72 | 184.91 | 194.55 | 620 | 598.95 | 199.65 | 209.68 |
| 576 | 555.70 | 185.23 | 194.88 | 621 | 599.93 | 199.98 | 210.01 |
| 577 | 556.69 | 185.56 | 195.22 | 622 | 600.91 | 200.30 | 210.35 |
| 578 | 557.67 | 185.89 | 195.56 | 623 | 601.90 | 200.63 | 210.68 |
| 579 | 558.65 | 186.22 | 195.89 | 624 | 602.88 | 200.96 | 211.02 |
| 580 | 559.63 | 186.54 | 196.23 | 625 | 603.86 | 201.29 | 211.36 |
| 581 | 560.62 | 186.87 | 196.56 | 626 | 604.85 | 201.62 | 211.69 |
| 582 | 561.60 | 187.20 | 196.90 | 627 | 605.83 | 201.94 | 212.03 |
| 583 | 562.58 | 187.53 | 197.24 | 628 | 606.81 | 202.27 | 212.36 |
| 584 | 563.56 | 187.85 | 197.57 | 629 | 607.80 | 202.60 | 212.70 |
| 585 | 564.55 | 188.18 | 197.91 | 630 | 608.78 | 202.93 | 213.04 |

Table 2.  CONSTANTS FOR p′ (LONG FORMULA), AN APPROXIMATION OF THE LOWER BINOMIAL CONFIDENCE LIMIT $\underline{p}$; c/n ≤ .5, n ≥ 20, c ≤ 1,000 (*Continued*)

| c | m | a | b | c | m | a | b |
|---|---|---|---|---|---|---|---|
| 631 | 609.76 | 203.25 | 213.37 | 676 | 654.02 | 218.01 | 228.50 |
| 632 | 610.74 | 203.58 | 213.71 | 677 | 655.00 | 218.33 | 228.83 |
| 633 | 611.73 | 203.91 | 214.05 | 678 | 655.99 | 218.66 | 229.17 |
| 634 | 612.71 | 204.24 | 214.38 | 679 | 656.97 | 218.99 | 229.50 |
| 635 | 613.69 | 204.56 | 214.72 | 680 | 657.96 | 219.32 | 229.84 |
| 636 | 614.68 | 204.89 | 215.05 | 681 | 658.94 | 219.65 | 230.18 |
| 637 | 615.66 | 205.22 | 215.39 | 682 | 659.92 | 219.97 | 230.51 |
| 638 | 616.64 | 205.55 | 215.73 | 683 | 660.91 | 220.30 | 230.85 |
| 639 | 617.63 | 205.88 | 216.06 | 684 | 661.89 | 220.63 | 231.18 |
| 640 | 618.61 | 206.20 | 216.40 | 685 | 662.88 | 220.96 | 231.52 |
| 641 | 619.59 | 206.53 | 216.73 | 686 | 663.86 | 221.29 | 231.86 |
| 642 | 620.58 | 206.86 | 217.07 | 687 | 664.84 | 221.61 | 232.19 |
| 643 | 621.56 | 207.19 | 217.41 | 688 | 665.83 | 221.94 | 232.53 |
| 644 | 622.54 | 207.51 | 217.74 | 689 | 666.81 | 222.27 | 232.86 |
| 645 | 623.53 | 207.84 | 218.08 | 690 | 667.80 | 222.60 | 233.20 |
| 646 | 624.51 | 208.17 | 218.41 | 691 | 668.78 | 222.93 | 233.54 |
| 647 | 625.50 | 208.50 | 218.75 | 692 | 669.76 | 223.25 | 233.87 |
| 648 | 626.48 | 208.83 | 219.09 | 693 | 670.75 | 223.58 | 234.21 |
| 649 | 627.46 | 209.15 | 219.42 | 694 | 671.73 | 223.91 | 234.54 |
| 650 | 628.45 | 209.48 | 219.76 | 695 | 672.72 | 224.24 | 234.88 |
| 651 | 629.43 | 209.81 | 220.10 | 696 | 673.70 | 224.57 | 235.22 |
| 652 | 630.41 | 210.14 | 220.43 | 697 | 674.68 | 224.89 | 235.55 |
| 653 | 631.40 | 210.47 | 220.77 | 698 | 675.67 | 225.22 | 235.89 |
| 654 | 632.38 | 210.79 | 221.10 | 699 | 676.65 | 225.55 | 236.22 |
| 655 | 633.36 | 211.12 | 221.44 | 700 | 677.64 | 225.88 | 236.56 |
| 656 | 634.35 | 211.45 | 221.78 | 701 | 678.62 | 226.21 | 236.90 |
| 657 | 635.33 | 211.78 | 222.11 | 702 | 679.60 | 226.53 | 237.23 |
| 658 | 636.31 | 212.10 | 222.45 | 703 | 680.59 | 226.86 | 237.57 |
| 659 | 637.30 | 212.43 | 222.78 | 704 | 681.57 | 227.19 | 237.90 |
| 660 | 638.28 | 212.76 | 223.12 | 705 | 682.56 | 227.52 | 238.24 |
| 661 | 639.26 | 213.09 | 223.46 | 706 | 683.54 | 227.85 | 238.58 |
| 662 | 640.25 | 213.42 | 223.79 | 707 | 684.52 | 228.17 | 238.91 |
| 663 | 641.23 | 213.74 | 224.13 | 708 | 685.51 | 228.50 | 239.25 |
| 664 | 642.22 | 214.07 | 224.46 | 709 | 686.49 | 228.83 | 239.58 |
| 665 | 643.20 | 214.40 | 224.80 | 710 | 687.48 | 229.16 | 239.92 |
| 666 | 644.18 | 214.73 | 225.14 | 711 | 688.46 | 229.49 | 240.26 |
| 667 | 645.17 | 215.06 | 225.47 | 712 | 689.45 | 229.82 | 240.59 |
| 668 | 646.15 | 215.38 | 225.81 | 713 | 690.43 | 230.14 | 240.93 |
| 669 | 647.13 | 215.71 | 226.14 | 714 | 691.41 | 230.47 | 241.26 |
| 670 | 648.12 | 216.04 | 226.48 | 715 | 692.40 | 230.80 | 241.60 |
| 671 | 649.10 | 216.37 | 226.82 | 716 | 693.38 | 231.13 | 241.94 |
| 672 | 650.09 | 216.70 | 227.15 | 717 | 694.37 | 231.46 | 242.27 |
| 673 | 651.07 | 217.02 | 227.49 | 718 | 695.35 | 231.78 | 242.61 |
| 674 | 652.05 | 217.35 | 227.82 | 719 | 696.34 | 232.11 | 242.94 |
| 675 | 653.04 | 217.68 | 228.16 | 720 | 697.32 | 232.44 | 243.28 |

Table 2. CONSTANTS FOR p′ (LONG FORMULA), AN APPROXIMATION OF
THE LOWER BINOMIAL CONFIDENCE LIMIT p; c/n ≤ .5, n ≥ 20,
c ≤ 1,000 (*Continued*)

Section 6

$\gamma_1 = 80\%$

$\gamma_2 = 60\%$

| c | m | a | b | c | m | a | b |
|---|---|---|---|---|---|---|---|
| 721 | 698.30 | 232.77 | 243.62 | 766 | 742.61 | 247.54 | 258.73 |
| 722 | 699.29 | 233.10 | 243.95 | 767 | 743.59 | 247.86 | 259.07 |
| 723 | 700.27 | 233.42 | 244.29 | 768 | 744.58 | 248.19 | 259.40 |
| 724 | 701.26 | 233.75 | 244.62 | 769 | 745.56 | 248.52 | 259.74 |
| 725 | 702.24 | 234.08 | 244.96 | 770 | 746.55 | 248.85 | 260.08 |
| 726 | 703.23 | 234.41 | 245.30 | 771 | 747.53 | 249.18 | 260.41 |
| 727 | 704.21 | 234.74 | 245.63 | 772 | 748.52 | 249.51 | 260.75 |
| 728 | 705.19 | 235.06 | 245.97 | 773 | 749.50 | 249.83 | 261.08 |
| 729 | 706.18 | 235.39 | 246.30 | 774 | 750.49 | 250.16 | 261.42 |
| 730 | 707.16 | 235.72 | 246.64 | 775 | 751.47 | 250.49 | 261.75 |
| 731 | 708.15 | 236.05 | 246.98 | 776 | 752.46 | 250.82 | 262.09 |
| 732 | 709.13 | 236.38 | 247.31 | 777 | 753.44 | 251.15 | 262.43 |
| 733 | 710.12 | 236.71 | 247.65 | 778 | 754.43 | 251.48 | 262.76 |
| 734 | 711.10 | 237.03 | 247.98 | 779 | 755.41 | 251.80 | 263.10 |
| 735 | 712.09 | 237.36 | 248.32 | 780 | 756.40 | 252.13 | 263.43 |
| 736 | 713.07 | 237.69 | 248.65 | 781 | 757.38 | 252.46 | 263.77 |
| 737 | 714.05 | 238.02 | 248.99 | 782 | 758.37 | 252.79 | 264.11 |
| 738 | 715.04 | 238.35 | 249.33 | 783 | 759.35 | 253.12 | 264.44 |
| 739 | 716.02 | 238.67 | 249.66 | 784 | 760.34 | 253.45 | 264.78 |
| 740 | 717.01 | 239.00 | 250.00 | 785 | 761.32 | 253.77 | 265.11 |
| 741 | 717.99 | 239.33 | 250.33 | 786 | 762.31 | 254.10 | 265.45 |
| 742 | 718.98 | 239.66 | 250.67 | 787 | 763.29 | 254.43 | 265.78 |
| 743 | 719.96 | 239.99 | 251.01 | 788 | 764.28 | 254.76 | 266.12 |
| 744 | 720.95 | 240.32 | 251.34 | 789 | 765.26 | 255.09 | 266.46 |
| 745 | 721.93 | 240.64 | 251.68 | 790 | 766.25 | 255.42 | 266.79 |
| 746 | 722.92 | 240.97 | 252.01 | 791 | 767.23 | 255.74 | 267.13 |
| 747 | 723.90 | 241.30 | 252.35 | 792 | 768.22 | 256.07 | 267.46 |
| 748 | 724.88 | 241.63 | 252.69 | 793 | 769.20 | 256.40 | 267.80 |
| 749 | 725.87 | 241.96 | 253.02 | 794 | 770.19 | 256.73 | 268.14 |
| 750 | 726.85 | 242.28 | 253.36 | 795 | 771.17 | 257.06 | 268.47 |
| 751 | 727.84 | 242.61 | 253.69 | 796 | 772.16 | 257.39 | 268.81 |
| 752 | 728.82 | 242.94 | 254.03 | 797 | 773.14 | 257.71 | 269.14 |
| 753 | 729.81 | 243.27 | 254.37 | 798 | 774.13 | 258.04 | 269.48 |
| 754 | 730.79 | 243.60 | 254.70 | 799 | 775.11 | 258.37 | 269.81 |
| 755 | 731.78 | 243.93 | 255.04 | 800 | 776.10 | 258.70 | 270.15 |
| 756 | 732.76 | 244.25 | 255.37 | 801 | 777.08 | 259.03 | 270.49 |
| 757 | 733.75 | 244.58 | 255.71 | 802 | 778.07 | 259.36 | 270.82 |
| 758 | 734.73 | 244.91 | 256.04 | 803 | 779.05 | 259.68 | 271.16 |
| 759 | 735.72 | 245.24 | 256.38 | 804 | 780.04 | 260.01 | 271.49 |
| 760 | 736.70 | 245.57 | 256.72 | 805 | 781.02 | 260.34 | 271.83 |
| 761 | 737.69 | 245.90 | 257.05 | 806 | 782.01 | 260.67 | 272.17 |
| 762 | 738.67 | 246.22 | 257.39 | 807 | 782.99 | 261.00 | 272.50 |
| 763 | 739.66 | 246.55 | 257.72 | 808 | 783.98 | 261.33 | 272.84 |
| 764 | 740.64 | 246.88 | 258.06 | 809 | 784.96 | 261.65 | 273.17 |
| 765 | 741.62 | 247.21 | 258.40 | 810 | 785.95 | 261.98 | 273.51 |

Table 2. **CONSTANTS FOR $\underline{p}'$ (LONG FORMULA), AN APPROXIMATION OF THE LOWER BINOMIAL CONFIDENCE LIMIT $\underline{p}$; $c/n \leqslant .5$, $n \geqslant 20$, $c \leqslant 1,000$** (*Continued*)

| c | m | a | b | c | m | a | b |
|---|---|---|---|---|---|---|---|
| 811 | 786.94 | 262.31 | 273.84 | 856 | 831.28 | 277.09 | 288.95 |
| 812 | 787.92 | 262.64 | 274.18 | 857 | 832.26 | 277.42 | 289.29 |
| 813 | 788.91 | 262.97 | 274.52 | 858 | 833.25 | 277.75 | 289.63 |
| 814 | 789.89 | 263.30 | 274.85 | 859 | 834.24 | 278.08 | 289.96 |
| 815 | 790.88 | 263.63 | 275.19 | 860 | 835.22 | 278.41 | 290.30 |
| 816 | 791.86 | 263.95 | 275.52 | 861 | 836.21 | 278.74 | 290.63 |
| 817 | 792.85 | 264.28 | 275.86 | 862 | 837.19 | 279.06 | 290.97 |
| 818 | 793.83 | 264.61 | 276.19 | 863 | 838.18 | 279.39 | 291.30 |
| 819 | 794.82 | 264.94 | 276.53 | 864 | 839.16 | 279.72 | 291.64 |
| 820 | 795.80 | 265.27 | 276.87 | 865 | 840.15 | 280.05 | 291.98 |
| 821 | 796.79 | 265.60 | 277.20 | 866 | 841.14 | 280.38 | 292.31 |
| 822 | 797.77 | 265.92 | 277.54 | 867 | 842.12 | 280.71 | 292.65 |
| 823 | 798.76 | 266.25 | 277.87 | 868 | 843.11 | 281.04 | 292.98 |
| 824 | 799.74 | 266.58 | 278.21 | 869 | 844.09 | 281.36 | 293.32 |
| 825 | 800.73 | 266.91 | 278.55 | 870 | 845.08 | 281.69 | 293.65 |
| 826 | 801.71 | 267.24 | 278.88 | 871 | 846.06 | 282.02 | 293.99 |
| 827 | 802.70 | 267.57 | 279.22 | 872 | 847.05 | 282.35 | 294.33 |
| 828 | 803.69 | 267.90 | 279.55 | 873 | 848.04 | 282.68 | 294.66 |
| 829 | 804.67 | 268.22 | 279.89 | 874 | 849.02 | 283.01 | 295.00 |
| 830 | 805.66 | 268.55 | 280.22 | 875 | 850.01 | 283.34 | 295.33 |
| 831 | 806.64 | 268.88 | 280.56 | 876 | 850.99 | 283.66 | 295.67 |
| 832 | 807.63 | 269.21 | 280.90 | 877 | 851.98 | 283.99 | 296.00 |
| 833 | 808.61 | 269.54 | 281.23 | 878 | 852.96 | 284.32 | 296.34 |
| 834 | 809.60 | 269.87 | 281.57 | 879 | 853.95 | 284.65 | 296.67 |
| 835 | 810.58 | 270.19 | 281.90 | 880 | 854.94 | 284.98 | 297.01 |
| 836 | 811.57 | 270.52 | 282.24 | 881 | 855.92 | 285.31 | 297.35 |
| 837 | 812.55 | 270.85 | 282.57 | 882 | 856.91 | 285.64 | 297.68 |
| 838 | 813.54 | 271.18 | 282.91 | 883 | 857.89 | 285.96 | 298.02 |
| 839 | 814.52 | 271.51 | 283.25 | 884 | 858.88 | 286.29 | 298.35 |
| 840 | 815.51 | 271.84 | 283.58 | 885 | 859.87 | 286.62 | 298.69 |
| 841 | 816.50 | 272.17 | 283.92 | 886 | 860.85 | 286.95 | 299.02 |
| 842 | 817.48 | 272.49 | 284.25 | 887 | 861.84 | 287.28 | 299.36 |
| 843 | 818.47 | 272.82 | 284.59 | 888 | 862.82 | 287.61 | 299.70 |
| 844 | 819.45 | 273.15 | 284.92 | 889 | 863.81 | 287.94 | 300.03 |
| 845 | 820.44 | 273.48 | 285.26 | 890 | 864.79 | 288.26 | 300.37 |
| 846 | 821.42 | 273.81 | 285.60 | 891 | 865.78 | 288.59 | 300.70 |
| 847 | 822.41 | 274.14 | 285.93 | 892 | 866.77 | 288.92 | 301.04 |
| 848 | 823.39 | 274.46 | 286.27 | 893 | 867.75 | 289.25 | 301.37 |
| 849 | 824.38 | 274.79 | 286.60 | 894 | 868.74 | 289.58 | 301.71 |
| 850 | 825.37 | 275.12 | 286.94 | 895 | 869.72 | 289.91 | 302.05 |
| 851 | 826.35 | 275.45 | 287.27 | 896 | 870.71 | 290.24 | 302.38 |
| 852 | 827.34 | 275.78 | 287.61 | 897 | 871.70 | 290.57 | 302.72 |
| 853 | 828.32 | 276.11 | 287.95 | 898 | 872.68 | 290.89 | 303.05 |
| 854 | 829.31 | 276.44 | 288.28 | 899 | 873.67 | 291.22 | 303.39 |
| 855 | 830.29 | 276.76 | 288.62 | 900 | 874.65 | 291.55 | 303.72 |

Table 2. CONSTANTS FOR p′ (LONG FORMULA), AN APPROXIMATION OF THE LOWER BINOMIAL CONFIDENCE LIMIT p; c/n ⩽ .5, n ⩾ 20, c ⩽ 1,000 (*Continued*)

Section 6

$\gamma_1 = 80\%$
$\gamma_2 = 60\%$

| c | m | a | b | c | m | a | b |
|---|---|---|---|---|---|---|---|
| 901 | 875.64 | 291.88 | 304.06 | 946 | 920.02 | 306.67 | 319.16 |
| 902 | 876.63 | 292.21 | 304.40 | 947 | 921.00 | 307.00 | 319.50 |
| 903 | 877.61 | 292.54 | 304.73 | 948 | 921.99 | 307.33 | 319.84 |
| 904 | 878.60 | 292.87 | 305.07 | 949 | 922.98 | 307.66 | 320.17 |
| 905 | 879.58 | 293.19 | 305.40 | 950 | 923.96 | 307.99 | 320.51 |
| 906 | 880.57 | 293.52 | 305.74 | 951 | 924.95 | 308.32 | 320.84 |
| 907 | 881.56 | 293.85 | 306.07 | 952 | 925.93 | 308.64 | 321.18 |
| 908 | 882.54 | 294.18 | 306.41 | 953 | 926.92 | 308.97 | 321.51 |
| 909 | 883.53 | 294.51 | 306.75 | 954 | 927.91 | 309.30 | 321.85 |
| 910 | 884.51 | 294.84 | 307.08 | 955 | 928.89 | 309.63 | 322.18 |
| 911 | 885.50 | 295.17 | 307.42 | 956 | 929.88 | 309.96 | 322.52 |
| 912 | 886.49 | 295.50 | 307.75 | 957 | 930.87 | 310.29 | 322.86 |
| 913 | 887.47 | 295.82 | 308.09 | 958 | 931.85 | 310.62 | 323.19 |
| 914 | 888.46 | 296.15 | 308.42 | 959 | 932.84 | 310.95 | 323.53 |
| 915 | 889.44 | 296.48 | 308.76 | 960 | 933.83 | 311.28 | 323.86 |
| 916 | 890.43 | 296.81 | 309.09 | 961 | 934.81 | 311.60 | 324.20 |
| 917 | 891.42 | 297.14 | 309.43 | 962 | 935.80 | 311.93 | 324.53 |
| 918 | 892.40 | 297.47 | 309.77 | 963 | 936.79 | 312.26 | 324.87 |
| 919 | 893.39 | 297.80 | 310.10 | 964 | 937.77 | 312.59 | 325.20 |
| 920 | 894.38 | 298.13 | 310.44 | 965 | 938.76 | 312.92 | 325.54 |
| 921 | 895.36 | 298.45 | 310.77 | 966 | 939.74 | 313.25 | 325.88 |
| 922 | 896.35 | 298.78 | 311.11 | 967 | 940.73 | 313.58 | 326.21 |
| 923 | 897.33 | 299.11 | 311.44 | 968 | 941.72 | 313.91 | 326.55 |
| 924 | 898.32 | 299.44 | 311.78 | 969 | 942.70 | 314.23 | 326.88 |
| 925 | 899.31 | 299.77 | 312.12 | 970 | 943.69 | 314.56 | 327.22 |
| 926 | 900.29 | 300.10 | 312.45 | 971 | 944.68 | 314.89 | 327.55 |
| 927 | 901.28 | 300.43 | 312.79 | 972 | 945.66 | 315.22 | 327.89 |
| 928 | 902.26 | 300.75 | 313.12 | 973 | 946.65 | 315.55 | 328.22 |
| 929 | 903.25 | 301.08 | 313.46 | 974 | 947.64 | 315.88 | 328.56 |
| 930 | 904.24 | 301.41 | 313.79 | 975 | 948.62 | 316.21 | 328.90 |
| 931 | 905.22 | 301.74 | 314.13 | 976 | 949.61 | 316.54 | 329.23 |
| 932 | 906.21 | 302.07 | 314.47 | 977 | 950.60 | 316.87 | 329.57 |
| 933 | 907.20 | 302.40 | 314.80 | 978 | 951.58 | 317.19 | 329.90 |
| 934 | 908.18 | 302.73 | 315.14 | 979 | 952.57 | 317.52 | 330.24 |
| 935 | 909.17 | 303.06 | 315.47 | 980 | 953.56 | 317.85 | 330.57 |
| 936 | 910.15 | 303.38 | 315.81 | 981 | 954.54 | 318.18 | 330.91 |
| 937 | 911.14 | 303.71 | 316.14 | 982 | 955.53 | 318.51 | 331.25 |
| 938 | 912.13 | 304.04 | 316.48 | 983 | 956.52 | 318.84 | 331.58 |
| 939 | 913.11 | 304.37 | 316.81 | 984 | 957.50 | 319.17 | 331.92 |
| 940 | 914.10 | 304.70 | 317.15 | 985 | 958.49 | 319.50 | 332.25 |
| 941 | 915.09 | 305.03 | 317.49 | 986 | 959.48 | 319.83 | 332.59 |
| 942 | 916.07 | 305.36 | 317.82 | 987 | 960.46 | 320.15 | 332.92 |
| 943 | 917.06 | 305.69 | 318.16 | 988 | 961.45 | 320.48 | 333.26 |
| 944 | 918.04 | 306.01 | 318.49 | 989 | 962.44 | 320.81 | 333.59 |
| 945 | 919.03 | 306.34 | 318.83 | 990 | 963.42 | 321.14 | 333.93 |

Table 2.  CONSTANTS FOR $\underline{p}'$ (LONG FORMULA), AN APPROXIMATION OF
THE LOWER BINOMIAL CONFIDENCE LIMIT $\underline{p}$; $c/n \leqslant .5$, $n \geqslant 20$,
$c \leqslant 1,000$ (*Continued*)

| c | m | a | b | c | m | a | b |
|---|---|---|---|---|---|---|---|
| 991 | 964.41 | 321.47 | 334.27 | 996 | 969.34 | 323.11 | 335.94 |
| 992 | 965.40 | 321.80 | 334.60 | 997 | 970.33 | 323.44 | 336.28 |
| 993 | 966.38 | 322.13 | 334.94 | 998 | 971.31 | 323.77 | 336.61 |
| 994 | 967.37 | 322.46 | 335.27 | 999 | 972.30 | 324.10 | 336.95 |
| 995 | 968.35 | 322.78 | 335.61 | 1000 | 973.29 | 324.43 | 337.29 |

# Table 3

CONSTANTS FOR $\bar{p}'$ (SHORT FORMULA), AN APPROXIMATION OF THE UPPER BINOMIAL CONFIDENCE LIMIT $\bar{p}$; $c/n \leqslant .1$, $n \geqslant 40$, $c \leqslant 1{,}000$

$$\text{Approximation Formula: } \bar{p}' = \frac{\bar{m}}{n + \bar{k}}$$

n     sample size.

c     number of events (items having a designated characteristic) in the sample.

$\bar{p}'$     approximation of the upper binomial confidence limit $\bar{p}$; $\bar{p}'$ has relative accuracy of at least .999.

$\bar{m}$     upper confidence limit for the parameter m of a Poisson distribution; each $\bar{m}$ is for a given c and a given $\gamma_1$ or $\gamma_2$.

$\bar{k}$     approximation constant for a given c and a given $\gamma_1$ or $\gamma_2$.

$\gamma_1$     confidence level for a one-sided (upper or lower) confidence limit.

$\gamma_2$     confidence level for two-sided (upper and lower) confidence limits. ($\gamma_2 = 2\gamma_1 - 100\%$.)

For $c > 1{,}000$, calculate $\bar{k}$ from the extension formula in Table 7–C; and calculate $\bar{m}$ from the extension formula in Table 7–E.

(Note. For the purpose of acceptance sampling in conjunction with Tables 8 and 13, c in Table 3 may be read as acceptance number $\dot{c}$.)

| | | |
|---|---|---|
| **Section 1** | $\gamma_1 = 99.5\%$ | $\gamma_2 = 99\%$ |
| **Section 2** | $\gamma_1 = 99\%$ | $\gamma_2 = 98\%$ |
| **Section 3** | $\gamma_1 = 97.5\%$ | $\gamma_2 = 95\%$ |
| **Section 4** | $\gamma_1 = 95\%$ | $\gamma_2 = 90\%$ |
| **Section 5** | $\gamma_1 = 90\%$ | $\gamma_2 = 80\%$ |
| **Section 6** | $\gamma_1 = 80\%$ | $\gamma_2 = 60\%$ |

**Table 3.** **CONSTANTS FOR $\overline{p}'$ (SHORT FORMULA), AN APPROXIMATION OF THE UPPER BINOMIAL CONFIDENCE LIMIT $\overline{p}$; $c/n \leqslant .1$, $n \geqslant 40$, $c \leqslant 1,000$**

| c | $\overline{m}$ | $\overline{k}$ | c | $\overline{m}$ | $\overline{k}$ | c | $\overline{m}$ | $\overline{k}$ |
|---|---|---|---|---|---|---|---|---|
| 0 | 5.2983 | 2.693 | 45 | 65.341 | 10.47 | 90 | 117.45 | 13.72 |
| 1 | 7.4302 | 3.301 | 46 | 66.530 | 10.59 | 91 | 118.58 | 13.79 |
| 2 | 9.2738 | 3.761 | 47 | 67.717 | 10.68 | 92 | 119.71 | 13.86 |
| 3 | 10.978 | 4.154 | 48 | 68.902 | 10.80 | 93 | 120.85 | 13.93 |
| 4 | 12.594 | 4.508 | 49 | 70.085 | 10.89 | 94 | 121.98 | 13.99 |
| 5 | 14.150 | 4.782 | 50 | 71.266 | 10.98 | 95 | 123.11 | 14.06 |
| 6 | 15.660 | 5.031 | 51 | 72.446 | 10.73 | 96 | 124.24 | 14.12 |
| 7 | 17.134 | 5.270 | 52 | 73.624 | 10.82 | 97 | 125.37 | 14.19 |
| 8 | 18.578 | 5.492 | 53 | 74.800 | 10.90 | 98 | 126.50 | 14.25 |
| 9 | 19.998 | 5.700 | 54 | 75.974 | 10.99 | 99 | 127.63 | 14.32 |
| 10 | 21.398 | 5.91 | 55 | 77.147 | 11.08 | 100 | 128.76 | 14.38 |
| 11 | 22.779 | 6.09 | 56 | 78.319 | 11.16 | 101 | 129.89 | 14.45 |
| 12 | 24.145 | 6.28 | 57 | 79.489 | 11.25 | 102 | 131.02 | 14.51 |
| 13 | 25.497 | 6.46 | 58 | 80.657 | 11.33 | 103 | 132.14 | 14.57 |
| 14 | 26.836 | 6.63 | 59 | 81.824 | 11.42 | 104 | 133.27 | 14.64 |
| 15 | 28.164 | 6.79 | 60 | 82.990 | 11.50 | 105 | 134.39 | 14.70 |
| 16 | 29.482 | 6.95 | 61 | 84.154 | 11.58 | 106 | 135.52 | 14.76 |
| 17 | 30.791 | 7.12 | 62 | 85.317 | 11.66 | 107 | 136.64 | 14.82 |
| 18 | 32.091 | 7.27 | 63 | 86.479 | 11.74 | 108 | 137.77 | 14.89 |
| 19 | 33.383 | 7.42 | 64 | 87.639 | 11.82 | 109 | 138.89 | 14.95 |
| 20 | 34.668 | 7.56 | 65 | 88.799 | 11.90 | 110 | 140.01 | 15.01 |
| 21 | 35.946 | 7.70 | 66 | 89.957 | 11.98 | 111 | 141.13 | 15.07 |
| 22 | 37.218 | 7.83 | 67 | 91.113 | 12.06 | 112 | 142.26 | 15.13 |
| 23 | 38.484 | 7.97 | 68 | 92.269 | 12.14 | 113 | 143.38 | 15.19 |
| 24 | 39.745 | 8.10 | 69 | 93.424 | 12.21 | 114 | 144.50 | 15.25 |
| 25 | 41.000 | 8.24 | 70 | 94.577 | 12.29 | 115 | 145.62 | 15.31 |
| 26 | 42.251 | 8.37 | 71 | 95.729 | 12.37 | 116 | 146.74 | 15.37 |
| 27 | 43.497 | 8.49 | 72 | 96.881 | 12.44 | 117 | 147.86 | 15.43 |
| 28 | 44.738 | 8.60 | 73 | 98.031 | 12.52 | 118 | 148.97 | 15.49 |
| 29 | 45.976 | 8.74 | 74 | 99.180 | 12.59 | 119 | 150.09 | 15.55 |
| 30 | 47.209 | 8.85 | 75 | 100.33 | 12.67 | 120 | 151.21 | 15.61 |
| 31 | 48.439 | 8.97 | 76 | 101.48 | 12.74 | 121 | 152.33 | 15.66 |
| 32 | 49.665 | 9.07 | 77 | 102.62 | 12.81 | 122 | 153.44 | 15.72 |
| 33 | 50.888 | 9.19 | 78 | 103.77 | 12.89 | 123 | 154.56 | 15.78 |
| 34 | 52.108 | 9.30 | 79 | 104.91 | 12.96 | 124 | 155.67 | 15.84 |
| 35 | 53.324 | 9.41 | 80 | 106.06 | 13.03 | 125 | 156.79 | 15.90 |
| 36 | 54.537 | 9.53 | 81 | 107.20 | 13.10 | 126 | 157.90 | 15.95 |
| 37 | 55.748 | 9.64 | 82 | 108.34 | 13.17 | 127 | 159.02 | 16.01 |
| 38 | 56.956 | 9.73 | 83 | 109.48 | 13.24 | 128 | 160.13 | 16.07 |
| 39 | 58.161 | 9.84 | 84 | 110.62 | 13.31 | 129 | 161.24 | 16.12 |
| 40 | 59.363 | 9.95 | 85 | 111.76 | 13.38 | 130 | 162.36 | 16.18 |
| 41 | 60.563 | 10.05 | 86 | 112.90 | 13.45 | 131 | 163.47 | 16.24 |
| 42 | 61.761 | 10.17 | 87 | 114.04 | 13.52 | 132 | 164.58 | 16.29 |
| 43 | 62.957 | 10.27 | 88 | 115.17 | 13.59 | 133 | 165.69 | 16.35 |
| 44 | 64.150 | 10.38 | 89 | 116.31 | 13.66 | 134 | 166.80 | 16.40 |

$\gamma_1 = 99.5\%$

$\gamma_2 = 99\%$

**Table 3.  CONSTANTS FOR $\overline{p}'$ (SHORT FORMULA), AN APPROXIMATION OF THE UPPER BINOMIAL CONFIDENCE LIMIT $\overline{p}$; $c/n \leqslant .1$, $n \geqslant 40$, $c \leqslant 1,000$ (_Continued_)**

| c | $\overline{m}$ | $\overline{k}$ | c | $\overline{m}$ | $\overline{k}$ | c | $\overline{m}$ | $\overline{k}$ |
|---|---|---|---|---|---|---|---|---|
| 135 | 167.91 | 16.46 | 180 | 217.53 | 18.77 | 225 | 266.60 | 20.8 |
| 136 | 169.02 | 16.51 | 181 | 218.63 | 18.81 | 226 | 267.69 | 20.8 |
| 137 | 170.13 | 16.57 | 182 | 219.72 | 18.86 | 227 | 268.77 | 20.8 |
| 138 | 171.24 | 16.62 | 183 | 220.82 | 18.91 | 228 | 269.86 | 20.9 |
| 139 | 172.35 | 16.68 | 184 | 221.91 | 18.96 | 229 | 270.94 | 20.9 |
| 140 | 173.46 | 16.73 | 185 | 223.01 | 19.00 | 230 | 272.03 | 21.0 |
| 141 | 174.57 | 16.79 | 186 | 224.10 | 19.05 | 231 | 273.11 | 21.0 |
| 142 | 175.68 | 16.84 | 187 | 225.20 | 19.10 | 232 | 274.20 | 21.1 |
| 143 | 176.78 | 16.89 | 188 | 226.29 | 19.14 | 233 | 275.28 | 21.1 |
| 144 | 177.89 | 16.95 | 189 | 227.38 | 19.19 | 234 | 276.36 | 21.1 |
| 145 | 179.00 | 17.00 | 190 | 228.48 | 19.24 | 235 | 277.45 | 21.2 |
| 146 | 180.11 | 17.05 | 191 | 229.57 | 19.28 | 236 | 278.53 | 21.2 |
| 147 | 181.21 | 17.11 | 192 | 230.66 | 19.33 | 237 | 279.62 | 21.3 |
| 148 | 182.32 | 17.16 | 193 | 231.76 | 19.38 | 238 | 280.70 | 21.3 |
| 149 | 183.42 | 17.21 | 194 | 232.85 | 19.42 | 239 | 281.78 | 21.3 |
| 150 | 184.53 | 17.27 | 195 | 233.94 | 19.47 | 240 | 282.87 | 21.4 |
| 151 | 185.63 | 17.32 | 196 | 235.03 | 19.52 | 241 | 283.95 | 21.4 |
| 152 | 186.74 | 17.37 | 197 | 236.12 | 19.56 | 242 | 285.03 | 21.5 |
| 153 | 187.84 | 17.42 | 198 | 237.21 | 19.61 | 243 | 286.11 | 21.5 |
| 154 | 188.94 | 17.47 | 199 | 238.31 | 19.65 | 244 | 287.20 | 21.6 |
| 155 | 190.05 | 17.53 | 200 | 239.40 | 19.70 | 245 | 288.28 | 21.6 |
| 156 | 191.15 | 17.58 | 201 | 240.49 | 19.74 | 246 | 289.36 | 21.6 |
| 157 | 192.25 | 17.63 | 202 | 241.58 | 19.79 | 247 | 290.44 | 21.7 |
| 158 | 193.36 | 17.68 | 203 | 242.67 | 19.83 | 248 | 291.52 | 21.7 |
| 159 | 194.46 | 17.73 | 204 | 243.76 | 19.88 | 249 | 292.61 | 21.8 |
| 160 | 195.56 | 17.78 | 205 | 244.85 | 19.92 | 250 | 293.69 | 21.8 |
| 161 | 196.66 | 17.83 | 206 | 245.94 | 19.97 | 251 | 294.77 | 21.8 |
| 162 | 197.76 | 17.88 | 207 | 247.03 | 20.01 | 252 | 295.85 | 21.9 |
| 163 | 198.86 | 17.93 | 208 | 248.12 | 20.06 | 253 | 296.93 | 21.9 |
| 164 | 199.97 | 17.98 | 209 | 249.21 | 20.10 | 254 | 298.01 | 22.0 |
| 165 | 201.07 | 18.03 | 210 | 250.29 | 20.15 | 255 | 299.09 | 22.0 |
| 166 | 202.16 | 18.08 | 211 | 251.38 | 20.19 | 256 | 300.17 | 22.0 |
| 167 | 203.26 | 18.13 | 212 | 252.47 | 20.24 | 257 | 301.25 | 22.1 |
| 168 | 204.36 | 18.18 | 213 | 253.56 | 20.28 | 258 | 302.33 | 22.1 |
| 169 | 205.46 | 18.23 | 214 | 254.65 | 20.32 | 259 | 303.41 | 22.2 |
| 170 | 206.56 | 18.28 | 215 | 255.73 | 20.37 | 260 | 304.49 | 22.2 |
| 171 | 207.66 | 18.33 | 216 | 256.82 | 20.41 | 261 | 305.57 | 22.2 |
| 172 | 208.76 | 18.38 | 217 | 257.91 | 20.45 | 262 | 306.65 | 22.3 |
| 173 | 209.86 | 18.43 | 218 | 259.00 | 20.50 | 263 | 307.73 | 22.3 |
| 174 | 210.95 | 18.48 | 219 | 260.08 | 20.54 | 264 | 308.81 | 22.4 |
| 175 | 212.05 | 18.53 | 220 | 261.17 | 20.59 | 265 | 309.89 | 22.4 |
| 176 | 213.15 | 18.57 | 221 | 262.26 | 20.63 | 266 | 310.97 | 22.4 |
| 177 | 214.24 | 18.62 | 222 | 263.34 | 20.67 | 267 | 312.05 | 22.5 |
| 178 | 215.34 | 18.67 | 223 | 264.43 | 20.71 | 268 | 313.12 | 22.5 |
| 179 | 216.44 | 18.72 | 224 | 265.52 | 20.76 | 269 | 314.20 | 22.6 |

Table 3. CONSTANTS FOR $\bar{p}'$ (SHORT FORMULA), AN APPROXIMATION OF THE UPPER BINOMIAL CONFIDENCE LIMIT $\bar{p}$; $c/n \leqslant .1$, $n \geqslant 40$, $c \leqslant 1,000$ (*Continued*)

Section 1

$\gamma_1 = 99.5\%$
$\gamma_2 = 99\%$

| c | $\bar{m}$ | $\bar{k}$ | c | $\bar{m}$ | $\bar{k}$ | c | $\bar{m}$ | $\bar{k}$ |
|---|---|---|---|---|---|---|---|---|
| 270 | 315.28 | 22.64 | 315 | 363.67 | 24.33 | 360 | 411.82 | 25.91 |
| 271 | 316.36 | 22.68 | 316 | 364.74 | 24.37 | 361 | 412.89 | 25.94 |
| 272 | 317.44 | 22.72 | 317 | 365.81 | 24.41 | 362 | 413.95 | 25.98 |
| 273 | 318.52 | 22.76 | 318 | 366.88 | 24.44 | 363 | 415.02 | 26.01 |
| 274 | 319.59 | 22.80 | 319 | 367.96 | 24.48 | 364 | 416.09 | 26.04 |
| 275 | 320.67 | 22.84 | 320 | 369.03 | 24.51 | 365 | 417.16 | 26.08 |
| 276 | 321.75 | 22.87 | 321 | 370.10 | 24.55 | 366 | 418.22 | 26.11 |
| 277 | 322.83 | 22.91 | 322 | 371.17 | 24.59 | 367 | 419.29 | 26.15 |
| 278 | 323.90 | 22.95 | 323 | 372.24 | 24.62 | 368 | 420.36 | 26.18 |
| 279 | 324.98 | 22.99 | 324 | 373.31 | 24.66 | 369 | 421.42 | 26.21 |
| 280 | 326.06 | 23.03 | 325 | 374.39 | 24.69 | 370 | 422.49 | 26.25 |
| 281 | 327.13 | 23.07 | 326 | 375.46 | 24.73 | 371 | 423.56 | 26.28 |
| 282 | 328.21 | 23.10 | 327 | 376.53 | 24.76 | 372 | 424.63 | 26.31 |
| 283 | 329.29 | 23.14 | 328 | 377.60 | 24.80 | 373 | 425.69 | 26.35 |
| 284 | 330.36 | 23.18 | 329 | 378.67 | 24.83 | 374 | 426.76 | 26.38 |
| 285 | 331.44 | 23.22 | 330 | 379.74 | 24.87 | 375 | 427.82 | 26.41 |
| 286 | 332.52 | 23.26 | 331 | 380.81 | 24.91 | 376 | 428.89 | 26.45 |
| 287 | 333.59 | 23.30 | 332 | 381.88 | 24.94 | 377 | 429.96 | 26.48 |
| 288 | 334.67 | 23.33 | 333 | 382.95 | 24.98 | 378 | 431.02 | 26.51 |
| 289 | 335.74 | 23.37 | 334 | 384.02 | 25.01 | 379 | 432.09 | 26.54 |
| 290 | 336.82 | 23.41 | 335 | 385.09 | 25.05 | 380 | 433.16 | 26.58 |
| 291 | 337.89 | 23.45 | 336 | 386.16 | 25.08 | 381 | 434.22 | 26.61 |
| 292 | 338.97 | 23.48 | 337 | 387.23 | 25.12 | 382 | 435.29 | 26.64 |
| 293 | 340.04 | 23.52 | 338 | 388.30 | 25.15 | 383 | 436.35 | 26.68 |
| 294 | 341.12 | 23.56 | 339 | 389.37 | 25.19 | 384 | 437.42 | 26.71 |
| 295 | 342.19 | 23.60 | 340 | 390.44 | 25.22 | 385 | 438.48 | 26.74 |
| 296 | 343.27 | 23.63 | 341 | 391.51 | 25.26 | 386 | 439.55 | 26.78 |
| 297 | 344.34 | 23.67 | 342 | 392.58 | 25.29 | 387 | 440.62 | 26.81 |
| 298 | 345.42 | 23.71 | 343 | 393.65 | 25.33 | 388 | 441.68 | 26.84 |
| 299 | 346.49 | 23.75 | 344 | 394.72 | 25.36 | 389 | 442.75 | 26.87 |
| 300 | 347.57 | 23.78 | 345 | 395.79 | 25.40 | 390 | 443.81 | 26.91 |
| 301 | 348.64 | 23.82 | 346 | 396.86 | 25.43 | 391 | 444.88 | 26.94 |
| 302 | 349.71 | 23.86 | 347 | 397.93 | 25.46 | 392 | 445.94 | 26.97 |
| 303 | 350.79 | 23.89 | 348 | 399.00 | 25.50 | 393 | 447.01 | 27.00 |
| 304 | 351.86 | 23.93 | 349 | 400.07 | 25.53 | 394 | 448.07 | 27.04 |
| 305 | 352.94 | 23.97 | 350 | 401.14 | 25.57 | 395 | 449.14 | 27.07 |
| 306 | 354.01 | 24.00 | 351 | 402.20 | 25.60 | 396 | 450.20 | 27.10 |
| 307 | 355.08 | 24.04 | 352 | 403.27 | 25.64 | 397 | 451.27 | 27.13 |
| 308 | 356.16 | 24.08 | 353 | 404.34 | 25.67 | 398 | 452.33 | 27.16 |
| 309 | 357.23 | 24.11 | 354 | 405.41 | 25.71 | 399 | 453.39 | 27.20 |
| 310 | 358.30 | 24.15 | 355 | 406.48 | 25.74 | 400 | 454.46 | 27.23 |
| 311 | 359.38 | 24.19 | 356 | 407.55 | 25.77 | 401 | 455.52 | 27.26 |
| 312 | 360.45 | 24.22 | 357 | 408.61 | 25.81 | 402 | 456.59 | 27.29 |
| 313 | 361.52 | 24.26 | 358 | 409.68 | 25.84 | 403 | 457.65 | 27.33 |
| 314 | 362.59 | 24.30 | 359 | 410.75 | 25.88 | 404 | 458.72 | 27.36 |

Table 3. CONSTANTS FOR $\overline{p}'$ (SHORT FORMULA), AN APPROXIMATION OF THE UPPER BINOMIAL CONFIDENCE LIMIT $\overline{p}$; $c/n \leqslant .1$, $n \geqslant 40$, $c \leqslant 1,000$ (*Continued*)

| c | $\overline{m}$ | $\overline{k}$ | c | $\overline{m}$ | $\overline{k}$ | c | $\overline{m}$ | $\overline{k}$ |
|---|---|---|---|---|---|---|---|---|
| 405 | 459.78 | 27.39 | 450 | 507.58 | 28.79 | 495 | 555.24 | 30.12 |
| 406 | 460.84 | 27.42 | 451 | 508.64 | 28.82 | 496 | 556.30 | 30.15 |
| 407 | 461.91 | 27.45 | 452 | 509.70 | 28.85 | 497 | 557.36 | 30.18 |
| 408 | 462.97 | 27.49 | 453 | 510.76 | 28.88 | 498 | 558.42 | 30.21 |
| 409 | 464.03 | 27.52 | 454 | 511.82 | 28.91 | 499 | 559.47 | 30.24 |
| 410 | 465.10 | 27.55 | 455 | 512.88 | 28.94 | 500 | 560.53 | 30.27 |
| 411 | 466.16 | 27.58 | 456 | 513.94 | 28.97 | 501 | 561.59 | 30.29 |
| 412 | 467.22 | 27.61 | 457 | 515.00 | 29.00 | 502 | 562.65 | 30.32 |
| 413 | 468.29 | 27.64 | 458 | 516.06 | 29.03 | 503 | 563.70 | 30.35 |
| 414 | 469.35 | 27.68 | 459 | 517.12 | 29.06 | 504 | 564.76 | 30.38 |
| 415 | 470.41 | 27.71 | 460 | 518.18 | 29.09 | 505 | 565.82 | 30.41 |
| 416 | 471.48 | 27.74 | 461 | 519.24 | 29.12 | 506 | 566.88 | 30.44 |
| 417 | 472.54 | 27.77 | 462 | 520.30 | 29.15 | 507 | 567.93 | 30.47 |
| 418 | 473.60 | 27.80 | 463 | 521.36 | 29.18 | 508 | 568.99 | 30.50 |
| 419 | 474.67 | 27.83 | 464 | 522.42 | 29.21 | 509 | 570.05 | 30.52 |
| 420 | 475.73 | 27.86 | 465 | 523.48 | 29.24 | 510 | 571.10 | 30.55 |
| 421 | 476.79 | 27.90 | 466 | 524.54 | 29.27 | 511 | 572.16 | 30.58 |
| 422 | 477.85 | 27.93 | 467 | 525.60 | 29.30 | 512 | 573.22 | 30.61 |
| 423 | 478.92 | 27.96 | 468 | 526.66 | 29.33 | 513 | 574.28 | 30.64 |
| 424 | 479.98 | 27.99 | 469 | 527.72 | 29.36 | 514 | 575.33 | 30.67 |
| 425 | 481.04 | 28.02 | 470 | 528.78 | 29.39 | 515 | 576.39 | 30.69 |
| 426 | 482.10 | 28.05 | 471 | 529.84 | 29.42 | 516 | 577.45 | 30.72 |
| 427 | 483.17 | 28.08 | 472 | 530.90 | 29.45 | 517 | 578.50 | 30.75 |
| 428 | 484.23 | 28.11 | 473 | 531.96 | 29.48 | 518 | 579.56 | 30.78 |
| 429 | 485.29 | 28.15 | 474 | 533.02 | 29.51 | 519 | 580.62 | 30.81 |
| 430 | 486.35 | 28.18 | 475 | 534.08 | 29.54 | 520 | 581.67 | 30.84 |
| 431 | 487.42 | 28.21 | 476 | 535.13 | 29.57 | 521 | 582.73 | 30.86 |
| 432 | 488.48 | 28.24 | 477 | 536.19 | 29.60 | 522 | 583.78 | 30.89 |
| 433 | 489.54 | 28.27 | 478 | 537.25 | 29.63 | 523 | 584.84 | 30.92 |
| 434 | 490.60 | 28.30 | 479 | 538.31 | 29.66 | 524 | 585.90 | 30.95 |
| 435 | 491.66 | 28.33 | 480 | 539.37 | 29.68 | 525 | 586.95 | 30.98 |
| 436 | 492.72 | 28.36 | 481 | 540.43 | 29.71 | 526 | 588.01 | 31.00 |
| 437 | 493.79 | 28.39 | 482 | 541.49 | 29.74 | 527 | 589.07 | 31.03 |
| 438 | 494.85 | 28.42 | 483 | 542.55 | 29.77 | 528 | 590.12 | 31.06 |
| 439 | 495.91 | 28.45 | 484 | 543.60 | 29.80 | 529 | 591.18 | 31.09 |
| 440 | 496.97 | 28.49 | 485 | 544.66 | 29.83 | 530 | 592.23 | 31.12 |
| 441 | 498.03 | 28.52 | 486 | 545.72 | 29.86 | 531 | 593.29 | 31.14 |
| 442 | 499.09 | 28.55 | 487 | 546.78 | 29.89 | 532 | 594.35 | 31.17 |
| 443 | 500.15 | 28.58 | 488 | 547.84 | 29.92 | 533 | 595.40 | 31.20 |
| 444 | 501.21 | 28.61 | 489 | 548.90 | 29.95 | 534 | 596.46 | 31.23 |
| 445 | 502.28 | 28.64 | 490 | 549.95 | 29.98 | 535 | 597.51 | 31.26 |
| 446 | 503.34 | 28.67 | 491 | 551.01 | 30.01 | 536 | 598.57 | 31.28 |
| 447 | 504.40 | 28.70 | 492 | 552.07 | 30.04 | 537 | 599.62 | 31.31 |
| 448 | 505.46 | 28.73 | 493 | 553.13 | 30.06 | 538 | 600.68 | 31.34 |
| 449 | 506.52 | 28.76 | 494 | 554.19 | 30.09 | 539 | 601.73 | 31.37 |

**Table 3. CONSTANTS FOR $\overline{p}'$ (SHORT FORMULA), AN APPROXIMATION OF THE UPPER BINOMIAL CONFIDENCE LIMIT $\overline{p}$; $c/n \leqslant .1$, $n \geqslant 40$, $c \leqslant 1,000$ (Continued)**

| c | $\overline{m}$ | $\overline{k}$ | c | $\overline{m}$ | $\overline{k}$ | c | $\overline{m}$ | $\overline{k}$ |
|---|---|---|---|---|---|---|---|---|
| 540 | 602.79 | 31.39 | 585 | 650.23 | 32.62 | 630 | 697.58 | 33.79 |
| 541 | 603.85 | 31.42 | 586 | 651.28 | 32.64 | 631 | 698.63 | 33.82 |
| 542 | 604.90 | 31.45 | 587 | 652.34 | 32.67 | 632 | 699.68 | 33.84 |
| 543 | 605.96 | 31.48 | 588 | 653.39 | 32.70 | 633 | 700.74 | 33.87 |
| 544 | 607.01 | 31.51 | 589 | 654.44 | 32.72 | 634 | 701.79 | 33.89 |
| 545 | 608.07 | 31.53 | 590 | 655.50 | 32.75 | 635 | 702.84 | 33.92 |
| 546 | 609.12 | 31.56 | 591 | 656.55 | 32.78 | 636 | 703.89 | 33.94 |
| 547 | 610.18 | 31.59 | 592 | 657.60 | 32.80 | 637 | 704.94 | 33.97 |
| 548 | 611.23 | 31.62 | 593 | 658.66 | 32.83 | 638 | 705.99 | 34.00 |
| 549 | 612.29 | 31.64 | 594 | 659.71 | 32.85 | 639 | 707.04 | 34.02 |
| 550 | 613.34 | 31.67 | 595 | 660.76 | 32.88 | 640 | 708.09 | 34.05 |
| 551 | 614.40 | 31.70 | 596 | 661.81 | 32.91 | 641 | 709.14 | 34.07 |
| 552 | 615.45 | 31.73 | 597 | 662.87 | 32.93 | 642 | 710.19 | 34.10 |
| 553 | 616.51 | 31.75 | 598 | 663.92 | 32.96 | 643 | 711.24 | 34.12 |
| 554 | 617.56 | 31.78 | 599 | 664.97 | 32.99 | 644 | 712.30 | 34.15 |
| 555 | 618.61 | 31.81 | 600 | 666.02 | 33.01 | 645 | 713.35 | 34.17 |
| 556 | 619.67 | 31.83 | 601 | 667.08 | 33.04 | 646 | 714.40 | 34.20 |
| 557 | 620.72 | 31.86 | 602 | 668.13 | 33.06 | 647 | 715.45 | 34.22 |
| 558 | 621.78 | 31.89 | 603 | 669.18 | 33.09 | 648 | 716.50 | 34.25 |
| 559 | 622.83 | 31.92 | 604 | 670.23 | 33.12 | 649 | 717.55 | 34.27 |
| 560 | 623.89 | 31.94 | 605 | 671.29 | 33.14 | 650 | 718.60 | 34.30 |
| 561 | 624.94 | 31.97 | 606 | 672.34 | 33.17 | 651 | 719.65 | 34.32 |
| 562 | 626.00 | 32.00 | 607 | 673.39 | 33.20 | 652 | 720.70 | 34.35 |
| 563 | 627.05 | 32.03 | 608 | 674.44 | 33.22 | 653 | 721.75 | 34.38 |
| 564 | 628.10 | 32.05 | 609 | 675.50 | 33.25 | 654 | 722.80 | 34.40 |
| 565 | 629.16 | 32.08 | 610 | 676.55 | 33.27 | 655 | 723.85 | 34.43 |
| 566 | 630.21 | 32.11 | 611 | 677.60 | 33.30 | 656 | 724.90 | 34.45 |
| 567 | 631.27 | 32.13 | 612 | 678.65 | 33.33 | 657 | 725.95 | 34.48 |
| 568 | 632.32 | 32.16 | 613 | 679.70 | 33.35 | 658 | 727.00 | 34.50 |
| 569 | 633.37 | 32.19 | 614 | 680.76 | 33.38 | 659 | 728.05 | 34.53 |
| 570 | 634.43 | 32.21 | 615 | 681.81 | 33.40 | 660 | 729.10 | 34.55 |
| 571 | 635.48 | 32.24 | 616 | 682.86 | 33.43 | 661 | 730.15 | 34.58 |
| 572 | 636.54 | 32.27 | 617 | 683.91 | 33.46 | 662 | 731.20 | 34.60 |
| 573 | 637.59 | 32.29 | 618 | 684.96 | 33.48 | 663 | 732.25 | 34.63 |
| 574 | 638.64 | 32.32 | 619 | 686.02 | 33.51 | 664 | 733.30 | 34.65 |
| 575 | 639.70 | 32.35 | 620 | 687.07 | 33.53 | 665 | 734.35 | 34.68 |
| 576 | 640.75 | 32.38 | 621 | 688.12 | 33.56 | 666 | 735.40 | 34.70 |
| 577 | 641.80 | 32.40 | 622 | 689.17 | 33.59 | 667 | 736.45 | 34.73 |
| 578 | 642.86 | 32.43 | 623 | 690.22 | 33.61 | 668 | 737.50 | 34.75 |
| 579 | 643.91 | 32.46 | 624 | 691.27 | 33.64 | 669 | 738.55 | 34.78 |
| 580 | 644.97 | 32.48 | 625 | 692.32 | 33.66 | 670 | 739.60 | 34.80 |
| 581 | 646.02 | 32.51 | 626 | 693.38 | 33.69 | 671 | 740.65 | 34.83 |
| 582 | 647.07 | 32.54 | 627 | 694.43 | 33.71 | 672 | 741.70 | 34.85 |
| 583 | 648.13 | 32.56 | 628 | 695.48 | 33.74 | 673 | 742.75 | 34.87 |
| 584 | 649.18 | 32.59 | 629 | 696.53 | 33.77 | 674 | 743.80 | 34.90 |

Table 3. CONSTANTS FOR $\bar{p}'$ (SHORT FORMULA), AN APPROXIMATION OF THE UPPER BINOMIAL CONFIDENCE LIMIT $\bar{p}$; $c/n \leqslant .1$, $n \geqslant 40$, $c \leqslant 1,000$ (*Continued*)

| c | $\overline{m}$ | $\overline{k}$ | c | $\overline{m}$ | $\overline{k}$ | c | $\overline{m}$ | $\overline{k}$ |
|---|---|---|---|---|---|---|---|---|
| 675 | 744.85 | 34.92 | 720 | 792.04 | 36.02 | 765 | 839.17 | 37.08 |
| 676 | 745.90 | 34.95 | 721 | 793.09 | 36.05 | 766 | 840.21 | 37.11 |
| 677 | 746.95 | 34.97 | 722 | 794.14 | 36.07 | 767 | 841.26 | 37.13 |
| 678 | 748.00 | 35.00 | 723 | 795.19 | 36.09 | 768 | 842.31 | 37.15 |
| 679 | 749.05 | 35.02 | 724 | 796.23 | 36.12 | 769 | 843.35 | 37.18 |
| 680 | 750.10 | 35.05 | 725 | 797.28 | 36.14 | 770 | 844.40 | 37.20 |
| 681 | 751.15 | 35.07 | 726 | 798.33 | 36.16 | 771 | 845.45 | 37.22 |
| 682 | 752.19 | 35.10 | 727 | 799.38 | 36.19 | 772 | 846.49 | 37.25 |
| 683 | 753.24 | 35.12 | 728 | 800.42 | 36.21 | 773 | 847.54 | 37.27 |
| 684 | 754.29 | 35.15 | 729 | 801.47 | 36.24 | 774 | 848.59 | 37.29 |
| 685 | 755.34 | 35.17 | 730 | 802.52 | 36.26 | 775 | 849.63 | 37.32 |
| 686 | 756.39 | 35.20 | 731 | 803.57 | 36.28 | 776 | 850.68 | 37.34 |
| 687 | 757.44 | 35.22 | 732 | 804.62 | 36.31 | 777 | 851.72 | 37.36 |
| 688 | 758.49 | 35.24 | 733 | 805.66 | 36.33 | 778 | 852.77 | 37.39 |
| 689 | 759.54 | 35.27 | 734 | 806.71 | 36.36 | 779 | 853.82 | 37.41 |
| 690 | 760.59 | 35.29 | 735 | 807.76 | 36.38 | 780 | 854.86 | 37.43 |
| 691 | 761.64 | 35.32 | 736 | 808.81 | 36.40 | 781 | 855.91 | 37.45 |
| 692 | 762.69 | 35.34 | 737 | 809.85 | 36.43 | 782 | 856.95 | 37.48 |
| 693 | 763.73 | 35.37 | 738 | 810.90 | 36.45 | 783 | 858.00 | 37.50 |
| 694 | 764.78 | 35.39 | 739 | 811.95 | 36.47 | 784 | 859.05 | 37.52 |
| 695 | 765.83 | 35.42 | 740 | 812.99 | 36.50 | 785 | 860.09 | 37.55 |
| 696 | 766.88 | 35.44 | 741 | 814.04 | 36.52 | 786 | 861.14 | 37.57 |
| 697 | 767.93 | 35.47 | 742 | 815.09 | 36.54 | 787 | 862.18 | 37.59 |
| 698 | 768.98 | 35.49 | 743 | 816.14 | 36.57 | 788 | 863.23 | 37.62 |
| 699 | 770.03 | 35.51 | 744 | 817.18 | 36.59 | 789 | 864.28 | 37.64 |
| 700 | 771.08 | 35.54 | 745 | 818.23 | 36.62 | 790 | 865.32 | 37.66 |
| 701 | 772.12 | 35.56 | 746 | 819.28 | 36.64 | 791 | 866.37 | 37.68 |
| 702 | 773.17 | 35.59 | 747 | 820.33 | 36.66 | 792 | 867.41 | 37.71 |
| 703 | 774.22 | 35.61 | 748 | 821.37 | 36.69 | 793 | 868.46 | 37.73 |
| 704 | 775.27 | 35.64 | 749 | 822.42 | 36.71 | 794 | 869.50 | 37.75 |
| 705 | 776.32 | 35.66 | 750 | 823.47 | 36.73 | 795 | 870.55 | 37.78 |
| 706 | 777.37 | 35.68 | 751 | 824.51 | 36.76 | 796 | 871.60 | 37.80 |
| 707 | 778.42 | 35.71 | 752 | 825.56 | 36.78 | 797 | 872.64 | 37.82 |
| 708 | 779.46 | 35.73 | 753 | 826.61 | 36.80 | 798 | 873.69 | 37.84 |
| 709 | 780.51 | 35.76 | 754 | 827.65 | 36.83 | 799 | 874.73 | 37.87 |
| 710 | 781.56 | 35.78 | 755 | 828.70 | 36.85 | 800 | 875.78 | 37.89 |
| 711 | 782.61 | 35.80 | 756 | 829.75 | 36.87 | 801 | 876.82 | 37.91 |
| 712 | 783.66 | 35.83 | 757 | 830.79 | 36.90 | 802 | 877.87 | 37.93 |
| 713 | 784.71 | 35.85 | 758 | 831.84 | 36.92 | 803 | 878.91 | 37.96 |
| 714 | 785.75 | 35.88 | 759 | 832.89 | 36.94 | 804 | 879.96 | 37.98 |
| 715 | 786.80 | 35.90 | 760 | 833.93 | 36.97 | 805 | 881.01 | 38.00 |
| 716 | 787.85 | 35.93 | 761 | 834.98 | 36.99 | 806 | 882.05 | 38.03 |
| 717 | 788.90 | 35.95 | 762 | 836.03 | 37.01 | 807 | 883.10 | 38.05 |
| 718 | 789.95 | 35.97 | 763 | 837.07 | 37.04 | 808 | 884.14 | 38.07 |
| 719 | 790.99 | 36.00 | 764 | 838.12 | 37.06 | 809 | 885.19 | 38.09 |

Table 3. **CONSTANTS FOR $\overline{p}'$ (SHORT FORMULA), AN APPROXIMATION OF** Section 1
**THE UPPER BINOMIAL CONFIDENCE LIMIT** $\overline{p}$; c/n $\leqslant$ .1, n $\geqslant$ 40, $\gamma_1$ = 99.5%
c $\leqslant$ 1,000 (*Continued*) $\gamma_2$ = 99%

| c | $\overline{m}$ | $\overline{k}$ | c | $\overline{m}$ | $\overline{k}$ | c | $\overline{m}$ | $\overline{k}$ |
|---|---|---|---|---|---|---|---|---|
| 810 | 886.23 | 38.12 | 855 | 933.24 | 39.12 | 900 | 980.20 | 40.10 |
| 811 | 887.28 | 38.14 | 856 | 934.28 | 39.14 | 901 | 981.24 | 40.12 |
| 812 | 888.32 | 38.16 | 857 | 935.33 | 39.16 | 902 | 982.28 | 40.14 |
| 813 | 889.37 | 38.18 | 858 | 936.37 | 39.19 | 903 | 983.32 | 40.16 |
| 814 | 890.41 | 38.21 | 859 | 937.42 | 39.21 | 904 | 984.37 | 40.18 |
| 815 | 891.46 | 38.23 | 860 | 938.46 | 39.23 | 905 | 985.41 | 40.20 |
| 816 | 892.50 | 38.25 | 861 | 939.50 | 39.25 | 906 | 986.45 | 40.23 |
| 817 | 893.55 | 38.27 | 862 | 940.55 | 39.27 | 907 | 987.49 | 40.25 |
| 818 | 894.59 | 38.30 | 863 | 941.59 | 39.30 | 908 | 988.54 | 40.27 |
| 819 | 895.64 | 38.32 | 864 | 942.63 | 39.32 | 909 | 989.58 | 40.29 |
| 820 | 896.68 | 38.34 | 865 | 943.68 | 39.34 | 910 | 990.62 | 40.31 |
| 821 | 897.73 | 38.36 | 866 | 944.72 | 39.36 | 911 | 991.67 | 40.33 |
| 822 | 898.77 | 38.39 | 867 | 945.77 | 39.38 | 912 | 992.71 | 40.35 |
| 823 | 899.82 | 38.41 | 868 | 946.81 | 39.40 | 913 | 993.75 | 40.38 |
| 824 | 900.86 | 38.43 | 869 | 947.85 | 39.43 | 914 | 994.79 | 40.40 |
| 825 | 901.91 | 38.45 | 870 | 948.90 | 39.45 | 915 | 995.84 | 40.42 |
| 826 | 902.95 | 38.48 | 871 | 949.94 | 39.47 | 916 | 996.88 | 40.44 |
| 827 | 904.00 | 38.50 | 872 | 950.98 | 39.49 | 917 | 997.92 | 40.46 |
| 828 | 905.04 | 38.52 | 873 | 952.03 | 39.51 | 918 | 998.96 | 40.48 |
| 829 | 906.09 | 38.54 | 874 | 953.07 | 39.54 | 919 | 1000.01 | 40.50 |
| 830 | 907.13 | 38.57 | 875 | 954.11 | 39.56 | 920 | 1001.05 | 40.52 |
| 831 | 908.18 | 38.59 | 876 | 955.16 | 39.58 | 921 | 1002.09 | 40.55 |
| 832 | 909.22 | 38.61 | 877 | 956.20 | 39.60 | 922 | 1003.13 | 40.57 |
| 833 | 910.26 | 38.63 | 878 | 957.25 | 39.62 | 923 | 1004.18 | 40.59 |
| 834 | 911.31 | 38.65 | 879 | 958.29 | 39.64 | 924 | 1005.22 | 40.61 |
| 835 | 912.35 | 38.68 | 880 | 959.33 | 39.67 | 925 | 1006.26 | 40.63 |
| 836 | 913.40 | 38.70 | 881 | 960.38 | 39.69 | 926 | 1007.30 | 40.65 |
| 837 | 914.44 | 38.72 | 882 | 961.42 | 39.71 | 927 | 1008.35 | 40.67 |
| 838 | 915.49 | 38.74 | 883 | 962.46 | 39.73 | 928 | 1009.39 | 40.69 |
| 839 | 916.53 | 38.77 | 884 | 963.51 | 39.75 | 929 | 1010.43 | 40.71 |
| 840 | 917.58 | 38.79 | 885 | 964.55 | 39.77 | 930 | 1011.47 | 40.74 |
| 841 | 918.62 | 38.81 | 886 | 965.59 | 39.80 | 931 | 1012.51 | 40.76 |
| 842 | 919.67 | 38.83 | 887 | 966.64 | 39.82 | 932 | 1013.56 | 40.78 |
| 843 | 920.71 | 38.85 | 888 | 967.68 | 39.84 | 933 | 1014.60 | 40.80 |
| 844 | 921.75 | 38.88 | 889 | 968.72 | 39.86 | 934 | 1015.64 | 40.82 |
| 845 | 922.80 | 38.90 | 890 | 969.76 | 39.88 | 935 | 1016.68 | 40.84 |
| 846 | 923.84 | 38.92 | 891 | 970.81 | 39.90 | 936 | 1017.72 | 40.86 |
| 847 | 924.89 | 38.94 | 892 | 971.85 | 39.93 | 937 | 1018.77 | 40.88 |
| 848 | 925.93 | 38.97 | 893 | 972.89 | 39.95 | 938 | 1019.81 | 40.90 |
| 849 | 926.98 | 38.99 | 894 | 973.94 | 39.97 | 939 | 1020.85 | 40.93 |
| 850 | 928.02 | 39.01 | 895 | 974.98 | 39.99 | 940 | 1021.89 | 40.95 |
| 851 | 929.06 | 39.03 | 896 | 976.02 | 40.01 | 941 | 1022.93 | 40.97 |
| 852 | 930.11 | 39.05 | 897 | 977.07 | 40.03 | 942 | 1023.98 | 40.99 |
| 853 | 931.15 | 39.08 | 898 | 978.11 | 40.05 | 943 | 1025.02 | 41.01 |
| 854 | 932.20 | 39.10 | 899 | 979.15 | 40.08 | 944 | 1026.06 | 41.03 |

Table 3. CONSTANTS FOR $\bar{p}'$ (SHORT FORMULA), AN APPROXIMATION OF
THE UPPER BINOMIAL CONFIDENCE LIMIT $\bar{p}$; $c/n \leqslant .1$, $n \geqslant 40$,
$c \leqslant 1,000$ (*Continued*)

| c | $\bar{m}$ | $\bar{k}$ | c | $\bar{m}$ | $\bar{k}$ | c | $\bar{m}$ | $\bar{k}$ |
|---|---|---|---|---|---|---|---|---|
| 945 | 1027.10 | 41.05 | 965 | 1047.94 | 41.47 | 985 | 1068.76 | 41.88 |
| 946 | 1028.14 | 41.07 | 966 | 1048.98 | 41.49 | 986 | 1069.80 | 41.90 |
| 947 | 1029.19 | 41.09 | 967 | 1050.02 | 41.51 | 987 | 1070.84 | 41.92 |
| 948 | 1030.23 | 41.11 | 968 | 1051.06 | 41.53 | 988 | 1071.88 | 41.94 |
| 949 | 1031.27 | 41.13 | 969 | 1052.10 | 41.55 | 989 | 1072.92 | 41.96 |
| 950 | 1032.31 | 41.16 | 970 | 1053.14 | 41.57 | 990 | 1073.96 | 41.98 |
| 951 | 1033.35 | 41.18 | 971 | 1054.18 | 41.59 | 991 | 1075.01 | 42.00 |
| 952 | 1034.40 | 41.20 | 972 | 1055.22 | 41.61 | 992 | 1076.05 | 42.02 |
| 953 | 1035.44 | 41.22 | 973 | 1056.27 | 41.63 | 993 | 1077.09 | 42.04 |
| 954 | 1036.48 | 41.24 | 974 | 1057.31 | 41.65 | 994 | 1078.13 | 42.06 |
| 955 | 1037.52 | 41.26 | 975 | 1058.35 | 41.67 | 995 | 1079.17 | 42.08 |
| 956 | 1038.56 | 41.28 | 976 | 1059.39 | 41.69 | 996 | 1080.21 | 42.10 |
| 957 | 1039.60 | 41.30 | 977 | 1060.43 | 41.72 | 997 | 1081.25 | 42.13 |
| 958 | 1040.64 | 41.32 | 978 | 1061.47 | 41.74 | 998 | 1082.29 | 42.15 |
| 959 | 1041.69 | 41.34 | 979 | 1062.51 | 41.76 | 999 | 1083.33 | 42.17 |
| 960 | 1042.73 | 41.36 | 980 | 1063.55 | 41.78 | 1000 | 1084.37 | 42.19 |
| 961 | 1043.77 | 41.38 | 981 | 1064.60 | 41.80 | | | |
| 962 | 1044.81 | 41.41 | 982 | 1065.64 | 41.82 | | | |
| 963 | 1045.85 | 41.43 | 983 | 1066.68 | 41.84 | | | |
| 964 | 1046.89 | 41.45 | 984 | 1067.72 | 41.86 | | | |

# Table 3. CONSTANTS FOR p̄′ (SHORT FORMULA), AN APPROXIMATION OF THE UPPER BINOMIAL CONFIDENCE LIMIT p̄; $c/n \leqslant .1$, $n \geqslant 40$, $c \leqslant 1,000$ (*Continued*)

$\gamma_1 = 99\%$
$\gamma_2 = 98\%$

| c | m̄ | k̄ | c | m̄ | k̄ | c | m̄ | k̄ |
|---|------|------|---|------|------|---|------|------|
| 0 | 4.6052 | 2.337 | 45 | 63.231 | 9.39 | 90 | 114.65 | 12.33 |
| 1 | 6.6384 | 2.887 | 46 | 64.402 | 9.47 | 91 | 115.77 | 12.39 |
| 2 | 8.4060 | 3.303 | 47 | 65.571 | 9.56 | 92 | 116.89 | 12.45 |
| 3 | 10.045 | 3.657 | 48 | 66.738 | 9.64 | 93 | 118.01 | 12.51 |
| 4 | 11.605 | 3.975 | 49 | 67.904 | 9.75 | 94 | 119.13 | 12.57 |
| 5 | 13.109 | 4.225 | 50 | 69.067 | 9.83 | 95 | 120.25 | 12.63 |
| 6 | 14.571 | 4.451 | 51 | 70.230 | 9.62 | 96 | 121.37 | 12.69 |
| 7 | 16.000 | 4.669 | 52 | 71.390 | 9.70 | 97 | 122.49 | 12.75 |
| 8 | 17.403 | 4.867 | 53 | 72.550 | 9.78 | 98 | 123.61 | 12.81 |
| 9 | 18.783 | 5.058 | 54 | 73.707 | 9.86 | 99 | 124.72 | 12.87 |
| 10 | 20.145 | 5.24 | 55 | 74.864 | 9.94 | 100 | 125.84 | 12.92 |
| 11 | 21.490 | 5.42 | 56 | 76.019 | 10.02 | 101 | 126.95 | 12.98 |
| 12 | 22.821 | 5.58 | 57 | 77.172 | 10.09 | 102 | 128.07 | 13.04 |
| 13 | 24.139 | 5.75 | 58 | 78.324 | 10.17 | 103 | 129.18 | 13.10 |
| 14 | 25.446 | 5.91 | 59 | 79.475 | 10.25 | 104 | 130.30 | 13.15 |
| 15 | 26.743 | 6.05 | 60 | 80.625 | 10.32 | 105 | 131.41 | 13.21 |
| 16 | 28.030 | 6.20 | 61 | 81.773 | 10.39 | 106 | 132.52 | 13.27 |
| 17 | 29.310 | 6.34 | 62 | 82.921 | 10.47 | 107 | 133.64 | 13.32 |
| 18 | 30.581 | 6.48 | 63 | 84.067 | 10.54 | 108 | 134.75 | 13.38 |
| 19 | 31.845 | 6.61 | 64 | 85.212 | 10.61 | 109 | 135.86 | 13.43 |
| 20 | 33.103 | 6.73 | 65 | 86.356 | 10.68 | 110 | 136.97 | 13.49 |
| 21 | 34.355 | 6.86 | 66 | 87.498 | 10.76 | 111 | 138.08 | 13.54 |
| 22 | 35.601 | 6.99 | 67 | 88.640 | 10.83 | 112 | 139.19 | 13.60 |
| 23 | 36.841 | 7.12 | 68 | 89.781 | 10.90 | 113 | 140.30 | 13.65 |
| 24 | 38.077 | 7.25 | 69 | 90.920 | 10.97 | 114 | 141.41 | 13.71 |
| 25 | 39.308 | 7.34 | 70 | 92.059 | 11.04 | 115 | 142.52 | 13.76 |
| 26 | 40.534 | 7.47 | 71 | 93.197 | 11.10 | 116 | 143.62 | 13.82 |
| 27 | 41.757 | 7.58 | 72 | 94.333 | 11.17 | 117 | 144.73 | 13.87 |
| 28 | 42.975 | 7.68 | 73 | 95.469 | 11.24 | 118 | 145.84 | 13.92 |
| 29 | 44.190 | 7.79 | 74 | 96.604 | 11.31 | 119 | 146.94 | 13.98 |
| 30 | 45.401 | 7.90 | 75 | 97.738 | 11.38 | 120 | 148.05 | 14.03 |
| 31 | 46.608 | 8.02 | 76 | 98.871 | 11.44 | 121 | 149.16 | 14.08 |
| 32 | 47.813 | 8.13 | 77 | 100.00 | 11.51 | 122 | 150.26 | 14.14 |
| 33 | 49.014 | 8.22 | 78 | 101.13 | 11.57 | 123 | 151.37 | 14.19 |
| 34 | 50.213 | 8.31 | 79 | 102.27 | 11.64 | 124 | 152.47 | 14.24 |
| 35 | 51.408 | 8.44 | 80 | 103.40 | 11.70 | 125 | 153.57 | 14.29 |
| 36 | 52.601 | 8.52 | 81 | 104.52 | 11.77 | 126 | 154.68 | 14.34 |
| 37 | 53.791 | 8.61 | 82 | 105.65 | 11.83 | 127 | 155.78 | 14.39 |
| 38 | 54.979 | 8.71 | 83 | 106.78 | 11.90 | 128 | 156.88 | 14.45 |
| 39 | 56.165 | 8.80 | 84 | 107.91 | 11.96 | 129 | 157.99 | 14.50 |
| 40 | 57.348 | 8.90 | 85 | 109.03 | 12.02 | 130 | 159.09 | 14.55 |
| 41 | 58.529 | 8.99 | 86 | 110.16 | 12.08 | 131 | 160.19 | 14.60 |
| 42 | 59.707 | 9.10 | 87 | 111.28 | 12.15 | 132 | 161.29 | 14.65 |
| 43 | 60.884 | 9.19 | 88 | 112.41 | 12.21 | 133 | 162.39 | 14.70 |
| 44 | 62.058 | 9.30 | 89 | 113.53 | 12.27 | 134 | 163.49 | 14.75 |

Table 3.  CONSTANTS FOR $\overline{p}'$ (SHORT FORMULA), AN APPROXIMATION OF THE UPPER BINOMIAL CONFIDENCE LIMIT $\overline{p}$; $c/n \leq .1$, $n \geq 40$, $c \leq 1,000$ (*Continued*)

| c | $\overline{m}$ | $\overline{k}$ | c | $\overline{m}$ | $\overline{k}$ | c | $\overline{m}$ | $\overline{k}$ |
|---|---|---|---|---|---|---|---|---|
| 135 | 164.59 | 14.80 | 180 | 213.77 | 16.88 | 225 | 262.44 | 18.72 |
| 136 | 165.69 | 14.85 | 181 | 214.85 | 16.93 | 226 | 263.52 | 18.76 |
| 137 | 166.79 | 14.90 | 182 | 215.94 | 16.97 | 227 | 264.60 | 18.80 |
| 138 | 167.89 | 14.95 | 183 | 217.03 | 17.01 | 228 | 265.67 | 18.84 |
| 139 | 168.99 | 15.00 | 184 | 218.11 | 17.06 | 229 | 266.75 | 18.88 |
| 140 | 170.09 | 15.05 | 185 | 219.20 | 17.10 | 230 | 267.83 | 18.91 |
| 141 | 171.18 | 15.10 | 186 | 220.28 | 17.14 | 231 | 268.90 | 18.95 |
| 142 | 172.28 | 15.14 | 187 | 221.37 | 17.18 | 232 | 269.98 | 18.99 |
| 143 | 173.38 | 15.19 | 188 | 222.45 | 17.23 | 233 | 271.06 | 19.03 |
| 144 | 174.47 | 15.24 | 189 | 223.54 | 17.27 | 234 | 272.13 | 19.07 |
| 145 | 175.57 | 15.29 | 190 | 224.62 | 17.31 | 235 | 273.21 | 19.10 |
| 146 | 176.67 | 15.34 | 191 | 225.70 | 17.35 | 236 | 274.28 | 19.14 |
| 147 | 177.76 | 15.39 | 192 | 226.79 | 17.39 | 237 | 275.36 | 19.18 |
| 148 | 178.86 | 15.43 | 193 | 227.87 | 17.44 | 238 | 276.43 | 19.22 |
| 149 | 179.95 | 15.48 | 194 | 228.96 | 17.48 | 239 | 277.51 | 19.25 |
| 150 | 181.05 | 15.53 | 195 | 230.04 | 17.52 | 240 | 278.58 | 19.29 |
| 151 | 182.14 | 15.58 | 196 | 231.12 | 17.56 | 241 | 279.66 | 19.33 |
| 152 | 183.24 | 15.62 | 197 | 232.20 | 17.60 | 242 | 280.73 | 19.37 |
| 153 | 184.33 | 15.67 | 198 | 233.29 | 17.64 | 243 | 281.81 | 19.40 |
| 154 | 185.42 | 15.72 | 199 | 234.37 | 17.68 | 244 | 282.88 | 19.44 |
| 155 | 186.52 | 15.76 | 200 | 235.45 | 17.73 | 245 | 283.96 | 19.48 |
| 156 | 187.61 | 15.81 | 201 | 236.53 | 17.77 | 246 | 285.03 | 19.52 |
| 157 | 188.70 | 15.86 | 202 | 237.62 | 17.81 | 247 | 286.11 | 19.55 |
| 158 | 189.80 | 15.90 | 203 | 238.70 | 17.85 | 248 | 287.18 | 19.59 |
| 159 | 190.89 | 15.95 | 204 | 239.78 | 17.89 | 249 | 288.25 | 19.63 |
| 160 | 191.98 | 15.99 | 205 | 240.86 | 17.93 | 250 | 289.33 | 19.66 |
| 161 | 193.08 | 16.04 | 206 | 241.94 | 17.97 | 251 | 290.40 | 19.70 |
| 162 | 194.17 | 16.09 | 207 | 243.02 | 18.01 | 252 | 291.47 | 19.74 |
| 163 | 195.26 | 16.13 | 208 | 244.10 | 18.05 | 253 | 292.55 | 19.77 |
| 164 | 196.35 | 16.18 | 209 | 245.18 | 18.09 | 254 | 293.62 | 19.81 |
| 165 | 197.44 | 16.22 | 210 | 246.26 | 18.13 | 255 | 294.69 | 19.85 |
| 166 | 198.53 | 16.27 | 211 | 247.34 | 18.17 | 256 | 295.76 | 19.88 |
| 167 | 199.62 | 16.31 | 212 | 248.42 | 18.21 | 257 | 296.84 | 19.92 |
| 168 | 200.71 | 16.36 | 213 | 249.50 | 18.25 | 258 | 297.91 | 19.95 |
| 169 | 201.80 | 16.40 | 214 | 250.58 | 18.29 | 259 | 298.98 | 19.99 |
| 170 | 202.89 | 16.45 | 215 | 251.66 | 18.33 | 260 | 300.05 | 20.03 |
| 171 | 203.98 | 16.49 | 216 | 252.74 | 18.37 | 261 | 301.13 | 20.06 |
| 172 | 205.07 | 16.53 | 217 | 253.82 | 18.41 | 262 | 302.20 | 20.10 |
| 173 | 206.16 | 16.58 | 218 | 254.90 | 18.45 | 263 | 303.27 | 20.13 |
| 174 | 207.24 | 16.62 | 219 | 255.98 | 18.49 | 264 | 304.34 | 20.17 |
| 175 | 208.33 | 16.67 | 220 | 257.05 | 18.53 | 265 | 305.41 | 20.21 |
| 176 | 209.42 | 16.71 | 221 | 258.13 | 18.57 | 266 | 306.48 | 20.24 |
| 177 | 210.51 | 16.75 | 222 | 259.21 | 18.60 | 267 | 307.55 | 20.28 |
| 178 | 211.59 | 16.80 | 223 | 260.29 | 18.64 | 268 | 308.62 | 20.31 |
| 179 | 212.68 | 16.84 | 224 | 261.37 | 18.68 | 269 | 309.70 | 20.35 |

Table 3.  CONSTANTS FOR $\bar{p}'$ (SHORT FORMULA), AN APPROXIMATION OF
THE UPPER BINOMIAL CONFIDENCE LIMIT $\bar{p}$; $c/n \leqslant .1$, $n \geqslant 40$,
$c \leqslant 1,000$ (*Continued*)

| c | $\bar{m}$ | $\bar{k}$ | c | $\bar{m}$ | $\bar{k}$ | c | $\bar{m}$ | $\bar{k}$ |
|---|---|---|---|---|---|---|---|---|
| 270 | 310.77 | 20.38 | 315 | 358.82 | 21.91 | 360 | 406.67 | 23.34 |
| 271 | 311.84 | 20.42 | 316 | 359.89 | 21.94 | 361 | 407.73 | 23.37 |
| 272 | 312.91 | 20.45 | 317 | 360.95 | 21.98 | 362 | 408.79 | 23.40 |
| 273 | 313.98 | 20.49 | 318 | 362.02 | 22.01 | 363 | 409.85 | 23.43 |
| 274 | 315.05 | 20.52 | 319 | 363.08 | 22.04 | 364 | 410.91 | 23.46 |
| 275 | 316.12 | 20.56 | 320 | 364.15 | 22.07 | 365 | 411.98 | 23.49 |
| 276 | 317.19 | 20.59 | 321 | 365.21 | 22.11 | 366 | 413.04 | 23.52 |
| 277 | 318.26 | 20.63 | 322 | 366.28 | 22.14 | 367 | 414.10 | 23.55 |
| 278 | 319.33 | 20.66 | 323 | 367.34 | 22.17 | 368 | 415.16 | 23.58 |
| 279 | 320.40 | 20.70 | 324 | 368.41 | 22.20 | 369 | 416.22 | 23.61 |
| 280 | 321.47 | 20.73 | 325 | 369.47 | 22.24 | 370 | 417.28 | 23.64 |
| 281 | 322.54 | 20.77 | 326 | 370.54 | 22.27 | 371 | 418.34 | 23.67 |
| 282 | 323.60 | 20.80 | 327 | 371.60 | 22.30 | 372 | 419.40 | 23.70 |
| 283 | 324.67 | 20.84 | 328 | 372.67 | 22.33 | 373 | 420.46 | 23.73 |
| 284 | 325.74 | 20.87 | 329 | 373.73 | 22.36 | 374 | 421.52 | 23.76 |
| 285 | 326.81 | 20.91 | 330 | 374.79 | 22.40 | 375 | 422.58 | 23.79 |
| 286 | 327.88 | 20.94 | 331 | 375.86 | 22.43 | 376 | 423.64 | 23.82 |
| 287 | 328.95 | 20.97 | 332 | 376.92 | 22.46 | 377 | 424.70 | 23.85 |
| 288 | 330.02 | 21.01 | 333 | 377.99 | 22.49 | 378 | 425.76 | 23.88 |
| 289 | 331.09 | 21.04 | 334 | 379.05 | 22.52 | 379 | 426.82 | 23.91 |
| 290 | 332.15 | 21.08 | 335 | 380.11 | 22.56 | 380 | 427.88 | 23.94 |
| 291 | 333.22 | 21.11 | 336 | 381.18 | 22.59 | 381 | 428.94 | 23.97 |
| 292 | 334.29 | 21.15 | 337 | 382.24 | 22.62 | 382 | 430.00 | 24.00 |
| 293 | 335.36 | 21.18 | 338 | 383.30 | 22.65 | 383 | 431.06 | 24.03 |
| 294 | 336.43 | 21.21 | 339 | 384.37 | 22.68 | 384 | 432.12 | 24.06 |
| 295 | 337.49 | 21.25 | 340 | 385.43 | 22.71 | 385 | 433.18 | 24.09 |
| 296 | 338.56 | 21.28 | 341 | 386.49 | 22.75 | 386 | 434.23 | 24.12 |
| 297 | 339.63 | 21.31 | 342 | 387.55 | 22.78 | 387 | 435.29 | 24.15 |
| 298 | 340.70 | 21.35 | 343 | 388.62 | 22.81 | 388 | 436.35 | 24.18 |
| 299 | 341.76 | 21.38 | 344 | 389.68 | 22.84 | 389 | 437.41 | 24.21 |
| 300 | 342.83 | 21.42 | 345 | 390.74 | 22.87 | 390 | 438.47 | 24.24 |
| 301 | 343.90 | 21.45 | 346 | 391.80 | 22.90 | 391 | 439.53 | 24.26 |
| 302 | 344.96 | 21.48 | 347 | 392.87 | 22.93 | 392 | 440.59 | 24.29 |
| 303 | 346.03 | 21.52 | 348 | 393.93 | 22.96 | 393 | 441.65 | 24.32 |
| 304 | 347.10 | 21.55 | 349 | 394.99 | 23.00 | 394 | 442.70 | 24.35 |
| 305 | 348.16 | 21.58 | 350 | 396.05 | 23.03 | 395 | 443.76 | 24.38 |
| 306 | 349.23 | 21.62 | 351 | 397.12 | 23.06 | 396 | 444.82 | 24.41 |
| 307 | 350.30 | 21.65 | 352 | 398.18 | 23.09 | 397 | 445.88 | 24.44 |
| 308 | 351.36 | 21.68 | 353 | 399.24 | 23.12 | 398 | 446.94 | 24.47 |
| 309 | 352.43 | 21.71 | 354 | 400.30 | 23.15 | 399 | 448.00 | 24.50 |
| 310 | 353.50 | 21.75 | 355 | 401.36 | 23.18 | 400 | 449.05 | 24.53 |
| 311 | 354.56 | 21.78 | 356 | 402.42 | 23.21 | 401 | 450.11 | 24.56 |
| 312 | 355.63 | 21.81 | 357 | 403.49 | 23.24 | 402 | 451.17 | 24.59 |
| 313 | 356.69 | 21.85 | 358 | 404.55 | 23.27 | 403 | 452.23 | 24.61 |
| 314 | 357.76 | 21.88 | 359 | 405.61 | 23.30 | 404 | 453.29 | 24.64 |

Table 3.  CONSTANTS FOR $\overline{p}'$ (SHORT FORMULA), AN APPROXIMATION OF
THE UPPER BINOMIAL CONFIDENCE LIMIT $\overline{p}$; $c/n \leqslant .1$, $n \geqslant 40$,
$c \leqslant 1,000$ (*Continued*)

| c | $\overline{m}$ | $\overline{k}$ | c | $\overline{m}$ | $\overline{k}$ | c | $\overline{m}$ | $\overline{k}$ |
|---|---|---|---|---|---|---|---|---|
| 405 | 454.34 | 24.67 | 450 | 501.87 | 25.94 | 495 | 549.28 | 27.14 |
| 406 | 455.40 | 24.70 | 451 | 502.93 | 25.96 | 496 | 550.33 | 27.17 |
| 407 | 456.46 | 24.73 | 452 | 503.98 | 25.99 | 497 | 551.38 | 27.19 |
| 408 | 457.52 | 24.76 | 453 | 505.04 | 26.02 | 498 | 552.44 | 27.22 |
| 409 | 458.57 | 24.79 | 454 | 506.09 | 26.05 | 499 | 553.49 | 27.24 |
| 410 | 459.63 | 24.82 | 455 | 507.15 | 26.07 | 500 | 554.54 | 27.27 |
| 411 | 460.69 | 24.84 | 456 | 508.20 | 26.10 | 501 | 555.59 | 27.30 |
| 412 | 461.75 | 24.87 | 457 | 509.26 | 26.13 | 502 | 556.64 | 27.32 |
| 413 | 462.80 | 24.90 | 458 | 510.31 | 26.15 | 503 | 557.70 | 27.35 |
| 414 | 463.86 | 24.93 | 459 | 511.36 | 26.18 | 504 | 558.75 | 27.37 |
| 415 | 464.92 | 24.96 | 460 | 512.42 | 26.21 | 505 | 559.80 | 27.40 |
| 416 | 465.97 | 24.99 | 461 | 513.47 | 26.24 | 506 | 560.85 | 27.43 |
| 417 | 467.03 | 25.02 | 462 | 514.53 | 26.26 | 507 | 561.90 | 27.45 |
| 418 | 468.09 | 25.04 | 463 | 515.58 | 26.29 | 508 | 562.95 | 27.48 |
| 419 | 469.15 | 25.07 | 464 | 516.63 | 26.32 | 509 | 564.01 | 27.50 |
| 420 | 470.20 | 25.10 | 465 | 517.69 | 26.34 | 510 | 565.06 | 27.53 |
| 421 | 471.26 | 25.13 | 466 | 518.74 | 26.37 | 511 | 566.11 | 27.55 |
| 422 | 472.32 | 25.16 | 467 | 519.80 | 26.40 | 512 | 567.16 | 27.58 |
| 423 | 473.37 | 25.19 | 468 | 520.85 | 26.42 | 513 | 568.21 | 27.61 |
| 424 | 474.43 | 25.21 | 469 | 521.90 | 26.45 | 514 | 569.26 | 27.63 |
| 425 | 475.48 | 25.24 | 470 | 522.96 | 26.48 | 515 | 570.31 | 27.66 |
| 426 | 476.54 | 25.27 | 471 | 524.01 | 26.51 | 516 | 571.37 | 27.68 |
| 427 | 477.60 | 25.30 | 472 | 525.06 | 26.53 | 517 | 572.42 | 27.71 |
| 428 | 478.65 | 25.33 | 473 | 526.12 | 26.56 | 518 | 573.47 | 27.73 |
| 429 | 479.71 | 25.35 | 474 | 527.17 | 26.59 | 519 | 574.52 | 27.76 |
| 430 | 480.77 | 25.38 | 475 | 528.22 | 26.61 | 520 | 575.57 | 27.78 |
| 431 | 481.82 | 25.41 | 476 | 529.28 | 26.64 | 521 | 576.62 | 27.81 |
| 432 | 482.88 | 25.44 | 477 | 530.33 | 26.67 | 522 | 577.67 | 27.84 |
| 433 | 483.93 | 25.47 | 478 | 531.38 | 26.69 | 523 | 578.72 | 27.86 |
| 434 | 484.99 | 25.49 | 479 | 532.44 | 26.72 | 524 | 579.77 | 27.89 |
| 435 | 486.05 | 25.52 | 480 | 533.49 | 26.75 | 525 | 580.82 | 27.91 |
| 436 | 487.10 | 25.55 | 481 | 534.54 | 26.77 | 526 | 581.87 | 27.94 |
| 437 | 488.16 | 25.58 | 482 | 535.60 | 26.80 | 527 | 582.92 | 27.96 |
| 438 | 489.21 | 25.61 | 483 | 536.65 | 26.82 | 528 | 583.98 | 27.99 |
| 439 | 490.27 | 25.63 | 484 | 537.70 | 26.85 | 529 | 585.03 | 28.01 |
| 440 | 491.32 | 25.66 | 485 | 538.75 | 26.88 | 530 | 586.08 | 28.04 |
| 441 | 492.38 | 25.69 | 486 | 539.81 | 26.90 | 531 | 587.13 | 28.06 |
| 442 | 493.43 | 25.72 | 487 | 540.86 | 26.93 | 532 | 588.18 | 28.09 |
| 443 | 494.49 | 25.74 | 488 | 541.91 | 26.96 | 533 | 589.23 | 28.11 |
| 444 | 495.54 | 25.77 | 489 | 542.97 | 26.98 | 534 | 590.28 | 28.14 |
| 445 | 496.60 | 25.80 | 490 | 544.02 | 27.01 | 535 | 591.33 | 28.16 |
| 446 | 497.65 | 25.83 | 491 | 545.07 | 27.04 | 536 | 592.38 | 28.19 |
| 447 | 498.71 | 25.85 | 492 | 546.12 | 27.06 | 537 | 593.43 | 28.21 |
| 448 | 499.76 | 25.88 | 493 | 547.18 | 27.09 | 538 | 594.48 | 28.24 |
| 449 | 500.82 | 25.91 | 494 | 548.23 | 27.11 | 539 | 595.53 | 28.26 |

Table 3. CONSTANTS FOR $\bar{p}'$ (SHORT FORMULA), AN APPROXIMATION OF THE UPPER BINOMIAL CONFIDENCE LIMIT $\bar{p}$; $c/n \leqslant .1$, $n \geqslant 40$, $c \leqslant 1,000$ (*Continued*)

Section 2

$\gamma_1 = 99\%$
$\gamma_2 = 98\%$

| c | $\bar{m}$ | $\bar{k}$ | c | $\bar{m}$ | $\bar{k}$ | c | $\bar{m}$ | $\bar{k}$ |
|---|---|---|---|---|---|---|---|---|
| 540 | 596.58 | 28.29 | 585 | 643.78 | 29.39 | 630 | 690.91 | 30.45 |
| 541 | 597.63 | 28.31 | 586 | 644.83 | 29.42 | 631 | 691.95 | 30.48 |
| 542 | 598.68 | 28.34 | 587 | 645.88 | 29.44 | 632 | 693.00 | 30.50 |
| 543 | 599.73 | 28.36 | 588 | 646.93 | 29.46 | 633 | 694.05 | 30.52 |
| 544 | 600.78 | 28.39 | 589 | 647.98 | 29.49 | 634 | 695.09 | 30.55 |
| 545 | 601.83 | 28.41 | 590 | 649.02 | 29.51 | 635 | 696.14 | 30.57 |
| 546 | 602.88 | 28.44 | 591 | 650.07 | 29.54 | 636 | 697.18 | 30.59 |
| 547 | 603.93 | 28.46 | 592 | 651.12 | 29.56 | 637 | 698.23 | 30.61 |
| 548 | 604.98 | 28.49 | 593 | 652.17 | 29.58 | 638 | 699.28 | 30.64 |
| 549 | 606.03 | 28.51 | 594 | 653.22 | 29.61 | 639 | 700.32 | 30.66 |
| 550 | 607.08 | 28.54 | 595 | 654.26 | 29.63 | 640 | 701.37 | 30.68 |
| 551 | 608.13 | 28.56 | 596 | 655.31 | 29.66 | 641 | 702.41 | 30.71 |
| 552 | 609.18 | 28.59 | 597 | 656.36 | 29.68 | 642 | 703.46 | 30.73 |
| 553 | 610.23 | 28.61 | 598 | 657.41 | 29.70 | 643 | 704.51 | 30.75 |
| 554 | 611.27 | 28.64 | 599 | 658.45 | 29.73 | 644 | 705.55 | 30.78 |
| 555 | 612.32 | 28.66 | 600 | 659.50 | 29.75 | 645 | 706.60 | 30.80 |
| 556 | 613.37 | 28.69 | 601 | 660.55 | 29.77 | 646 | 707.64 | 30.82 |
| 557 | 614.42 | 28.71 | 602 | 661.60 | 29.80 | 647 | 708.69 | 30.84 |
| 558 | 615.47 | 28.74 | 603 | 662.64 | 29.82 | 648 | 709.73 | 30.87 |
| 559 | 616.52 | 28.76 | 604 | 663.69 | 29.84 | 649 | 710.78 | 30.89 |
| 560 | 617.57 | 28.78 | 605 | 664.74 | 29.87 | 650 | 711.83 | 30.91 |
| 561 | 618.62 | 28.81 | 606 | 665.78 | 29.89 | 651 | 712.87 | 30.94 |
| 562 | 619.67 | 28.83 | 607 | 666.83 | 29.92 | 652 | 713.92 | 30.96 |
| 563 | 620.72 | 28.86 | 608 | 667.88 | 29.94 | 653 | 714.96 | 30.98 |
| 564 | 621.77 | 28.88 | 609 | 668.93 | 29.96 | 654 | 716.01 | 31.00 |
| 565 | 622.81 | 28.91 | 610 | 669.97 | 29.99 | 655 | 717.05 | 31.03 |
| 566 | 623.86 | 28.93 | 611 | 671.02 | 30.01 | 656 | 718.10 | 31.05 |
| 567 | 624.91 | 28.96 | 612 | 672.07 | 30.03 | 657 | 719.14 | 31.07 |
| 568 | 625.96 | 28.98 | 613 | 673.11 | 30.06 | 658 | 720.19 | 31.09 |
| 569 | 627.01 | 29.01 | 614 | 674.16 | 30.08 | 659 | 721.23 | 31.12 |
| 570 | 628.06 | 29.03 | 615 | 675.21 | 30.10 | 660 | 722.28 | 31.14 |
| 571 | 629.11 | 29.05 | 616 | 676.25 | 30.13 | 661 | 723.32 | 31.16 |
| 572 | 630.16 | 29.08 | 617 | 677.30 | 30.15 | 662 | 724.37 | 31.18 |
| 573 | 631.20 | 29.10 | 618 | 678.35 | 30.17 | 663 | 725.42 | 31.21 |
| 574 | 632.25 | 29.13 | 619 | 679.39 | 30.20 | 664 | 726.46 | 31.23 |
| 575 | 633.30 | 29.15 | 620 | 680.44 | 30.22 | 665 | 727.51 | 31.25 |
| 576 | 634.35 | 29.18 | 621 | 681.49 | 30.24 | 666 | 728.55 | 31.28 |
| 577 | 635.40 | 29.20 | 622 | 682.53 | 30.27 | 667 | 729.60 | 31.30 |
| 578 | 636.45 | 29.22 | 623 | 683.58 | 30.29 | 668 | 730.64 | 31.32 |
| 579 | 637.50 | 29.25 | 624 | 684.63 | 30.31 | 669 | 731.69 | 31.34 |
| 580 | 638.54 | 29.27 | 625 | 685.67 | 30.34 | 670 | 732.73 | 31.37 |
| 581 | 639.59 | 29.30 | 626 | 686.72 | 30.36 | 671 | 733.78 | 31.39 |
| 582 | 640.64 | 29.32 | 627 | 687.77 | 30.38 | 672 | 734.82 | 31.41 |
| 583 | 641.69 | 29.34 | 628 | 688.81 | 30.41 | 673 | 735.86 | 31.43 |
| 584 | 642.74 | 29.37 | 629 | 689.86 | 30.43 | 674 | 736.91 | 31.45 |

Table 3. CONSTANTS FOR $\overline{p}'$ (SHORT FORMULA), AN APPROXIMATION OF THE UPPER BINOMIAL CONFIDENCE LIMIT $\overline{p}$; $c/n \leqslant .1$, $n \geqslant 40$, $c \leqslant 1,000$ (*Continued*)

| c | $\overline{m}$ | $\overline{k}$ | c | $\overline{m}$ | $\overline{k}$ | c | $\overline{m}$ | $\overline{k}$ |
|---|---|---|---|---|---|---|---|---|
| 675 | 737.95 | 31.48 | 720 | 784.94 | 32.47 | 765 | 831.85 | 33.43 |
| 676 | 739.00 | 31.50 | 721 | 785.98 | 32.49 | 766 | 832.90 | 33.45 |
| 677 | 740.04 | 31.52 | 722 | 787.02 | 32.51 | 767 | 833.94 | 33.47 |
| 678 | 741.09 | 31.54 | 723 | 788.06 | 32.53 | 768 | 834.98 | 33.49 |
| 679 | 742.13 | 31.57 | 724 | 789.11 | 32.55 | 769 | 836.02 | 33.51 |
| 680 | 743.18 | 31.59 | 725 | 790.15 | 32.58 | 770 | 837.06 | 33.53 |
| 681 | 744.22 | 31.61 | 726 | 791.19 | 32.60 | 771 | 838.11 | 33.55 |
| 682 | 745.27 | 31.63 | 727 | 792.24 | 32.62 | 772 | 839.15 | 33.57 |
| 683 | 746.31 | 31.66 | 728 | 793.28 | 32.64 | 773 | 840.19 | 33.60 |
| 684 | 747.36 | 31.68 | 729 | 794.32 | 32.66 | 774 | 841.23 | 33.62 |
| 685 | 748.40 | 31.70 | 730 | 795.37 | 32.68 | 775 | 842.27 | 33.64 |
| 686 | 749.44 | 31.72 | 731 | 796.41 | 32.70 | 776 | 843.32 | 33.66 |
| 687 | 750.49 | 31.74 | 732 | 797.45 | 32.73 | 777 | 844.36 | 33.68 |
| 688 | 751.53 | 31.77 | 733 | 798.50 | 32.75 | 778 | 845.40 | 33.70 |
| 689 | 752.58 | 31.79 | 734 | 799.54 | 32.77 | 779 | 846.44 | 33.72 |
| 690 | 753.62 | 31.81 | 735 | 800.58 | 32.79 | 780 | 847.48 | 33.74 |
| 691 | 754.67 | 31.83 | 736 | 801.62 | 32.81 | 781 | 848.52 | 33.76 |
| 692 | 755.71 | 31.86 | 737 | 802.67 | 32.83 | 782 | 849.57 | 33.78 |
| 693 | 756.75 | 31.88 | 738 | 803.71 | 32.85 | 783 | 850.61 | 33.80 |
| 694 | 757.80 | 31.90 | 739 | 804.75 | 32.88 | 784 | 851.65 | 33.82 |
| 695 | 758.84 | 31.92 | 740 | 805.80 | 32.90 | 785 | 852.69 | 33.84 |
| 696 | 759.89 | 31.94 | 741 | 806.84 | 32.92 | 786 | 853.73 | 33.87 |
| 697 | 760.93 | 31.97 | 742 | 807.88 | 32.94 | 787 | 854.77 | 33.89 |
| 698 | 761.97 | 31.99 | 743 | 808.92 | 32.96 | 788 | 855.81 | 33.91 |
| 699 | 763.02 | 32.01 | 744 | 809.97 | 32.98 | 789 | 856.86 | 33.93 |
| 700 | 764.06 | 32.03 | 745 | 811.01 | 33.00 | 790 | 857.90 | 33.95 |
| 701 | 765.11 | 32.05 | 746 | 812.05 | 33.03 | 791 | 858.94 | 33.97 |
| 702 | 766.15 | 32.08 | 747 | 813.09 | 33.05 | 792 | 859.98 | 33.99 |
| 703 | 767.19 | 32.10 | 748 | 814.14 | 33.07 | 793 | 861.02 | 34.01 |
| 704 | 768.24 | 32.12 | 749 | 815.18 | 33.09 | 794 | 862.06 | 34.03 |
| 705 | 769.28 | 32.14 | 750 | 816.22 | 33.11 | 795 | 863.10 | 34.05 |
| 706 | 770.33 | 32.16 | 751 | 817.26 | 33.13 | 796 | 864.14 | 34.07 |
| 707 | 771.37 | 32.18 | 752 | 818.31 | 33.15 | 797 | 865.19 | 34.09 |
| 708 | 772.41 | 32.21 | 753 | 819.35 | 33.17 | 798 | 866.23 | 34.11 |
| 709 | 773.46 | 32.23 | 754 | 820.39 | 33.20 | 799 | 867.27 | 34.13 |
| 710 | 774.50 | 32.25 | 755 | 821.43 | 33.22 | 800 | 868.31 | 34.15 |
| 711 | 775.54 | 32.27 | 756 | 822.48 | 33.24 | 801 | 869.35 | 34.18 |
| 712 | 776.59 | 32.29 | 757 | 823.52 | 33.26 | 802 | 870.39 | 34.20 |
| 713 | 777.63 | 32.32 | 758 | 824.56 | 33.28 | 803 | 871.43 | 34.22 |
| 714 | 778.67 | 32.34 | 759 | 825.60 | 33.30 | 804 | 872.47 | 34.24 |
| 715 | 779.72 | 32.36 | 760 | 826.64 | 33.32 | 805 | 873.51 | 34.26 |
| 716 | 780.76 | 32.38 | 761 | 827.69 | 33.34 | 806 | 874.56 | 34.28 |
| 717 | 781.80 | 32.40 | 762 | 828.73 | 33.36 | 807 | 875.60 | 34.30 |
| 718 | 782.85 | 32.42 | 763 | 829.77 | 33.39 | 808 | 876.64 | 34.32 |
| 719 | 783.89 | 32.45 | 764 | 830.81 | 33.41 | 809 | 877.68 | 34.34 |

Table 3. CONSTANTS FOR $\overline{p}'$ (SHORT FORMULA), AN APPROXIMATION OF THE UPPER BINOMIAL CONFIDENCE LIMIT $\overline{p}$; $c/n \leqslant .1$, $n \geqslant 40$, $c \leqslant 1{,}000$ (*Continued*)

| c | $\overline{m}$ | $\overline{k}$ | c | $\overline{m}$ | $\overline{k}$ | c | $\overline{m}$ | $\overline{k}$ |
|---|---|---|---|---|---|---|---|---|
| 810 | 878.72 | 34.36 | 855 | 925.53 | 35.27 | 900 | 972.30 | 36.15 |
| 811 | 879.76 | 34.38 | 856 | 926.57 | 35.29 | 901 | 973.34 | 36.17 |
| 812 | 880.80 | 34.40 | 857 | 927.61 | 35.31 | 902 | 974.38 | 36.19 |
| 813 | 881.84 | 34.42 | 858 | 928.65 | 35.33 | 903 | 975.41 | 36.21 |
| 814 | 882.88 | 34.44 | 859 | 929.69 | 35.35 | 904 | 976.45 | 36.23 |
| 815 | 883.92 | 34.46 | 860 | 930.73 | 35.37 | 905 | 977.49 | 36.25 |
| 816 | 884.96 | 34.48 | 861 | 931.77 | 35.39 | 906 | 978.53 | 36.27 |
| 817 | 886.00 | 34.50 | 862 | 932.81 | 35.40 | 907 | 979.57 | 36.28 |
| 818 | 887.05 | 34.52 | 863 | 933.85 | 35.42 | 908 | 980.61 | 36.30 |
| 819 | 888.09 | 34.54 | 864 | 934.89 | 35.44 | 909 | 981.65 | 36.32 |
| 820 | 889.13 | 34.56 | 865 | 935.93 | 35.46 | 910 | 982.68 | 36.34 |
| 821 | 890.17 | 34.58 | 866 | 936.97 | 35.48 | 911 | 983.72 | 36.36 |
| 822 | 891.21 | 34.60 | 867 | 938.01 | 35.50 | 912 | 984.76 | 36.38 |
| 823 | 892.25 | 34.62 | 868 | 939.05 | 35.52 | 913 | 985.80 | 36.40 |
| 824 | 893.29 | 34.64 | 869 | 940.09 | 35.54 | 914 | 986.84 | 36.42 |
| 825 | 894.33 | 34.66 | 870 | 941.13 | 35.56 | 915 | 987.88 | 36.44 |
| 826 | 895.37 | 34.68 | 871 | 942.17 | 35.58 | 916 | 988.92 | 36.46 |
| 827 | 896.41 | 34.70 | 872 | 943.20 | 35.60 | 917 | 989.95 | 36.48 |
| 828 | 897.45 | 34.73 | 873 | 944.24 | 35.62 | 918 | 990.99 | 36.50 |
| 829 | 898.49 | 34.75 | 874 | 945.28 | 35.64 | 919 | 992.03 | 36.52 |
| 830 | 899.53 | 34.77 | 875 | 946.32 | 35.66 | 920 | 993.07 | 36.53 |
| 831 | 900.57 | 34.79 | 876 | 947.36 | 35.68 | 921 | 994.11 | 36.55 |
| 832 | 901.61 | 34.81 | 877 | 948.40 | 35.70 | 922 | 995.15 | 36.57 |
| 833 | 902.65 | 34.83 | 878 | 949.44 | 35.72 | 923 | 996.18 | 36.59 |
| 834 | 903.69 | 34.85 | 879 | 950.48 | 35.74 | 924 | 997.22 | 36.61 |
| 835 | 904.73 | 34.87 | 880 | 951.52 | 35.76 | 925 | 998.26 | 36.63 |
| 836 | 905.77 | 34.89 | 881 | 952.56 | 35.78 | 926 | 999.30 | 36.65 |
| 837 | 906.81 | 34.91 | 882 | 953.60 | 35.80 | 927 | 1000.34 | 36.67 |
| 838 | 907.85 | 34.93 | 883 | 954.64 | 35.82 | 928 | 1001.38 | 36.69 |
| 839 | 908.89 | 34.95 | 884 | 955.68 | 35.84 | 929 | 1002.41 | 36.71 |
| 840 | 909.93 | 34.97 | 885 | 956.71 | 35.86 | 930 | 1003.45 | 36.73 |
| 841 | 910.97 | 34.99 | 886 | 957.75 | 35.88 | 931 | 1004.49 | 36.74 |
| 842 | 912.01 | 35.01 | 887 | 958.79 | 35.90 | 932 | 1005.53 | 36.76 |
| 843 | 913.05 | 35.03 | 888 | 959.83 | 35.92 | 933 | 1006.57 | 36.78 |
| 844 | 914.09 | 35.05 | 889 | 960.87 | 35.94 | 934 | 1007.60 | 36.80 |
| 845 | 915.13 | 35.07 | 890 | 961.91 | 35.95 | 935 | 1008.64 | 36.82 |
| 846 | 916.17 | 35.09 | 891 | 962.95 | 35.97 | 936 | 1009.68 | 36.84 |
| 847 | 917.21 | 35.11 | 892 | 963.99 | 35.99 | 937 | 1010.72 | 36.86 |
| 848 | 918.25 | 35.13 | 893 | 965.03 | 36.01 | 938 | 1011.76 | 36.88 |
| 849 | 919.29 | 35.15 | 894 | 966.07 | 36.03 | 939 | 1012.79 | 36.90 |
| 850 | 920.33 | 35.17 | 895 | 967.10 | 36.05 | 940 | 1013.83 | 36.92 |
| 851 | 921.37 | 35.19 | 896 | 968.14 | 36.07 | 941 | 1014.87 | 36.93 |
| 852 | 922.41 | 35.21 | 897 | 969.18 | 36.09 | 942 | 1015.91 | 36.95 |
| 853 | 923.45 | 35.23 | 898 | 970.22 | 36.11 | 943 | 1016.95 | 36.97 |
| 854 | 924.49 | 35.25 | 899 | 971.26 | 36.13 | 944 | 1017.98 | 36.99 |

Table 3.  CONSTANTS FOR $\overline{p}'$ (SHORT FORMULA), AN APPROXIMATION OF
THE UPPER BINOMIAL CONFIDENCE LIMIT $\overline{p}$; $c/n \leqslant .1$, $n \geqslant 40$,
$c \leqslant 1,000$ (*Continued*)

| c | $\overline{m}$ | $\overline{k}$ | c | $\overline{m}$ | $\overline{k}$ | c | $\overline{m}$ | $\overline{k}$ |
|---|---|---|---|---|---|---|---|---|
| 945 | 1019.02 | 37.01 | 965 | 1039.77 | 37.39 | 985 | 1060.52 | 37.7( |
| 946 | 1020.06 | 37.03 | 966 | 1040.81 | 37.41 | 986 | 1061.55 | 37.7& |
| 947 | 1021.10 | 37.05 | 967 | 1041.85 | 37.42 | 987 | 1062.59 | 37.8( |
| 948 | 1022.13 | 37.07 | 968 | 1042.89 | 37.44 | 988 | 1063.63 | 37.8 |
| 949 | 1023.17 | 37.09 | 969 | 1043.92 | 37.46 | 989 | 1064.67 | 37.8: |
| 950 | 1024.21 | 37.10 | 970 | 1044.96 | 37.48 | 990 | 1065.70 | 37.8' |
| 951 | 1025.25 | 37.12 | 971 | 1046.00 | 37.50 | 991 | 1066.74 | 37.8' |
| 952 | 1026.29 | 37.14 | 972 | 1047.03 | 37.52 | 992 | 1067.78 | 37.8( |
| 953 | 1027.32 | 37.16 | 973 | 1048.07 | 37.54 | 993 | 1068.81 | 37.9 |
| 954 | 1028.36 | 37.18 | 974 | 1049.11 | 37.55 | 994 | 1069.85 | 37.9: |
| 955 | 1029.40 | 37.20 | 975 | 1050.15 | 37.57 | 995 | 1070.89 | 37.9₄ |
| 956 | 1030.44 | 37.22 | 976 | 1051.18 | 37.59 | 996 | 1071.92 | 37.9( |
| 957 | 1031.47 | 37.24 | 977 | 1052.22 | 37.61 | 997 | 1072.96 | 37.9' |
| 958 | 1032.51 | 37.26 | 978 | 1053.26 | 37.63 | 998 | 1074.00 | 38.0( |
| 959 | 1033.55 | 37.27 | 979 | 1054.30 | 37.65 | 999 | 1075.03 | 38.0 |
| 960 | 1034.59 | 37.29 | 980 | 1055.33 | 37.67 | 1000 | 1076.07 | 38.0₄ |
| 961 | 1035.62 | 37.31 | 981 | 1056.37 | 37.68 | | | |
| 962 | 1036.66 | 37.33 | 982 | 1057.41 | 37.70 | | | |
| 963 | 1037.70 | 37.35 | 983 | 1058.44 | 37.72 | | | |
| 964 | 1038.74 | 37.37 | 984 | 1059.48 | 37.74 | | | |

Table 3. CONSTANTS FOR $\overline{p}'$ (SHORT FORMULA), AN APPROXIMATION OF Section 3
THE UPPER BINOMIAL CONFIDENCE LIMIT $\overline{p}$; $c/n \leqslant .1$, $n \geqslant 40$, $\gamma_1 = 97.5\%$
$c \leqslant 1{,}000$ (*Continued*) $\gamma_2 = 95\%$

| c | $\overline{m}$ | $\overline{k}$ | c | $\overline{m}$ | $\overline{k}$ | c | $\overline{m}$ | $\overline{k}$ |
|---|---|---|---|---|---|---|---|---|
| 0 | 3.6889 | 1.868 | 45 | 60.214 | 7.83 | 90 | 110.63 | 10.32 |
| 1 | 5.5717 | 2.331 | 46 | 61.358 | 7.88 | 91 | 111.73 | 10.37 |
| 2 | 7.2247 | 2.681 | 47 | 62.500 | 7.96 | 92 | 112.83 | 10.42 |
| 3 | 8.7673 | 2.981 | 48 | 63.641 | 8.07 | 93 | 113.93 | 10.48 |
| 4 | 10.242 | 3.247 | 49 | 64.781 | 8.12 | 94 | 115.03 | 10.53 |
| 5 | 11.668 | 3.461 | 50 | 65.919 | 8.20 | 95 | 116.13 | 10.58 |
| 6 | 13.059 | 3.652 | 51 | 67.056 | 8.04 | 96 | 117.23 | 10.63 |
| 7 | 14.423 | 3.834 | 52 | 68.191 | 8.11 | 97 | 118.33 | 10.68 |
| 8 | 15.763 | 4.006 | 53 | 69.326 | 8.18 | 98 | 119.43 | 10.72 |
| 9 | 17.085 | 4.168 | 54 | 70.459 | 8.24 | 99 | 120.53 | 10.77 |
| 10 | 18.390 | 4.33 | 55 | 71.590 | 8.31 | 100 | 121.63 | 10.82 |
| 11 | 19.682 | 4.48 | 56 | 72.721 | 8.37 | 101 | 122.72 | 10.87 |
| 12 | 20.962 | 4.62 | 57 | 73.850 | 8.44 | 102 | 123.82 | 10.92 |
| 13 | 22.230 | 4.76 | 58 | 74.979 | 8.50 | 103 | 124.92 | 10.97 |
| 14 | 23.490 | 4.89 | 59 | 76.106 | 8.56 | 104 | 126.01 | 11.02 |
| 15 | 24.740 | 5.01 | 60 | 77.232 | 8.63 | 105 | 127.11 | 11.06 |
| 16 | 25.983 | 5.14 | 61 | 78.357 | 8.69 | 106 | 128.20 | 11.11 |
| 17 | 27.219 | 5.25 | 62 | 79.481 | 8.75 | 107 | 129.30 | 11.16 |
| 18 | 28.448 | 5.36 | 63 | 80.605 | 8.81 | 108 | 130.39 | 11.21 |
| 19 | 29.671 | 5.49 | 64 | 81.727 | 8.87 | 109 | 131.49 | 11.25 |
| 20 | 30.888 | 5.59 | 65 | 82.848 | 8.94 | 110 | 132.58 | 11.30 |
| 21 | 32.101 | 5.71 | 66 | 83.968 | 9.00 | 111 | 133.67 | 11.34 |
| 22 | 33.308 | 5.80 | 67 | 85.088 | 9.05 | 112 | 134.77 | 11.39 |
| 23 | 34.511 | 5.92 | 68 | 86.206 | 9.11 | 113 | 135.86 | 11.44 |
| 24 | 35.710 | 6.00 | 69 | 87.324 | 9.17 | 114 | 136.95 | 11.48 |
| 25 | 36.905 | 6.12 | 70 | 88.441 | 9.23 | 115 | 138.04 | 11.53 |
| 26 | 38.096 | 6.21 | 71 | 89.557 | 9.29 | 116 | 139.13 | 11.57 |
| 27 | 39.284 | 6.30 | 72 | 90.672 | 9.35 | 117 | 140.22 | 11.62 |
| 28 | 40.468 | 6.39 | 73 | 91.787 | 9.40 | 118 | 141.31 | 11.66 |
| 29 | 41.649 | 6.48 | 74 | 92.900 | 9.46 | 119 | 142.40 | 11.71 |
| 30 | 42.827 | 6.56 | 75 | 94.013 | 9.52 | 120 | 143.49 | 11.75 |
| 31 | 44.002 | 6.68 | 76 | 95.126 | 9.57 | 121 | 144.58 | 11.80 |
| 32 | 45.174 | 6.76 | 77 | 96.237 | 9.63 | 122 | 145.67 | 11.84 |
| 33 | 46.344 | 6.84 | 78 | 97.348 | 9.68 | 123 | 146.76 | 11.89 |
| 34 | 47.512 | 6.92 | 79 | 98.458 | 9.74 | 124 | 147.84 | 11.93 |
| 35 | 48.677 | 7.01 | 80 | 99.567 | 9.79 | 125 | 148.93 | 11.97 |
| 36 | 49.839 | 7.11 | 81 | 100.68 | 9.85 | 126 | 150.02 | 12.02 |
| 37 | 51.000 | 7.18 | 82 | 101.78 | 9.90 | 127 | 151.11 | 12.06 |
| 38 | 52.158 | 7.27 | 83 | 102.89 | 9.96 | 128 | 152.19 | 12.10 |
| 39 | 53.315 | 7.34 | 84 | 104.00 | 10.01 | 129 | 153.28 | 12.15 |
| 40 | 54.469 | 7.39 | 85 | 105.10 | 10.06 | 130 | 154.36 | 12.19 |
| 41 | 55.621 | 7.49 | 86 | 106.21 | 10.11 | 131 | 155.45 | 12.23 |
| 42 | 56.772 | 7.57 | 87 | 107.31 | 10.17 | 132 | 156.54 | 12.28 |
| 43 | 57.921 | 7.67 | 88 | 108.42 | 10.22 | 133 | 157.62 | 12.32 |
| 44 | 59.068 | 7.72 | 89 | 109.52 | 10.27 | 134 | 158.70 | 12.36 |

Table 3. **CONSTANTS FOR** $\bar{p}'$ **(SHORT FORMULA), AN APPROXIMATION OF THE UPPER BINOMIAL CONFIDENCE LIMIT** $\bar{p}$; $c/n \leqslant .1$, $n \geqslant 40$, $c \leqslant 1,000$ *(Continued)*

| c | $\bar{m}$ | $\bar{k}$ | c | $\bar{m}$ | $\bar{k}$ | c | $\bar{m}$ | $\bar{k}$ |
|---|---|---|---|---|---|---|---|---|
| 135 | 159.79 | 12.40 | 180 | 208.32 | 14.16 | 225 | 256.41 | 15.71 |
| 136 | 160.87 | 12.44 | 181 | 209.39 | 14.19 | 226 | 257.48 | 15.74 |
| 137 | 161.96 | 12.49 | 182 | 210.46 | 14.23 | 227 | 258.54 | 15.77 |
| 138 | 163.04 | 12.53 | 183 | 211.53 | 14.27 | 228 | 259.61 | 15.80 |
| 139 | 164.12 | 12.57 | 184 | 212.61 | 14.30 | 229 | 260.67 | 15.84 |
| 140 | 165.21 | 12.61 | 185 | 213.68 | 14.34 | 230 | 261.74 | 15.87 |
| 141 | 166.29 | 12.65 | 186 | 214.75 | 14.37 | 231 | 262.80 | 15.90 |
| 142 | 167.37 | 12.69 | 187 | 215.82 | 14.41 | 232 | 263.87 | 15.93 |
| 143 | 168.45 | 12.73 | 188 | 216.89 | 14.45 | 233 | 264.93 | 15.96 |
| 144 | 169.53 | 12.77 | 189 | 217.96 | 14.48 | 234 | 265.99 | 16.00 |
| 145 | 170.62 | 12.81 | 190 | 219.03 | 14.52 | 235 | 267.06 | 16.03 |
| 146 | 171.70 | 12.86 | 191 | 220.11 | 14.55 | 236 | 268.12 | 16.06 |
| 147 | 172.78 | 12.90 | 192 | 221.18 | 14.59 | 237 | 269.18 | 16.09 |
| 148 | 173.86 | 12.94 | 193 | 222.25 | 14.62 | 238 | 270.25 | 16.12 |
| 149 | 174.94 | 12.98 | 194 | 223.32 | 14.66 | 239 | 271.31 | 16.16 |
| 150 | 176.02 | 13.02 | 195 | 224.39 | 14.69 | 240 | 272.37 | 16.19 |
| 151 | 177.10 | 13.06 | 196 | 225.46 | 14.73 | 241 | 273.44 | 16.22 |
| 152 | 178.18 | 13.10 | 197 | 226.53 | 14.76 | 242 | 274.50 | 16.25 |
| 153 | 179.26 | 13.14 | 198 | 227.60 | 14.80 | 243 | 275.56 | 16.28 |
| 154 | 180.33 | 13.17 | 199 | 228.67 | 14.83 | 244 | 276.63 | 16.31 |
| 155 | 181.41 | 13.21 | 200 | 229.73 | 14.87 | 245 | 277.69 | 16.34 |
| 156 | 182.49 | 13.25 | 201 | 230.80 | 14.90 | 246 | 278.75 | 16.38 |
| 157 | 183.57 | 13.29 | 202 | 231.87 | 14.94 | 247 | 279.81 | 16.41 |
| 158 | 184.65 | 13.33 | 203 | 232.94 | 14.97 | 248 | 280.88 | 16.44 |
| 159 | 185.73 | 13.37 | 204 | 234.01 | 15.01 | 249 | 281.94 | 16.47 |
| 160 | 186.80 | 13.41 | 205 | 235.08 | 15.04 | 250 | 283.00 | 16.50 |
| 161 | 187.89 | 13.45 | 206 | 236.15 | 15.07 | 251 | 284.06 | 16.53 |
| 162 | 188.97 | 13.49 | 207 | 237.21 | 15.11 | 252 | 285.12 | 16.56 |
| 163 | 190.05 | 13.52 | 208 | 238.28 | 15.14 | 253 | 286.18 | 16.59 |
| 164 | 191.12 | 13.56 | 209 | 239.35 | 15.18 | 254 | 287.25 | 16.62 |
| 165 | 192.20 | 13.60 | 210 | 240.42 | 15.21 | 255 | 288.31 | 16.65 |
| 166 | 193.28 | 13.64 | 211 | 241.49 | 15.24 | 256 | 289.37 | 16.68 |
| 167 | 194.35 | 13.68 | 212 | 242.55 | 15.28 | 257 | 290.43 | 16.71 |
| 168 | 195.43 | 13.71 | 213 | 243.62 | 15.31 | 258 | 291.49 | 16.75 |
| 169 | 196.50 | 13.75 | 214 | 244.69 | 15.34 | 259 | 292.55 | 16.78 |
| 170 | 197.58 | 13.79 | 215 | 245.75 | 15.38 | 260 | 293.61 | 16.81 |
| 171 | 198.65 | 13.83 | 216 | 246.82 | 15.41 | 261 | 294.67 | 16.84 |
| 172 | 199.73 | 13.86 | 217 | 247.89 | 15.44 | 262 | 295.73 | 16.87 |
| 173 | 200.80 | 13.90 | 218 | 248.95 | 15.48 | 263 | 296.79 | 16.90 |
| 174 | 201.88 | 13.94 | 219 | 250.02 | 15.51 | 264 | 297.85 | 16.93 |
| 175 | 202.95 | 13.97 | 220 | 251.08 | 15.54 | 265 | 298.91 | 16.96 |
| 176 | 204.02 | 14.01 | 221 | 252.15 | 15.58 | 266 | 299.97 | 16.99 |
| 177 | 205.10 | 14.05 | 222 | 253.22 | 15.61 | 267 | 301.03 | 17.02 |
| 178 | 206.17 | 14.09 | 223 | 254.28 | 15.64 | 268 | 302.09 | 17.05 |
| 179 | 207.24 | 14.12 | 224 | 255.35 | 15.67 | 269 | 303.15 | 17.08 |

Table 3. **CONSTANTS FOR $\bar{p}'$ (SHORT FORMULA), AN APPROXIMATION OF THE UPPER BINOMIAL CONFIDENCE LIMIT** $\bar{p}$; c/n ≤ .1, n ≥ 40, c ≤ 1,000 (*Continued*)

| c | $\bar{m}$ | $\bar{k}$ | c | $\bar{m}$ | $\bar{k}$ | c | $\bar{m}$ | $\bar{k}$ |
|---|---|---|---|---|---|---|---|---|
| 270 | 304.21 | 17.11 | 315 | 351.79 | 18.39 | 360 | 399.19 | 19.59 |
| 271 | 305.27 | 17.14 | 316 | 352.84 | 18.42 | 361 | 400.24 | 19.62 |
| 272 | 306.33 | 17.17 | 317 | 353.90 | 18.45 | 362 | 401.29 | 19.65 |
| 273 | 307.39 | 17.20 | 318 | 354.95 | 18.48 | 363 | 402.34 | 19.67 |
| 274 | 308.45 | 17.23 | 319 | 356.01 | 18.50 | 364 | 403.39 | 19.70 |
| 275 | 309.51 | 17.25 | 320 | 357.06 | 18.53 | 365 | 404.44 | 19.72 |
| 276 | 310.57 | 17.28 | 321 | 358.12 | 18.56 | 366 | 405.50 | 19.75 |
| 277 | 311.63 | 17.31 | 322 | 359.17 | 18.59 | 367 | 406.55 | 19.77 |
| 278 | 312.69 | 17.34 | 323 | 360.23 | 18.61 | 368 | 407.60 | 19.80 |
| 279 | 313.74 | 17.37 | 324 | 361.28 | 18.64 | 369 | 408.65 | 19.82 |
| 280 | 314.80 | 17.40 | 325 | 362.34 | 18.67 | 370 | 409.70 | 19.85 |
| 281 | 315.86 | 17.43 | 326 | 363.39 | 18.70 | 371 | 410.75 | 19.88 |
| 282 | 316.92 | 17.46 | 327 | 364.44 | 18.72 | 372 | 411.80 | 19.90 |
| 283 | 317.98 | 17.49 | 328 | 365.50 | 18.75 | 373 | 412.85 | 19.93 |
| 284 | 319.04 | 17.52 | 329 | 366.55 | 18.78 | 374 | 413.90 | 19.95 |
| 285 | 320.09 | 17.55 | 330 | 367.61 | 18.80 | 375 | 414.95 | 19.98 |
| 286 | 321.15 | 17.58 | 331 | 368.66 | 18.83 | 376 | 416.00 | 20.00 |
| 287 | 322.21 | 17.60 | 332 | 369.71 | 18.86 | 377 | 417.05 | 20.03 |
| 288 | 323.27 | 17.63 | 333 | 370.77 | 18.88 | 378 | 418.10 | 20.05 |
| 289 | 324.32 | 17.66 | 334 | 371.82 | 18.91 | 379 | 419.15 | 20.08 |
| 290 | 325.38 | 17.69 | 335 | 372.87 | 18.94 | 380 | 420.20 | 20.10 |
| 291 | 326.44 | 17.72 | 336 | 373.93 | 18.96 | 381 | 421.26 | 20.13 |
| 292 | 327.50 | 17.75 | 337 | 374.98 | 18.99 | 382 | 422.31 | 20.15 |
| 293 | 328.55 | 17.78 | 338 | 376.03 | 19.02 | 383 | 423.36 | 20.18 |
| 294 | 329.61 | 17.81 | 339 | 377.09 | 19.04 | 384 | 424.41 | 20.20 |
| 295 | 330.67 | 17.83 | 340 | 378.14 | 19.07 | 385 | 425.46 | 20.23 |
| 296 | 331.73 | 17.86 | 341 | 379.19 | 19.10 | 386 | 426.50 | 20.25 |
| 297 | 332.78 | 17.89 | 342 | 380.25 | 19.12 | 387 | 427.55 | 20.28 |
| 298 | 333.84 | 17.92 | 343 | 381.30 | 19.15 | 388 | 428.60 | 20.30 |
| 299 | 334.90 | 17.95 | 344 | 382.35 | 19.18 | 389 | 429.65 | 20.33 |
| 300 | 335.95 | 17.98 | 345 | 383.41 | 19.20 | 390 | 430.70 | 20.35 |
| 301 | 337.01 | 18.00 | 346 | 384.46 | 19.23 | 391 | 431.75 | 20.38 |
| 302 | 338.06 | 18.03 | 347 | 385.51 | 19.26 | 392 | 432.80 | 20.40 |
| 303 | 339.12 | 18.06 | 348 | 386.56 | 19.28 | 393 | 433.85 | 20.43 |
| 304 | 340.18 | 18.09 | 349 | 387.62 | 19.31 | 394 | 434.90 | 20.45 |
| 305 | 341.23 | 18.12 | 350 | 388.67 | 19.33 | 395 | 435.95 | 20.48 |
| 306 | 342.29 | 18.14 | 351 | 389.72 | 19.36 | 396 | 437.00 | 20.50 |
| 307 | 343.35 | 18.17 | 352 | 390.77 | 19.39 | 397 | 438.05 | 20.52 |
| 308 | 344.40 | 18.20 | 353 | 391.82 | 19.41 | 398 | 439.10 | 20.55 |
| 309 | 345.46 | 18.23 | 354 | 392.88 | 19.44 | 399 | 440.15 | 20.57 |
| 310 | 346.51 | 18.26 | 355 | 393.93 | 19.46 | 400 | 441.20 | 20.60 |
| 311 | 347.57 | 18.28 | 356 | 394.98 | 19.49 | 401 | 442.25 | 20.62 |
| 312 | 348.62 | 18.31 | 357 | 396.03 | 19.52 | 402 | 443.29 | 20.65 |
| 313 | 349.68 | 18.34 | 358 | 397.08 | 19.54 | 403 | 444.34 | 20.67 |
| 314 | 350.73 | 18.37 | 359 | 398.14 | 19.57 | 404 | 445.39 | 20.70 |

Table 3. CONSTANTS FOR $\overline{p}'$ (SHORT FORMULA), AN APPROXIMATION OF THE UPPER BINOMIAL CONFIDENCE LIMIT $\overline{p}$; $c/n \leqslant .1$, $n \geqslant 40$, $c \leqslant 1,000$ (*Continued*)

| c | $\overline{m}$ | $\overline{k}$ | c | $\overline{m}$ | $\overline{k}$ | c | $\overline{m}$ | $\overline{k}$ |
|---|---|---|---|---|---|---|---|---|
| 405 | 446.44 | 20.72 | 450 | 493.57 | 21.79 | 495 | 540.60 | 22.80 |
| 406 | 447.49 | 20.74 | 451 | 494.62 | 21.81 | 496 | 541.64 | 22.82 |
| 407 | 448.54 | 20.77 | 452 | 495.66 | 21.83 | 497 | 542.69 | 22.84 |
| 408 | 449.59 | 20.79 | 453 | 496.71 | 21.85 | 498 | 543.73 | 22.87 |
| 409 | 450.63 | 20.82 | 454 | 497.76 | 21.88 | 499 | 544.77 | 22.89 |
| 410 | 451.68 | 20.84 | 455 | 498.80 | 21.90 | 500 | 545.82 | 22.91 |
| 411 | 452.73 | 20.87 | 456 | 499.85 | 21.92 | 501 | 546.86 | 22.93 |
| 412 | 453.78 | 20.89 | 457 | 500.89 | 21.95 | 502 | 547.91 | 22.95 |
| 413 | 454.83 | 20.91 | 458 | 501.94 | 21.97 | 503 | 548.95 | 22.97 |
| 414 | 455.88 | 20.94 | 459 | 502.98 | 21.99 | 504 | 549.99 | 23.00 |
| 415 | 456.92 | 20.96 | 460 | 504.03 | 22.02 | 505 | 551.04 | 23.02 |
| 416 | 457.97 | 20.99 | 461 | 505.08 | 22.04 | 506 | 552.08 | 23.04 |
| 417 | 459.02 | 21.01 | 462 | 506.12 | 22.06 | 507 | 553.12 | 23.06 |
| 418 | 460.07 | 21.03 | 463 | 507.17 | 22.08 | 508 | 554.17 | 23.08 |
| 419 | 461.12 | 21.06 | 464 | 508.21 | 22.11 | 509 | 555.21 | 23.11 |
| 420 | 462.16 | 21.08 | 465 | 509.26 | 22.13 | 510 | 556.25 | 23.13 |
| 421 | 463.21 | 21.11 | 466 | 510.30 | 22.15 | 511 | 557.30 | 23.15 |
| 422 | 464.26 | 21.13 | 467 | 511.35 | 22.17 | 512 | 558.34 | 23.17 |
| 423 | 465.31 | 21.15 | 468 | 512.39 | 22.20 | 513 | 559.38 | 23.19 |
| 424 | 466.35 | 21.18 | 469 | 513.44 | 22.22 | 514 | 560.43 | 23.21 |
| 425 | 467.40 | 21.20 | 470 | 514.48 | 22.24 | 515 | 561.47 | 23.23 |
| 426 | 468.45 | 21.22 | 471 | 515.53 | 22.26 | 516 | 562.51 | 23.26 |
| 427 | 469.50 | 21.25 | 472 | 516.57 | 22.29 | 517 | 563.56 | 23.28 |
| 428 | 470.54 | 21.27 | 473 | 517.62 | 22.31 | 518 | 564.60 | 23.30 |
| 429 | 471.59 | 21.30 | 474 | 518.66 | 22.33 | 519 | 565.64 | 23.32 |
| 430 | 472.64 | 21.32 | 475 | 519.71 | 22.35 | 520 | 566.68 | 23.34 |
| 431 | 473.69 | 21.34 | 476 | 520.75 | 22.38 | 521 | 567.73 | 23.36 |
| 432 | 474.73 | 21.37 | 477 | 521.80 | 22.40 | 522 | 568.77 | 23.39 |
| 433 | 475.78 | 21.39 | 478 | 522.84 | 22.42 | 523 | 569.81 | 23.41 |
| 434 | 476.83 | 21.41 | 479 | 523.89 | 22.44 | 524 | 570.86 | 23.43 |
| 435 | 477.87 | 21.44 | 480 | 524.93 | 22.47 | 525 | 571.90 | 23.45 |
| 436 | 478.92 | 21.46 | 481 | 525.98 | 22.49 | 526 | 572.94 | 23.47 |
| 437 | 479.97 | 21.48 | 482 | 527.02 | 22.51 | 527 | 573.98 | 23.49 |
| 438 | 481.01 | 21.51 | 483 | 528.07 | 22.53 | 528 | 575.03 | 23.51 |
| 439 | 482.06 | 21.53 | 484 | 529.11 | 22.56 | 529 | 576.07 | 23.53 |
| 440 | 483.11 | 21.55 | 485 | 530.16 | 22.58 | 530 | 577.11 | 23.56 |
| 441 | 484.15 | 21.58 | 486 | 531.20 | 22.60 | 531 | 578.15 | 23.58 |
| 442 | 485.20 | 21.60 | 487 | 532.25 | 22.62 | 532 | 579.20 | 23.60 |
| 443 | 486.25 | 21.62 | 488 | 533.29 | 22.64 | 533 | 580.24 | 23.62 |
| 444 | 487.29 | 21.65 | 489 | 534.33 | 22.67 | 534 | 581.28 | 23.64 |
| 445 | 488.34 | 21.67 | 490 | 535.38 | 22.69 | 535 | 582.32 | 23.66 |
| 446 | 489.39 | 21.69 | 491 | 536.42 | 22.71 | 536 | 583.37 | 23.68 |
| 447 | 490.43 | 21.72 | 492 | 537.47 | 22.73 | 537 | 584.41 | 23.70 |
| 448 | 491.48 | 21.74 | 493 | 538.51 | 22.76 | 538 | 585.45 | 23.73 |
| 449 | 492.53 | 21.76 | 494 | 539.55 | 22.78 | 539 | 586.49 | 23.75 |

Table 3.  CONSTANTS FOR $\overline{p}'$ (SHORT FORMULA), AN APPROXIMATION OF     **Section 3**
THE UPPER BINOMIAL CONFIDENCE LIMIT $\overline{p}$; $c/n \leqslant .1$, $n \geqslant 40$,     $\gamma_1 = 97.5\%$
c $\leqslant$ 1,000 (*Continued*)     $\gamma_2 = 95\%$

| c | $\overline{m}$ | $\overline{k}$ | c | $\overline{m}$ | $\overline{k}$ | c | $\overline{m}$ | $\overline{k}$ |
|---|---|---|---|---|---|---|---|---|
| 540 | 587.54 | 23.77 | 585 | 634.39 | 24.70 | 630 | 681.18 | 25.59 |
| 541 | 588.58 | 23.79 | 586 | 635.43 | 24.72 | 631 | 682.22 | 25.61 |
| 542 | 589.62 | 23.81 | 587 | 636.47 | 24.74 | 632 | 683.26 | 25.63 |
| 543 | 590.66 | 23.83 | 588 | 637.52 | 24.76 | 633 | 684.30 | 25.65 |
| 544 | 591.70 | 23.85 | 589 | 638.56 | 24.78 | 634 | 685.34 | 25.67 |
| 545 | 592.75 | 23.87 | 590 | 639.60 | 24.80 | 635 | 686.38 | 25.69 |
| 546 | 593.79 | 23.89 | 591 | 640.64 | 24.82 | 636 | 687.42 | 25.71 |
| 547 | 594.83 | 23.91 | 592 | 641.68 | 24.84 | 637 | 688.45 | 25.73 |
| 548 | 595.87 | 23.94 | 593 | 642.72 | 24.86 | 638 | 689.49 | 25.75 |
| 549 | 596.91 | 23.96 | 594 | 643.76 | 24.88 | 639 | 690.53 | 25.77 |
| 550 | 597.95 | 23.98 | 595 | 644.80 | 24.90 | 640 | 691.57 | 25.79 |
| 551 | 599.00 | 24.00 | 596 | 645.84 | 24.92 | 641 | 692.61 | 25.80 |
| 552 | 600.04 | 24.02 | 597 | 646.88 | 24.94 | 642 | 693.65 | 25.82 |
| 553 | 601.08 | 24.04 | 598 | 647.92 | 24.96 | 643 | 694.69 | 25.84 |
| 554 | 602.12 | 24.06 | 599 | 648.96 | 24.98 | 644 | 695.72 | 25.86 |
| 555 | 603.16 | 24.08 | 600 | 650.00 | 25.00 | 645 | 696.76 | 25.88 |
| 556 | 604.20 | 24.10 | 601 | 651.04 | 25.02 | 646 | 697.80 | 25.90 |
| 557 | 605.25 | 24.12 | 602 | 652.08 | 25.04 | 647 | 698.84 | 25.92 |
| 558 | 606.29 | 24.14 | 603 | 653.12 | 25.06 | 648 | 699.88 | 25.94 |
| 559 | 607.33 | 24.16 | 604 | 654.16 | 25.08 | 649 | 700.92 | 25.96 |
| 560 | 608.37 | 24.19 | 605 | 655.20 | 25.10 | 650 | 701.96 | 25.98 |
| 561 | 609.41 | 24.21 | 606 | 656.24 | 25.12 | 651 | 702.99 | 26.00 |
| 562 | 610.45 | 24.23 | 607 | 657.28 | 25.14 | 652 | 704.03 | 26.02 |
| 563 | 611.49 | 24.25 | 608 | 658.32 | 25.16 | 653 | 705.07 | 26.04 |
| 564 | 612.54 | 24.27 | 609 | 659.36 | 25.18 | 654 | 706.11 | 26.05 |
| 565 | 613.58 | 24.29 | 610 | 660.40 | 25.20 | 655 | 707.15 | 26.07 |
| 566 | 614.62 | 24.31 | 611 | 661.43 | 25.22 | 656 | 708.19 | 26.09 |
| 567 | 615.66 | 24.33 | 612 | 662.47 | 25.24 | 657 | 709.22 | 26.11 |
| 568 | 616.70 | 24.35 | 613 | 663.51 | 25.26 | 658 | 710.26 | 26.13 |
| 569 | 617.74 | 24.37 | 614 | 664.55 | 25.28 | 659 | 711.30 | 26.15 |
| 570 | 618.78 | 24.39 | 615 | 665.59 | 25.30 | 660 | 712.34 | 26.17 |
| 571 | 619.82 | 24.41 | 616 | 666.63 | 25.32 | 661 | 713.38 | 26.19 |
| 572 | 620.86 | 24.43 | 617 | 667.67 | 25.34 | 662 | 714.41 | 26.21 |
| 573 | 621.91 | 24.45 | 618 | 668.71 | 25.36 | 663 | 715.45 | 26.23 |
| 574 | 622.95 | 24.47 | 619 | 669.75 | 25.38 | 664 | 716.49 | 26.25 |
| 575 | 623.99 | 24.49 | 620 | 670.79 | 25.40 | 665 | 717.53 | 26.26 |
| 576 | 625.03 | 24.51 | 621 | 671.83 | 25.41 | 666 | 718.57 | 26.28 |
| 577 | 626.07 | 24.53 | 622 | 672.87 | 25.43 | 667 | 719.60 | 26.30 |
| 578 | 627.11 | 24.55 | 623 | 673.91 | 25.45 | 668 | 720.64 | 26.32 |
| 579 | 628.15 | 24.58 | 624 | 674.95 | 25.47 | 669 | 721.68 | 26.34 |
| 580 | 629.19 | 24.60 | 625 | 675.99 | 25.49 | 670 | 722.72 | 26.36 |
| 581 | 630.23 | 24.62 | 626 | 677.03 | 25.51 | 671 | 723.76 | 26.38 |
| 582 | 631.27 | 24.64 | 627 | 678.06 | 25.53 | 672 | 724.79 | 26.40 |
| 583 | 632.31 | 24.66 | 628 | 679.10 | 25.55 | 673 | 725.83 | 26.42 |
| 584 | 633.35 | 24.68 | 629 | 680.14 | 25.57 | 674 | 726.87 | 26.43 |

Table 3. CONSTANTS FOR $\overline{p}'$ (SHORT FORMULA), AN APPROXIMATION OF THE UPPER BINOMIAL CONFIDENCE LIMIT $\overline{p}$; $c/n \leqslant .1$, $n \geqslant 40$, $c \leqslant 1,000$ (Continued)

| c | $\overline{m}$ | $\overline{k}$ | c | $\overline{m}$ | $\overline{k}$ | c | $\overline{m}$ | $\overline{k}$ |
|---|---|---|---|---|---|---|---|---|
| 675 | 727.91 | 26.45 | 720 | 774.58 | 27.29 | 765 | 821.19 | 28.10 |
| 676 | 728.94 | 26.47 | 721 | 775.61 | 27.31 | 766 | 822.23 | 28.11 |
| 677 | 729.98 | 26.49 | 722 | 776.65 | 27.32 | 767 | 823.26 | 28.13 |
| 678 | 731.02 | 26.51 | 723 | 777.69 | 27.34 | 768 | 824.30 | 28.15 |
| 679 | 732.06 | 26.53 | 724 | 778.72 | 27.36 | 769 | 825.33 | 28.17 |
| 680 | 733.10 | 26.55 | 725 | 779.76 | 27.38 | 770 | 826.37 | 28.19 |
| 681 | 734.13 | 26.57 | 726 | 780.79 | 27.40 | 771 | 827.41 | 28.20 |
| 682 | 735.17 | 26.59 | 727 | 781.83 | 27.42 | 772 | 828.44 | 28.22 |
| 683 | 736.21 | 26.60 | 728 | 782.87 | 27.43 | 773 | 829.48 | 28.24 |
| 684 | 737.25 | 26.62 | 729 | 783.90 | 27.45 | 774 | 830.51 | 28.26 |
| 685 | 738.28 | 26.64 | 730 | 784.94 | 27.47 | 775 | 831.55 | 28.27 |
| 686 | 739.32 | 26.66 | 731 | 785.98 | 27.49 | 776 | 832.58 | 28.29 |
| 687 | 740.36 | 26.68 | 732 | 787.01 | 27.51 | 777 | 833.62 | 28.31 |
| 688 | 741.39 | 26.70 | 733 | 788.05 | 27.52 | 778 | 834.65 | 28.33 |
| 689 | 742.43 | 26.72 | 734 | 789.08 | 27.54 | 779 | 835.69 | 28.34 |
| 690 | 743.47 | 26.73 | 735 | 790.12 | 27.56 | 780 | 836.72 | 28.36 |
| 691 | 744.51 | 26.75 | 736 | 791.16 | 27.58 | 781 | 837.76 | 28.38 |
| 692 | 745.54 | 26.77 | 737 | 792.19 | 27.60 | 782 | 838.79 | 28.40 |
| 693 | 746.58 | 26.79 | 738 | 793.23 | 27.61 | 783 | 839.83 | 28.41 |
| 694 | 747.62 | 26.81 | 739 | 794.26 | 27.63 | 784 | 840.86 | 28.43 |
| 695 | 748.66 | 26.83 | 740 | 795.30 | 27.65 | 785 | 841.90 | 28.45 |
| 696 | 749.69 | 26.85 | 741 | 796.34 | 27.67 | 786 | 842.93 | 28.47 |
| 697 | 750.73 | 26.86 | 742 | 797.37 | 27.69 | 787 | 843.97 | 28.48 |
| 698 | 751.77 | 26.88 | 743 | 798.41 | 27.70 | 788 | 845.00 | 28.50 |
| 699 | 752.80 | 26.90 | 744 | 799.44 | 27.72 | 789 | 846.04 | 28.52 |
| 700 | 753.84 | 26.92 | 745 | 800.48 | 27.74 | 790 | 847.07 | 28.54 |
| 701 | 754.88 | 26.94 | 746 | 801.52 | 27.76 | 791 | 848.11 | 28.55 |
| 702 | 755.91 | 26.96 | 747 | 802.55 | 27.78 | 792 | 849.14 | 28.57 |
| 703 | 756.95 | 26.98 | 748 | 803.59 | 27.79 | 793 | 850.18 | 28.59 |
| 704 | 757.99 | 26.99 | 749 | 804.62 | 27.81 | 794 | 851.21 | 28.61 |
| 705 | 759.03 | 27.01 | 750 | 805.66 | 27.83 | 795 | 852.25 | 28.62 |
| 706 | 760.06 | 27.03 | 751 | 806.70 | 27.85 | 796 | 853.28 | 28.64 |
| 707 | 761.10 | 27.05 | 752 | 807.73 | 27.87 | 797 | 854.31 | 28.66 |
| 708 | 762.14 | 27.07 | 753 | 808.77 | 27.88 | 798 | 855.35 | 28.67 |
| 709 | 763.17 | 27.09 | 754 | 809.80 | 27.90 | 799 | 856.38 | 28.69 |
| 710 | 764.21 | 27.10 | 755 | 810.84 | 27.92 | 800 | 857.42 | 28.71 |
| 711 | 765.25 | 27.12 | 756 | 811.87 | 27.94 | 801 | 858.45 | 28.73 |
| 712 | 766.28 | 27.14 | 757 | 812.91 | 27.95 | 802 | 859.49 | 28.74 |
| 713 | 767.32 | 27.16 | 758 | 813.95 | 27.97 | 803 | 860.52 | 28.76 |
| 714 | 768.36 | 27.18 | 759 | 814.98 | 27.99 | 804 | 861.56 | 28.78 |
| 715 | 769.39 | 27.20 | 760 | 816.02 | 28.01 | 805 | 862.59 | 28.80 |
| 716 | 770.43 | 27.21 | 761 | 817.05 | 28.03 | 806 | 863.63 | 28.81 |
| 717 | 771.47 | 27.23 | 762 | 818.09 | 28.04 | 807 | 864.66 | 28.83 |
| 718 | 772.50 | 27.25 | 763 | 819.12 | 28.06 | 808 | 865.70 | 28.85 |
| 719 | 773.54 | 27.27 | 764 | 820.16 | 28.08 | 809 | 866.73 | 28.86 |

Table 3.  CONSTANTS FOR $\overline{p}'$ (SHORT FORMULA), AN APPROXIMATION OF
THE UPPER BINOMIAL CONFIDENCE LIMIT $\overline{p}$; $c/n \leqslant .1$, $n \geqslant 40$,
$c \leqslant 1,000$ (*Continued*)

| c | $\overline{m}$ | $\overline{k}$ | c | $\overline{m}$ | $\overline{k}$ | c | $\overline{m}$ | $\overline{k}$ |
|---|---|---|---|---|---|---|---|---|
| 810 | 867.76 | 28.88 | 855 | 914.29 | 29.65 | 900 | 960.78 | 30.39 |
| 811 | 868.80 | 28.90 | 856 | 915.33 | 29.66 | 901 | 961.81 | 30.41 |
| 812 | 869.83 | 28.92 | 857 | 916.36 | 29.68 | 902 | 962.85 | 30.42 |
| 813 | 870.87 | 28.93 | 858 | 917.39 | 29.70 | 903 | 963.88 | 30.44 |
| 814 | 871.90 | 28.95 | 859 | 918.43 | 29.71 | 904 | 964.91 | 30.46 |
| 815 | 872.94 | 28.97 | 860 | 919.46 | 29.73 | 905 | 965.94 | 30.47 |
| 816 | 873.97 | 28.99 | 861 | 920.49 | 29.75 | 906 | 966.98 | 30.49 |
| 817 | 875.00 | 29.00 | 862 | 921.53 | 29.76 | 907 | 968.01 | 30.50 |
| 818 | 876.04 | 29.02 | 863 | 922.56 | 29.78 | 908 | 969.04 | 30.52 |
| 819 | 877.07 | 29.04 | 864 | 923.59 | 29.80 | 909 | 970.07 | 30.54 |
| 820 | 878.11 | 29.05 | 865 | 924.63 | 29.81 | 910 | 971.11 | 30.55 |
| 821 | 879.14 | 29.07 | 866 | 925.66 | 29.83 | 911 | 972.14 | 30.57 |
| 822 | 880.18 | 29.09 | 867 | 926.69 | 29.85 | 912 | 973.17 | 30.59 |
| 823 | 881.21 | 29.10 | 868 | 927.73 | 29.86 | 913 | 974.20 | 30.60 |
| 824 | 882.24 | 29.12 | 869 | 928.76 | 29.88 | 914 | 975.24 | 30.62 |
| 825 | 883.28 | 29.14 | 870 | 929.79 | 29.90 | 915 | 976.27 | 30.63 |
| 826 | 884.31 | 29.16 | 871 | 930.83 | 29.91 | 916 | 977.30 | 30.65 |
| 827 | 885.35 | 29.17 | 872 | 931.86 | 29.93 | 917 | 978.33 | 30.67 |
| 828 | 886.38 | 29.19 | 873 | 932.89 | 29.95 | 918 | 979.36 | 30.68 |
| 829 | 887.41 | 29.21 | 874 | 933.92 | 29.96 | 919 | 980.40 | 30.70 |
| 830 | 888.45 | 29.22 | 875 | 934.96 | 29.98 | 920 | 981.43 | 30.71 |
| 831 | 889.48 | 29.24 | 876 | 935.99 | 30.00 | 921 | 982.46 | 30.73 |
| 832 | 890.52 | 29.26 | 877 | 937.02 | 30.01 | 922 | 983.49 | 30.75 |
| 833 | 891.55 | 29.28 | 878 | 938.06 | 30.03 | 923 | 984.53 | 30.76 |
| 834 | 892.58 | 29.29 | 879 | 939.09 | 30.05 | 924 | 985.56 | 30.78 |
| 835 | 893.62 | 29.31 | 880 | 940.12 | 30.06 | 925 | 986.59 | 30.80 |
| 836 | 894.65 | 29.33 | 881 | 941.16 | 30.08 | 926 | 987.62 | 30.81 |
| 837 | 895.69 | 29.34 | 882 | 942.19 | 30.09 | 927 | 988.65 | 30.83 |
| 838 | 896.72 | 29.36 | 883 | 943.22 | 30.11 | 928 | 989.69 | 30.84 |
| 839 | 897.75 | 29.38 | 884 | 944.26 | 30.13 | 929 | 990.72 | 30.86 |
| 840 | 898.79 | 29.39 | 885 | 945.29 | 30.14 | 930 | 991.75 | 30.88 |
| 841 | 899.82 | 29.41 | 886 | 946.32 | 30.16 | 931 | 992.78 | 30.89 |
| 842 | 900.85 | 29.43 | 887 | 947.35 | 30.18 | 932 | 993.82 | 30.91 |
| 843 | 901.89 | 29.44 | 888 | 948.39 | 30.19 | 933 | 994.85 | 30.92 |
| 844 | 902.92 | 29.46 | 889 | 949.42 | 30.21 | 934 | 995.88 | 30.94 |
| 845 | 903.96 | 29.48 | 890 | 950.45 | 30.23 | 935 | 996.91 | 30.96 |
| 846 | 904.99 | 29.49 | 891 | 951.49 | 30.24 | 936 | 997.94 | 30.97 |
| 847 | 906.02 | 29.51 | 892 | 952.52 | 30.26 | 937 | 998.98 | 30.99 |
| 848 | 907.06 | 29.53 | 893 | 953.55 | 30.28 | 938 | 1000.01 | 31.00 |
| 849 | 908.09 | 29.55 | 894 | 954.58 | 30.29 | 939 | 1001.04 | 31.02 |
| 850 | 909.12 | 29.56 | 895 | 955.62 | 30.31 | 940 | 1002.07 | 31.04 |
| 851 | 910.16 | 29.58 | 896 | 956.65 | 30.32 | 941 | 1003.10 | 31.05 |
| 852 | 911.19 | 29.60 | 897 | 957.68 | 30.34 | 942 | 1004.14 | 31.07 |
| 853 | 912.22 | 29.61 | 898 | 958.71 | 30.36 | 943 | 1005.17 | 31.08 |
| 854 | 913.26 | 29.63 | 899 | 959.75 | 30.37 | 944 | 1006.20 | 31.10 |

Table 3.  **CONSTANTS FOR $\overline{p}'$ (SHORT FORMULA), AN APPROXIMATION OF THE UPPER BINOMIAL CONFIDENCE LIMIT $\overline{p}$; $c/n \leqslant .1$, $n \geqslant 40$, $c \leqslant 1,000$ (*Continued*)**

| c | $\overline{m}$ | $\overline{k}$ | c | $\overline{m}$ | $\overline{k}$ | c | $\overline{m}$ | $\overline{k}$ |
|---|---|---|---|---|---|---|---|---|
| 945 | 1007.23 | 31.12 | 965 | 1027.87 | 31.43 | 985 | 1048.49 | 31.75 |
| 946 | 1008.26 | 31.13 | 966 | 1028.90 | 31.45 | 986 | 1049.52 | 31.76 |
| 947 | 1009.29 | 31.15 | 967 | 1029.93 | 31.46 | 987 | 1050.55 | 31.78 |
| 948 | 1010.33 | 31.16 | 968 | 1030.96 | 31.48 | 988 | 1051.59 | 31.79 |
| 949 | 1011.36 | 31.18 | 969 | 1031.99 | 31.50 | 989 | 1052.62 | 31.81 |
| 950 | 1012.39 | 31.20 | 970 | 1033.02 | 31.51 | 990 | 1053.65 | 31.82 |
| 951 | 1013.42 | 31.21 | 971 | 1034.05 | 31.53 | 991 | 1054.68 | 31.84 |
| 952 | 1014.45 | 31.23 | 972 | 1035.09 | 31.54 | 992 | 1055.71 | 31.86 |
| 953 | 1015.49 | 31.24 | 973 | 1036.12 | 31.56 | 993 | 1056.74 | 31.87 |
| 954 | 1016.52 | 31.26 | 974 | 1037.15 | 31.57 | 994 | 1057.77 | 31.89 |
| 955 | 1017.55 | 31.27 | 975 | 1038.18 | 31.59 | 995 | 1058.80 | 31.90 |
| 956 | 1018.58 | 31.29 | 976 | 1039.21 | 31.61 | 996 | 1059.83 | 31.92 |
| 957 | 1019.61 | 31.31 | 977 | 1040.24 | 31.62 | 997 | 1060.87 | 31.93 |
| 958 | 1020.64 | 31.32 | 978 | 1041.27 | 31.64 | 998 | 1061.90 | 31.95 |
| 959 | 1021.68 | 31.34 | 979 | 1042.30 | 31.65 | 999 | 1062.93 | 31.96 |
| 960 | 1022.71 | 31.35 | 980 | 1043.34 | 31.67 | 1000 | 1063.96 | 31.98 |
| 961 | 1023.74 | 31.37 | 981 | 1044.37 | 31.68 | | | |
| 962 | 1024.77 | 31.39 | 982 | 1045.40 | 31.70 | | | |
| 963 | 1025.80 | 31.40 | 983 | 1046.43 | 31.71 | | | |
| 964 | 1026.83 | 31.42 | 984 | 1047.46 | 31.73 | | | |

# Table 3. CONSTANTS FOR $\overline{p}'$ (SHORT FORMULA), AN APPROXIMATION OF THE UPPER BINOMIAL CONFIDENCE LIMIT $\overline{p}$; $c/n \leqslant .1$, $n \geqslant 40$, $c \leqslant 1,000$ (Continued)

| c | $\overline{m}$ | $\overline{k}$ | c | $\overline{m}$ | $\overline{k}$ | c | $\overline{m}$ | $\overline{k}$ |
|---|---|---|---|---|---|---|---|---|
| 0 | 2.9957 | 1.508 | 45 | 57.695 | 6.50 | 90 | 107.24 | 8.63 |
| 1 | 4.7439 | 1.902 | 46 | 58.816 | 6.60 | 91 | 108.32 | 8.67 |
| 2 | 6.2958 | 2.200 | 47 | 59.936 | 6.64 | 92 | 109.41 | 8.72 |
| 3 | 7.7537 | 2.446 | 48 | 61.054 | 6.71 | 93 | 110.50 | 8.76 |
| 4 | 9.1535 | 2.670 | 49 | 62.171 | 6.77 | 94 | 111.58 | 8.80 |
| 5 | 10.513 | 2.851 | 50 | 63.287 | 6.82 | 95 | 112.66 | 8.84 |
| 6 | 11.842 | 3.013 | 51 | 64.402 | 6.72 | 96 | 113.75 | 8.88 |
| 7 | 13.148 | 3.167 | 52 | 65.516 | 6.77 | 97 | 114.83 | 8.93 |
| 8 | 14.435 | 3.311 | 53 | 66.629 | 6.83 | 98 | 115.91 | 8.97 |
| 9 | 15.705 | 3.454 | 54 | 67.740 | 6.88 | 99 | 117.00 | 9.01 |
| 10 | 16.962 | 3.58 | 55 | 68.851 | 6.94 | 100 | 118.08 | 9.05 |
| 11 | 18.208 | 3.71 | 56 | 69.961 | 6.99 | 101 | 119.16 | 9.09 |
| 12 | 19.443 | 3.83 | 57 | 71.069 | 7.05 | 102 | 120.24 | 9.13 |
| 13 | 20.669 | 3.93 | 58 | 72.177 | 7.10 | 103 | 121.32 | 9.17 |
| 14 | 21.887 | 4.05 | 59 | 73.284 | 7.15 | 104 | 122.40 | 9.21 |
| 15 | 23.097 | 4.15 | 60 | 74.390 | 7.21 | 105 | 123.48 | 9.25 |
| 16 | 24.301 | 4.26 | 61 | 75.495 | 7.26 | 106 | 124.56 | 9.29 |
| 17 | 25.499 | 4.37 | 62 | 76.599 | 7.31 | 107 | 125.64 | 9.33 |
| 18 | 26.692 | 4.47 | 63 | 77.703 | 7.36 | 108 | 126.72 | 9.37 |
| 19 | 27.879 | 4.55 | 64 | 78.805 | 7.42 | 109 | 127.80 | 9.41 |
| 20 | 29.062 | 4.64 | 65 | 79.907 | 7.47 | 110 | 128.88 | 9.45 |
| 21 | 30.240 | 4.74 | 66 | 81.008 | 7.52 | 111 | 129.96 | 9.49 |
| 22 | 31.415 | 4.81 | 67 | 82.108 | 7.57 | 112 | 131.04 | 9.53 |
| 23 | 32.585 | 4.91 | 68 | 83.208 | 7.62 | 113 | 132.11 | 9.57 |
| 24 | 33.752 | 5.01 | 69 | 84.307 | 7.67 | 114 | 133.19 | 9.60 |
| 25 | 34.916 | 5.07 | 70 | 85.405 | 7.71 | 115 | 134.27 | 9.64 |
| 26 | 36.077 | 5.16 | 71 | 86.502 | 7.76 | 116 | 135.34 | 9.68 |
| 27 | 37.234 | 5.26 | 72 | 87.599 | 7.81 | 117 | 136.42 | 9.72 |
| 28 | 38.389 | 5.32 | 73 | 88.695 | 7.86 | 118 | 137.49 | 9.76 |
| 29 | 39.541 | 5.41 | 74 | 89.791 | 7.91 | 119 | 138.57 | 9.79 |
| 30 | 40.691 | 5.46 | 75 | 90.885 | 7.95 | 120 | 139.64 | 9.83 |
| 31 | 41.838 | 5.55 | 76 | 91.980 | 8.00 | 121 | 140.72 | 9.87 |
| 32 | 42.982 | 5.61 | 77 | 93.073 | 8.05 | 122 | 141.79 | 9.91 |
| 33 | 44.125 | 5.69 | 78 | 94.166 | 8.09 | 123 | 142.87 | 9.94 |
| 34 | 45.266 | 5.77 | 79 | 95.258 | 8.14 | 124 | 143.94 | 9.98 |
| 35 | 46.404 | 5.83 | 80 | 96.350 | 8.19 | 125 | 145.01 | 10.02 |
| 36 | 47.541 | 5.90 | 81 | 97.442 | 8.23 | 126 | 146.09 | 10.05 |
| 37 | 48.676 | 5.99 | 82 | 98.532 | 8.28 | 127 | 147.16 | 10.09 |
| 38 | 49.808 | 6.06 | 83 | 99.622 | 8.32 | 128 | 148.23 | 10.13 |
| 39 | 50.940 | 6.12 | 84 | 100.71 | 8.37 | 129 | 149.31 | 10.16 |
| 40 | 52.070 | 6.17 | 85 | 101.80 | 8.41 | 130 | 150.38 | 10.20 |
| 41 | 53.198 | 6.26 | 86 | 102.89 | 8.46 | 131 | 151.45 | 10.23 |
| 42 | 54.324 | 6.33 | 87 | 103.98 | 8.50 | 132 | 152.52 | 10.27 |
| 43 | 55.449 | 6.38 | 88 | 105.07 | 8.54 | 133 | 153.59 | 10.30 |
| 44 | 56.573 | 6.47 | 89 | 106.15 | 8.59 | 134 | 154.66 | 10.34 |

Table 3.  CONSTANTS FOR $\overline{p}'$ (SHORT FORMULA), AN APPROXIMATION OF THE UPPER BINOMIAL CONFIDENCE LIMIT $\overline{p}$; $c/n \leqslant .1$, $n \geqslant 40$, $c \leqslant 1{,}000$ (*Continued*)

| c | $\overline{m}$ | $\overline{k}$ | c | $\overline{m}$ | $\overline{k}$ | c | $\overline{m}$ | $\overline{k}$ |
|---|---|---|---|---|---|---|---|---|
| 135 | 155.73 | 10.38 | 180 | 203.70 | 11.85 | 225 | 251.30 | 13.15 |
| 136 | 156.80 | 10.41 | 181 | 204.76 | 11.88 | 226 | 252.35 | 13.18 |
| 137 | 157.87 | 10.45 | 182 | 205.82 | 11.91 | 227 | 253.41 | 13.20 |
| 138 | 158.94 | 10.48 | 183 | 206.88 | 11.94 | 228 | 254.46 | 13.23 |
| 139 | 160.01 | 10.52 | 184 | 207.94 | 11.97 | 229 | 255.51 | 13.26 |
| 140 | 161.08 | 10.55 | 185 | 209.00 | 12.00 | 230 | 256.57 | 13.28 |
| 141 | 162.15 | 10.58 | 186 | 210.06 | 12.03 | 231 | 257.62 | 13.31 |
| 142 | 163.22 | 10.62 | 187 | 211.12 | 12.06 | 232 | 258.68 | 13.34 |
| 143 | 164.29 | 10.65 | 188 | 212.18 | 12.09 | 233 | 259.73 | 13.37 |
| 144 | 165.36 | 10.69 | 189 | 213.24 | 12.12 | 234 | 260.78 | 13.39 |
| 145 | 166.43 | 10.72 | 190 | 214.30 | 12.15 | 235 | 261.84 | 13.42 |
| 146 | 167.50 | 10.76 | 191 | 215.36 | 12.18 | 236 | 262.89 | 13.45 |
| 147 | 168.56 | 10.79 | 192 | 216.42 | 12.21 | 237 | 263.94 | 13.47 |
| 148 | 169.63 | 10.82 | 193 | 217.48 | 12.24 | 238 | 265.00 | 13.50 |
| 149 | 170.70 | 10.86 | 194 | 218.54 | 12.27 | 239 | 266.05 | 13.53 |
| 150 | 171.76 | 10.89 | 195 | 219.60 | 12.30 | 240 | 267.10 | 13.55 |
| 151 | 172.83 | 10.92 | 196 | 220.66 | 12.33 | 241 | 268.16 | 13.58 |
| 152 | 173.90 | 10.96 | 197 | 221.71 | 12.36 | 242 | 269.21 | 13.60 |
| 153 | 174.96 | 10.99 | 198 | 222.77 | 12.39 | 243 | 270.26 | 13.63 |
| 154 | 176.03 | 11.02 | 199 | 223.83 | 12.42 | 244 | 271.32 | 13.66 |
| 155 | 177.10 | 11.06 | 200 | 224.89 | 12.44 | 245 | 272.37 | 13.68 |
| 156 | 178.16 | 11.09 | 201 | 225.95 | 12.47 | 246 | 273.42 | 13.71 |
| 157 | 179.23 | 11.12 | 202 | 227.00 | 12.50 | 247 | 274.47 | 13.74 |
| 158 | 180.29 | 11.15 | 203 | 228.06 | 12.53 | 248 | 275.52 | 13.76 |
| 159 | 181.36 | 11.19 | 204 | 229.12 | 12.56 | 249 | 276.58 | 13.79 |
| 160 | 182.42 | 11.22 | 205 | 230.18 | 12.59 | 250 | 277.63 | 13.81 |
| 161 | 183.50 | 11.25 | 206 | 231.23 | 12.62 | 251 | 278.68 | 13.84 |
| 162 | 184.57 | 11.28 | 207 | 232.29 | 12.65 | 252 | 279.73 | 13.87 |
| 163 | 185.63 | 11.32 | 208 | 233.35 | 12.67 | 253 | 280.78 | 13.89 |
| 164 | 186.70 | 11.35 | 209 | 234.41 | 12.70 | 254 | 281.84 | 13.92 |
| 165 | 187.76 | 11.38 | 210 | 235.46 | 12.73 | 255 | 282.89 | 13.94 |
| 166 | 188.83 | 11.41 | 211 | 236.52 | 12.76 | 256 | 283.94 | 13.97 |
| 167 | 189.89 | 11.44 | 212 | 237.58 | 12.79 | 257 | 284.99 | 13.99 |
| 168 | 190.95 | 11.48 | 213 | 238.63 | 12.82 | 258 | 286.04 | 14.02 |
| 169 | 192.02 | 11.51 | 214 | 239.69 | 12.84 | 259 | 287.09 | 14.05 |
| 170 | 193.08 | 11.54 | 215 | 240.74 | 12.87 | 260 | 288.14 | 14.07 |
| 171 | 194.14 | 11.57 | 216 | 241.80 | 12.90 | 261 | 289.19 | 14.10 |
| 172 | 195.20 | 11.60 | 217 | 242.86 | 12.93 | 262 | 290.24 | 14.12 |
| 173 | 196.27 | 11.63 | 218 | 243.91 | 12.96 | 263 | 291.29 | 14.15 |
| 174 | 197.33 | 11.66 | 219 | 244.97 | 12.98 | 264 | 292.35 | 14.17 |
| 175 | 198.39 | 11.70 | 220 | 246.02 | 13.01 | 265 | 293.40 | 14.20 |
| 176 | 199.45 | 11.73 | 221 | 247.08 | 13.04 | 266 | 294.45 | 14.22 |
| 177 | 200.51 | 11.76 | 222 | 248.13 | 13.07 | 267 | 295.50 | 14.25 |
| 178 | 201.58 | 11.79 | 223 | 249.19 | 13.09 | 268 | 296.55 | 14.27 |
| 179 | 202.64 | 11.82 | 224 | 250.24 | 13.12 | 269 | 297.60 | 14.30 |

Table 3. CONSTANTS FOR $\overline{p}'$ (SHORT FORMULA), AN APPROXIMATION OF THE UPPER BINOMIAL CONFIDENCE LIMIT $\overline{p}$; $c/n \leqslant .1$, $n \geqslant 40$, $c \leqslant 1,000$ (*Continued*)

| c | $\overline{m}$ | $\overline{k}$ | c | $\overline{m}$ | $\overline{k}$ | c | $\overline{m}$ | $\overline{k}$ |
|---|---|---|---|---|---|---|---|---|
| 270 | 298.65 | 14.32 | 315 | 345.81 | 15.40 | 360 | 392.82 | 16.41 |
| 271 | 299.70 | 14.35 | 316 | 346.86 | 15.43 | 361 | 393.86 | 16.43 |
| 272 | 300.75 | 14.37 | 317 | 347.90 | 15.45 | 362 | 394.91 | 16.45 |
| 273 | 301.80 | 14.40 | 318 | 348.95 | 15.47 | 363 | 395.95 | 16.48 |
| 274 | 302.85 | 14.42 | 319 | 349.99 | 15.50 | 364 | 396.99 | 16.50 |
| 275 | 303.90 | 14.45 | 320 | 351.04 | 15.52 | 365 | 398.04 | 16.52 |
| 276 | 304.95 | 14.47 | 321 | 352.09 | 15.54 | 366 | 399.08 | 16.54 |
| 277 | 305.99 | 14.50 | 322 | 353.13 | 15.57 | 367 | 400.12 | 16.56 |
| 278 | 307.04 | 14.52 | 323 | 354.18 | 15.59 | 368 | 401.17 | 16.58 |
| 279 | 308.09 | 14.55 | 324 | 355.22 | 15.61 | 369 | 402.21 | 16.60 |
| 280 | 309.14 | 14.57 | 325 | 356.27 | 15.63 | 370 | 403.25 | 16.63 |
| 281 | 310.19 | 14.60 | 326 | 357.31 | 15.66 | 371 | 404.29 | 16.65 |
| 282 | 311.24 | 14.62 | 327 | 358.36 | 15.68 | 372 | 405.34 | 16.67 |
| 283 | 312.29 | 14.64 | 328 | 359.40 | 15.70 | 373 | 406.38 | 16.69 |
| 284 | 313.34 | 14.67 | 329 | 360.45 | 15.72 | 374 | 407.42 | 16.71 |
| 285 | 314.39 | 14.69 | 330 | 361.49 | 15.75 | 375 | 408.46 | 16.73 |
| 286 | 315.43 | 14.72 | 331 | 362.54 | 15.77 | 376 | 409.51 | 16.75 |
| 287 | 316.48 | 14.74 | 332 | 363.59 | 15.79 | 377 | 410.55 | 16.77 |
| 288 | 317.53 | 14.77 | 333 | 364.63 | 15.82 | 378 | 411.59 | 16.80 |
| 289 | 318.58 | 14.79 | 334 | 365.68 | 15.84 | 379 | 412.63 | 16.82 |
| 290 | 319.63 | 14.81 | 335 | 366.72 | 15.86 | 380 | 413.68 | 16.84 |
| 291 | 320.68 | 14.84 | 336 | 367.76 | 15.88 | 381 | 414.72 | 16.86 |
| 292 | 321.72 | 14.86 | 337 | 368.81 | 15.90 | 382 | 415.76 | 16.88 |
| 293 | 322.77 | 14.89 | 338 | 369.85 | 15.93 | 383 | 416.80 | 16.90 |
| 294 | 323.82 | 14.91 | 339 | 370.90 | 15.95 | 384 | 417.84 | 16.92 |
| 295 | 324.87 | 14.93 | 340 | 371.94 | 15.97 | 385 | 418.89 | 16.94 |
| 296 | 325.92 | 14.96 | 341 | 372.99 | 15.99 | 386 | 419.93 | 16.96 |
| 297 | 326.96 | 14.98 | 342 | 374.03 | 16.02 | 387 | 420.97 | 16.98 |
| 298 | 328.01 | 15.01 | 343 | 375.08 | 16.04 | 388 | 422.01 | 17.01 |
| 299 | 329.06 | 15.03 | 344 | 376.12 | 16.06 | 389 | 423.05 | 17.03 |
| 300 | 330.11 | 15.05 | 345 | 377.17 | 16.08 | 390 | 424.09 | 17.05 |
| 301 | 331.15 | 15.08 | 346 | 378.21 | 16.10 | 391 | 425.14 | 17.07 |
| 302 | 332.20 | 15.10 | 347 | 379.25 | 16.13 | 392 | 426.18 | 17.09 |
| 303 | 333.25 | 15.12 | 348 | 380.30 | 16.15 | 393 | 427.22 | 17.11 |
| 304 | 334.30 | 15.15 | 349 | 381.34 | 16.17 | 394 | 428.26 | 17.13 |
| 305 | 335.34 | 15.17 | 350 | 382.39 | 16.19 | 395 | 429.30 | 17.15 |
| 306 | 336.39 | 15.19 | 351 | 383.43 | 16.21 | 396 | 430.34 | 17.17 |
| 307 | 337.44 | 15.22 | 352 | 384.47 | 16.24 | 397 | 431.38 | 17.19 |
| 308 | 338.48 | 15.24 | 353 | 385.52 | 16.26 | 398 | 432.43 | 17.21 |
| 309 | 339.53 | 15.26 | 354 | 386.56 | 16.28 | 399 | 433.47 | 17.23 |
| 310 | 340.58 | 15.29 | 355 | 387.60 | 16.30 | 400 | 434.51 | 17.25 |
| 311 | 341.62 | 15.31 | 356 | 388.65 | 16.32 | 401 | 435.55 | 17.27 |
| 312 | 342.67 | 15.33 | 357 | 389.69 | 16.35 | 402 | 436.59 | 17.29 |
| 313 | 343.72 | 15.36 | 358 | 390.73 | 16.37 | 403 | 437.63 | 17.32 |
| 314 | 344.76 | 15.38 | 359 | 391.78 | 16.39 | 404 | 438.67 | 17.34 |

Table 3.  CONSTANTS FOR $\overline{p}'$ (SHORT FORMULA), AN APPROXIMATION OF
THE UPPER BINOMIAL CONFIDENCE LIMIT $\overline{p}$; c/n $\leqslant$ .1, n $\geqslant$ 40,
c $\leqslant$ 1,000 (*Continued*)

| c | $\overline{m}$ | $\overline{k}$ | c | $\overline{m}$ | $\overline{k}$ | c | $\overline{m}$ | $\overline{k}$ |
|---|---|---|---|---|---|---|---|---|
| 405 | 439.71 | 17.36 | 450 | 486.50 | 18.25 | 495 | 533.20 | 19.10 |
| 406 | 440.75 | 17.38 | 451 | 487.54 | 18.27 | 496 | 534.24 | 19.12 |
| 407 | 441.79 | 17.40 | 452 | 488.58 | 18.29 | 497 | 535.28 | 19.14 |
| 408 | 442.83 | 17.42 | 453 | 489.62 | 18.31 | 498 | 536.31 | 19.16 |
| 409 | 443.88 | 17.44 | 454 | 490.66 | 18.33 | 499 | 537.35 | 19.17 |
| 410 | 444.92 | 17.46 | 455 | 491.69 | 18.35 | 500 | 538.39 | 19.19 |
| 411 | 445.96 | 17.48 | 456 | 492.73 | 18.37 | 501 | 539.42 | 19.21 |
| 412 | 447.00 | 17.50 | 457 | 493.77 | 18.39 | 502 | 540.46 | 19.23 |
| 413 | 448.04 | 17.52 | 458 | 494.81 | 18.40 | 503 | 541.50 | 19.25 |
| 414 | 449.08 | 17.54 | 459 | 495.85 | 18.42 | 504 | 542.53 | 19.27 |
| 415 | 450.12 | 17.56 | 460 | 496.89 | 18.44 | 505 | 543.57 | 19.28 |
| 416 | 451.16 | 17.58 | 461 | 497.92 | 18.46 | 506 | 544.61 | 19.30 |
| 417 | 452.20 | 17.60 | 462 | 498.96 | 18.48 | 507 | 545.64 | 19.32 |
| 418 | 453.24 | 17.62 | 463 | 500.00 | 18.50 | 508 | 546.68 | 19.34 |
| 419 | 454.28 | 17.64 | 464 | 501.04 | 18.52 | 509 | 547.72 | 19.36 |
| 420 | 455.32 | 17.66 | 465 | 502.08 | 18.54 | 510 | 548.75 | 19.38 |
| 421 | 456.36 | 17.68 | 466 | 503.12 | 18.56 | 511 | 549.79 | 19.39 |
| 422 | 457.40 | 17.70 | 467 | 504.15 | 18.58 | 512 | 550.82 | 19.41 |
| 423 | 458.44 | 17.72 | 468 | 505.19 | 18.60 | 513 | 551.86 | 19.43 |
| 424 | 459.48 | 17.74 | 469 | 506.23 | 18.61 | 514 | 552.90 | 19.45 |
| 425 | 460.52 | 17.76 | 470 | 507.27 | 18.63 | 515 | 553.93 | 19.47 |
| 426 | 461.56 | 17.78 | 471 | 508.30 | 18.65 | 516 | 554.97 | 19.48 |
| 427 | 462.60 | 17.80 | 472 | 509.34 | 18.67 | 517 | 556.01 | 19.50 |
| 428 | 463.64 | 17.82 | 473 | 510.38 | 18.69 | 518 | 557.04 | 19.52 |
| 429 | 464.68 | 17.84 | 474 | 511.42 | 18.71 | 519 | 558.08 | 19.54 |
| 430 | 465.72 | 17.86 | 475 | 512.46 | 18.73 | 520 | 559.11 | 19.56 |
| 431 | 466.76 | 17.88 | 476 | 513.49 | 18.75 | 521 | 560.15 | 19.58 |
| 432 | 467.80 | 17.90 | 477 | 514.53 | 18.77 | 522 | 561.19 | 19.59 |
| 433 | 468.84 | 17.92 | 478 | 515.57 | 18.78 | 523 | 562.22 | 19.61 |
| 434 | 469.88 | 17.94 | 479 | 516.61 | 18.80 | 524 | 563.26 | 19.63 |
| 435 | 470.91 | 17.96 | 480 | 517.64 | 18.82 | 525 | 564.29 | 19.65 |
| 436 | 471.95 | 17.98 | 481 | 518.68 | 18.84 | 526 | 565.33 | 19.66 |
| 437 | 472.99 | 18.00 | 482 | 519.72 | 18.86 | 527 | 566.37 | 19.68 |
| 438 | 474.03 | 18.02 | 483 | 520.76 | 18.88 | 528 | 567.40 | 19.70 |
| 439 | 475.07 | 18.04 | 484 | 521.79 | 18.90 | 529 | 568.44 | 19.72 |
| 440 | 476.11 | 18.06 | 485 | 522.83 | 18.92 | 530 | 569.47 | 19.74 |
| 441 | 477.15 | 18.08 | 486 | 523.87 | 18.93 | 531 | 570.51 | 19.75 |
| 442 | 478.19 | 18.09 | 487 | 524.91 | 18.95 | 532 | 571.54 | 19.77 |
| 443 | 479.23 | 18.11 | 488 | 525.94 | 18.97 | 533 | 572.58 | 19.79 |
| 444 | 480.27 | 18.13 | 489 | 526.98 | 18.99 | 534 | 573.62 | 19.81 |
| 445 | 481.31 | 18.15 | 490 | 528.02 | 19.01 | 535 | 574.65 | 19.83 |
| 446 | 482.35 | 18.17 | 491 | 529.05 | 19.03 | 536 | 575.69 | 19.84 |
| 447 | 483.38 | 18.19 | 492 | 530.09 | 19.05 | 537 | 576.72 | 19.86 |
| 448 | 484.42 | 18.21 | 493 | 531.13 | 19.06 | 538 | 577.76 | 19.88 |
| 449 | 485.46 | 18.23 | 494 | 532.17 | 19.08 | 539 | 578.79 | 19.90 |

**Table 3.** **CONSTANTS FOR $\overline{p}'$ (SHORT FORMULA), AN APPROXIMATION OF THE UPPER BINOMIAL CONFIDENCE LIMIT $\overline{p}$; $c/n \leq .1$, $n \geq 40$, $c \leq 1,000$ (*Continued*)**

Section 4

$\gamma_1 = 95\%$

$\gamma_2 = 90\%$

| c | $\overline{m}$ | $\overline{k}$ | c | $\overline{m}$ | $\overline{k}$ | c | $\overline{m}$ | $\overline{k}$ |
|---|---|---|---|---|---|---|---|---|
| 540 | 579.83 | 19.91 | 585 | 626.39 | 20.69 | 630 | 672.89 | 21.44 |
| 541 | 580.86 | 19.93 | 586 | 627.42 | 20.71 | 631 | 673.92 | 21.46 |
| 542 | 581.90 | 19.95 | 587 | 628.46 | 20.73 | 632 | 674.95 | 21.48 |
| 543 | 582.93 | 19.97 | 588 | 629.49 | 20.74 | 633 | 675.99 | 21.49 |
| 544 | 583.97 | 19.98 | 589 | 630.52 | 20.76 | 634 | 677.02 | 21.51 |
| 545 | 585.00 | 20.00 | 590 | 631.56 | 20.78 | 635 | 678.05 | 21.53 |
| 546 | 586.04 | 20.02 | 591 | 632.59 | 20.80 | 636 | 679.08 | 21.54 |
| 547 | 587.07 | 20.04 | 592 | 633.62 | 20.81 | 637 | 680.12 | 21.56 |
| 548 | 588.11 | 20.05 | 593 | 634.66 | 20.83 | 638 | 681.15 | 21.57 |
| 549 | 589.14 | 20.07 | 594 | 635.69 | 20.85 | 639 | 682.18 | 21.59 |
| 550 | 590.18 | 20.09 | 595 | 636.73 | 20.86 | 640 | 683.21 | 21.61 |
| 551 | 591.21 | 20.11 | 596 | 637.76 | 20.88 | 641 | 684.25 | 21.62 |
| 552 | 592.25 | 20.12 | 597 | 638.79 | 20.90 | 642 | 685.28 | 21.64 |
| 553 | 593.28 | 20.14 | 598 | 639.83 | 20.91 | 643 | 686.31 | 21.66 |
| 554 | 594.32 | 20.16 | 599 | 640.86 | 20.93 | 644 | 687.34 | 21.67 |
| 555 | 595.35 | 20.18 | 600 | 641.89 | 20.95 | 645 | 688.38 | 21.69 |
| 556 | 596.39 | 20.19 | 601 | 642.93 | 20.96 | 646 | 689.41 | 21.70 |
| 557 | 597.42 | 20.21 | 602 | 643.96 | 20.98 | 647 | 690.44 | 21.72 |
| 558 | 598.46 | 20.23 | 603 | 644.99 | 21.00 | 648 | 691.47 | 21.74 |
| 559 | 599.49 | 20.25 | 604 | 646.03 | 21.01 | 649 | 692.51 | 21.75 |
| 560 | 600.53 | 20.26 | 605 | 647.06 | 21.03 | 650 | 693.54 | 21.77 |
| 561 | 601.56 | 20.28 | 606 | 648.09 | 21.05 | 651 | 694.57 | 21.78 |
| 562 | 602.60 | 20.30 | 607 | 649.13 | 21.06 | 652 | 695.60 | 21.80 |
| 563 | 603.63 | 20.32 | 608 | 650.16 | 21.08 | 653 | 696.63 | 21.82 |
| 564 | 604.67 | 20.33 | 609 | 651.19 | 21.10 | 654 | 697.67 | 21.83 |
| 565 | 605.70 | 20.35 | 610 | 652.23 | 21.11 | 655 | 698.70 | 21.85 |
| 566 | 606.74 | 20.37 | 611 | 653.26 | 21.13 | 656 | 699.73 | 21.87 |
| 567 | 607.77 | 20.39 | 612 | 654.29 | 21.15 | 657 | 700.76 | 21.88 |
| 568 | 608.81 | 20.40 | 613 | 655.33 | 21.16 | 658 | 701.79 | 21.90 |
| 569 | 609.84 | 20.42 | 614 | 656.36 | 21.18 | 659 | 702.83 | 21.91 |
| 570 | 610.87 | 20.44 | 615 | 657.39 | 21.20 | 660 | 703.86 | 21.93 |
| 571 | 611.91 | 20.45 | 616 | 658.43 | 21.21 | 661 | 704.89 | 21.95 |
| 572 | 612.94 | 20.47 | 617 | 659.46 | 21.23 | 662 | 705.92 | 21.96 |
| 573 | 613.98 | 20.49 | 618 | 660.49 | 21.25 | 663 | 706.95 | 21.98 |
| 574 | 615.01 | 20.51 | 619 | 661.53 | 21.26 | 664 | 707.99 | 21.99 |
| 575 | 616.05 | 20.52 | 620 | 662.56 | 21.28 | 665 | 709.02 | 22.01 |
| 576 | 617.08 | 20.54 | 621 | 663.59 | 21.30 | 666 | 710.05 | 22.03 |
| 577 | 618.11 | 20.56 | 622 | 664.63 | 21.31 | 667 | 711.08 | 22.04 |
| 578 | 619.15 | 20.57 | 623 | 665.66 | 21.33 | 668 | 712.11 | 22.06 |
| 579 | 620.18 | 20.59 | 624 | 666.69 | 21.35 | 669 | 713.15 | 22.07 |
| 580 | 621.22 | 20.61 | 625 | 667.72 | 21.36 | 670 | 714.18 | 22.09 |
| 581 | 622.25 | 20.63 | 626 | 668.76 | 21.38 | 671 | 715.21 | 22.10 |
| 582 | 623.29 | 20.64 | 627 | 669.79 | 21.39 | 672 | 716.24 | 22.12 |
| 583 | 624.32 | 20.66 | 628 | 670.82 | 21.41 | 673 | 717.27 | 22.14 |
| 584 | 625.35 | 20.68 | 629 | 671.86 | 21.43 | 674 | 718.30 | 22.15 |

Table 3.   CONSTANTS FOR $\overline{p}'$ (SHORT FORMULA), AN APPROXIMATION OF THE UPPER BINOMIAL CONFIDENCE LIMIT $\overline{p}$; $c/n \leqslant .1$, $n \geqslant 40$, $c \leqslant 1,000$ (*Continued*)

| c | $\overline{m}$ | $\overline{k}$ | c | $\overline{m}$ | $\overline{k}$ | c | $\overline{m}$ | $\overline{k}$ |
|---|---|---|---|---|---|---|---|---|
| 675 | 719.34 | 22.17 | 720 | 765.74 | 22.87 | 765 | 812.09 | 23.55 |
| 676 | 720.37 | 22.18 | 721 | 766.77 | 22.88 | 766 | 813.12 | 23.56 |
| 677 | 721.40 | 22.20 | 722 | 767.80 | 22.90 | 767 | 814.15 | 23.58 |
| 678 | 722.43 | 22.22 | 723 | 768.83 | 22.91 | 768 | 815.18 | 23.59 |
| 679 | 723.46 | 22.23 | 724 | 769.86 | 22.93 | 769 | 816.21 | 23.61 |
| 680 | 724.49 | 22.25 | 725 | 770.89 | 22.94 | 770 | 817.24 | 23.62 |
| 681 | 725.53 | 22.26 | 726 | 771.92 | 22.96 | 771 | 818.27 | 23.64 |
| 682 | 726.56 | 22.28 | 727 | 772.95 | 22.98 | 772 | 819.30 | 23.65 |
| 683 | 727.59 | 22.29 | 728 | 773.98 | 22.99 | 773 | 820.33 | 23.67 |
| 684 | 728.62 | 22.31 | 729 | 775.01 | 23.01 | 774 | 821.36 | 23.68 |
| 685 | 729.65 | 22.33 | 730 | 776.04 | 23.02 | 775 | 822.39 | 23.70 |
| 686 | 730.68 | 22.34 | 731 | 777.07 | 23.04 | 776 | 823.42 | 23.71 |
| 687 | 731.71 | 22.36 | 732 | 778.10 | 23.05 | 777 | 824.45 | 23.72 |
| 688 | 732.75 | 22.37 | 733 | 779.13 | 23.07 | 778 | 825.48 | 23.74 |
| 689 | 733.78 | 22.39 | 734 | 780.16 | 23.08 | 779 | 826.51 | 23.75 |
| 690 | 734.81 | 22.40 | 735 | 781.19 | 23.10 | 780 | 827.54 | 23.77 |
| 691 | 735.84 | 22.42 | 736 | 782.22 | 23.11 | 781 | 828.57 | 23.78 |
| 692 | 736.87 | 22.44 | 737 | 783.25 | 23.13 | 782 | 829.60 | 23.80 |
| 693 | 737.90 | 22.45 | 738 | 784.28 | 23.14 | 783 | 830.63 | 23.81 |
| 694 | 738.93 | 22.47 | 739 | 785.31 | 23.16 | 784 | 831.66 | 23.83 |
| 695 | 739.96 | 22.48 | 740 | 786.34 | 23.17 | 785 | 832.68 | 23.84 |
| 696 | 741.00 | 22.50 | 741 | 787.38 | 23.19 | 786 | 833.71 | 23.86 |
| 697 | 742.03 | 22.51 | 742 | 788.41 | 23.20 | 787 | 834.74 | 23.87 |
| 698 | 743.06 | 22.53 | 743 | 789.44 | 23.22 | 788 | 835.77 | 23.89 |
| 699 | 744.09 | 22.54 | 744 | 790.47 | 23.23 | 789 | 836.80 | 23.90 |
| 700 | 745.12 | 22.56 | 745 | 791.50 | 23.25 | 790 | 837.83 | 23.92 |
| 701 | 746.15 | 22.58 | 746 | 792.53 | 23.26 | 791 | 838.86 | 23.93 |
| 702 | 747.18 | 22.59 | 747 | 793.56 | 23.28 | 792 | 839.89 | 23.94 |
| 703 | 748.21 | 22.61 | 748 | 794.59 | 23.29 | 793 | 840.92 | 23.96 |
| 704 | 749.24 | 22.62 | 749 | 795.62 | 23.31 | 794 | 841.95 | 23.97 |
| 705 | 750.27 | 22.64 | 750 | 796.65 | 23.32 | 795 | 842.98 | 23.99 |
| 706 | 751.31 | 22.65 | 751 | 797.68 | 23.34 | 796 | 844.01 | 24.00 |
| 707 | 752.34 | 22.67 | 752 | 798.71 | 23.35 | 797 | 845.04 | 24.02 |
| 708 | 753.37 | 22.68 | 753 | 799.74 | 23.37 | 798 | 846.06 | 24.03 |
| 709 | 754.40 | 22.70 | 754 | 800.77 | 23.38 | 799 | 847.09 | 24.05 |
| 710 | 755.43 | 22.71 | 755 | 801.80 | 23.40 | 800 | 848.12 | 24.06 |
| 711 | 756.46 | 22.73 | 756 | 802.83 | 23.41 | 801 | 849.15 | 24.08 |
| 712 | 757.49 | 22.75 | 757 | 803.86 | 23.43 | 802 | 850.18 | 24.09 |
| 713 | 758.52 | 22.76 | 758 | 804.89 | 23.44 | 803 | 851.21 | 24.10 |
| 714 | 759.55 | 22.78 | 759 | 805.92 | 23.46 | 804 | 852.24 | 24.12 |
| 715 | 760.58 | 22.79 | 760 | 806.95 | 23.47 | 805 | 853.27 | 24.13 |
| 716 | 761.61 | 22.81 | 761 | 807.97 | 23.49 | 806 | 854.30 | 24.15 |
| 717 | 762.64 | 22.82 | 762 | 809.00 | 23.50 | 807 | 855.33 | 24.16 |
| 718 | 763.68 | 22.84 | 763 | 810.03 | 23.52 | 808 | 856.35 | 24.18 |
| 719 | 764.71 | 22.85 | 764 | 811.06 | 23.53 | 809 | 857.38 | 24.19 |

**Table 3. CONSTANTS FOR $\overline{p}'$ (SHORT FORMULA), AN APPROXIMATION OF THE UPPER BINOMIAL CONFIDENCE LIMIT $\overline{p}$; $c/n \leqslant .1$, $n \geqslant 40$, $c \leqslant 1,000$ (Continued)**

| c | $\overline{m}$ | $\overline{k}$ | c | $\overline{m}$ | $\overline{k}$ | c | $\overline{m}$ | $\overline{k}$ |
|---|---|---|---|---|---|---|---|---|
| 810 | 858.41 | 24.21 | 855 | 904.69 | 24.85 | 900 | 950.94 | 25.47 |
| 811 | 859.44 | 24.22 | 856 | 905.72 | 24.86 | 901 | 951.97 | 25.49 |
| 812 | 860.47 | 24.23 | 857 | 906.75 | 24.88 | 902 | 953.00 | 25.50 |
| 813 | 861.50 | 24.25 | 858 | 907.78 | 24.89 | 903 | 954.03 | 25.51 |
| 814 | 862.53 | 24.26 | 859 | 908.81 | 24.90 | 904 | 955.05 | 25.53 |
| 815 | 863.56 | 24.28 | 860 | 909.83 | 24.92 | 905 | 956.08 | 25.54 |
| 816 | 864.58 | 24.29 | 861 | 910.86 | 24.93 | 906 | 957.11 | 25.55 |
| 817 | 865.61 | 24.31 | 862 | 911.89 | 24.95 | 907 | 958.13 | 25.57 |
| 818 | 866.64 | 24.32 | 863 | 912.92 | 24.96 | 908 | 959.16 | 25.58 |
| 819 | 867.67 | 24.34 | 864 | 913.95 | 24.97 | 909 | 960.19 | 25.59 |
| 820 | 868.70 | 24.35 | 865 | 914.97 | 24.99 | 910 | 961.22 | 25.61 |
| 821 | 869.73 | 24.36 | 866 | 916.00 | 25.00 | 911 | 962.24 | 25.62 |
| 822 | 870.76 | 24.38 | 867 | 917.03 | 25.02 | 912 | 963.27 | 25.64 |
| 823 | 871.79 | 24.39 | 868 | 918.06 | 25.03 | 913 | 964.30 | 25.65 |
| 824 | 872.81 | 24.41 | 869 | 919.09 | 25.04 | 914 | 965.32 | 25.66 |
| 825 | 873.84 | 24.42 | 870 | 920.11 | 25.06 | 915 | 966.35 | 25.68 |
| 826 | 874.87 | 24.44 | 871 | 921.14 | 25.07 | 916 | 967.38 | 25.69 |
| 827 | 875.90 | 24.45 | 872 | 922.17 | 25.08 | 917 | 968.41 | 25.70 |
| 828 | 876.93 | 24.46 | 873 | 923.20 | 25.10 | 918 | 969.43 | 25.72 |
| 829 | 877.96 | 24.48 | 874 | 924.23 | 25.11 | 919 | 970.46 | 25.73 |
| 830 | 878.99 | 24.49 | 875 | 925.25 | 25.13 | 920 | 971.49 | 25.74 |
| 831 | 880.01 | 24.51 | 876 | 926.28 | 25.14 | 921 | 972.51 | 25.76 |
| 832 | 881.04 | 24.52 | 877 | 927.31 | 25.15 | 922 | 973.54 | 25.77 |
| 833 | 882.07 | 24.54 | 878 | 928.34 | 25.17 | 923 | 974.57 | 25.78 |
| 834 | 883.10 | 24.55 | 879 | 929.36 | 25.18 | 924 | 975.60 | 25.80 |
| 835 | 884.13 | 24.56 | 880 | 930.39 | 25.20 | 925 | 976.62 | 25.81 |
| 836 | 885.16 | 24.58 | 881 | 931.42 | 25.21 | 926 | 977.65 | 25.83 |
| 837 | 886.19 | 24.59 | 882 | 932.45 | 25.22 | 927 | 978.68 | 25.84 |
| 838 | 887.21 | 24.61 | 883 | 933.47 | 25.24 | 928 | 979.70 | 25.85 |
| 839 | 888.24 | 24.62 | 884 | 934.50 | 25.25 | 929 | 980.73 | 25.87 |
| 840 | 889.27 | 24.64 | 885 | 935.53 | 25.27 | 930 | 981.76 | 25.88 |
| 841 | 890.30 | 24.65 | 886 | 936.56 | 25.28 | 931 | 982.79 | 25.89 |
| 842 | 891.33 | 24.66 | 887 | 937.59 | 25.29 | 932 | 983.81 | 25.91 |
| 843 | 892.36 | 24.68 | 888 | 938.61 | 25.31 | 933 | 984.84 | 25.92 |
| 844 | 893.38 | 24.69 | 889 | 939.64 | 25.32 | 934 | 985.87 | 25.93 |
| 845 | 894.41 | 24.71 | 890 | 940.67 | 25.33 | 935 | 986.89 | 25.95 |
| 846 | 895.44 | 24.72 | 891 | 941.70 | 25.35 | 936 | 987.92 | 25.96 |
| 847 | 896.47 | 24.73 | 892 | 942.72 | 25.36 | 937 | 988.95 | 25.97 |
| 848 | 897.50 | 24.75 | 893 | 943.75 | 25.38 | 938 | 989.97 | 25.99 |
| 849 | 898.53 | 24.76 | 894 | 944.78 | 25.39 | 939 | 991.00 | 26.00 |
| 850 | 899.55 | 24.78 | 895 | 945.81 | 25.40 | 940 | 992.03 | 26.01 |
| 851 | 900.58 | 24.79 | 896 | 946.83 | 25.42 | 941 | 993.05 | 26.03 |
| 852 | 901.61 | 24.80 | 897 | 947.86 | 25.43 | 942 | 994.08 | 26.04 |
| 853 | 902.64 | 24.82 | 898 | 948.89 | 25.44 | 943 | 995.11 | 26.05 |
| 854 | 903.67 | 24.83 | 899 | 949.92 | 25.46 | 944 | 996.13 | 26.07 |

Table 3. CONSTANTS FOR $\overline{p}'$ (SHORT FORMULA), AN APPROXIMATION OF
THE UPPER BINOMIAL CONFIDENCE LIMIT $\overline{p}$; $c/n \leqslant .1$, $n \geqslant 40$,
$c \leqslant 1,000$ (*Continued*)

| c | $\overline{m}$ | $\overline{k}$ | c | $\overline{m}$ | $\overline{k}$ | c | $\overline{m}$ | $\overline{k}$ |
|---|---|---|---|---|---|---|---|---|
| 945 | 997.16 | 26.08 | 965 | 1017.69 | 26.35 | 985 | 1038.22 | 26 |
| 946 | 998.19 | 26.09 | 966 | 1018.72 | 26.36 | 986 | 1039.25 | 26 |
| 947 | 999.21 | 26.11 | 967 | 1019.75 | 26.37 | 987 | 1040.27 | 26 |
| 948 | 1000.24 | 26.12 | 968 | 1020.77 | 26.39 | 988 | 1041.30 | 26 |
| 949 | 1001.27 | 26.13 | 969 | 1021.80 | 26.40 | 989 | 1042.32 | 26 |
| 950 | 1002.29 | 26.15 | 970 | 1022.82 | 26.41 | 990 | 1043.35 | 26 |
| 951 | 1003.32 | 26.16 | 971 | 1023.85 | 26.43 | 991 | 1044.38 | 26 |
| 952 | 1004.35 | 26.17 | 972 | 1024.88 | 26.44 | 992 | 1045.40 | 26 |
| 953 | 1005.37 | 26.19 | 973 | 1025.90 | 26.45 | 993 | 1046.43 | 26 |
| 954 | 1006.40 | 26.20 | 974 | 1026.93 | 26.47 | 994 | 1047.45 | 26 |
| 955 | 1007.43 | 26.21 | 975 | 1027.96 | 26.48 | 995 | 1048.48 | 26 |
| 956 | 1008.45 | 26.23 | 976 | 1028.98 | 26.49 | 996 | 1049.51 | 26 |
| 957 | 1009.48 | 26.24 | 977 | 1030.01 | 26.50 | 997 | 1050.53 | 26 |
| 958 | 1010.51 | 26.25 | 978 | 1031.04 | 26.52 | 998 | 1051.56 | 26 |
| 959 | 1011.53 | 26.27 | 979 | 1032.06 | 26.53 | 999 | 1052.58 | 26 |
| 960 | 1012.56 | 26.28 | 980 | 1033.09 | 26.54 | 1000 | 1053.61 | 26 |
| 961 | 1013.59 | 26.29 | 981 | 1034.11 | 26.56 | | | |
| 962 | 1014.61 | 26.31 | 982 | 1035.14 | 26.57 | | | |
| 963 | 1015.64 | 26.32 | 983 | 1036.17 | 26.58 | | | |
| 964 | 1016.67 | 26.33 | 984 | 1037.19 | 26.60 | | | |

# Table 3. CONSTANTS FOR $\overline{p}'$ (SHORT FORMULA), AN APPROXIMATION OF THE UPPER BINOMIAL CONFIDENCE LIMIT $\overline{p}$; c/n ≤ .1, n ≥ 40, c ≤ 1,000 (*Continued*)

| c | $\overline{m}$ | $\overline{k}$ | c | $\overline{m}$ | $\overline{k}$ | c | $\overline{m}$ | $\overline{k}$ |
|---|---|---|---|---|---|---|---|---|
| 0 | 2.3026 | 1.157 | 45 | 54.878 | 5.04 | 90 | 103.42 | 6.72 |
| 1 | 3.8897 | 1.465 | 46 | 55.972 | 5.12 | 91 | 104.49 | 6.75 |
| 2 | 5.3223 | 1.693 | 47 | 57.066 | 5.18 | 92 | 105.55 | 6.79 |
| 3 | 6.6808 | 1.886 | 48 | 58.158 | 5.23 | 93 | 106.62 | 6.82 |
| 4 | 7.9936 | 2.059 | 49 | 59.249 | 5.28 | 94 | 107.69 | 6.85 |
| 5 | 9.2747 | 2.202 | 50 | 60.340 | 5.31 | 95 | 108.75 | 6.89 |
| 6 | 10.532 | 2.328 | 51 | 61.429 | 5.23 | 96 | 109.82 | 6.92 |
| 7 | 11.771 | 2.451 | 52 | 62.518 | 5.27 | 97 | 110.88 | 6.95 |
| 8 | 12.995 | 2.563 | 53 | 63.606 | 5.32 | 98 | 111.95 | 6.98 |
| 9 | 14.206 | 2.675 | 54 | 64.693 | 5.36 | 99 | 113.01 | 7.02 |
| 10 | 15.407 | 2.77 | 55 | 65.779 | 5.40 | 100 | 114.07 | 7.05 |
| 11 | 16.598 | 2.87 | 56 | 66.865 | 5.44 | 101 | 115.14 | 7.08 |
| 12 | 17.782 | 2.96 | 57 | 67.949 | 5.49 | 102 | 116.20 | 7.11 |
| 13 | 18.958 | 3.05 | 58 | 69.033 | 5.53 | 103 | 117.26 | 7.14 |
| 14 | 20.128 | 3.14 | 59 | 70.117 | 5.57 | 104 | 118.33 | 7.17 |
| 15 | 21.292 | 3.23 | 60 | 71.199 | 5.61 | 105 | 119.39 | 7.20 |
| 16 | 22.452 | 3.31 | 61 | 72.281 | 5.65 | 106 | 120.45 | 7.24 |
| 17 | 23.606 | 3.38 | 62 | 73.362 | 5.69 | 107 | 121.51 | 7.27 |
| 18 | 24.756 | 3.46 | 63 | 74.443 | 5.73 | 108 | 122.58 | 7.30 |
| 19 | 25.903 | 3.53 | 64 | 75.523 | 5.77 | 109 | 123.64 | 7.33 |
| 20 | 27.045 | 3.61 | 65 | 76.602 | 5.81 | 110 | 124.70 | 7.36 |
| 21 | 28.184 | 3.69 | 66 | 77.681 | 5.85 | 111 | 125.76 | 7.39 |
| 22 | 29.320 | 3.74 | 67 | 78.759 | 5.89 | 112 | 126.82 | 7.42 |
| 23 | 30.453 | 3.81 | 68 | 79.837 | 5.93 | 113 | 127.88 | 7.45 |
| 24 | 31.584 | 3.89 | 69 | 80.914 | 5.97 | 114 | 128.94 | 7.48 |
| 25 | 32.711 | 3.96 | 70 | 81.990 | 6.01 | 115 | 130.00 | 7.51 |
| 26 | 33.836 | 4.02 | 71 | 83.066 | 6.04 | 116 | 131.06 | 7.54 |
| 27 | 34.959 | 4.07 | 72 | 84.142 | 6.08 | 117 | 132.12 | 7.57 |
| 28 | 36.080 | 4.14 | 73 | 85.216 | 6.12 | 118 | 133.18 | 7.60 |
| 29 | 37.199 | 4.20 | 74 | 86.291 | 6.16 | 119 | 134.24 | 7.63 |
| 30 | 38.315 | 4.24 | 75 | 87.365 | 6.19 | 120 | 135.29 | 7.66 |
| 31 | 39.430 | 4.32 | 76 | 88.438 | 6.23 | 121 | 136.35 | 7.68 |
| 32 | 40.543 | 4.38 | 77 | 89.511 | 6.27 | 122 | 137.41 | 7.71 |
| 33 | 41.654 | 4.41 | 78 | 90.584 | 6.30 | 123 | 138.47 | 7.74 |
| 34 | 42.764 | 4.48 | 79 | 91.656 | 6.34 | 124 | 139.53 | 7.77 |
| 35 | 43.872 | 4.54 | 80 | 92.727 | 6.37 | 125 | 140.58 | 7.80 |
| 36 | 44.978 | 4.60 | 81 | 93.798 | 6.41 | 126 | 141.64 | 7.83 |
| 37 | 46.083 | 4.66 | 82 | 94.869 | 6.45 | 127 | 142.70 | 7.86 |
| 38 | 47.187 | 4.69 | 83 | 95.939 | 6.48 | 128 | 143.75 | 7.89 |
| 39 | 48.289 | 4.74 | 84 | 97.009 | 6.51 | 129 | 144.81 | 7.91 |
| 40 | 49.390 | 4.81 | 85 | 98.078 | 6.55 | 130 | 145.87 | 7.94 |
| 41 | 50.490 | 4.85 | 86 | 99.147 | 6.58 | 131 | 146.92 | 7.97 |
| 42 | 51.589 | 4.93 | 87 | 100.22 | 6.62 | 132 | 147.98 | 8.00 |
| 43 | 52.686 | 4.95 | 88 | 101.28 | 6.65 | 133 | 149.03 | 8.02 |
| 44 | 53.783 | 5.00 | 89 | 102.35 | 6.69 | 134 | 150.09 | 8.05 |

Table 3.   CONSTANTS FOR $\bar{p}'$ (SHORT FORMULA), AN APPROXIMATION OF
THE UPPER BINOMIAL CONFIDENCE LIMIT $\bar{p}$; $c/n \leqslant .1$, $n \geqslant 40$,
$c \leqslant 1,000$ (*Continued*)

| c | $\bar{m}$ | $\bar{k}$ | c | $\bar{m}$ | $\bar{k}$ | c | $\bar{m}$ | $\bar{k}$ |
|---|---|---|---|---|---|---|---|---|
| 135 | 151.14 | 8.08 | 180 | 198.46 | 9.23 | 225 | 245.48 | 10.24 |
| 136 | 152.20 | 8.11 | 181 | 199.50 | 9.25 | 226 | 246.52 | 10.26 |
| 137 | 153.25 | 8.13 | 182 | 200.55 | 9.28 | 227 | 247.57 | 10.28 |
| 138 | 154.31 | 8.16 | 183 | 201.60 | 9.30 | 228 | 248.61 | 10.30 |
| 139 | 155.36 | 8.19 | 184 | 202.65 | 9.32 | 229 | 249.65 | 10.33 |
| 140 | 156.42 | 8.22 | 185 | 203.69 | 9.35 | 230 | 250.69 | 10.35 |
| 141 | 157.47 | 8.24 | 186 | 204.74 | 9.37 | 231 | 251.73 | 10.37 |
| 142 | 158.52 | 8.27 | 187 | 205.79 | 9.39 | 232 | 252.78 | 10.39 |
| 143 | 159.58 | 8.30 | 188 | 206.83 | 9.42 | 233 | 253.82 | 10.41 |
| 144 | 160.63 | 8.32 | 189 | 207.88 | 9.44 | 234 | 254.86 | 10.43 |
| 145 | 161.68 | 8.35 | 190 | 208.93 | 9.46 | 235 | 255.90 | 10.45 |
| 146 | 162.74 | 8.38 | 191 | 209.97 | 9.49 | 236 | 256.94 | 10.47 |
| 147 | 163.79 | 8.40 | 192 | 211.02 | 9.51 | 237 | 257.99 | 10.49 |
| 148 | 164.84 | 8.43 | 193 | 212.06 | 9.53 | 238 | 259.03 | 10.51 |
| 149 | 165.89 | 8.46 | 194 | 213.11 | 9.56 | 239 | 260.07 | 10.53 |
| 150 | 166.95 | 8.48 | 195 | 214.16 | 9.58 | 240 | 261.11 | 10.55 |
| 151 | 168.00 | 8.51 | 196 | 215.20 | 9.60 | 241 | 262.15 | 10.58 |
| 152 | 169.05 | 8.53 | 197 | 216.25 | 9.62 | 242 | 263.19 | 10.60 |
| 153 | 170.10 | 8.56 | 198 | 217.29 | 9.65 | 243 | 264.23 | 10.62 |
| 154 | 171.15 | 8.58 | 199 | 218.34 | 9.67 | 244 | 265.27 | 10.64 |
| 155 | 172.21 | 8.61 | 200 | 219.38 | 9.69 | 245 | 266.32 | 10.66 |
| 156 | 173.26 | 8.64 | 201 | 220.43 | 9.71 | 246 | 267.36 | 10.68 |
| 157 | 174.31 | 8.66 | 202 | 221.47 | 9.74 | 247 | 268.40 | 10.70 |
| 158 | 175.36 | 8.69 | 203 | 222.52 | 9.76 | 248 | 269.44 | 10.72 |
| 159 | 176.41 | 8.71 | 204 | 223.56 | 9.78 | 249 | 270.48 | 10.74 |
| 160 | 177.46 | 8.74 | 205 | 224.61 | 9.80 | 250 | 271.52 | 10.76 |
| 161 | 178.53 | 8.76 | 206 | 225.65 | 9.83 | 251 | 272.56 | 10.78 |
| 162 | 179.58 | 8.79 | 207 | 226.70 | 9.85 | 252 | 273.60 | 10.80 |
| 163 | 180.63 | 8.81 | 208 | 227.74 | 9.87 | 253 | 274.64 | 10.82 |
| 164 | 181.68 | 8.84 | 209 | 228.79 | 9.89 | 254 | 275.68 | 10.84 |
| 165 | 182.73 | 8.86 | 210 | 229.83 | 9.92 | 255 | 276.72 | 10.86 |
| 166 | 183.78 | 8.89 | 211 | 230.87 | 9.94 | 256 | 277.76 | 10.88 |
| 167 | 184.83 | 8.91 | 212 | 231.92 | 9.96 | 257 | 278.80 | 10.90 |
| 168 | 185.87 | 8.94 | 213 | 232.96 | 9.98 | 258 | 279.84 | 10.91 |
| 169 | 186.92 | 8.96 | 214 | 234.01 | 10.00 | 259 | 280.88 | 10.94 |
| 170 | 187.97 | 8.99 | 215 | 235.05 | 10.02 | 260 | 281.92 | 10.9 |
| 171 | 189.02 | 9.01 | 216 | 236.09 | 10.05 | 261 | 282.96 | 10.9 |
| 172 | 190.07 | 9.04 | 217 | 237.14 | 10.07 | 262 | 284.00 | 11.0 |
| 173 | 191.12 | 9.06 | 218 | 238.18 | 10.09 | 263 | 285.04 | 11.0 |
| 174 | 192.17 | 9.08 | 219 | 239.22 | 10.11 | 264 | 286.08 | 11.0 |
| 175 | 193.22 | 9.11 | 220 | 240.27 | 10.13 | 265 | 287.12 | 11.0 |
| 176 | 194.26 | 9.13 | 221 | 241.31 | 10.15 | 266 | 288.16 | 11.0 |
| 177 | 195.31 | 9.16 | 222 | 242.35 | 10.18 | 267 | 289.19 | 11.1 |
| 178 | 196.36 | 9.18 | 223 | 243.40 | 10.20 | 268 | 290.23 | 11.1 |
| 179 | 197.41 | 9.20 | 224 | 244.44 | 10.22 | 269 | 291.27 | 11.1 |

# Table 3. CONSTANTS FOR $\overline{p}'$ (SHORT FORMULA), AN APPROXIMATION OF THE UPPER BINOMIAL CONFIDENCE LIMIT $\overline{p}$; $c/n \leqslant .1$, $n \geqslant 40$, $c \leqslant 1,000$ (*Continued*)

| c | $\overline{m}$ | $\overline{k}$ | c | $\overline{m}$ | $\overline{k}$ | c | $\overline{m}$ | $\overline{k}$ |
|-----|--------|-------|-----|--------|-------|-----|--------|-------|
| 270 | 292.31 | 11.16 | 315 | 339.00 | 12.00 | 360 | 385.56 | 12.78 |
| 271 | 293.35 | 11.18 | 316 | 340.03 | 12.02 | 361 | 386.60 | 12.80 |
| 272 | 294.39 | 11.19 | 317 | 341.07 | 12.03 | 362 | 387.63 | 12.82 |
| 273 | 295.43 | 11.21 | 318 | 342.10 | 12.05 | 363 | 388.67 | 12.83 |
| 274 | 296.47 | 11.23 | 319 | 343.14 | 12.07 | 364 | 389.70 | 12.85 |
| 275 | 297.51 | 11.25 | 320 | 344.18 | 12.09 | 365 | 390.73 | 12.87 |
| 276 | 298.54 | 11.27 | 321 | 345.21 | 12.11 | 366 | 391.77 | 12.88 |
| 277 | 299.58 | 11.29 | 322 | 346.25 | 12.12 | 367 | 392.80 | 12.90 |
| 278 | 300.62 | 11.31 | 323 | 347.28 | 12.14 | 368 | 393.83 | 12.92 |
| 279 | 301.66 | 11.33 | 324 | 348.32 | 12.16 | 369 | 394.87 | 12.93 |
| 280 | 302.70 | 11.35 | 325 | 349.35 | 12.18 | 370 | 395.90 | 12.95 |
| 281 | 303.74 | 11.37 | 326 | 350.39 | 12.19 | 371 | 396.93 | 12.97 |
| 282 | 304.77 | 11.39 | 327 | 351.42 | 12.21 | 372 | 397.97 | 12.98 |
| 283 | 305.81 | 11.41 | 328 | 352.46 | 12.23 | 373 | 399.00 | 13.00 |
| 284 | 306.85 | 11.42 | 329 | 353.50 | 12.25 | 374 | 400.03 | 13.02 |
| 285 | 307.89 | 11.44 | 330 | 354.53 | 12.27 | 375 | 401.07 | 13.03 |
| 286 | 308.93 | 11.46 | 331 | 355.57 | 12.28 | 376 | 402.10 | 13.05 |
| 287 | 309.96 | 11.48 | 332 | 356.60 | 12.30 | 377 | 403.13 | 13.07 |
| 288 | 311.00 | 11.50 | 333 | 357.64 | 12.32 | 378 | 404.16 | 13.08 |
| 289 | 312.04 | 11.52 | 334 | 358.67 | 12.34 | 379 | 405.20 | 13.10 |
| 290 | 313.08 | 11.54 | 335 | 359.71 | 12.35 | 380 | 406.23 | 13.11 |
| 291 | 314.11 | 11.56 | 336 | 360.74 | 12.37 | 381 | 407.26 | 13.13 |
| 292 | 315.15 | 11.58 | 337 | 361.78 | 12.39 | 382 | 408.30 | 13.15 |
| 293 | 316.19 | 11.59 | 338 | 362.81 | 12.41 | 383 | 409.33 | 13.16 |
| 294 | 317.23 | 11.61 | 339 | 363.85 | 12.42 | 384 | 410.36 | 13.18 |
| 295 | 318.26 | 11.63 | 340 | 364.88 | 12.44 | 385 | 411.39 | 13.20 |
| 296 | 319.30 | 11.65 | 341 | 365.91 | 12.46 | 386 | 412.43 | 13.21 |
| 297 | 320.34 | 11.67 | 342 | 366.95 | 12.47 | 387 | 413.46 | 13.23 |
| 298 | 321.38 | 11.69 | 343 | 367.98 | 12.49 | 388 | 414.49 | 13.25 |
| 299 | 322.41 | 11.71 | 344 | 369.02 | 12.51 | 389 | 415.52 | 13.26 |
| 300 | 323.45 | 11.72 | 345 | 370.05 | 12.53 | 390 | 416.56 | 13.28 |
| 301 | 324.49 | 11.74 | 346 | 371.09 | 12.54 | 391 | 417.59 | 13.29 |
| 302 | 325.52 | 11.76 | 347 | 372.12 | 12.56 | 392 | 418.62 | 13.31 |
| 303 | 326.56 | 11.78 | 348 | 373.16 | 12.58 | 393 | 419.65 | 13.33 |
| 304 | 327.60 | 11.80 | 349 | 374.19 | 12.60 | 394 | 420.69 | 13.34 |
| 305 | 328.63 | 11.82 | 350 | 375.22 | 12.61 | 395 | 421.72 | 13.36 |
| 306 | 329.67 | 11.83 | 351 | 376.26 | 12.63 | 396 | 422.75 | 13.37 |
| 307 | 330.71 | 11.85 | 352 | 377.29 | 12.65 | 397 | 423.78 | 13.39 |
| 308 | 331.74 | 11.87 | 353 | 378.33 | 12.66 | 398 | 424.81 | 13.41 |
| 309 | 332.78 | 11.89 | 354 | 379.36 | 12.68 | 399 | 425.85 | 13.42 |
| 310 | 333.82 | 11.91 | 355 | 380.40 | 12.70 | 400 | 426.88 | 13.44 |
| 311 | 334.85 | 11.93 | 356 | 381.43 | 12.71 | 401 | 427.91 | 13.46 |
| 312 | 335.89 | 11.94 | 357 | 382.46 | 12.73 | 402 | 428.94 | 13.47 |
| 313 | 336.92 | 11.96 | 358 | 383.50 | 12.75 | 403 | 429.97 | 13.49 |
| 314 | 337.96 | 11.98 | 359 | 384.53 | 12.77 | 404 | 431.01 | 13.50 |

Table 3. CONSTANTS FOR $\overline{p}'$ (SHORT FORMULA), AN APPROXIMATION OF THE UPPER BINOMIAL CONFIDENCE LIMIT $\overline{p}$; $c/n \leqslant .1$, $n \geqslant 40$, $c \leqslant 1{,}000$ (*Continued*)

| c | $\overline{m}$ | $\overline{k}$ | c | $\overline{m}$ | $\overline{k}$ | c | $\overline{m}$ | $\overline{k}$ |
|---|---|---|---|---|---|---|---|---|
| 405 | 432.04 | 13.52 | 450 | 478.43 | 14.22 | 495 | 524.76 | 14.88 |
| 406 | 433.07 | 13.53 | 451 | 479.46 | 14.23 | 496 | 525.79 | 14.89 |
| 407 | 434.10 | 13.55 | 452 | 480.49 | 14.25 | 497 | 526.81 | 14.91 |
| 408 | 435.13 | 13.57 | 453 | 481.52 | 14.26 | 498 | 527.84 | 14.92 |
| 409 | 436.16 | 13.58 | 454 | 482.55 | 14.28 | 499 | 528.87 | 14.94 |
| 410 | 437.20 | 13.60 | 455 | 483.58 | 14.29 | 500 | 529.90 | 14.95 |
| 411 | 438.23 | 13.61 | 456 | 484.61 | 14.31 | 501 | 530.93 | 14.96 |
| 412 | 439.26 | 13.63 | 457 | 485.64 | 14.32 | 502 | 531.96 | 14.98 |
| 413 | 440.29 | 13.65 | 458 | 486.67 | 14.34 | 503 | 532.99 | 14.99 |
| 414 | 441.32 | 13.66 | 459 | 487.70 | 14.35 | 504 | 534.01 | 15.01 |
| 415 | 442.35 | 13.68 | 460 | 488.73 | 14.37 | 505 | 535.04 | 15.02 |
| 416 | 443.39 | 13.69 | 461 | 489.76 | 14.38 | 506 | 536.07 | 15.04 |
| 417 | 444.42 | 13.71 | 462 | 490.79 | 14.40 | 507 | 537.10 | 15.05 |
| 418 | 445.45 | 13.72 | 463 | 491.82 | 14.41 | 508 | 538.13 | 15.06 |
| 419 | 446.48 | 13.74 | 464 | 492.85 | 14.43 | 509 | 539.16 | 15.08 |
| 420 | 447.51 | 13.76 | 465 | 493.88 | 14.44 | 510 | 540.19 | 15.09 |
| 421 | 448.54 | 13.77 | 466 | 494.91 | 14.45 | 511 | 541.21 | 15.11 |
| 422 | 449.57 | 13.79 | 467 | 495.94 | 14.47 | 512 | 542.24 | 15.12 |
| 423 | 450.60 | 13.80 | 468 | 496.97 | 14.48 | 513 | 543.27 | 15.13 |
| 424 | 451.63 | 13.82 | 469 | 498.00 | 14.50 | 514 | 544.30 | 15.15 |
| 425 | 452.67 | 13.83 | 470 | 499.03 | 14.51 | 515 | 545.33 | 15.16 |
| 426 | 453.70 | 13.85 | 471 | 500.06 | 14.53 | 516 | 546.35 | 15.18 |
| 427 | 454.73 | 13.86 | 472 | 501.09 | 14.54 | 517 | 547.38 | 15.19 |
| 428 | 455.76 | 13.88 | 473 | 502.12 | 14.56 | 518 | 548.41 | 15.21 |
| 429 | 456.79 | 13.89 | 474 | 503.15 | 14.57 | 519 | 549.44 | 15.22 |
| 430 | 457.82 | 13.91 | 475 | 504.18 | 14.59 | 520 | 550.47 | 15.23 |
| 431 | 458.85 | 13.93 | 476 | 505.20 | 14.60 | 521 | 551.50 | 15.25 |
| 432 | 459.88 | 13.94 | 477 | 506.23 | 14.62 | 522 | 552.52 | 15.26 |
| 433 | 460.91 | 13.96 | 478 | 507.26 | 14.63 | 523 | 553.55 | 15.28 |
| 434 | 461.94 | 13.97 | 479 | 508.29 | 14.65 | 524 | 554.58 | 15.29 |
| 435 | 462.97 | 13.99 | 480 | 509.32 | 14.66 | 525 | 555.61 | 15.30 |
| 436 | 464.01 | 14.00 | 481 | 510.35 | 14.68 | 526 | 556.64 | 15.32 |
| 437 | 465.04 | 14.02 | 482 | 511.38 | 14.69 | 527 | 557.66 | 15.33 |
| 438 | 466.07 | 14.03 | 483 | 512.41 | 14.70 | 528 | 558.69 | 15.35 |
| 439 | 467.10 | 14.05 | 484 | 513.44 | 14.72 | 529 | 559.72 | 15.36 |
| 440 | 468.13 | 14.06 | 485 | 514.47 | 14.73 | 530 | 560.75 | 15.37 |
| 441 | 469.16 | 14.08 | 486 | 515.50 | 14.75 | 531 | 561.77 | 15.39 |
| 442 | 470.19 | 14.09 | 487 | 516.53 | 14.76 | 532 | 562.80 | 15.40 |
| 443 | 471.22 | 14.11 | 488 | 517.55 | 14.78 | 533 | 563.83 | 15.41 |
| 444 | 472.25 | 14.12 | 489 | 518.58 | 14.79 | 534 | 564.86 | 15.43 |
| 445 | 473.28 | 14.14 | 490 | 519.61 | 14.81 | 535 | 565.89 | 15.44 |
| 446 | 474.31 | 14.16 | 491 | 520.64 | 14.82 | 536 | 566.91 | 15.46 |
| 447 | 475.34 | 14.17 | 492 | 521.67 | 14.84 | 537 | 567.94 | 15.47 |
| 448 | 476.37 | 14.19 | 493 | 522.70 | 14.85 | 538 | 568.97 | 15.48 |
| 449 | 477.40 | 14.20 | 494 | 523.73 | 14.86 | 539 | 570.00 | 15.50 |

Table 3.   CONSTANTS FOR $\bar{p}'$ (SHORT FORMULA), AN APPROXIMATION OF
THE UPPER BINOMIAL CONFIDENCE LIMIT $\bar{p}$; $c/n \leqslant .1$, $n \geqslant 40$,
$c \leqslant 1,000$ (*Continued*)

Section 5

$\gamma_1 = 90\%$

$\gamma_2 = 80\%$

| c | $\bar{m}$ | $\bar{k}$ | c | $\bar{m}$ | $\bar{k}$ | c | $\bar{m}$ | $\bar{k}$ |
|---|---|---|---|---|---|---|---|---|
| 540 | 571.02 | 15.51 | 585 | 617.24 | 16.12 | 630 | 663.41 | 16.70 |
| 541 | 572.05 | 15.53 | 586 | 618.26 | 16.13 | 631 | 664.43 | 16.72 |
| 542 | 573.08 | 15.54 | 587 | 619.29 | 16.15 | 632 | 665.46 | 16.73 |
| 543 | 574.11 | 15.55 | 588 | 620.32 | 16.16 | 633 | 666.48 | 16.74 |
| 544 | 575.13 | 15.57 | 589 | 621.34 | 16.17 | 634 | 667.51 | 16.75 |
| 545 | 576.16 | 15.58 | 590 | 622.37 | 16.19 | 635 | 668.53 | 16.77 |
| 546 | 577.19 | 15.59 | 591 | 623.40 | 16.20 | 636 | 669.56 | 16.78 |
| 547 | 578.22 | 15.61 | 592 | 624.42 | 16.21 | 637 | 670.59 | 16.79 |
| 548 | 579.24 | 15.62 | 593 | 625.45 | 16.22 | 638 | 671.61 | 16.81 |
| 549 | 580.27 | 15.64 | 594 | 626.48 | 16.24 | 639 | 672.64 | 16.82 |
| 550 | 581.30 | 15.65 | 595 | 627.50 | 16.25 | 640 | 673.66 | 16.83 |
| 551 | 582.32 | 15.66 | 596 | 628.53 | 16.26 | 641 | 674.69 | 16.84 |
| 552 | 583.35 | 15.68 | 597 | 629.55 | 16.28 | 642 | 675.71 | 16.86 |
| 553 | 584.38 | 15.69 | 598 | 630.58 | 16.29 | 643 | 676.74 | 16.87 |
| 554 | 585.41 | 15.70 | 599 | 631.61 | 16.30 | 644 | 677.76 | 16.88 |
| 555 | 586.43 | 15.72 | 600 | 632.63 | 16.32 | 645 | 678.79 | 16.89 |
| 556 | 587.46 | 15.73 | 601 | 633.66 | 16.33 | 646 | 679.81 | 16.91 |
| 557 | 588.49 | 15.74 | 602 | 634.69 | 16.34 | 647 | 680.84 | 16.92 |
| 558 | 589.52 | 15.76 | 603 | 635.71 | 16.36 | 648 | 681.86 | 16.93 |
| 559 | 590.54 | 15.77 | 604 | 636.74 | 16.37 | 649 | 682.89 | 16.94 |
| 560 | 591.57 | 15.78 | 605 | 637.76 | 16.38 | 650 | 683.91 | 16.96 |
| 561 | 592.60 | 15.80 | 606 | 638.79 | 16.39 | 651 | 684.94 | 16.97 |
| 562 | 593.62 | 15.81 | 607 | 639.82 | 16.41 | 652 | 685.96 | 16.98 |
| 563 | 594.65 | 15.83 | 608 | 640.84 | 16.42 | 653 | 686.99 | 16.99 |
| 564 | 595.68 | 15.84 | 609 | 641.87 | 16.43 | 654 | 688.01 | 17.01 |
| 565 | 596.70 | 15.85 | 610 | 642.89 | 16.45 | 655 | 689.04 | 17.02 |
| 566 | 597.73 | 15.87 | 611 | 643.92 | 16.46 | 656 | 690.06 | 17.03 |
| 567 | 598.76 | 15.88 | 612 | 644.94 | 16.47 | 657 | 691.09 | 17.04 |
| 568 | 599.78 | 15.89 | 613 | 645.97 | 16.49 | 658 | 692.11 | 17.06 |
| 569 | 600.81 | 15.91 | 614 | 647.00 | 16.50 | 659 | 693.14 | 17.07 |
| 570 | 601.84 | 15.92 | 615 | 648.02 | 16.51 | 660 | 694.16 | 17.08 |
| 571 | 602.87 | 15.93 | 616 | 649.05 | 16.52 | 661 | 695.19 | 17.09 |
| 572 | 603.89 | 15.95 | 617 | 650.07 | 16.54 | 662 | 696.21 | 17.11 |
| 573 | 604.92 | 15.96 | 618 | 651.10 | 16.55 | 663 | 697.24 | 17.12 |
| 574 | 605.95 | 15.97 | 619 | 652.13 | 16.56 | 664 | 698.26 | 17.13 |
| 575 | 606.97 | 15.99 | 620 | 653.15 | 16.58 | 665 | 699.29 | 17.14 |
| 576 | 608.00 | 16.00 | 621 | 654.18 | 16.59 | 666 | 700.31 | 17.16 |
| 577 | 609.03 | 16.01 | 622 | 655.20 | 16.60 | 667 | 701.34 | 17.17 |
| 578 | 610.05 | 16.03 | 623 | 656.23 | 16.61 | 668 | 702.36 | 17.18 |
| 579 | 611.08 | 16.04 | 624 | 657.25 | 16.63 | 669 | 703.39 | 17.19 |
| 580 | 612.11 | 16.05 | 625 | 658.28 | 16.64 | 670 | 704.41 | 17.21 |
| 581 | 613.13 | 16.07 | 626 | 659.31 | 16.65 | 671 | 705.44 | 17.22 |
| 582 | 614.16 | 16.08 | 627 | 660.33 | 16.67 | 672 | 706.46 | 17.23 |
| 583 | 615.19 | 16.09 | 628 | 661.36 | 16.68 | 673 | 707.49 | 17.24 |
| 584 | 616.21 | 16.11 | 629 | 662.38 | 16.69 | 674 | 708.51 | 17.26 |

Table 3.  CONSTANTS FOR $\overline{p}'$ (SHORT FORMULA), AN APPROXIMATION OF THE UPPER BINOMIAL CONFIDENCE LIMIT $\overline{p}$; $c/n \leqslant .1$, $n \geqslant 40$, $c \leqslant 1,000$ (*Continued*)

| c | $\overline{m}$ | $\overline{k}$ | c | $\overline{m}$ | $\overline{k}$ | c | $\overline{m}$ | $\overline{k}$ |
|---|---|---|---|---|---|---|---|---|
| 675 | 709.54 | 17.27 | 720 | 755.63 | 17.81 | 765 | 801.68 | 18.34 |
| 676 | 710.56 | 17.28 | 721 | 756.65 | 17.83 | 766 | 802.71 | 18.35 |
| 677 | 711.58 | 17.29 | 722 | 757.67 | 17.84 | 767 | 803.73 | 18.37 |
| 678 | 712.61 | 17.30 | 723 | 758.70 | 17.85 | 768 | 804.75 | 18.38 |
| 679 | 713.63 | 17.32 | 724 | 759.72 | 17.86 | 769 | 805.78 | 18.39 |
| 680 | 714.66 | 17.33 | 725 | 760.75 | 17.87 | 770 | 806.80 | 18.40 |
| 681 | 715.68 | 17.34 | 726 | 761.77 | 17.88 | 771 | 807.82 | 18.41 |
| 682 | 716.71 | 17.35 | 727 | 762.79 | 17.90 | 772 | 808.85 | 18.42 |
| 683 | 717.73 | 17.37 | 728 | 763.82 | 17.91 | 773 | 809.87 | 18.43 |
| 684 | 718.76 | 17.38 | 729 | 764.84 | 17.92 | 774 | 810.89 | 18.45 |
| 685 | 719.78 | 17.39 | 730 | 765.86 | 17.93 | 775 | 811.92 | 18.46 |
| 686 | 720.81 | 17.40 | 731 | 766.89 | 17.94 | 776 | 812.94 | 18.47 |
| 687 | 721.83 | 17.42 | 732 | 767.91 | 17.96 | 777 | 813.96 | 18.48 |
| 688 | 722.85 | 17.43 | 733 | 768.94 | 17.97 | 778 | 814.98 | 18.49 |
| 689 | 723.88 | 17.44 | 734 | 769.96 | 17.98 | 779 | 816.01 | 18.50 |
| 690 | 724.90 | 17.45 | 735 | 770.98 | 17.99 | 780 | 817.03 | 18.52 |
| 691 | 725.93 | 17.46 | 736 | 772.01 | 18.00 | 781 | 818.05 | 18.53 |
| 692 | 726.95 | 17.48 | 737 | 773.03 | 18.02 | 782 | 819.08 | 18.54 |
| 693 | 727.98 | 17.49 | 738 | 774.05 | 18.03 | 783 | 820.10 | 18.55 |
| 694 | 729.00 | 17.50 | 739 | 775.08 | 18.04 | 784 | 821.12 | 18.56 |
| 695 | 730.03 | 17.51 | 740 | 776.10 | 18.05 | 785 | 822.14 | 18.57 |
| 696 | 731.05 | 17.52 | 741 | 777.12 | 18.06 | 786 | 823.17 | 18.58 |
| 697 | 732.07 | 17.54 | 742 | 778.15 | 18.07 | 787 | 824.19 | 18.60 |
| 698 | 733.10 | 17.55 | 743 | 779.17 | 18.09 | 788 | 825.21 | 18.61 |
| 699 | 734.12 | 17.56 | 744 | 780.19 | 18.10 | 789 | 826.24 | 18.62 |
| 700 | 735.15 | 17.57 | 745 | 781.22 | 18.11 | 790 | 827.26 | 18.63 |
| 701 | 736.17 | 17.59 | 746 | 782.24 | 18.12 | 791 | 828.28 | 18.64 |
| 702 | 737.19 | 17.60 | 747 | 783.27 | 18.13 | 792 | 829.30 | 18.65 |
| 703 | 738.22 | 17.61 | 748 | 784.29 | 18.14 | 793 | 830.33 | 18.66 |
| 704 | 739.24 | 17.62 | 749 | 785.31 | 18.16 | 794 | 831.35 | 18.67 |
| 705 | 740.27 | 17.63 | 750 | 786.34 | 18.17 | 795 | 832.37 | 18.69 |
| 706 | 741.29 | 17.65 | 751 | 787.36 | 18.18 | 796 | 833.40 | 18.70 |
| 707 | 742.32 | 17.66 | 752 | 788.38 | 18.19 | 797 | 834.42 | 18.71 |
| 708 | 743.34 | 17.67 | 753 | 789.41 | 18.20 | 798 | 835.44 | 18.72 |
| 709 | 744.36 | 17.68 | 754 | 790.43 | 18.21 | 799 | 836.46 | 18.73 |
| 710 | 745.39 | 17.69 | 755 | 791.45 | 18.23 | 800 | 837.49 | 18.74 |
| 711 | 746.41 | 17.71 | 756 | 792.48 | 18.24 | 801 | 838.51 | 18.75 |
| 712 | 747.44 | 17.72 | 757 | 793.50 | 18.25 | 802 | 839.53 | 18.77 |
| 713 | 748.46 | 17.73 | 758 | 794.52 | 18.26 | 803 | 840.55 | 18.78 |
| 714 | 749.48 | 17.74 | 759 | 795.55 | 18.27 | 804 | 841.58 | 18.79 |
| 715 | 750.51 | 17.75 | 760 | 796.57 | 18.28 | 805 | 842.60 | 18.80 |
| 716 | 751.53 | 17.77 | 761 | 797.59 | 18.30 | 806 | 843.62 | 18.81 |
| 717 | 752.56 | 17.78 | 762 | 798.61 | 18.31 | 807 | 844.64 | 18.82 |
| 718 | 753.58 | 17.79 | 763 | 799.64 | 18.32 | 808 | 845.67 | 18.83 |
| 719 | 754.60 | 17.80 | 764 | 800.66 | 18.33 | 809 | 846.69 | 18.84 |

# Table 3. CONSTANTS FOR $\overline{p}'$ (SHORT FORMULA), AN APPROXIMATION OF THE UPPER BINOMIAL CONFIDENCE LIMIT $\overline{p}$; $c/n \leqslant .1$, $n \geqslant 40$, $c \leqslant 1,000$ (Continued)

| c | $\overline{m}$ | $\overline{k}$ | c | $\overline{m}$ | $\overline{k}$ | c | $\overline{m}$ | $\overline{k}$ |
|---|---|---|---|---|---|---|---|---|
| 810 | 847.71 | 18.86 | 855 | 893.71 | 19.36 | 900 | 939.68 | 19.84 |
| 811 | 848.73 | 18.87 | 856 | 894.73 | 19.37 | 901 | 940.70 | 19.85 |
| 812 | 849.76 | 18.88 | 857 | 895.75 | 19.38 | 902 | 941.73 | 19.86 |
| 813 | 850.78 | 18.89 | 858 | 896.78 | 19.39 | 903 | 942.75 | 19.87 |
| 814 | 851.80 | 18.90 | 859 | 897.80 | 19.40 | 904 | 943.77 | 19.88 |
| 815 | 852.82 | 18.91 | 860 | 898.82 | 19.41 | 905 | 944.79 | 19.90 |
| 816 | 853.85 | 18.92 | 861 | 899.84 | 19.42 | 906 | 945.81 | 19.91 |
| 817 | 854.87 | 18.93 | 862 | 900.86 | 19.43 | 907 | 946.83 | 19.92 |
| 818 | 855.89 | 18.95 | 863 | 901.89 | 19.44 | 908 | 947.85 | 19.93 |
| 819 | 856.91 | 18.96 | 864 | 902.91 | 19.45 | 909 | 948.88 | 19.94 |
| 820 | 857.94 | 18.97 | 865 | 903.93 | 19.46 | 910 | 949.90 | 19.95 |
| 821 | 858.96 | 18.98 | 866 | 904.95 | 19.48 | 911 | 950.92 | 19.96 |
| 822 | 859.98 | 18.99 | 867 | 905.97 | 19.49 | 912 | 951.94 | 19.97 |
| 823 | 861.00 | 19.00 | 868 | 906.99 | 19.50 | 913 | 952.96 | 19.98 |
| 824 | 862.03 | 19.01 | 869 | 908.02 | 19.51 | 914 | 953.98 | 19.99 |
| 825 | 863.05 | 19.02 | 870 | 909.04 | 19.52 | 915 | 955.00 | 20.00 |
| 826 | 864.07 | 19.03 | 871 | 910.06 | 19.53 | 916 | 956.02 | 20.01 |
| 827 | 865.09 | 19.05 | 872 | 911.08 | 19.54 | 917 | 957.04 | 20.02 |
| 828 | 866.11 | 19.06 | 873 | 912.10 | 19.55 | 918 | 958.07 | 20.03 |
| 829 | 867.14 | 19.07 | 874 | 913.12 | 19.56 | 919 | 959.09 | 20.04 |
| 830 | 868.16 | 19.08 | 875 | 914.15 | 19.57 | 920 | 960.11 | 20.05 |
| 831 | 869.18 | 19.09 | 876 | 915.17 | 19.58 | 921 | 961.13 | 20.06 |
| 832 | 870.20 | 19.10 | 877 | 916.19 | 19.59 | 922 | 962.15 | 20.08 |
| 833 | 871.23 | 19.11 | 878 | 917.21 | 19.61 | 923 | 963.17 | 20.09 |
| 834 | 872.25 | 19.12 | 879 | 918.23 | 19.62 | 924 | 964.19 | 20.10 |
| 835 | 873.27 | 19.13 | 880 | 919.25 | 19.63 | 925 | 965.21 | 20.11 |
| 836 | 874.29 | 19.15 | 881 | 920.28 | 19.64 | 926 | 966.23 | 20.12 |
| 837 | 875.31 | 19.16 | 882 | 921.30 | 19.65 | 927 | 967.26 | 20.13 |
| 838 | 876.34 | 19.17 | 883 | 922.32 | 19.66 | 928 | 968.28 | 20.14 |
| 839 | 877.36 | 19.18 | 884 | 923.34 | 19.67 | 929 | 969.30 | 20.15 |
| 840 | 878.38 | 19.19 | 885 | 924.36 | 19.68 | 930 | 970.32 | 20.16 |
| 841 | 879.40 | 19.20 | 886 | 925.38 | 19.69 | 931 | 971.34 | 20.17 |
| 842 | 880.42 | 19.21 | 887 | 926.40 | 19.70 | 932 | 972.36 | 20.18 |
| 843 | 881.45 | 19.22 | 888 | 927.43 | 19.71 | 933 | 973.38 | 20.19 |
| 844 | 882.47 | 19.23 | 889 | 928.45 | 19.72 | 934 | 974.40 | 20.20 |
| 845 | 883.49 | 19.25 | 890 | 929.47 | 19.73 | 935 | 975.42 | 20.21 |
| 846 | 884.51 | 19.26 | 891 | 930.49 | 19.75 | 936 | 976.44 | 20.22 |
| 847 | 885.53 | 19.27 | 892 | 931.51 | 19.76 | 937 | 977.47 | 20.23 |
| 848 | 886.56 | 19.28 | 893 | 932.53 | 19.77 | 938 | 978.49 | 20.24 |
| 849 | 887.58 | 19.29 | 894 | 933.56 | 19.78 | 939 | 979.51 | 20.25 |
| 850 | 888.60 | 19.30 | 895 | 934.58 | 19.79 | 940 | 980.53 | 20.26 |
| 851 | 889.62 | 19.31 | 896 | 935.60 | 19.80 | 941 | 981.55 | 20.27 |
| 852 | 890.64 | 19.32 | 897 | 936.62 | 19.81 | 942 | 982.57 | 20.28 |
| 853 | 891.67 | 19.33 | 898 | 937.64 | 19.82 | 943 | 983.59 | 20.30 |
| 854 | 892.69 | 19.34 | 899 | 938.66 | 19.83 | 944 | 984.61 | 20.31 |

Table 3. **CONSTANTS FOR $\bar{p}'$ (SHORT FORMULA), AN APPROXIMATION OF THE UPPER BINOMIAL CONFIDENCE LIMIT $\bar{p}$; $c/n \leqslant .1$, $n \geqslant 40$, $c \leqslant 1,000$** (*Continued*)

| c | $\bar{m}$ | $\bar{k}$ | c | $\bar{m}$ | $\bar{k}$ | c | $\bar{m}$ | $\bar{k}$ |
|---|---|---|---|---|---|---|---|---|
| 945 | 985.63 | 20.32 | 965 | 1006.05 | 20.52 | 985 | 1026.46 | 20.73 |
| 946 | 986.65 | 20.33 | 966 | 1007.07 | 20.53 | 986 | 1027.48 | 20.74 |
| 947 | 987.67 | 20.34 | 967 | 1008.09 | 20.54 | 987 | 1028.50 | 20.75 |
| 948 | 988.69 | 20.35 | 968 | 1009.11 | 20.55 | 988 | 1029.52 | 20.76 |
| 949 | 989.72 | 20.36 | 969 | 1010.13 | 20.56 | 989 | 1030.54 | 20.77 |
| 950 | 990.74 | 20.37 | 970 | 1011.15 | 20.57 | 990 | 1031.56 | 20.78 |
| 951 | 991.76 | 20.38 | 971 | 1012.17 | 20.59 | 991 | 1032.58 | 20.79 |
| 952 | 992.78 | 20.39 | 972 | 1013.19 | 20.60 | 992 | 1033.60 | 20.80 |
| 953 | 993.80 | 20.40 | 973 | 1014.21 | 20.61 | 993 | 1034.62 | 20.81 |
| 954 | 994.82 | 20.41 | 974 | 1015.23 | 20.62 | 994 | 1035.64 | 20.82 |
| 955 | 995.84 | 20.42 | 975 | 1016.25 | 20.63 | 995 | 1036.66 | 20.83 |
| 956 | 996.86 | 20.43 | 976 | 1017.27 | 20.64 | 996 | 1037.68 | 20.84 |
| 957 | 997.88 | 20.44 | 977 | 1018.29 | 20.65 | 997 | 1038.70 | 20.85 |
| 958 | 998.90 | 20.45 | 978 | 1019.31 | 20.66 | 998 | 1039.72 | 20.86 |
| 959 | 999.92 | 20.46 | 979 | 1020.33 | 20.67 | 999 | 1040.74 | 20.87 |
| 960 | 1000.94 | 20.47 | 980 | 1021.35 | 20.68 | 1000 | 1041.76 | 20.88 |
| 961 | 1001.96 | 20.48 | 981 | 1022.38 | 20.69 | | | |
| 962 | 1002.98 | 20.49 | 982 | 1023.40 | 20.70 | | | |
| 963 | 1004.01 | 20.50 | 983 | 1024.42 | 20.71 | | | |
| 964 | 1005.03 | 20.51 | 984 | 1025.44 | 20.72 | | | |

Table 3.  CONSTANTS FOR $\overline{p}'$ (SHORT FORMULA), AN APPROXIMATION OF THE UPPER BINOMIAL CONFIDENCE LIMIT $\overline{p}$; c/n $\leqslant$ .1, n $\geqslant$ 40, c $\leqslant$ 1,000 (*Continued*)

| c | $\overline{m}$ | $\overline{k}$ | c | $\overline{m}$ | $\overline{k}$ | c | $\overline{m}$ | $\overline{k}$ |
|---|---|---|---|---|---|---|---|---|
| 0 | 1.6094 | 0.803 | 45 | 51.590 | 3.36 | 90 | 98.916 | 4.47 |
| 1 | 2.9943 | 1.010 | 46 | 52.652 | 3.41 | 91 | 99.960 | 4.49 |
| 2 | 4.2790 | 1.154 | 47 | 53.713 | 3.44 | 92 | 101.00 | 4.51 |
| 3 | 5.5151 | 1.284 | 48 | 54.774 | 3.48 | 93 | 102.05 | 4.53 |
| 4 | 6.7210 | 1.396 | 49 | 55.834 | 3.50 | 94 | 103.09 | 4.55 |
| 5 | 7.9060 | 1.487 | 50 | 56.893 | 3.52 | 95 | 104.13 | 4.57 |
| 6 | 9.0754 | 1.573 | 51 | 57.952 | 3.49 | 96 | 105.18 | 4.60 |
| 7 | 10.233 | 1.655 | 52 | 59.010 | 3.51 | 97 | 106.22 | 4.62 |
| 8 | 11.380 | 1.732 | 53 | 60.068 | 3.54 | 98 | 107.26 | 4.64 |
| 9 | 12.519 | 1.796 | 54 | 61.125 | 3.57 | 99 | 108.30 | 4.66 |
| 10 | 13.651 | 1.87 | 55 | 62.182 | 3.60 | 100 | 109.35 | 4.68 |
| 11 | 14.777 | 1.93 | 56 | 63.238 | 3.63 | 101 | 110.39 | 4.70 |
| 12 | 15.897 | 2.00 | 57 | 64.294 | 3.66 | 102 | 111.43 | 4.72 |
| 13 | 17.013 | 2.05 | 58 | 65.349 | 3.68 | 103 | 112.47 | 4.74 |
| 14 | 18.125 | 2.11 | 59 | 66.403 | 3.71 | 104 | 113.51 | 4.76 |
| 15 | 19.233 | 2.15 | 60 | 67.458 | 3.74 | 105 | 114.55 | 4.78 |
| 16 | 20.338 | 2.21 | 61 | 68.511 | 3.76 | 106 | 115.59 | 4.80 |
| 17 | 21.439 | 2.27 | 62 | 69.565 | 3.79 | 107 | 116.64 | 4.82 |
| 18 | 22.538 | 2.31 | 63 | 70.618 | 3.82 | 108 | 117.68 | 4.84 |
| 19 | 23.634 | 2.36 | 64 | 71.670 | 3.84 | 109 | 118.72 | 4.86 |
| 20 | 24.728 | 2.40 | 65 | 72.722 | 3.87 | 110 | 119.76 | 4.88 |
| 21 | 25.819 | 2.46 | 66 | 73.774 | 3.90 | 111 | 120.80 | 4.90 |
| 22 | 26.909 | 2.50 | 67 | 74.826 | 3.92 | 112 | 121.84 | 4.92 |
| 23 | 27.996 | 2.55 | 68 | 75.877 | 3.95 | 113 | 122.88 | 4.94 |
| 24 | 29.082 | 2.59 | 69 | 76.927 | 3.97 | 114 | 123.91 | 4.96 |
| 25 | 30.166 | 2.65 | 70 | 77.977 | 4.00 | 115 | 124.95 | 4.98 |
| 26 | 31.248 | 2.69 | 71 | 79.027 | 4.02 | 116 | 125.99 | 5.00 |
| 27 | 32.329 | 2.73 | 72 | 80.077 | 4.05 | 117 | 127.03 | 5.02 |
| 28 | 33.408 | 2.78 | 73 | 81.126 | 4.07 | 118 | 128.07 | 5.04 |
| 29 | 34.486 | 2.82 | 74 | 82.175 | 4.10 | 119 | 129.11 | 5.06 |
| 30 | 35.563 | 2.84 | 75 | 83.223 | 4.12 | 120 | 130.15 | 5.08 |
| 31 | 36.638 | 2.89 | 76 | 84.272 | 4.14 | 121 | 131.19 | 5.10 |
| 32 | 37.712 | 2.91 | 77 | 85.320 | 4.17 | 122 | 132.22 | 5.12 |
| 33 | 38.785 | 2.95 | 78 | 86.367 | 4.19 | 123 | 133.26 | 5.14 |
| 34 | 39.857 | 2.98 | 79 | 87.414 | 4.22 | 124 | 134.30 | 5.16 |
| 35 | 40.928 | 3.04 | 80 | 88.461 | 4.24 | 125 | 135.34 | 5.17 |
| 36 | 41.998 | 3.06 | 81 | 89.508 | 4.26 | 126 | 136.37 | 5.19 |
| 37 | 43.067 | 3.09 | 82 | 90.555 | 4.29 | 127 | 137.41 | 5.21 |
| 38 | 44.135 | 3.12 | 83 | 91.601 | 4.31 | 128 | 138.45 | 5.23 |
| 39 | 45.203 | 3.16 | 84 | 92.647 | 4.33 | 129 | 139.49 | 5.25 |
| 40 | 46.269 | 3.20 | 85 | 93.692 | 4.35 | 130 | 140.52 | 5.27 |
| 41 | 47.335 | 3.24 | 86 | 94.738 | 4.38 | 131 | 141.56 | 5.29 |
| 42 | 48.400 | 3.27 | 87 | 95.783 | 4.40 | 132 | 142.60 | 5.30 |
| 43 | 49.464 | 3.29 | 88 | 96.827 | 4.42 | 133 | 143.63 | 5.32 |
| 44 | 50.527 | 3.36 | 89 | 97.872 | 4.44 | 134 | 144.67 | 5.34 |

Table 3.  CONSTANTS FOR $\overline{p}'$ (SHORT FORMULA), AN APPROXIMATION OF THE UPPER BINOMIAL CONFIDENCE LIMIT $\overline{p}$; $c/n \leqslant .1$, $n \geqslant 40$, $c \leqslant 1,000$ (*Continued*)

| c | $\overline{m}$ | $\overline{k}$ | c | $\overline{m}$ | $\overline{k}$ | c | $\overline{m}$ | $\overline{k}$ |
|---|---|---|---|---|---|---|---|---|
| 135 | 145.71 | 5.36 | 180 | 192.23 | 6.11 | 225 | 238.56 | 6.78 |
| 136 | 146.74 | 5.38 | 181 | 193.26 | 6.13 | 226 | 239.58 | 6.79 |
| 137 | 147.78 | 5.39 | 182 | 194.29 | 6.14 | 227 | 240.61 | 6.81 |
| 138 | 148.81 | 5.41 | 183 | 195.32 | 6.16 | 228 | 241.64 | 6.82 |
| 139 | 149.85 | 5.43 | 184 | 196.35 | 6.18 | 229 | 242.67 | 6.83 |
| 140 | 150.88 | 5.45 | 185 | 197.38 | 6.19 | 230 | 243.69 | 6.85 |
| 141 | 151.92 | 5.47 | 186 | 198.41 | 6.21 | 231 | 244.72 | 6.86 |
| 142 | 152.96 | 5.48 | 187 | 199.44 | 6.22 | 232 | 245.75 | 6.87 |
| 143 | 153.99 | 5.50 | 188 | 200.47 | 6.24 | 233 | 246.78 | 6.89 |
| 144 | 155.03 | 5.52 | 189 | 201.50 | 6.25 | 234 | 247.80 | 6.90 |
| 145 | 156.06 | 5.54 | 190 | 202.53 | 6.27 | 235 | 248.83 | 6.92 |
| 146 | 157.10 | 5.55 | 191 | 203.56 | 6.28 | 236 | 249.86 | 6.93 |
| 147 | 158.13 | 5.57 | 192 | 204.59 | 6.30 | 237 | 250.89 | 6.94 |
| 148 | 159.16 | 5.59 | 193 | 205.63 | 6.31 | 238 | 251.91 | 6.96 |
| 149 | 160.20 | 5.61 | 194 | 206.66 | 6.33 | 239 | 252.94 | 6.97 |
| 150 | 161.23 | 5.62 | 195 | 207.69 | 6.34 | 240 | 253.97 | 6.98 |
| 151 | 162.27 | 5.64 | 196 | 208.72 | 6.36 | 241 | 255.00 | 7.00 |
| 152 | 163.30 | 5.66 | 197 | 209.75 | 6.37 | 242 | 256.02 | 7.01 |
| 153 | 164.34 | 5.67 | 198 | 210.78 | 6.39 | 243 | 257.05 | 7.02 |
| 154 | 165.37 | 5.69 | 199 | 211.81 | 6.40 | 244 | 258.08 | 7.04 |
| 155 | 166.40 | 5.71 | 200 | 212.83 | 6.42 | 245 | 259.10 | 7.05 |
| 156 | 167.44 | 5.72 | 201 | 213.86 | 6.43 | 246 | 260.13 | 7.06 |
| 157 | 168.47 | 5.74 | 202 | 214.89 | 6.45 | 247 | 261.16 | 7.08 |
| 158 | 169.50 | 5.76 | 203 | 215.92 | 6.46 | 248 | 262.18 | 7.09 |
| 159 | 170.54 | 5.77 | 204 | 216.95 | 6.48 | 249 | 263.21 | 7.10 |
| 160 | 171.57 | 5.79 | 205 | 217.98 | 6.49 | 250 | 264.24 | 7.12 |
| 161 | 172.61 | 5.81 | 206 | 219.01 | 6.51 | 251 | 265.26 | 7.13 |
| 162 | 173.65 | 5.82 | 207 | 220.04 | 6.52 | 252 | 266.29 | 7.14 |
| 163 | 174.68 | 5.84 | 208 | 221.07 | 6.53 | 253 | 267.32 | 7.16 |
| 164 | 175.71 | 5.86 | 209 | 222.10 | 6.55 | 254 | 268.34 | 7.17 |
| 165 | 176.75 | 5.87 | 210 | 223.13 | 6.56 | 255 | 269.37 | 7.18 |
| 166 | 177.78 | 5.89 | 211 | 224.16 | 6.58 | 256 | 270.39 | 7.20 |
| 167 | 178.81 | 5.91 | 212 | 225.19 | 6.59 | 257 | 271.42 | 7.21 |
| 168 | 179.84 | 5.92 | 213 | 226.21 | 6.61 | 258 | 272.45 | 7.22 |
| 169 | 180.88 | 5.94 | 214 | 227.24 | 6.62 | 259 | 273.47 | 7.24 |
| 170 | 181.91 | 5.95 | 215 | 228.27 | 6.64 | 260 | 274.50 | 7.25 |
| 171 | 182.94 | 5.97 | 216 | 229.30 | 6.65 | 261 | 275.53 | 7.26 |
| 172 | 183.97 | 5.99 | 217 | 230.33 | 6.66 | 262 | 276.55 | 7.28 |
| 173 | 185.00 | 6.00 | 218 | 231.36 | 6.68 | 263 | 277.58 | 7.29 |
| 174 | 186.04 | 6.02 | 219 | 232.39 | 6.69 | 264 | 278.60 | 7.30 |
| 175 | 187.07 | 6.03 | 220 | 233.41 | 6.71 | 265 | 279.63 | 7.31 |
| 176 | 188.10 | 6.05 | 221 | 234.44 | 6.72 | 266 | 280.65 | 7.33 |
| 177 | 189.13 | 6.07 | 222 | 235.47 | 6.74 | 267 | 281.68 | 7.34 |
| 178 | 190.16 | 6.08 | 223 | 236.50 | 6.75 | 268 | 282.71 | 7.35 |
| 179 | 191.19 | 6.10 | 224 | 237.53 | 6.76 | 269 | 283.73 | 7.37 |

**Table 3.   CONSTANTS FOR $\bar{p}'$ (SHORT FORMULA), AN APPROXIMATION OF THE UPPER BINOMIAL CONFIDENCE LIMIT $\bar{p}$; $c/n \leqslant .1$, $n \geqslant 40$, $c \leqslant 1,000$ (*Continued*)**

| c | $\bar{m}$ | $\bar{k}$ | c | $\bar{m}$ | $\bar{k}$ | c | $\bar{m}$ | $\bar{k}$ |
|---|---|---|---|---|---|---|---|---|
| 270 | 284.76 | 7.38 | 315 | 330.86 | 7.93 | 360 | 376.89 | 8.45 |
| 271 | 285.78 | 7.39 | 316 | 331.89 | 7.94 | 361 | 377.92 | 8.46 |
| 272 | 286.81 | 7.40 | 317 | 332.91 | 7.96 | 362 | 378.94 | 8.47 |
| 273 | 287.83 | 7.42 | 318 | 333.93 | 7.97 | 363 | 379.96 | 8.48 |
| 274 | 288.86 | 7.43 | 319 | 334.96 | 7.98 | 364 | 380.98 | 8.49 |
| 275 | 289.88 | 7.44 | 320 | 335.98 | 7.99 | 365 | 382.00 | 8.50 |
| 276 | 290.91 | 7.46 | 321 | 337.01 | 8.00 | 366 | 383.03 | 8.51 |
| 277 | 291.94 | 7.47 | 322 | 338.03 | 8.01 | 367 | 384.05 | 8.52 |
| 278 | 292.96 | 7.48 | 323 | 339.05 | 8.03 | 368 | 385.07 | 8.53 |
| 279 | 293.99 | 7.49 | 324 | 340.08 | 8.04 | 369 | 386.09 | 8.55 |
| 280 | 295.01 | 7.51 | 325 | 341.10 | 8.05 | 370 | 387.11 | 8.56 |
| 281 | 296.04 | 7.52 | 326 | 342.12 | 8.06 | 371 | 388.14 | 8.57 |
| 282 | 297.06 | 7.53 | 327 | 343.15 | 8.07 | 372 | 389.16 | 8.58 |
| 283 | 298.09 | 7.54 | 328 | 344.17 | 8.08 | 373 | 390.18 | 8.59 |
| 284 | 299.11 | 7.56 | 329 | 345.19 | 8.10 | 374 | 391.20 | 8.60 |
| 285 | 300.14 | 7.57 | 330 | 346.21 | 8.11 | 375 | 392.22 | 8.61 |
| 286 | 301.16 | 7.58 | 331 | 347.24 | 8.12 | 376 | 393.24 | 8.62 |
| 287 | 302.19 | 7.59 | 332 | 348.26 | 8.13 | 377 | 394.27 | 8.63 |
| 288 | 303.21 | 7.61 | 333 | 349.28 | 8.14 | 378 | 395.29 | 8.64 |
| 289 | 304.24 | 7.62 | 334 | 350.31 | 8.15 | 379 | 396.31 | 8.65 |
| 290 | 305.26 | 7.63 | 335 | 351.33 | 8.16 | 380 | 397.33 | 8.67 |
| 291 | 306.28 | 7.64 | 336 | 352.35 | 8.18 | 381 | 398.35 | 8.68 |
| 292 | 307.31 | 7.65 | 337 | 353.38 | 8.19 | 382 | 399.37 | 8.69 |
| 293 | 308.33 | 7.67 | 338 | 354.40 | 8.20 | 383 | 400.40 | 8.70 |
| 294 | 309.36 | 7.68 | 339 | 355.42 | 8.21 | 384 | 401.42 | 8.71 |
| 295 | 310.38 | 7.69 | 340 | 356.44 | 8.22 | 385 | 402.44 | 8.72 |
| 296 | 311.41 | 7.70 | 341 | 357.47 | 8.23 | 386 | 403.46 | 8.73 |
| 297 | 312.43 | 7.72 | 342 | 358.49 | 8.24 | 387 | 404.48 | 8.74 |
| 298 | 313.46 | 7.73 | 343 | 359.51 | 8.26 | 388 | 405.50 | 8.75 |
| 299 | 314.48 | 7.74 | 344 | 360.54 | 8.27 | 389 | 406.52 | 8.76 |
| 300 | 315.50 | 7.75 | 345 | 361.56 | 8.28 | 390 | 407.54 | 8.77 |
| 301 | 316.53 | 7.76 | 346 | 362.58 | 8.29 | 391 | 408.57 | 8.78 |
| 302 | 317.55 | 7.78 | 347 | 363.60 | 8.30 | 392 | 409.59 | 8.79 |
| 303 | 318.58 | 7.79 | 348 | 364.63 | 8.31 | 393 | 410.61 | 8.80 |
| 304 | 319.60 | 7.80 | 349 | 365.65 | 8.32 | 394 | 411.63 | 8.81 |
| 305 | 320.63 | 7.81 | 350 | 366.67 | 8.34 | 395 | 412.65 | 8.83 |
| 306 | 321.65 | 7.82 | 351 | 367.69 | 8.35 | 396 | 413.67 | 8.84 |
| 307 | 322.67 | 7.84 | 352 | 368.72 | 8.36 | 397 | 414.69 | 8.85 |
| 308 | 323.70 | 7.85 | 353 | 369.74 | 8.37 | 398 | 415.71 | 8.86 |
| 309 | 324.72 | 7.86 | 354 | 370.76 | 8.38 | 399 | 416.74 | 8.87 |
| 310 | 325.74 | 7.87 | 355 | 371.78 | 8.39 | 400 | 417.76 | 8.88 |
| 311 | 326.77 | 7.88 | 356 | 372.80 | 8.40 | 401 | 418.78 | 8.89 |
| 312 | 327.79 | 7.90 | 357 | 373.83 | 8.41 | 402 | 419.80 | 8.90 |
| 313 | 328.82 | 7.91 | 358 | 374.85 | 8.42 | 403 | 420.82 | 8.91 |
| 314 | 329.84 | 7.92 | 359 | 375.87 | 8.44 | 404 | 421.84 | 8.92 |

Table 3.  CONSTANTS FOR $\overline{p}'$ (SHORT FORMULA), AN APPROXIMATION OF THE UPPER BINOMIAL CONFIDENCE LIMIT $\overline{p}$; $c/n \leqslant .1$, $n \geqslant 40$, $c \leqslant 1,000$ (*Continued*)

| c | $\overline{m}$ | $\overline{k}$ | c | $\overline{m}$ | $\overline{k}$ | c | $\overline{m}$ | $\overline{k}$ |
|---|---|---|---|---|---|---|---|---|
| 405 | 422.86 | 8.93 | 450 | 468.78 | 9.39 | 495 | 514.65 | 9.8 |
| 406 | 423.88 | 8.94 | 451 | 469.80 | 9.40 | 496 | 515.67 | 9.8 |
| 407 | 424.90 | 8.95 | 452 | 470.82 | 9.41 | 497 | 516.68 | 9.8 |
| 408 | 425.92 | 8.96 | 453 | 471.84 | 9.42 | 498 | 517.70 | 9.8 |
| 409 | 426.94 | 8.97 | 454 | 472.86 | 9.43 | 499 | 518.72 | 9.8 |
| 410 | 427.97 | 8.98 | 455 | 473.87 | 9.44 | 500 | 519.74 | 9.8 |
| 411 | 428.99 | 8.99 | 456 | 474.89 | 9.45 | 501 | 520.76 | 9.8 |
| 412 | 430.01 | 9.00 | 457 | 475.91 | 9.46 | 502 | 521.78 | 9.8 |
| 413 | 431.03 | 9.01 | 458 | 476.93 | 9.47 | 503 | 522.80 | 9.9 |
| 414 | 432.05 | 9.02 | 459 | 477.95 | 9.48 | 504 | 523.82 | 9.9 |
| 415 | 433.07 | 9.03 | 460 | 478.97 | 9.49 | 505 | 524.83 | 9.9 |
| 416 | 434.09 | 9.04 | 461 | 479.99 | 9.50 | 506 | 525.85 | 9.9 |
| 417 | 435.11 | 9.05 | 462 | 481.01 | 9.51 | 507 | 526.87 | 9.9 |
| 418 | 436.13 | 9.07 | 463 | 482.03 | 9.52 | 508 | 527.89 | 9.9 |
| 419 | 437.15 | 9.08 | 464 | 483.05 | 9.53 | 509 | 528.91 | 9.9 |
| 420 | 438.17 | 9.09 | 465 | 484.07 | 9.54 | 510 | 529.93 | 9.9 |
| 421 | 439.19 | 9.10 | 466 | 485.09 | 9.55 | 511 | 530.95 | 9.9 |
| 422 | 440.21 | 9.11 | 467 | 486.11 | 9.55 | 512 | 531.96 | 9.9 |
| 423 | 441.23 | 9.12 | 468 | 487.13 | 9.56 | 513 | 532.98 | 9.9 |
| 424 | 442.25 | 9.13 | 469 | 488.15 | 9.57 | 514 | 534.00 | 10.0 |
| 425 | 443.27 | 9.14 | 470 | 489.17 | 9.58 | 515 | 535.02 | 10.0 |
| 426 | 444.29 | 9.15 | 471 | 490.19 | 9.59 | 516 | 536.04 | 10.0 |
| 427 | 445.31 | 9.16 | 472 | 491.21 | 9.60 | 517 | 537.06 | 10.0 |
| 428 | 446.33 | 9.17 | 473 | 492.23 | 9.61 | 518 | 538.08 | 10.0 |
| 429 | 447.35 | 9.18 | 474 | 493.25 | 9.62 | 519 | 539.09 | 10.0 |
| 430 | 448.38 | 9.19 | 475 | 494.26 | 9.63 | 520 | 540.11 | 10.0 |
| 431 | 449.40 | 9.20 | 476 | 495.28 | 9.64 | 521 | 541.13 | 10.0 |
| 432 | 450.42 | 9.21 | 477 | 496.30 | 9.65 | 522 | 542.15 | 10.0 |
| 433 | 451.44 | 9.22 | 478 | 497.32 | 9.66 | 523 | 543.17 | 10.0 |
| 434 | 452.46 | 9.23 | 479 | 498.34 | 9.67 | 524 | 544.19 | 10.0 |
| 435 | 453.48 | 9.24 | 480 | 499.36 | 9.68 | 525 | 545.20 | 10.1 |
| 436 | 454.50 | 9.25 | 481 | 500.38 | 9.69 | 526 | 546.22 | 10.1 |
| 437 | 455.52 | 9.26 | 482 | 501.40 | 9.70 | 527 | 547.24 | 10.1 |
| 438 | 456.54 | 9.27 | 483 | 502.42 | 9.71 | 528 | 548.26 | 10.1 |
| 439 | 457.56 | 9.28 | 484 | 503.44 | 9.72 | 529 | 549.28 | 10.1 |
| 440 | 458.58 | 9.29 | 485 | 504.46 | 9.73 | 530 | 550.30 | 10.1 |
| 441 | 459.60 | 9.30 | 486 | 505.48 | 9.74 | 531 | 551.31 | 10.1 |
| 442 | 460.62 | 9.31 | 487 | 506.49 | 9.75 | 532 | 552.33 | 10.1 |
| 443 | 461.64 | 9.32 | 488 | 507.51 | 9.76 | 533 | 553.35 | 10.1 |
| 444 | 462.66 | 9.33 | 489 | 508.53 | 9.77 | 534 | 554.37 | 10.1 |
| 445 | 463.68 | 9.34 | 490 | 509.55 | 9.78 | 535 | 555.39 | 10.1 |
| 446 | 464.70 | 9.35 | 491 | 510.57 | 9.79 | 536 | 556.41 | 10.2 |
| 447 | 465.72 | 9.36 | 492 | 511.59 | 9.79 | 537 | 557.42 | 10.2 |
| 448 | 466.74 | 9.37 | 493 | 512.61 | 9.80 | 538 | 558.44 | 10.2 |
| 449 | 467.76 | 9.38 | 494 | 513.63 | 9.81 | 539 | 559.46 | 10.2 |

Table 3. **CONSTANTS FOR** $\overline{p}'$ **(SHORT FORMULA), AN APPROXIMATION OF** Section 6
**THE UPPER BINOMIAL CONFIDENCE LIMIT** $\overline{p}$; $c/n \leqslant .1$, $n \geqslant 40$, $\gamma_1 = 80\%$
$c \leqslant 1,000$ (*Continued*) $\gamma_2 = 60\%$

| c | $\overline{m}$ | $\overline{k}$ | c | $\overline{m}$ | $\overline{k}$ | c | $\overline{m}$ | $\overline{k}$ |
|---|---|---|---|---|---|---|---|---|
| 540 | 560.48 | 10.24 | 585 | 606.28 | 10.64 | 630 | 652.04 | 11.02 |
| 541 | 561.50 | 10.25 | 586 | 607.29 | 10.65 | 631 | 653.06 | 11.03 |
| 542 | 562.51 | 10.26 | 587 | 608.31 | 10.66 | 632 | 654.08 | 11.04 |
| 543 | 563.53 | 10.27 | 588 | 609.33 | 10.66 | 633 | 655.09 | 11.05 |
| 544 | 564.55 | 10.28 | 589 | 610.35 | 10.67 | 634 | 656.11 | 11.06 |
| 545 | 565.57 | 10.28 | 590 | 611.36 | 10.68 | 635 | 657.13 | 11.06 |
| 546 | 566.59 | 10.29 | 591 | 612.38 | 10.69 | 636 | 658.14 | 11.07 |
| 547 | 567.60 | 10.30 | 592 | 613.40 | 10.70 | 637 | 659.16 | 11.08 |
| 548 | 568.62 | 10.31 | 593 | 614.41 | 10.71 | 638 | 660.18 | 11.09 |
| 549 | 569.64 | 10.32 | 594 | 615.43 | 10.72 | 639 | 661.19 | 11.10 |
| 550 | 570.66 | 10.33 | 595 | 616.45 | 10.72 | 640 | 662.21 | 11.11 |
| 551 | 571.68 | 10.34 | 596 | 617.47 | 10.73 | 641 | 663.23 | 11.11 |
| 552 | 572.69 | 10.35 | 597 | 618.48 | 10.74 | 642 | 664.24 | 11.12 |
| 553 | 573.71 | 10.36 | 598 | 619.50 | 10.75 | 643 | 665.26 | 11.13 |
| 554 | 574.73 | 10.36 | 599 | 620.52 | 10.76 | 644 | 666.28 | 11.14 |
| 555 | 575.75 | 10.37 | 600 | 621.54 | 10.77 | 645 | 667.29 | 11.15 |
| 556 | 576.77 | 10.38 | 601 | 622.55 | 10.78 | 646 | 668.31 | 11.16 |
| 557 | 577.78 | 10.39 | 602 | 623.57 | 10.78 | 647 | 669.33 | 11.16 |
| 558 | 578.80 | 10.40 | 603 | 624.59 | 10.79 | 648 | 670.34 | 11.17 |
| 559 | 579.82 | 10.41 | 604 | 625.60 | 10.80 | 649 | 671.36 | 11.18 |
| 560 | 580.84 | 10.42 | 605 | 626.62 | 10.81 | 650 | 672.38 | 11.19 |
| 561 | 581.85 | 10.43 | 606 | 627.64 | 10.82 | 651 | 673.39 | 11.20 |
| 562 | 582.87 | 10.44 | 607 | 628.66 | 10.83 | 652 | 674.41 | 11.20 |
| 563 | 583.89 | 10.45 | 608 | 629.67 | 10.84 | 653 | 675.43 | 11.21 |
| 564 | 584.91 | 10.45 | 609 | 630.69 | 10.84 | 654 | 676.44 | 11.22 |
| 565 | 585.93 | 10.46 | 610 | 631.71 | 10.85 | 655 | 677.46 | 11.23 |
| 566 | 586.94 | 10.47 | 611 | 632.72 | 10.86 | 656 | 678.48 | 11.24 |
| 567 | 587.96 | 10.48 | 612 | 633.74 | 10.87 | 657 | 679.49 | 11.25 |
| 568 | 588.98 | 10.49 | 613 | 634.76 | 10.88 | 658 | 680.51 | 11.25 |
| 569 | 590.00 | 10.50 | 614 | 635.77 | 10.89 | 659 | 681.52 | 11.26 |
| 570 | 591.01 | 10.51 | 615 | 636.79 | 10.90 | 660 | 682.54 | 11.27 |
| 571 | 592.03 | 10.52 | 616 | 637.81 | 10.90 | 661 | 683.56 | 11.28 |
| 572 | 593.05 | 10.52 | 617 | 638.83 | 10.91 | 662 | 684.57 | 11.29 |
| 573 | 594.07 | 10.53 | 618 | 639.84 | 10.92 | 663 | 685.59 | 11.29 |
| 574 | 595.08 | 10.54 | 619 | 640.86 | 10.93 | 664 | 686.61 | 11.30 |
| 575 | 596.10 | 10.55 | 620 | 641.88 | 10.94 | 665 | 687.62 | 11.31 |
| 576 | 597.12 | 10.56 | 621 | 642.89 | 10.95 | 666 | 688.64 | 11.32 |
| 577 | 598.14 | 10.57 | 622 | 643.91 | 10.95 | 667 | 689.66 | 11.33 |
| 578 | 599.15 | 10.58 | 623 | 644.93 | 10.96 | 668 | 690.67 | 11.34 |
| 579 | 600.17 | 10.59 | 624 | 645.94 | 10.97 | 669 | 691.69 | 11.34 |
| 580 | 601.19 | 10.59 | 625 | 646.96 | 10.98 | 670 | 692.70 | 11.35 |
| 581 | 602.21 | 10.60 | 626 | 647.98 | 10.99 | 671 | 693.72 | 11.36 |
| 582 | 603.22 | 10.61 | 627 | 648.99 | 11.00 | 672 | 694.74 | 11.37 |
| 583 | 604.24 | 10.62 | 628 | 650.01 | 11.01 | 673 | 695.75 | 11.38 |
| 584 | 605.26 | 10.63 | 629 | 651.03 | 11.01 | 674 | 696.77 | 11.38 |

Table 3.  CONSTANTS FOR $\overline{p}'$ (SHORT FORMULA), AN APPROXIMATION OF THE UPPER BINOMIAL CONFIDENCE LIMIT $\overline{p}$; $c/n \leqslant .1$, $n \geqslant 40$, $c \leqslant 1,000$ (*Continued*)

| c | $\overline{m}$ | $\overline{k}$ | c | $\overline{m}$ | $\overline{k}$ | c | $\overline{m}$ | $\overline{k}$ |
|---|---|---|---|---|---|---|---|---|
| 675 | 697.78 | 11.39 | 720 | 743.50 | 11.75 | 765 | 789.20 | 12.10 |
| 676 | 698.80 | 11.40 | 721 | 744.52 | 11.76 | 766 | 790.21 | 12.11 |
| 677 | 699.82 | 11.41 | 722 | 745.53 | 11.77 | 767 | 791.23 | 12.11 |
| 678 | 700.83 | 11.42 | 723 | 746.55 | 11.77 | 768 | 792.24 | 12.12 |
| 679 | 701.85 | 11.42 | 724 | 747.56 | 11.78 | 769 | 793.26 | 12.13 |
| 680 | 702.87 | 11.43 | 725 | 748.58 | 11.79 | 770 | 794.27 | 12.14 |
| 681 | 703.88 | 11.44 | 726 | 749.60 | 11.80 | 771 | 795.29 | 12.14 |
| 682 | 704.90 | 11.45 | 727 | 750.61 | 11.81 | 772 | 796.30 | 12.15 |
| 683 | 705.91 | 11.46 | 728 | 751.63 | 11.81 | 773 | 797.32 | 12.16 |
| 684 | 706.93 | 11.47 | 729 | 752.64 | 11.82 | 774 | 798.33 | 12.17 |
| 685 | 707.95 | 11.47 | 730 | 753.66 | 11.83 | 775 | 799.35 | 12.17 |
| 686 | 708.96 | 11.48 | 731 | 754.67 | 11.84 | 776 | 800.36 | 12.18 |
| 687 | 709.98 | 11.49 | 732 | 755.69 | 11.84 | 777 | 801.38 | 12.19 |
| 688 | 710.99 | 11.50 | 733 | 756.70 | 11.85 | 778 | 802.39 | 12.20 |
| 689 | 712.01 | 11.51 | 734 | 757.72 | 11.86 | 779 | 803.41 | 12.20 |
| 690 | 713.03 | 11.51 | 735 | 758.74 | 11.87 | 780 | 804.42 | 12.21 |
| 691 | 714.04 | 11.52 | 736 | 759.75 | 11.88 | 781 | 805.44 | 12.22 |
| 692 | 715.06 | 11.53 | 737 | 760.77 | 11.88 | 782 | 806.45 | 12.23 |
| 693 | 716.07 | 11.54 | 738 | 761.78 | 11.89 | 783 | 807.47 | 12.23 |
| 694 | 717.09 | 11.55 | 739 | 762.80 | 11.90 | 784 | 808.48 | 12.24 |
| 695 | 718.11 | 11.55 | 740 | 763.81 | 11.91 | 785 | 809.50 | 12.25 |
| 696 | 719.12 | 11.56 | 741 | 764.83 | 11.91 | 786 | 810.51 | 12.26 |
| 697 | 720.14 | 11.57 | 742 | 765.84 | 11.92 | 787 | 811.53 | 12.26 |
| 698 | 721.15 | 11.58 | 743 | 766.86 | 11.93 | 788 | 812.54 | 12.27 |
| 699 | 722.17 | 11.58 | 744 | 767.87 | 11.94 | 789 | 813.56 | 12.28 |
| 700 | 723.19 | 11.59 | 745 | 768.89 | 11.94 | 790 | 814.57 | 12.29 |
| 701 | 724.20 | 11.60 | 746 | 769.91 | 11.95 | 791 | 815.59 | 12.29 |
| 702 | 725.22 | 11.61 | 747 | 770.92 | 11.96 | 792 | 816.60 | 12.30 |
| 703 | 726.23 | 11.62 | 748 | 771.94 | 11.97 | 793 | 817.62 | 12.31 |
| 704 | 727.25 | 11.62 | 749 | 772.95 | 11.98 | 794 | 818.63 | 12.32 |
| 705 | 728.27 | 11.63 | 750 | 773.97 | 11.98 | 795 | 819.65 | 12.32 |
| 706 | 729.28 | 11.64 | 751 | 774.98 | 11.99 | 796 | 820.66 | 12.33 |
| 707 | 730.30 | 11.65 | 752 | 776.00 | 12.00 | 797 | 821.68 | 12.34 |
| 708 | 731.31 | 11.66 | 753 | 777.01 | 12.01 | 798 | 822.69 | 12.35 |
| 709 | 732.33 | 11.66 | 754 | 778.03 | 12.01 | 799 | 823.71 | 12.35 |
| 710 | 733.34 | 11.67 | 755 | 779.04 | 12.02 | 800 | 824.72 | 12.36 |
| 711 | 734.36 | 11.68 | 756 | 780.06 | 12.03 | 801 | 825.74 | 12.37 |
| 712 | 735.38 | 11.69 | 757 | 781.07 | 12.04 | 802 | 826.75 | 12.38 |
| 713 | 736.39 | 11.70 | 758 | 782.09 | 12.04 | 803 | 827.77 | 12.38 |
| 714 | 737.41 | 11.70 | 759 | 783.10 | 12.05 | 804 | 828.78 | 12.39 |
| 715 | 738.42 | 11.71 | 760 | 784.12 | 12.06 | 805 | 829.80 | 12.40 |
| 716 | 739.44 | 11.72 | 761 | 785.14 | 12.07 | 806 | 830.81 | 12.41 |
| 717 | 740.45 | 11.73 | 762 | 786.15 | 12.08 | 807 | 831.83 | 12.41 |
| 718 | 741.47 | 11.74 | 763 | 787.17 | 12.08 | 808 | 832.84 | 12.42 |
| 719 | 742.49 | 11.74 | 764 | 788.18 | 12.09 | 809 | 833.86 | 12.43 |

Table 3.   CONSTANTS FOR $\overline{p}'$ (SHORT FORMULA), AN APPROXIMATION OF
THE UPPER BINOMIAL CONFIDENCE LIMIT $\overline{p}$; $c/n \leqslant .1$, $n \geqslant 40$,
$c \leqslant 1{,}000$ (*Continued*)

Section 6
$\gamma_1 = 80\%$
$\gamma_2 = 60\%$

| c | $\overline{m}$ | $\overline{k}$ | c | $\overline{m}$ | $\overline{k}$ | c | $\overline{m}$ | $\overline{k}$ |
|---|---|---|---|---|---|---|---|---|
| 810 | 834.87 | 12.44 | 855 | 880.53 | 12.76 | 900 | 926.17 | 13.08 |
| 811 | 835.89 | 12.44 | 856 | 881.54 | 12.77 | 901 | 927.18 | 13.09 |
| 812 | 836.90 | 12.45 | 857 | 882.56 | 12.78 | 902 | 928.19 | 13.10 |
| 813 | 837.91 | 12.46 | 858 | 883.57 | 12.78 | 903 | 929.21 | 13.10 |
| 814 | 838.93 | 12.46 | 859 | 884.58 | 12.79 | 904 | 930.22 | 13.11 |
| 815 | 839.94 | 12.47 | 860 | 885.60 | 12.80 | 905 | 931.24 | 13.12 |
| 816 | 840.96 | 12.48 | 861 | 886.61 | 12.81 | 906 | 932.25 | 13.12 |
| 817 | 841.97 | 12.49 | 862 | 887.63 | 12.81 | 907 | 933.26 | 13.13 |
| 818 | 842.99 | 12.49 | 863 | 888.64 | 12.82 | 908 | 934.28 | 13.14 |
| 819 | 844.00 | 12.50 | 864 | 889.66 | 12.83 | 909 | 935.29 | 13.15 |
| 820 | 845.02 | 12.51 | 865 | 890.67 | 12.83 | 910 | 936.31 | 13.15 |
| 821 | 846.03 | 12.52 | 866 | 891.68 | 12.84 | 911 | 937.32 | 13.16 |
| 822 | 847.05 | 12.52 | 867 | 892.70 | 12.85 | 912 | 938.33 | 13.17 |
| 823 | 848.06 | 12.53 | 868 | 893.71 | 12.86 | 913 | 939.35 | 13.17 |
| 824 | 849.08 | 12.54 | 869 | 894.73 | 12.86 | 914 | 940.36 | 13.18 |
| 825 | 850.09 | 12.55 | 870 | 895.74 | 12.87 | 915 | 941.37 | 13.19 |
| 826 | 851.11 | 12.55 | 871 | 896.76 | 12.88 | 916 | 942.39 | 13.19 |
| 827 | 852.12 | 12.56 | 872 | 897.77 | 12.88 | 917 | 943.40 | 13.20 |
| 828 | 853.13 | 12.57 | 873 | 898.78 | 12.89 | 918 | 944.42 | 13.21 |
| 829 | 854.15 | 12.57 | 874 | 899.80 | 12.90 | 919 | 945.43 | 13.22 |
| 830 | 855.16 | 12.58 | 875 | 900.81 | 12.91 | 920 | 946.44 | 13.22 |
| 831 | 856.18 | 12.59 | 876 | 901.83 | 12.91 | 921 | 947.46 | 13.23 |
| 832 | 857.19 | 12.60 | 877 | 902.84 | 12.92 | 922 | 948.47 | 13.24 |
| 833 | 858.21 | 12.60 | 878 | 903.85 | 12.93 | 923 | 949.49 | 13.24 |
| 834 | 859.22 | 12.61 | 879 | 904.87 | 12.93 | 924 | 950.50 | 13.25 |
| 835 | 860.24 | 12.62 | 880 | 905.88 | 12.94 | 925 | 951.51 | 13.26 |
| 836 | 861.25 | 12.63 | 881 | 906.90 | 12.95 | 926 | 952.53 | 13.26 |
| 837 | 862.27 | 12.63 | 882 | 907.91 | 12.96 | 927 | 953.54 | 13.27 |
| 838 | 863.28 | 12.64 | 883 | 908.93 | 12.96 | 928 | 954.55 | 13.28 |
| 839 | 864.30 | 12.65 | 884 | 909.94 | 12.97 | 929 | 955.57 | 13.28 |
| 840 | 865.31 | 12.65 | 885 | 910.95 | 12.98 | 930 | 956.58 | 13.29 |
| 841 | 866.32 | 12.66 | 886 | 911.97 | 12.98 | 931 | 957.60 | 13.30 |
| 842 | 867.34 | 12.67 | 887 | 912.98 | 12.99 | 932 | 958.61 | 13.30 |
| 843 | 868.35 | 12.68 | 888 | 914.00 | 13.00 | 933 | 959.62 | 13.31 |
| 844 | 869.37 | 12.68 | 889 | 915.01 | 13.01 | 934 | 960.64 | 13.32 |
| 845 | 870.38 | 12.69 | 890 | 916.02 | 13.01 | 935 | 961.65 | 13.33 |
| 846 | 871.40 | 12.70 | 891 | 917.04 | 13.02 | 936 | 962.67 | 13.33 |
| 847 | 872.41 | 12.71 | 892 | 918.05 | 13.03 | 937 | 963.68 | 13.34 |
| 848 | 873.43 | 12.71 | 893 | 919.07 | 13.03 | 938 | 964.69 | 13.35 |
| 849 | 874.44 | 12.72 | 894 | 920.08 | 13.04 | 939 | 965.71 | 13.35 |
| 850 | 875.45 | 12.73 | 895 | 921.10 | 13.05 | 940 | 966.72 | 13.36 |
| 851 | 876.47 | 12.73 | 896 | 922.11 | 13.05 | 941 | 967.73 | 13.37 |
| 852 | 877.48 | 12.74 | 897 | 923.12 | 13.06 | 942 | 968.75 | 13.37 |
| 853 | 878.50 | 12.75 | 898 | 924.14 | 13.07 | 943 | 969.76 | 13.38 |
| 854 | 879.51 | 12.76 | 899 | 925.15 | 13.08 | 944 | 970.77 | 13.39 |

Table 3.  **CONSTANTS FOR $\bar{p}'$ (SHORT FORMULA), AN APPROXIMATION OF THE UPPER BINOMIAL CONFIDENCE LIMIT $\bar{p}$; $c/n \leqslant .1$, $n \geqslant 40$, $c \leqslant 1,000$ (*Continued*)**

| c | $\bar{m}$ | $\bar{k}$ | c | $\bar{m}$ | $\bar{k}$ | c | $\bar{m}$ | $\bar{k}$ |
|---|---|---|---|---|---|---|---|---|
| 945 | 971.79 | 13.39 | 965 | 992.06 | 13.53 | 985 | 1012.33 | 13.6 |
| 946 | 972.80 | 13.40 | 966 | 993.07 | 13.54 | 986 | 1013.34 | 13.6 |
| 947 | 973.82 | 13.41 | 967 | 994.09 | 13.54 | 987 | 1014.36 | 13.6 |
| 948 | 974.83 | 13.41 | 968 | 995.10 | 13.55 | 988 | 1015.37 | 13.6 |
| 949 | 975.84 | 13.42 | 969 | 996.11 | 13.56 | 989 | 1016.38 | 13.6 |
| 950 | 976.86 | 13.43 | 970 | 997.13 | 13.56 | 990 | 1017.40 | 13.7 |
| 951 | 977.87 | 13.44 | 971 | 998.14 | 13.57 | 991 | 1018.41 | 13.7 |
| 952 | 978.88 | 13.44 | 972 | 999.16 | 13.58 | 992 | 1019.42 | 13.7 |
| 953 | 979.90 | 13.45 | 973 | 1000.17 | 13.58 | 993 | 1020.44 | 13.7 |
| 954 | 980.91 | 13.46 | 974 | 1001.18 | 13.59 | 994 | 1021.45 | 13.7 |
| 955 | 981.92 | 13.46 | 975 | 1002.20 | 13.60 | 995 | 1022.46 | 13.7 |
| 956 | 982.94 | 13.47 | 976 | 1003.21 | 13.60 | 996 | 1023.48 | 13.7 |
| 957 | 983.95 | 13.48 | 977 | 1004.22 | 13.61 | 997 | 1024.49 | 13.7 |
| 958 | 984.97 | 13.48 | 978 | 1005.24 | 13.62 | 998 | 1025.50 | 13.7 |
| 959 | 985.98 | 13.49 | 979 | 1006.25 | 13.62 | 999 | 1026.52 | 13.7 |
| 960 | 986.99 | 13.50 | 980 | 1007.26 | 13.63 | 1000 | 1027.53 | 13.7 |
| 961 | 988.01 | 13.50 | 981 | 1008.28 | 13.64 | | | |
| 962 | 989.02 | 13.51 | 982 | 1009.29 | 13.64 | | | |
| 963 | 990.03 | 13.52 | 983 | 1010.30 | 13.65 | | | |
| 964 | 991.05 | 13.52 | 984 | 1011.32 | 13.66 | | | |

# Table 4

CONSTANTS FOR $\underline{p}'$ (SHORT FORMULA), AN APPROXIMATION OF
THE LOWER BINOMIAL CONFIDENCE LIMIT $\underline{p}$; $c/n \leqslant .125$, $n \geqslant 16$,
$c \leqslant 1,000$

Approximation Formula: $\underline{p}' = \dfrac{\underline{m}}{n - \underline{k}}$

n     sample size.

c     number of events (items having a designated characteristic) in the sample.

$\underline{p}'$     approximation of the lower binomial confidence limit $\underline{p}$; $\underline{p}'$ has relative accuracy of at least .999.

$\underline{m}$     lower confidence limit for the parameter m of a Poisson distribution; each $\underline{m}$ is for a given c and a given $\gamma_1$ or $\gamma_2$.

$\underline{k}$     approximation constant for a given c and a given $\gamma_1$ or $\gamma_2$.

$\gamma_1$     confidence level for a one-sided (upper or lower) confidence limit.

$\gamma_2$     confidence level for two-sided (upper and lower) confidence limits. ($\gamma_2 = 2\gamma_1 - 100\%$.)

For $c > 1,000$, calculate $\underline{k}$ from the extension formula in Table 7–D; and calculate $\underline{m}$ from the extension formula in Table 7–E.

(**Note.** For the purpose of acceptance sampling in conjunction with Table 9, c and $\underline{m}$ in Table 4 may respectively be read as rejection number $\dot{c}_1$ and $\underline{m}_1$.)

Section 1   $\gamma_1 = 99.5\%$   $\gamma_2 = 99\%$
Section 2   $\gamma_1 = 99\%$     $\gamma_2 = 98\%$
Section 3   $\gamma_1 = 97.5\%$   $\gamma_2 = 95\%$
Section 4   $\gamma_1 = 95\%$     $\gamma_2 = 90\%$
Section 5   $\gamma_1 = 90\%$     $\gamma_2 = 80\%$
Section 6   $\gamma_1 = 80\%$     $\gamma_2 = 60\%$

Table 4. CONSTANTS FOR $p'$ (SHORT FORMULA), AN APPROXIMATION OF THE LOWER BINOMIAL CONFIDENCE LIMIT $\underline{p}$; $c/n \leqslant .125$, $n \geqslant 16$, $c \leqslant 1,000$

| c | m | k | c | m | k | c | m | k |
|---|---|---|---|---|---|---|---|---|
| 1 | .0050126 | -0.003 | 46 | 30.407 | 7.30 | 91 | 68.307 | 10.85 |
| 2 | .10350 | 0.448 | 47 | 31.219 | 7.39 | 92 | 69.172 | 10.91 |
| 3 | .33786 | 0.831 | 48 | 32.032 | 7.48 | 93 | 70.039 | 10.98 |
| 4 | .67221 | 1.164 | 49 | 32.847 | 7.58 | 94 | 70.905 | 11.05 |
| 5 | 1.0779 | 1.461 | 50 | 33.664 | 7.67 | 95 | 71.773 | 11.11 |
| 6 | 1.5369 | 1.732 | 51 | 34.483 | 7.76 | 96 | 72.641 | 11.18 |
| 7 | 2.0373 | 1.981 | 52 | 35.303 | 7.85 | 97 | 73.510 | 11.25 |
| 8 | 2.5711 | 2.214 | 53 | 36.125 | 7.94 | 98 | 74.380 | 11.31 |
| 9 | 3.1324 | 2.434 | 54 | 36.949 | 8.02 | 99 | 75.250 | 11.38 |
| 10 | 3.7169 | 2.64 | 55 | 37.775 | 8.11 | 100 | 76.121 | 11.44 |
| 11 | 4.3214 | 2.84 | 56 | 38.602 | 8.20 | 101 | 76.992 | 11.50 |
| 12 | 4.9431 | 3.03 | 57 | 39.431 | 8.28 | 102 | 77.864 | 11.57 |
| 13 | 5.5801 | 3.21 | 58 | 40.261 | 8.37 | 103 | 78.737 | 11.63 |
| 14 | 6.2307 | 3.38 | 59 | 41.093 | 8.45 | 104 | 79.611 | 11.69 |
| 15 | 6.8934 | 3.55 | 60 | 41.926 | 8.54 | 105 | 80.485 | 11.76 |
| 16 | 7.5670 | 3.72 | 61 | 42.760 | 8.62 | 106 | 81.359 | 11.82 |
| 17 | 8.2507 | 3.87 | 62 | 43.596 | 8.70 | 107 | 82.235 | 11.88 |
| 18 | 8.9434 | 4.03 | 63 | 44.433 | 8.78 | 108 | 83.110 | 11.95 |
| 19 | 9.6445 | 4.18 | 64 | 45.272 | 8.86 | 109 | 83.987 | 12.01 |
| 20 | 10.353 | 4.32 | 65 | 46.111 | 8.94 | 110 | 84.864 | 12.07 |
| 21 | 11.069 | 4.47 | 66 | 46.952 | 9.02 | 111 | 85.741 | 12.13 |
| 22 | 11.792 | 4.60 | 67 | 47.794 | 9.10 | 112 | 86.619 | 12.19 |
| 23 | 12.521 | 4.74 | 68 | 48.637 | 9.18 | 113 | 87.498 | 12.25 |
| 24 | 13.255 | 4.87 | 69 | 49.482 | 9.26 | 114 | 88.377 | 12.31 |
| 25 | 13.995 | 5.00 | 70 | 50.328 | 9.34 | 115 | 89.256 | 12.37 |
| 26 | 14.741 | 5.13 | 71 | 51.174 | 9.41 | 116 | 90.137 | 12.43 |
| 27 | 15.491 | 5.25 | 72 | 52.022 | 9.49 | 117 | 91.017 | 12.49 |
| 28 | 16.245 | 5.38 | 73 | 52.871 | 9.56 | 118 | 91.898 | 12.55 |
| 29 | 17.004 | 5.50 | 74 | 53.721 | 9.64 | 119 | 92.780 | 12.61 |
| 30 | 17.767 | 5.62 | 75 | 54.571 | 9.71 | 120 | 93.662 | 12.67 |
| 31 | 18.534 | 5.73 | 76 | 55.423 | 9.79 | 121 | 94.545 | 12.73 |
| 32 | 19.305 | 5.85 | 77 | 56.276 | 9.86 | 122 | 95.428 | 12.79 |
| 33 | 20.079 | 5.96 | 78 | 57.130 | 9.94 | 123 | 96.312 | 12.84 |
| 34 | 20.857 | 6.07 | 79 | 57.984 | 10.01 | 124 | 97.196 | 12.90 |
| 35 | 21.638 | 6.18 | 80 | 58.840 | 10.08 | 125 | 98.081 | 12.96 |
| 36 | 22.422 | 6.29 | 81 | 59.696 | 10.15 | 126 | 98.966 | 13.02 |
| 37 | 23.209 | 6.40 | 82 | 60.554 | 10.22 | 127 | 99.851 | 13.07 |
| 38 | 23.998 | 6.50 | 83 | 61.412 | 10.29 | 128 | 100.74 | 13.13 |
| 39 | 24.791 | 6.60 | 84 | 62.271 | 10.36 | 129 | 101.62 | 13.19 |
| 40 | 25.586 | 6.71 | 85 | 63.131 | 10.43 | 130 | 102.51 | 13.25 |
| 41 | 26.384 | 6.81 | 86 | 63.992 | 10.50 | 131 | 103.40 | 13.30 |
| 42 | 27.184 | 6.91 | 87 | 64.853 | 10.57 | 132 | 104.29 | 13.36 |
| 43 | 27.986 | 7.01 | 88 | 65.715 | 10.64 | 133 | 105.17 | 13.41 |
| 44 | 28.791 | 7.10 | 89 | 66.579 | 10.71 | 134 | 106.06 | 13.47 |
| 45 | 29.598 | 7.20 | 90 | 67.442 | 10.78 | 135 | 106.95 | 13.52 |

Table 4.  CONSTANTS FOR p′ (SHORT FORMULA), AN APPROXIMATION OF
THE LOWER BINOMIAL CONFIDENCE LIMIT p̲; c/n ≤ .125, n ≥ 16,
c ≤ 1,000 (*Continued*)

| c | m̲ | k̲ | c | m̲ | k̲ | c | m̲ | k̲ |
|---|---|---|---|---|---|---|---|---|
| 136 | 107.84 | 13.58 | 181 | 148.22 | 15.89 | 226 | 189.16 | 17.92 |
| 137 | 108.73 | 13.64 | 182 | 149.13 | 15.94 | 227 | 190.07 | 17.97 |
| 138 | 109.62 | 13.69 | 183 | 150.03 | 15.98 | 228 | 190.98 | 18.01 |
| 139 | 110.51 | 13.74 | 184 | 150.94 | 16.03 | 229 | 191.90 | 18.05 |
| 140 | 111.40 | 13.80 | 185 | 151.84 | 16.08 | 230 | 192.81 | 18.09 |
| 141 | 112.29 | 13.85 | 186 | 152.75 | 16.13 | 231 | 193.73 | 18.14 |
| 142 | 113.18 | 13.91 | 187 | 153.65 | 16.17 | 232 | 194.64 | 18.18 |
| 143 | 114.08 | 13.96 | 188 | 154.56 | 16.22 | 233 | 195.56 | 18.22 |
| 144 | 114.97 | 14.02 | 189 | 155.47 | 16.27 | 234 | 196.48 | 18.26 |
| 145 | 115.86 | 14.07 | 190 | 156.37 | 16.31 | 235 | 197.39 | 18.30 |
| 146 | 116.76 | 14.12 | 191 | 157.28 | 16.36 | 236 | 198.31 | 18.35 |
| 147 | 117.65 | 14.18 | 192 | 158.19 | 16.41 | 237 | 199.22 | 18.39 |
| 148 | 118.54 | 14.23 | 193 | 159.09 | 16.45 | 238 | 200.14 | 18.43 |
| 149 | 119.44 | 14.28 | 194 | 160.00 | 16.50 | 239 | 201.06 | 18.47 |
| 150 | 120.33 | 14.33 | 195 | 160.91 | 16.55 | 240 | 201.97 | 18.51 |
| 151 | 121.23 | 14.39 | 196 | 161.82 | 16.59 | 241 | 202.89 | 18.55 |
| 152 | 122.12 | 14.44 | 197 | 162.73 | 16.64 | 242 | 203.81 | 18.60 |
| 153 | 123.02 | 14.49 | 198 | 163.63 | 16.68 | 243 | 204.73 | 18.64 |
| 154 | 123.91 | 14.54 | 199 | 164.54 | 16.73 | 244 | 205.64 | 18.68 |
| 155 | 124.81 | 14.60 | 200 | 165.45 | 16.77 | 245 | 206.56 | 18.72 |
| 156 | 125.71 | 14.65 | 201 | 166.36 | 16.82 | 246 | 207.48 | 18.76 |
| 157 | 126.60 | 14.70 | 202 | 167.27 | 16.87 | 247 | 208.40 | 18.80 |
| 158 | 127.50 | 14.75 | 203 | 168.18 | 16.91 | 248 | 209.31 | 18.84 |
| 159 | 128.40 | 14.80 | 204 | 169.09 | 16.96 | 249 | 210.23 | 18.88 |
| 160 | 129.30 | 14.85 | 205 | 170.00 | 17.00 | 250 | 211.15 | 18.92 |
| 161 | 130.20 | 14.90 | 206 | 170.91 | 17.05 | 251 | 212.07 | 18.97 |
| 162 | 131.09 | 14.95 | 207 | 171.82 | 17.09 | 252 | 212.99 | 19.01 |
| 163 | 131.99 | 15.00 | 208 | 172.73 | 17.14 | 253 | 213.91 | 19.05 |
| 164 | 132.89 | 15.05 | 209 | 173.64 | 17.18 | 254 | 214.83 | 19.09 |
| 165 | 133.79 | 15.10 | 210 | 174.55 | 17.22 | 255 | 215.75 | 19.13 |
| 166 | 134.69 | 15.15 | 211 | 175.46 | 17.27 | 256 | 216.67 | 19.17 |
| 167 | 135.59 | 15.20 | 212 | 176.37 | 17.31 | 257 | 217.59 | 19.21 |
| 168 | 136.49 | 15.25 | 213 | 177.29 | 17.36 | 258 | 218.50 | 19.25 |
| 169 | 137.39 | 15.30 | 214 | 178.20 | 17.40 | 259 | 219.42 | 19.29 |
| 170 | 138.29 | 15.35 | 215 | 179.11 | 17.45 | 260 | 220.34 | 19.33 |
| 171 | 139.20 | 15.40 | 216 | 180.02 | 17.49 | 261 | 221.26 | 19.37 |
| 172 | 140.10 | 15.45 | 217 | 180.93 | 17.53 | 262 | 222.19 | 19.41 |
| 173 | 141.00 | 15.50 | 218 | 181.85 | 17.58 | 263 | 223.11 | 19.45 |
| 174 | 141.90 | 15.55 | 219 | 182.76 | 17.62 | 264 | 224.03 | 19.49 |
| 175 | 142.80 | 15.60 | 220 | 183.67 | 17.66 | 265 | 224.95 | 19.53 |
| 176 | 143.71 | 15.65 | 221 | 184.59 | 17.71 | 266 | 225.87 | 19.57 |
| 177 | 144.61 | 15.70 | 222 | 185.50 | 17.75 | 267 | 226.79 | 19.61 |
| 178 | 145.51 | 15.74 | 223 | 186.41 | 17.79 | 268 | 227.71 | 19.64 |
| 179 | 146.42 | 15.79 | 224 | 187.33 | 17.84 | 269 | 228.63 | 19.68 |
| 180 | 147.32 | 15.84 | 225 | 188.24 | 17.88 | 270 | 229.55 | 19.72 |

Table 4. CONSTANTS FOR p' (SHORT FORMULA), AN APPROXIMATION OF Section 1
THE LOWER BINOMIAL CONFIDENCE LIMIT p; c/n ≤ .125, n ≥ 16, $\gamma_1 = 99.5\%$
c ≤ 1,000 (*Continued*) $\gamma_2 = 99\%$

| c | m | k | c | m | k | c | m | k |
|---|---|---|---|---|---|---|---|---|
| 271 | 230.48 | 19.76 | 316 | 272.09 | 21.46 | 361 | 313.94 | 23.03 |
| 272 | 231.40 | 19.80 | 317 | 273.02 | 21.49 | 362 | 314.87 | 23.06 |
| 273 | 232.32 | 19.84 | 318 | 273.95 | 21.53 | 363 | 315.80 | 23.10 |
| 274 | 233.24 | 19.88 | 319 | 274.87 | 21.56 | 364 | 316.74 | 23.13 |
| 275 | 234.16 | 19.92 | 320 | 275.80 | 21.60 | 365 | 317.67 | 23.17 |
| 276 | 235.09 | 19.96 | 321 | 276.73 | 21.64 | 366 | 318.60 | 23.20 |
| 277 | 236.01 | 20.00 | 322 | 277.66 | 21.67 | 367 | 319.53 | 23.23 |
| 278 | 236.93 | 20.03 | 323 | 278.59 | 21.71 | 368 | 320.47 | 23.27 |
| 279 | 237.85 | 20.07 | 324 | 279.51 | 21.74 | 369 | 321.40 | 23.30 |
| 280 | 238.78 | 20.11 | 325 | 280.44 | 21.78 | 370 | 322.33 | 23.33 |
| 281 | 239.70 | 20.15 | 326 | 281.37 | 21.81 | 371 | 323.26 | 23.37 |
| 282 | 240.62 | 20.19 | 327 | 282.30 | 21.85 | 372 | 324.20 | 23.40 |
| 283 | 241.55 | 20.23 | 328 | 283.23 | 21.89 | 373 | 325.13 | 23.43 |
| 284 | 242.47 | 20.26 | 329 | 284.16 | 21.92 | 374 | 326.06 | 23.47 |
| 285 | 243.39 | 20.30 | 330 | 285.09 | 21.96 | 375 | 327.00 | 23.50 |
| 286 | 244.32 | 20.34 | 331 | 286.02 | 21.99 | 376 | 327.93 | 23.53 |
| 287 | 245.24 | 20.38 | 332 | 286.94 | 22.03 | 377 | 328.87 | 23.57 |
| 288 | 246.17 | 20.42 | 333 | 287.87 | 22.06 | 378 | 329.80 | 23.60 |
| 289 | 247.09 | 20.46 | 334 | 288.80 | 22.10 | 379 | 330.73 | 23.63 |
| 290 | 248.01 | 20.49 | 335 | 289.73 | 22.13 | 380 | 331.67 | 23.67 |
| 291 | 248.94 | 20.53 | 336 | 290.66 | 22.17 | 381 | 332.60 | 23.70 |
| 292 | 249.86 | 20.57 | 337 | 291.59 | 22.20 | 382 | 333.53 | 23.73 |
| 293 | 250.79 | 20.61 | 338 | 292.52 | 22.24 | 383 | 334.47 | 23.77 |
| 294 | 251.71 | 20.64 | 339 | 293.45 | 22.27 | 384 | 335.40 | 23.80 |
| 295 | 252.64 | 20.68 | 340 | 294.38 | 22.31 | 385 | 336.34 | 23.83 |
| 296 | 253.56 | 20.72 | 341 | 295.31 | 22.34 | 386 | 337.27 | 23.86 |
| 297 | 254.49 | 20.76 | 342 | 296.24 | 22.38 | 387 | 338.21 | 23.90 |
| 298 | 255.41 | 20.79 | 343 | 297.17 | 22.41 | 388 | 339.14 | 23.93 |
| 299 | 256.34 | 20.83 | 344 | 298.10 | 22.45 | 389 | 340.08 | 23.96 |
| 300 | 257.26 | 20.87 | 345 | 299.03 | 22.48 | 390 | 341.01 | 23.99 |
| 301 | 258.19 | 20.91 | 346 | 299.97 | 22.52 | 391 | 341.95 | 24.03 |
| 302 | 259.12 | 20.94 | 347 | 300.90 | 22.55 | 392 | 342.88 | 24.06 |
| 303 | 260.04 | 20.98 | 348 | 301.83 | 22.59 | 393 | 343.82 | 24.09 |
| 304 | 260.97 | 21.02 | 349 | 302.76 | 22.62 | 394 | 344.75 | 24.12 |
| 305 | 261.89 | 21.05 | 350 | 303.69 | 22.66 | 395 | 345.69 | 24.16 |
| 306 | 262.82 | 21.09 | 351 | 304.62 | 22.69 | 396 | 346.62 | 24.19 |
| 307 | 263.75 | 21.13 | 352 | 305.55 | 22.72 | 397 | 347.56 | 24.22 |
| 308 | 264.67 | 21.16 | 353 | 306.48 | 22.76 | 398 | 348.49 | 24.25 |
| 309 | 265.60 | 21.20 | 354 | 307.41 | 22.79 | 399 | 349.43 | 24.29 |
| 310 | 266.53 | 21.24 | 355 | 308.35 | 22.83 | 400 | 350.36 | 24.32 |
| 311 | 267.45 | 21.27 | 356 | 309.28 | 22.86 | 401 | 351.30 | 24.35 |
| 312 | 268.38 | 21.31 | 357 | 310.21 | 22.90 | 402 | 352.23 | 24.38 |
| 313 | 269.31 | 21.35 | 358 | 311.14 | 22.93 | 403 | 353.17 | 24.42 |
| 314 | 270.23 | 21.38 | 359 | 312.07 | 22.96 | 404 | 354.11 | 24.45 |
| 315 | 271.16 | 21.42 | 360 | 313.01 | 23.00 | 405 | 355.04 | 24.48 |

Table 4.  CONSTANTS FOR $\underline{p}'$ (SHORT FORMULA), AN APPROXIMATION OF THE LOWER BINOMIAL CONFIDENCE LIMIT $\underline{p}$; $c/n \leqslant .125$, $n \geqslant 16$, $c \leqslant 1,000$ (*Continued*)

| c | m | k | c | m | k | c | m | k |
|---|---|---|---|---|---|---|---|---|
| 406 | 355.98 | 24.51 | 451 | 398.18 | 25.91 | 496 | 440.51 | 27.24 |
| 407 | 356.91 | 24.54 | 452 | 399.12 | 25.94 | 497 | 441.45 | 27.27 |
| 408 | 357.85 | 24.58 | 453 | 400.06 | 25.97 | 498 | 442.40 | 27.30 |
| 409 | 358.79 | 24.61 | 454 | 401.00 | 26.00 | 499 | 443.34 | 27.33 |
| 410 | 359.72 | 24.64 | 455 | 401.93 | 26.03 | 500 | 444.28 | 27.36 |
| 411 | 360.66 | 24.67 | 456 | 402.87 | 26.06 | 501 | 445.22 | 27.39 |
| 412 | 361.60 | 24.70 | 457 | 403.81 | 26.09 | 502 | 446.17 | 27.42 |
| 413 | 362.53 | 24.73 | 458 | 404.75 | 26.12 | 503 | 447.11 | 27.45 |
| 414 | 363.47 | 24.77 | 459 | 405.69 | 26.15 | 504 | 448.05 | 27.47 |
| 415 | 364.41 | 24.80 | 460 | 406.63 | 26.18 | 505 | 448.99 | 27.50 |
| 416 | 365.34 | 24.83 | 461 | 407.57 | 26.21 | 506 | 449.94 | 27.53 |
| 417 | 366.28 | 24.86 | 462 | 408.51 | 26.24 | 507 | 450.88 | 27.56 |
| 418 | 367.22 | 24.89 | 463 | 409.45 | 26.27 | 508 | 451.82 | 27.59 |
| 419 | 368.15 | 24.92 | 464 | 410.39 | 26.30 | 509 | 452.77 | 27.62 |
| 420 | 369.09 | 24.95 | 465 | 411.33 | 26.33 | 510 | 453.71 | 27.65 |
| 421 | 370.03 | 24.99 | 466 | 412.27 | 26.36 | 511 | 454.65 | 27.67 |
| 422 | 370.96 | 25.02 | 467 | 413.21 | 26.39 | 512 | 455.59 | 27.70 |
| 423 | 371.90 | 25.05 | 468 | 414.16 | 26.42 | 513 | 456.54 | 27.73 |
| 424 | 372.84 | 25.08 | 469 | 415.10 | 26.45 | 514 | 457.48 | 27.76 |
| 425 | 373.78 | 25.11 | 470 | 416.04 | 26.48 | 515 | 458.42 | 27.79 |
| 426 | 374.71 | 25.14 | 471 | 416.98 | 26.51 | 516 | 459.37 | 27.82 |
| 427 | 375.65 | 25.17 | 472 | 417.92 | 26.54 | 517 | 460.31 | 27.84 |
| 428 | 376.59 | 25.21 | 473 | 418.86 | 26.57 | 518 | 461.25 | 27.87 |
| 429 | 377.53 | 25.24 | 474 | 419.80 | 26.60 | 519 | 462.20 | 27.90 |
| 430 | 378.47 | 25.27 | 475 | 420.74 | 26.63 | 520 | 463.14 | 27.93 |
| 431 | 379.40 | 25.30 | 476 | 421.68 | 26.66 | 521 | 464.08 | 27.96 |
| 432 | 380.34 | 25.33 | 477 | 422.62 | 26.69 | 522 | 465.03 | 27.99 |
| 433 | 381.28 | 25.36 | 478 | 423.56 | 26.72 | 523 | 465.97 | 28.01 |
| 434 | 382.22 | 25.39 | 479 | 424.50 | 26.75 | 524 | 466.92 | 28.04 |
| 435 | 383.16 | 25.42 | 480 | 425.45 | 26.78 | 525 | 467.86 | 28.07 |
| 436 | 384.09 | 25.45 | 481 | 426.39 | 26.81 | 526 | 468.80 | 28.10 |
| 437 | 385.03 | 25.48 | 482 | 427.33 | 26.84 | 527 | 469.75 | 28.13 |
| 438 | 385.97 | 25.51 | 483 | 428.27 | 26.87 | 528 | 470.69 | 28.15 |
| 439 | 386.91 | 25.55 | 484 | 429.21 | 26.89 | 529 | 471.63 | 28.18 |
| 440 | 387.85 | 25.58 | 485 | 430.15 | 26.92 | 530 | 472.58 | 28.21 |
| 441 | 388.79 | 25.61 | 486 | 431.09 | 26.95 | 531 | 473.52 | 28.24 |
| 442 | 389.73 | 25.64 | 487 | 432.04 | 26.98 | 532 | 474.47 | 28.27 |
| 443 | 390.66 | 25.67 | 488 | 432.98 | 27.01 | 533 | 475.41 | 28.29 |
| 444 | 391.60 | 25.70 | 489 | 433.92 | 27.04 | 534 | 476.36 | 28.32 |
| 445 | 392.54 | 25.73 | 490 | 434.86 | 27.07 | 535 | 477.30 | 28.35 |
| 446 | 393.48 | 25.76 | 491 | 435.80 | 27.10 | 536 | 478.24 | 28.38 |
| 447 | 394.42 | 25.79 | 492 | 436.74 | 27.13 | 537 | 479.19 | 28.41 |
| 448 | 395.36 | 25.82 | 493 | 437.69 | 27.16 | 538 | 480.13 | 28.43 |
| 449 | 396.30 | 25.85 | 494 | 438.63 | 27.19 | 539 | 481.08 | 28.46 |
| 450 | 397.24 | 25.88 | 495 | 439.57 | 27.21 | 540 | 482.02 | 28.49 |

Table 4.  CONSTANTS FOR $p'$ (SHORT FORMULA), AN APPROXIMATION OF
THE LOWER BINOMIAL CONFIDENCE LIMIT $\underline{p}$; $c/n \leqslant .125$, $n \geqslant 16$,
$c \leqslant 1{,}000$ (*Continued*)

| c | m | k | c | m | k | c | m | k |
|---|---|---|---|---|---|---|---|---|
| 541 | 482.97 | 28.52 | 586 | 525.52 | 29.74 | 631 | 568.17 | 30.91 |
| 542 | 483.91 | 28.54 | 587 | 526.47 | 29.76 | 632 | 569.12 | 30.94 |
| 543 | 484.86 | 28.57 | 588 | 527.42 | 29.79 | 633 | 570.07 | 30.96 |
| 544 | 485.80 | 28.60 | 589 | 528.37 | 29.82 | 634 | 571.02 | 30.99 |
| 545 | 486.75 | 28.63 | 590 | 529.31 | 29.84 | 635 | 571.97 | 31.01 |
| 546 | 487.69 | 28.65 | 591 | 530.26 | 29.87 | 636 | 572.92 | 31.04 |
| 547 | 488.64 | 28.68 | 592 | 531.21 | 29.90 | 637 | 573.87 | 31.07 |
| 548 | 489.58 | 28.71 | 593 | 532.15 | 29.92 | 638 | 574.82 | 31.09 |
| 549 | 490.53 | 28.74 | 594 | 533.10 | 29.95 | 639 | 575.77 | 31.12 |
| 550 | 491.47 | 28.76 | 595 | 534.05 | 29.98 | 640 | 576.72 | 31.14 |
| 551 | 492.42 | 28.79 | 596 | 535.00 | 30.00 | 641 | 577.66 | 31.17 |
| 552 | 493.36 | 28.82 | 597 | 535.94 | 30.03 | 642 | 578.61 | 31.19 |
| 553 | 494.31 | 28.85 | 598 | 536.89 | 30.06 | 643 | 579.56 | 31.22 |
| 554 | 495.25 | 28.87 | 599 | 537.84 | 30.08 | 644 | 580.51 | 31.24 |
| 555 | 496.20 | 28.90 | 600 | 538.78 | 30.11 | 645 | 581.46 | 31.27 |
| 556 | 497.14 | 28.93 | 601 | 539.73 | 30.13 | 646 | 582.41 | 31.29 |
| 557 | 498.09 | 28.96 | 602 | 540.68 | 30.16 | 647 | 583.36 | 31.32 |
| 558 | 499.03 | 28.98 | 603 | 541.63 | 30.19 | 648 | 584.31 | 31.35 |
| 559 | 499.98 | 29.01 | 604 | 542.57 | 30.21 | 649 | 585.26 | 31.37 |
| 560 | 500.92 | 29.04 | 605 | 543.52 | 30.24 | 650 | 586.21 | 31.40 |
| 561 | 501.87 | 29.07 | 606 | 544.47 | 30.27 | 651 | 587.16 | 31.42 |
| 562 | 502.81 | 29.09 | 607 | 545.42 | 30.29 | 652 | 588.11 | 31.45 |
| 563 | 503.76 | 29.12 | 608 | 546.37 | 30.32 | 653 | 589.06 | 31.47 |
| 564 | 504.71 | 29.15 | 609 | 547.31 | 30.34 | 654 | 590.01 | 31.50 |
| 565 | 505.65 | 29.17 | 610 | 548.26 | 30.37 | 655 | 590.96 | 31.52 |
| 566 | 506.60 | 29.20 | 611 | 549.21 | 30.40 | 656 | 591.91 | 31.55 |
| 567 | 507.54 | 29.23 | 612 | 550.16 | 30.42 | 657 | 592.86 | 31.57 |
| 568 | 508.49 | 29.26 | 613 | 551.10 | 30.45 | 658 | 593.81 | 31.60 |
| 569 | 509.44 | 29.28 | 614 | 552.05 | 30.47 | 659 | 594.75 | 31.62 |
| 570 | 510.38 | 29.31 | 615 | 553.00 | 30.50 | 660 | 595.70 | 31.65 |
| 571 | 511.33 | 29.34 | 616 | 553.95 | 30.53 | 661 | 596.65 | 31.67 |
| 572 | 512.27 | 29.36 | 617 | 554.90 | 30.55 | 662 | 597.60 | 31.70 |
| 573 | 513.22 | 29.39 | 618 | 555.84 | 30.58 | 663 | 598.55 | 31.72 |
| 574 | 514.17 | 29.42 | 619 | 556.79 | 30.60 | 664 | 599.50 | 31.75 |
| 575 | 515.11 | 29.44 | 620 | 557.74 | 30.63 | 665 | 600.45 | 31.77 |
| 576 | 516.06 | 29.47 | 621 | 558.69 | 30.66 | 666 | 601.40 | 31.80 |
| 577 | 517.01 | 29.50 | 622 | 559.64 | 30.68 | 667 | 602.35 | 31.82 |
| 578 | 517.95 | 29.52 | 623 | 560.59 | 30.71 | 668 | 603.30 | 31.85 |
| 579 | 518.90 | 29.55 | 624 | 561.53 | 30.73 | 669 | 604.26 | 31.87 |
| 580 | 519.84 | 29.58 | 625 | 562.48 | 30.76 | 670 | 605.21 | 31.90 |
| 581 | 520.79 | 29.60 | 626 | 563.43 | 30.78 | 671 | 606.16 | 31.92 |
| 582 | 521.74 | 29.63 | 627 | 564.38 | 30.81 | 672 | 607.11 | 31.95 |
| 583 | 522.68 | 29.66 | 628 | 565.33 | 30.84 | 673 | 608.06 | 31.97 |
| 584 | 523.63 | 29.68 | 629 | 566.28 | 30.86 | 674 | 609.01 | 32.00 |
| 585 | 524.58 | 29.71 | 630 | 567.23 | 30.89 | 675 | 609.96 | 32.02 |

Table 4. CONSTANTS FOR p′ (SHORT FORMULA), AN APPROXIMATION OF THE LOWER BINOMIAL CONFIDENCE LIMIT p; c/n ⩽ .125, n ⩾ 16, c ⩽ 1,000 (*Continued*)

| c | m | k | c | m | k | c | m | k |
|---|---|---|---|---|---|---|---|---|
| 676 | 610.91 | 32.05 | 721 | 653.71 | 33.14 | 766 | 696.59 | 34.21 |
| 677 | 611.86 | 32.07 | 722 | 654.67 | 33.17 | 767 | 697.54 | 34.23 |
| 678 | 612.81 | 32.10 | 723 | 655.62 | 33.19 | 768 | 698.50 | 34.25 |
| 679 | 613.76 | 32.12 | 724 | 656.57 | 33.21 | 769 | 699.45 | 34.28 |
| 680 | 614.71 | 32.15 | 725 | 657.52 | 33.24 | 770 | 700.40 | 34.30 |
| 681 | 615.66 | 32.17 | 726 | 658.47 | 33.26 | 771 | 701.36 | 34.32 |
| 682 | 616.61 | 32.19 | 727 | 659.43 | 33.29 | 772 | 702.31 | 34.35 |
| 683 | 617.56 | 32.22 | 728 | 660.38 | 33.31 | 773 | 703.26 | 34.37 |
| 684 | 618.51 | 32.24 | 729 | 661.33 | 33.33 | 774 | 704.22 | 34.39 |
| 685 | 619.46 | 32.27 | 730 | 662.28 | 33.36 | 775 | 705.17 | 34.41 |
| 686 | 620.41 | 32.29 | 731 | 663.24 | 33.38 | 776 | 706.12 | 34.44 |
| 687 | 621.36 | 32.32 | 732 | 664.19 | 33.41 | 777 | 707.08 | 34.46 |
| 688 | 622.32 | 32.34 | 733 | 665.14 | 33.43 | 778 | 708.03 | 34.48 |
| 689 | 623.27 | 32.37 | 734 | 666.09 | 33.45 | 779 | 708.99 | 34.51 |
| 690 | 624.22 | 32.39 | 735 | 667.05 | 33.48 | 780 | 709.94 | 34.53 |
| 691 | 625.17 | 32.42 | 736 | 668.00 | 33.50 | 781 | 710.89 | 34.55 |
| 692 | 626.12 | 32.44 | 737 | 668.95 | 33.52 | 782 | 711.85 | 34.58 |
| 693 | 627.07 | 32.46 | 738 | 669.90 | 33.55 | 783 | 712.80 | 34.60 |
| 694 | 628.02 | 32.49 | 739 | 670.86 | 33.57 | 784 | 713.76 | 34.62 |
| 695 | 628.97 | 32.51 | 740 | 671.81 | 33.60 | 785 | 714.71 | 34.65 |
| 696 | 629.92 | 32.54 | 741 | 672.76 | 33.62 | 786 | 715.66 | 34.67 |
| 697 | 630.88 | 32.56 | 742 | 673.71 | 33.64 | 787 | 716.62 | 34.69 |
| 698 | 631.83 | 32.59 | 743 | 674.67 | 33.67 | 788 | 717.57 | 34.71 |
| 699 | 632.78 | 32.61 | 744 | 675.62 | 33.69 | 789 | 718.53 | 34.74 |
| 700 | 633.73 | 32.64 | 745 | 676.57 | 33.71 | 790 | 719.48 | 34.76 |
| 701 | 634.68 | 32.66 | 746 | 677.53 | 33.74 | 791 | 720.43 | 34.78 |
| 702 | 635.63 | 32.68 | 747 | 678.48 | 33.76 | 792 | 721.39 | 34.81 |
| 703 | 636.58 | 32.71 | 748 | 679.43 | 33.78 | 793 | 722.34 | 34.83 |
| 704 | 637.53 | 32.73 | 749 | 680.38 | 33.81 | 794 | 723.30 | 34.85 |
| 705 | 638.49 | 32.76 | 750 | 681.34 | 33.83 | 795 | 724.25 | 34.87 |
| 706 | 639.44 | 32.78 | 751 | 682.29 | 33.85 | 796 | 725.21 | 34.90 |
| 707 | 640.39 | 32.81 | 752 | 683.24 | 33.88 | 797 | 726.16 | 34.92 |
| 708 | 641.34 | 32.83 | 753 | 684.20 | 33.90 | 798 | 727.11 | 34.94 |
| 709 | 642.29 | 32.85 | 754 | 685.15 | 33.93 | 799 | 728.07 | 34.97 |
| 710 | 643.24 | 32.88 | 755 | 686.10 | 33.95 | 800 | 729.02 | 34.99 |
| 711 | 644.20 | 32.90 | 756 | 687.06 | 33.97 | 801 | 729.98 | 35.01 |
| 712 | 645.15 | 32.93 | 757 | 688.01 | 34.00 | 802 | 730.93 | 35.03 |
| 713 | 646.10 | 32.95 | 758 | 688.96 | 34.02 | 803 | 731.89 | 35.06 |
| 714 | 647.05 | 32.97 | 759 | 689.92 | 34.04 | 804 | 732.84 | 35.08 |
| 715 | 648.00 | 33.00 | 760 | 690.87 | 34.07 | 805 | 733.80 | 35.10 |
| 716 | 648.95 | 33.02 | 761 | 691.82 | 34.09 | 806 | 734.75 | 35.12 |
| 717 | 649.91 | 33.05 | 762 | 692.77 | 34.11 | 807 | 735.71 | 35.15 |
| 718 | 650.86 | 33.07 | 763 | 693.73 | 34.14 | 808 | 736.66 | 35.17 |
| 719 | 651.81 | 33.09 | 764 | 694.68 | 34.16 | 809 | 737.61 | 35.19 |
| 720 | 652.76 | 33.12 | 765 | 695.64 | 34.18 | 810 | 738.57 | 35.22 |

Table 4.  CONSTANTS FOR p′ (SHORT FORMULA), AN APPROXIMATION OF
THE LOWER BINOMIAL CONFIDENCE LIMIT p; c/n ⩽ .125, n ⩾ 16,
c ⩽ 1,000 (*Continued*)

Section 1

$\gamma_1$ = 99.5%

$\gamma_2$ = 99%

| c | m | k | c | m | k | c | m | k |
|---|---|---|---|---|---|---|---|---|
| 811 | 739.52 | 35.24 | 856 | 782.52 | 36.24 | 901 | 825.56 | 37.22 |
| 812 | 740.48 | 35.26 | 857 | 783.47 | 36.26 | 902 | 826.52 | 37.24 |
| 813 | 741.43 | 35.28 | 858 | 784.43 | 36.29 | 903 | 827.48 | 37.26 |
| 814 | 742.39 | 35.31 | 859 | 785.38 | 36.31 | 904 | 828.43 | 37.28 |
| 815 | 743.34 | 35.33 | 860 | 786.34 | 36.33 | 905 | 829.39 | 37.31 |
| 816 | 744.30 | 35.35 | 861 | 787.30 | 36.35 | 906 | 830.35 | 37.33 |
| 817 | 745.25 | 35.37 | 862 | 788.25 | 36.37 | 907 | 831.30 | 37.35 |
| 818 | 746.21 | 35.40 | 863 | 789.21 | 36.40 | 908 | 832.26 | 37.37 |
| 819 | 747.16 | 35.42 | 864 | 790.17 | 36.42 | 909 | 833.22 | 37.39 |
| 820 | 748.12 | 35.44 | 865 | 791.12 | 36.44 | 910 | 834.18 | 37.41 |
| 821 | 749.07 | 35.46 | 866 | 792.08 | 36.46 | 911 | 835.13 | 37.43 |
| 822 | 750.03 | 35.49 | 867 | 793.03 | 36.48 | 912 | 836.09 | 37.45 |
| 823 | 750.98 | 35.51 | 868 | 793.99 | 36.50 | 913 | 837.05 | 37.48 |
| 824 | 751.94 | 35.53 | 869 | 794.95 | 36.53 | 914 | 838.01 | 37.50 |
| 825 | 752.89 | 35.55 | 870 | 795.90 | 36.55 | 915 | 838.96 | 37.52 |
| 826 | 753.85 | 35.58 | 871 | 796.86 | 36.57 | 916 | 839.92 | 37.54 |
| 827 | 754.80 | 35.60 | 872 | 797.82 | 36.59 | 917 | 840.88 | 37.56 |
| 828 | 755.76 | 35.62 | 873 | 798.77 | 36.61 | 918 | 841.84 | 37.58 |
| 829 | 756.71 | 35.64 | 874 | 799.73 | 36.64 | 919 | 842.79 | 37.60 |
| 830 | 757.67 | 35.66 | 875 | 800.69 | 36.66 | 920 | 843.75 | 37.62 |
| 831 | 758.63 | 35.69 | 876 | 801.64 | 36.68 | 921 | 844.71 | 37.65 |
| 832 | 759.58 | 35.71 | 877 | 802.60 | 36.70 | 922 | 845.67 | 37.67 |
| 833 | 760.54 | 35.73 | 878 | 803.55 | 36.72 | 923 | 846.62 | 37.69 |
| 834 | 761.49 | 35.75 | 879 | 804.51 | 36.74 | 924 | 847.58 | 37.71 |
| 835 | 762.45 | 35.78 | 880 | 805.47 | 36.77 | 925 | 848.54 | 37.73 |
| 836 | 763.40 | 35.80 | 881 | 806.42 | 36.79 | 926 | 849.50 | 37.75 |
| 837 | 764.36 | 35.82 | 882 | 807.38 | 36.81 | 927 | 850.45 | 37.77 |
| 838 | 765.31 | 35.84 | 883 | 808.34 | 36.83 | 928 | 851.41 | 37.79 |
| 839 | 766.27 | 35.87 | 884 | 809.29 | 36.85 | 929 | 852.37 | 37.82 |
| 840 | 767.22 | 35.89 | 885 | 810.25 | 36.87 | 930 | 853.33 | 37.84 |
| 841 | 768.18 | 35.91 | 886 | 811.21 | 36.90 | 931 | 854.28 | 37.86 |
| 842 | 769.14 | 35.93 | 887 | 812.16 | 36.92 | 932 | 855.24 | 37.88 |
| 843 | 770.09 | 35.95 | 888 | 813.12 | 36.94 | 933 | 856.20 | 37.90 |
| 844 | 771.05 | 35.98 | 889 | 814.08 | 36.96 | 934 | 857.16 | 37.92 |
| 845 | 772.00 | 36.00 | 890 | 815.03 | 36.98 | 935 | 858.12 | 37.94 |
| 846 | 772.96 | 36.02 | 891 | 815.99 | 37.00 | 936 | 859.07 | 37.96 |
| 847 | 773.91 | 36.04 | 892 | 816.95 | 37.03 | 937 | 860.03 | 37.98 |
| 848 | 774.87 | 36.07 | 893 | 817.91 | 37.05 | 938 | 860.99 | 38.01 |
| 849 | 775.83 | 36.09 | 894 | 818.86 | 37.07 | 939 | 861.95 | 38.03 |
| 850 | 776.78 | 36.11 | 895 | 819.82 | 37.09 | 940 | 862.91 | 38.05 |
| 851 | 777.74 | 36.13 | 896 | 820.78 | 37.11 | 941 | 863.86 | 38.07 |
| 852 | 778.69 | 36.15 | 897 | 821.73 | 37.13 | 942 | 864.82 | 38.09 |
| 853 | 779.65 | 36.18 | 898 | 822.69 | 37.15 | 943 | 865.78 | 38.11 |
| 854 | 780.60 | 36.20 | 899 | 823.65 | 37.18 | 944 | 866.74 | 38.13 |
| 855 | 781.56 | 36.22 | 900 | 824.60 | 37.20 | 945 | 867.70 | 38.15 |

Table 4. CONSTANTS FOR p′ (SHORT FORMULA), AN APPROXIMATION OF THE LOWER BINOMIAL CONFIDENCE LIMIT p̲; c/n ⩽ .125, n ⩾ 16, c ⩽ 1,000 *(Continued)*

| c | m | k | c | m | k | c | m | k |
|---|---|---|---|---|---|---|---|---|
| 946 | 868.65 | 38.17 | 966 | 887.82 | 38.59 | 986 | 907.00 | 39.00 |
| 947 | 869.61 | 38.19 | 967 | 888.78 | 38.61 | 987 | 907.96 | 39.02 |
| 948 | 870.57 | 38.21 | 968 | 889.74 | 38.63 | 988 | 908.91 | 39.04 |
| 949 | 871.53 | 38.24 | 969 | 890.70 | 38.65 | 989 | 909.87 | 39.06 |
| 950 | 872.49 | 38.26 | 970 | 891.66 | 38.67 | 990 | 910.83 | 39.08 |
| 951 | 873.44 | 38.28 | 971 | 892.61 | 38.69 | 991 | 911.79 | 39.10 |
| 952 | 874.40 | 38.30 | 972 | 893.57 | 38.71 | 992 | 912.75 | 39.12 |
| 953 | 875.36 | 38.32 | 973 | 894.53 | 38.73 | 993 | 913.71 | 39.15 |
| 954 | 876.32 | 38.34 | 974 | 895.49 | 38.75 | 994 | 914.67 | 39.17 |
| 955 | 877.28 | 38.36 | 975 | 896.45 | 38.78 | 995 | 915.63 | 39.19 |
| 956 | 878.24 | 38.38 | 976 | 897.41 | 38.80 | 996 | 916.59 | 39.21 |
| 957 | 879.19 | 38.40 | 977 | 898.37 | 38.82 | 997 | 917.55 | 39.23 |
| 958 | 880.15 | 38.42 | 978 | 899.33 | 38.84 | 998 | 918.51 | 39.25 |
| 959 | 881.11 | 38.44 | 979 | 900.28 | 38.86 | 999 | 919.47 | 39.27 |
| 960 | 882.07 | 38.47 | 980 | 901.24 | 38.88 | 1000 | 920.42 | 39.29 |
| 961 | 883.03 | 38.49 | 981 | 902.20 | 38.90 | | | |
| 962 | 883.99 | 38.51 | 982 | 903.16 | 38.92 | | | |
| 963 | 884.95 | 38.53 | 983 | 904.12 | 38.94 | | | |
| 964 | 885.90 | 38.55 | 984 | 905.08 | 38.96 | | | |
| 965 | 886.86 | 38.57 | 985 | 906.04 | 38.98 | | | |

# Table 4. CONSTANTS FOR p′ (SHORT FORMULA), AN APPROXIMATION OF THE LOWER BINOMIAL CONFIDENCE LIMIT p; c/n ⩽ .125, n ⩾ 16, c ⩽ 1,000 (*Continued*)

| c | m | k | c | m | k | c | m | k |
|---|---|---|---|---|---|---|---|---|
| 1 | .010051 | -0.005 | 46 | 31.705 | 6.65 | 91 | 70.288 | 9.86 |
| 2 | .14856 | 0.426 | 47 | 32.534 | 6.73 | 92 | 71.166 | 9.92 |
| 3 | .43605 | 0.782 | 48 | 33.365 | 6.82 | 93 | 72.045 | 9.98 |
| 4 | .82325 | 1.088 | 49 | 34.198 | 6.90 | 94 | 72.925 | 10.04 |
| 5 | 1.2791 | 1.360 | 50 | 35.032 | 6.98 | 95 | 73.805 | 10.10 |
| 6 | 1.7853 | 1.607 | 51 | 35.869 | 7.07 | 96 | 74.686 | 10.16 |
| 7 | 2.3302 | 1.835 | 52 | 36.707 | 7.15 | 97 | 75.568 | 10.22 |
| 8 | 2.9061 | 2.047 | 53 | 37.546 | 7.23 | 98 | 76.450 | 10.28 |
| 9 | 3.5075 | 2.246 | 54 | 38.387 | 7.31 | 99 | 77.333 | 10.34 |
| 10 | 4.1302 | 2.43 | 55 | 39.229 | 7.39 | 100 | 78.216 | 10.40 |
| 11 | 4.7713 | 2.61 | 56 | 40.073 | 7.47 | 101 | 79.100 | 10.45 |
| 12 | 5.4282 | 2.79 | 57 | 40.918 | 7.55 | 102 | 79.985 | 10.51 |
| 13 | 6.0991 | 2.95 | 58 | 41.765 | 7.62 | 103 | 80.870 | 10.57 |
| 14 | 6.7824 | 3.11 | 59 | 42.613 | 7.70 | 104 | 81.755 | 10.63 |
| 15 | 7.4768 | 3.26 | 60 | 43.462 | 7.77 | 105 | 82.642 | 10.68 |
| 16 | 8.1811 | 3.41 | 61 | 44.312 | 7.85 | 106 | 83.528 | 10.74 |
| 17 | 8.8946 | 3.55 | 62 | 45.164 | 7.92 | 107 | 84.416 | 10.80 |
| 18 | 9.6164 | 3.69 | 63 | 46.016 | 8.00 | 108 | 85.303 | 10.85 |
| 19 | 10.346 | 3.83 | 64 | 46.870 | 8.07 | 109 | 86.192 | 10.91 |
| 20 | 11.082 | 3.96 | 65 | 47.726 | 8.14 | 110 | 87.080 | 10.96 |
| 21 | 11.825 | 4.09 | 66 | 48.582 | 8.21 | 111 | 87.970 | 11.02 |
| 22 | 12.574 | 4.21 | 67 | 49.439 | 8.29 | 112 | 88.860 | 11.07 |
| 23 | 13.329 | 4.34 | 68 | 50.298 | 8.36 | 113 | 89.750 | 11.13 |
| 24 | 14.089 | 4.46 | 69 | 51.157 | 8.43 | 114 | 90.641 | 11.18 |
| 25 | 14.853 | 4.57 | 70 | 52.017 | 8.50 | 115 | 91.532 | 11.24 |
| 26 | 15.623 | 4.69 | 71 | 52.879 | 8.57 | 116 | 92.424 | 11.29 |
| 27 | 16.397 | 4.80 | 72 | 53.741 | 8.63 | 117 | 93.316 | 11.35 |
| 28 | 17.175 | 4.91 | 73 | 54.605 | 8.70 | 118 | 94.209 | 11.40 |
| 29 | 17.957 | 5.02 | 74 | 55.469 | 8.77 | 119 | 95.102 | 11.45 |
| 30 | 18.742 | 5.13 | 75 | 56.334 | 8.84 | 120 | 95.995 | 11.51 |
| 31 | 19.532 | 5.23 | 76 | 57.200 | 8.90 | 121 | 96.889 | 11.56 |
| 32 | 20.324 | 5.34 | 77 | 58.067 | 8.97 | 122 | 97.784 | 11.61 |
| 33 | 21.120 | 5.44 | 78 | 58.935 | 9.04 | 123 | 98.679 | 11.66 |
| 34 | 21.919 | 5.54 | 79 | 59.804 | 9.10 | 124 | 99.574 | 11.72 |
| 35 | 22.721 | 5.64 | 80 | 60.673 | 9.17 | 125 | 100.47 | 11.77 |
| 36 | 23.526 | 5.74 | 81 | 61.543 | 9.23 | 126 | 101.37 | 11.82 |
| 37 | 24.333 | 5.83 | 82 | 62.414 | 9.30 | 127 | 102.26 | 11.87 |
| 38 | 25.143 | 5.93 | 83 | 63.286 | 9.36 | 128 | 103.16 | 11.92 |
| 39 | 25.955 | 6.02 | 84 | 64.159 | 9.43 | 129 | 104.06 | 11.98 |
| 40 | 26.770 | 6.11 | 85 | 65.032 | 9.49 | 130 | 104.95 | 12.03 |
| 41 | 27.587 | 6.21 | 86 | 65.907 | 9.55 | 131 | 105.85 | 12.08 |
| 42 | 28.407 | 6.30 | 87 | 66.781 | 9.61 | 132 | 106.75 | 12.13 |
| 43 | 29.228 | 6.39 | 88 | 67.657 | 9.68 | 133 | 107.65 | 12.18 |
| 44 | 30.052 | 6.47 | 89 | 68.533 | 9.74 | 134 | 108.55 | 12.23 |
| 45 | 30.877 | 6.56 | 90 | 69.410 | 9.80 | 135 | 109.45 | 12.28 |

Table 4.  CONSTANTS FOR p′ (SHORT FORMULA), AN APPROXIMATION OF THE LOWER BINOMIAL CONFIDENCE LIMIT p; c/n ≤ .125, n ≥ 16, c ≤ 1,000 (*Continued*)

| c | m | k | c | m | k | c | m | k |
|---|---|---|---|---|---|---|---|---|
| 136 | 110.35 | 12.33 | 181 | 151.17 | 14.41 | 226 | 192.50 | 16.25 |
| 137 | 111.25 | 12.38 | 182 | 152.09 | 14.46 | 227 | 193.42 | 16.29 |
| 138 | 112.15 | 12.43 | 183 | 153.00 | 14.50 | 228 | 194.34 | 16.33 |
| 139 | 113.05 | 12.48 | 184 | 153.92 | 14.54 | 229 | 195.27 | 16.37 |
| 140 | 113.95 | 12.53 | 185 | 154.83 | 14.59 | 230 | 196.19 | 16.40 |
| 141 | 114.85 | 12.58 | 186 | 155.74 | 14.63 | 231 | 197.11 | 16.44 |
| 142 | 115.76 | 12.63 | 187 | 156.66 | 14.67 | 232 | 198.04 | 16.48 |
| 143 | 116.66 | 12.67 | 188 | 157.57 | 14.71 | 233 | 198.96 | 16.52 |
| 144 | 117.56 | 12.72 | 189 | 158.49 | 14.76 | 234 | 199.89 | 16.56 |
| 145 | 118.47 | 12.77 | 190 | 159.40 | 14.80 | 235 | 200.81 | 16.60 |
| 146 | 119.37 | 12.82 | 191 | 160.32 | 14.84 | 236 | 201.73 | 16.63 |
| 147 | 120.27 | 12.87 | 192 | 161.24 | 14.88 | 237 | 202.66 | 16.67 |
| 148 | 121.18 | 12.92 | 193 | 162.15 | 14.92 | 238 | 203.58 | 16.71 |
| 149 | 122.08 | 12.96 | 194 | 163.07 | 14.97 | 239 | 204.51 | 16.75 |
| 150 | 122.99 | 13.01 | 195 | 163.99 | 15.01 | 240 | 205.43 | 16.78 |
| 151 | 123.89 | 13.06 | 196 | 164.90 | 15.05 | 241 | 206.36 | 16.82 |
| 152 | 124.80 | 13.10 | 197 | 165.82 | 15.09 | 242 | 207.28 | 16.86 |
| 153 | 125.70 | 13.15 | 198 | 166.74 | 15.13 | 243 | 208.21 | 16.90 |
| 154 | 126.61 | 13.20 | 199 | 167.65 | 15.17 | 244 | 209.13 | 16.93 |
| 155 | 127.52 | 13.25 | 200 | 168.57 | 15.21 | 245 | 210.06 | 16.97 |
| 156 | 128.42 | 13.29 | 201 | 169.49 | 15.26 | 246 | 210.98 | 17.01 |
| 157 | 129.33 | 13.34 | 202 | 170.41 | 15.30 | 247 | 211.91 | 17.05 |
| 158 | 130.24 | 13.39 | 203 | 171.33 | 15.34 | 248 | 212.84 | 17.08 |
| 159 | 131.14 | 13.43 | 204 | 172.24 | 15.38 | 249 | 213.76 | 17.12 |
| 160 | 132.05 | 13.48 | 205 | 173.16 | 15.42 | 250 | 214.69 | 17.16 |
| 161 | 132.95 | 13.52 | 206 | 174.08 | 15.46 | 251 | 215.62 | 17.19 |
| 162 | 133.86 | 13.57 | 207 | 175.00 | 15.50 | 252 | 216.54 | 17.23 |
| 163 | 134.77 | 13.61 | 208 | 175.92 | 15.54 | 253 | 217.47 | 17.27 |
| 164 | 135.68 | 13.66 | 209 | 176.84 | 15.58 | 254 | 218.40 | 17.30 |
| 165 | 136.59 | 13.71 | 210 | 177.76 | 15.62 | 255 | 219.32 | 17.34 |
| 166 | 137.50 | 13.75 | 211 | 178.68 | 15.66 | 256 | 220.25 | 17.38 |
| 167 | 138.41 | 13.80 | 212 | 179.60 | 15.70 | 257 | 221.18 | 17.41 |
| 168 | 139.32 | 13.84 | 213 | 180.52 | 15.74 | 258 | 222.10 | 17.45 |
| 169 | 140.23 | 13.89 | 214 | 181.44 | 15.78 | 259 | 223.03 | 17.48 |
| 170 | 141.14 | 13.93 | 215 | 182.36 | 15.82 | 260 | 223.96 | 17.52 |
| 171 | 142.05 | 13.97 | 216 | 183.28 | 15.86 | 261 | 224.89 | 17.56 |
| 172 | 142.96 | 14.02 | 217 | 184.20 | 15.90 | 262 | 225.82 | 17.59 |
| 173 | 143.87 | 14.06 | 218 | 185.12 | 15.94 | 263 | 226.74 | 17.63 |
| 174 | 144.78 | 14.11 | 219 | 186.04 | 15.98 | 264 | 227.67 | 17.66 |
| 175 | 145.70 | 14.15 | 220 | 186.97 | 16.02 | 265 | 228.60 | 17.70 |
| 176 | 146.61 | 14.20 | 221 | 187.89 | 16.06 | 266 | 229.53 | 17.74 |
| 177 | 147.52 | 14.24 | 222 | 188.81 | 16.10 | 267 | 230.46 | 17.77 |
| 178 | 148.43 | 14.28 | 223 | 189.73 | 16.13 | 268 | 231.39 | 17.81 |
| 179 | 149.35 | 14.33 | 224 | 190.65 | 16.17 | 269 | 232.32 | 17.84 |
| 180 | 150.26 | 14.37 | 225 | 191.58 | 16.21 | 270 | 233.25 | 17.88 |

# Table 4.  CONSTANTS FOR $\underline{p}'$ (SHORT FORMULA), AN APPROXIMATION OF THE LOWER BINOMIAL CONFIDENCE LIMIT $\underline{p}$; $c/n \leqslant .125$, $n \geqslant 16$, $c \leqslant 1,000$ (*Continued*)

| c | m | k | c | m | k | c | m | k |
|---|---|---|---|---|---|---|---|---|
| 271 | 234.17 | 17.91 | 316 | 276.12 | 19.44 | 361 | 318.27 | 20.86 |
| 272 | 235.10 | 17.95 | 317 | 277.05 | 19.47 | 362 | 319.21 | 20.90 |
| 273 | 236.03 | 17.98 | 318 | 277.99 | 19.51 | 363 | 320.15 | 20.93 |
| 274 | 236.96 | 18.02 | 319 | 278.92 | 19.54 | 364 | 321.09 | 20.96 |
| 275 | 237.89 | 18.05 | 320 | 279.86 | 19.57 | 365 | 322.03 | 20.99 |
| 276 | 238.82 | 18.09 | 321 | 280.79 | 19.60 | 366 | 322.97 | 21.02 |
| 277 | 239.75 | 18.12 | 322 | 281.73 | 19.64 | 367 | 323.91 | 21.05 |
| 278 | 240.68 | 18.16 | 323 | 282.66 | 19.67 | 368 | 324.84 | 21.08 |
| 279 | 241.61 | 18.19 | 324 | 283.60 | 19.70 | 369 | 325.78 | 21.11 |
| 280 | 242.54 | 18.23 | 325 | 284.53 | 19.73 | 370 | 326.72 | 21.14 |
| 281 | 243.47 | 18.26 | 326 | 285.47 | 19.77 | 371 | 327.66 | 21.17 |
| 282 | 244.41 | 18.30 | 327 | 286.40 | 19.80 | 372 | 328.60 | 21.20 |
| 283 | 245.34 | 18.33 | 328 | 287.34 | 19.83 | 373 | 329.54 | 21.23 |
| 284 | 246.27 | 18.37 | 329 | 288.28 | 19.86 | 374 | 330.48 | 21.26 |
| 285 | 247.20 | 18.40 | 330 | 289.21 | 19.89 | 375 | 331.42 | 21.29 |
| 286 | 248.13 | 18.44 | 331 | 290.15 | 19.93 | 376 | 332.36 | 21.32 |
| 287 | 249.06 | 18.47 | 332 | 291.08 | 19.96 | 377 | 333.30 | 21.35 |
| 288 | 249.99 | 18.50 | 333 | 292.02 | 19.99 | 378 | 334.24 | 21.38 |
| 289 | 250.92 | 18.54 | 334 | 292.96 | 20.02 | 379 | 335.18 | 21.41 |
| 290 | 251.86 | 18.57 | 335 | 293.89 | 20.05 | 380 | 336.12 | 21.44 |
| 291 | 252.79 | 18.61 | 336 | 294.83 | 20.09 | 381 | 337.06 | 21.47 |
| 292 | 253.72 | 18.64 | 337 | 295.77 | 20.12 | 382 | 338.00 | 21.50 |
| 293 | 254.65 | 18.67 | 338 | 296.70 | 20.15 | 383 | 338.94 | 21.53 |
| 294 | 255.58 | 18.71 | 339 | 297.64 | 20.18 | 384 | 339.88 | 21.56 |
| 295 | 256.52 | 18.74 | 340 | 298.58 | 20.21 | 385 | 340.83 | 21.59 |
| 296 | 257.45 | 18.78 | 341 | 299.51 | 20.24 | 386 | 341.77 | 21.62 |
| 297 | 258.38 | 18.81 | 342 | 300.45 | 20.28 | 387 | 342.71 | 21.65 |
| 298 | 259.31 | 18.84 | 343 | 301.39 | 20.31 | 388 | 343.65 | 21.68 |
| 299 | 260.25 | 18.88 | 344 | 302.32 | 20.34 | 389 | 344.59 | 21.71 |
| 300 | 261.18 | 18.91 | 345 | 303.26 | 20.37 | 390 | 345.53 | 21.74 |
| 301 | 262.11 | 18.94 | 346 | 304.20 | 20.40 | 391 | 346.47 | 21.76 |
| 302 | 263.04 | 18.98 | 347 | 305.14 | 20.43 | 392 | 347.41 | 21.79 |
| 303 | 263.98 | 19.01 | 348 | 306.07 | 20.46 | 393 | 348.35 | 21.82 |
| 304 | 264.91 | 19.04 | 349 | 307.01 | 20.49 | 394 | 349.29 | 21.85 |
| 305 | 265.84 | 19.08 | 350 | 307.95 | 20.53 | 395 | 350.24 | 21.88 |
| 306 | 266.78 | 19.11 | 351 | 308.89 | 20.56 | 396 | 351.18 | 21.91 |
| 307 | 267.71 | 19.14 | 352 | 309.83 | 20.59 | 397 | 352.12 | 21.94 |
| 308 | 268.64 | 19.18 | 353 | 310.76 | 20.62 | 398 | 353.06 | 21.97 |
| 309 | 269.58 | 19.21 | 354 | 311.70 | 20.65 | 399 | 354.00 | 22.00 |
| 310 | 270.51 | 19.24 | 355 | 312.64 | 20.68 | 400 | 354.94 | 22.03 |
| 311 | 271.45 | 19.28 | 356 | 313.58 | 20.71 | 401 | 355.89 | 22.06 |
| 312 | 272.38 | 19.31 | 357 | 314.52 | 20.74 | 402 | 356.83 | 22.09 |
| 313 | 273.31 | 19.34 | 358 | 315.45 | 20.77 | 403 | 357.77 | 22.11 |
| 314 | 274.25 | 19.38 | 359 | 316.39 | 20.80 | 404 | 358.71 | 22.14 |
| 315 | 275.18 | 19.41 | 360 | 317.33 | 20.83 | 405 | 359.65 | 22.17 |

**Table 4. CONSTANTS FOR p′ (SHORT FORMULA), AN APPROXIMATION OF THE LOWER BINOMIAL CONFIDENCE LIMIT p̲; c/n ⩽ .125, n ⩾ 16, c ⩽ 1,000 (*Continued*)**

| c | m | k | c | m | k | c | m | k |
|---|---|---|---|---|---|---|---|---|
| 406 | 360.60 | 22.20 | 451 | 403.07 | 23.47 | 496 | 445.66 | 24.67 |
| 407 | 361.54 | 22.23 | 452 | 404.01 | 23.49 | 497 | 446.61 | 24.70 |
| 408 | 362.48 | 22.26 | 453 | 404.96 | 23.52 | 498 | 447.56 | 24.72 |
| 409 | 363.42 | 22.29 | 454 | 405.90 | 23.55 | 499 | 448.51 | 24.75 |
| 410 | 364.37 | 22.32 | 455 | 406.85 | 23.58 | 500 | 449.45 | 24.77 |
| 411 | 365.31 | 22.35 | 456 | 407.79 | 23.60 | 501 | 450.40 | 24.80 |
| 412 | 366.25 | 22.37 | 457 | 408.74 | 23.63 | 502 | 451.35 | 24.83 |
| 413 | 367.19 | 22.40 | 458 | 409.69 | 23.66 | 503 | 452.30 | 24.85 |
| 414 | 368.14 | 22.43 | 459 | 410.63 | 23.68 | 504 | 453.25 | 24.88 |
| 415 | 369.08 | 22.46 | 460 | 411.58 | 23.71 | 505 | 454.19 | 24.90 |
| 416 | 370.02 | 22.49 | 461 | 412.52 | 23.74 | 506 | 455.14 | 24.93 |
| 417 | 370.97 | 22.52 | 462 | 413.47 | 23.77 | 507 | 456.09 | 24.95 |
| 418 | 371.91 | 22.55 | 463 | 414.41 | 23.79 | 508 | 457.04 | 24.98 |
| 419 | 372.85 | 22.57 | 464 | 415.36 | 23.82 | 509 | 457.99 | 25.01 |
| 420 | 373.80 | 22.60 | 465 | 416.31 | 23.85 | 510 | 458.94 | 25.03 |
| 421 | 374.74 | 22.63 | 466 | 417.25 | 23.87 | 511 | 459.88 | 25.06 |
| 422 | 375.68 | 22.66 | 467 | 418.20 | 23.90 | 512 | 460.83 | 25.08 |
| 423 | 376.63 | 22.69 | 468 | 419.15 | 23.93 | 513 | 461.78 | 25.11 |
| 424 | 377.57 | 22.72 | 469 | 420.09 | 23.95 | 514 | 462.73 | 25.14 |
| 425 | 378.51 | 22.74 | 470 | 421.04 | 23.98 | 515 | 463.68 | 25.16 |
| 426 | 379.46 | 22.77 | 471 | 421.98 | 24.01 | 516 | 464.63 | 25.19 |
| 427 | 380.40 | 22.80 | 472 | 422.93 | 24.03 | 517 | 465.58 | 25.21 |
| 428 | 381.34 | 22.83 | 473 | 423.88 | 24.06 | 518 | 466.52 | 25.24 |
| 429 | 382.29 | 22.86 | 474 | 424.82 | 24.09 | 519 | 467.47 | 25.26 |
| 430 | 383.23 | 22.88 | 475 | 425.77 | 24.11 | 520 | 468.42 | 25.29 |
| 431 | 384.18 | 22.91 | 476 | 426.72 | 24.14 | 521 | 469.37 | 25.31 |
| 432 | 385.12 | 22.94 | 477 | 427.66 | 24.17 | 522 | 470.32 | 25.34 |
| 433 | 386.06 | 22.97 | 478 | 428.61 | 24.19 | 523 | 471.27 | 25.37 |
| 434 | 387.01 | 23.00 | 479 | 429.56 | 24.22 | 524 | 472.22 | 25.39 |
| 435 | 387.95 | 23.02 | 480 | 430.50 | 24.25 | 525 | 473.17 | 25.42 |
| 436 | 388.90 | 23.05 | 481 | 431.45 | 24.27 | 526 | 474.12 | 25.44 |
| 437 | 389.84 | 23.08 | 482 | 432.40 | 24.30 | 527 | 475.07 | 25.47 |
| 438 | 390.78 | 23.11 | 483 | 433.34 | 24.33 | 528 | 476.02 | 25.49 |
| 439 | 391.73 | 23.14 | 484 | 434.29 | 24.35 | 529 | 476.97 | 25.52 |
| 440 | 392.67 | 23.16 | 485 | 435.24 | 24.38 | 530 | 477.92 | 25.54 |
| 441 | 393.62 | 23.19 | 486 | 436.19 | 24.41 | 531 | 478.86 | 25.57 |
| 442 | 394.56 | 23.22 | 487 | 437.13 | 24.43 | 532 | 479.81 | 25.59 |
| 443 | 395.51 | 23.25 | 488 | 438.08 | 24.46 | 533 | 480.76 | 25.62 |
| 444 | 396.45 | 23.27 | 489 | 439.03 | 24.49 | 534 | 481.71 | 25.64 |
| 445 | 397.40 | 23.30 | 490 | 439.98 | 24.51 | 535 | 482.66 | 25.67 |
| 446 | 398.34 | 23.33 | 491 | 440.92 | 24.54 | 536 | 483.61 | 25.69 |
| 447 | 399.29 | 23.36 | 492 | 441.87 | 24.56 | 537 | 484.56 | 25.72 |
| 448 | 400.23 | 23.38 | 493 | 442.82 | 24.59 | 538 | 485.51 | 25.74 |
| 449 | 401.18 | 23.41 | 494 | 443.77 | 24.62 | 539 | 486.46 | 25.77 |
| 450 | 402.12 | 23.44 | 495 | 444.71 | 24.64 | 540 | 487.41 | 25.79 |

| c | m | k | c | m | k | c | m | k |
|---|---|---|---|---|---|---|---|---|
| 541 | 488.36 | 25.82 | 586 | 531.16 | 26.92 | 631 | 574.03 | 27.98 |
| 542 | 489.31 | 25.84 | 587 | 532.11 | 26.95 | 632 | 574.99 | 28.01 |
| 543 | 490.26 | 25.87 | 588 | 533.06 | 26.97 | 633 | 575.94 | 28.03 |
| 544 | 491.21 | 25.89 | 589 | 534.01 | 26.99 | 634 | 576.90 | 28.05 |
| 545 | 492.16 | 25.92 | 590 | 534.96 | 27.02 | 635 | 577.85 | 28.08 |
| 546 | 493.11 | 25.94 | 591 | 535.92 | 27.04 | 636 | 578.80 | 28.10 |
| 547 | 494.06 | 25.97 | 592 | 536.87 | 27.07 | 637 | 579.76 | 28.12 |
| 548 | 495.01 | 25.99 | 593 | 537.82 | 27.09 | 638 | 580.71 | 28.14 |
| 549 | 495.96 | 26.02 | 594 | 538.77 | 27.11 | 639 | 581.67 | 28.17 |
| 550 | 496.91 | 26.04 | 595 | 539.73 | 27.14 | 640 | 582.62 | 28.19 |
| 551 | 497.86 | 26.07 | 596 | 540.68 | 27.16 | 641 | 583.57 | 28.21 |
| 552 | 498.81 | 26.09 | 597 | 541.63 | 27.18 | 642 | 584.53 | 28.24 |
| 553 | 499.77 | 26.12 | 598 | 542.58 | 27.21 | 643 | 585.48 | 28.26 |
| 554 | 500.72 | 26.14 | 599 | 543.54 | 27.23 | 644 | 586.44 | 28.28 |
| 555 | 501.67 | 26.17 | 600 | 544.49 | 27.26 | 645 | 587.39 | 28.31 |
| 556 | 502.62 | 26.19 | 601 | 545.44 | 27.28 | 646 | 588.34 | 28.33 |
| 557 | 503.57 | 26.22 | 602 | 546.39 | 27.30 | 647 | 589.30 | 28.35 |
| 558 | 504.52 | 26.24 | 603 | 547.35 | 27.33 | 648 | 590.25 | 28.37 |
| 559 | 505.47 | 26.27 | 604 | 548.30 | 27.35 | 649 | 591.21 | 28.40 |
| 560 | 506.42 | 26.29 | 605 | 549.25 | 27.37 | 650 | 592.16 | 28.42 |
| 561 | 507.37 | 26.31 | 606 | 550.20 | 27.40 | 651 | 593.12 | 28.44 |
| 562 | 508.32 | 26.34 | 607 | 551.16 | 27.42 | 652 | 594.07 | 28.46 |
| 563 | 509.27 | 26.36 | 608 | 552.11 | 27.45 | 653 | 595.02 | 28.49 |
| 564 | 510.22 | 26.39 | 609 | 553.06 | 27.47 | 654 | 595.98 | 28.51 |
| 565 | 511.18 | 26.41 | 610 | 554.02 | 27.49 | 655 | 596.93 | 28.53 |
| 566 | 512.13 | 26.44 | 611 | 554.97 | 27.52 | 656 | 597.89 | 28.56 |
| 567 | 513.08 | 26.46 | 612 | 555.92 | 27.54 | 657 | 598.84 | 28.58 |
| 568 | 514.03 | 26.49 | 613 | 556.87 | 27.56 | 658 | 599.80 | 28.60 |
| 569 | 514.98 | 26.51 | 614 | 557.83 | 27.59 | 659 | 600.75 | 28.62 |
| 570 | 515.93 | 26.53 | 615 | 558.78 | 27.61 | 660 | 601.71 | 28.65 |
| 571 | 516.88 | 26.56 | 616 | 559.73 | 27.63 | 661 | 602.66 | 28.67 |
| 572 | 517.83 | 26.58 | 617 | 560.69 | 27.66 | 662 | 603.62 | 28.69 |
| 573 | 518.78 | 26.61 | 618 | 561.64 | 27.68 | 663 | 604.57 | 28.71 |
| 574 | 519.74 | 26.63 | 619 | 562.59 | 27.70 | 664 | 605.53 | 28.74 |
| 575 | 520.69 | 26.66 | 620 | 563.55 | 27.73 | 665 | 606.48 | 28.76 |
| 576 | 521.64 | 26.68 | 621 | 564.50 | 27.75 | 666 | 607.44 | 28.78 |
| 577 | 522.59 | 26.70 | 622 | 565.45 | 27.77 | 667 | 608.39 | 28.80 |
| 578 | 523.54 | 26.73 | 623 | 566.41 | 27.80 | 668 | 609.35 | 28.83 |
| 579 | 524.49 | 26.75 | 624 | 567.36 | 27.82 | 669 | 610.30 | 28.85 |
| 580 | 525.45 | 26.78 | 625 | 568.31 | 27.84 | 670 | 611.26 | 28.87 |
| 581 | 526.40 | 26.80 | 626 | 569.27 | 27.87 | 671 | 612.21 | 28.89 |
| 582 | 527.35 | 26.83 | 627 | 570.22 | 27.89 | 672 | 613.17 | 28.92 |
| 583 | 528.30 | 26.85 | 628 | 571.17 | 27.91 | 673 | 614.12 | 28.94 |
| 584 | 529.25 | 26.87 | 629 | 572.13 | 27.94 | 674 | 615.08 | 28.96 |
| 585 | 530.20 | 26.90 | 630 | 573.08 | 27.96 | 675 | 616.03 | 28.98 |

Table 4. CONSTANTS FOR $p'$ (SHORT FORMULA), AN APPROXIMATION OF THE LOWER BINOMIAL CONFIDENCE LIMIT $\underline{p}$; $c/n \leqslant .125$, $n \geqslant 1$ $c \leqslant 1,000$ (*Continued*)

| c | m | k | c | m | k | c | m | k |
|---|---|---|---|---|---|---|---|---|
| 676 | 616.99 | 29.01 | 721 | 660.01 | 30.00 | 766 | 703.09 | 30.96 |
| 677 | 617.94 | 29.03 | 722 | 660.96 | 30.02 | 767 | 704.04 | 30.98 |
| 678 | 618.90 | 29.05 | 723 | 661.92 | 30.04 | 768 | 705.00 | 31.00 |
| 679 | 619.85 | 29.07 | 724 | 662.88 | 30.06 | 769 | 705.96 | 31.02 |
| 680 | 620.81 | 29.10 | 725 | 663.83 | 30.08 | 770 | 706.92 | 31.04 |
| 681 | 621.76 | 29.12 | 726 | 664.79 | 30.11 | 771 | 707.88 | 31.06 |
| 682 | 622.72 | 29.14 | 727 | 665.75 | 30.13 | 772 | 708.83 | 31.08 |
| 683 | 623.67 | 29.16 | 728 | 666.70 | 30.15 | 773 | 709.79 | 31.10 |
| 684 | 624.63 | 29.19 | 729 | 667.66 | 30.17 | 774 | 710.75 | 31.12 |
| 685 | 625.59 | 29.21 | 730 | 668.62 | 30.19 | 775 | 711.71 | 31.15 |
| 686 | 626.54 | 29.23 | 731 | 669.57 | 30.21 | 776 | 712.67 | 31.17 |
| 687 | 627.50 | 29.25 | 732 | 670.53 | 30.23 | 777 | 713.63 | 31.19 |
| 688 | 628.45 | 29.27 | 733 | 671.49 | 30.26 | 778 | 714.58 | 31.21 |
| 689 | 629.41 | 29.30 | 734 | 672.45 | 30.28 | 779 | 715.54 | 31.23 |
| 690 | 630.36 | 29.32 | 735 | 673.40 | 30.30 | 780 | 716.50 | 31.25 |
| 691 | 631.32 | 29.34 | 736 | 674.36 | 30.32 | 781 | 717.46 | 31.27 |
| 692 | 632.28 | 29.36 | 737 | 675.32 | 30.34 | 782 | 718.42 | 31.29 |
| 693 | 633.23 | 29.38 | 738 | 676.27 | 30.36 | 783 | 719.38 | 31.31 |
| 694 | 634.19 | 29.41 | 739 | 677.23 | 30.38 | 784 | 720.33 | 31.33 |
| 695 | 635.14 | 29.43 | 740 | 678.19 | 30.41 | 785 | 721.29 | 31.35 |
| 696 | 636.10 | 29.45 | 741 | 679.15 | 30.43 | 786 | 722.25 | 31.37 |
| 697 | 637.05 | 29.47 | 742 | 680.10 | 30.45 | 787 | 723.21 | 31.40 |
| 698 | 638.01 | 29.49 | 743 | 681.06 | 30.47 | 788 | 724.17 | 31.42 |
| 699 | 638.97 | 29.52 | 744 | 682.02 | 30.49 | 789 | 725.13 | 31.44 |
| 700 | 639.92 | 29.54 | 745 | 682.97 | 30.51 | 790 | 726.09 | 31.46 |
| 701 | 640.88 | 29.56 | 746 | 683.93 | 30.53 | 791 | 727.04 | 31.48 |
| 702 | 641.83 | 29.58 | 747 | 684.89 | 30.56 | 792 | 728.00 | 31.50 |
| 703 | 642.79 | 29.60 | 748 | 685.85 | 30.58 | 793 | 728.96 | 31.52 |
| 704 | 643.75 | 29.63 | 749 | 686.80 | 30.60 | 794 | 729.92 | 31.54 |
| 705 | 644.70 | 29.65 | 750 | 687.76 | 30.62 | 795 | 730.88 | 31.56 |
| 706 | 645.66 | 29.67 | 751 | 688.72 | 30.64 | 796 | 731.84 | 31.58 |
| 707 | 646.62 | 29.69 | 752 | 689.68 | 30.66 | 797 | 732.80 | 31.60 |
| 708 | 647.57 | 29.71 | 753 | 690.63 | 30.68 | 798 | 733.76 | 31.62 |
| 709 | 648.53 | 29.74 | 754 | 691.59 | 30.70 | 799 | 734.71 | 31.64 |
| 710 | 649.48 | 29.76 | 755 | 692.55 | 30.72 | 800 | 735.67 | 31.66 |
| 711 | 650.44 | 29.78 | 756 | 693.51 | 30.75 | 801 | 736.63 | 31.68 |
| 712 | 651.40 | 29.80 | 757 | 694.47 | 30.77 | 802 | 737.59 | 31.70 |
| 713 | 652.35 | 29.82 | 758 | 695.42 | 30.79 | 803 | 738.55 | 31.73 |
| 714 | 653.31 | 29.84 | 759 | 696.38 | 30.81 | 804 | 739.51 | 31.75 |
| 715 | 654.27 | 29.87 | 760 | 697.34 | 30.83 | 805 | 740.47 | 31.77 |
| 716 | 655.22 | 29.89 | 761 | 698.30 | 30.85 | 806 | 741.43 | 31.79 |
| 717 | 656.18 | 29.91 | 762 | 699.25 | 30.87 | 807 | 742.39 | 31.81 |
| 718 | 657.14 | 29.93 | 763 | 700.21 | 30.89 | 808 | 743.34 | 31.83 |
| 719 | 658.09 | 29.95 | 764 | 701.17 | 30.91 | 809 | 744.30 | 31.85 |
| 720 | 659.05 | 29.98 | 765 | 702.13 | 30.94 | 810 | 745.26 | 31.87 |

Table 4. CONSTANTS FOR $p'$ (SHORT FORMULA), AN APPROXIMATION OF     Section 2
THE LOWER BINOMIAL CONFIDENCE LIMIT $\underline{p}$; $c/n \leqslant .125$, $n \geqslant 16$,    $\gamma_1 = 99\%$
$c \leqslant 1,000$ (Continued)            $\gamma_2 = 98\%$

| c | m | k | c | m | k | c | m | k |
|---|---|---|---|---|---|---|---|---|
| 811 | 746.22 | 31.89 | 856 | 789.41 | 32.80 | 901 | 832.64 | 33.68 |
| 812 | 747.18 | 31.91 | 857 | 790.37 | 32.82 | 902 | 833.60 | 33.70 |
| 813 | 748.14 | 31.93 | 858 | 791.33 | 32.84 | 903 | 834.57 | 33.72 |
| 814 | 749.10 | 31.95 | 859 | 792.29 | 32.86 | 904 | 835.53 | 33.74 |
| 815 | 750.06 | 31.97 | 860 | 793.25 | 32.88 | 905 | 836.49 | 33.76 |
| 816 | 751.02 | 31.99 | 861 | 794.21 | 32.89 | 906 | 837.45 | 33.78 |
| 817 | 751.98 | 32.01 | 862 | 795.17 | 32.91 | 907 | 838.41 | 33.79 |
| 818 | 752.94 | 32.03 | 863 | 796.13 | 32.93 | 908 | 839.37 | 33.81 |
| 819 | 753.90 | 32.05 | 864 | 797.09 | 32.95 | 909 | 840.33 | 33.83 |
| 820 | 754.86 | 32.07 | 865 | 798.05 | 32.97 | 910 | 841.29 | 33.85 |
| 821 | 755.81 | 32.09 | 866 | 799.01 | 32.99 | 911 | 842.26 | 33.87 |
| 822 | 756.77 | 32.11 | 867 | 799.97 | 33.01 | 912 | 843.22 | 33.89 |
| 823 | 757.73 | 32.13 | 868 | 800.93 | 33.03 | 913 | 844.18 | 33.91 |
| 824 | 758.69 | 32.15 | 869 | 801.89 | 33.05 | 914 | 845.14 | 33.93 |
| 825 | 759.65 | 32.17 | 870 | 802.85 | 33.07 | 915 | 846.10 | 33.95 |
| 826 | 760.61 | 32.19 | 871 | 803.82 | 33.09 | 916 | 847.06 | 33.97 |
| 827 | 761.57 | 32.21 | 872 | 804.78 | 33.11 | 917 | 848.03 | 33.99 |
| 828 | 762.53 | 32.23 | 873 | 805.74 | 33.13 | 918 | 848.99 | 34.01 |
| 829 | 763.49 | 32.25 | 874 | 806.70 | 33.15 | 919 | 849.95 | 34.03 |
| 830 | 764.45 | 32.27 | 875 | 807.66 | 33.17 | 920 | 850.91 | 34.04 |
| 831 | 765.41 | 32.29 | 876 | 808.62 | 33.19 | 921 | 851.87 | 34.06 |
| 832 | 766.37 | 32.32 | 877 | 809.58 | 33.21 | 922 | 852.83 | 34.08 |
| 833 | 767.33 | 32.34 | 878 | 810.54 | 33.23 | 923 | 853.80 | 34.10 |
| 834 | 768.29 | 32.36 | 879 | 811.50 | 33.25 | 924 | 854.76 | 34.12 |
| 835 | 769.25 | 32.38 | 880 | 812.46 | 33.27 | 925 | 855.72 | 34.14 |
| 836 | 770.21 | 32.40 | 881 | 813.42 | 33.29 | 926 | 856.68 | 34.16 |
| 837 | 771.17 | 32.42 | 882 | 814.38 | 33.31 | 927 | 857.64 | 34.18 |
| 838 | 772.13 | 32.44 | 883 | 815.34 | 33.33 | 928 | 858.60 | 34.20 |
| 839 | 773.09 | 32.46 | 884 | 816.30 | 33.35 | 929 | 859.57 | 34.22 |
| 840 | 774.05 | 32.48 | 885 | 817.27 | 33.37 | 930 | 860.53 | 34.24 |
| 841 | 775.01 | 32.50 | 886 | 818.23 | 33.39 | 931 | 861.49 | 34.26 |
| 842 | 775.97 | 32.52 | 887 | 819.19 | 33.41 | 932 | 862.45 | 34.27 |
| 843 | 776.93 | 32.54 | 888 | 820.15 | 33.43 | 933 | 863.41 | 34.29 |
| 844 | 777.89 | 32.56 | 889 | 821.11 | 33.45 | 934 | 864.38 | 34.31 |
| 845 | 778.85 | 32.58 | 890 | 822.07 | 33.46 | 935 | 865.34 | 34.33 |
| 846 | 779.81 | 32.60 | 891 | 823.03 | 33.48 | 936 | 866.30 | 34.35 |
| 847 | 780.77 | 32.62 | 892 | 823.99 | 33.50 | 937 | 867.26 | 34.37 |
| 848 | 781.73 | 32.64 | 893 | 824.95 | 33.52 | 938 | 868.22 | 34.39 |
| 849 | 782.69 | 32.66 | 894 | 825.91 | 33.54 | 939 | 869.19 | 34.41 |
| 850 | 783.65 | 32.68 | 895 | 826.88 | 33.56 | 940 | 870.15 | 34.43 |
| 851 | 784.61 | 32.70 | 896 | 827.84 | 33.58 | 941 | 871.11 | 34.45 |
| 852 | 785.57 | 32.72 | 897 | 828.80 | 33.60 | 942 | 872.07 | 34.46 |
| 853 | 786.53 | 32.74 | 898 | 829.76 | 33.62 | 943 | 873.03 | 34.48 |
| 854 | 787.49 | 32.76 | 899 | 830.72 | 33.64 | 944 | 874.00 | 34.50 |
| 855 | 788.45 | 32.78 | 900 | 831.68 | 33.66 | 945 | 874.96 | 34.52 |

Table 4.  CONSTANTS FOR $\underline{p}'$ (SHORT FORMULA), AN APPROXIMATION OF THE LOWER BINOMIAL CONFIDENCE LIMIT $\underline{p}$; $c/n \leqslant .125$, $n \geqslant$ $c \leqslant 1,000$ (*Continued*)

| c | m | k | c | m | k | c | m | k |
|---|---|---|---|---|---|---|---|---|
| 946 | 875.92 | 34.54 | 966 | 895.17 | 34.92 | 986 | 914.42 | 35.2 |
| 947 | 876.88 | 34.56 | 967 | 896.13 | 34.93 | 987 | 915.39 | 35.3 |
| 948 | 877.84 | 34.58 | 968 | 897.09 | 34.95 | 988 | 916.35 | 35.3 |
| 949 | 878.81 | 34.60 | 969 | 898.06 | 34.97 | 989 | 917.31 | 35.3 |
| 950 | 879.77 | 34.62 | 970 | 899.02 | 34.99 | 990 | 918.28 | 35.3 |
| 951 | 880.73 | 34.63 | 971 | 899.98 | 35.01 | 991 | 919.24 | 35.3 |
| 952 | 881.69 | 34.65 | 972 | 900.94 | 35.03 | 992 | 920.20 | 35.4 |
| 953 | 882.66 | 34.67 | 973 | 901.91 | 35.05 | 993 | 921.16 | 35.4 |
| 954 | 883.62 | 34.69 | 974 | 902.87 | 35.07 | 994 | 922.13 | 35.4 |
| 955 | 884.58 | 34.71 | 975 | 903.83 | 35.08 | 995 | 923.09 | 35.4 |
| 956 | 885.54 | 34.73 | 976 | 904.79 | 35.10 | 996 | 924.05 | 35.4 |
| 957 | 886.51 | 34.75 | 977 | 905.76 | 35.12 | 997 | 925.02 | 35.4 |
| 958 | 887.47 | 34.77 | 978 | 906.72 | 35.14 | 998 | 925.98 | 35.5 |
| 959 | 888.43 | 34.78 | 979 | 907.68 | 35.16 | 999 | 926.94 | 35.5 |
| 960 | 889.39 | 34.80 | 980 | 908.65 | 35.18 | 1000 | 927.91 | 35.5 |
| 961 | 890.36 | 34.82 | 981 | 909.61 | 35.20 | | | |
| 962 | 891.32 | 34.84 | 982 | 910.57 | 35.21 | | | |
| 963 | 892.28 | 34.86 | 983 | 911.53 | 35.23 | | | |
| 964 | 893.24 | 34.88 | 984 | 912.50 | 35.25 | | | |
| 965 | 894.21 | 34.90 | 985 | 913.46 | 35.27 | | | |

Table 4. CONSTANTS FOR $p'$ (SHORT FORMULA), AN APPROXIMATION OF THE LOWER BINOMIAL CONFIDENCE LIMIT $\underline{p}$; $c/n \leqslant .125$, $n \geqslant 16$, $c \leqslant 1,000$ (*Continued*)

Section 3

$\gamma_1 = 97.5\%$
$\gamma_2 = 95\%$

| c | m | k | c | m | k | c | m | k |
|---|---|---|---|---|---|---|---|---|
| 1 | .025318 | -0.013 | 46 | 33.678 | 5.66 | 91 | 73.268 | 8.38 |
| 2 | .24221 | 0.379 | 47 | 34.534 | 5.73 | 92 | 74.165 | 8.43 |
| 3 | .61867 | 0.691 | 48 | 35.391 | 5.80 | 93 | 75.063 | 8.48 |
| 4 | 1.0899 | 0.955 | 49 | 36.250 | 5.87 | 94 | 75.962 | 8.53 |
| 5 | 1.6235 | 1.188 | 50 | 37.111 | 5.94 | 95 | 76.861 | 8.58 |
| 6 | 2.2019 | 1.399 | 51 | 37.973 | 6.03 | 96 | 77.761 | 8.63 |
| 7 | 2.8144 | 1.593 | 52 | 38.836 | 6.09 | 97 | 78.661 | 8.68 |
| 8 | 3.4538 | 1.773 | 53 | 39.701 | 6.16 | 98 | 79.561 | 8.73 |
| 9 | 4.1154 | 1.942 | 54 | 40.566 | 6.23 | 99 | 80.463 | 8.78 |
| 10 | 4.7954 | 2.10 | 55 | 41.434 | 6.29 | 100 | 81.364 | 8.83 |
| 11 | 5.4912 | 2.25 | 56 | 42.302 | 6.36 | 101 | 82.266 | 8.88 |
| 12 | 6.2006 | 2.40 | 57 | 43.171 | 6.43 | 102 | 83.169 | 8.92 |
| 13 | 6.9220 | 2.54 | 58 | 44.042 | 6.49 | 103 | 84.072 | 8.97 |
| 14 | 7.6540 | 2.67 | 59 | 44.914 | 6.55 | 104 | 84.976 | 9.02 |
| 15 | 8.3954 | 2.80 | 60 | 45.786 | 6.62 | 105 | 85.880 | 9.07 |
| 16 | 9.1454 | 2.93 | 61 | 46.660 | 6.68 | 106 | 86.784 | 9.12 |
| 17 | 9.9032 | 3.05 | 62 | 47.535 | 6.74 | 107 | 87.689 | 9.16 |
| 18 | 10.668 | 3.17 | 63 | 48.411 | 6.80 | 108 | 88.595 | 9.21 |
| 19 | 11.439 | 3.28 | 64 | 49.288 | 6.87 | 109 | 89.501 | 9.26 |
| 20 | 12.217 | 3.39 | 65 | 50.166 | 6.93 | 110 | 90.407 | 9.30 |
| 21 | 12.999 | 3.50 | 66 | 51.045 | 6.99 | 111 | 91.314 | 9.35 |
| 22 | 13.787 | 3.61 | 67 | 51.924 | 7.05 | 112 | 92.221 | 9.40 |
| 23 | 14.580 | 3.71 | 68 | 52.805 | 7.11 | 113 | 93.128 | 9.44 |
| 24 | 15.377 | 3.81 | 69 | 53.686 | 7.17 | 114 | 94.036 | 9.49 |
| 25 | 16.179 | 3.91 | 70 | 54.569 | 7.23 | 115 | 94.945 | 9.54 |
| 26 | 16.984 | 4.01 | 71 | 55.452 | 7.28 | 116 | 95.853 | 9.58 |
| 27 | 17.793 | 4.10 | 72 | 56.336 | 7.34 | 117 | 96.762 | 9.63 |
| 28 | 18.606 | 4.20 | 73 | 57.221 | 7.40 | 118 | 97.672 | 9.67 |
| 29 | 19.422 | 4.29 | 74 | 58.106 | 7.46 | 119 | 98.582 | 9.72 |
| 30 | 20.241 | 4.38 | 75 | 58.993 | 7.51 | 120 | 99.492 | 9.76 |
| 31 | 21.063 | 4.47 | 76 | 59.880 | 7.57 | 121 | 100.40 | 9.81 |
| 32 | 21.888 | 4.56 | 77 | 60.767 | 7.63 | 122 | 101.31 | 9.85 |
| 33 | 22.716 | 4.64 | 78 | 61.656 | 7.68 | 123 | 102.23 | 9.90 |
| 34 | 23.546 | 4.73 | 79 | 62.545 | 7.74 | 124 | 103.14 | 9.94 |
| 35 | 24.379 | 4.81 | 80 | 63.435 | 7.79 | 125 | 104.05 | 9.98 |
| 36 | 25.214 | 4.89 | 81 | 64.326 | 7.85 | 126 | 104.96 | 10.03 |
| 37 | 26.051 | 4.97 | 82 | 65.217 | 7.90 | 127 | 105.87 | 10.07 |
| 38 | 26.891 | 5.05 | 83 | 66.109 | 7.95 | 128 | 106.79 | 10.11 |
| 39 | 27.733 | 5.13 | 84 | 67.002 | 8.01 | 129 | 107.70 | 10.16 |
| 40 | 28.577 | 5.21 | 85 | 67.895 | 8.06 | 130 | 108.61 | 10.20 |
| 41 | 29.422 | 5.29 | 86 | 68.789 | 8.11 | 131 | 109.53 | 10.24 |
| 42 | 30.270 | 5.37 | 87 | 69.684 | 8.17 | 132 | 110.44 | 10.29 |
| 43 | 31.119 | 5.44 | 88 | 70.579 | 8.22 | 133 | 111.36 | 10.33 |
| 44 | 31.970 | 5.51 | 89 | 71.475 | 8.27 | 134 | 112.27 | 10.37 |
| 45 | 32.823 | 5.59 | 90 | 72.371 | 8.32 | 135 | 113.19 | 10.41 |

**Table 4. CONSTANTS FOR p′ (SHORT FORMULA), AN APPROXIMATION OF THE LOWER BINOMIAL CONFIDENCE LIMIT p; c/n ≤ .125, n ≥ 16, c ≤ 1,000** (*Continued*)

| c | m | k | c | m | k | c | m | k |
|---|---|---|---|---|---|---|---|---|
| 136 | 114.10 | 10.46 | 181 | 155.58 | 12.21 | 226 | 197.48 | 13.76 |
| 137 | 115.02 | 10.50 | 182 | 156.51 | 12.25 | 227 | 198.42 | 13.79 |
| 138 | 115.94 | 10.54 | 183 | 157.43 | 12.28 | 228 | 199.35 | 13.82 |
| 139 | 116.85 | 10.58 | 184 | 158.36 | 12.32 | 229 | 200.29 | 13.86 |
| 140 | 117.77 | 10.62 | 185 | 159.29 | 12.36 | 230 | 201.22 | 13.89 |
| 141 | 118.69 | 10.66 | 186 | 160.22 | 12.39 | 231 | 202.16 | 13.92 |
| 142 | 119.61 | 10.70 | 187 | 161.14 | 12.43 | 232 | 203.09 | 13.95 |
| 143 | 120.52 | 10.75 | 188 | 162.07 | 12.46 | 233 | 204.03 | 13.99 |
| 144 | 121.44 | 10.79 | 189 | 163.00 | 12.50 | 234 | 204.96 | 14.02 |
| 145 | 122.36 | 10.83 | 190 | 163.93 | 12.53 | 235 | 205.90 | 14.05 |
| 146 | 123.28 | 10.87 | 191 | 164.86 | 12.57 | 236 | 206.84 | 14.08 |
| 147 | 124.20 | 10.91 | 192 | 165.79 | 12.61 | 237 | 207.77 | 14.11 |
| 148 | 125.12 | 10.95 | 193 | 166.72 | 12.64 | 238 | 208.71 | 14.15 |
| 149 | 126.04 | 10.99 | 194 | 167.65 | 12.68 | 239 | 209.65 | 14.18 |
| 150 | 126.96 | 11.03 | 195 | 168.58 | 12.71 | 240 | 210.58 | 14.21 |
| 151 | 127.88 | 11.07 | 196 | 169.51 | 12.75 | 241 | 211.52 | 14.24 |
| 152 | 128.80 | 11.11 | 197 | 170.44 | 12.78 | 242 | 212.46 | 14.27 |
| 153 | 129.72 | 11.15 | 198 | 171.37 | 12.82 | 243 | 213.39 | 14.30 |
| 154 | 130.64 | 11.19 | 199 | 172.30 | 12.85 | 244 | 214.33 | 14.33 |
| 155 | 131.56 | 11.23 | 200 | 173.23 | 12.89 | 245 | 215.27 | 14.37 |
| 156 | 132.48 | 11.27 | 201 | 174.16 | 12.92 | 246 | 216.21 | 14.40 |
| 157 | 133.40 | 11.31 | 202 | 175.09 | 12.95 | 247 | 217.14 | 14.43 |
| 158 | 134.32 | 11.34 | 203 | 176.02 | 12.99 | 248 | 218.08 | 14.46 |
| 159 | 135.25 | 11.38 | 204 | 176.95 | 13.02 | 249 | 219.02 | 14.49 |
| 160 | 136.17 | 11.42 | 205 | 177.88 | 13.06 | 250 | 219.96 | 14.52 |
| 161 | 137,08 | 11.46 | 206 | 178.82 | 13.09 | 251 | 220.89 | 14.55 |
| 162 | 138.00 | 11.50 | 207 | 179.75 | 13.13 | 252 | 221.83 | 14.58 |
| 163 | 138.92 | 11.54 | 208 | 180.68 | 13.16 | 253 | 222.77 | 14.61 |
| 164 | 139.85 | 11.58 | 209 | 181.61 | 13.19 | 254 | 223.71 | 14.65 |
| 165 | 140.77 | 11.61 | 210 | 182.54 | 13.23 | 255 | 224.65 | 14.68 |
| 166 | 141.69 | 11.65 | 211 | 183.48 | 13.26 | 256 | 225.59 | 14.71 |
| 167 | 142.62 | 11.69 | 212 | 184.41 | 13.30 | 257 | 226.53 | 14.74 |
| 168 | 143.54 | 11.73 | 213 | 185.34 | 13.33 | 258 | 227.46 | 14.77 |
| 169 | 144.47 | 11.77 | 214 | 186.27 | 13.36 | 259 | 228.40 | 14.80 |
| 170 | 145.39 | 11.80 | 215 | 187.21 | 13.40 | 260 | 229.34 | 14.83 |
| 171 | 146.32 | 11.84 | 216 | 188.14 | 13.43 | 261 | 230.28 | 14.86 |
| 172 | 147.24 | 11.88 | 217 | 189.07 | 13.46 | 262 | 231.22 | 14.89 |
| 173 | 148.17 | 11.92 | 218 | 190.01 | 13.50 | 263 | 232.16 | 14.92 |
| 174 | 149.09 | 11.95 | 219 | 190.94 | 13.53 | 264 | 233.10 | 14.95 |
| 175 | 150.02 | 11.99 | 220 | 191.88 | 13.56 | 265 | 234.04 | 14.98 |
| 176 | 150.94 | 12.03 | 221 | 192.81 | 13.60 | 266 | 234.98 | 15.01 |
| 177 | 151.87 | 12.06 | 222 | 193.74 | 13.63 | 267 | 235.92 | 15.04 |
| 178 | 152.80 | 12.10 | 223 | 194.68 | 13.66 | 268 | 236.86 | 15.07 |
| 179 | 153.72 | 12.14 | 224 | 195.61 | 13.69 | 269 | 237.80 | 15.10 |
| 180 | 154.65 | 12.17 | 225 | 196.55 | 13.73 | 270 | 238.74 | 15.13 |

**Table 4.** **CONSTANTS FOR** $\underline{p}'$ **(SHORT FORMULA), AN APPROXIMATION OF**  **Section 3**
**THE LOWER BINOMIAL CONFIDENCE LIMIT** $\underline{p}$; $c/n \leqslant .125$, $n \geqslant 16$,  $\gamma_1 = 97.5\%$
$c \leqslant 1,000$ (*Continued*)  $\gamma_2 = 95\%$

| c | m | k | c | m | k | c | m | k |
|---|---|---|---|---|---|---|---|---|
| 271 | 239.68 | 15.16 | 316 | 282.11 | 16.45 | 361 | 324.71 | 17.65 |
| 272 | 240.62 | 15.19 | 317 | 283.05 | 16.47 | 362 | 325.66 | 17.67 |
| 273 | 241.56 | 15.22 | 318 | 284.00 | 16.50 | 363 | 326.60 | 17.70 |
| 274 | 242.50 | 15.25 | 319 | 284.94 | 16.53 | 364 | 327.55 | 17.72 |
| 275 | 243.44 | 15.28 | 320 | 285.89 | 16.56 | 365 | 328.50 | 17.75 |
| 276 | 244.39 | 15.31 | 321 | 286.83 | 16.58 | 366 | 329.45 | 17.77 |
| 277 | 245.33 | 15.34 | 322 | 287.78 | 16.61 | 367 | 330.40 | 17.80 |
| 278 | 246.27 | 15.37 | 323 | 288.72 | 16.64 | 368 | 331.35 | 17.83 |
| 279 | 247.21 | 15.40 | 324 | 289.67 | 16.67 | 369 | 332.30 | 17.85 |
| 280 | 248.15 | 15.42 | 325 | 290.61 | 16.69 | 370 | 333.25 | 17.88 |
| 281 | 249.09 | 15.45 | 326 | 291.56 | 16.72 | 371 | 334.19 | 17.90 |
| 282 | 250.03 | 15.48 | 327 | 292.50 | 16.75 | 372 | 335.14 | 17.93 |
| 283 | 250.97 | 15.51 | 328 | 293.45 | 16.77 | 373 | 336.09 | 17.95 |
| 284 | 251.92 | 15.54 | 329 | 294.40 | 16.80 | 374 | 337.04 | 17.98 |
| 285 | 252.86 | 15.57 | 330 | 295.34 | 16.83 | 375 | 337.99 | 18.00 |
| 286 | 253.80 | 15.60 | 331 | 296.29 | 16.86 | 376 | 338.94 | 18.03 |
| 287 | 254.74 | 15.63 | 332 | 297.23 | 16.88 | 377 | 339.89 | 18.05 |
| 288 | 255.68 | 15.66 | 333 | 298.18 | 16.91 | 378 | 340.84 | 18.08 |
| 289 | 256.63 | 15.69 | 334 | 299.13 | 16.94 | 379 | 341.79 | 18.10 |
| 290 | 257.57 | 15.72 | 335 | 300.07 | 16.96 | 380 | 342.74 | 18.13 |
| 291 | 258.51 | 15.74 | 336 | 301.02 | 16.99 | 381 | 343.69 | 18.16 |
| 292 | 259.45 | 15.77 | 337 | 301.97 | 17.02 | 382 | 344.64 | 18.18 |
| 293 | 260.40 | 15.80 | 338 | 302.91 | 17.04 | 383 | 345.59 | 18.21 |
| 294 | 261.34 | 15.83 | 339 | 303.86 | 17.07 | 384 | 346.54 | 18.23 |
| 295 | 262.28 | 15.86 | 340 | 304.81 | 17.10 | 385 | 347.49 | 18.26 |
| 296 | 263.23 | 15.89 | 341 | 305.75 | 17.12 | 386 | 348.44 | 18.28 |
| 297 | 264.17 | 15.92 | 342 | 306.70 | 17.15 | 387 | 349.39 | 18.31 |
| 298 | 265.11 | 15.94 | 343 | 307.65 | 17.18 | 388 | 350.34 | 18.33 |
| 299 | 266.06 | 15.97 | 344 | 308.59 | 17.20 | 389 | 351.29 | 18.36 |
| 300 | 267.00 | 16.00 | 345 | 309.54 | 17.23 | 390 | 352.24 | 18.38 |
| 301 | 267.94 | 16.03 | 346 | 310.49 | 17.26 | 391 | 353.19 | 18.40 |
| 302 | 268.89 | 16.06 | 347 | 311.44 | 17.28 | 392 | 354.14 | 18.43 |
| 303 | 269.83 | 16.09 | 348 | 312.38 | 17.31 | 393 | 355.09 | 18.45 |
| 304 | 270.77 | 16.11 | 349 | 313.33 | 17.33 | 394 | 356.04 | 18.48 |
| 305 | 271.72 | 16.14 | 350 | 314.28 | 17.36 | 395 | 356.99 | 18.50 |
| 306 | 272.66 | 16.17 | 351 | 315.23 | 17.39 | 396 | 357.94 | 18.53 |
| 307 | 273.61 | 16.20 | 352 | 316.17 | 17.41 | 397 | 358.89 | 18.55 |
| 308 | 274.55 | 16.23 | 353 | 317.12 | 17.44 | 398 | 359.85 | 18.58 |
| 309 | 275.49 | 16.25 | 354 | 318.07 | 17.46 | 399 | 360.80 | 18.60 |
| 310 | 276.44 | 16.28 | 355 | 319.02 | 17.49 | 400 | 361.75 | 18.63 |
| 311 | 277.38 | 16.31 | 356 | 319.97 | 17.52 | 401 | 362.70 | 18.65 |
| 312 | 278.33 | 16.34 | 357 | 320.91 | 17.54 | 402 | 363.65 | 18.68 |
| 313 | 279.27 | 16.36 | 358 | 321.86 | 17.57 | 403 | 364.60 | 18.70 |
| 314 | 280.22 | 16.39 | 359 | 322.81 | 17.59 | 404 | 365.55 | 18.72 |
| 315 | 281.16 | 16.42 | 360 | 323.76 | 17.62 | 405 | 366.50 | 18.75 |

Table 4.  CONSTANTS FOR $\underline{p}'$ (SHORT FORMULA), AN APPROXIMATION OF THE LOWER BINOMIAL CONFIDENCE LIMIT $\underline{p}$; $c/n \leqslant .125$, $n \geqslant 16$
$c \leqslant 1,000$ (*Continued*)

| c | m | k | c | m | k | c | m | k |
|---|---|---|---|---|---|---|---|---|
| 406 | 367.45 | 18.77 | 451 | 410.32 | 19.84 | 496 | 453.30 | 20.85 |
| 407 | 368.41 | 18.80 | 452 | 411.28 | 19.86 | 497 | 454.25 | 20.87 |
| 408 | 369.36 | 18.82 | 453 | 412.23 | 19.88 | 498 | 455.21 | 20.90 |
| 409 | 370.31 | 18.85 | 454 | 413.18 | 19.91 | 499 | 456.16 | 20.92 |
| 410 | 371.26 | 18.87 | 455 | 414.14 | 19.93 | 500 | 457.12 | 20.94 |
| 411 | 372.21 | 18.89 | 456 | 415.09 | 19.95 | 501 | 458.08 | 20.96 |
| 412 | 373.16 | 18.92 | 457 | 416.05 | 19.98 | 502 | 459.03 | 20.98 |
| 413 | 374.12 | 18.94 | 458 | 417.00 | 20.00 | 503 | 459.99 | 21.01 |
| 414 | 375.07 | 18.97 | 459 | 417.96 | 20.02 | 504 | 460.95 | 21.03 |
| 415 | 376.02 | 18.99 | 460 | 418.91 | 20.05 | 505 | 461.90 | 21.05 |
| 416 | 376.97 | 19.01 | 461 | 419.86 | 20.07 | 506 | 462.86 | 21.07 |
| 417 | 377.92 | 19.04 | 462 | 420.82 | 20.09 | 507 | 463.81 | 21.09 |
| 418 | 378.87 | 19.06 | 463 | 421.77 | 20.11 | 508 | 464.77 | 21.11 |
| 419 | 379.83 | 19.09 | 464 | 422.73 | 20.14 | 509 | 465.73 | 21.14 |
| 420 | 380.78 | 19.11 | 465 | 423.68 | 20.16 | 510 | 466.68 | 21.16 |
| 421 | 381.73 | 19.13 | 466 | 424.64 | 20.18 | 511 | 467.64 | 21.18 |
| 422 | 382.68 | 19.16 | 467 | 425.59 | 20.20 | 512 | 468.60 | 21.20 |
| 423 | 383.64 | 19.18 | 468 | 426.55 | 20.23 | 513 | 469.55 | 21.22 |
| 424 | 384.59 | 19.21 | 469 | 427.50 | 20.25 | 514 | 470.51 | 21.24 |
| 425 | 385.54 | 19.23 | 470 | 428.46 | 20.27 | 515 | 471.47 | 21.27 |
| 426 | 386.49 | 19.25 | 471 | 429.41 | 20.29 | 516 | 472.42 | 21.29 |
| 427 | 387.45 | 19.28 | 472 | 430.37 | 20.32 | 517 | 473.38 | 21.31 |
| 428 | 388.40 | 19.30 | 473 | 431.32 | 20.34 | 518 | 474.34 | 21.33 |
| 429 | 389.35 | 19.32 | 474 | 432.27 | 20.36 | 519 | 475.30 | 21.35 |
| 430 | 390.30 | 19.35 | 475 | 433.23 | 20.39 | 520 | 476.25 | 21.37 |
| 431 | 391.26 | 19.37 | 476 | 434.19 | 20.41 | 521 | 477.21 | 21.40 |
| 432 | 392.21 | 19.40 | 477 | 435.14 | 20.43 | 522 | 478.17 | 21.42 |
| 433 | 393.16 | 19.42 | 478 | 436.10 | 20.45 | 523 | 479.12 | 21.44 |
| 434 | 394.12 | 19.44 | 479 | 437.05 | 20.47 | 524 | 480.08 | 21.46 |
| 435 | 395.07 | 19.47 | 480 | 438.01 | 20.50 | 525 | 481.04 | 21.48 |
| 436 | 396.02 | 19.49 | 481 | 438.96 | 20.52 | 526 | 482.00 | 21.50 |
| 437 | 396.97 | 19.51 | 482 | 439.92 | 20.54 | 527 | 482.95 | 21.52 |
| 438 | 397.93 | 19.54 | 483 | 440.87 | 20.56 | 528 | 483.91 | 21.55 |
| 439 | 398.88 | 19.56 | 484 | 441.83 | 20.59 | 529 | 484.87 | 21.57 |
| 440 | 399.83 | 19.58 | 485 | 442.78 | 20.61 | 530 | 485.82 | 21.59 |
| 441 | 400.79 | 19.61 | 486 | 443.74 | 20.63 | 531 | 486.78 | 21.61 |
| 442 | 401.74 | 19.63 | 487 | 444.69 | 20.65 | 532 | 487.74 | 21.63 |
| 443 | 402.69 | 19.65 | 488 | 445.65 | 20.68 | 533 | 488.70 | 21.65 |
| 444 | 403.65 | 19.68 | 489 | 446.61 | 20.70 | 534 | 489.65 | 21.67 |
| 445 | 404.60 | 19.70 | 490 | 447.56 | 20.72 | 535 | 490.61 | 21.69 |
| 446 | 405.55 | 19.72 | 491 | 448.52 | 20.74 | 536 | 491.57 | 21.72 |
| 447 | 406.51 | 19.75 | 492 | 449.47 | 20.76 | 537 | 492.53 | 21.74 |
| 448 | 407.46 | 19.77 | 493 | 450.43 | 20.79 | 538 | 493.49 | 21.76 |
| 449 | 408.42 | 19.79 | 494 | 451.38 | 20.81 | 539 | 494.44 | 21.78 |
| 450 | 409.37 | 19.82 | 495 | 452.34 | 20.83 | 540 | 495.40 | 21.80 |

## Table 4. CONSTANTS FOR p′ (SHORT FORMULA), AN APPROXIMATION OF THE LOWER BINOMIAL CONFIDENCE LIMIT p; c/n ≤ .125, n ≥ 16, c ≤ 1,000 (*Continued*)

$\gamma_1 = 97.5\%$
$\gamma_2 = 95\%$

| c | m | k | c | m | k | c | m | k |
|---|---|---|---|---|---|---|---|---|
| 541 | 496.36 | 21.82 | 586 | 539.50 | 22.75 | 631 | 582.71 | 23.64 |
| 542 | 497.32 | 21.84 | 587 | 540.46 | 22.77 | 632 | 583.67 | 23.66 |
| 543 | 498.27 | 21.86 | 588 | 541.42 | 22.79 | 633 | 584.63 | 23.68 |
| 544 | 499.23 | 21.88 | 589 | 542.38 | 22.81 | 634 | 585.60 | 23.70 |
| 545 | 500.19 | 21.90 | 590 | 543.34 | 22.83 | 635 | 586.56 | 23.72 |
| 546 | 501.15 | 21.93 | 591 | 544.30 | 22.85 | 636 | 587.52 | 23.74 |
| 547 | 502.11 | 21.95 | 592 | 545.26 | 22.87 | 637 | 588.48 | 23.76 |
| 548 | 503.06 | 21.97 | 593 | 546.22 | 22.89 | 638 | 589.44 | 23.78 |
| 549 | 504.02 | 21.99 | 594 | 547.18 | 22.91 | 639 | 590.40 | 23.80 |
| 550 | 504.98 | 22.01 | 595 | 548.14 | 22.93 | 640 | 591.36 | 23.82 |
| 551 | 505.94 | 22.03 | 596 | 549.10 | 22.95 | 641 | 592.32 | 23.84 |
| 552 | 506.90 | 22.05 | 597 | 550.06 | 22.97 | 642 | 593.29 | 23.86 |
| 553 | 507.86 | 22.07 | 598 | 551.02 | 22.99 | 643 | 594.25 | 23.88 |
| 554 | 508.81 | 22.09 | 599 | 551.98 | 23.01 | 644 | 595.21 | 23.90 |
| 555 | 509.77 | 22.11 | 600 | 552.94 | 23.03 | 645 | 596.17 | 23.92 |
| 556 | 510.73 | 22.13 | 601 | 553.90 | 23.05 | 646 | 597.13 | 23.93 |
| 557 | 511.69 | 22.16 | 602 | 554.86 | 23.07 | 647 | 598.09 | 23.95 |
| 558 | 512.65 | 22.18 | 603 | 555.82 | 23.09 | 648 | 599.05 | 23.97 |
| 559 | 513.61 | 22.20 | 604 | 556.78 | 23.11 | 649 | 600.02 | 23.99 |
| 560 | 514.57 | 22.22 | 605 | 557.74 | 23.13 | 650 | 600.98 | 24.01 |
| 561 | 515.52 | 22.24 | 606 | 558.70 | 23.15 | 651 | 601.94 | 24.03 |
| 562 | 516.48 | 22.26 | 607 | 559.66 | 23.17 | 652 | 602.90 | 24.05 |
| 563 | 517.44 | 22.28 | 608 | 560.62 | 23.19 | 653 | 603.86 | 24.07 |
| 564 | 518.40 | 22.30 | 609 | 561.58 | 23.21 | 654 | 604.82 | 24.09 |
| 565 | 519.36 | 22.32 | 610 | 562.54 | 23.23 | 655 | 605.78 | 24.11 |
| 566 | 520.32 | 22.34 | 611 | 563.50 | 23.25 | 656 | 606.75 | 24.13 |
| 567 | 521.28 | 22.36 | 612 | 564.46 | 23.27 | 657 | 607.71 | 24.15 |
| 568 | 522.23 | 22.38 | 613 | 565.42 | 23.29 | 658 | 608.67 | 24.16 |
| 569 | 523.19 | 22.40 | 614 | 566.38 | 23.31 | 659 | 609.63 | 24.18 |
| 570 | 524.15 | 22.42 | 615 | 567.34 | 23.33 | 660 | 610.59 | 24.20 |
| 571 | 525.11 | 22.44 | 616 | 568.30 | 23.35 | 661 | 611.56 | 24.22 |
| 572 | 526.07 | 22.46 | 617 | 569.26 | 23.37 | 662 | 612.52 | 24.24 |
| 573 | 527.03 | 22.49 | 618 | 570.22 | 23.39 | 663 | 613.48 | 24.26 |
| 574 | 527.99 | 22.51 | 619 | 571.18 | 23.41 | 664 | 614.44 | 24.28 |
| 575 | 528.95 | 22.53 | 620 | 572.14 | 23.43 | 665 | 615.40 | 24.30 |
| 576 | 529.91 | 22.55 | 621 | 573.10 | 23.45 | 666 | 616.37 | 24.32 |
| 577 | 530.87 | 22.57 | 622 | 574.06 | 23.47 | 667 | 617.33 | 24.34 |
| 578 | 531.83 | 22.59 | 623 | 575.03 | 23.49 | 668 | 618.29 | 24.36 |
| 579 | 532.78 | 22.61 | 624 | 575.99 | 23.51 | 669 | 619.25 | 24.37 |
| 580 | 533.74 | 22.63 | 625 | 576.95 | 23.53 | 670 | 620.21 | 24.39 |
| 581 | 534.70 | 22.65 | 626 | 577.91 | 23.55 | 671 | 621.18 | 24.41 |
| 582 | 535.66 | 22.67 | 627 | 578.87 | 23.57 | 672 | 622.14 | 24.43 |
| 583 | 536.62 | 22.69 | 628 | 579.83 | 23.59 | 673 | 623.10 | 24.45 |
| 584 | 537.58 | 22.71 | 629 | 580.79 | 23.60 | 674 | 624.06 | 24.47 |
| 585 | 538.54 | 22.73 | 630 | 581.75 | 23.62 | 675 | 625.02 | 24.49 |

Table 4.  CONSTANTS FOR $\underline{p}'$ (SHORT FORMULA), AN APPROXIMATION OF
THE LOWER BINOMIAL CONFIDENCE LIMIT $\underline{p}$; $c/n \leqslant .125$, $n \geqslant 16$,
$c \leqslant 1,000$ (*Continued*)

| c | $\underline{m}$ | $\underline{k}$ | c | $\underline{m}$ | $\underline{k}$ | c | $\underline{m}$ | $\underline{k}$ |
|---|---|---|---|---|---|---|---|---|
| 676 | 625.99 | 24.51 | 721 | 669.32 | 25.34 | 766 | 712.70 | 26.15 |
| 677 | 626.95 | 24.53 | 722 | 670.28 | 25.36 | 767 | 713.67 | 26.17 |
| 678 | 627.91 | 24.54 | 723 | 671.25 | 25.38 | 768 | 714.63 | 26.18 |
| 679 | 628.87 | 24.56 | 724 | 672.21 | 25.40 | 769 | 715.59 | 26.20 |
| 680 | 629.84 | 24.58 | 725 | 673.17 | 25.41 | 770 | 716.56 | 26.22 |
| 681 | 630.80 | 24.60 | 726 | 674.14 | 25.43 | 771 | 717.52 | 26.24 |
| 682 | 631.76 | 24.62 | 727 | 675.10 | 25.45 | 772 | 718.49 | 26.26 |
| 683 | 632.72 | 24.64 | 728 | 676.06 | 25.47 | 773 | 719.45 | 26.27 |
| 684 | 633.69 | 24.66 | 729 | 677.03 | 25.49 | 774 | 720.42 | 26.29 |
| 685 | 634.65 | 24.68 | 730 | 677.99 | 25.50 | 775 | 721.38 | 26.31 |
| 686 | 635.61 | 24.69 | 731 | 678.95 | 25.52 | 776 | 722.35 | 26.33 |
| 687 | 636.57 | 24.71 | 732 | 679.92 | 25.54 | 777 | 723.31 | 26.34 |
| 688 | 637.54 | 24.73 | 733 | 680.88 | 25.56 | 778 | 724.28 | 26.36 |
| 689 | 638.50 | 24.75 | 734 | 681.85 | 25.58 | 779 | 725.24 | 26.38 |
| 690 | 639.46 | 24.77 | 735 | 682.81 | 25.60 | 780 | 726.21 | 26.40 |
| 691 | 640.42 | 24.79 | 736 | 683.77 | 25.61 | 781 | 727.17 | 26.41 |
| 692 | 641.39 | 24.81 | 737 | 684.74 | 25.63 | 782 | 728.14 | 26.43 |
| 693 | 642.35 | 24.82 | 738 | 685.70 | 25.65 | 783 | 729.10 | 26.45 |
| 694 | 643.31 | 24.84 | 739 | 686.67 | 25.67 | 784 | 730.07 | 26.47 |
| 695 | 644.28 | 24.86 | 740 | 687.63 | 25.69 | 785 | 731.03 | 26.48 |
| 696 | 645.24 | 24.88 | 741 | 688.59 | 25.70 | 786 | 732.00 | 26.50 |
| 697 | 646.20 | 24.90 | 742 | 689.56 | 25.72 | 787 | 732.96 | 26.52 |
| 698 | 647.16 | 24.92 | 743 | 690.52 | 25.74 | 788 | 733.93 | 26.54 |
| 699 | 648.13 | 24.94 | 744 | 691.49 | 25.76 | 789 | 734.89 | 26.55 |
| 700 | 649.09 | 24.95 | 745 | 692.45 | 25.78 | 790 | 735.86 | 26.57 |
| 701 | 650.05 | 24.97 | 746 | 693.41 | 25.79 | 791 | 736.82 | 26.59 |
| 702 | 651.02 | 24.99 | 747 | 694.38 | 25.81 | 792 | 737.79 | 26.61 |
| 703 | 651.98 | 25.01 | 748 | 695.34 | 25.83 | 793 | 738.75 | 26.62 |
| 704 | 652.94 | 25.03 | 749 | 696.31 | 25.85 | 794 | 739.72 | 26.64 |
| 705 | 653.91 | 25.05 | 750 | 697.27 | 25.86 | 795 | 740.68 | 26.66 |
| 706 | 654.87 | 25.07 | 751 | 698.23 | 25.88 | 796 | 741.65 | 26.68 |
| 707 | 655.83 | 25.08 | 752 | 699.20 | 25.90 | 797 | 742.61 | 26.69 |
| 708 | 656.79 | 25.10 | 753 | 700.16 | 25.92 | 798 | 743.58 | 26.71 |
| 709 | 657.76 | 25.12 | 754 | 701.13 | 25.94 | 799 | 744.54 | 26.73 |
| 710 | 658.72 | 25.14 | 755 | 702.09 | 25.95 | 800 | 745.51 | 26.74 |
| 711 | 659.68 | 25.16 | 756 | 703.06 | 25.97 | 801 | 746.48 | 26.76 |
| 712 | 660.65 | 25.18 | 757 | 704.02 | 25.99 | 802 | 747.44 | 26.78 |
| 713 | 661.61 | 25.19 | 758 | 704.98 | 26.01 | 803 | 748.41 | 26.80 |
| 714 | 662.57 | 25.21 | 759 | 705.95 | 26.03 | 804 | 749.37 | 26.81 |
| 715 | 663.54 | 25.23 | 760 | 706.91 | 26.04 | 805 | 750.34 | 26.83 |
| 716 | 664.50 | 25.25 | 761 | 707.88 | 26.06 | 806 | 751.30 | 26.85 |
| 717 | 665.46 | 25.27 | 762 | 708.84 | 26.08 | 807 | 752.27 | 26.87 |
| 718 | 666.43 | 25.29 | 763 | 709.81 | 26.10 | 808 | 753.23 | 26.88 |
| 719 | 667.39 | 25.30 | 764 | 710.77 | 26.11 | 809 | 754.20 | 26.90 |
| 720 | 668.35 | 25.32 | 765 | 711.74 | 26.13 | 810 | 755.16 | 26.92 |

Table 4. CONSTANTS FOR p′ (SHORT FORMULA), AN APPROXIMATION OF    Section 3
THE LOWER BINOMIAL CONFIDENCE LIMIT p; c/n ≤ .125, n ≥ 16,    $\gamma_1 = 97.5\%$
c ≤ 1,000 (Continued)    $\gamma_2 = 95\%$

| c | m | k | c | m | k | c | m | k |
|---|---|---|---|---|---|---|---|---|
| 811 | 756.13 | 26.93 | 856 | 799.60 | 27.70 | 901 | 843.11 | 28.44 |
| 812 | 757.10 | 26.95 | 857 | 800.57 | 27.72 | 902 | 844.08 | 28.46 |
| 813 | 758.06 | 26.97 | 858 | 801.54 | 27.73 | 903 | 845.05 | 28.48 |
| 814 | 759.03 | 26.99 | 859 | 802.50 | 27.75 | 904 | 846.02 | 28.49 |
| 815 | 759.99 | 27.00 | 860 | 803.47 | 27.77 | 905 | 846.98 | 28.51 |
| 816 | 760.96 | 27.02 | 861 | 804.44 | 27.78 | 906 | 847.95 | 28.52 |
| 817 | 761.92 | 27.04 | 862 | 805.40 | 27.80 | 907 | 848.92 | 28.54 |
| 818 | 762.89 | 27.06 | 863 | 806.37 | 27.82 | 908 | 849.89 | 28.56 |
| 819 | 763.86 | 27.07 | 864 | 807.34 | 27.83 | 909 | 850.85 | 28.57 |
| 820 | 764.82 | 27.09 | 865 | 808.30 | 27.85 | 910 | 851.82 | 28.59 |
| 821 | 765.79 | 27.11 | 866 | 809.27 | 27.87 | 911 | 852.79 | 28.61 |
| 822 | 766.75 | 27.12 | 867 | 810.24 | 27.88 | 912 | 853.76 | 28.62 |
| 823 | 767.72 | 27.14 | 868 | 811.20 | 27.90 | 913 | 854.72 | 28.64 |
| 824 | 768.68 | 27.16 | 869 | 812.17 | 27.92 | 914 | 855.69 | 28.65 |
| 825 | 769.65 | 27.17 | 870 | 813.14 | 27.93 | 915 | 856.66 | 28.67 |
| 826 | 770.62 | 27.19 | 871 | 814.10 | 27.95 | 916 | 857.63 | 28.69 |
| 827 | 771.58 | 27.21 | 872 | 815.07 | 27.97 | 917 | 858.59 | 28.70 |
| 828 | 772.55 | 27.23 | 873 | 816.04 | 27.98 | 918 | 859.56 | 28.72 |
| 829 | 773.51 | 27.24 | 874 | 817.00 | 28.00 | 919 | 860.53 | 28.74 |
| 830 | 774.48 | 27.26 | 875 | 817.97 | 28.02 | 920 | 861.50 | 28.75 |
| 831 | 775.45 | 27.28 | 876 | 818.94 | 28.03 | 921 | 862.47 | 28.77 |
| 832 | 776.41 | 27.29 | 877 | 819.90 | 28.05 | 922 | 863.43 | 28.78 |
| 833 | 777.38 | 27.31 | 878 | 820.87 | 28.06 | 923 | 864.40 | 28.80 |
| 834 | 778.34 | 27.33 | 879 | 821.84 | 28.08 | 924 | 865.37 | 28.82 |
| 835 | 779.31 | 27.34 | 880 | 822.80 | 28.10 | 925 | 866.34 | 28.83 |
| 836 | 780.28 | 27.36 | 881 | 823.77 | 28.11 | 926 | 867.30 | 28.85 |
| 837 | 781.24 | 27.38 | 882 | 824.74 | 28.13 | 927 | 868.27 | 28.86 |
| 838 | 782.21 | 27.40 | 883 | 825.71 | 28.15 | 928 | 869.24 | 28.88 |
| 839 | 783.17 | 27.41 | 884 | 826.67 | 28.16 | 929 | 870.21 | 28.90 |
| 840 | 784.14 | 27.43 | 885 | 827.64 | 28.18 | 930 | 871.18 | 28.91 |
| 841 | 785.11 | 27.45 | 886 | 828.61 | 28.20 | 931 | 872.14 | 28.93 |
| 842 | 786.07 | 27.46 | 887 | 829.57 | 28.21 | 932 | 873.11 | 28.94 |
| 843 | 787.04 | 27.48 | 888 | 830.54 | 28.23 | 933 | 874.08 | 28.96 |
| 844 | 788.01 | 27.50 | 889 | 831.51 | 28.25 | 934 | 875.05 | 28.98 |
| 845 | 788.97 | 27.51 | 890 | 832.47 | 28.26 | 935 | 876.01 | 28.99 |
| 846 | 789.94 | 27.53 | 891 | 833.44 | 28.28 | 936 | 876.98 | 29.01 |
| 847 | 790.90 | 27.55 | 892 | 834.41 | 28.30 | 937 | 877.95 | 29.02 |
| 848 | 791.87 | 27.56 | 893 | 835.38 | 28.31 | 938 | 878.92 | 29.04 |
| 849 | 792.84 | 27.58 | 894 | 836.34 | 28.33 | 939 | 879.89 | 29.06 |
| 850 | 793.80 | 27.60 | 895 | 837.31 | 28.34 | 940 | 880.85 | 29.07 |
| 851 | 794.77 | 27.61 | 896 | 838.28 | 28.36 | 941 | 881.82 | 29.09 |
| 852 | 795.74 | 27.63 | 897 | 839.25 | 28.38 | 942 | 882.79 | 29.10 |
| 853 | 796.70 | 27.65 | 898 | 840.21 | 28.39 | 943 | 883.76 | 29.12 |
| 854 | 797.67 | 27.67 | 899 | 841.18 | 28.41 | 944 | 884.73 | 29.14 |
| 855 | 798.64 | 27.68 | 900 | 842.15 | 28.43 | 945 | 885.70 | 29.15 |

Table 4. **CONSTANTS FOR $\underline{p}'$ (SHORT FORMULA), AN APPROXIMATION OF THE LOWER BINOMIAL CONFIDENCE LIMIT $\underline{p}$; $c/n \leqslant .125$, $n \geqslant 16$, $c \leqslant 1,000$** (*Continued*)

| c | m | k | c | m | k | c | m | k |
|---|---|---|---|---|---|---|---|---|
| 946 | 886.66 | 29.17 | 966 | 906.03 | 29.49 | 986 | 925.40 | 29.80 |
| 947 | 887.63 | 29.18 | 967 | 907.00 | 29.50 | 987 | 926.37 | 29.81 |
| 948 | 888.60 | 29.20 | 968 | 907.97 | 29.52 | 988 | 927.34 | 29.83 |
| 949 | 889.57 | 29.22 | 969 | 908.93 | 29.53 | 989 | 928.31 | 29.85 |
| 950 | 890.54 | 29.23 | 970 | 909.90 | 29.55 | 990 | 929.28 | 29.86 |
| 951 | 891.50 | 29.25 | 971 | 910.87 | 29.56 | 991 | 930.25 | 29.88 |
| 952 | 892.47 | 29.26 | 972 | 911.84 | 29.58 | 992 | 931.21 | 29.89 |
| 953 | 893.44 | 29.28 | 973 | 912.81 | 29.60 | 993 | 932.18 | 29.91 |
| 954 | 894.41 | 29.30 | 974 | 913.78 | 29.61 | 994 | 933.15 | 29.92 |
| 955 | 895.38 | 29.31 | 975 | 914.75 | 29.63 | 995 | 934.12 | 29.94 |
| 956 | 896.35 | 29.33 | 976 | 915.71 | 29.64 | 996 | 935.09 | 29.95 |
| 957 | 897.31 | 29.34 | 977 | 916.68 | 29.66 | 997 | 936.06 | 29.97 |
| 958 | 898.28 | 29.36 | 978 | 917.65 | 29.67 | 998 | 937.03 | 29.99 |
| 959 | 899.25 | 29.37 | 979 | 918.62 | 29.69 | 999 | 938.00 | 30.00 |
| 960 | 900.22 | 29.39 | 980 | 919.59 | 29.71 | 1000 | 938.97 | 30.02 |
| 961 | 901.19 | 29.41 | 981 | 920.56 | 29.72 | | | |
| 962 | 902.16 | 29.42 | 982 | 921.53 | 29.74 | | | |
| 963 | 903.12 | 29.44 | 983 | 922.50 | 29.75 | | | |
| 964 | 904.09 | 29.45 | 984 | 923.46 | 29.77 | | | |
| 965 | 905.06 | 29.47 | 985 | 924.43 | 29.78 | | | |

# Table 4. CONSTANTS FOR $p'$ (SHORT FORMULA), AN APPROXIMATION OF THE LOWER BINOMIAL CONFIDENCE LIMIT $\underline{p}$; $c/n \leqslant .125$, $n \geqslant 16$, $c \leqslant 1,000$ (Continued)

| c | m | k | c | m | k | c | m | k |
|---|-----|-----|----|--------|------|-----|---------|------|
| 1 | 0.051294 | -0.026 | 46 | 35.441 | 4.78 | 91 | 75.898 | 7.06 |
| 2 | 0.35536 | 0.322 | 47 | 36.320 | 4.84 | 92 | 76.812 | 7.10 |
| 3 | 0.81769 | 0.591 | 48 | 37.200 | 4.90 | 93 | 77.726 | 7.15 |
| 4 | 1.3663 | 0.817 | 49 | 38.082 | 4.96 | 94 | 78.641 | 7.19 |
| 5 | 1.9702 | 1.015 | 50 | 38.965 | 5.02 | 95 | 79.557 | 7.23 |
| 6 | 2.6130 | 1.193 | 51 | 39.849 | 5.09 | 96 | 80.472 | 7.27 |
| 7 | 3.2853 | 1.357 | 52 | 40.734 | 5.15 | 97 | 81.388 | 7.32 |
| 8 | 3.9808 | 1.510 | 53 | 41.620 | 5.20 | 98 | 82.305 | 7.36 |
| 9 | 4.6952 | 1.652 | 54 | 42.507 | 5.26 | 99 | 83.222 | 7.40 |
| 10 | 5.4254 | 1.79 | 55 | 43.396 | 5.32 | 100 | 84.139 | 7.44 |
| 11 | 6.1690 | 1.92 | 56 | 44.285 | 5.37 | 101 | 85.057 | 7.48 |
| 12 | 6.9242 | 2.04 | 57 | 45.176 | 5.43 | 102 | 85.976 | 7.52 |
| 13 | 7.6896 | 2.16 | 58 | 46.067 | 5.48 | 103 | 86.894 | 7.56 |
| 14 | 8.4640 | 2.27 | 59 | 46.959 | 5.53 | 104 | 87.813 | 7.60 |
| 15 | 9.2464 | 2.38 | 60 | 47.852 | 5.59 | 105 | 88.733 | 7.64 |
| 16 | 10.036 | 2.48 | 61 | 48.746 | 5.64 | 106 | 89.653 | 7.68 |
| 17 | 10.832 | 2.58 | 62 | 49.641 | 5.69 | 107 | 90.573 | 7.72 |
| 18 | 11.634 | 2.68 | 63 | 50.537 | 5.74 | 108 | 91.493 | 7.76 |
| 19 | 12.442 | 2.78 | 64 | 51.434 | 5.80 | 109 | 92.414 | 7.80 |
| 20 | 13.255 | 2.87 | 65 | 52.331 | 5.85 | 110 | 93.336 | 7.84 |
| 21 | 14.072 | 2.96 | 66 | 53.230 | 5.90 | 111 | 94.257 | 7.88 |
| 22 | 14.894 | 3.05 | 67 | 54.129 | 5.95 | 112 | 95.179 | 7.92 |
| 23 | 15.720 | 3.14 | 68 | 55.028 | 6.00 | 113 | 96.102 | 7.96 |
| 24 | 16.549 | 3.23 | 69 | 55.929 | 6.05 | 114 | 97.025 | 8.00 |
| 25 | 17.382 | 3.31 | 70 | 56.830 | 6.10 | 115 | 97.948 | 8.04 |
| 26 | 18.219 | 3.39 | 71 | 57.732 | 6.15 | 116 | 98.871 | 8.07 |
| 27 | 19.058 | 3.47 | 72 | 58.634 | 6.19 | 117 | 99.795 | 8.11 |
| 28 | 19.901 | 3.55 | 73 | 59.538 | 6.24 | 118 | 100.72 | 8.15 |
| 29 | 20.746 | 3.63 | 74 | 60.442 | 6.29 | 119 | 101.64 | 8.19 |
| 30 | 21.594 | 3.70 | 75 | 61.346 | 6.34 | 120 | 102.57 | 8.23 |
| 31 | 22.445 | 3.78 | 76 | 62.251 | 6.39 | 121 | 103.49 | 8.26 |
| 32 | 23.297 | 3.85 | 77 | 63.157 | 6.43 | 122 | 104.42 | 8.30 |
| 33 | 24.153 | 3.92 | 78 | 64.064 | 6.48 | 123 | 105.34 | 8.34 |
| 34 | 25.010 | 3.99 | 79 | 64.971 | 6.53 | 124 | 106.27 | 8.37 |
| 35 | 25.870 | 4.07 | 80 | 65.878 | 6.57 | 125 | 107.20 | 8.41 |
| 36 | 26.731 | 4.13 | 81 | 66.786 | 6.62 | 126 | 108.12 | 8.45 |
| 37 | 27.595 | 4.20 | 82 | 67.695 | 6.66 | 127 | 109.05 | 8.48 |
| 38 | 28.460 | 4.27 | 83 | 68.605 | 6.71 | 128 | 109.98 | 8.52 |
| 39 | 29.327 | 4.34 | 84 | 69.514 | 6.75 | 129 | 110.90 | 8.56 |
| 40 | 30.196 | 4.40 | 85 | 70.425 | 6.80 | 130 | 111.83 | 8.59 |
| 41 | 31.066 | 4.47 | 86 | 71.336 | 6.84 | 131 | 112.76 | 8.63 |
| 42 | 31.938 | 4.53 | 87 | 72.247 | 6.89 | 132 | 113.69 | 8.66 |
| 43 | 32.812 | 4.59 | 88 | 73.159 | 6.93 | 133 | 114.62 | 8.70 |
| 44 | 33.687 | 4.66 | 89 | 74.072 | 6.97 | 134 | 115.54 | 8.74 |
| 45 | 34.563 | 4.72 | 90 | 74.985 | 7.02 | 135 | 116.47 | 8.77 |

Table 4. CONSTANTS FOR p' (SHORT FORMULA), AN APPROXIMATION OF
THE LOWER BINOMIAL CONFIDENCE LIMIT p; c/n ≤ .125, n ≥ 16,
c ≤ 1,000 (*Continued*)

| c | m | k | c | m | k | c | m | k |
|---|---|---|---|---|---|---|---|---|
| 136 | 117.40 | 8.81 | 181 | 159.44 | 10.28 | 226 | 201.84 | 11.58 |
| 137 | 118.33 | 8.84 | 182 | 160.38 | 10.31 | 227 | 202.79 | 11.61 |
| 138 | 119.26 | 8.88 | 183 | 161.32 | 10.34 | 228 | 203.73 | 11.63 |
| 139 | 120.19 | 8.91 | 184 | 162.26 | 10.37 | 229 | 204.68 | 11.66 |
| 140 | 121.12 | 8.95 | 185 | 163.20 | 10.40 | 230 | 205.62 | 11.69 |
| 141 | 122.05 | 8.98 | 186 | 164.14 | 10.43 | 231 | 206.57 | 11.72 |
| 142 | 122.98 | 9.02 | 187 | 165.07 | 10.46 | 232 | 207.51 | 11.74 |
| 143 | 123.92 | 9.05 | 188 | 166.01 | 10.49 | 233 | 208.46 | 11.77 |
| 144 | 124.85 | 9.09 | 189 | 166.95 | 10.52 | 234 | 209.41 | 11.80 |
| 145 | 125.78 | 9.12 | 190 | 167.90 | 10.55 | 235 | 210.35 | 11.82 |
| 146 | 126.71 | 9.15 | 191 | 168.84 | 10.58 | 236 | 211.30 | 11.85 |
| 147 | 127.64 | 9.19 | 192 | 169.78 | 10.61 | 237 | 212.25 | 11.88 |
| 148 | 128.57 | 9.22 | 193 | 170.72 | 10.64 | 238 | 213.19 | 11.90 |
| 149 | 129.51 | 9.26 | 194 | 171.66 | 10.67 | 239 | 214.14 | 11.93 |
| 150 | 130.44 | 9.29 | 195 | 172.60 | 10.70 | 240 | 215.09 | 11.96 |
| 151 | 131.37 | 9.32 | 196 | 173.54 | 10.73 | 241 | 216.03 | 11.98 |
| 152 | 132.31 | 9.36 | 197 | 174.48 | 10.76 | 242 | 216.98 | 12.01 |
| 153 | 133.24 | 9.39 | 198 | 175.42 | 10.79 | 243 | 217.93 | 12.04 |
| 154 | 134.17 | 9.42 | 199 | 176.36 | 10.82 | 244 | 218.87 | 12.06 |
| 155 | 135.11 | 9.46 | 200 | 177.31 | 10.85 | 245 | 219.82 | 12.09 |
| 156 | 136.04 | 9.49 | 201 | 178.25 | 10.88 | 246 | 220.77 | 12.12 |
| 157 | 136.97 | 9.52 | 202 | 179.19 | 10.90 | 247 | 221.72 | 12.14 |
| 158 | 137.91 | 9.55 | 203 | 180.13 | 10.93 | 248 | 222.66 | 12.17 |
| 159 | 138.84 | 9.59 | 204 | 181.07 | 10.96 | 249 | 223.61 | 12.19 |
| 160 | 139.78 | 9.62 | 205 | 182.02 | 10.99 | 250 | 224.56 | 12.22 |
| 161 | 140.70 | 9.65 | 206 | 182.96 | 11.02 | 251 | 225.51 | 12.25 |
| 162 | 141.63 | 9.68 | 207 | 183.90 | 11.05 | 252 | 226.46 | 12.27 |
| 163 | 142.57 | 9.72 | 208 | 184.85 | 11.08 | 253 | 227.40 | 12.30 |
| 164 | 143.50 | 9.75 | 209 | 185.79 | 11.11 | 254 | 228.35 | 12.32 |
| 165 | 144.44 | 9.78 | 210 | 186.73 | 11.13 | 255 | 229.30 | 12.35 |
| 166 | 145.38 | 9.81 | 211 | 187.67 | 11.16 | 256 | 230.25 | 12.37 |
| 167 | 146.31 | 9.84 | 212 | 188.62 | 11.19 | 257 | 231.20 | 12.40 |
| 168 | 147.25 | 9.88 | 213 | 189.56 | 11.22 | 258 | 232.15 | 12.43 |
| 169 | 148.18 | 9.91 | 214 | 190.51 | 11.25 | 259 | 233.10 | 12.45 |
| 170 | 149.12 | 9.94 | 215 | 191.45 | 11.28 | 260 | 234.05 | 12.48 |
| 171 | 150.06 | 9.97 | 216 | 192.39 | 11.30 | 261 | 234.99 | 12.50 |
| 172 | 151.00 | 10.00 | 217 | 193.34 | 11.33 | 262 | 235.94 | 12.53 |
| 173 | 151.93 | 10.03 | 218 | 194.28 | 11.36 | 263 | 236.89 | 12.55 |
| 174 | 152.87 | 10.06 | 219 | 195.23 | 11.39 | 264 | 237.84 | 12.58 |
| 175 | 153.81 | 10.10 | 220 | 196.17 | 11.41 | 265 | 238.79 | 12.60 |
| 176 | 154.75 | 10.13 | 221 | 197.12 | 11.44 | 266 | 239.74 | 12.63 |
| 177 | 155.68 | 10.16 | 222 | 198.06 | 11.47 | 267 | 240.69 | 12.65 |
| 178 | 156.62 | 10.19 | 223 | 199.00 | 11.50 | 268 | 241.64 | 12.68 |
| 179 | 157.56 | 10.22 | 224 | 199.95 | 11.53 | 269 | 242.59 | 12.70 |
| 180 | 158.50 | 10.25 | 225 | 200.89 | 11.55 | 270 | 243.54 | 12.73 |

Table 4. CONSTANTS FOR p′ (SHORT FORMULA), AN APPROXIMATION OF
THE LOWER BINOMIAL CONFIDENCE LIMIT p; c/n ≤ .125, n ≥ 16,
c ≤ 1,000 (*Continued*)

Section 4
$\gamma_1$ = 95%
$\gamma_2$ = 90%

| c | m | k | c | m | k | c | m | k |
|---|---|---|---|---|---|---|---|---|
| 271 | 244.49 | 12.75 | 316 | 287.33 | 13.84 | 361 | 330.32 | 14.84 |
| 272 | 245.44 | 12.78 | 317 | 288.28 | 13.86 | 362 | 331.27 | 14.86 |
| 273 | 246.39 | 12.80 | 318 | 289.24 | 13.88 | 363 | 332.23 | 14.89 |
| 274 | 247.34 | 12.83 | 319 | 290.19 | 13.91 | 364 | 333.19 | 14.91 |
| 275 | 248.29 | 12.85 | 320 | 291.14 | 13.93 | 365 | 334.14 | 14.93 |
| 276 | 249.24 | 12.88 | 321 | 292.10 | 13.95 | 366 | 335.10 | 14.95 |
| 277 | 250.19 | 12.90 | 322 | 293.05 | 13.97 | 367 | 336.06 | 14.97 |
| 278 | 251.14 | 12.93 | 323 | 294.01 | 14.00 | 368 | 337.01 | 14.99 |
| 279 | 252.09 | 12.95 | 324 | 294.96 | 14.02 | 369 | 337.97 | 15.01 |
| 280 | 253.04 | 12.98 | 325 | 295.91 | 14.04 | 370 | 338.93 | 15.04 |
| 281 | 253.99 | 13.00 | 326 | 296.87 | 14.07 | 371 | 339.89 | 15.06 |
| 282 | 254.95 | 13.03 | 327 | 297.82 | 14.09 | 372 | 340.84 | 15.08 |
| 283 | 255.90 | 13.05 | 328 | 298.78 | 14.11 | 373 | 341.80 | 15.10 |
| 284 | 256.85 | 13.08 | 329 | 299.73 | 14.13 | 374 | 342.76 | 15.12 |
| 285 | 257.80 | 13.10 | 330 | 300.69 | 14.16 | 375 | 343.72 | 15.14 |
| 286 | 258.75 | 13.12 | 331 | 301.64 | 14.18 | 376 | 344.67 | 15.16 |
| 287 | 259.70 | 13.15 | 332 | 302.60 | 14.20 | 377 | 345.63 | 15.18 |
| 288 | 260.65 | 13.17 | 333 | 303.55 | 14.22 | 378 | 346.59 | 15.21 |
| 289 | 261.61 | 13.20 | 334 | 304.51 | 14.25 | 379 | 347.55 | 15.23 |
| 290 | 262.56 | 13.22 | 335 | 305.46 | 14.27 | 380 | 348.50 | 15.25 |
| 291 | 263.51 | 13.25 | 336 | 306.42 | 14.29 | 381 | 349.46 | 15.27 |
| 292 | 264.46 | 13.27 | 337 | 307.37 | 14.31 | 382 | 350.42 | 15.29 |
| 293 | 265.41 | 13.29 | 338 | 308.33 | 14.34 | 383 | 351.38 | 15.31 |
| 294 | 266.36 | 13.32 | 339 | 309.28 | 14.36 | 384 | 352.34 | 15.33 |
| 295 | 267.32 | 13.34 | 340 | 310.24 | 14.38 | 385 | 353.29 | 15.35 |
| 296 | 268.27 | 13.37 | 341 | 311.19 | 14.40 | 386 | 354.25 | 15.37 |
| 297 | 269.22 | 13.39 | 342 | 312.15 | 14.43 | 387 | 355.21 | 15.40 |
| 298 | 270.17 | 13.41 | 343 | 313.10 | 14.45 | 388 | 356.17 | 15.42 |
| 299 | 271.13 | 13.44 | 344 | 314.06 | 14.47 | 389 | 357.13 | 15.44 |
| 300 | 272.08 | 13.46 | 345 | 315.02 | 14.49 | 390 | 358.08 | 15.46 |
| 301 | 273.03 | 13.48 | 346 | 315.97 | 14.51 | 391 | 359.04 | 15.48 |
| 302 | 273.98 | 13.51 | 347 | 316.93 | 14.54 | 392 | 360.00 | 15.50 |
| 303 | 274.94 | 13.53 | 348 | 317.88 | 14.56 | 393 | 360.96 | 15.52 |
| 304 | 275.89 | 13.56 | 349 | 318.84 | 14.58 | 394 | 361.92 | 15.54 |
| 305 | 276.84 | 13.58 | 350 | 319.80 | 14.60 | 395 | 362.88 | 15.56 |
| 306 | 277.79 | 13.60 | 351 | 320.75 | 14.62 | 396 | 363.84 | 15.58 |
| 307 | 278.75 | 13.63 | 352 | 321.71 | 14.65 | 397 | 364.79 | 15.60 |
| 308 | 279.70 | 13.65 | 353 | 322.66 | 14.67 | 398 | 365.75 | 15.62 |
| 309 | 280.65 | 13.67 | 354 | 323.62 | 14.69 | 399 | 366.71 | 15.64 |
| 310 | 281.61 | 13.70 | 355 | 324.58 | 14.71 | 400 | 367.67 | 15.66 |
| 311 | 282.56 | 13.72 | 356 | 325.53 | 14.73 | 401 | 368.63 | 15.69 |
| 312 | 283.51 | 13.74 | 357 | 326.49 | 14.76 | 402 | 369.59 | 15.71 |
| 313 | 284.47 | 13.77 | 358 | 327.45 | 14.78 | 403 | 370.55 | 15.73 |
| 314 | 285.42 | 13.79 | 359 | 328.40 | 14.80 | 404 | 371.51 | 15.75 |
| 315 | 286.37 | 13.81 | 360 | 329.36 | 14.82 | 405 | 372.47 | 15.77 |

Table 4. CONSTANTS FOR $\underline{p}'$ (SHORT FORMULA), AN APPROXIMATION OF THE LOWER BINOMIAL CONFIDENCE LIMIT $\underline{p}$; $c/n \leqslant .125$, $n \geqslant 16$, $c \leqslant 1{,}000$ (*Continued*)

| c | m | k | c | m | k | c | m | k |
|---|---|---|---|---|---|---|---|---|
| 406 | 373.42 | 15.79 | 451 | 416.64 | 16.68 | 496 | 459.93 | 17.53 |
| 407 | 374.38 | 15.81 | 452 | 417.60 | 16.70 | 497 | 460.90 | 17.55 |
| 408 | 375.34 | 15.83 | 453 | 418.56 | 16.72 | 498 | 461.86 | 17.57 |
| 409 | 376.30 | 15.85 | 454 | 419.52 | 16.74 | 499 | 462.82 | 17.59 |
| 410 | 377.26 | 15.87 | 455 | 420.48 | 16.76 | 500 | 463.79 | 17.61 |
| 411 | 378.22 | 15.89 | 456 | 421.44 | 16.78 | 501 | 464.75 | 17.62 |
| 412 | 379.18 | 15.91 | 457 | 422.40 | 16.80 | 502 | 465.71 | 17.64 |
| 413 | 380.14 | 15.93 | 458 | 423.37 | 16.82 | 503 | 466.68 | 17.66 |
| 414 | 381.10 | 15.95 | 459 | 424.33 | 16.84 | 504 | 467.64 | 17.68 |
| 415 | 382.06 | 15.97 | 460 | 425.29 | 16.86 | 505 | 468.60 | 17.70 |
| 416 | 383.02 | 15.99 | 461 | 426.25 | 16.87 | 506 | 469.57 | 17.72 |
| 417 | 383.98 | 16.01 | 462 | 427.21 | 16.89 | 507 | 470.53 | 17.73 |
| 418 | 384.94 | 16.03 | 463 | 428.17 | 16.91 | 508 | 471.49 | 17.75 |
| 419 | 385.90 | 16.05 | 464 | 429.14 | 16.93 | 509 | 472.46 | 17.77 |
| 420 | 386.86 | 16.07 | 465 | 430.10 | 16.95 | 510 | 473.42 | 17.79 |
| 421 | 387.82 | 16.09 | 466 | 431.06 | 16.97 | 511 | 474.39 | 17.81 |
| 422 | 388.78 | 16.11 | 467 | 432.02 | 16.99 | 512 | 475.35 | 17.83 |
| 423 | 389.74 | 16.13 | 468 | 432.98 | 17.01 | 513 | 476.31 | 17.84 |
| 424 | 390.70 | 16.15 | 469 | 433.95 | 17.03 | 514 | 477.28 | 17.86 |
| 425 | 391.66 | 16.17 | 470 | 434.91 | 17.05 | 515 | 478.24 | 17.88 |
| 426 | 392.62 | 16.19 | 471 | 435.87 | 17.07 | 516 | 479.20 | 17.90 |
| 427 | 393.58 | 16.21 | 472 | 436.83 | 17.08 | 517 | 480.17 | 17.92 |
| 428 | 394.54 | 16.23 | 473 | 437.79 | 17.10 | 518 | 481.13 | 17.93 |
| 429 | 395.50 | 16.25 | 474 | 438.76 | 17.12 | 519 | 482.10 | 17.95 |
| 430 | 396.46 | 16.27 | 475 | 439.72 | 17.14 | 520 | 483.06 | 17.97 |
| 431 | 397.42 | 16.29 | 476 | 440.68 | 17.16 | 521 | 484.02 | 17.99 |
| 432 | 398.38 | 16.31 | 477 | 441.64 | 17.18 | 522 | 484.99 | 18.01 |
| 433 | 399.34 | 16.33 | 478 | 442.61 | 17.20 | 523 | 485.95 | 18.02 |
| 434 | 400.30 | 16.35 | 479 | 443.57 | 17.22 | 524 | 486.91 | 18.04 |
| 435 | 401.26 | 16.37 | 480 | 444.53 | 17.23 | 525 | 487.88 | 18.06 |
| 436 | 402.22 | 16.39 | 481 | 445.49 | 17.25 | 526 | 488.84 | 18.08 |
| 437 | 403.18 | 16.41 | 482 | 446.46 | 17.27 | 527 | 489.81 | 18.10 |
| 438 | 404.14 | 16.43 | 483 | 447.42 | 17.29 | 528 | 490.77 | 18.11 |
| 439 | 405.10 | 16.45 | 484 | 448.38 | 17.31 | 529 | 491.74 | 18.13 |
| 440 | 406.06 | 16.47 | 485 | 449.34 | 17.33 | 530 | 492.70 | 18.15 |
| 441 | 407.03 | 16.49 | 486 | 450.31 | 17.35 | 531 | 493.66 | 18.17 |
| 442 | 407.99 | 16.51 | 487 | 451.27 | 17.37 | 532 | 494.63 | 18.19 |
| 443 | 408.95 | 16.53 | 488 | 452.23 | 17.38 | 533 | 495.59 | 18.20 |
| 444 | 409.91 | 16.55 | 489 | 453.19 | 17.40 | 534 | 496.56 | 18.22 |
| 445 | 410.87 | 16.57 | 490 | 454.16 | 17.42 | 535 | 497.52 | 18.24 |
| 446 | 411.83 | 16.58 | 491 | 455.12 | 17.44 | 536 | 498.49 | 18.26 |
| 447 | 412.79 | 16.60 | 492 | 456.08 | 17.46 | 537 | 499.45 | 18.27 |
| 448 | 413.75 | 16.62 | 493 | 457.05 | 17.48 | 538 | 500.42 | 18.29 |
| 449 | 414.71 | 16.64 | 494 | 458.01 | 17.50 | 539 | 501.38 | 18.31 |
| 450 | 415.67 | 16.66 | 495 | 458.97 | 17.51 | 540 | 502.34 | 18.33 |

Table 4. CONSTANTS FOR p' (SHORT FORMULA), AN APPROXIMATION OF
THE LOWER BINOMIAL CONFIDENCE LIMIT p; c/n ≤ .125, n ≥ 16,
c ≤ 1,000 (*Continued*)

Section 4
$\gamma_1 = 95\%$
$\gamma_2 = 90\%$

| c | m | k | c | m | k | c | m | k |
|---|---|---|---|---|---|---|---|---|
| 541 | 503.31 | 18.35 | 586 | 546.75 | 19.13 | 631 | 590.25 | 19.88 |
| 542 | 504.27 | 18.36 | 587 | 547.72 | 19.14 | 632 | 591.22 | 19.89 |
| 543 | 505.24 | 18.38 | 588 | 548.68 | 19.16 | 633 | 592.18 | 19.91 |
| 544 | 506.20 | 18.40 | 589 | 549.65 | 19.18 | 634 | 593.15 | 19.92 |
| 545 | 507.17 | 18.42 | 590 | 550.61 | 19.19 | 635 | 594.12 | 19.94 |
| 546 | 508.13 | 18.43 | 591 | 551.58 | 19.21 | 636 | 595.09 | 19.96 |
| 547 | 509.10 | 18.45 | 592 | 552.55 | 19.23 | 637 | 596.05 | 19.97 |
| 548 | 510.06 | 18.47 | 593 | 553.51 | 19.24 | 638 | 597.02 | 19.99 |
| 549 | 511.03 | 18.49 | 594 | 554.48 | 19.26 | 639 | 597.99 | 20.01 |
| 550 | 511.99 | 18.50 | 595 | 555.45 | 19 28 | 640 | 598.96 | 20.02 |
| 551 | 512.96 | 18.52 | 596 | 556.41 | 1%.29 | 641 | 599.92 | 20.04 |
| 552 | 513.92 | 18.54 | 597 | 557.38 | 19.31 | 642 | 600.89 | 20.05 |
| 553 | 514.89 | 18.56 | 598 | 558.34 | 19.33 | 643 | 601.86 | 20.07 |
| 554 | 515.85 | 18.57 | 599 | 559.31 | 19.34 | 644 | 602.83 | 20.09 |
| 555 | 516.82 | 18.59 | 600 | 560.28 | 19.36 | 645 | 603.79 | 20.10 |
| 556 | 517.78 | 18.61 | 601 | 561.24 | 19.38 | 646 | 604.76 | 20.12 |
| 557 | 518.75 | 18.63 | 602 | 562.21 | 19.40 | 647 | 605.73 | 20.14 |
| 558 | 519.71 | 18.64 | 603 | 563.18 | 19.41 | 648 | 606.70 | 20.15 |
| 559 | 520.68 | 18.66 | 604 | 564.14 | 19.43 | 649 | 607.66 | 20.17 |
| 560 | 521.64 | 18.68 | 605 | 565.11 | 19.45 | 650 | 608.63 | 20.18 |
| 561 | 522.61 | 18.70 | 606 | 566.08 | 19.46 | 651 | 609.60 | 20.20 |
| 562 | 523.57 | 18.71 | 607 | 567.04 | 19.48 | 652 | 610.57 | 20.22 |
| 563 | 524.54 | 18.73 | 608 | 568.01 | 19.50 | 653 | 611.53 | 20.23 |
| 564 | 525.50 | 18.75 | 609 | 568.98 | 19.51 | 654 | 612.50 | 20.25 |
| 565 | 526.47 | 18.77 | 610 | 569.94 | 19.53 | 655 | 613.47 | 20.26 |
| 566 | 527.44 | 18.78 | 611 | 570.91 | 19.55 | 656 | 614.44 | 20.28 |
| 567 | 528.40 | 18.80 | 612 | 571.88 | 19.56 | 657 | 615.41 | 20.30 |
| 568 | 529.37 | 18.82 | 613 | 572.84 | 19.58 | 658 | 616.37 | 20.31 |
| 569 | 530.33 | 18.83 | 614 | 573.81 | 19.60 | 659 | 617.34 | 20.33 |
| 570 | 531.30 | 18.85 | 615 | 574.78 | 19.61 | 660 | 618.31 | 20.34 |
| 571 | 532.26 | 18.87 | 616 | 575.74 | 19.63 | 661 | 619.28 | 20.36 |
| 572 | 533.23 | 18.89 | 617 | 576.71 | 19.65 | 662 | 620.25 | 20.38 |
| 573 | 534.19 | 18.90 | 618 | 577.68 | 19.66 | 663 | 621.21 | 20.39 |
| 574 | 535.16 | 18.92 | 619 | 578.64 | 19.68 | 664 | 622.18 | 20.41 |
| 575 | 536.13 | 18.94 | 620 | 579.61 | 19.69 | 665 | 623.15 | 20.42 |
| 576 | 537.09 | 18.95 | 621 | 580.58 | 19.71 | 666 | 624.12 | 20.44 |
| 577 | 538.06 | 18.97 | 622 | 581.54 | 19.73 | 667 | 625.09 | 20.46 |
| 578 | 539.02 | 18.99 | 623 | 582.51 | 19.74 | 668 | 626.05 | 20.47 |
| 579 | 539.99 | 19.01 | 624 | 583.48 | 19.76 | 669 | 627.02 | 20.49 |
| 580 | 540.95 | 19.02 | 625 | 584.45 | 19.78 | 670 | 627.99 | 20.50 |
| 581 | 541.92 | 19.04 | 626 | 585.41 | 19.79 | 671 | 628.96 | 20.52 |
| 582 | 542.89 | 19.06 | 627 | 586.38 | 19.81 | 672 | 629.93 | 20.54 |
| 583 | 543.85 | 19.07 | 628 | 587.35 | 19.83 | 673 | 630.90 | 20.55 |
| 584 | 544.82 | 19.09 | 629 | 588.31 | 19.84 | 674 | 631.86 | 20.57 |
| 585 | 545.78 | 19.11 | 630 | 589.28 | 19.86 | 675 | 632.83 | 20.58 |

Table 4.  CONSTANTS FOR $\underline{p}'$ (SHORT FORMULA), AN APPROXIMATION OF
THE LOWER BINOMIAL CONFIDENCE LIMIT $\underline{p}$; c/n $\leqslant$ .125, n $\geqslant$ 1
c $\leqslant$ 1,000 (*Continued*)

| c | m | k | c | m | k | c | m | k |
|---|---|---|---|---|---|---|---|---|
| 676 | 633.80 | 20.60 | 721 | 677.40 | 21.30 | 766 | 721.04 | 21.98 |
| 677 | 634.77 | 20.62 | 722 | 678.37 | 21.32 | 767 | 722.01 | 21.99 |
| 678 | 635.74 | 20.63 | 723 | 679.34 | 21.33 | 768 | 722.98 | 22.01 |
| 679 | 636.71 | 20.65 | 724 | 680.31 | 21.35 | 769 | 723.95 | 22.02 |
| 680 | 637.67 | 20.66 | 725 | 681.28 | 21.36 | 770 | 724.92 | 22.04 |
| 681 | 638.64 | 20.68 | 726 | 682.25 | 21.38 | 771 | 725.89 | 22.05 |
| 682 | 639.61 | 20.69 | 727 | 683.22 | 21.39 | 772 | 726.87 | 22.07 |
| 683 | 640.58 | 20.71 | 728 | 684.19 | 21.41 | 773 | 727.84 | 22.08 |
| 684 | 641.55 | 20.73 | 729 | 685.16 | 21.42 | 774 | 728.81 | 22.10 |
| 685 | 642.52 | 20.74 | 730 | 686.13 | 21.44 | 775 | 729.78 | 22.11 |
| 686 | 643.49 | 20.76 | 731 | 687.10 | 21.45 | 776 | 730.75 | 22.13 |
| 687 | 644.45 | 20.77 | 732 | 688.06 | 21.47 | 777 | 731.72 | 22.14 |
| 688 | 645.42 | 20.79 | 733 | 689.03 | 21.48 | 778 | 732.69 | 22.16 |
| 689 | 646.39 | 20.80 | 734 | 690.00 | 21.50 | 779 | 733.66 | 22.17 |
| 690 | 647.36 | 20.82 | 735 | 690.97 | 21.51 | 780 | 734.63 | 22.19 |
| 691 | 648.33 | 20.84 | 736 | 691.94 | 21.53 | 781 | 735.60 | 22.20 |
| 692 | 649.30 | 20.85 | 737 | 692.91 | 21.54 | 782 | 736.57 | 22.21 |
| 693 | 650.27 | 20.87 | 738 | 693.88 | 21.56 | 783 | 737.54 | 22.23 |
| 694 | 651.24 | 20.88 | 739 | 694.85 | 21.57 | 784 | 738.51 | 22.24 |
| 695 | 652.20 | 20.90 | 740 | 695.82 | 21.59 | 785 | 739.48 | 22.26 |
| 696 | 653.17 | 20.91 | 741 | 696.79 | 21.60 | 786 | 740.45 | 22.27 |
| 697 | 654.14 | 20.93 | 742 | 697.76 | 21.62 | 787 | 741.42 | 22.29 |
| 698 | 655.11 | 20.94 | 743 | 698.73 | 21.63 | 788 | 742.39 | 22.30 |
| 699 | 656.08 | 20.96 | 744 | 699.70 | 21.65 | 789 | 743.36 | 22.32 |
| 700 | 657.05 | 20.98 | 745 | 700.67 | 21.66 | 790 | 744.34 | 22.33 |
| 701 | 658.02 | 20.99 | 746 | 701.64 | 21.68 | 791 | 745.31 | 22.35 |
| 702 | 658.99 | 21.01 | 747 | 702.61 | 21.69 | 792 | 746.28 | 22.36 |
| 703 | 659.96 | 21.02 | 748 | 703.58 | 21.71 | 793 | 747.25 | 22.38 |
| 704 | 660.92 | 21.04 | 749 | 704.55 | 21.72 | 794 | 748.22 | 22.39 |
| 705 | 661.89 | 21.05 | 750 | 705.52 | 21.74 | 795 | 749.19 | 22.41 |
| 706 | 662.86 | 21.07 | 751 | 706.49 | 21.75 | 796 | 750.16 | 22.42 |
| 707 | 663.83 | 21.08 | 752 | 707.46 | 21.77 | 797 | 751.13 | 22.43 |
| 708 | 664.80 | 21.10 | 753 | 708.43 | 21.78 | 798 | 752.10 | 22.45 |
| 709 | 665.77 | 21.12 | 754 | 709.40 | 21.80 | 799 | 753.07 | 22.46 |
| 710 | 666.74 | 21.13 | 755 | 710.37 | 21.81 | 800 | 754.04 | 22.48 |
| 711 | 667.71 | 21.15 | 756 | 711.34 | 21.83 | 801 | 755.01 | 22.49 |
| 712 | 668.68 | 21.16 | 757 | 712.31 | 21.84 | 802 | 755.99 | 22.51 |
| 713 | 669.65 | 21.18 | 758 | 713.28 | 21.86 | 803 | 756.96 | 22.52 |
| 714 | 670.62 | 21.19 | 759 | 714.25 | 21.87 | 804 | 757.93 | 22.54 |
| 715 | 671.58 | 21.21 | 760 | 715.22 | 21.89 | 805 | 758.90 | 22.55 |
| 716 | 672.55 | 21.22 | 761 | 716.19 | 21.90 | 806 | 759.87 | 22.57 |
| 717 | 673.52 | 21.24 | 762 | 717.16 | 21.92 | 807 | 760.84 | 22.58 |
| 718 | 674.49 | 21.25 | 763 | 718.13 | 21.93 | 808 | 761.81 | 22.59 |
| 719 | 675.46 | 21.27 | 764 | 719.10 | 21.95 | 809 | 762.78 | 22.61 |
| 720 | 676.43 | 21.28 | 765 | 720.07 | 21.96 | 810 | 763.75 | 22.62 |

Table 4.  CONSTANTS FOR p' (SHORT FORMULA), AN APPROXIMATION OF
THE LOWER BINOMIAL CONFIDENCE LIMIT p; c/n ≤ .125, n ≥ 16,
c ≤ 1,000 (*Continued*)

Section 4

$\gamma_1 = 95\%$
$\gamma_2 = 90\%$

| c | m | k | c | m | k | c | m | k |
|---|---|---|---|---|---|---|---|---|
| 811 | 764.72 | 22.64 | 856 | 808.44 | 23.28 | 901 | 852.19 | 23.90 |
| 812 | 765.70 | 22.65 | 857 | 809.41 | 23.29 | 902 | 853.17 | 23.92 |
| 813 | 766.67 | 22.67 | 858 | 810.39 | 23.31 | 903 | 854.14 | 23.93 |
| 814 | 767.64 | 22.68 | 859 | 811.36 | 23.32 | 904 | 855.11 | 23.94 |
| 815 | 768.61 | 22.70 | 860 | 812.33 | 23.33 | 905 | 856.08 | 23.96 |
| 816 | 769.58 | 22.71 | 861 | 813.30 | 23.35 | 906 | 857.06 | 23.97 |
| 817 | 770.55 | 22.72 | 862 | 814.27 | 23.36 | 907 | 858.03 | 23.99 |
| 818 | 771.52 | 22.74 | 863 | 815.25 | 23.38 | 908 | 859.00 | 24.00 |
| 819 | 772.49 | 22.75 | 864 | 816.22 | 23.39 | 909 | 859.98 | 24.01 |
| 820 | 773.47 | 22.77 | 865 | 817.19 | 23.40 | 910 | 860.95 | 24.03 |
| 821 | 774.44 | 22.78 | 866 | 818.16 | 23.42 | 911 | 861.92 | 24.04 |
| 822 | 775.41 | 22.80 | 867 | 819.13 | 23.43 | 912 | 862.89 | 24.05 |
| 823 | 776.38 | 22.81 | 868 | 820.11 | 23.45 | 913 | 863.87 | 24.07 |
| 824 | 777.35 | 22.82 | 869 | 821.08 | 23.46 | 914 | 864.84 | 24.08 |
| 825 | 778.32 | 22.84 | 870 | 822.05 | 23.47 | 915 | 865.81 | 24.09 |
| 826 | 779.29 | 22.85 | 871 | 823.02 | 23.49 | 916 | 866.78 | 24.11 |
| 827 | 780.27 | 22.87 | 872 | 824.00 | 23.50 | 917 | 867.76 | 24.12 |
| 828 | 781.24 | 22.88 | 873 | 824.97 | 23.52 | 918 | 868.73 | 24.13 |
| 829 | 782.21 | 22.90 | 874 | 825.94 | 23.53 | 919 | 869.70 | 24.15 |
| 830 | 783.18 | 22.91 | 875 | 826.91 | 23.54 | 920 | 870.68 | 24.16 |
| 831 | 784.15 | 22.92 | 876 | 827.88 | 23.56 | 921 | 871.65 | 24.18 |
| 832 | 785.12 | 22.94 | 877 | 828.86 | 23.57 | 922 | 872.62 | 24.19 |
| 833 | 786.09 | 22.95 | 878 | 829.83 | 23.59 | 923 | 873.59 | 24.20 |
| 834 | 787.07 | 22.97 | 879 | 830.80 | 23.60 | 924 | 874.57 | 24.22 |
| 835 | 788.04 | 22.98 | 880 | 831.77 | 23.61 | 925 | 875.54 | 24.23 |
| 836 | 789.01 | 23.00 | 881 | 832.75 | 23.63 | 926 | 876.51 | 24.24 |
| 837 | 789.98 | 23.01 | 882 | 833.72 | 23.64 | 927 | 877.49 | 24.26 |
| 838 | 790.95 | 23.02 | 883 | 834.69 | 23.66 | 928 | 878.46 | 24.27 |
| 839 | 791.92 | 23.04 | 884 | 835.66 | 23.67 | 929 | 879.43 | 24.28 |
| 840 | 792.89 | 23.05 | 885 | 836.63 | 23.68 | 930 | 880.41 | 24.30 |
| 841 | 793.87 | 23.07 | 886 | 837.61 | 23.70 | 931 | 881.38 | 24.31 |
| 842 | 794.84 | 23.08 | 887 | 838.58 | 23.71 | 932 | 882.35 | 24.32 |
| 843 | 795.81 | 23.10 | 888 | 839.55 | 23.72 | 933 | 883.32 | 24.34 |
| 844 | 796.78 | 23.11 | 889 | 840.52 | 23.74 | 934 | 884.30 | 24.35 |
| 845 | 797.75 | 23.12 | 890 | 841.50 | 23.75 | 935 | 885.27 | 24.36 |
| 846 | 798.72 | 23.14 | 891 | 842.47 | 23.77 | 936 | 886.24 | 24.38 |
| 847 | 799.70 | 23.15 | 892 | 843.44 | 23.78 | 937 | 887.22 | 24.39 |
| 848 | 800.67 | 23.17 | 893 | 844.41 | 23.79 | 938 | 888.19 | 24.40 |
| 849 | 801.64 | 23.18 | 894 | 845.39 | 23.81 | 939 | 889.16 | 24.42 |
| 850 | 802.61 | 23.19 | 895 | 846.36 | 23.82 | 940 | 890.14 | 24.43 |
| 851 | 803.58 | 23.21 | 896 | 847.33 | 23.83 | 941 | 891.11 | 24.45 |
| 852 | 804.56 | 23.22 | 897 | 848.30 | 23.85 | 942 | 892.08 | 24.46 |
| 853 | 805.53 | 23.24 | 898 | 849.28 | 23.86 | 943 | 893.06 | 24.47 |
| 854 | 806.50 | 23.25 | 899 | 850.25 | 23.88 | 944 | 894.03 | 24.49 |
| 855 | 807.47 | 23.26 | 900 | 851.22 | 23.89 | 945 | 895.00 | 24.50 |

Table 4. CONSTANTS FOR p' (SHORT FORMULA), AN APPROXIMATION OF
THE LOWER BINOMIAL CONFIDENCE LIMIT p; c/n ≤ .125, n ≥ 10
c ≤ 1,000 (*Continued*)

| c | m | k | c | m | k | c | m | k |
|---|---|---|---|---|---|---|---|---|
| 946 | 895.98 | 24.51 | 966 | 915.44 | 24.78 | 986 | 934.92 | 25.04 |
| 947 | 896.95 | 24.53 | 967 | 916.42 | 24.79 | 987 | 935.89 | 25.05 |
| 948 | 897.92 | 24.54 | 968 | 917.39 | 24.80 | 988 | 936.87 | 25.07 |
| 949 | 898.90 | 24.55 | 969 | 918.36 | 24.82 | 989 | 937.84 | 25.08 |
| 950 | 899.87 | 24.57 | 970 | 919.34 | 24.83 | 990 | 938.81 | 25.09 |
| 951 | 900.84 | 24.58 | 971 | 920.31 | 24.84 | 991 | 939.79 | 25.11 |
| 952 | 901.82 | 24.59 | 972 | 921.29 | 24.86 | 992 | 940.76 | 25.12 |
| 953 | 902.79 | 24.61 | 973 | 922.26 | 24.87 | 993 | 941.73 | 25.13 |
| 954 | 903.76 | 24.62 | 974 | 923.23 | 24.88 | 994 | 942.71 | 25.15 |
| 955 | 904.74 | 24.63 | 975 | 924.21 | 24.90 | 995 | 943.68 | 25.16 |
| 956 | 905.71 | 24.65 | 976 | 925.18 | 24.91 | 996 | 944.66 | 25.17 |
| 957 | 906.68 | 24.66 | 977 | 926.15 | 24.92 | 997 | 945.63 | 25.18 |
| 958 | 907.66 | 24.67 | 978 | 927.13 | 24.94 | 998 | 946.60 | 25.20 |
| 959 | 908.63 | 24.69 | 979 | 928.10 | 24.95 | 999 | 947.58 | 25.21 |
| 960 | 909.60 | 24.70 | 980 | 929.07 | 24.96 | 1000 | 948.55 | 25.22 |
| 961 | 910.58 | 24.71 | 981 | 930.05 | 24.98 | | | |
| 962 | 911.55 | 24.72 | 982 | 931.02 | 24.99 | | | |
| 963 | 912.52 | 24.74 | 983 | 932.00 | 25.00 | | | |
| 964 | 913.50 | 24.75 | 984 | 932.97 | 25.02 | | | |
| 965 | 914.47 | 24.76 | 985 | 933.94 | 25.03 | | | |

Table 4. **CONSTANTS FOR** $\underline{p}'$ **(SHORT FORMULA), AN APPROXIMATION OF** **THE LOWER BINOMIAL CONFIDENCE LIMIT** $\underline{p}$; $c/n \leqslant .125$, $n \geqslant 16$, $c \leqslant 1,000$ (*Continued*)

Section 5

$\gamma_1 = 90\%$
$\gamma_2 = 80\%$

| c | $\underline{m}$ | $\underline{k}$ | c | $\underline{m}$ | $\underline{k}$ | c | $\underline{m}$ | $\underline{k}$ |
|---|---|---|---|---|---|---|---|---|
| 1 | .10536 | −0.053 | 46 | 37.550 | 3.72 | 91 | 79.009 | 5.51 |
| 2 | .53181 | 0.234 | 47 | 38.456 | 3.77 | 92 | 79.942 | 5.54 |
| 3 | 1.1021 | 0.449 | 48 | 39.363 | 3.82 | 93 | 80.875 | 5.57 |
| 4 | 1.7448 | 0.628 | 49 | 40.270 | 3.86 | 94 | 81.809 | 5.61 |
| 5 | 2.4326 | 0.784 | 50 | 41.179 | 3.91 | 95 | 82.743 | 5.64 |
| 6 | 3.1519 | 0.924 | 51 | 42.089 | 3.97 | 96 | 83.677 | 5.67 |
| 7 | 3.8948 | 1.053 | 52 | 42.999 | 4.01 | 97 | 84.612 | 5.70 |
| 8 | 4.6561 | 1.172 | 53 | 43.910 | 4.06 | 98 | 85.547 | 5.74 |
| 9 | 5.4325 | 1.284 | 54 | 44.823 | 4.10 | 99 | 86.482 | 5.77 |
| 10 | 6.2213 | 1.39 | 55 | 45.736 | 4.15 | 100 | 87.418 | 5.80 |
| 11 | 7.0208 | 1.49 | 56 | 46.649 | 4.19 | 101 | 88.354 | 5.83 |
| 12 | 7.8294 | 1.59 | 57 | 47.564 | 4.23 | 102 | 89.290 | 5.86 |
| 13 | 8.6460 | 1.68 | 58 | 48.479 | 4.27 | 103 | 90.227 | 5.90 |
| 14 | 9.4696 | 1.77 | 59 | 49.395 | 4.32 | 104 | 91.164 | 5.93 |
| 15 | 10.300 | 1.85 | 60 | 50.312 | 4.36 | 105 | 92.101 | 5.96 |
| 16 | 11.135 | 1.93 | 61 | 51.229 | 4.40 | 106 | 93.038 | 5.99 |
| 17 | 11.976 | 2.01 | 62 | 52.148 | 4.44 | 107 | 93.976 | 6.02 |
| 18 | 12.822 | 2.09 | 63 | 53.066 | 4.48 | 108 | 94.914 | 6.05 |
| 19 | 13.671 | 2.16 | 64 | 53.986 | 4.52 | 109 | 95.853 | 6.08 |
| 20 | 14.525 | 2.24 | 65 | 54.906 | 4.56 | 110 | 96.792 | 6.11 |
| 21 | 15.383 | 2.31 | 66 | 55.826 | 4.60 | 111 | 97.730 | 6.14 |
| 22 | 16.244 | 2.38 | 67 | 56.748 | 4.64 | 112 | 98.670 | 6.17 |
| 23 | 17.108 | 2.45 | 68 | 57.669 | 4.68 | 113 | 99.609 | 6.20 |
| 24 | 17.975 | 2.51 | 69 | 58.592 | 4.72 | 114 | 100.55 | 6.23 |
| 25 | 18.844 | 2.58 | 70 | 59.515 | 4.75 | 115 | 101.49 | 6.26 |
| 26 | 19.717 | 2.64 | 71 | 60.438 | 4.79 | 116 | 102.43 | 6.29 |
| 27 | 20.592 | 2.70 | 72 | 61.362 | 4.83 | 117 | 103.37 | 6.32 |
| 28 | 21.469 | 2.77 | 73 | 62.287 | 4.87 | 118 | 104.31 | 6.35 |
| 29 | 22.348 | 2.83 | 74 | 63.212 | 4.91 | 119 | 105.25 | 6.38 |
| 30 | 23.229 | 2.89 | 75 | 64.138 | 4.94 | 120 | 106.19 | 6.41 |
| 31 | 24.113 | 2.94 | 76 | 65.064 | 4.98 | 121 | 107.13 | 6.44 |
| 32 | 24.998 | 3.00 | 77 | 65.990 | 5.02 | 122 | 108.08 | 6.47 |
| 33 | 25.885 | 3.06 | 78 | 66.918 | 5.05 | 123 | 109.02 | 6.50 |
| 34 | 26.774 | 3.11 | 79 | 67.845 | 5.09 | 124 | 109.96 | 6.53 |
| 35 | 27.664 | 3.17 | 80 | 68.773 | 5.12 | 125 | 110.90 | 6.56 |
| 36 | 28.556 | 3.22 | 81 | 69.702 | 5.16 | 126 | 111.85 | 6.59 |
| 37 | 29.450 | 3.28 | 82 | 70.630 | 5.20 | 127 | 112.79 | 6.61 |
| 38 | 30.345 | 3.33 | 83 | 71.560 | 5.23 | 128 | 113.73 | 6.64 |
| 39 | 31.241 | 3.38 | 84 | 72.490 | 5.27 | 129 | 114.68 | 6.67 |
| 40 | 32.139 | 3.43 | 85 | 73.420 | 5.30 | 130 | 115.62 | 6.70 |
| 41 | 33.038 | 3.48 | 86 | 74.350 | 5.34 | 131 | 116.56 | 6.73 |
| 42 | 33.938 | 3.53 | 87 | 75.281 | 5.37 | 132 | 117.51 | 6.76 |
| 43 | 34.839 | 3.58 | 88 | 76.213 | 5.40 | 133 | 118.45 | 6.78 |
| 44 | 35.742 | 3.63 | 89 | 77.144 | 5.44 | 134 | 119.40 | 6.81 |
| 45 | 36.646 | 3.68 | 90 | 78.077 | 5.47 | 135 | 120.34 | 6.84 |

**Table 4. CONSTANTS FOR $\underline{p}'$ (SHORT FORMULA), AN APPROXIMATION OF THE LOWER BINOMIAL CONFIDENCE LIMIT $\underline{p}$; $c/n \leqslant .125$, $n \geqslant 16$, $c \leqslant 1,000$ (Continued)**

| c | m | k | c | m | k | c | m | k |
|---|---|---|---|---|---|---|---|---|
| 136 | 121.29 | 6.87 | 181 | 163.97 | 8.01 | 226 | 206.95 | 9.03 |
| 137 | 122.23 | 6.89 | 182 | 164.92 | 8.04 | 227 | 207.90 | 9.05 |
| 138 | 123.18 | 6.92 | 183 | 165.88 | 8.06 | 228 | 208.86 | 9.07 |
| 139 | 124.12 | 6.95 | 184 | 166.83 | 8.09 | 229 | 209.82 | 9.09 |
| 140 | 125.07 | 6.97 | 185 | 167.78 | 8.11 | 230 | 210.78 | 9.11 |
| 141 | 126.01 | 7.00 | 186 | 168.74 | 8.13 | 231 | 211.74 | 9.13 |
| 142 | 126.96 | 7.03 | 187 | 169.69 | 8.16 | 232 | 212.69 | 9.15 |
| 143 | 127.91 | 7.06 | 188 | 170.64 | 8.18 | 233 | 213.65 | 9.17 |
| 144 | 128.85 | 7.08 | 189 | 171.60 | 8.20 | 234 | 214.61 | 9.20 |
| 145 | 129.80 | 7.11 | 190 | 172.55 | 8.23 | 235 | 215.57 | 9.22 |
| 146 | 130.75 | 7.14 | 191 | 173.50 | 8.25 | 236 | 216.53 | 9.24 |
| 147 | 131.69 | 7.16 | 192 | 174.46 | 8.27 | 237 | 217.48 | 9.26 |
| 148 | 132.64 | 7.19 | 193 | 175.41 | 8.30 | 238 | 218.44 | 9.28 |
| 149 | 133.59 | 7.21 | 194 | 176.36 | 8.32 | 239 | 219.40 | 9.30 |
| 150 | 134.53 | 7.24 | 195 | 177.32 | 8.34 | 240 | 220.36 | 9.32 |
| 151 | 135.48 | 7.27 | 196 | 178.27 | 8.36 | 241 | 221.32 | 9.34 |
| 152 | 136.43 | 7.29 | 197 | 179.23 | 8.39 | 242 | 222.28 | 9.36 |
| 153 | 137.38 | 7.32 | 198 | 180.18 | 8.41 | 243 | 223.24 | 9.38 |
| 154 | 138.33 | 7.35 | 199 | 181.13 | 8.43 | 244 | 224.19 | 9.40 |
| 155 | 139.27 | 7.37 | 200 | 182.09 | 8.46 | 245 | 225.15 | 9.42 |
| 156 | 140.22 | 7.40 | 201 | 183.04 | 8.48 | 246 | 226.11 | 9.44 |
| 157 | 141.17 | 7.42 | 202 | 184.00 | 8.50 | 247 | 227.07 | 9.46 |
| 158 | 142.12 | 7.45 | 203 | 184.95 | 8.52 | 248 | 228.03 | 9.48 |
| 159 | 143.07 | 7.47 | 204 | 185.91 | 8.55 | 249 | 228.99 | 9.50 |
| 160 | 144.02 | 7.50 | 205 | 186.86 | 8.57 | 250 | 229.95 | 9.52 |
| 161 | 144.95 | 7.52 | 206 | 187.82 | 8.59 | 251 | 230.91 | 9.55 |
| 162 | 145.90 | 7.55 | 207 | 188.78 | 8.61 | 252 | 231.87 | 9.57 |
| 163 | 146.85 | 7.57 | 208 | 189.73 | 8.63 | 253 | 232.83 | 9.59 |
| 164 | 147.80 | 7.60 | 209 | 190.69 | 8.66 | 254 | 233.79 | 9.61 |
| 165 | 148.75 | 7.62 | 210 | 191.64 | 8.68 | 255 | 234.75 | 9.63 |
| 166 | 149.70 | 7.65 | 211 | 192.60 | 8.70 | 256 | 235.71 | 9.65 |
| 167 | 150.65 | 7.67 | 212 | 193.55 | 8.72 | 257 | 236.67 | 9.67 |
| 168 | 151.60 | 7.70 | 213 | 194.51 | 8.75 | 258 | 237.63 | 9.69 |
| 169 | 152.55 | 7.72 | 214 | 195.47 | 8.77 | 259 | 238.59 | 9.71 |
| 170 | 153.50 | 7.75 | 215 | 196.42 | 8.79 | 260 | 239.55 | 9.73 |
| 171 | 154.45 | 7.77 | 216 | 197.38 | 8.81 | 261 | 240.51 | 9.75 |
| 172 | 155.41 | 7.80 | 217 | 198.33 | 8.83 | 262 | 241.47 | 9.77 |
| 173 | 156.36 | 7.82 | 218 | 199.29 | 8.85 | 263 | 242.43 | 9.78 |
| 174 | 157.31 | 7.85 | 219 | 200.25 | 8.88 | 264 | 243.39 | 9.80 |
| 175 | 158.26 | 7.87 | 220 | 201.20 | 8.90 | 265 | 244.35 | 9.82 |
| 176 | 159.21 | 7.89 | 221 | 202.16 | 8.92 | 266 | 245.31 | 9.84 |
| 177 | 160.16 | 7.92 | 222 | 203.12 | 8.94 | 267 | 246.27 | 9.86 |
| 178 | 161.12 | 7.94 | 223 | 204.08 | 8.96 | 268 | 247.23 | 9.88 |
| 179 | 162.07 | 7.97 | 224 | 205.03 | 8.98 | 269 | 248.19 | 9.90 |
| 180 | 163.02 | 7.99 | 225 | 205.99 | 9.00 | 270 | 249.16 | 9.92 |

Table 4. CONSTANTS FOR p′ (SHORT FORMULA), AN APPROXIMATION OF    Section 5
THE LOWER BINOMIAL CONFIDENCE LIMIT p; c/n ≤ .125, n ≥ 16,    $\gamma_1$ = 90%
c ≤ 1,000 (Continued)    $\gamma_2$ = 80%

| c | m | k | c | m | k | c | m | k |
|---|---|---|---|---|---|---|---|---|
| 271 | 250.12 | 9.94 | 316 | 293.43 | 10.78 | 361 | 336.86 | 11.57 |
| 272 | 251.08 | 9.96 | 317 | 294.40 | 10.80 | 362 | 337.83 | 11.59 |
| 273 | 252.04 | 9.98 | 318 | 295.36 | 10.82 | 363 | 338.80 | 11.60 |
| 274 | 253.00 | 10.00 | 319 | 296.32 | 10.84 | 364 | 339.76 | 11.62 |
| 275 | 253.96 | 10.02 | 320 | 297.29 | 10.86 | 365 | 340.73 | 11.64 |
| 276 | 254.92 | 10.04 | 321 | 298.25 | 10.87 | 366 | 341.70 | 11.65 |
| 277 | 255.88 | 10.06 | 322 | 299.22 | 10.89 | 367 | 342.66 | 11.67 |
| 278 | 256.85 | 10.08 | 323 | 300.18 | 10.91 | 368 | 343.63 | 11.69 |
| 279 | 257.81 | 10.10 | 324 | 301.15 | 10.93 | 369 | 344.60 | 11.70 |
| 280 | 258.77 | 10.12 | 325 | 302.11 | 10.95 | 370 | 345.56 | 11.72 |
| 281 | 259.73 | 10.13 | 326 | 303.07 | 10.96 | 371 | 346.53 | 11.74 |
| 282 | 260.69 | 10.15 | 327 | 304.04 | 10.98 | 372 | 347.50 | 11.75 |
| 283 | 261.65 | 10.17 | 328 | 305.00 | 11.00 | 373 | 348.46 | 11.77 |
| 284 | 262.62 | 10.19 | 329 | 305.97 | 11.02 | 374 | 349.43 | 11.79 |
| 285 | 263.58 | 10.21 | 330 | 306.93 | 11.03 | 375 | 350.40 | 11.80 |
| 286 | 264.54 | 10.23 | 331 | 307.90 | 11.05 | 376 | 351.36 | 11.82 |
| 287 | 265.50 | 10.25 | 332 | 308.86 | 11.07 | 377 | 352.33 | 11.84 |
| 288 | 266.46 | 10.27 | 333 | 309.83 | 11.09 | 378 | 353.30 | 11.85 |
| 289 | 267.43 | 10.29 | 334 | 310.79 | 11.10 | 379 | 354.26 | 11.87 |
| 290 | 268.39 | 10.31 | 335 | 311.76 | 11.12 | 380 | 355.23 | 11.88 |
| 291 | 269.35 | 10.32 | 336 | 312.72 | 11.14 | 381 | 356.20 | 11.90 |
| 292 | 270.31 | 10.34 | 337 | 313.69 | 11.16 | 382 | 357.17 | 11.92 |
| 293 | 271.28 | 10.36 | 338 | 314.65 | 11.17 | 383 | 358.13 | 11.93 |
| 294 | 272.24 | 10.38 | 339 | 315.62 | 11.19 | 384 | 359.10 | 11.95 |
| 295 | 273.20 | 10.40 | 340 | 316.58 | 11.21 | 385 | 360.07 | 11.97 |
| 296 | 274.16 | 10.42 | 341 | 317.55 | 11.23 | 386 | 361.03 | 11.98 |
| 297 | 275.13 | 10.44 | 342 | 318.51 | 11.24 | 387 | 362.00 | 12.00 |
| 298 | 276.09 | 10.45 | 343 | 319.48 | 11.26 | 388 | 362.97 | 12.02 |
| 299 | 277.05 | 10.47 | 344 | 320.44 | 11.28 | 389 | 363.94 | 12.03 |
| 300 | 278.02 | 10.49 | 345 | 321.41 | 11.30 | 390 | 364.90 | 12.05 |
| 301 | 278.98 | 10.51 | 346 | 322.37 | 11.31 | 391 | 365.87 | 12.06 |
| 302 | 279.94 | 10.53 | 347 | 323.34 | 11.33 | 392 | 366.84 | 12.08 |
| 303 | 280.91 | 10.55 | 348 | 324.31 | 11.35 | 393 | 367.81 | 12.10 |
| 304 | 281.87 | 10.57 | 349 | 325.27 | 11.36 | 394 | 368.78 | 12.11 |
| 305 | 282.83 | 10.58 | 350 | 326.24 | 11.38 | 395 | 369.74 | 12.13 |
| 306 | 283.80 | 10.60 | 351 | 327.20 | 11.40 | 396 | 370.71 | 12.14 |
| 307 | 284.76 | 10.62 | 352 | 328.17 | 11.42 | 397 | 371.68 | 12.16 |
| 308 | 285.72 | 10.64 | 353 | 329.14 | 11.43 | 398 | 372.65 | 12.18 |
| 309 | 286.69 | 10.66 | 354 | 330.10 | 11.45 | 399 | 373.61 | 12.19 |
| 310 | 287.65 | 10.68 | 355 | 331.07 | 11.47 | 400 | 374.58 | 12.21 |
| 311 | 288.61 | 10.69 | 356 | 332.03 | 11.48 | 401 | 375.55 | 12.22 |
| 312 | 289.58 | 10.71 | 357 | 333.00 | 11.50 | 402 | 376.52 | 12.24 |
| 313 | 290.54 | 10.73 | 358 | 333.97 | 11.52 | 403 | 377.49 | 12.26 |
| 314 | 291.50 | 10.75 | 359 | 334.93 | 11.53 | 404 | 378.45 | 12.27 |
| 315 | 292.47 | 10.77 | 360 | 335.90 | 11.55 | 405 | 379.42 | 12.29 |

Table 4.  CONSTANTS FOR $\underline{p}'$ (SHORT FORMULA), AN APPROXIMATION OF THE LOWER BINOMIAL CONFIDENCE LIMIT $\underline{p}$; $c/n \leqslant .125$, $n \geqslant 16$, $c \leqslant 1{,}000$ (*Continued*)

| c | m | k | c | m | k | c | m | k |
|---|---|---|---|---|---|---|---|---|
| 406 | 380.39 | 12.30 | 451 | 424.00 | 13.00 | 496 | 467.67 | 13.6 |
| 407 | 381.36 | 12.32 | 452 | 424.97 | 13.02 | 497 | 468.64 | 13.6 |
| 408 | 382.33 | 12.34 | 453 | 425.94 | 13.03 | 498 | 469.61 | 13.6 |
| 409 | 383.30 | 12.35 | 454 | 426.91 | 13.05 | 499 | 470.59 | 13.7 |
| 410 | 384.26 | 12.37 | 455 | 427.88 | 13.06 | 500 | 471.56 | 13.7 |
| 411 | 385.23 | 12.38 | 456 | 428.85 | 13.08 | 501 | 472.53 | 13.7 |
| 412 | 386.20 | 12.40 | 457 | 429.82 | 13.09 | 502 | 473.50 | 13.7 |
| 413 | 387.17 | 12.42 | 458 | 430.79 | 13.11 | 503 | 474.47 | 13.7 |
| 414 | 388.14 | 12.43 | 459 | 431.76 | 13.12 | 504 | 475.44 | 13.7 |
| 415 | 389.11 | 12.45 | 460 | 432.73 | 13.14 | 505 | 476.41 | 13.7 |
| 416 | 390.07 | 12.46 | 461 | 433.70 | 13.15 | 506 | 477.39 | 13.8 |
| 417 | 391.04 | 12.48 | 462 | 434.67 | 13.17 | 507 | 478.36 | 13.8 |
| 418 | 392.01 | 12.49 | 463 | 435.64 | 13.18 | 508 | 479.33 | 13.8 |
| 419 | 392.98 | 12.51 | 464 | 436.61 | 13.20 | 509 | 480.30 | 13.8 |
| 420 | 393.95 | 12.53 | 465 | 437.58 | 13.21 | 510 | 481.27 | 13.8 |
| 421 | 394.92 | 12.54 | 466 | 438.55 | 13.23 | 511 | 482.24 | 13.8 |
| 422 | 395.89 | 12.56 | 467 | 439.52 | 13.24 | 512 | 483.21 | 13.8 |
| 423 | 396.86 | 12.57 | 468 | 440.49 | 13.26 | 513 | 484.19 | 13.9 |
| 424 | 397.82 | 12.59 | 469 | 441.46 | 13.27 | 514 | 485.16 | 13.9 |
| 425 | 398.79 | 12.60 | 470 | 442.43 | 13.29 | 515 | 486.13 | 13.9 |
| 426 | 399.76 | 12.62 | 471 | 443.40 | 13.30 | 516 | 487.10 | 13.9 |
| 427 | 400.73 | 12.63 | 472 | 444.37 | 13.31 | 517 | 488.07 | 13.9 |
| 428 | 401.70 | 12.65 | 473 | 445.34 | 13.33 | 518 | 489.05 | 13.9 |
| 429 | 402.67 | 12.67 | 474 | 446.31 | 13.34 | 519 | 490.02 | 13.9 |
| 430 | 403.64 | 12.68 | 475 | 447.28 | 13.36 | 520 | 490.99 | 14.0 |
| 431 | 404.61 | 12.70 | 476 | 448.25 | 13.37 | 521 | 491.96 | 14.0 |
| 432 | 405.58 | 12.71 | 477 | 449.22 | 13.39 | 522 | 492.93 | 14.0 |
| 433 | 406.55 | 12.73 | 478 | 450.19 | 13.40 | 523 | 493.90 | 14.0 |
| 434 | 407.51 | 12.74 | 479 | 451.16 | 13.42 | 524 | 494.88 | 14.00 |
| 435 | 408.48 | 12.76 | 480 | 452.14 | 13.43 | 525 | 495.85 | 14.08 |
| 436 | 409.45 | 12.77 | 481 | 453.11 | 13.45 | 526 | 496.82 | 14.0 |
| 437 | 410.42 | 12.79 | 482 | 454.08 | 13.46 | 527 | 497.79 | 14.1 |
| 438 | 411.39 | 12.80 | 483 | 455.05 | 13.48 | 528 | 498.77 | 14.1 |
| 439 | 412.36 | 12.82 | 484 | 456.02 | 13.49 | 529 | 499.74 | 14.1 |
| 440 | 413.33 | 12.83 | 485 | 456.99 | 13.51 | 530 | 500.71 | 14.1 |
| 441 | 414.30 | 12.85 | 486 | 457.96 | 13.52 | 531 | 501.68 | 14.1 |
| 442 | 415.27 | 12.87 | 487 | 458.93 | 13.53 | 532 | 502.65 | 14.1 |
| 443 | 416.24 | 12.88 | 488 | 459.90 | 13.55 | 533 | 503.63 | 14.1 |
| 444 | 417.21 | 12.90 | 489 | 460.87 | 13.56 | 534 | 504.60 | 14.20 |
| 445 | 418.18 | 12.91 | 490 | 461.84 | 13.58 | 535 | 505.57 | 14.2 |
| 446 | 419.15 | 12.93 | 491 | 462.82 | 13.59 | 536 | 506.54 | 14.2 |
| 447 | 420.12 | 12.94 | 492 | 463.79 | 13.61 | 537 | 507.52 | 14.2 |
| 448 | 421.09 | 12.96 | 493 | 464.76 | 13.62 | 538 | 508.49 | 14.2 |
| 449 | 422.06 | 12.97 | 494 | 465.73 | 13.64 | 539 | 509.46 | 14.2 |
| 450 | 423.03 | 12.99 | 495 | 466.70 | 13.65 | 540 | 510.43 | 14.2 |

**Table 4. CONSTANTS FOR p′ (SHORT FORMULA), AN APPROXIMATION OF THE LOWER BINOMIAL CONFIDENCE LIMIT p; c/n ⩽ .125, n ⩾ 16, c ⩽ 1,000 (Continued)**

| c | m | k | c | m | k | c | m | k |
|---|---|---|---|---|---|---|---|---|
| 541 | 511.40 | 14.30 | 586 | 555.19 | 14.91 | 631 | 599.02 | 15.49 |
| 542 | 512.38 | 14.31 | 587 | 556.16 | 14.92 | 632 | 600.00 | 15.50 |
| 543 | 513.35 | 14.33 | 588 | 557.14 | 14.93 | 633 | 600.97 | 15.52 |
| 544 | 514.32 | 14.34 | 589 | 558.11 | 14.94 | 634 | 601.94 | 15.53 |
| 545 | 515.29 | 14.35 | 590 | 559.08 | 14.96 | 635 | 602.92 | 15.54 |
| 546 | 516.27 | 14.37 | 591 | 560.06 | 14.97 | 636 | 603.89 | 15.55 |
| 547 | 517.24 | 14.38 | 592 | 561.03 | 14.98 | 637 | 604.87 | 15.57 |
| 548 | 518.21 | 14.39 | 593 | 562.01 | 15.00 | 638 | 605.84 | 15.58 |
| 549 | 519.19 | 14.41 | 594 | 562.98 | 15.01 | 639 | 606.82 | 15.59 |
| 550 | 520.16 | 14.42 | 595 | 563.95 | 15.02 | 640 | 607.79 | 15.60 |
| 551 | 521.13 | 14.43 | 596 | 564.93 | 15.04 | 641 | 608.77 | 15.62 |
| 552 | 522.10 | 14.45 | 597 | 565.90 | 15.05 | 642 | 609.74 | 15.63 |
| 553 | 523.08 | 14.46 | 598 | 566.87 | 15.06 | 643 | 610.72 | 15.64 |
| 554 | 524.05 | 14.48 | 599 | 567.85 | 15.08 | 644 | 611.69 | 15.65 |
| 555 | 525.02 | 14.49 | 600 | 568.82 | 15.09 | 645 | 612.67 | 15.67 |
| 556 | 525.99 | 14.50 | 601 | 569.80 | 15.10 | 646 | 613.64 | 15.68 |
| 557 | 526.97 | 14.52 | 602 | 570.77 | 15.12 | 647 | 614.61 | 15.69 |
| 558 | 527.94 | 14.53 | 603 | 571.74 | 15.13 | 648 | 615.59 | 15.71 |
| 559 | 528.91 | 14.54 | 604 | 572.72 | 15.14 | 649 | 616.56 | 15.72 |
| 560 | 529.89 | 14.56 | 605 | 573.69 | 15.15 | 650 | 617.54 | 15.73 |
| 561 | 530.86 | 14.57 | 606 | 574.66 | 15.17 | 651 | 618.51 | 15.74 |
| 562 | 531.83 | 14.58 | 607 | 575.64 | 15.18 | 652 | 619.49 | 15.76 |
| 563 | 532.80 | 14.60 | 608 | 576.61 | 15.19 | 653 | 620.46 | 15.77 |
| 564 | 533.78 | 14.61 | 609 | 577.59 | 15.21 | 654 | 621.44 | 15.78 |
| 565 | 534.75 | 14.62 | 610 | 578.56 | 15.22 | 655 | 622.41 | 15.79 |
| 566 | 535.72 | 14.64 | 611 | 579.53 | 15.23 | 656 | 623.39 | 15.81 |
| 567 | 536.70 | 14.65 | 612 | 580.51 | 15.25 | 657 | 624.36 | 15.82 |
| 568 | 537.67 | 14.66 | 613 | 581.48 | 15.26 | 658 | 625.34 | 15.83 |
| 569 | 538.64 | 14.68 | 614 | 582.46 | 15.27 | 659 | 626.31 | 15.84 |
| 570 | 539.62 | 14.69 | 615 | 583.43 | 15.28 | 660 | 627.29 | 15.86 |
| 571 | 540.59 | 14.71 | 616 | 584.41 | 15.30 | 661 | 628.26 | 15.87 |
| 572 | 541.56 | 14.72 | 617 | 585.38 | 15.31 | 662 | 629.24 | 15.88 |
| 573 | 542.54 | 14.73 | 618 | 586.35 | 15.32 | 663 | 630.21 | 15.89 |
| 574 | 543.51 | 14.75 | 619 | 587.33 | 15.34 | 664 | 631.19 | 15.91 |
| 575 | 544;48 | 14.76 | 620 | 588.30 | 15.35 | 665 | 632.16 | 15.92 |
| 576 | 545.46 | 14.77 | 621 | 589.28 | 15.36 | 666 | 633.14 | 15.93 |
| 577 | 546.43 | 14.79 | 622 | 590.25 | 15.37 | 667 | 634.11 | 15.94 |
| 578 | 547.40 | 14.80 | 623 | 591.23 | 15.39 | 668 | 635.09 | 15.95 |
| 579 | 548.38 | 14.81 | 624 | 592.20 | 15.40 | 669 | 636.07 | 15.97 |
| 580 | 549.35 | 14.83 | 625 | 593.17 | 15.41 | 670 | 637.04 | 15.98 |
| 581 | 550.32 | 14.84 | 626 | 594.15 | 15.43 | 671 | 638.02 | 15.99 |
| 582 | 551.30 | 14.85 | 627 | 595.12 | 15.44 | 672 | 638.99 | 16.00 |
| 583 | 552.27 | 14.87 | 628 | 596.10 | 15.45 | 673 | 639.97 | 16.02 |
| 584 | 553.24 | 14.88 | 629 | 597.07 | 15.46 | 674 | 640.94 | 16.03 |
| 585 | 554.22 | 14.89 | 630 | 598.05 | 15.48 | 675 | 641.92 | 16.04 |

Table 4.  CONSTANTS FOR p' (SHORT FORMULA), AN APPROXIMATION OF
THE LOWER BINOMIAL CONFIDENCE LIMIT p; c/n ≤ .125, n ≥ 16,
c ≤ 1,000 (*Continued*)

| c | m | k | c | m | k | c | m | k |
|---|---|---|---|---|---|---|---|---|
| 676 | 642.89 | 16.05 | 721 | 686.80 | 16.60 | 766 | 730.74 | 17.13 |
| 677 | 643.87 | 16.07 | 722 | 687.78 | 16.61 | 767 | 731.72 | 17.14 |
| 678 | 644.84 | 16.08 | 723 | 688.75 | 16.62 | 768 | 732.70 | 17.15 |
| 679 | 645.82 | 16.09 | 724 | 689.73 | 16.64 | 769 | 733.67 | 17.16 |
| 680 | 646.79 | 16.10 | 725 | 690.71 | 16.65 | 770 | 734.65 | 17.17 |
| 681 | 647.77 | 16.12 | 726 | 691.68 | 16.66 | 771 | 735.63 | 17.19 |
| 682 | 648.74 | 16.13 | 727 | 692.66 | 16.67 | 772 | 736.60 | 17.20 |
| 683 | 649.72 | 16.14 | 728 | 693.63 | 16.68 | 773 | 737.58 | 17.21 |
| 684 | 650.70 | 16.15 | 729 | 694.61 | 16.69 | 774 | 738.56 | 17.22 |
| 685 | 651.67 | 16.16 | 730 | 695.59 | 16.71 | 775 | 739.54 | 17.23 |
| 686 | 652.65 | 16.18 | 731 | 696.56 | 16.72 | 776 | 740.51 | 17.24 |
| 687 | 653.62 | 16.19 | 732 | 697.54 | 16.73 | 777 | 741.49 | 17.26 |
| 688 | 654.60 | 16.20 | 733 | 698.52 | 16.74 | 778 | 742.47 | 17.27 |
| 689 | 655.57 | 16.21 | 734 | 699.49 | 16.75 | 779 | 743.44 | 17.28 |
| 690 | 656.55 | 16.23 | 735 | 700.47 | 16.77 | 780 | 744.42 | 17.29 |
| 691 | 657.52 | 16.24 | 736 | 701.45 | 16.78 | 781 | 745.40 | 17.30 |
| 692 | 658.50 | 16.25 | 737 | 702.42 | 16.79 | 782 | 746.38 | 17.31 |
| 693 | 659.48 | 16.26 | 738 | 703.40 | 16.80 | 783 | 747.35 | 17.32 |
| 694 | 660.45 | 16.27 | 739 | 704.37 | 16.81 | 784 | 748.33 | 17.34 |
| 695 | 661.43 | 16.29 | 740 | 705.35 | 16.82 | 785 | 749.31 | 17.35 |
| 696 | 662.40 | 16.30 | 741 | 706.33 | 16.84 | 786 | 750.28 | 17.36 |
| 697 | 663.38 | 16.31 | 742 | 707.30 | 16.85 | 787 | 751.26 | 17.37 |
| 698 | 664.35 | 16.32 | 743 | 708.28 | 16.86 | 788 | 752.24 | 17.38 |
| 699 | 665.33 | 16.33 | 744 | 709.26 | 16.87 | 789 | 753.21 | 17.39 |
| 700 | 666.31 | 16.35 | 745 | 710.23 | 16.88 | 790 | 754.19 | 17.40 |
| 701 | 667.28 | 16.36 | 746 | 711.21 | 16.90 | 791 | 755.17 | 17.42 |
| 702 | 668.26 | 16.37 | 747 | 712.19 | 16.91 | 792 | 756.15 | 17.43 |
| 703 | 669.23 | 16.38 | 748 | 713.16 | 16.92 | 793 | 757.12 | 17.44 |
| 704 | 670.21 | 16.40 | 749 | 714.14 | 16.93 | 794 | 758.10 | 17.45 |
| 705 | 671.19 | 16.41 | 750 | 715.12 | 16.94 | 795 | 759.08 | 17.46 |
| 706 | 672.16 | 16.42 | 751 | 716.09 | 16.95 | 796 | 760.06 | 17.47 |
| 707 | 673.14 | 16.43 | 752 | 717.07 | 16.97 | 797 | 761.03 | 17.48 |
| 708 | 674.11 | 16.44 | 753 | 718.05 | 16.98 | 798 | 762.01 | 17.49 |
| 709 | 675.09 | 16.46 | 754 | 719.02 | 16.99 | 799 | 762.99 | 17.51 |
| 710 | 676.06 | 16.47 | 755 | 720.00 | 17.00 | 800 | 763.96 | 17.52 |
| 711 | 677.04 | 16.48 | 756 | 720.98 | 17.01 | 801 | 764.94 | 17.53 |
| 712 | 678.02 | 16.49 | 757 | 721.95 | 17.02 | 802 | 765.92 | 17.54 |
| 713 | 678.99 | 16.50 | 758 | 722.93 | 17.04 | 803 | 766.90 | 17.55 |
| 714 | 679.97 | 16.52 | 759 | 723.91 | 17.05 | 804 | 767.87 | 17.56 |
| 715 | 680.94 | 16.53 | 760 | 724.88 | 17.06 | 805 | 768.85 | 17.57 |
| 716 | 681.92 | 16.54 | 761 | 725.86 | 17.07 | 806 | 769.83 | 17.59 |
| 717 | 682.90 | 16.55 | 762 | 726.84 | 17.08 | 807 | 770.81 | 17.60 |
| 718 | 683.87 | 16.56 | 763 | 727.81 | 17.09 | 808 | 771.78 | 17.61 |
| 719 | 684.85 | 16.58 | 764 | 728.79 | 17.11 | 809 | 772.76 | 17.62 |
| 720 | 685.83 | 16.59 | 765 | 729.77 | 17.12 | 810 | 773.74 | 17.63 |

# Table 4. CONSTANTS FOR $p'$ (SHORT FORMULA), AN APPROXIMATION OF THE LOWER BINOMIAL CONFIDENCE LIMIT $\underline{p}$; $c/n \leqslant .125$, $n \geqslant 16$, $c \leqslant 1,000$ (Continued)

| c | $\underline{m}$ | $\underline{k}$ | c | $\underline{m}$ | $\underline{k}$ | c | $\underline{m}$ | $\underline{k}$ |
|---|---|---|---|---|---|---|---|---|
| 811 | 774.72 | 17.64 | 856 | 818.72 | 18.14 | 901 | 862.74 | 18.63 |
| 812 | 775.69 | 17.65 | 857 | 819.70 | 18.15 | 902 | 863.72 | 18.64 |
| 813 | 776.67 | 17.66 | 858 | 820.67 | 18.16 | 903 | 864.70 | 18.65 |
| 814 | 777.65 | 17.68 | 859 | 821.65 | 18.17 | 904 | 865.68 | 18.66 |
| 815 | 778.63 | 17.69 | 860 | 822.63 | 18.18 | 905 | 866.66 | 18.67 |
| 816 | 779.60 | 17.70 | 861 | 823.61 | 18.20 | 906 | 867.64 | 18.68 |
| 817 | 780.58 | 17.71 | 862 | 824.59 | 18.21 | 907 | 868.62 | 18.69 |
| 818 | 781.56 | 17.72 | 863 | 825.56 | 18.22 | 908 | 869.60 | 18.70 |
| 819 | 782.54 | 17.73 | 864 | 826.54 | 18.23 | 909 | 870.57 | 18.71 |
| 820 | 783.51 | 17.74 | 865 | 827.52 | 18.24 | 910 | 871.55 | 18.72 |
| 821 | 784.49 | 17.75 | 866 | 828.50 | 18.25 | 911 | 872.53 | 18.73 |
| 822 | 785.47 | 17.77 | 867 | 829.48 | 18.26 | 912 | 873.51 | 18.74 |
| 823 | 786.45 | 17.78 | 868 | 830.46 | 18.27 | 913 | 874.49 | 18.76 |
| 824 | 787.43 | 17.79 | 869 | 831.43 | 18.28 | 914 | 875.47 | 18.77 |
| 825 | 788.40 | 17.80 | 870 | 832.41 | 18.29 | 915 | 876.45 | 18.78 |
| 826 | 789.38 | 17.81 | 871 | 833.39 | 18.30 | 916 | 877.43 | 18.79 |
| 827 | 790.36 | 17.82 | 872 | 834.37 | 18.32 | 917 | 878.40 | 18.80 |
| 828 | 791.34 | 17.83 | 873 | 835.35 | 18.33 | 918 | 879.38 | 18.81 |
| 829 | 792.31 | 17.84 | 874 | 836.33 | 18.34 | 919 | 880.36 | 18.82 |
| 830 | 793.29 | 17.85 | 875 | 837.30 | 18.35 | 920 | 881.34 | 18.83 |
| 831 | 794.27 | 17.87 | 876 | 838.28 | 18.36 | 921 | 882.32 | 18.84 |
| 832 | 795.25 | 17.88 | 877 | 839.26 | 18.37 | 922 | 883.30 | 18.85 |
| 833 | 796.22 | 17.89 | 878 | 840.24 | 18.38 | 923 | 884.28 | 18.86 |
| 834 | 797.20 | 17.90 | 879 | 841.22 | 18.39 | 924 | 885.26 | 18.87 |
| 835 | 798.18 | 17.91 | 880 | 842.20 | 18.40 | 925 | 886.24 | 18.88 |
| 836 | 799.16 | 17.92 | 881 | 843.17 | 18.41 | 926 | 887.21 | 18.89 |
| 837 | 800.14 | 17.93 | 882 | 844.15 | 18.42 | 927 | 888.19 | 18.90 |
| 838 | 801.11 | 17.94 | 883 | 845.13 | 18.43 | 928 | 889.17 | 18.91 |
| 839 | 802.09 | 17.95 | 884 | 846.11 | 18.45 | 929 | 890.15 | 18.92 |
| 840 | 803.07 | 17.97 | 885 | 847.09 | 18.46 | 930 | 891.13 | 18.93 |
| 841 | 804.05 | 17.98 | 886 | 848.07 | 18.47 | 931 | 892.11 | 18.95 |
| 842 | 805.03 | 17.99 | 887 | 849.04 | 18.48 | 932 | 893.09 | 18.96 |
| 843 | 806.00 | 18.00 | 888 | 850.02 | 18.49 | 933 | 894.07 | 18.97 |
| 844 | 806.98 | 18.01 | 889 | 851.00 | 18.50 | 934 | 895.05 | 18.98 |
| 845 | 807.96 | 18.02 | 890 | 851.98 | 18.51 | 935 | 896.03 | 18.99 |
| 846 | 808.94 | 18.03 | 891 | 852.96 | 18.52 | 936 | 897.00 | 19.00 |
| 847 | 809.92 | 18.04 | 892 | 853.94 | 18.53 | 937 | 897.98 | 19.01 |
| 848 | 810.89 | 18.05 | 893 | 854.92 | 18.54 | 938 | 898.96 | 19.02 |
| 849 | 811.87 | 18.06 | 894 | 855.89 | 18.55 | 939 | 899.94 | 19.03 |
| 850 | 812.85 | 18.08 | 895 | 856.87 | 18.56 | 940 | 900.92 | 19.04 |
| 851 | 813.83 | 18.09 | 896 | 857.85 | 18.57 | 941 | 901.90 | 19.05 |
| 852 | 814.81 | 18.10 | 897 | 858.83 | 18.58 | 942 | 902.88 | 19.06 |
| 853 | 815.78 | 18.11 | 898 | 859.81 | 18.60 | 943 | 903.86 | 19.07 |
| 854 | 816.76 | 18.12 | 899 | 860.79 | 18.61 | 944 | 904.84 | 19.08 |
| 855 | 817.74 | 18.13 | 900 | 861.77 | 18.62 | 945 | 905.82 | 19.09 |

Table 4.  CONSTANTS FOR $\underline{p}'$ (SHORT FORMULA), AN APPROXIMATION OF
THE LOWER BINOMIAL CONFIDENCE LIMIT $\underline{p}$; $c/n \leqslant .125$, $n \geqslant 16$,
$c \leqslant 1,000$ (*Continued*)

| c | $\underline{m}$ | $\underline{k}$ | c | $\underline{m}$ | $\underline{k}$ | c | $\underline{m}$ | $\underline{k}$ |
|---|---|---|---|---|---|---|---|---|
| 946 | 906.80 | 19.10 | 966 | 926.38 | 19.31 | 986 | 945.97 | 19.51 |
| 947 | 907.77 | 19.11 | 967 | 927.36 | 19.32 | 987 | 946.95 | 19.52 |
| 948 | 908.75 | 19.12 | 968 | 928.34 | 19.33 | 988 | 947.93 | 19.53 |
| 949 | 909.73 | 19.13 | 969 | 929.32 | 19.34 | 989 | 948.91 | 19.55 |
| 950 | 910.71 | 19.14 | 970 | 930.30 | 19.35 | 990 | 949.89 | 19.56 |
| 951 | 911.69 | 19.15 | 971 | 931.28 | 19.36 | 991 | 950.87 | 19.57 |
| 952 | 912.67 | 19.16 | 972 | 932.26 | 19.37 | 992 | 951.85 | 19.58 |
| 953 | 913.65 | 19.17 | 973 | 933.24 | 19.38 | 993 | 952.83 | 19.59 |
| 954 | 914.63 | 19.19 | 974 | 934.22 | 19.39 | 994 | 953.81 | 19.60 |
| 955 | 915.61 | 19.20 | 975 | 935.20 | 19.40 | 995 | 954.79 | 19.61 |
| 956 | 916.59 | 19.21 | 976 | 936.18 | 19.41 | 996 | 955.77 | 19.62 |
| 957 | 917.57 | 19.22 | 977 | 937.16 | 19.42 | 997 | 956.75 | 19.63 |
| 958 | 918.55 | 19.23 | 978 | 938.13 | 19.43 | 998 | 957.73 | 19.64 |
| 959 | 919.53 | 19.24 | 979 | 939.11 | 19.44 | 999 | 958.71 | 19.65 |
| 960 | 920.51 | 19.25 | 980 | 940.09 | 19.45 | 1000 | 959.69 | 19.66 |
| 961 | 921.48 | 19.26 | 981 | 941.07 | 19.46 | | | |
| 962 | 922.46 | 19.27 | 982 | 942.05 | 19.47 | | | |
| 963 | 923.44 | 19.28 | 983 | 943.03 | 19.48 | | | |
| 964 | 924.42 | 19.29 | 984 | 944.01 | 19.49 | | | |
| 965 | 925.40 | 19.30 | 985 | 944.99 | 19.50 | | | |

**Table 4.   CONSTANTS FOR $p'$ (SHORT FORMULA), AN APPROXIMATION OF THE LOWER BINOMIAL CONFIDENCE LIMIT $\underline{p}$; $c/n \leqslant .125$, $n \geqslant 16$, $c \leqslant 1,000$ (Continued)**

| c | m | k | c | m | k | c | m | k |
|---|---|---|---|---|---|---|---|---|
| 1 | .22314 | −0.112 | 46 | 40.217 | 2.39 | 91 | 82.890 | 3.56 |
| 2 | .82439 | 0.088 | 47 | 41.155 | 2.42 | 92 | 83.846 | 3.58 |
| 3 | 1.5351 | 0.232 | 48 | 42.093 | 2.45 | 93 | 84.802 | 3.61 |
| 4 | 2.2968 | 0.352 | 49 | 43.033 | 2.48 | 94 | 85.759 | 3.63 |
| 5 | 3.0895 | 0.455 | 50 | 43.973 | 2.51 | 95 | 86.715 | 3.65 |
| 6 | 3.9037 | 0.548 | 51 | 44.913 | 2.55 | 96 | 87.672 | 3.67 |
| 7 | 4.7337 | 0.633 | 52 | 45.854 | 2.58 | 97 | 88.629 | 3.69 |
| 8 | 5.5761 | 0.712 | 53 | 46.796 | 2.61 | 98 | 89.586 | 3.71 |
| 9 | 6.4285 | 0.786 | 54 | 47.738 | 2.64 | 99 | 90.544 | 3.74 |
| 10 | 7.2892 | 0.86 | 55 | 48.681 | 2.67 | 100 | 91.502 | 3.76 |
| 11 | 8.1570 | 0.92 | 56 | 49.625 | 2.70 | 101 | 92.460 | 3.78 |
| 12 | 9.0309 | 0.98 | 57 | 50.569 | 2.73 | 102 | 93.418 | 3.80 |
| 13 | 9.9101 | 1.04 | 58 | 51.513 | 2.75 | 103 | 94.376 | 3.82 |
| 14 | 10.794 | 1.10 | 59 | 52.458 | 2.78 | 104 | 95.335 | 3.84 |
| 15 | 11.682 | 1.16 | 60 | 53.403 | 2.81 | 105 | 96.293 | 3.86 |
| 16 | 12.574 | 1.21 | 61 | 54.349 | 2.84 | 106 | 97.252 | 3.88 |
| 17 | 13.469 | 1.27 | 62 | 55.295 | 2.86 | 107 | 98.212 | 3.90 |
| 18 | 14.368 | 1.32 | 63 | 56.242 | 2.89 | 108 | 99.171 | 3.92 |
| 19 | 15.269 | 1.37 | 64 | 57.189 | 2.92 | 109 | 100.13 | 3.94 |
| 20 | 16.173 | 1.41 | 65 | 58.136 | 2.94 | 110 | 101.09 | 3.96 |
| 21 | 17.079 | 1.46 | 66 | 59.084 | 2.97 | 111 | 102.05 | 3.98 |
| 22 | 17.987 | 1.51 | 67 | 60.032 | 2.99 | 112 | 103.01 | 4.00 |
| 23 | 18.898 | 1.55 | 68 | 60.981 | 3.02 | 113 | 103.97 | 4.02 |
| 24 | 19.810 | 1.59 | 69 | 61.930 | 3.04 | 114 | 104.93 | 4.04 |
| 25 | 20.725 | 1.64 | 70 | 62.879 | 3.07 | 115 | 105.89 | 4.06 |
| 26 | 21.641 | 1.68 | 71 | 63.829 | 3.09 | 116 | 106.85 | 4.08 |
| 27 | 22.558 | 1.72 | 72 | 64.779 | 3.12 | 117 | 107.81 | 4.10 |
| 28 | 23.478 | 1.76 | 73 | 65.730 | 3.14 | 118 | 108.77 | 4.12 |
| 29 | 24.398 | 1.80 | 74 | 66.680 | 3.17 | 119 | 109.74 | 4.14 |
| 30 | 25.320 | 1.84 | 75 | 67.632 | 3.19 | 120 | 110.70 | 4.16 |
| 31 | 26.244 | 1.88 | 76 | 68.583 | 3.22 | 121 | 111.66 | 4.18 |
| 32 | 27.168 | 1.92 | 77 | 69.535 | 3.24 | 122 | 112.62 | 4.20 |
| 33 | 28.094 | 1.95 | 78 | 70.487 | 3.27 | 123 | 113.58 | 4.22 |
| 34 | 29.021 | 1.99 | 79 | 71.438 | 3.29 | 124 | 114.54 | 4.23 |
| 35 | 29.949 | 2.03 | 80 | 72.393 | 3.31 | 125 | 115.51 | 4.25 |
| 36 | 30.878 | 2.06 | 81 | 73.345 | 3.34 | 126 | 116.47 | 4.27 |
| 37 | 31.808 | 2.10 | 82 | 74.298 | 3.36 | 127 | 117.43 | 4.29 |
| 38 | 32.739 | 2.13 | 83 | 75.252 | 3.38 | 128 | 118.39 | 4.31 |
| 39 | 33.671 | 2.16 | 84 | 76.206 | 3.41 | 129 | 119.36 | 4.33 |
| 40 | 34.603 | 2.20 | 85 | 77.160 | 3.43 | 130 | 120.32 | 4.35 |
| 41 | 35.537 | 2.23 | 86 | 78.114 | 3.45 | 131 | 121.28 | 4.37 |
| 42 | 36.471 | 2.26 | 87 | 79.069 | 3.47 | 132 | 122.25 | 4.38 |
| 43 | 37.407 | 2.30 | 88 | 80.024 | 3.50 | 133 | 123.21 | 4.40 |
| 44 | 38.343 | 2.33 | 89 | 80.979 | 3.52 | 134 | 124.17 | 4.42 |
| 45 | 39.279 | 2.36 | 90 | 81.934 | 3.54 | 135 | 125.14 | 4.44 |

Table 4.  CONSTANTS FOR $p'$ (SHORT FORMULA), AN APPROXIMATION OF THE LOWER BINOMIAL CONFIDENCE LIMIT $\underline{p}$; $c/n \leqslant .125$, $n \geqslant 16$, $c \leqslant 1,000$ (*Continued*)

| c | m | k | c | m | k | c | m | k |
|---|---|---|---|---|---|---|---|---|
| 136 | 126.10 | 4.46 | 181 | 169.58 | 5.21 | 226 | 213.25 | 5.8 |
| 137 | 127.06 | 4.47 | 182 | 170.55 | 5.23 | 227 | 214.22 | 5.8 |
| 138 | 128.03 | 4.49 | 183 | 171.52 | 5.24 | 228 | 215.19 | 5.9 |
| 139 | 128.99 | 4.51 | 184 | 172.49 | 5.26 | 229 | 216.17 | 5.9 |
| 140 | 129.96 | 4.53 | 185 | 173.46 | 5.27 | 230 | 217.14 | 5.9 |
| 141 | 130.92 | 4.55 | 186 | 174.42 | 5.29 | 231 | 218.11 | 5.9 |
| 142 | 131.89 | 4.56 | 187 | 175.39 | 5.30 | 232 | 219.08 | 5.9 |
| 143 | 132.85 | 4.58 | 188 | 176.36 | 5.32 | 233 | 220.06 | 5.9 |
| 144 | 133.82 | 4.60 | 189 | 177.33 | 5.33 | 234 | 221.03 | 5.9 |
| 145 | 134.78 | 4.62 | 190 | 178.30 | 5.35 | 235 | 222.00 | 6.0 |
| 146 | 135.75 | 4.63 | 191 | 179.27 | 5.36 | 236 | 222.97 | 6.0 |
| 147 | 136.71 | 4.65 | 192 | 180.24 | 5.38 | 237 | 223.95 | 6.0 |
| 148 | 137.68 | 4.67 | 193 | 181.21 | 5.39 | 238 | 224.92 | 6.0 |
| 149 | 138.64 | 4.69 | 194 | 182.18 | 5.41 | 239 | 225.89 | 6.0 |
| 150 | 139.61 | 4.70 | 195 | 183.15 | 5.42 | 240 | 226.86 | 6.0 |
| 151 | 140.57 | 4.72 | 196 | 184.12 | 5.44 | 241 | 227.84 | 6.0 |
| 152 | 141.54 | 4.74 | 197 | 185.09 | 5.45 | 242 | 228.81 | 6.0 |
| 153 | 142.50 | 4.75 | 198 | 186.06 | 5.47 | 243 | 229.78 | 6.1 |
| 154 | 143.47 | 4.77 | 199 | 187.03 | 5.48 | 244 | 230.76 | 6.1 |
| 155 | 144.44 | 4.79 | 200 | 188.00 | 5.50 | 245 | 231.73 | 6.1 |
| 156 | 145.40 | 4.80 | 201 | 188.97 | 5.51 | 246 | 232.70 | 6.1 |
| 157 | 146.37 | 4.82 | 202 | 189.94 | 5.53 | 247 | 233.68 | 6.1 |
| 158 | 147.34 | 4.84 | 203 | 190.91 | 5.54 | 248 | 234.65 | 6.1 |
| 159 | 148.30 | 4.85 | 204 | 191.88 | 5.56 | 249 | 235.62 | 6.1 |
| 160 | 149.27 | 4.87 | 205 | 192.85 | 5.57 | 250 | 236.60 | 6.2 |
| 161 | 150.22 | 4.89 | 206 | 193.82 | 5.59 | 251 | 237.57 | 6.2 |
| 162 | 151.19 | 4.90 | 207 | 194.79 | 5.60 | 252 | 238.54 | 6.2 |
| 163 | 152.16 | 4.92 | 208 | 195.76 | 5.62 | 253 | 239.52 | 6.2 |
| 164 | 153.12 | 4.94 | 209 | 196.74 | 5.63 | 254 | 240.49 | 6.2 |
| 165 | 154.09 | 4.95 | 210 | 197.71 | 5.65 | 255 | 241.46 | 6.2 |
| 166 | 155.06 | 4.97 | 211 | 198.68 | 5.66 | 256 | 242.44 | 6.2 |
| 167 | 156.03 | 4.99 | 212 | 199.65 | 5.68 | 257 | 243.41 | 6.2 |
| 168 | 156.99 | 5.00 | 213 | 200.62 | 5.69 | 258 | 244.38 | 6.3 |
| 169 | 157.96 | 5.02 | 214 | 201.59 | 5.70 | 259 | 245.36 | 6.3 |
| 170 | 158.93 | 5.04 | 215 | 202.56 | 5.72 | 260 | 246.33 | 6.3 |
| 171 | 159.90 | 5.05 | 216 | 203.53 | 5.73 | 261 | 247.31 | 6.3 |
| 172 | 160.87 | 5.07 | 217 | 204.50 | 5.75 | 262 | 248.28 | 6.3 |
| 173 | 161.83 | 5.08 | 218 | 205.48 | 5.76 | 263 | 249.25 | 6.3 |
| 174 | 162.80 | 5.10 | 219 | 206.45 | 5.78 | 264 | 250.23 | 6.3 |
| 175 | 163.77 | 5.12 | 220 | 207.42 | 5.79 | 265 | 251.20 | 6.4 |
| 176 | 164.74 | 5.13 | 221 | 208.39 | 5.80 | 266 | 252.18 | 6.4 |
| 177 | 165.71 | 5.15 | 222 | 209.36 | 5.82 | 267 | 253.15 | 6.4 |
| 178 | 166.67 | 5.16 | 223 | 210.33 | 5.83 | 268 | 254.12 | 6.4 |
| 179 | 167.64 | 5.18 | 224 | 211.31 | 5.85 | 269 | 255.10 | 6.4 |
| 180 | 168.61 | 5.19 | 225 | 212.28 | 5.86 | 270 | 256.07 | 6.4 |

Table 4. CONSTANTS FOR $p'$ (SHORT FORMULA), AN APPROXIMATION OF
THE LOWER BINOMIAL CONFIDENCE LIMIT $\underline{p}$; $c/n \leqslant .125$, $n \geqslant 16$,
$c \leqslant 1,000$ (Continued)

Section 6

$\gamma_1 = 80\%$
$\gamma_2 = 60\%$

| c | m | k | c | m | k | c | m | k |
|---|---|---|---|---|---|---|---|---|
| 271 | 257.05 | 6.48 | 316 | 300.94 | 7.03 | 361 | 344.91 | 7.54 |
| 272 | 258.02 | 6.49 | 317 | 301.92 | 7.04 | 362 | 345.89 | 7.56 |
| 273 | 259.00 | 6.50 | 318 | 302.89 | 7.05 | 363 | 346.87 | 7.57 |
| 274 | 259.97 | 6.51 | 319 | 303.87 | 7.06 | 364 | 347.85 | 7.58 |
| 275 | 260.95 | 6.53 | 320 | 304.85 | 7.08 | 365 | 348.82 | 7.59 |
| 276 | 261.92 | 6.54 | 321 | 305.82 | 7.09 | 366 | 349.80 | 7.60 |
| 277 | 262.90 | 6.55 | 322 | 306.80 | 7.10 | 367 | 350.78 | 7.61 |
| 278 | 263.87 | 6.56 | 323 | 307.78 | 7.11 | 368 | 351.76 | 7.62 |
| 279 | 264.84 | 6.58 | 324 | 308.75 | 7.12 | 369 | 352.74 | 7.63 |
| 280 | 265.82 | 6.59 | 325 | 309.73 | 7.13 | 370 | 353.71 | 7.64 |
| 281 | 266.79 | 6.60 | 326 | 310.71 | 7.15 | 371 | 354.69 | 7.65 |
| 282 | 267.77 | 6.62 | 327 | 311.68 | 7.16 | 372 | 355.67 | 7.66 |
| 283 | 268.74 | 6.63 | 328 | 312.66 | 7.17 | 373 | 356.65 | 7.68 |
| 284 | 269.72 | 6.64 | 329 | 313.64 | 7.18 | 374 | 357.63 | 7.69 |
| 285 | 270.69 | 6.65 | 330 | 314.61 | 7.19 | 375 | 358.60 | 7.70 |
| 286 | 271.67 | 6.67 | 331 | 315.59 | 7.20 | 376 | 359.58 | 7.71 |
| 287 | 272.64 | 6.68 | 332 | 316.57 | 7.22 | 377 | 360.56 | 7.72 |
| 288 | 273.62 | 6.69 | 333 | 317.54 | 7.23 | 378 | 361.54 | 7.73 |
| 289 | 274.60 | 6.70 | 334 | 318.52 | 7.24 | 379 | 362.52 | 7.74 |
| 290 | 275.57 | 6.71 | 335 | 319.50 | 7.25 | 380 | 363.50 | 7.75 |
| 291 | 276.55 | 6.73 | 336 | 320.48 | 7.26 | 381 | 364.47 | 7.76 |
| 292 | 277.52 | 6.74 | 337 | 321.45 | 7.27 | 382 | 365.45 | 7.77 |
| 293 | 278.50 | 6.75 | 338 | 322.43 | 7.29 | 383 | 366.43 | 7.78 |
| 294 | 279.47 | 6.76 | 339 | 323.41 | 7.30 | 384 | 367.41 | 7.79 |
| 295 | 280.45 | 6.78 | 340 | 324.38 | 7.31 | 385 | 368.39 | 7.81 |
| 296 | 281.42 | 6.79 | 341 | 325.36 | 7.32 | 386 | 369.37 | 7.82 |
| 297 | 282.40 | 6.80 | 342 | 326.34 | 7.33 | 387 | 370.35 | 7.83 |
| 298 | 283.37 | 6.81 | 343 | 327.32 | 7.34 | 388 | 371.32 | 7.84 |
| 299 | 284.35 | 6.83 | 344 | 328.29 | 7.35 | 389 | 372.30 | 7.85 |
| 300 | 285.33 | 6.84 | 345 | 329.27 | 7.36 | 390 | 373.28 | 7.86 |
| 301 | 286.30 | 6.85 | 346 | 330.25 | 7.38 | 391 | 374.26 | 7.87 |
| 302 | 287.28 | 6.86 | 347 | 331.23 | 7.39 | 392 | 375.24 | 7.88 |
| 303 | 288.25 | 6.87 | 348 | 332.20 | 7.40 | 393 | 376.22 | 7.89 |
| 304 | 289.23 | 6.89 | 349 | 333.18 | 7.41 | 394 | 377.20 | 7.90 |
| 305 | 290.20 | 6.90 | 350 | 334.16 | 7.42 | 395 | 378.18 | 7.91 |
| 306 | 291.18 | 6.91 | 351 | 335.14 | 7.43 | 396 | 379.15 | 7.92 |
| 307 | 292.16 | 6.92 | 352 | 336.11 | 7.44 | 397 | 380.13 | 7.93 |
| 308 | 293.13 | 6.93 | 353 | 337.09 | 7.45 | 398 | 381.11 | 7.94 |
| 309 | 294.11 | 6.95 | 354 | 338.07 | 7.47 | 399 | 382.09 | 7.95 |
| 310 | 295.08 | 6.96 | 355 | 339.05 | 7.48 | 400 | 383.07 | 7.96 |
| 311 | 296.06 | 6.97 | 356 | 340.02 | 7.49 | 401 | 384.05 | 7.98 |
| 312 | 297.04 | 6.98 | 357 | 341.00 | 7.50 | 402 | 385.03 | 7.99 |
| 313 | 298.01 | 6.99 | 358 | 341.98 | 7.51 | 403 | 386.01 | 8.00 |
| 314 | 298.99 | 7.01 | 359 | 342.96 | 7.52 | 404 | 386.99 | 8.01 |
| 315 | 299.97 | 7.02 | 360 | 343.93 | 7.53 | 405 | 387.97 | 8.02 |

Table 4. CONSTANTS FOR p′ (SHORT FORMULA), AN APPROXIMATION OF THE LOWER BINOMIAL CONFIDENCE LIMIT p; c/n ≤ .125, n ≥ 16, c ≤ 1,000 (*Continued*)

| c | m | k | c | m | k | c | m | k |
|---|---|---|---|---|---|---|---|---|
| 406 | 388.94 | 8.03 | 451 | 433.03 | 8.49 | 496 | 477.16 | 8.92 |
| 407 | 389.92 | 8.04 | 452 | 434.01 | 8.50 | 497 | 478.14 | 8.93 |
| 408 | 390.90 | 8.05 | 453 | 434.99 | 8.51 | 498 | 479.12 | 8.94 |
| 409 | 391.88 | 8.06 | 454 | 435.97 | 8.51 | 499 | 480.10 | 8.95 |
| 410 | 392.86 | 8.07 | 455 | 436.95 | 8.52 | 500 | 481.08 | 8.96 |
| 411 | 393.84 | 8.08 | 456 | 437.93 | 8.53 | 501 | 482.06 | 8.97 |
| 412 | 394.82 | 8.09 | 457 | 438.91 | 8.54 | 502 | 483.05 | 8.98 |
| 413 | 395.80 | 8.10 | 458 | 439.89 | 8.55 | 503 | 484.03 | 8.99 |
| 414 | 396.78 | 8.11 | 459 | 440.87 | 8.56 | 504 | 485.01 | 9.00 |
| 415 | 397.76 | 8.12 | 460 | 441.85 | 8.57 | 505 | 485.99 | 9.01 |
| 416 | 398.74 | 8.13 | 461 | 442.83 | 8.58 | 506 | 486.97 | 9.01 |
| 417 | 399.72 | 8.14 | 462 | 443.81 | 8.59 | 507 | 487.95 | 9.02 |
| 418 | 400.70 | 8.15 | 463 | 444.79 | 8.60 | 508 | 488.93 | 9.03 |
| 419 | 401.68 | 8.16 | 464 | 445.77 | 8.61 | 509 | 489.91 | 9.04 |
| 420 | 402.65 | 8.17 | 465 | 446.75 | 8.62 | 510 | 490.90 | 9.05 |
| 421 | 403.63 | 8.18 | 466 | 447.73 | 8.63 | 511 | 491.88 | 9.06 |
| 422 | 404.61 | 8.19 | 467 | 448.72 | 8.64 | 512 | 492.86 | 9.07 |
| 423 | 405.59 | 8.20 | 468 | 449.70 | 8.65 | 513 | 493.84 | 9.08 |
| 424 | 406.57 | 8.21 | 469 | 450.68 | 8.66 | 514 | 494.82 | 9.09 |
| 425 | 407.55 | 8.22 | 470 | 451.66 | 8.67 | 515 | 495.80 | 9.10 |
| 426 | 408.53 | 8.23 | 471 | 452.64 | 8.68 | 516 | 496.78 | 9.11 |
| 427 | 409.51 | 8.24 | 472 | 453.62 | 8.69 | 517 | 497.77 | 9.12 |
| 428 | 410.49 | 8.25 | 473 | 454.60 | 8.70 | 518 | 498.75 | 9.13 |
| 429 | 411.47 | 8.26 | 474 | 455.58 | 8.71 | 519 | 499.73 | 9.14 |
| 430 | 412.45 | 8.27 | 475 | 456.56 | 8.72 | 520 | 500.71 | 9.14 |
| 431 | 413.43 | 8.28 | 476 | 457.54 | 8.73 | 521 | 501.69 | 9.15 |
| 432 | 414.41 | 8.29 | 477 | 458.52 | 8.74 | 522 | 502.67 | 9.16 |
| 433 | 415.39 | 8.31 | 478 | 459.50 | 8.75 | 523 | 503.66 | 9.17 |
| 434 | 416.37 | 8.32 | 479 | 460.48 | 8.76 | 524 | 504.64 | 9.18 |
| 435 | 417.35 | 8.33 | 480 | 461.46 | 8.77 | 525 | 505.62 | 9.19 |
| 436 | 418.33 | 8.34 | 481 | 462.44 | 8.78 | 526 | 506.60 | 9.20 |
| 437 | 419.31 | 8.35 | 482 | 463.43 | 8.79 | 527 | 507.58 | 9.21 |
| 438 | 420.29 | 8.36 | 483 | 464.41 | 8.80 | 528 | 508.56 | 9.22 |
| 439 | 421.27 | 8.37 | 484 | 465.39 | 8.81 | 529 | 509.55 | 9.23 |
| 440 | 422.25 | 8.38 | 485 | 466.37 | 8.82 | 530 | 510.53 | 9.24 |
| 441 | 423.23 | 8.39 | 486 | 467.35 | 8.83 | 531 | 511.51 | 9.25 |
| 442 | 424.21 | 8.40 | 487 | 468.33 | 8.84 | 532 | 512.49 | 9.25 |
| 443 | 425.19 | 8.41 | 488 | 469.31 | 8.84 | 533 | 513.47 | 9.26 |
| 444 | 426.17 | 8.42 | 489 | 470.29 | 8.85 | 534 | 514.45 | 9.27 |
| 445 | 427.15 | 8.43 | 490 | 471.27 | 8.86 | 535 | 515.44 | 9.28 |
| 446 | 428.13 | 8.44 | 491 | 472.25 | 8.87 | 536 | 516.42 | 9.29 |
| 447 | 429.11 | 8.45 | 492 | 473.23 | 8.88 | 537 | 517.40 | 9.30 |
| 448 | 430.09 | 8.46 | 493 | 474.22 | 8.89 | 538 | 518.38 | 9.31 |
| 449 | 431.07 | 8.47 | 494 | 475.20 | 8.90 | 539 | 519.36 | 9.32 |
| 450 | 432.05 | 8.48 | 495 | 476.18 | 8.91 | 540 | 520.35 | 9.33 |

Table 4.   CONSTANTS FOR p′ (SHORT FORMULA), AN APPROXIMATION OF   Section 6
THE LOWER BINOMIAL CONFIDENCE LIMIT p̲; c/n ⩽ .125, n ⩾ 16,   $\gamma_1$ = 80%
c ⩽ 1,000 (Continued)   $\gamma_2$ = 60%

| c | m̲ | k̲ | c | m̲ | k̲ | c | m̲ | k̲ |
|---|---|---|---|---|---|---|---|---|
| 541 | 521.33 | 9.34 | 586 | 565.53 | 9.74 | 631 | 609.76 | 10.12 |
| 542 | 522.31 | 9.35 | 587 | 566.51 | 9.74 | 632 | 610.74 | 10.13 |
| 543 | 523.29 | 9.35 | 588 | 567.49 | 9.75 | 633 | 611.73 | 10.14 |
| 544 | 524.27 | 9.36 | 589 | 568.48 | 9.76 | 634 | 612.71 | 10.14 |
| 545 | 525.25 | 9.37 | 590 | 569.46 | 9.77 | 635 | 613.69 | 10.15 |
| 546 | 526.24 | 9.38 | 591 | 570.44 | 9.78 | 636 | 614.68 | 10.16 |
| 547 | 527.22 | 9.39 | 592 | 571.43 | 9.79 | 637 | 615.66 | 10.17 |
| 548 | 528.20 | 9.40 | 593 | 572.41 | 9.80 | 638 | 616.64 | 10.18 |
| 549 | 529.18 | 9.41 | 594 | 573.39 | 9.80 | 639 | 617.63 | 10.19 |
| 550 | 530.17 | 9.42 | 595 | 574.37 | 9.81 | 640 | 618.61 | 10.19 |
| 551 | 531.15 | 9.43 | 596 | 575.36 | 9.82 | 641 | 619.59 | 10.20 |
| 552 | 532.13 | 9.44 | 597 | 576.34 | 9.83 | 642 | 620.58 | 10.21 |
| 553 | 533.11 | 9.44 | 598 | 577.32 | 9.84 | 643 | 621.56 | 10.22 |
| 554 | 534.09 | 9.45 | 599 | 578.30 | 9.85 | 644 | 622.54 | 10.23 |
| 555 | 535.08 | 9.46 | 600 | 579.29 | 9.86 | 645 | 623.53 | 10.24 |
| 556 | 536.06 | 9.47 | 601 | 580.27 | 9.86 | 646 | 624.51 | 10.24 |
| 557 | 537.04 | 9.48 | 602 | 581.25 | 9.87 | 647 | 625.50 | 10.25 |
| 558 | 538.02 | 9.49 | 603 | 582.24 | 9.88 | 648 | 626.48 | 10.26 |
| 559 | 539.00 | 9.50 | 604 | 583.22 | 9.89 | 649 | 627.46 | 10.27 |
| 560 | 539.99 | 9.51 | 605 | 584.20 | 9.90 | 650 | 628.45 | 10.28 |
| 561 | 540.97 | 9.52 | 606 | 585.18 | 9.91 | 651 | 629.43 | 10.29 |
| 562 | 541.95 | 9.52 | 607 | 586.17 | 9.92 | 652 | 630.41 | 10.29 |
| 563 | 542.93 | 9.53 | 608 | 587.15 | 9.92 | 653 | 631.40 | 10.30 |
| 564 | 543.92 | 9.54 | 609 | 588.13 | 9.93 | 654 | 632.38 | 10.31 |
| 565 | 544.90 | 9.55 | 610 | 589.12 | 9.94 | 655 | 633.36 | 10.32 |
| 566 | 545.88 | 9.56 | 611 | 590.10 | 9.95 | 656 | 634.35 | 10.33 |
| 567 | 546.86 | 9.57 | 612 | 591.08 | 9.96 | 657 | 635.33 | 10.33 |
| 568 | 547.84 | 9.58 | 613 | 592.07 | 9.97 | 658 | 636.31 | 10.34 |
| 569 | 548.83 | 9.59 | 614 | 593.05 | 9.98 | 659 | 637.30 | 10.35 |
| 570 | 549.81 | 9.60 | 615 | 594.03 | 9.98 | 660 | 638.28 | 10.36 |
| 571 | 550.79 | 9.60 | 616 | 595.01 | 9.99 | 661 | 639.26 | 10.37 |
| 572 | 551.77 | 9.61 | 617 | 596.00 | 10.00 | 662 | 640.25 | 10.38 |
| 573 | 552.76 | 9.62 | 618 | 596.98 | 10.01 | 663 | 641.23 | 10.38 |
| 574 | 553.74 | 9.63 | 619 | 597.96 | 10.02 | 664 | 642.22 | 10.39 |
| 575 | 554.72 | 9.64 | 620 | 598.95 | 10.03 | 665 | 643.20 | 10.40 |
| 576 | 555.70 | 9.65 | 621 | 599.93 | 10.04 | 666 | 644.18 | 10.41 |
| 577 | 556.69 | 9.66 | 622 | 600.91 | 10.04 | 667 | 645.17 | 10.42 |
| 578 | 557.67 | 9.67 | 623 | 601.90 | 10.05 | 668 | 646.15 | 10.42 |
| 579 | 558.65 | 9.67 | 624 | 602.88 | 10.06 | 669 | 647.13 | 10.43 |
| 580 | 559.63 | 9.68 | 625 | 603.86 | 10.07 | 670 | 648.12 | 10.44 |
| 581 | 560.62 | 9.69 | 626 | 604.85 | 10.08 | 671 | 649.10 | 10.45 |
| 582 | 561.60 | 9.70 | 627 | 605.83 | 10.09 | 672 | 650.09 | 10.46 |
| 583 | 562.58 | 9.71 | 628 | 606.81 | 10.09 | 673 | 651.07 | 10.47 |
| 584 | 563.56 | 9.72 | 629 | 607.80 | 10.10 | 674 | 652.05 | 10.47 |
| 585 | 564.55 | 9.73 | 630 | 608.78 | 10.11 | 675 | 653.04 | 10.48 |

Table 4.   CONSTANTS FOR p′ (SHORT FORMULA), AN APPROXIMATION OF
THE LOWER BINOMIAL CONFIDENCE LIMIT p̲; c/n ≤ .125, n ≥ 16,
c ≤ 1,000 (*Continued*)

| c | m | k | c | m | k | c | m | k |
|---|---|---|---|---|---|---|---|---|
| 676 | 654.02 | 10.49 | 721 | 698.30 | 10.85 | 766 | 742.61 | 11.20 |
| 677 | 655.00 | 10.50 | 722 | 699.29 | 10.86 | 767 | 743.59 | 11.20 |
| 678 | 655.99 | 10.51 | 723 | 700.27 | 10.86 | 768 | 744.58 | 11.21 |
| 679 | 656.97 | 10.51 | 724 | 701.26 | 10.87 | 769 | 745.56 | 11.22 |
| 680 | 657.96 | 10.52 | 725 | 702.24 | 10.88 | 770 | 746.55 | 11.23 |
| 681 | 658.94 | 10.53 | 726 | 703.23 | 10.89 | 771 | 747.53 | 11.23 |
| 682 | 659.92 | 10.54 | 727 | 704.21 | 10.89 | 772 | 748.52 | 11.24 |
| 683 | 660.91 | 10.55 | 728 | 705.19 | 10.90 | 773 | 749.50 | 11.25 |
| 684 | 661.89 | 10.55 | 729 | 706.18 | 10.91 | 774 | 750.49 | 11.26 |
| 685 | 662.88 | 10.56 | 730 | 707.16 | 10.92 | 775 | 751.47 | 11.26 |
| 686 | 663.86 | 10.57 | 731 | 708.15 | 10.93 | 776 | 752.46 | 11.27 |
| 687 | 664.84 | 10.58 | 732 | 709.13 | 10.93 | 777 | 753.44 | 11.28 |
| 688 | 665.83 | 10.59 | 733 | 710.12 | 10.94 | 778 | 754.43 | 11.29 |
| 689 | 666.81 | 10.59 | 734 | 711.10 | 10.95 | 779 | 755.41 | 11.29 |
| 690 | 667.80 | 10.60 | 735 | 712.09 | 10.96 | 780 | 756.40 | 11.30 |
| 691 | 668.78 | 10.61 | 736 | 713.07 | 10.96 | 781 | 757.38 | 11.31 |
| 692 | 669.76 | 10.62 | 737 | 714.05 | 10.97 | 782 | 758.37 | 11.32 |
| 693 | 670.75 | 10.63 | 738 | 715.04 | 10.98 | 783 | 759.35 | 11.32 |
| 694 | 671.73 | 10.63 | 739 | 716.02 | 10.99 | 784 | 760.34 | 11.33 |
| 695 | 672.72 | 10.64 | 740 | 717.01 | 11.00 | 785 | 761.32 | 11.34 |
| 696 | 673.70 | 10.65 | 741 | 717.99 | 11.00 | 786 | 762.31 | 11.35 |
| 697 | 674.68 | 10.66 | 742 | 718.98 | 11.01 | 787 | 763.29 | 11.35 |
| 698 | 675.67 | 10.67 | 743 | 719.96 | 11.02 | 788 | 764.28 | 11.36 |
| 699 | 676.65 | 10.67 | 744 | 720.95 | 11.03 | 789 | 765.26 | 11.37 |
| 700 | 677.64 | 10.68 | 745 | 721.93 | 11.03 | 790 | 766.25 | 11.38 |
| 701 | 678.62 | 10.69 | 746 | 722.92 | 11.04 | 791 | 767.23 | 11.38 |
| 702 | 679.60 | 10.70 | 747 | 723.90 | 11.05 | 792 | 768.22 | 11.39 |
| 703 | 680.59 | 10.71 | 748 | 724.88 | 11.06 | 793 | 769.20 | 11.40 |
| 704 | 681.57 | 10.71 | 749 | 725.87 | 11.07 | 794 | 770.19 | 11.41 |
| 705 | 682.56 | 10.72 | 750 | 726.85 | 11.07 | 795 | 771.17 | 11.41 |
| 706 | 683.54 | 10.73 | 751 | 727.84 | 11.08 | 796 | 772.16 | 11.42 |
| 707 | 684.52 | 10.74 | 752 | 728.82 | 11.09 | 797 | 773.14 | 11.43 |
| 708 | 685.51 | 10.75 | 753 | 729.81 | 11.10 | 798 | 774.13 | 11.44 |
| 709 | 686.49 | 10.75 | 754 | 730.79 | 11.10 | 799 | 775.11 | 11.44 |
| 710 | 687.48 | 10.76 | 755 | 731.78 | 11.11 | 800 | 776.10 | 11.45 |
| 711 | 688.46 | 10.77 | 756 | 732.76 | 11.12 | 801 | 777.08 | 11.46 |
| 712 | 689.45 | 10.78 | 757 | 733.75 | 11.13 | 802 | 778.07 | 11.47 |
| 713 | 690.43 | 10.79 | 758 | 734.73 | 11.13 | 803 | 779.05 | 11.47 |
| 714 | 691.41 | 10.79 | 759 | 735.72 | 11.14 | 804 | 780.04 | 11.48 |
| 715 | 692.40 | 10.80 | 760 | 736.70 | 11.15 | 805 | 781.02 | 11.49 |
| 716 | 693.38 | 10.81 | 761 | 737.69 | 11.16 | 806 | 782.01 | 11.50 |
| 717 | 694.37 | 10.82 | 762 | 738.67 | 11.16 | 807 | 782.99 | 11.50 |
| 718 | 695.35 | 10.82 | 763 | 739.66 | 11.17 | 808 | 783.98 | 11.51 |
| 719 | 696.34 | 10.83 | 764 | 740.64 | 11.18 | 809 | 784.96 | 11.52 |
| 720 | 697.32 | 10.84 | 765 | 741.62 | 11.19 | 810 | 785.95 | 11.53 |

Table 4. CONSTANTS FOR p' (SHORT FORMULA), AN APPROXIMATION OF THE LOWER BINOMIAL CONFIDENCE LIMIT p; c/n ≤ .125, n ≥ 16, c ≤ 1,000 (*Continued*)

Section 6

$\gamma_1 = 80\%$
$\gamma_2 = 60\%$

| c | m | k | c | m | k | c | m | k |
|---|---|---|---|---|---|---|---|---|
| 811 | 786.94 | 11.53 | 856 | 831.28 | 11.86 | 901 | 875.64 | 12.18 |
| 812 | 787.92 | 11.54 | 857 | 832.26 | 11.87 | 902 | 876.63 | 12.19 |
| 813 | 788.91 | 11.55 | 858 | 833.25 | 11.87 | 903 | 877.61 | 12.19 |
| 814 | 789.89 | 11.55 | 859 | 834.24 | 11.88 | 904 | 878.60 | 12.20 |
| 815 | 790.88 | 11.56 | 860 | 835.22 | 11.89 | 905 | 879.58 | 12.21 |
| 816 | 791.86 | 11.57 | 861 | 836.21 | 11.90 | 906 | 880.57 | 12.21 |
| 817 | 792.85 | 11.58 | 862 | 837.19 | 11.90 | 907 | 881.56 | 12.22 |
| 818 | 793.83 | 11.58 | 863 | 838.18 | 11.91 | 908 | 882.54 | 12.23 |
| 819 | 794.82 | 11.59 | 864 | 839.16 | 11.92 | 909 | 883.53 | 12.24 |
| 820 | 795.80 | 11.60 | 865 | 840.15 | 11.93 | 910 | 884.51 | 12.24 |
| 821 | 796.79 | 11.61 | 866 | 841.14 | 11.93 | 911 | 885.50 | 12.25 |
| 822 | 797.77 | 11.61 | 867 | 842.12 | 11.94 | 912 | 886.49 | 12.26 |
| 823 | 798.76 | 11.62 | 868 | 843.11 | 11.95 | 913 | 887.47 | 12.26 |
| 824 | 799.74 | 11.63 | 869 | 844.09 | 11.95 | 914 | 888.46 | 12.27 |
| 825 | 800.73 | 11.64 | 870 | 845.08 | 11.96 | 915 | 889.44 | 12.28 |
| 826 | 801.71 | 11.64 | 871 | 846.06 | 11.97 | 916 | 890.43 | 12.28 |
| 827 | 802.70 | 11.65 | 872 | 847.05 | 11.98 | 917 | 891.42 | 12.29 |
| 828 | 803.69 | 11.66 | 873 | 848.04 | 11.98 | 918 | 892.40 | 12.30 |
| 829 | 804.67 | 11.66 | 874 | 849.02 | 11.99 | 919 | 893.39 | 12.31 |
| 830 | 805.66 | 11.67 | 875 | 850.01 | 12.00 | 920 | 894.38 | 12.31 |
| 831 | 806.64 | 11.68 | 876 | 850.99 | 12.00 | 921 | 895.36 | 12.32 |
| 832 | 807.63 | 11.69 | 877 | 851.98 | 12.01 | 922 | 896.35 | 12.33 |
| 833 | 808.61 | 11.69 | 878 | 852.96 | 12.02 | 923 | 897.33 | 12.33 |
| 834 | 809.60 | 11.70 | 879 | 853.95 | 12.02 | 924 | 898.32 | 12.34 |
| 835 | 810.58 | 11.71 | 880 | 854.94 | 12.03 | 925 | 899.31 | 12.35 |
| 836 | 811.57 | 11.72 | 881 | 855.92 | 12.04 | 926 | 900.29 | 12.35 |
| 837 | 812.55 | 11.72 | 882 | 856.91 | 12.05 | 927 | 901.28 | 12.36 |
| 838 | 813.54 | 11.73 | 883 | 857.89 | 12.05 | 928 | 902.26 | 12.37 |
| 839 | 814.52 | 1'.74 | 884 | 858.88 | 12.06 | 929 | 903.25 | 12.37 |
| 840 | 815.51 | 11.74 | 885 | 859.87 | 12.07 | 930 | 904.24 | 12.38 |
| 841 | 816.50 | 11.75 | 886 | 860.85 | 12.07 | 931 | 905.22 | 12.39 |
| 842 | 817.48 | 11.76 | 887 | 861.84 | 12.08 | 932 | 906.21 | 12.40 |
| 843 | 818.47 | 11.77 | 888 | 862.82 | 12.09 | 933 | 907.20 | 12.40 |
| 844 | 819.45 | 11.77 | 889 | 863.81 | 12.10 | 934 | 908.18 | 12.41 |
| 845 | 820.44 | 11.78 | 890 | 864.79 | 12.10 | 935 | 909.17 | 12.42 |
| 846 | 821.42 | 11.79 | 891 | 865.78 | 12.11 | 936 | 910.15 | 12.42 |
| 847 | 822.41 | 11.80 | 892 | 866.77 | 12.12 | 937 | 911.14 | 12.43 |
| 848 | 823.39 | 11.80 | 893 | 867.75 | 12.12 | 938 | 912.13 | 12.44 |
| 849 | 824.38 | 11.81 | 894 | 868.74 | 12.13 | 939 | 913.11 | 12.44 |
| 850 | 825.37 | 11.82 | 895 | 869.72 | 12.14 | 940 | 914.10 | 12.45 |
| 851 | 826.35 | 11.82 | 896 | 870.71 | 12.14 | 941 | 915.09 | 12.46 |
| 852 | 827.34 | 11.83 | 897 | 871.70 | 12.15 | 942 | 916.07 | 12.46 |
| 853 | 828.32 | 11.84 | 898 | 872.68 | 12.16 | 943 | 917.06 | 12.47 |
| 854 | 829.31 | 11.85 | 899 | 873.67 | 12.17 | 944 | 918.04 | 12.48 |
| 855 | 830.29 | 11.85 | 900 | 874.65 | 12.17 | 945 | 919.03 | 12.48 |

Table 4.  CONSTANTS FOR $p'$ (SHORT FORMULA), AN APPROXIMATION OF THE LOWER BINOMIAL CONFIDENCE LIMIT $\underline{p}$; $c/n \leqslant .125$, $n \geqslant 1\,$ $c \leqslant 1,000$ (*Continued*)

| c | $\underline{m}$ | $\underline{k}$ | c | $\underline{m}$ | $\underline{k}$ | c | $\underline{m}$ | $\underline{k}$ |
|---|---|---|---|---|---|---|---|---|
| 946 | 920.02 | 12.49 | 966 | 939.74 | 12.63 | 986 | 959.48 | 12. |
| 947 | 921.00 | 12.50 | 967 | 940.73 | 12.63 | 987 | 960.46 | 12. |
| 948 | 921.99 | 12.51 | 968 | 941.72 | 12.64 | 988 | 961.45 | 12. |
| 949 | 922.98 | 12.51 | 969 | 942.70 | 12.65 | 989 | 962.44 | 12. |
| 950 | 923.96 | 12.52 | 970 | 943.69 | 12.65 | 990 | 963.42 | 12. |
| 951 | 924.95 | 12.53 | 971 | 944.68 | 12.66 | 991 | 964.41 | 12. |
| 952 | 925.93 | 12.53 | 972 | 945.66 | 12.67 | 992 | 965.40 | 12. |
| 953 | 926.92 | 12.54 | 973 | 946.65 | 12.67 | 993 | 966.38 | 12. |
| 954 | 927.91 | 12.55 | 974 | 947.64 | 12.68 | 994 | 967.37 | 12. |
| 955 | 928.89 | 12.55 | 975 | 948.62 | 12.69 | 995 | 968.35 | 12. |
| 956 | 929.88 | 12.56 | 976 | 949.61 | 12.70 | 996 | 969.34 | 12. |
| 957 | 930.87 | 12.57 | 977 | 950.60 | 12.70 | 997 | 970.33 | 12. |
| 958 | 931.85 | 12.57 | 978 | 951.58 | 12.71 | 998 | 971.31 | 12. |
| 959 | 932.84 | 12.58 | 979 | 952.57 | 12.72 | 999 | 972.30 | 12. |
| 960 | 933.83 | 12.59 | 980 | 953.56 | 12.72 | 1000 | 973.29 | 12. |
| 961 | 934.81 | 12.59 | 981 | 954.54 | 12.73 | | | |
| 962 | 935.80 | 12.60 | 982 | 955.53 | 12.74 | | | |
| 963 | 936.79 | 12.61 | 983 | 956.52 | 12.74 | | | |
| 964 | 937.77 | 12.61 | 984 | 957.50 | 12.75 | | | |
| 965 | 938.76 | 12.62 | 985 | 958.49 | 12.76 | | | |

# Table 5

**UPPER BINOMIAL CONFIDENCE LIMIT** $\bar{p}$ **FOR SMALL SAMPLES;** $c/n \leqslant .5$ **(APPROXIMATELY),** $n \leqslant 20$

Enter the line for n and c, and the column for $\gamma_1$ or $\gamma_2$, to find $\bar{p}$ in the body of the table.

n  sample size.

c  number of events (items having a designated characteristic) in the sample.

$\bar{p}$  exact upper binomial confidence limit.

$\gamma_1$  confidence level for a one-sided (upper or lower) confidence limit.

$\gamma_2$  confidence level for two-sided (upper and lower) confidence limits. ($\gamma_2 = 2\gamma_1 - 100\%$.)

| | | | | Confidence Level | | | |
|---|---|---|---|---|---|---|---|
| | | $\gamma_1$: 99.5% | 99% | 97.5% | 95% | 90% | 80% |
| | | $\gamma_2$: 99% | 98% | 95% | 90% | 80% | 60% |
| n | c | | | | $\bar{p}$ | | |
| 1 | 0 | .99500 | .99000 | .97500 | .95000 | .90000 | .80000 |
| 2 | 0 | .92929 | .90000 | .84189 | .77639 | .68377 | .55279 |
| | 1 | .99750 | .99499 | .98742 | .97468 | .94868 | .89443 |
| 3 | 0 | .82900 | .78456 | .70760 | .63160 | .53584 | .41520 |
| | 1 | .95860 | .94110 | .90570 | .86465 | .80420 | .71286 |
| | 2 | .99833 | .99666 | .99160 | .98305 | .96549 | .92832 |
| 4 | 0 | .73409 | .68377 | .60236 | .52713 | .43766 | .33126 |
| | 1 | .88912 | .85913 | .80588 | .75140 | .67954 | .58245 |
| | 2 | .97056 | .95800 | .93241 | .90239 | .85744 | .78768 |
| 5 | 0 | .65343 | .60189 | .52182 | .45072 | .36904 | .27522 |
| | 1 | .81490 | .77793 | .71642 | .65741 | .58389 | .49019 |
| | 2 | .91717 | .89436 | .85337 | .81074 | .75336 | .67340 |
| | 3 | .97712 | .96732 | .94726 | .92356 | .88777 | .83139 |
| 6 | 0 | .58648 | .53584 | .45926 | .39304 | .31871 | .23528 |
| | 1 | .74601 | .70569 | .64123 | .58180 | .51032 | .42245 |
| | 2 | .85640 | .82693 | .77722 | .72866 | .66681 | .58539 |
| | 3 | .93372 | .91527 | .88188 | .84684 | .79909 | .73135 |
| 7 | 0 | .53088 | .48205 | .40962 | .34816 | .28031 | .20540 |
| | 1 | .68491 | .64336 | .57872 | .52070 | .45256 | .37086 |
| | 2 | .79703 | .76368 | .70958 | .65874 | .59618 | .51676 |
| | 3 | .88230 | .85773 | .81595 | .77468 | .72140 | .64991 |
| | 4 | .94470 | .92920 | .90101 | .87124 | .83036 | .77167 |
| 8 | 0 | .48433 | .43766 | .36942 | .31234 | .25011 | .18223 |
| | 1 | .63152 | .58994 | .52651 | .47068 | .40625 | .33037 |
| | 2 | .74217 | .70677 | .65086 | .59969 | .53822 | .46210 |
| | 3 | .83030 | .80180 | .75514 | .71076 | .65538 | .58366 |
| | 4 | .90013 | .87905 | .84299 | .80710 | .76034 | .69677 |
| 9 | 0 | .44495 | .40052 | .33627 | .28313 | .22574 | .16375 |
| | 1 | .58497 | .54403 | .48250 | .42914 | .36836 | .29777 |
| | 2 | .69261 | .65631 | .60009 | .54964 | .49008 | .41768 |
| | 3 | .78086 | .74997 | .70070 | .65506 | .59942 | .52914 |
| | 4 | .85394 | .82903 | .78799 | .74863 | .69903 | .63391 |
| | 5 | .91321 | .89474 | .86300 | .83125 | .78960 | .73245 |
| 10 | 0 | .41130 | .36904 | .30850 | .25887 | .20567 | .14866 |
| | 1 | .54429 | .50435 | .44502 | .39416 | .33685 | .27099 |
| | 2 | .64820 | .61174 | .55610 | .50690 | .44960 | .38094 |
| | 3 | .73511 | .70288 | .65245 | .60662 | .55173 | .48366 |
| | 4 | .80908 | .78166 | .73762 | .69646 | .64578 | .58087 |
| | 5 | .87169 | .84956 | .81291 | .77756 | .73268 | .67317 |
| 11 | 0 | .38225 | .34207 | .28491 | .23840 | .18887 | .13611 |
| | 1 | .50856 | .46982 | .41278 | .36436 | .31024 | .24860 |
| | 2 | .60850 | .57232 | .51776 | .47009 | .41516 | .35007 |

# Table 5. UPPER BINOMIAL CONFIDENCE LIMIT $\bar{p}$ FOR SMALL SAMPLES; c/n ⩽ .5 (APPROXIMATELY), n ⩽ 20 (*Continued*)

| | | Confidence Level | | | | | |
|---|---|---|---|---|---|---|---|
| | $\gamma_1$: | 99.5% | 99% | 97.5% | 95% | 90% | 80% |
| | $\gamma_2$: | 99% | 98% | 95% | 90% | 80% | 60% |
| n | c | | | | $\bar{p}$ | | |
| 11 | 3 | .69328 | .66042 | .60974 | .56437 | .51076 | .44522 |
| | 4 | .76680 | .73780 | .69210 | .65019 | .59947 | .53569 |
| | 5 | .83069 | .80602 | .76621 | .72875 | .68228 | .62213 |
| | 6 | .88553 | .86561 | .83251 | .80042 | .75947 | .70474 |
| 12 | 0 | .35695 | .31871 | .26465 | .22092 | .17460 | .12551 |
| | 1 | .47703 | .43954 | .38480 | .33868 | .28750 | .22962 |
| | 2 | .57295 | .53734 | .48414 | .43811 | .38552 | .32378 |
| | 3 | .65522 | .62219 | .57186 | .52733 | .47527 | .41235 |
| | 4 | .72752 | .69760 | .65112 | .60914 | .55900 | .49685 |
| | 5 | .79147 | .76511 | .72333 | .68476 | .63772 | .57794 |
| | 6 | .84781 | .82539 | .78906 | .75470 | .71183 | .65589 |
| 13 | 0 | .33473 | .29830 | .24705 | .20582 | .16232 | .11645 |
| | 1 | .44902 | .41283 | .36030 | .31634 | .26784 | .21332 |
| | 2 | .54104 | .50617 | .45447 | .41010 | .35978 | .30114 |
| | 3 | .62064 | .58776 | .53813 | .49465 | .44426 | .38394 |
| | 4 | .69128 | .66090 | .61426 | .57262 | .52343 | .46314 |
| | 5 | .75457 | .72711 | .68422 | .64520 | .59824 | .53939 |
| | 6 | .81130 | .78712 | .74865 | .71295 | .66914 | .61300 |
| | 7 | .86173 | .84118 | .80777 | .77604 | .73627 | .68404 |
| 14 | 0 | .31508 | .28031 | .23164 | .19264 | .15166 | .10860 |
| | 1 | .42403 | .38910 | .33868 | .29673 | .25067 | .19917 |
| | 2 | .51231 | .47826 | .42813 | .38539 | .33721 | .28144 |
| | 3 | .58918 | .55667 | .50798 | .46566 | .41698 | .35915 |
| | 4 | .65794 | .62743 | .58104 | .54001 | .49197 | .43364 |
| | 5 | .72014 | .69203 | .64862 | .60959 | .56311 | .50554 |
| | 6 | .77657 | .75120 | .71139 | .67497 | .63087 | .57514 |
| | 7 | .82760 | .80528 | .76964 | .73642 | .69545 | .64259 |
| 15 | 0 | .29758 | .26436 | .21802 | .18104 | .14230 | .10174 |
| | 1 | .40159 | .36789 | .31948 | .27940 | .23557 | .18678 |
| | 2 | .48633 | .45317 | .40460 | .36344 | .31729 | .26415 |
| | 3 | .56053 | .52851 | .48089 | .43978 | .39279 | .33735 |
| | 4 | .62731 | .59689 | .55100 | .51075 | .46397 | .40763 |
| | 5 | .68816 | .65971 | .61620 | .57744 | .53171 | .47559 |
| | 6 | .74387 | .71771 | .67713 | .64043 | .59647 | .54154 |
| | 7 | .79486 | .77127 | .73414 | .70001 | .65848 | .60564 |
| | 8 | .84127 | .82054 | .78733 | .75627 | .71782 | .66793 |
| 16 | 0 | .28190 | .25011 | .20591 | .17075 | .13404 | .095696 |
| | 1 | .38136 | .34884 | .30232 | .26396 | .22217 | .17583 |
| | 2 | .46276 | .43049 | .38348 | .34383 | .29956 | .24885 |
| | 3 | .53436 | .50294 | .45646 | .41657 | .37122 | .31802 |
| | 4 | .59913 | .56897 | .52377 | .48440 | .43892 | .38452 |
| | 5 | .65849 | .62995 | .58662 | .54835 | .50351 | .44893 |
| | 6 | .71323 | .68659 | .64565 | .60899 | .56544 | .51156 |

# Table 5. UPPER BINOMIAL CONFIDENCE LIMIT p̄ FOR SMALL SAMPLES; c/n ≤ .5 (APPROXIMATELY), n ≤ 20 (*Continued*)

| | | Confidence Level | | | | | |
|---|---|---|---|---|---|---|---|
| | γ₁: | 99.5% | 99% | 97.5% | 95% | 90% | 80% |
| | γ₂: | 99% | 98% | 95% | 90% | 80% | 60% |
| n | c | | | | p̄ | | |
| 16 | 7 | .76377 | .73931 | .70122 | .66663 | .62496 | .57255 |
| | 8 | .81031 | .78828 | .75349 | .72140 | .68217 | .63199 |
| 17 | 0 | .26777 | .23730 | .19506 | .16157 | .12667 | .090330 |
| | 1 | .36303 | .33163 | .28689 | .25012 | .21021 | .16610 |
| | 2 | .44129 | .40992 | .36441 | .32619 | .28370 | .23522 |
| | 3 | .51040 | .47962 | .43432 | .39564 | .35187 | .30077 |
| | 4 | .57318 | .54339 | .49899 | .46055 | .41639 | .36387 |
| | 5 | .63099 | .60251 | .55958 | .52192 | .47807 | .42507 |
| | 6 | .68459 | .65771 | .61672 | .58029 | .53735 | .48465 |
| | 7 | .73442 | .70938 | .67075 | .63599 | .59449 | .54278 |
| | 8 | .78072 | .75775 | .72188 | .68917 | .64961 | .59956 |
| | 9 | .82356 | .80289 | .77017 | .73989 | .70274 | .65500 |
| 18 | 0 | .25499 | .22574 | .18530 | .15332 | .12008 | .085532 |
| | 1 | .34635 | .31602 | .27294 | .23766 | .19947 | .15738 |
| | 2 | .42167 | .39119 | .34712 | .31026 | .26942 | .22300 |
| | 3 | .48841 | .45830 | .41418 | .37668 | .33441 | .28529 |
| | 4 | .54924 | .51989 | .47637 | .43888 | .39602 | .34531 |
| | 5 | .60548 | .57720 | .53480 | .49783 | .45502 | .40358 |
| | 6 | .65786 | .63091 | .59007 | .55405 | .51184 | .46038 |
| | 7 | .70682 | .68142 | .64255 | .60784 | .56672 | .51588 |
| | 8 | .75261 | .72899 | .69243 | .65940 | .61980 | .57018 |
| | 9 | .79535 | .77370 | .73981 | .70880 | .67115 | .62331 |
| 19 | 0 | .24335 | .21524 | .17647 | .14587 | .11413 | .081219 |
| | 1 | .33111 | .30180 | .26028 | .22637 | .18977 | .14954 |
| | 2 | .40368 | .37405 | .33138 | .29580 | .25651 | .21198 |
| | 3 | .46816 | .43873 | .39578 | .35943 | .31859 | .27131 |
| | 4 | .52711 | .49825 | .45565 | .41912 | .37753 | .32853 |
| | 5 | .58179 | .55379 | .51203 | .47580 | .43405 | .38413 |
| | 6 | .63291 | .60601 | .56550 | .52997 | .48856 | .43839 |
| | 7 | .68090 | .65532 | .61642 | .58194 | .54132 | .49146 |
| | 8 | .72601 | .70195 | .66500 | .63188 | .59246 | .54346 |
| | 9 | .76840 | .74605 | .71136 | .67991 | .64207 | .59442 |
| | 10 | .80811 | .78765 | .75553 | .72605 | .69017 | .64436 |
| 20 | 0 | .23273 | .20567 | .16843 | .13911 | .10875 | .077319 |
| | 1 | .31714 | .28879 | .24873 | .21611 | .18096 | .14243 |
| | 2 | .38713 | .35834 | .31698 | .28262 | .24477 | .20200 |
| | 3 | .44947 | .42073 | .37893 | .34366 | .30419 | .25864 |
| | 4 | .50661 | .47828 | .43661 | .40103 | .36066 | .31330 |
| | 5 | .55976 | .53211 | .49105 | .45558 | .41489 | .36646 |
| | 6 | .60961 | .58286 | .54279 | .50782 | .46727 | .41838 |
| | 7 | .65657 | .63094 | .59219 | .55803 | .51803 | .46922 |
| | 8 | .70091 | .67658 | .63946 | .60642 | .56733 | .51907 |
| | 9 | .74277 | .71992 | .68472 | .65307 | .61525 | .56800 |
| | 10 | .78225 | .76104 | .72804 | .69805 | .66183 | .61603 |

# Table 6

**LOWER BINOMIAL CONFIDENCE LIMIT $\underline{p}$ FOR SMALL SAMPLES; $c/n \leqslant .5$ (APPROXIMATELY), $n \leqslant 20$**

Enter the line for n and c, and the column for $\gamma_1$ or $\gamma_2$, to find $\underline{p}$ in the body of the table.

| | |
|---|---|
| n | sample size. |
| c | number of events (items having a designated characteristic) in the sample. |
| $\underline{p}$ | exact lower binomial confidence limit. |
| $\gamma_1$ | confidence level for a one-sided (upper or lower) confidence limit. |
| $\gamma_2$ | confidence level for two-sided (upper and lower) confidence limits. ($\gamma_2 = 2\gamma_1 - 100\%$.) |

## Table 6. LOWER BINOMIAL CONFIDENCE LIMIT p̲ FOR SMALL SAMPLES; c/n ⩽ .5 (APPROXIMATELY), n ⩽ 20

| | | Confidence Level | | | | | |
|---|---|---|---|---|---|---|---|
| | $\gamma_1$: | 99.5% | 99% | 97.5% | 95% | 90% | 80% |
| | $\gamma_2$: | 99% | 98% | 95% | 90% | 80% | 60% |
| n | c | | | | p̲ | | |
| 1 | 1 | .0050000 | .010000 | .025000 | .050000 | .10000 | .20000 |
| 2 | 1 | .0025031 | .0050126 | .012579 | .025321 | .051317 | .10557 |
| 3 | 1 | .0016695 | .0033445 | .0084038 | .016952 | .034511 | .071682 |
| | 2 | .041400 | .058903 | .094299 | .13535 | .19580 | .28714 |
| 4 | 1 | .0012524 | .0025094 | .0063095 | .012741 | .025996 | .054258 |
| | 2 | .029445 | .041999 | .067586 | .097611 | .14256 | .21232 |
| 5 | 1 | .0010020 | .0020080 | .0050508 | .010206 | .020852 | .043648 |
| | 2 | .022881 | .032682 | .052745 | .076440 | .11224 | .16861 |
| | 3 | .082829 | .10564 | .14663 | .18926 | .24664 | .32660 |
| 6 | 1 | .00083507 | .0016737 | .0042107 | .0085124 | .017407 | .036508 |
| | 2 | .018721 | .026763 | .043272 | .062850 | .092595 | .13988 |
| | 3 | .066279 | .084730 | .11812 | .15316 | .20091 | .26865 |
| 7 | 1 | .00071582 | .0014347 | .0036103 | .0073008 | .014939 | .031375 |
| | 2 | .015844 | .022665 | .036693 | .053376 | .078823 | .11954 |
| | 3 | .055299 | .070804 | .098988 | .12876 | .16964 | .22833 |
| | 4 | .11770 | .14227 | .18405 | .22532 | .27860 | .35009 |
| 8 | 1 | .00062637 | .0012555 | .0031597 | .0063912 | .013084 | .027508 |
| | 2 | .013736 | .019658 | .031854 | .046389 | .068626 | .10437 |
| | 3 | .047464 | .060840 | .085233 | .11111 | .14685 | .19860 |
| | 4 | .099867 | .12095 | .15701 | .19290 | .23966 | .30323 |
| 9 | 1 | .00055679 | .0011161 | .0028091 | .0056830 | .011638 | .024489 |
| | 2 | .012124 | .017356 | .028145 | .041023 | .060769 | .092627 |
| | 3 | .041585 | .053348 | .074855 | .097747 | .12950 | .17575 |
| | 4 | .086788 | .10526 | .13700 | .16875 | .21040 | .26755 |
| | 5 | .14606 | .17097 | .21201 | .25137 | .30097 | .36609 |
| 10 | 1 | .00050113 | .0010045 | .0025286 | .0051162 | .010481 | .022067 |
| | 2 | .010851 | .015538 | .025211 | .036771 | .054529 | .083260 |
| | 3 | .037007 | .047507 | .066740 | .087264 | .11583 | .15763 |
| | 4 | .076768 | .093214 | .12155 | .15003 | .18756 | .23944 |
| | 5 | .12831 | .15044 | .18709 | .22244 | .26732 | .32683 |
| 11 | 1 | .00045558 | .00091325 | .0022990 | .0046522 | .0095325 | .020081 |
| | 2 | .0098197 | .014065 | .022831 | .033319 | .049452 | .075615 |
| | 3 | .033341 | .042823 | .060218 | .078820 | .10477 | .14292 |
| | 4 | .068839 | .083660 | .10926 | .13508 | .16923 | .21671 |
| | 5 | .11447 | .13439 | .16749 | .19958 | .24053 | .29526 |
| | 6 | .16931 | .19398 | .23379 | .27125 | .31772 | .37787 |
| 12 | 1 | .00041762 | .00083718 | .0021076 | .0042653 | .0087416 | .018423 |
| | 2 | .0089679 | .012847 | .020863 | .030460 | .045241 | .069257 |
| | 3 | .030337 | .038982 | .054861 | .071870 | .095653 | .13072 |
| | 4 | .062405 | .075895 | .099246 | .12285 | .15419 | .19795 |

## Table 6. LOWER BINOMIAL CONFIDENCE LIMIT p FOR SMALL SAMPLES; c/n ⩽ .5 (APPROXIMATELY), n ⩽ 20 (*Continued*)

| | | \(\gamma_1\): 99.5% | 99% | 97.5% | 95% | 90% | 80% |
|---|---|---|---|---|---|---|---|
| | | \(\gamma_2\): 99% | 98% | 95% | 90% | 80% | 60% |
| n | c | | | | p | | |
| 12 | 5 | .10336 | .12147 | .15165 | .18102 | .21868 | .26931 |
| | 6 | .15219 | .17461 | .21094 | .24530 | .28817 | .34411 |
| 13 | 1 | .00038551 | .00077280 | .0019456 | .0039379 | .0080719 | .017018 |
| | 2 | .0082522 | .011824 | .019207 | .028053 | .041691 | .063886 |
| | 3 | .027832 | .035775 | .050381 | .066050 | .087996 | .12044 |
| | 4 | .057076 | .069455 | .090920 | .11267 | .14161 | .18218 |
| | 5 | .094229 | .11083 | .13858 | .16566 | .20050 | .24758 |
| | 6 | .13827 | .15882 | .19223 | .22396 | .26373 | .31596 |
| | 7 | .18870 | .21288 | .25135 | .28705 | .33086 | .38700 |
| 14 | 1 | .00035797 | .00071762 | .0018068 | .0036571 | .0074975 | .015812 |
| | 2 | .0076424 | .010952 | .017795 | .025999 | .038658 | .059289 |
| | 3 | .025709 | .033057 | .046579 | .061103 | .081477 | .11166 |
| | 4 | .052590 | .064028 | .083889 | .10405 | .13094 | .16875 |
| | 5 | .086595 | .10193 | .12760 | .15272 | .18513 | .22912 |
| | 6 | .12671 | .14568 | .17661 | .20607 | .24316 | .29211 |
| | 7 | .17240 | .19472 | .23036 | .26358 | .30455 | .35741 |
| 15 | 1 | .00033411 | .00066980 | .0016864 | .0034137 | .0069994 | .014766 |
| | 2 | .0071165 | .010199 | .016576 | .024226 | .036037 | .055309 |
| | 3 | .023888 | .030723 | .043312 | .056847 | .075859 | .10408 |
| | 4 | .048759 | .059390 | .077872 | .096658 | .12177 | .15717 |
| | 5 | .080113 | .094356 | .11824 | .14166 | .17197 | .21323 |
| | 6 | .11696 | .13458 | .16336 | .19086 | .22559 | .27164 |
| | 7 | .15873 | .17946 | .21267 | .24373 | .28218 | .33207 |
| | 8 | .20514 | .22873 | .26586 | .29999 | .34152 | .39436 |
| 16 | 1 | .00031323 | .00062795 | .0015811 | .0032007 | .0065634 | .013850 |
| | 2 | .0066584 | .0095436 | .015514 | .022679 | .033749 | .051830 |
| | 3 | .022308 | .028698 | .040474 | .053146 | .070966 | .097464 |
| | 4 | .045451 | .055381 | .072662 | .090252 | .11380 | .17408 |
| | 5 | .074540 | .087838 | .11017 | .13211 | .16056 | .19941 |
| | 6 | .10862 | .12506 | .15198 | .17777 | .21041 | .25387 |
| | 7 | .14710 | .16646 | .19753 | .22669 | .26292 | .31012 |
| | 8 | .18969 | .21172 | .24651 | .27860 | .31783 | .36801 |
| 17 | 1 | .00029481 | .00059102 | .0014882 | .0030127 | .0061785 | .013040 |
| | 2 | .0062557 | .0089672 | .014579 | .021318 | .031734 | .048763 |
| | 3 | .020924 | .026923 | .037985 | .049898 | .066668 | .091639 |
| | 4 | .042564 | .051880 | .068108 | .084645 | .10682 | .13821 |
| | 5 | .069695 | .082166 | .10314 | .12377 | .15058 | .18728 |
| | 6 | .10139 | .11681 | .14210 | .16636 | .19716 | .23829 |
| | 7 | .13708 | .15523 | .18444 | .21191 | .24614 | .29092 |
| | 8 | .17644 | .19711 | .22983 | .26011 | .29726 | .34500 |
| | 9 | .21928 | .24225 | .27812 | .31083 | .35039 | .40044 |
| 18 | 1 | .00027844 | .00055820 | .0014056 | .0028456 | .0058363 | .012320 |
| | 2 | .0058990 | .0084565 | .013751 | .020111 | .029946 | .046039 |

**Table 6.** **LOWER BINOMIAL CONFIDENCE LIMIT p FOR SMALL SAMPLES;**
c/n ≤ .5 **(APPROXIMATELY),** n ≤ 20 (*Continued*)

| | | | | Confidence Level | | | |
|---|---|---|---|---|---|---|---|
| | $\gamma_1$: | 99.5% | 99% | 97.5% | 95% | 90% | 80% |
| | $\gamma_2$: | 99% | 98% | 95% | 90% | 80% | 60% |
| n | c | | | | p | | |
| 18 | 3 | .019703 | .025356 | .035785 | .047025 | .062860 | .086472 |
| | 4 | .040023 | .048797 | .064092 | .079695 | .10064 | .13035 |
| | 5 | .065444 | .077185 | .096949 | .11643 | .14177 | .17655 |
| | 6 | .095073 | .10959 | .13343 | .15634 | .18549 | .22452 |
| | 7 | .12835 | .14544 | .17299 | .19895 | .23139 | .27397 |
| | 8 | .16495 | .18441 | .21530 | .24396 | .27922 | .32473 |
| | 9 | .20465 | .22630 | .26019 | .29120 | .32885 | .37669 |
| 19 | 1 | .00026378 | .00052883 | .0013316 | .0026960 | .0055299 | .011676 |
| | 2 | .0055808 | .0080009 | .013012 | .019033 | .028349 | .043603 |
| | 3 | .018616 | .023961 | .033826 | .044465 | .059465 | .081857 |
| | 4 | .037769 | .046061 | .060525 | .075294 | .095142 | .12334 |
| | 5 | .061684 | .072776 | .091466 | .10991 | .13394 | .16698 |
| | 6 | .089501 | .10321 | .12576 | .14747 | .17513 | .21226 |
| | 7 | .12068 | .13683 | .16289 | .18750 | .21832 | .25890 |
| | 8 | .15488 | .17327 | .20252 | .22972 | .26327 | .30673 |
| | 9 | .19189 | .21235 | .24447 | .27395 | .30983 | .35564 |
| | 10 | .23160 | .25395 | .28864 | .32009 | .35793 | .40558 |
| 20 | 1 | .00025060 | .00050239 | .0012651 | .0025614 | .0052542 | .011095 |
| | 2 | .0052951 | .0075919 | .012349 | .018065 | .026914 | .041412 |
| | 3 | .017643 | .022711 | .032071 | .042169 | .056418 | .077710 |
| | 4 | .035756 | .043615 | .057334 | .071354 | .090213 | .11705 |
| | 5 | .058334 | .068845 | .086571 | .10408 | .12693 | .15840 |
| | 6 | .084550 | .097542 | .11893 | .13955 | .16587 | .20128 |
| | 7 | .11388 | .12918 | .15391 | .17731 | .20666 | .24541 |
| | 8 | .14598 | .16342 | .19119 | .21707 | .24906 | .29063 |
| | 9 | .18065 | .20005 | .23058 | .25865 | .29293 | .33684 |
| | 10 | .21775 | .23896 | .27196 | .30195 | .33817 | .38397 |

# Table 7

FORMULAS FOR APPROXIMATION CONSTANTS

To obtain $\bar{p}'$ and $\underline{p}'$ from the formulas in Tables 1–4, various constants ($\bar{a}$, $\bar{b}$, $\underline{a}$, $\underline{b}$, $\bar{k}$, $\underline{k}$, $\bar{m}$, and $\underline{m}$) are required, as given in those tables for $c \leqslant 1,000$. The constants are based on a given $c$ and a given $\gamma_1$ or $\gamma_2$. Alternatively, or for $c > 1,000$, the constants may be calculated from the formulas in Tables 7–A to 7–E.

To designate a calculated (rather than tabular) constant, it is marked with a prime. Hence Tables 7–A to 7–E permit calculation of $\bar{a}'$, $\bar{b}'$, $\underline{a}'$, $\underline{b}'$, $\bar{k}'$, $\underline{k}'$, $\bar{m}'$, and $\underline{m}'$. For $c > 160$, the calculated and tabular constants are the same; for $c \leqslant 160$, they may differ slightly.

$\bar{p}'$, $\underline{p}'$     approximations of the upper and lower binomial confidence limits, respectively; these approximations have relative accuracy of at least .999.

n     sample size.

c     number of events (items having a designated characteristic) in the sample.

$\gamma_1$     confidence level for a one-sided (upper or lower) confidence limit.

$\gamma_2$     confidence level for two-sided (upper and lower) confidence limits. ($\gamma_2 = 2\gamma_1 - 100\%$.)

Table 7–A.    Constants for $\bar{p}'$ (Long Formula, Table 1)

Table 7–B.    Constants for $\underline{p}'$ (Long Formula, Table 2)

Table 7–C.    Constants for $\bar{p}'$ (Short Formula, Table 3)

Table 7–D.    Constants for $\underline{p}'$ (Short Formula, Table 4)

Table 7–E.    Approximations of $\bar{m}$ and $\underline{m}$

**Table 7-A. CONSTANTS FOR $\bar{p}'$ (LONG FORMULA, TABLE 1); $c/n \leqslant .5$, $n \geqslant 20$**

1. Based on c and $\gamma_1$ or $\gamma_2$, obtain $\bar{m}$ from Table 1; or $\bar{m}'$ from the formula in Table 7-E provided $c \geqslant 50$.

2. Based on c and $\gamma_1$ or $\gamma_2$, obtain the auxiliary term f from the table below.

3. Based on c, $\bar{m}$, and f, calculate:

$$\bar{a}' = \frac{\bar{m}}{4} - f \quad \text{(Extension formula for } \bar{a} \text{ in Table 1)}$$

$$\bar{b}' = \frac{2c - \bar{m}}{4} - f \quad \text{(Extension formula for } \bar{b} \text{ in Table 1)}$$

4. Calculate:

$$\bar{p}' = \frac{\bar{m}}{n} \times \frac{n - \bar{a}'}{n - \bar{b}'} \quad \text{(Formula of Table 1)}$$

## Values of f

| Confidence Level | | | |
|---|---|---|---|
| $\gamma_1$ | $\gamma_2$ | c | f |
| 99.5% | 99% | 0-2 | .60 |
| | | 3-38 | .85 |
| 99% | 98% | 0-2 | .55 |
| | | 3-33 | .75 |
| 97.5% | 95% | 0-2 | .50 |
| | | 3-26 | .60 |
| 95% | 90% | 0-21 | .45 |
| 90% | 80% | 0-17 | .35 |
| 80% | 60% | 3-14 | .30 |

f = 0 for values of c not shown.

**Table 7-B. CONSTANTS FOR $\underline{p}'$ (LONG FORMULA, TABLE 2); $c/n \leqslant .5$, $n \geqslant 20$**

1. Based on c and $\gamma_1$ or $\gamma_2$, obtain $\underline{m}$ from Table 2; or $\underline{m}'$ from the formula in Table 7-E provided $c \geqslant 50$.

2. Based on c and $\gamma_1$ or $\gamma_2$, obtain the auxiliary term g from the table below.

3. Based on c, $\underline{m}$, and g, calculate:

$$\underline{a}' = \frac{m}{3} + g \qquad \text{(Extension formula for } \underline{a} \text{ in Table 2)}$$

$$\underline{b}' = \frac{3c - 3 - \underline{m}}{6} + g \qquad \text{(Extension formula for } \underline{b} \text{ in Table 2)}$$

4. Calculate:

$$\underline{p}' = \frac{m}{n} \times \frac{n - \underline{a}'}{n - \underline{b}'} \qquad \text{(Formula of Table 2)}$$

| | | Values of g | | |
|---|---|---|---|---|
| Confidence Level | | $c$ = 10 or More | | $c$ = 9 or Less |
| $\gamma_1$ | $\gamma_2$ | c | g | g |
| 99.5% | 99% | 10–57 | .75 | .075 x c |
| 99% | 98% | 10–45 | .70 | .070 x c |
| 97.5% | 95% | 10–35 | .65 | .065 x c |
| 95% | 90% | 10–28 | .60 | .060 x c |
| 90% | 80% | 10–23 | .60 | .060 x c |
| 80% | 60% | 10–19 | .75 | .075 x c |

g = 0 for values of c not shown.

**Table 7–C.   CONSTANTS FOR $\bar{p}'$ (SHORT FORMULA, TABLE 3); $c/n \leqslant .1$,**
                $n \geqslant 40$

1. Based on c and $\gamma_1$ or $\gamma_2$, obtain $\bar{m}$ from Table 1; or $\bar{m}'$ from the formula in Table 7–E provided $c \geqslant 50$.

2. Based on c and $\gamma_1$ or $\gamma_2$, obtain the auxiliary term d from the table below.

3. Based on c, $\bar{m}$, and d, calculate:

$$\bar{k}' = \frac{\bar{m} - c}{2} + d \quad \text{(Extension formula for } \bar{k} \text{ in Table 3)}$$

4. Calculate:

$$\bar{p}' = \frac{\bar{m}}{n + \bar{k}'} \quad \text{(Formula of Table 3)}$$

Values of d

| Confidence Level | | c = 4 to 50 | c = 3 or less |
|:---:|:---:|:---:|:---:|
| $\gamma_1$ | $\gamma_2$ | d | d |
| 99.5% | 99% | .211 | .042 × (c + 1) |
| 99% | 98% | .173 | .034 × (c + 1) |
| 97.5% | 95% | .126 | .025 × (c + 1) |
| 95% | 90% | .093 | .019 × (c + 1) |
| 90% | 80% | .062 | .012 × (c + 1) |
| 80% | 60% | .036 | .007 × (c + 1) |

d = 0 for c over 50.

**Table 7-D.  CONSTANTS FOR $\underline{p}'$ (SHORT FORMULA, TABLE 4);** $c/n \leqslant .125$, $n \geqslant 16$

1. Based on c and $\gamma_1$ or $\gamma_2$, obtain $\underline{m}$ from Table 2; or $\underline{m}'$ from the formula in Table 7-E provided $c \geqslant 50$.

2. Based on c and $\overline{m}$, calculate:

$$\underline{k}' = \frac{c - 1 - \underline{m}}{2} \qquad \text{(Extension formula for } \underline{k} \text{ in Table 4)}$$

3. Calculate:

$$\underline{p}' = \frac{m}{n - \underline{k}'} \qquad \text{(Formula of Table 4)}$$

# Table 7-E. APPROXIMATIONS OF $\bar{m}$ AND $\underline{m}$; $c \geqslant 50$

1. Based on $\gamma_1$ or $\gamma_2$, obtain the terms A and B from the table below.

2. Based on c, A, and B, calculate:

$$\bar{m}' = c + A \sqrt{c + 1} + B \qquad \text{(Extension formula for } \bar{m} \text{ in Tables 1 and 3)}$$

$$\underline{m}' = c - 1 - A \sqrt{c} + B \qquad \text{(Extension formula for } \underline{m} \text{ in Tables 2 and 4)}$$

**Note for Acceptance Sampling:**

To obtain $\bar{m}$ corresponding to acceptance number $\dot{c}$, read c as $\dot{c}$ in the first formula above; and use $\gamma_1 = 1 - \beta$ to obtain A and B. To obtain $\underline{m}_1$ corresponding to rejection number $\dot{c}_1$, read c as $\dot{c}_1$ and $\underline{m}$ as $\underline{m}_1$ in the second formula above; and use $\gamma_1 = 1 - \alpha$ to obtain A and B. Thus:

$$\bar{m}' = \dot{c} + A \sqrt{\dot{c} + 1} + B$$

$$\underline{m}'_1 = \dot{c}_1 - 1 - A \sqrt{\dot{c}_1} + B$$

## Values of A and B

| Confidence Level | | | |
|---|---|---|---|
| $\gamma_1$ | $\gamma_2$ | A | B |
| 99.5% | 99% | 2.5758 | 2.8783 |
| 99% | 98% | 2.3263 | 2.4706 |
| 97.5% | 95% | 1.9600 | 1.9472 |
| 95% | 90% | 1.6449 | 1.5685 |
| 90% | 80% | 1.2816 | 1.2141 |
| 80% | 60% | 0.84162 | 0.90278 |

# Table 8

**A. Proportion Sampling; Both $\bar{p}$ and $\underline{p}$, Or Only $\bar{p}$, Desired**

1. Calculate $\hat{\bar{p}} = \hat{p} + \hat{e}$. If $\hat{p} = 0$, use Table 12.

2. Calculate $\hat{Q} = \dfrac{\hat{\bar{p}}}{\hat{p}} \times \dfrac{2 - \hat{p}}{2 - \hat{\bar{p}}}$ .

3. In Table 8 and column for specified $\gamma_1$ or $\gamma_2$, find Q nearest to $\hat{Q}$; use $\gamma_2$ if both $\bar{p}$ and $\underline{p}$ are desired, and $\gamma_1$ if only $\bar{p}$ is desired. Find the value of c corresponding to Q. If necessary, interpolate linearly between values of Q bounding $\hat{Q}$ to find integer c.

4. Calculate $n = \dfrac{c}{\hat{p}}$.

5. Test n for Appropriate Sample Size. Using the tables for confidence limits, obtain $\bar{p}$ based on c, n, and $\gamma_1$ or $\gamma_2$. Calculate $\bar{e} = \bar{p} - c/n$. If $\bar{e}$ differs appreciably from $\hat{e}$, calculate $n' = n(\bar{e}/\hat{e})^2$.

    Q    $\bar{m}/c$, where $\bar{m}$ is the upper confidence limit for the parameter m of a Poisson distribution; $\bar{m}$ is based on c and $\gamma_1$ or $\gamma_2$.

    n    sample size.

    c    number of events (items having a designated characteristic) in the sample.

    $\bar{p}$    upper binomial confidence limit.

    $\underline{p}$    lower binomial confidence limit.

\*  $\hat{p}$    anticipated value of c/n.

    $\bar{e}$    upper error margin: $\bar{p} - c/n$.

\*  $\hat{e}$    anticipated error margin; specifically, $\hat{\bar{e}}$, the anticipated value of $\bar{e}$.

    $\hat{\bar{p}}$    anticipated upper confidence limit.

\*  Specifications for determining sample size.

**Table 8.   VALUES OF Q AND c (OR ċ) FOR DETERMINING SAMPLE
SIZE BY THE POISSON PROCEDURE; $\hat{p} \leqslant .25$ (BUT NOT 0)**
*(Continued)*

$\gamma_1$   confidence level for a one-sided confidence limit (only $\overline{p}$).

\*  *or*

$\gamma_2$   confidence level for two-sided confidence limits (both $\overline{p}$ and $\underline{p}$).
($\gamma_2 = 2\gamma_1 - 100\%$.)

**B.   Quasi-Form of Acceptance Sampling; $p_2$ and $\beta$ Specified**

1.  Calculate $\hat{Q} = \dfrac{p_2}{\overline{p}} \times \dfrac{2 - \hat{p}}{2 - p_2}$ .

2.  In Table 8 and column for specified $\beta$, find Q nearest to $\hat{Q}$. Find the value of ċ corresponding to Q; ċ is the acceptance number. If necessary, interpolate linearly between values of Q bounding $\hat{Q}$ to find integer ċ.

3.  In Table 1 or 7–E, for $\gamma_1 = 1 - \beta$, find $\overline{m}$ corresponding to ċ; read c as ċ.

4.  Calculate $n = \dfrac{\overline{m}}{p_2} - \dfrac{\overline{m} - ċ}{2}$ .

Q   $\overline{m}/c$, where $\overline{m}$ is the upper confidence limit for the parameter m of a Poisson distribution; $\overline{m}$ is based on c and $\gamma_1$.

n   sample size.

c   number of error (defective) items in the sample.

ċ   acceptance number.

\*  $p_2$   unsatisfactory error (defective) rate.

\*  $\hat{p}$   anticipated value of c/n.

\*  $\beta$   risk of accepting a lot (population) with error rate $p_2$.

$\gamma_1$   confidence level for a one-sided confidence limit; $\gamma_1 = 1 - \beta$.

\*  Specifications for determining sample size.

# Table 8. VALUES OF Q AND c (OR ċ) FOR DETERMINING SAMPLE SIZE BY THE POISSON PROCEDURE; $p \leqslant .25$ (BUT NOT 0) (*Continued*)

| | γ or β | | | | | |
|---|---|---|---|---|---|---|
| $\gamma_1$: | 99.5% | 99% | 97.5% | 95% | 90% | 80% |
| $\gamma_2$: | 99% | 98% | 95% | 90% | 80% | 60% |
| β: | 0.5% | 1% | 2.5% | 5% | 10% | 20% |
| c or ċ | | | | Q | | |
| 1 | 7.430 | 6.638 | 5.572 | 4.744 | 3.890 | 2.994 |
| 2 | 4.637 | 4.203 | 3.612 | 3.148 | 2.661 | 2.140 |
| 3 | 3.659 | 3.348 | 2.922 | 2.585 | 2.227 | 1.838 |
| 4 | 3.149 | 2.901 | 2.560 | 2.288 | 1.998 | 1.680 |
| 5 | 2.830 | 2.622 | 2.334 | 2.103 | 1.855 | 1.581 |
| 6 | 2.610 | 2.428 | 2.177 | 1.974 | 1.755 | 1.513 |
| 7 | 2.448 | 2.286 | 2.060 | 1.878 | 1.682 | 1.462 |
| 8 | 2.322 | 2.175 | 1.970 | 1.804 | 1.624 | 1.422 |
| 9 | 2.222 | 2.087 | 1.898 | 1.745 | 1.578 | 1.391 |
| 10 | 2.140 | 2.014 | 1.839 | 1.696 | 1.541 | 1.365 |
| 11 | 2.071 | 1.954 | 1.789 | 1.655 | 1.509 | 1.343 |
| 12 | 2.012 | 1.902 | 1.747 | 1.620 | 1.482 | 1.325 |
| 13 | 1.961 | 1.857 | 1.710 | 1.590 | 1.458 | 1.309 |
| 14 | 1.917 | 1.818 | 1.678 | 1.563 | 1.438 | 1.295 |
| 15 | 1.878 | 1.783 | 1.649 | 1.540 | 1.419 | 1.282 |
| 16 | 1.843 | 1.752 | 1.624 | 1.519 | 1.403 | 1.271 |
| 17 | 1.811 | 1.724 | 1.601 | 1.500 | 1.389 | 1.261 |
| 18 | 1.783 | 1.699 | 1.580 | 1.483 | 1.375 | 1.252 |
| 19 | 1.757 | 1.676 | 1.562 | 1.467 | 1.363 | 1.244 |
| 20 | 1.733 | 1.655 | 1.544 | 1.453 | 1.352 | 1.236 |
| 21 | 1.712 | 1.636 | 1.529 | 1.440 | 1.342 | 1.229 |
| 22 | 1.692 | 1.618 | 1.514 | 1.428 | 1.333 | 1.223 |
| 23 | 1.673 | 1.602 | 1.500 | 1.417 | 1.324 | 1.217 |
| 24 | 1.656 | 1.587 | 1.488 | 1.406 | 1.316 | 1.212 |
| 25 | 1.640 | 1.572 | 1.476 | 1.397 | 1.308 | 1.207 |
| 26 | 1.625 | 1.559 | 1.465 | 1.388 | 1.301 | 1.202 |
| 27 | 1.611 | 1.547 | 1.455 | 1.379 | 1.295 | 1.197 |
| 28 | 1.598 | 1.535 | 1.445 | 1.371 | 1.289 | 1.193 |
| 29 | 1.585 | 1.524 | 1.436 | 1.363 | 1.283 | 1.189 |
| 30 | 1.574 | 1.513 | 1.428 | 1.356 | 1.277 | 1.185 |
| 31 | 1.563 | 1.503 | 1.419 | 1.350 | 1.272 | 1.182 |
| 32 | 1.552 | 1.494 | 1.412 | 1.343 | 1.267 | 1.179 |
| 33 | 1.542 | 1.485 | 1.404 | 1.337 | 1.262 | 1.175 |
| 34 | 1.533 | 1.477 | 1.397 | 1.331 | 1.258 | 1.172 |
| 35 | 1.524 | 1.469 | 1.391 | 1.326 | 1.253 | 1.169 |
| 36 | 1.515 | 1.461 | 1.384 | 1.321 | 1.249 | 1.167 |
| 37 | 1.507 | 1.454 | 1.378 | 1.316 | 1.245 | 1.164 |
| 38 | 1.499 | 1.447 | 1.373 | 1.311 | 1.242 | 1.161 |
| 39 | 1.491 | 1.440 | 1.367 | 1.306 | 1.238 | 1.159 |
| 40 | 1.484 | 1.434 | 1.362 | 1.302 | 1.235 | 1.157 |

# Table 8. VALUES OF Q AND c (OR ċ) FOR DETERMINING SAMPLE SIZE BY THE POISSON PROCEDURE; p ≤ .25 (BUT NOT 0) *(Continued)*

| | γ or β | | | | | |
|---|---|---|---|---|---|---|
| $\gamma_1$: 99.5% | 99% | 97.5% | 95% | 90% | 80% |
| $\gamma_2$: 99% | 98% | 95% | 90% | 80% | 60% |
| β:  0.5% | 1% | 2.5% | 5% | 10% | 20% |
| c or ċ | | | Q | | | |
| 41 | 1.477 | 1.428 | 1.357 | 1.298 | 1.231 | 1.155 |
| 42 | 1.470 | 1.422 | 1.352 | 1.293 | 1.228 | 1.152 |
| 43 | 1.464 | 1.416 | 1.347 | 1.290 | 1.225 | 1.150 |
| 44 | 1.458 | 1.410 | 1.342 | 1.286 | 1.222 | 1.148 |
| 45 | 1.452 | 1.405 | 1.338 | 1.282 | 1.220 | 1.146 |
| 46 | 1.446 | 1.400 | 1.334 | 1.279 | 1.217 | 1.145 |
| 47 | 1.441 | 1.395 | 1.330 | 1.275 | 1.214 | 1.143 |
| 48 | 1.435 | 1.390 | 1.326 | 1.272 | 1.212 | 1.141 |
| 49 | 1.430 | 1.386 | 1.322 | 1.269 | 1.209 | 1.139 |
| 50 | 1.425 | 1.381 | 1.318 | 1.266 | 1.207 | 1.138 |
| 55 | 1.403 | 1.361 | 1.302 | 1.252 | 1.196 | 1.131 |
| 60 | 1.383 | 1.344 | 1.287 | 1.240 | 1.187 | 1.124 |
| 65 | 1.366 | 1.329 | 1.275 | 1.229 | 1.178 | 1.119 |
| 70 | 1.351 | 1.315 | 1.263 | 1.220 | 1.171 | 1.114 |
| 75 | 1.338 | 1.303 | 1.254 | 1.212 | 1.165 | 1.110 |
| 80 | 1.326 | 1.292 | 1.245 | 1.204 | 1.159 | 1.106 |
| 90 | 1.305 | 1.274 | 1.229 | 1.192 | 1.149 | 1.0991 |
| 100 | 1.288 | 1.258 | 1.216 | 1.181 | 1.141 | 1.0935 |
| 110 | 1.273 | 1.246 | 1.205 | 1.172 | 1.134 | 1.0887 |
| 120 | 1.260 | 1.234 | 1.196 | 1.164 | 1.127 | 1.0846 |
| 130 | 1.249 | 1.224 | 1.187 | 1.157 | 1.122 | 1.0809 |
| 150 | 1.230 | 1.207 | 1.173 | 1.145 | 1.113 | 1.0749 |
| 170 | 1.215 | 1.193 | 1.162 | 1.136 | 1.106 | 1.0700 |
| 200 | 1.197 | 1.177 | 1.149 | 1.124 | 1.0969 | 1.0642 |
| 250 | 1.175 | 1.157 | 1.132 | 1.111 | 1.0861 | 1.0569 |
| 300 | 1.159 | 1.143 | 1.120 | 1.100 | 1.0782 | 1.0517 |
| 350 | 1.146 | 1.132 | 1.110 | 1.0925 | 1.0721 | 1.0476 |
| 400 | 1.136 | 1.123 | 1.103 | 1.0863 | 1.0672 | 1.0444 |
| 450 | 1.128 | 1.115 | 1.0968 | 1.0811 | 1.0632 | 1.0417 |
| 500 | 1.121 | 1.109 | 1.0916 | 1.0768 | 1.0598 | 1.0395 |
| 600 | 1.110 | 1.0992 | 1.0833 | 1.0698 | 1.0544 | 1.0359 |
| 700 | 1.102 | 1.0915 | 1.0769 | 1.0645 | 1.0502 | 1.0331 |
| 800 | 1.0947 | 1.0854 | 1.0718 | 1.0602 | 1.0469 | 1.0309 |
| 900 | 1.0891 | 1.0803 | 1.0675 | 1.0566 | 1.0441 | 1.0291 |
| 1000 | 1.0844 | 1.0761 | 1.0640 | 1.0536 | 1.0418 | 1.0275 |
| 1100 | 1.0803 | 1.0724 | 1.0609 | 1.0510 | 1.0398 | 1.0262 |
| 1200 | 1.0768 | 1.0692 | 1.0582 | 1.0488 | 1.0380 | 1.0251 |
| 1500 | 1.0684 | 1.0617 | 1.0519 | 1.0435 | 1.0339 | 1.0223 |
| 2000 | 1.0591 | 1.0533 | 1.0448 | 1.0376 | 1.0293 | 1.0193 |
| 2500 | 1.0527 | 1.0475 | 1.0400 | 1.0335 | 1.0261 | 1.0172 |

# Table 8. VALUES OF Q AND c (OR ċ) FOR DETERMINING SAMPLE SIZE BY THE POISSON PROCEDURE; p ≤ .25 (BUT NOT 0) (*Continued*)

| | $\gamma$ or $\beta$ | | | | | |
|---|---|---|---|---|---|---|
| $\gamma_1$: 99.5% | 99% | 97.5% | 95% | 90% | 80% |
| $\gamma_2$: 99% | 98% | 95% | 90% | 80% | 60% |
| $\beta$: 0.5% | 1% | 2.5% | 5% | 10% | 20% |
| c or ċ | | | | Q | | |
| 3000 | 1.0480 | 1.0433 | 1.0364 | 1.0306 | 1.0238 | 1.0157 |
| 3500 | 1.0444 | 1.0400 | 1.0337 | 1.0283 | 1.0220 | 1.0145 |
| 4000 | 1.0415 | 1.0374 | 1.0315 | 1.0264 | 1.0206 | 1.0135 |
| 4500 | 1.0390 | 1.0352 | 1.0297 | 1.0249 | 1.0194 | 1.0127 |
| 5000 | 1.0370 | 1.0334 | 1.0281 | 1.0236 | 1.0184 | 1.0121 |
| 6000 | 1.0337 | 1.0304 | 1.0256 | 1.0215 | 1.0167 | 1.0110 |
| 7000 | 1.0312 | 1.0282 | 1.0237 | 1.0199 | 1.0155 | 1.0102 |
| 8000 | 1.0292 | 1.0263 | 1.0222 | 1.0186 | 1.0145 | 1.00952 |
| 9000 | 1.0275 | 1.0248 | 1.0209 | 1.0175 | 1.0136 | 1.00897 |
| 10000 | 1.0260 | 1.0235 | 1.0198 | 1.0166 | 1.0129 | 1.00851 |
| 11000 | 1.0248 | 1.0224 | 1.0189 | 1.0158 | 1.0123 | 1.00811 |
| 12000 | 1.0238 | 1.0214 | 1.0181 | 1.0151 | 1.0118 | 1.00776 |
| 15000 | 1.0212 | 1.0192 | 1.0161 | 1.0135 | 1.0105 | 1.00693 |
| 20000 | 1.0184 | 1.0166 | 1.0140 | 1.0117 | 1.00912 | 1.00600 |
| 25000 | 1.0164 | 1.0148 | 1.0125 | 1.0105 | 1.00815 | 1.00536 |
| 30000 | 1.0150 | 1.0135 | 1.0114 | 1.00955 | 1.00744 | 1.00489 |
| 35000 | 1.0139 | 1.0125 | 1.0105 | 1.00884 | 1.00689 | 1.00452 |
| 40000 | 1.0130 | 1.0117 | 1.00985 | 1.00826 | 1.00644 | 1.00423 |
| 45000 | 1.0122 | 1.0110 | 1.00928 | 1.00779 | 1.00607 | 1.00399 |
| 50000 | 1.0116 | 1.0105 | 1.00880 | 1.00739 | 1.00576 | 1.00378 |
| 60000 | 1.0106 | 1.00954 | 1.00803 | 1.00674 | 1.00525 | 1.00345 |
| 70000 | 1.00978 | 1.00883 | 1.00744 | 1.00624 | 1.00486 | 1.00319 |
| 80000 | 1.00914 | 1.00826 | 1.00695 | 1.00584 | 1.00455 | 1.00299 |
| 90000 | 1.00862 | 1.00778 | 1.00656 | 1.00550 | 1.00429 | 1.00282 |
| 100000 | 1.00817 | 1.00738 | 1.00622 | 1.00522 | 1.00406 | 1.00267 |
| 110000 | 1.00779 | 1.00704 | 1.00593 | 1.00497 | 1.00388 | 1.00255 |
| 120000 | 1.00746 | 1.00674 | 1.00567 | 1.00476 | 1.00371 | 1.00244 |
| 150000 | 1.00667 | 1.00602 | 1.00507 | 1.00426 | 1.00332 | 1.00218 |
| 200000 | 1.00577 | 1.00521 | 1.00439 | 1.00369 | 1.00287 | 1.00189 |
| 250000 | 1.00516 | 1.00466 | 1.00393 | 1.00330 | 1.00257 | 1.00169 |
| 300000 | 1.00471 | 1.00426 | 1.00358 | 1.00301 | 1.00234 | 1.00154 |
| 350000 | 1.00436 | 1.00394 | 1.00332 | 1.00278 | 1.00217 | 1.00143 |
| 400000 | 1.00408 | 1.00368 | 1.00310 | 1.00260 | 1.00203 | 1.00133 |
| 450000 | 1.00385 | 1.00347 | 1.00293 | 1.00246 | 1.00191 | 1.00126 |
| 500000 | 1.00365 | 1.00329 | 1.00278 | 1.00233 | 1.00181 | 1.00119 |
| 600000 | 1.00333 | 1.00301 | 1.00253 | 1.00213 | 1.00166 | 1.00109 |
| 700000 | 1.00308 | 1.00278 | 1.00235 | 1.00197 | 1.00153 | 1.00101 |
| 800000 | 1.00288 | 1.00260 | 1.00219 | 1.00184 | 1.00143 | 1.000942 |
| 900000 | 1.00272 | 1.00245 | 1.00207 | 1.00174 | 1.00135 | 1.000888 |
| 1000000 | 1.00258 | 1.00233 | 1.00196 | 1.00165 | 1.00128 | 1.000843 |

# Table 9

**VALUES OF R AND c (OR $\dot{c}_1$) FOR DETERMINING SAMPLE SIZE BY THE POISSON PROCEDURE; $\hat{p} \leqslant .25$**

## A. Proportion Sampling; Only $\underline{p}$ Desired

1. Calculate $\underline{\hat{p}} = \hat{p} - \hat{e}$.

2. Calculate $\hat{R} = \dfrac{\hat{p}}{\underline{\hat{p}}} \times \dfrac{2 - \hat{p}}{2 - \underline{p}}$ .

3. In Table 9 and column for specified $\gamma_1$, find R nearest to $\hat{R}$. Find the value of c corresponding to R. If necessary, interpolate linearly between values of R bounding $\hat{R}$ to find integer c.

4. Calculate $n = \dfrac{c}{\hat{\underline{p}}}$ .

5. Test n for Appropriate Sample Size. Using the tables for confidence limits, obtain $\underline{p}$ based on c, n, and $\gamma_1$. Calculate $\underline{e} = c/n - \underline{p}$. If $\underline{e}$ differs appreciably from $\hat{e}$, calculate $n' = n(\underline{e}/\hat{e})^2$.

    R    $\underline{m}/c$, where $\underline{m}$ is the lower confidence limit for the parameter m of a Poisson distribution; $\underline{m}$ is based on c and $\gamma_1$.

    n    sample size.

    c    number of events (items having a designated characteristic) in the sample.

    $\underline{p}$    lower binomial confidence limit.

\*  $\hat{p}$    anticipated value of c/n.

    $\underline{e}$    lower error margin: $c/n - \underline{p}$.

\*  $\hat{e}$    anticipated error margin; specifically $\hat{\underline{e}}$, the anticipated value of $\underline{e}$.

    $\underline{\hat{p}}$    anticipated lower confidence limit.

\*  $\gamma_1$    confidence level for a one-sided confidence limit (only $\underline{p}$).

\*   Specifications for determining sample size.

**B. Quasi-Form of Acceptance Sampling; $p_1$ and $\alpha$ Specified**

1. Calculate $\hat{R} = \dfrac{p_1}{\hat{p}} \times \dfrac{2 - \hat{p}}{2 - p_1}$ .

2. In Table 9 and column for specified $\alpha$, find R nearest to $\hat{R}$. Find the value of $\dot{c}_1$ corresponding to R; $\dot{c}_1$ is the rejection number. If necessary, interpolate linearly between values of R bounding $\hat{R}$ to find integer $\dot{c}_1$.

3. In Table 2 or 7–E, for $\gamma_1 = 1 - \alpha$, find $\underline{m}_1$ corresponding to $\dot{c}_1$; read c as $\dot{c}_1$, and $\underline{m}$ as $\underline{m}_1$.

4. Calculate $n = \dfrac{\underline{m}_1}{p_1} + \dfrac{c_1 - \underline{m}_1 - 1}{2}$ .

$\quad$ R $\quad$ $\underline{m}/c$, where $\underline{m}$ is the lower confidence limit for the parameter m of a Poisson distribution; $\underline{m}$ is based on c and $\gamma_1$.

$\quad$ n $\quad$ sample size.

$\quad$ c $\quad$ number of items in the sample with a designated error (or other characteristic).

$\quad$ $\dot{c}_1$ $\quad$ rejection (decision) number.

* $\quad$ $p_1$ $\quad$ satisfactory error rate (or other designated rate).

* $\quad$ $\hat{p}$ $\quad$ anticipated value of c/n.

* $\quad$ $\alpha$ $\quad$ risk of rejecting a lot (population) with rate $p_1$.

$\quad$ $\gamma_1$ $\quad$ confidence level for a one-sided confidence limit; $\gamma_1 = 1 - \alpha$.

* Specifications for determining sample size.

# Table 9. VALUES OF R AND c (OR $\dot{c}_1$) FOR DETERMINING SAMPLE SIZE BY THE POISSON PROCEDURE; $\hat{p} \leqslant .25$ (Continued)

| c or $\dot{c}_1$ | $\gamma_1$: 99.5%<br>$\alpha$: 0.5% | 99%<br>1% | 97.5%<br>2.5% | 95%<br>5% | 90%<br>10% | 80%<br>20% |
|---|---|---|---|---|---|---|
| | | | | R | | |
| 1 | 0.005 | 0.010 | 0.025 | 0.051 | 0.105 | 0.223 |
| 2 | 0.052 | 0.074 | 0.121 | 0.178 | 0.266 | 0.412 |
| 3 | 0.113 | 0.145 | 0.206 | 0.273 | 0.367 | 0.512 |
| 4 | 0.168 | 0.206 | 0.272 | 0.342 | 0.436 | 0.574 |
| 5 | 0.216 | 0.256 | 0.325 | 0.394 | 0.487 | 0.618 |
| 6 | 0.256 | 0.298 | 0.367 | 0.436 | 0.525 | 0.651 |
| 7 | 0.291 | 0.333 | 0.402 | 0.469 | 0.556 | 0.676 |
| 8 | 0.321 | 0.363 | 0.432 | 0.498 | 0.582 | 0.697 |
| 9 | 0.348 | 0.390 | 0.457 | 0.522 | 0.604 | 0.714 |
| 10 | 0.372 | 0.413 | 0.480 | 0.543 | 0.622 | 0.729 |
| 11 | 0.393 | 0.434 | 0.499 | 0.561 | 0.638 | 0.742 |
| 12 | 0.412 | 0.452 | 0.517 | 0.577 | 0.652 | 0.753 |
| 13 | 0.429 | 0.469 | 0.532 | 0.592 | 0.665 | 0.762 |
| 14 | 0.445 | 0.484 | 0.547 | 0.605 | 0.676 | 0.771 |
| 15 | 0.460 | 0.498 | 0.560 | 0.616 | 0.687 | 0.779 |
| 16 | 0.473 | 0.511 | 0.572 | 0.627 | 0.696 | 0.786 |
| 17 | 0.485 | 0.523 | 0.583 | 0.637 | 0.704 | 0.792 |
| 18 | 0.497 | 0.534 | 0.593 | 0.646 | 0.712 | 0.798 |
| 19 | 0.508 | 0.545 | 0.602 | 0.655 | 0.720 | 0.804 |
| 20 | 0.518 | 0.554 | 0.611 | 0.663 | 0.726 | 0.809 |
| 21 | 0.527 | 0.563 | 0.619 | 0.670 | 0.733 | 0.813 |
| 22 | 0.536 | 0.572 | 0.627 | 0.677 | 0.738 | 0.818 |
| 23 | 0.544 | 0.580 | 0.634 | 0.683 | 0.744 | 0.822 |
| 24 | 0.552 | 0.587 | 0.641 | 0.690 | 0.749 | 0.825 |
| 25 | 0.560 | 0.594 | 0.647 | 0.695 | 0.754 | 0.829 |
| 26 | 0.567 | 0.601 | 0.653 | 0.701 | 0.758 | 0.832 |
| 27 | 0.574 | 0.607 | 0.659 | 0.706 | 0.763 | 0.835 |
| 28 | 0.580 | 0.613 | 0.665 | 0.711 | 0.767 | 0.838 |
| 29 | 0.586 | 0.619 | 0.670 | 0.715 | 0.771 | 0.841 |
| 30 | 0.592 | 0.625 | 0.675 | 0.720 | 0.774 | 0.844 |
| 31 | 0.598 | 0.630 | 0.679 | 0.724 | 0.778 | 0.847 |
| 32 | 0.603 | 0.635 | 0.684 | 0.728 | 0.781 | 0.849 |
| 33 | 0.608 | 0.640 | 0.688 | 0.732 | 0.784 | 0.851 |
| 34 | 0.613 | 0.645 | 0.693 | 0.736 | 0.787 | 0.854 |
| 35 | 0.618 | 0.649 | 0.697 | 0.739 | 0.790 | 0.856 |
| 36 | 0.623 | 0.653 | 0.700 | 0.743 | 0.793 | 0.858 |
| 37 | 0.627 | 0.658 | 0.704 | 0.746 | 0.796 | 0.860 |
| 38 | 0.632 | 0.662 | 0.708 | 0.749 | 0.799 | 0.862 |
| 39 | 0.636 | 0.666 | 0.711 | 0.752 | 0.801 | 0.863 |
| 40 | 0.640 | 0.669 | 0.714 | 0.755 | 0.803 | 0.865 |

Table 9. VALUES OF R AND c (OR $\dot{c}_1$) FOR DETERMINING SAMPLE SIZE BY THE POISSON PROCEDURE; $\hat{p} \leqslant .25$ (*Continued*)

| | $\gamma_1$ or $\alpha$ | | | | | |
|---|---|---|---|---|---|---|
| $\gamma_1$: | 99.5% | 99% | 97.5% | 95% | 90% | 80% |
| $\alpha$: | 0.5% | 1% | 2.5% | 5% | 10% | 20% |
| c or $\dot{c}_1$ | | | | R | | |
| 41 | 0.644 | 0.673 | 0.718 | 0.758 | 0.806 | 0.867 |
| 42 | 0.647 | 0.676 | 0.721 | 0.760 | 0.808 | 0.868 |
| 43 | 0.651 | 0.680 | 0.724 | 0.763 | 0.810 | 0.870 |
| 44 | 0.654 | 0.683 | 0.727 | 0.766 | 0.812 | 0.871 |
| 45 | 0.658 | 0.686 | 0.729 | 0.768 | 0.814 | 0.873 |
| 46 | 0.661 | 0.689 | 0.732 | 0.770 | 0.816 | 0.874 |
| 47 | 0.664 | 0.692 | 0.735 | 0.773 | 0.818 | 0.876 |
| 48 | 0.667 | 0.695 | 0.737 | 0.775 | 0.820 | 0.877 |
| 49 | 0.670 | 0.698 | 0.740 | 0.777 | 0.822 | 0.878 |
| 50 | 0.673 | 0.701 | 0.742 | 0.779 | 0.824 | 0.879 |
| 55 | 0.687 | 0.713 | 0.753 | 0.789 | 0.832 | 0.885 |
| 60 | 0.699 | 0.724 | 0.763 | 0.798 | 0.839 | 0.890 |
| 65 | 0.709 | 0.734 | 0.772 | 0.805 | 0.845 | 0.894 |
| 70 | 0.719 | 0.743 | 0.780 | 0.812 | 0.850 | 0.898 |
| 75 | 0.728 | 0.751 | 0.787 | 0.818 | 0.855 | 0.9018 |
| 80 | 0.735 | 0.758 | 0.793 | 0.823 | 0.860 | 0.9049 |
| 90 | 0.749 | 0.771 | 0.804 | 0.833 | 0.868 | 0.9104 |
| 100 | 0.761 | 0.782 | 0.814 | 0.841 | 0.874 | 0.9150 |
| 110 | 0.771 | 0.792 | 0.822 | 0.849 | 0.880 | 0.9190 |
| 120 | 0.781 | 0.800 | 0.829 | 0.855 | 0.885 | 0.9225 |
| 130 | 0.789 | 0.807 | 0.835 | 0.860 | 0.889 | 0.9255 |
| 150 | 0.802 | 0.820 | 0.846 | 0.870 | 0.897 | 0.9307 |
| 170 | 0.813 | 0.830 | 0.855 | 0.877 | 0.9030 | 0.9349 |
| 200 | 0.827 | 0.843 | 0.866 | 0.887 | 0.9104 | 0.9400 |
| 250 | 0.845 | 0.859 | 0.880 | 0.898 | 0.9198 | 0.9464 |
| 300 | 0.858 | 0.871 | 0.890 | 0.9069 | 0.9267 | 0.9511 |
| 350 | 0.868 | 0.880 | 0.898 | 0.9137 | 0.9321 | 0.9547 |
| 400 | 0.876 | 0.887 | 0.9044 | 0.9192 | 0.9365 | 0.9577 |
| 450 | 0.883 | 0.894 | 0.9097 | 0.9237 | 0.9401 | 0.9601 |
| 500 | 0.889 | 0.899 | 0.9142 | 0.9276 | 0.9431 | 0.9622 |
| 600 | 0.898 | 0.9075 | 0.9216 | 0.9338 | 0.9480 | 0.9655 |
| 700 | 0.9053 | 0.9142 | 0.9273 | 0.9386 | 0.9519 | 0.9681 |
| 800 | 0.9113 | 0.9196 | 0.9319 | 0.9426 | 0.9550 | 0.9701 |
| 900 | 0.9162 | 0.9241 | 0.9357 | 0.9458 | 0.9575 | 0.9718 |
| 1000 | 0.9204 | 0.9279 | 0.9390 | 0.9486 | 0.9597 | 0.9733 |
| 1100 | 0.9240 | 0.9312 | 0.9418 | 0.9509 | 0.9616 | 0.9745 |
| 1200 | 0.9272 | 0.9341 | 0.9442 | 0.9530 | 0.9632 | 0.9756 |
| 1500 | 0.9347 | 0.9409 | 0.9500 | 0.9579 | 0.9671 | 0.9782 |
| 2000 | 0.9433 | 0.9487 | 0.9566 | 0.9635 | 0.9714 | 0.9811 |
| 2500 | 0.9492 | 0.9541 | 0.9612 | 0.9673 | 0.9745 | 0.9831 |

# Table 9. VALUES OF R AND c (OR $\dot{c}_1$) FOR DETERMINING SAMPLE SIZE BY THE POISSON PROCEDURE; $\hat{p} \leqslant .25$ (Continued)

| c or $\dot{c}_1$ | $\gamma_1$: 99.5%  $\alpha$: 0.5% | 99%  1% | 97.5%  2.5% | 95%  5% | 90%  10% | 80%  20% |
|---|---|---|---|---|---|---|
| | | | R | | | |
| 3000 | 0.9536 | 0.9580 | 0.9645 | 0.9702 | 0.9767 | 0.9846 |
| 3500 | 0.9570 | 0.9611 | 0.9671 | 0.9724 | 0.9784 | 0.9857 |
| 4000 | 0.9597 | 0.9636 | 0.9692 | 0.9741 | 0.9798 | 0.9867 |
| 4500 | 0.9620 | 0.9656 | 0.9710 | 0.9756 | 0.9809 | 0.9874 |
| 5000 | 0.9639 | 0.9674 | 0.9725 | 0.9769 | 0.9819 | 0.9881 |
| 6000 | 0.9671 | 0.9702 | 0.9749 | 0.9789 | 0.9835 | 0.9891 |
| 7000 | 0.9695 | 0.9724 | 0.9767 | 0.9804 | 0.9847 | 0.9899 |
| 8000 | 0.9714 | 0.9742 | 0.9782 | 0.9817 | 0.9857 | 0.99058 |
| 9000 | 0.9731 | 0.9756 | 0.9794 | 0.9827 | 0.9865 | 0.99112 |
| 10000 | 0.9744 | 0.9769 | 0.9805 | 0.9836 | 0.9872 | 0.99157 |
| 11000 | 0.9756 | 0.9780 | 0.9814 | 0.9844 | 0.9878 | 0.99197 |
| 12000 | 0.9766 | 0.9789 | 0.9822 | 0.9850 | 0.9883 | 0.99231 |
| 15000 | 0.9791 | 0.9811 | 0.9841 | 0.9866 | 0.9896 | 0.99312 |
| 20000 | 0.9819 | 0.9836 | 0.9862 | 0.9884 | 0.99095 | 0.99404 |
| 25000 | 0.9838 | 0.9853 | 0.9876 | 0.9896 | 0.99190 | 0.99467 |
| 30000 | 0.9852 | 0.9866 | 0.9887 | 0.99052 | 0.99261 | 0.99514 |
| 35000 | 0.9863 | 0.9876 | 0.9896 | 0.99122 | 0.99316 | 0.99550 |
| 40000 | 0.9872 | 0.9884 | 0.99022 | 0.99179 | 0.99360 | 0.99579 |
| 45000 | 0.9879 | 0.9891 | 0.99078 | 0.99226 | 0.99396 | 0.99603 |
| 50000 | 0.9885 | 0.9896 | 0.99125 | 0.99266 | 0.99427 | 0.99623 |
| 60000 | 0.9895 | 0.99053 | 0.99201 | 0.99329 | 0.99477 | 0.99656 |
| 70000 | 0.99029 | 0.99123 | 0.99261 | 0.99379 | 0.99516 | 0.99682 |
| 80000 | 0.99092 | 0.99179 | 0.99308 | 0.99419 | 0.99547 | 0.99702 |
| 90000 | 0.99143 | 0.99226 | 0.99348 | 0.99452 | 0.99573 | 0.99719 |
| 100000 | 0.99187 | 0.99266 | 0.99381 | 0.99480 | 0.99595 | 0.99734 |
| 110000 | 0.99225 | 0.99300 | 0.99410 | 0.99505 | 0.99614 | 0.99746 |
| 120000 | 0.99258 | 0.99330 | 0.99435 | 0.99526 | 0.99630 | 0.99757 |
| 150000 | 0.99336 | 0.99400 | 0.99495 | 0.99576 | 0.99669 | 0.99783 |
| 200000 | 0.99425 | 0.99481 | 0.99562 | 0.99632 | 0.99714 | 0.99812 |
| 250000 | 0.99486 | 0.99535 | 0.99608 | 0.99671 | 0.99744 | 0.99832 |
| 300000 | 0.99530 | 0.99576 | 0.99642 | 0.99700 | 0.99766 | 0.99846 |
| 350000 | 0.99565 | 0.99607 | 0.99669 | 0.99722 | 0.99783 | 0.99858 |
| 400000 | 0.99593 | 0.99633 | 0.99690 | 0.99740 | 0.99797 | 0.99867 |
| 450000 | 0.99616 | 0.99654 | 0.99708 | 0.99755 | 0.99809 | 0.99875 |
| 500000 | 0.99636 | 0.99671 | 0.99723 | 0.99767 | 0.99819 | 0.99881 |
| 600000 | 0.99668 | 0.99700 | 9.99747 | 0.99788 | 0.99835 | 0.99891 |
| 700000 | 0.99692 | 0.99722 | 0.99766 | 0.99803 | 0.99847 | 0.99899 |
| 800000 | 0.99712 | 0.99740 | 0.99781 | 0.99816 | 0.99857 | 0.999059 |
| 900000 | 0.99729 | 0.99755 | 0.99794 | 0.99827 | 0.99865 | 0.999113 |
| 1000000 | 0.99743 | 0.99768 | 0.99804 | 0.99836 | 0.99872 | 0.999158 |

# Table 10

FORMULAS FOR DETERMINING SAMPLE SIZE BY THE MODIFIED
NORMAL PROCEDURE; PROPORTION SAMPLING; $\hat{p} > .25$ (BUT NOT
OVER .5)

**A. Both $\bar{p}$ and $\underline{p}$, Or Only $\bar{p}$, Desired**

1. Calculate $\hat{\bar{p}} = \hat{p} + \hat{e}$.

2. Calculate $\hat{\bar{q}} = 1 - \hat{\bar{p}}$.

3. Calculate $n = \dfrac{A^2 \hat{\bar{p}} \hat{\bar{q}} + \hat{e}}{\hat{e}^2}$ .

4. Test n for Appropriate Sample Size. Calculate $c = n\hat{p}$. Using the tables for confidence limits, obtain $\bar{p}$ based on c, n, and $\gamma_1$ or $\gamma_2$; use $\gamma_2$ if both $\bar{p}$ and $\underline{p}$ are desired, and $\gamma_1$ if only $\bar{p}$ is desired. Calculate $\bar{e} = \bar{p} - c/n$. If $\bar{e}$ differs appreciably from $\hat{e}$, calculate $n' = n(\bar{e}/\hat{e})^2$.

**B. Only $\underline{p}$ Desired**

1. Calculate $\hat{\underline{p}} = \hat{p} - \hat{e}$.

2. Calculate $\hat{\underline{q}} = 1 - \hat{\underline{p}}$.

3. Calculate $n = \dfrac{A^2 \hat{\underline{p}} \hat{\underline{q}} + \hat{e}}{\hat{e}^2}$ .

4. Test n for Appropriate Sample Size. Calculate $c = n\hat{p}$. Using the tables for confidence limits, obtain $\underline{p}$ based on c, n, and $\gamma_1$. Calculate $\underline{e} = c/n - \underline{p}$. If $\underline{e}$ differs appreciably from $\hat{e}$, calculate $n' = n(\underline{e}/\hat{e})^2$.

| | |
|---|---|
| n | sample size. |
| c | number of events (items having a designated characteristic) in the sample. |
| $\bar{p}$ | upper binomial confidence limit. |
| $\underline{p}$ | lower binomial confidence limit. |
| * $\hat{p}$ | anticipated value of c/n. |
| $\bar{e}$ | upper error margin: $\bar{p}$ − c/n. |
| $\underline{e}$ | lower error margin: c/n − $\underline{p}$. |
| * $\hat{e}$ | anticipated error margin, specifically: $\hat{\bar{e}}$, the anticipated value of $\bar{e}$ (when both $\bar{p}$ and $\underline{p}$, or only $\bar{p}$, are desired); or $\hat{\underline{e}}$, the anticipated value of $\underline{e}$ (when only $\underline{p}$ is desired). |
| A | number of standard deviations, based on the normal distribution, for a given confidence level $\gamma_1$ or $\gamma_2$. Values of A and $A^2$ are given in the table below. |
| $\gamma_1$ | confidence level for a one-sided confidence limit (only $\bar{p}$ or only $\underline{p}$). |
| * *or* | |
| $\gamma_2$ | confidence level for two-sided confidence limits (both $\bar{p}$ and $\underline{p}$). |

\* Specifications for determining sample size.

## Values of A and $A^2$

| Confidence Level | | | |
|---|---|---|---|
| $\gamma_1$ | $\gamma_2$ | A | $A^2$ |
| 99.5% | 99% | 2.5758 | 6.6349 |
| 99% | 98% | 2.3263 | 5.4119 |
| 97.5% | 95% | 1.9600 | 3.8415 |
| 95% | 90% | 1.6449 | 2.7055 |
| 90% | 80% | 1.2816 | 1.6424 |
| 80% | 60% | 0.84162 | 0.70833 |

# Table 11

**FORMULAS FOR $c$, $\dot{c}$, AND $\dot{c}_1$, USED IN DETERMINING SAMPLE SIZE BY THE POISSON PROCEDURES IN TABLES 8 AND 9**

To determine sample size n by the Poisson Procedures in Tables 8 and 9, it is necessary to obtain the value of c (anticipated number of sample items having a designated characteristic) for proportion sampling; and of $\dot{c}$ and $\dot{c}_1$ (acceptance and rejection numbers) for the quasi-forms of acceptance sampling. Instead of obtaining c, $\dot{c}$, and $\dot{c}_1$ from Tables 8 and 9, they may be calculated from the following formulas. Except for obtaining c, $\dot{c}$, and $\dot{c}_1$ by formula, the sample size procedures remain the same as in Tables 8 and 9.

## A.  c Or $\dot{c}$ For Procedures in Table 8

$$c \text{ or } \dot{c} = \left[ \frac{A + \sqrt{A^2 + 4(\hat{Q} - 1)(\hat{Q} + B - 1)}}{2(\hat{Q} - 1)} \right]^2 - 1; \quad \hat{p} > 0$$

Round c or $\dot{c}$ to the nearest integer.

$\hat{Q}$ is obtained in the manner described in Table 8, based on the specifications for sample size. The values of A, $A^2$, and B, which depend on the specified confidence level or risk level, are given in the table below.

## B.  c Or $\dot{c}_1$ For Procedures in Table 9

$$c \text{ or } \dot{c}_1 = \left[ \frac{A + \sqrt{A^2 - 4(B - 1)(1 - \hat{R})}}{B(1 - \hat{R})} \right]^2$$

Round c or $\dot{c}_1$ to the nearest integer.

$\hat{R}$ is obtained in the manner described in Table 9, based on the specifications for sample size. The values of A, $A^2$, and B, which depend on the specified confidence level or risk level, are given in the table below.

If the term inside the radical is negative (which can happen in a limited
number of cases: when $\gamma_1$ = 99.5% or $\alpha$ = 0.5% and c or ċ$_1$ should be 1,
2, or 3; and when $\gamma_1$ = 99% or $\alpha$ = 1% and c or ċ$_1$ should be 1 or 2), do
not use this formula; instead use Table 9 to find c or ċ$_1$.

| | |
|---|---|
| A | number of standard deviations, based on the normal distribution, for a given confidence level $\gamma_1$ or $\gamma_2$. |
| B | $(A^2 + 2)/3$. |
| $\gamma_1$ | confidence level for a one-sided confidence limit (only the upper confidence limit $\bar{p}$ or only the lower confidence limit $\underline{p}$). |
| $\gamma_2$ | confidence level for two-sided confidence limits (both the upper confidence limit $\bar{p}$ and the lower confidence limit $\bar{p}$). |
| $\alpha$ | risk of rejecting a lot (population) with satisfactory error rate $p_1$. ($\gamma_1$ = 1 – $\alpha$.) |
| $\beta$ | risk of accepting a lot (population) with unsatisfactory error rate $p_2$. ($\gamma_1$ = 1 – $\beta$.) |

## Values of A, A$^2$, and B

| Confidence Level | | Risk Level | | | |
|---|---|---|---|---|---|
| $\gamma_1$ | $\gamma_2$ | $\alpha$ or $\beta$ | A | A$^2$ | B |
| 99.5% | 99% | 0.5% | 2.5758 | 6.6349 | 2.8783 |
| 99% | 98% | 1% | 2.3263 | 5.4119 | 2.4706 |
| 97.5% | 95% | 2.5% | 1.9600 | 3.8415 | 1.9472 |
| 95% | 90% | 5% | 1.6449 | 2.7055 | 1.5685 |
| 90% | 80% | 10% | 1.2816 | 1.6424 | 1.2141 |
| 80% | 60% | 20% | 0.84162 | 0.70833 | 0.90278 |

**Note.** Acceptance number ċ for acceptance sampling proper (where both $p_1$
and $p_2$ and both risks $\alpha$ and $\beta$ are specified) may also be obtained from
a formula, in this case a complex one. The formula is presented in Appendix
C, page 448.

# Table 12

**FORMULAS FOR SAMPLE SIZE;** $\hat{p} = 0$

For $\hat{p} \leqslant .25$ (but not 0), Table 8 applies to proportion sampling based on $\bar{p}$ and to the quasi-form of acceptance sampling based on $p_2$. In the special case $\hat{p} = 0$, the following procedures and formulas are used instead of those in Table 8.

## A. Proportion Sampling; $\bar{p}$ Desired

1. In the table below, find the value of $\bar{m}$ for $c = 0$ and $\gamma_1$.

2. Calculate $n = \dfrac{\bar{m}}{\hat{e}} - \dfrac{\bar{m}}{2}$.

    $n$    sample size.

    $c$    number of events (items having a designated characteristic) in the sample.

    $\bar{m}$    upper confidence limit for the parameter m of a Poisson distribution, based on c and $\gamma_1$.

    $\bar{p}$    upper binomial confidence limit.

\*  $\hat{p}$    anticipated value of c/n: 0.

    $\bar{e}$    upper error margin: $\bar{p} - c/n$; when $c/n = 0$, $\bar{e} = \bar{p}$.

\*  $\hat{e}$    anticipated error margin; specifically, $\hat{e}$, the anticipated value of $\bar{e}$.

\*  $\gamma_1$    confidence level for a one-sided confidence limit (only $\bar{p}$).

## B. Quasi-Form of Acceptance Sampling; $p_2$ and $\beta$ Specified

1. In the table below, find the value of $\bar{m}$ for $c = 0$ and $\beta$.

2. Calculate $n = \dfrac{\bar{m}}{p_2} - \dfrac{\bar{m}}{2}$.

\*  Specifications for determining sample size.

**Table 12. FORMULAS FOR SAMPLE SIZE; $\hat{p} = 0$** *(Continued)*

   n     sample size.

   c     number of error (defective) items in the sample.

   ċ     acceptance number: 0.

   $\overline{m}$     upper confidence limit for the parameter m of a Poisson distribution, based on ċ and $\gamma_1 = 1 - \beta$.

\*  $\hat{p}$     anticipated value of c/n: 0.

\*  $p_2$    unsatisfactory error (defective) rate.

\*  $\beta$     risk of accepting a lot (population) with error rate $p_2$.

## Values of $\overline{m}$ for c (or ċ) = 0

| Confidence Level | Risk Level | |
|:---:|:---:|:---:|
| $\gamma_1$ | $\beta$ | $\overline{m}$ |
| 99.5% | 0.5% | 5.2983 |
| 99% | 1% | 4.6052 |
| 97.5% | 2.5% | 3.6889 |
| 95% | 5% | 2.9957 |
| 90% | 10% | 2.3026 |
| 80% | 20% | 1.6094 |

\*  Specifications for determining sample size.

# Table 13

VALUES OF S AND $\dot{c}$ FOR DETERMINING SAMPLE SIZE FOR ACCEP–
TANCE SAMPLING; $p_1$ AND $p_2 \leqslant .25$

1.  Calculate $\hat{S} = \dfrac{p_2}{p_1} \times \dfrac{2 - p_1}{2 - p_2}$ .

2.  In Table 13, section for specified $\beta$ and column for specified $\alpha$, find S
    nearest to $\hat{S}$. Find the value of $\dot{c}$ corresponding to S; $\dot{c}$ is the acceptance
    number (and $\dot{c} + 1$, or $\dot{c}_1$, is the rejection number). If necessary, inter-
    polate linearly between values of S bounding $\hat{S}$ to find integer $\dot{c}$.
3.  Find the value of $\bar{m}$ corresponding to $\dot{c}$. If $\bar{m}$ is not given in Table 13
    (because $\dot{c}$ is obtained by interpolation), then in Table 1 or 7–E, for
    $\gamma_1 = 1 - \beta$, find $\bar{m}$ corresponding to $\dot{c}$; read c as $\dot{c}$.

4.  Calculate $n = \dfrac{\bar{m}}{p_2} - \dfrac{\bar{m} - \dot{c}}{2}$ .

| | |
|---|---|
| S | $\bar{m}/\underline{m}_1$. |
| $\bar{m}$ | upper confidence limit for the parameter m of a Poisson distri-bution; based on c and $\gamma_1 = 1 - \beta$. |
| $\underline{m}_1$ | lower confidence limit for the parameter m of a Poisson distri-bution; based on c + 1 and $\gamma_1 = 1 - \alpha$. |
| n | sample size. |
| c | number of error (defective) items in the sample. |
| $\dot{c}$ | acceptance number. |
| * $p_1$ | satisfactory error (defective) rate. |
| * $p_2$ | unsatisfactory error (defective) rate. |
| * $\alpha$ | risk of rejecting a lot (population) with error rate $p_1$. |
| * $\beta$ | risk of accepting a lot (population) with error rate $p_2$. |
| $\gamma_1$ | confidence level for a one-sided confidence limit; $\gamma_1 = 1 - \alpha$, and $\gamma_1 = 1 - \beta$. |

*   Specifications for determining sample size.

**Table 13. VALUES OF S AND ĉ FOR DETERMINING SAMPLE SIZE FOR ACCEPTANCE SAMPLING; $p_1$ AND $p_2 \leqslant .25$** *(Continued)*

Section 1    $\beta$ =   0.5%
Section 2    $\beta$ =   1%
Section 3    $\beta$ =   2.5%
Section 4    $\beta$ =   5%     $\alpha$ =   0.5%, 1%, 2.5%, 5%, 10%, 20%
Section 5    $\beta$ =   10%
Section 6    $\beta$ =   20%

Table 13. **VALUES OF S AND ċ FOR DETERMINING SAMPLE SIZE FOR ACCEPTANCE SAMPLING; $p_1$ AND $p_2 \leqslant .25$**

| | | $\alpha$ | | | | | |
|---|---|---|---|---|---|---|---|
| | | 0.5% | 1% | 2.5% | 5% | 10% | 20% |
| ċ | $\overline{m}$ | | | S | | | |
| 0 | 5.2983 | 1056.996 | 527.142 | 209.270 | 103.293 | 50.287 | 23.744 |
| 1 | 7.4302 | 71.792 | 50.016 | 30.677 | 20.909 | 13.971 | 9.013 |
| 2 | 9.2738 | 27.448 | 21.268 | 14.990 | 11.341 | 8.415 | 6.041 |
| 3 | 10.978 | 16.331 | 13.334 | 10.072 | 8.034 | 6.292 | 4.780 |
| 4 | 12.594 | 11.684 | 9.846 | 7.757 | 6.392 | 5.177 | 4.076 |
| 5 | 14.150 | 9.207 | 7.926 | 6.426 | 5.415 | 4.489 | 3.625 |
| 6 | 15.660 | 7.686 | 6.720 | 5.564 | 4.767 | 4.021 | 3.308 |
| 7 | 17.134 | 6.664 | 5.896 | 4.961 | 4.304 | 3.680 | 3.073 |
| 8 | 18.578 | 5.931 | 5.297 | 4.514 | 3.957 | 3.420 | 2.890 |
| 9 | 19.998 | 5.380 | 4.842 | 4.170 | 3.686 | 3.215 | 2.744 |
| 10 | 21.398 | 4.952 | 4.485 | 3.897 | 3.469 | 3.048 | 2.623 |
| 11 | 22.779 | 4.608 | 4.196 | 3.674 | 3.290 | 2.909 | 2.522 |
| 12 | 24.145 | 4.327 | 3.959 | 3.488 | 3.140 | 2.793 | 2.436 |
| 13 | 25.497 | 4.092 | 3.759 | 3.331 | 3.012 | 2.692 | 2.362 |
| 14 | 26.836 | 3.893 | 3.589 | 3.197 | 2.902 | 2.606 | 2.297 |
| 15 | 28.164 | 3.722 | 3.443 | 3.080 | 2.806 | 2.529 | 2.240 |
| 16 | 29.482 | 3.573 | 3.315 | 2.977 | 2.722 | 2.462 | 2.189 |
| 17 | 30.791 | 3.443 | 3.202 | 2.886 | 2.647 | 2.401 | 2.143 |
| 18 | 32.091 | 3.327 | 3.102 | 2.805 | 2.579 | 2.347 | 2.102 |
| 19 | 33.383 | 3.224 | 3.012 | 2.733 | 2.519 | 2.298 | 2.064 |
| 20 | 34.668 | 3.132 | 2.932 | 2.667 | 2.464 | 2.254 | 2.030 |
| 21 | 35.946 | 3.048 | 2.859 | 2.607 | 2.414 | 2.213 | 1.998 |
| 22 | 37.218 | 2.973 | 2.792 | 2.553 | 2.368 | 2.176 | 1.969 |
| 23 | 38.484 | 2.903 | 2.732 | 2.503 | 2.325 | 2.141 | 1.943 |
| 24 | 39.745 | 2.840 | 2.676 | 2.457 | 2.287 | 2.109 | 1.918 |
| 25 | 41.000 | 2.781 | 2.624 | 2.414 | 2.250 | 2.079 | 1.895 |
| 26 | 42.251 | 2.728 | 2.577 | 2.375 | 2.217 | 2.052 | 1.873 |
| 27 | 43.497 | 2.678 | 2.533 | 2.338 | 2.186 | 2.026 | 1.853 |
| 28 | 44.738 | 2.631 | 2.491 | 2.304 | 2.156 | 2.002 | 1.834 |
| 29 | 45.976 | 2.588 | 2.453 | 2.271 | 2.129 | 1.979 | 1.816 |
| 30 | 47.209 | 2.547 | 2.417 | 2.241 | 2.103 | 1.958 | 1.799 |
| 31 | 48.439 | 2.509 | 2.383 | 2.213 | 2.079 | 1.938 | 1.783 |
| 32 | 49.665 | 2.473 | 2.352 | 2.186 | 2.056 | 1.919 | 1.768 |
| 33 | 50.888 | 2.440 | 2.322 | 2.161 | 2.035 | 1.901 | 1.753 |
| 34 | 52.108 | 2.408 | 2.293 | 2.137 | 2.014 | 1.884 | 1.740 |
| 35 | 53.324 | 2.378 | 2.267 | 2.115 | 1.995 | 1.867 | 1.727 |
| 36 | 54.537 | 2.350 | 2.241 | 2.093 | 1.976 | 1.852 | 1.715 |
| 37 | 55.748 | 2.323 | 2.217 | 2.073 | 1.959 | 1.837 | 1.703 |
| 38 | 56.956 | 2.297 | 2.194 | 2.054 | 1.942 | 1.823 | 1.692 |
| 39 | 58.161 | 2.273 | 2.173 | 2.035 | 1.926 | 1.810 | 1.681 |

Table 13.  VALUES OF S AND ċ FOR DETERMINING SAMPLE SIZE FOR
ACCEPTANCE SAMPLING; $p_1$ AND $p_2$ ⩽ .25 (*Continued*)

| ċ | $\overline{m}$ | 0.5% | 1% | 2.5% | 5% | 10% | 20% |
|---|---|---|---|---|---|---|---|
| | | | | | S | | |
| 40 | 59.363 | 2.250 | 2.152 | 2.018 | 1.911 | 1.797 | 1.670 |
| 41 | 60.563 | 2.228 | 2.132 | 2.001 | 1.896 | 1.785 | 1.661 |
| 42 | 61.761 | 2.207 | 2.113 | 1.985 | 1.882 | 1.773 | 1.651 |
| 43 | 62.957 | 2.187 | 2.095 | 1.969 | 1.869 | 1.761 | 1.642 |
| 44 | 64.150 | 2.167 | 2.078 | 1.954 | 1.856 | 1.751 | 1.633 |
| 45 | 65.341 | 2.149 | 2.061 | 1.940 | 1.844 | 1.740 | 1.625 |
| 46 | 66.530 | 2.131 | 2.045 | 1.927 | 1.832 | 1.730 | 1.617 |
| 47 | 67.717 | 2.114 | 2.030 | 1.913 | 1.820 | 1.720 | 1.609 |
| 48 | 68.902 | 2.098 | 2.015 | 1.901 | 1.809 | 1.711 | 1.601 |
| 49 | 70.085 | 2.082 | 2.001 | 1.889 | 1.799 | 1.702 | 1.594 |
| 50 | 71.266 | 2.067 | 1.987 | 1.877 | 1.788 | 1.693 | 1.587 |
| 55 | 77.147 | 1.999 | 1.925 | 1.824 | 1.742 | 1.654 | 1.555 |
| 60 | 82.990 | 1.941 | 1.873 | 1.779 | 1.702 | 1.620 | 1.527 |
| 65 | 88.799 | 1.891 | 1.828 | 1.740 | 1.668 | 1.591 | 1.503 |
| 70 | 94.577 | 1.848 | 1.789 | 1.706 | 1.638 | 1.565 | 1.482 |
| 75 | 100.33 | 1.810 | 1.754 | 1.676 | 1.612 | 1.542 | 1.463 |
| 80 | 106.06 | 1.777 | 1.723 | 1.649 | 1.588 | 1.522 | 1.446 |
| 90 | 117.45 | 1.719 | 1.671 | 1.603 | 1.547 | 1.486 | 1.417 |
| 100 | 128.76 | 1.672 | 1.628 | 1.566 | 1.514 | 1.457 | 1.393 |
| 110 | 140.01 | 1.633 | 1.592 | 1.534 | 1.485 | 1.433 | 1.372 |
| 120 | 151.21 | 1.599 | 1.561 | 1.506 | 1.461 | 1.411 | 1.354 |
| 130 | 162.36 | 1.570 | 1.534 | 1.482 | 1.440 | 1.393 | 1.339 |
| 150 | 184.53 | 1.522 | 1.489 | 1.443 | 1.405 | 1.362 | 1.313 |
| 170 | 206.56 | 1.484 | 1.454 | 1.412 | 1.377 | 1.337 | 1.292 |
| 200 | 239.40 | 1.439 | 1.412 | 1.375 | 1.343 | 1.308 | 1.267 |
| 250 | 293.69 | 1.385 | 1.362 | 1.330 | 1.302 | 1.272 | 1.236 |
| 300 | 347.57 | 1.346 | 1.326 | 1.297 | 1.273 | 1.246 | 1.214 |
| 350 | 401.14 | 1.317 | 1.299 | 1.273 | 1.251 | 1.226 | 1.197 |
| 400 | 454.46 | 1.294 | 1.277 | 1.253 | 1.233 | 1.210 | 1.183 |
| 450 | 507.58 | 1.275 | 1.259 | 1.237 | 1.218 | 1.197 | 1.172 |
| 500 | 560.53 | 1.259 | 1.245 | 1.224 | 1.206 | 1.186 | 1.163 |
| 600 | 666.02 | 1.234 | 1.221 | 1.202 | 1.187 | 1.169 | 1.148 |
| 700 | 771.08 | 1.215 | 1.203 | 1.186 | 1.172 | 1.156 | 1.136 |
| 800 | 875.78 | 1.200 | 1.189 | 1.173 | 1.160 | 1.145 | 1.127 |
| 900 | 980.20 | 1.187 | 1.177 | 1.163 | 1.150 | 1.136 | 1.119 |
| 1000 | 1084.4 | 1.177 | 1.167 | 1.154 | 1.142 | 1.129 | 1.113 |
| 1100 | 1188.3 | 1.168 | 1.159 | 1.146 | 1.135 | 1.122 | 1.108 |
| 1200 | 1292.1 | 1.160 | 1.152 | 1.139 | 1.129 | 1.117 | 1.103 |
| 1500 | 1602.7 | 1.142 | 1.135 | 1.124 | 1.115 | 1.104 | 1.0915 |
| 2000 | 2118.1 | 1.122 | 1.116 | 1.106 | 1.0986 | 1.0896 | 1.0789 |

**Table 13.  VALUES OF S AND ċ FOR DETERMINING SAMPLE SIZE FOR ACCEPTANCE SAMPLING; $p_1$ AND $p_2 \leqslant .25$** (*Continued*)

| | | $\alpha$ | | | | | |
|---|---|---|---|---|---|---|---|
| | | 0.5% | 1% | 2.5% | 5% | 10% | 20% |
| ċ | $\overline{m}$ | | | S | | | |
| 2500 | 2631.7 | 1.109 | 1.103 | 1.0947 | 1.0878 | 1.0798 | 1.0703 |
| 3000 | 3144.0 | 1.0986 | 1.0935 | 1.0862 | 1.0799 | 1.0727 | 1.0640 |
| 3500 | 3655.3 | 1.0910 | 1.0863 | 1.0795 | 1.0737 | 1.0671 | 1.0592 |
| 4000 | 4165.8 | 1.0849 | 1.0805 | 1.0742 | 1.0688 | 1.0627 | 1.0553 |
| 4500 | 4675.7 | 1.0798 | 1.0758 | 1.0698 | 1.0648 | 1.0590 | 1.0520 |
| 5000 | 5185.0 | 1.0756 | 1.0717 | 1.0661 | 1.0614 | 1.0559 | 1.0493 |
| 6000 | 6202.4 | 1.0688 | 1.0653 | 1.0602 | 1.0559 | 1.0509 | 1.0449 |
| 7000 | 7218.4 | 1.0635 | 1.0603 | 1.0556 | 1.0516 | 1.0471 | 1.0415 |
| 8000 | 8233.3 | 1.0593 | 1.0563 | 1.0520 | 1.0482 | 1.0440 | 1.0388 |
| 9000 | 9247.3 | 1.0558 | 1.0530 | 1.0489 | 1.0454 | 1.0414 | 1.0366 |
| 10000 | 10260.5 | 1.0529 | 1.0502 | 1.0464 | 1.0430 | 1.0392 | 1.0347 |

# Table 13. VALUES OF S AND ċ FOR DETERMINING SAMPLE SIZE FOR ACCEPTANCE SAMPLING; $p_1$ AND $p_2 \leqslant .25$ (*Continued*)

| ċ | $\overline{m}$ | α 0.5% | 1% | 2.5% | 5% | 10% | 20% |
|---|---|---|---|---|---|---|---|
| | | | | S | | | |
| 0 | 4.6052 | 918.719 | 458.180 | 181.893 | 89.780 | 43.708 | 20.638 |
| 1 | 6.6384 | 64.142 | 44.686 | 27.408 | 18.681 | 12.483 | 8.052 |
| 2 | 8.4060 | 24.880 | 19.278 | 13.587 | 10.280 | 7.627 | 5.476 |
| 3 | 10.045 | 14.944 | 12.202 | 9.217 | 7.352 | 5.757 | 4.374 |
| 4 | 11.605 | 10.766 | 9.072 | 7.148 | 5.890 | 4.770 | 3.756 |
| 5 | 13.109 | 8.529 | 7.343 | 5.953 | 5.017 | 4.159 | 3.358 |
| 6 | 14.571 | 7.152 | 6.253 | 5.177 | 4.435 | 3.741 | 3.078 |
| 7 | 16.000 | 6.223 | 5.506 | 4.633 | 4.019 | 3.436 | 2.869 |
| 8 | 17.403 | 5.556 | 4.962 | 4.229 | 3.706 | 3.203 | 2.707 |
| 9 | 18.783 | 5.053 | 4.548 | 3.917 | 3.462 | 3.019 | 2.577 |
| 10 | 20.145 | 4.662 | 4.222 | 3.669 | 3.265 | 2.869 | 2.470 |
| 11 | 21.490 | 4.347 | 3.959 | 3.466 | 3.104 | 2.745 | 2.380 |
| 12 | 22.821 | 4.090 | 3.742 | 3.297 | 2.968 | 2.639 | 2.303 |
| 13 | 24.139 | 3.874 | 3.559 | 3.154 | 2.852 | 2.549 | 2.236 |
| 14 | 25.446 | 3.691 | 3.403 | 3.031 | 2.752 | 2.471 | 2.178 |
| 15 | 26.743 | 3.534 | 3.269 | 2.924 | 2.665 | 2.402 | 2.127 |
| 16 | 28.030 | 3.397 | 3.151 | 2.830 | 2.588 | 2.341 | 2.081 |
| 17 | 29.310 | 3.277 | 3.048 | 2.747 | 2.519 | 2.286 | 2.040 |
| 18 | 30.581 | 3.171 | 2.956 | 2.673 | 2.458 | 2.237 | 2.003 |
| 19 | 31.845 | 3.076 | 2.874 | 2.607 | 2.403 | 2.192 | 1.969 |
| 20 | 33.103 | 2.991 | 2.799 | 2.547 | 2.352 | 2.152 | 1.938 |
| 21 | 34.355 | 2.913 | 2.732 | 2.492 | 2.307 | 2.115 | 1.910 |
| 22 | 35.601 | 2.843 | 2.671 | 2.442 | 2.265 | 2.081 | 1.884 |
| 23 | 36.841 | 2.779 | 2.615 | 2.396 | 2.226 | 2.050 | 1.860 |
| 24 | 38.077 | 2.721 | 2.564 | 2.354 | 2.191 | 2.021 | 1.837 |
| 25 | 39.308 | 2.667 | 2.516 | 2.314 | 2.158 | 1.994 | 1.816 |
| 26 | 40.534 | 2.617 | 2.472 | 2.278 | 2.127 | 1.969 | 1.797 |
| 27 | 41.757 | 2.570 | 2.431 | 2.244 | 2.098 | 1.945 | 1.779 |
| 28 | 42.975 | 2.527 | 2.393 | 2.213 | 2.071 | 1.923 | 1.761 |
| 29 | 44.190 | 2.487 | 2.358 | 2.183 | 2.046 | 1.902 | 1.745 |
| 30 | 45.401 | 2.450 | 2.324 | 2.155 | 2.023 | 1.883 | 1.730 |
| 31 | 46.608 | 2.414 | 2.293 | 2.129 | 2.001 | 1.864 | 1.716 |
| 32 | 47.813 | 2.381 | 2.264 | 2.105 | 1.980 | 1.847 | 1.702 |
| 33 | 49.014 | 2.350 | 2.236 | 2.082 | 1.960 | 1.831 | 1.689 |
| 34 | 50.213 | 2.321 | 2.210 | 2.060 | 1.941 | 1.815 | 1.677 |
| 35 | 51.408 | 2.293 | 2.185 | 2.039 | 1.923 | 1.800 | 1.665 |
| 36 | 52.601 | 2.266 | 2.162 | 2.019 | 1.906 | 1.786 | 1.654 |
| 37 | 53.791 | 2.241 | 2.139 | 2.000 | 1.890 | 1.773 | 1.643 |
| 38 | 54.979 | 2.218 | 2.118 | 1.982 | 1.875 | 1.760 | 1.633 |
| 39 | 56.165 | 2.195 | 2.098 | 1.965 | 1.860 | 1.748 | 1.623 |

Table 13.  VALUES OF S AND ċ FOR DETERMINING SAMPLE SIZE FOR
ACCEPTANCE SAMPLING; $p_1$ AND $p_2$ ⩽ .25 (*Continued*)

| ċ | m̄ | α | | | | | |
|---|---|---|---|---|---|---|---|
| | | 0.5% | 1% | 2.5% | 5% | 10% | 20% |
| | | S | | | | | |
| 40 | 57.348 | 2.174 | 2.079 | 1.949 | 1.846 | 1.736 | 1.614 |
| 41 | 58.529 | 2.153 | 2.060 | 1.934 | 1.833 | 1.725 | 1.605 |
| 42 | 59.707 | 2.133 | 2.043 | 1.919 | 1.820 | 1.714 | 1.596 |
| 43 | 60.884 | 2.115 | 2.026 | 1.904 | 1.807 | 1.703 | 1.588 |
| 44 | 62.058 | 2.097 | 2.010 | 1.891 | 1.796 | 1.693 | 1.580 |
| 45 | 63.231 | 2.079 | 1.994 | 1.878 | 1.784 | 1.684 | 1.572 |
| 46 | 64.402 | 2.063 | 1.980 | 1.865 | 1.773 | 1.675 | 1.565 |
| 47 | 65.571 | 2.047 | 1.965 | 1.853 | 1.763 | 1.666 | 1.558 |
| 48 | 66.738 | 2.032 | 1.952 | 1.841 | 1.752 | 1.657 | 1.551 |
| 49 | 67.904 | 2.017 | 1.938 | 1.830 | 1.743 | 1.649 | 1.544 |
| 50 | 69.067 | 2.003 | 1.926 | 1.819 | 1.733 | 1.641 | 1.538 |
| 55 | 74.864 | 1.939 | 1.868 | 1.770 | 1.690 | 1.605 | 1.509 |
| 60 | 80.625 | 1.886 | 1.819 | 1.728 | 1.654 | 1.574 | 1.483 |
| 65 | 86.356 | 1.839 | 1.778 | 1.692 | 1.622 | 1.547 | 1.462 |
| 70 | 92.059 | 1.799 | 1.741 | 1.660 | 1.595 | 1.523 | 1.442 |
| 75 | 97.738 | 1.763 | 1.709 | 1.632 | 1.570 | 1.502 | 1.425 |
| 80 | 103.40 | 1.732 | 1.680 | 1.607 | 1.548 | 1.483 | 1.410 |
| 90 | 114.65 | 1.678 | 1.631 | 1.565 | 1.511 | 1.451 | 1.383 |
| 100 | 125.84 | 1.634 | 1.591 | 1.530 | 1.479 | 1.424 | 1.361 |
| 110 | 136.97 | 1.597 | 1.557 | 1.500 | 1.453 | 1.402 | 1.342 |
| 120 | 148.05 | 1.566 | 1.528 | 1.475 | 1.431 | 1.382 | 1.326 |
| 130 | 159.09 | 1.539 | 1.503 | 1.452 | 1.411 | 1.365 | 1.312 |
| 150 | 181.05 | 1.493 | 1.461 | 1.416 | 1.378 | 1.336 | 1.288 |
| 170 | 202.89 | 1.458 | 1.428 | 1.387 | 1.352 | 1.314 | 1.269 |
| 200 | 235.45 | 1.415 | 1.389 | 1.352 | 1.321 | 1.286 | 1.246 |
| 250 | 289.33 | 1.364 | 1.342 | 1.310 | 1.283 | 1.253 | 1.218 |
| 300 | 342.83 | 1.328 | 1.308 | 1.279 | 1.256 | 1.229 | 1.197 |
| 350 | 396.05 | 1.300 | 1.282 | 1.256 | 1.235 | 1.210 | 1.182 |
| 400 | 449.05 | 1.278 | 1.262 | 1.238 | 1.218 | 1.196 | 1.169 |
| 450 | 501.87 | 1.260 | 1.245 | 1.223 | 1.205 | 1.184 | 1.159 |
| 500 | 554.54 | 1.246 | 1.231 | 1.211 | 1.193 | 1.174 | 1.150 |
| 600 | 659.50 | 1.222 | 1.209 | 1.191 | 1.175 | 1.157 | 1.137 |
| 700 | 764.06 | 1.204 | 1.192 | 1.175 | 1.161 | 1.145 | 1.126 |
| 800 | 868.31 | 1.190 | 1.179 | 1.163 | 1.150 | 1.135 | 1.117 |
| 900 | 972.30 | 1.178 | 1.168 | 1.153 | 1.141 | 1.127 | 1.110 |
| 1000 | 1076.1 | 1.168 | 1.158 | 1.145 | 1.133 | 1.120 | 1.104 |
| 1100 | 1179.7 | 1.159 | 1.151 | 1.138 | 1.127 | 1.114 | 1.0994 |
| 1200 | 1283.1 | 1.152 | 1.144 | 1.131 | 1.121 | 1.109 | 1.0950 |
| 1500 | 1592.6 | 1.135 | 1.128 | 1.117 | 1.108 | 1.0972 | 1.0847 |
| 2000 | 2106.5 | 1.116 | 1.110 | 1.100 | 1.0926 | 1.0837 | 1.0730 |

**Table 13. VALUES OF S AND ċ FOR DETERMINING SAMPLE SIZE FOR ACCEPTANCE SAMPLING; $p_1$ AND $p_2 \leqslant .25$ (*Continued*)**

| ċ | $\overline{m}$ | $\alpha$ | | | | | |
|---|---|---|---|---|---|---|---|
| | | 0.5% | 1% | 2.5% | 5% | 10% | 20% |
| | | S | | | | | |
| 2500 | 2618.8 | 1.103 | 1.0975 | 1.0894 | 1.0825 | 1.0746 | 1.0651 |
| 3000 | 3129.9 | 1.0937 | 1.0887 | 1.0813 | 1.0750 | 1.0679 | 1.0593 |
| 3500 | 3640.1 | 1.0864 | 1.0818 | 1.0751 | 1.0693 | 1.0627 | 1.0548 |
| 4000 | 4149.6 | 1.0806 | 1.0763 | 1.0700 | 1.0647 | 1.0585 | 1.0512 |
| 4500 | 4658.5 | 1.0759 | 1.0718 | 1.0659 | 1.0609 | 1.0551 | 1.0482 |
| 5000 | 5167.0 | 1.0718 | 1.0680 | 1.0624 | 1.0577 | 1.0522 | 1.0457 |
| 6000 | 6182.7 | 1.0654 | 1.0619 | 1.0568 | 1.0525 | 1.0476 | 1.0416 |
| 7000 | 7197.1 | 1.0604 | 1.0572 | 1.0525 | 1.0485 | 1.0440 | 1.0385 |
| 8000 | 8210.6 | 1.0564 | 1.0534 | 1.0491 | 1.0453 | 1.0411 | 1.0360 |
| 9000 | 9223.2 | 1.0531 | 1.0503 | 1.0462 | 1.0427 | 1.0387 | 1.0339 |
| 10000 | 10235.1 | 1.0503 | 1.0476 | 1.0438 | 1.0405 | 1.0367 | 1.0321 |

**Table 13.** **VALUES OF S AND ċ FOR DETERMINING SAMPLE SIZE FOR** **Section 3**
**ACCEPTANCE SAMPLING;** $p_1$ AND $p_2 \leqslant .25$ *(Continued)*

$\beta = 2.5\%$

| ċ | $\overline{m}$ | 0.5% | 1% | 2.5% | 5% | 10% | 20% |
|---|---|---|---|---|---|---|---|
|   |   |      |    |      | S  |     |     |
| 0 | 3.6889 | 735.921 | 367.016 | 145.702 | 71.916 | 35.012 | 16.531 |
| 1 | 5.5717 | 53.835 | 37.506 | 23.003 | 15.679 | 10.477 | 6.759 |
| 2 | 7.2247 | 21.383 | 16.569 | 11.678 | 8.836 | 6.556 | 4.707 |
| 3 | 8.7673 | 13.043 | 10.650 | 8.044 | 6.417 | 5.025 | 3.817 |
| 4 | 10.242 | 9.501 | 8.007 | 6.308 | 5.198 | 4.210 | 3.315 |
| 5 | 11.668 | 7.592 | 6.536 | 5.299 | 4.465 | 3.702 | 2.989 |
| 6 | 13.059 | 6.410 | 5.604 | 4.640 | 3.975 | 3.353 | 2.759 |
| 7 | 14.423 | 5.610 | 4.963 | 4.176 | 3.623 | 3.098 | 2.587 |
| 8 | 15.763 | 5.032 | 4.494 | 3.830 | 3.357 | 2.902 | 2.452 |
| 9 | 17.085 | 4.596 | 4.137 | 3.563 | 3.149 | 2.746 | 2.344 |
| 10 | 18.390 | 4.256 | 3.854 | 3.349 | 2.981 | 2.619 | 2.255 |
| 11 | 19.682 | 3.982 | 3.626 | 3.174 | 2.843 | 2.514 | 2.179 |
| 12 | 20.962 | 3.756 | 3.437 | 3.028 | 2.726 | 2.424 | 2.115 |
| 13 | 22.230 | 3.568 | 3.278 | 2.904 | 2.626 | 2.348 | 2.060 |
| 14 | 23.490 | 3.408 | 3.142 | 2.798 | 2.540 | 2.281 | 2.011 |
| 15 | 24.740 | 3.269 | 3.024 | 2.705 | 2.465 | 2.222 | 1.968 |
| 16 | 25.983 | 3.149 | 2.921 | 2.624 | 2.399 | 2.170 | 1.929 |
| 17 | 27.219 | 3.043 | 2.830 | 2.551 | 2.340 | 2.123 | 1.894 |
| 18 | 28.448 | 2.950 | 2.750 | 2.487 | 2.286 | 2.081 | 1.863 |
| 19 | 29.671 | 2.866 | 2.677 | 2.429 | 2.239 | 2.043 | 1.835 |
| 20 | 30.888 | 2.790 | 2.612 | 2.376 | 2.195 | 2.008 | 1.809 |
| 21 | 32.101 | 2.722 | 2.553 | 2.328 | 2.155 | 1.976 | 1.785 |
| 22 | 33.308 | 2.660 | 2.499 | 2.285 | 2.119 | 1.947 | 1.763 |
| 23 | 34.511 | 2.604 | 2.450 | 2.244 | 2.085 | 1.920 | 1.742 |
| 24 | 35.710 | 2.552 | 2.404 | 2.207 | 2.054 | 1.895 | 1.723 |
| 25 | 36.905 | 2.504 | 2.362 | 2.173 | 2.026 | 1.872 | 1.705 |
| 26 | 38.096 | 2.459 | 2.323 | 2.141 | 1.999 | 1.850 | 1.689 |
| 27 | 39.284 | 2.418 | 2.287 | 2.111 | 1.974 | 1.830 | 1.673 |
| 28 | 40.468 | 2.380 | 2.254 | 2.084 | 1.951 | 1.811 | 1.659 |
| 29 | 41.649 | 2.344 | 2.222 | 2.058 | 1.929 | 1.793 | 1.645 |
| 30 | 42.827 | 2.311 | 2.193 | 2.033 | 1.908 | 1.776 | 1.632 |
| 31 | 44.002 | 2.279 | 2.165 | 2.010 | 1.889 | 1.760 | 1.620 |
| 32 | 45.174 | 2.250 | 2.139 | 1.989 | 1.870 | 1.745 | 1.608 |
| 33 | 46.344 | 2.222 | 2.114 | 1.968 | 1.853 | 1.731 | 1.597 |
| 34 | 47.512 | 2.196 | 2.091 | 1.949 | 1.837 | 1.717 | 1.586 |
| 35 | 48.677 | 2.171 | 2.069 | 1.931 | 1.821 | 1.705 | 1.576 |
| 36 | 49.839 | 2.147 | 2.048 | 1.913 | 1.806 | 1.692 | 1.567 |
| 37 | 51.000 | 2.125 | 2.028 | 1.897 | 1.792 | 1.681 | 1.558 |
| 38 | 52.158 | 2.104 | 2.010 | 1.881 | 1.779 | 1.670 | 1.549 |
| 39 | 53.315 | 2.084 | 1.992 | 1.866 | 1.766 | 1.659 | 1.541 |

Table 13.  VALUES OF S AND ċ FOR DETERMINING SAMPLE SIZE FOR
ACCEPTANCE SAMPLING; $p_1$ AND $p_2 \leqslant .25$ (*Continued*)

| ċ | $\overline{m}$ | 0.5% | 1% | 2.5% | 5% | 10% | 20% |
|---|---|---|---|---|---|---|---|
| | | | | S | | | |
| 40 | 54.469 | 2.064 | 1.974 | 1.851 | 1.753 | 1.649 | 1.533 |
| 41 | 55.621 | 2.046 | 1.958 | 1.838 | 1.742 | 1.639 | 1.525 |
| 42 | 56.772 | 2.029 | 1.942 | 1.824 | 1.730 | 1.630 | 1.518 |
| 43 | 57.921 | 2.012 | 1.927 | 1.812 | 1.719 | 1.621 | 1.511 |
| 44 | 59.068 | 1.996 | 1.913 | 1.800 | 1.709 | 1.612 | 1.504 |
| 45 | 60.214 | 1.980 | 1.899 | 1.788 | 1.699 | 1.604 | 1.497 |
| 46 | 61.358 | 1.965 | 1.886 | 1.777 | 1.689 | 1.596 | 1.491 |
| 47 | 62.500 | 1.951 | 1.873 | 1.766 | 1.680 | 1.588 | 1.485 |
| 48 | 63.641 | 1.938 | 1.861 | 1.756 | 1.671 | 1.580 | 1.479 |
| 49 | 64.781 | 1.924 | 1.849 | 1.746 | 1.663 | 1.573 | 1.473 |
| 50 | 65.919 | 1.912 | 1.838 | 1.736 | 1.654 | 1.566 | 1.468 |
| 55 | 71.590 | 1.855 | 1.786 | 1.692 | 1.617 | 1.535 | 1.443 |
| 60 | 77.232 | 1.806 | 1.743 | 1.655 | 1.584 | 1.508 | 1.421 |
| 65 | 82.848 | 1.765 | 1.705 | 1.623 | 1.556 | 1.484 | 1.402 |
| 70 | 88.441 | 1.728 | 1.673 | 1.595 | 1.532 | 1.463 | 1.386 |
| 75 | 94.013 | 1.696 | 1.644 | 1.570 | 1.510 | 1.445 | 1.371 |
| 80 | 99.567 | 1.668 | 1.618 | 1.548 | 1.491 | 1.428 | 1.358 |
| 90 | 110.63 | 1.620 | 1.574 | 1.510 | 1.458 | 1.400 | 1.335 |
| 100 | 121.63 | 1.580 | 1.538 | 1.478 | 1.430 | 1.377 | 1.315 |
| 110 | 132.58 | 1.546 | 1.507 | 1.452 | 1.407 | 1.357 | 1.299 |
| 120 | 143.49 | 1.518 | 1.481 | 1.429 | 1.386 | 1.339 | 1.285 |
| 130 | 154.36 | 1.493 | 1.458 | 1.409 | 1.369 | 1.324 | 1.273 |
| 150 | 176.02 | 1.452 | 1.421 | 1.376 | 1.340 | 1.299 | 1.252 |
| 170 | 197.58 | 1.419 | 1.391 | 1.350 | 1.317 | 1.279 | 1.236 |
| 200 | 229.73 | 1.381 | 1.355 | 1.319 | 1.289 | 1.255 | 1.216 |
| 250 | 283.00 | 1.334 | 1.313 | 1.281 | 1.255 | 1.226 | 1.191 |
| 300 | 335.95 | 1.301 | 1.282 | 1.254 | 1.230 | 1.204 | 1.173 |
| 350 | 388.67 | 1.276 | 1.258 | 1.233 | 1.212 | 1.188 | 1.160 |
| 400 | 441.20 | 1.256 | 1.240 | 1.216 | 1.197 | 1.175 | 1.149 |
| 450 | 493.57 | 1.240 | 1.225 | 1.203 | 1.185 | 1.164 | 1.140 |
| 500 | 545.82 | 1.226 | 1.212 | 1.192 | 1.174 | 1.155 | 1.132 |
| 600 | 650.00 | 1.204 | 1.192 | 1.173 | 1.158 | 1.141 | 1.120 |
| 700 | 753.84 | 1.188 | 1.176 | 1.160 | 1.146 | 1.130 | 1.111 |
| 800 | 857.42 | 1.175 | 1.164 | 1.149 | 1.136 | 1.121 | 1.103 |
| 900 | 960.78 | 1.164 | 1.154 | 1.140 | 1.127 | 1.114 | 1.0972 |
| 1000 | 1064.0 | 1.155 | 1.145 | 1.132 | 1.121 | 1.108 | 1.0921 |
| 1100 | 1167.0 | 1.147 | 1.138 | 1.125 | 1.115 | 1.102 | 1.0876 |
| 1200 | 1269.9 | 1.140 | 1.132 | 1.120 | 1.109 | 1.0977 | 1.0838 |
| 1500 | 1577.9 | 1.125 | 1.117 | 1.107 | 1.0974 | 1.0870 | 1.0746 |
| 2000 | 2089.6 | 1.107 | 1.101 | 1.0916 | 1.0838 | 1.0750 | 1.0644 |

**Table 13. VALUES OF S AND ċ FOR DETERMINING SAMPLE SIZE FOR ACCEPTANCE SAMPLING; $p_1$ AND $p_2$ $\leqslant$ .25 (*Continued*)**

| | | | | | α | | |
|---|---|---|---|---|---|---|---|
| | | 0.5% | 1% | 2.5% | 5% | 10% | 20% |
| ċ | $\overline{m}$ | | | | S | | |
| 2500 | 2600.0 | 1.0952 | 1.0896 | 1.0815 | 1.0747 | 1.0668 | 1.0574 |
| 3000 | 3109.3 | 1.0865 | 1.0815 | 1.0742 | 1.0680 | 1.0608 | 1.0523 |
| 3500 | 3617.9 | 1.0798 | 1.0752 | 1.0685 | 1.0628 | 1.0562 | 1.0483 |
| 4000 | 4125.9 | 1.0745 | 1.0702 | 1.0639 | 1.0586 | 1.0525 | 1.0452 |
| 4500 | 4633.4 | 1.0701 | 1.0660 | 1.0602 | 1.0552 | 1.0494 | 1.0425 |
| 5000 | 5140.6 | 1.0663 | 1.0625 | 1.0570 | 1.0523 | 1.0468 | 1.0403 |
| 6000 | 6153.8 | 1.0604 | 1.0569 | 1.0519 | 1.0476 | 1.0427 | 1.0367 |
| 7000 | 7165.9 | 1.0558 | 1.0526 | 1.0480 | 1.0440 | 1.0394 | 1.0340 |
| 8000 | 8177.3 | 1.0521 | 1.0491 | 1.0448 | 1.0411 | 1.0369 | 1.0318 |
| 9000 | 9187.9 | 1.0490 | 1.0462 | 1.0422 | 1.0387 | 1.0347 | 1.0299 |
| 10000 | 10198.0 | 1.0465 | 1.0438 | 1.0400 | 1.0367 | 1.0329 | 1.0284 |

| | | α | | | | | |
|---|---|---|---|---|---|---|---|
| | | 0.5% | 1% | 2.5% | 5% | 10% | 20% |
| ċ | m̄ | | | | S | | |
| 0 | 2.9957 | 597.640 | 298.053 | 118.324 | 58.403 | 28.433 | 13.425 |
| 1 | 4.7439 | 45.837 | 31.933 | 19.586 | 13.349 | 8.920 | 5.754 |
| 2 | 6.2958 | 18.634 | 14.438 | 10.176 | 7.699 | 5.713 | 4.101 |
| 3 | 7.7537 | 11.535 | 9.418 | 7.114 | 5.675 | 4.444 | 3.376 |
| 4 | 9.1535 | 8.492 | 7.156 | 5.638 | 4.646 | 3.763 | 2.963 |
| 5 | 10.513 | 6.840 | 5.889 | 4.775 | 4.023 | 3.335 | 2.693 |
| 6 | 11.842 | 5.813 | 5.082 | 4.208 | 3.605 | 3.041 | 2.502 |
| 7 | 13.148 | 5.114 | 4.524 | 3.807 | 3.303 | 2.824 | 2.358 |
| 8 | 14.435 | 4.608 | 4.115 | 3.507 | 3.074 | 2.657 | 2.245 |
| 9 | 15.705 | 4.225 | 3.803 | 3.275 | 2.895 | 2.524 | 2.155 |
| 10 | 16.962 | 3.925 | 3.555 | 3.089 | 2.750 | 2.416 | 2.079 |
| 11 | 18.208 | 3.683 | 3.354 | 2.936 | 2.630 | 2.326 | 2.016 |
| 12 | 19.443 | 3.484 | 3.188 | 2.809 | 2.528 | 2.249 | 1.962 |
| 13 | 20.669 | 3.317 | 3.047 | 2.700 | 2.442 | 2.183 | 1.915 |
| 14 | 21.887 | 3.175 | 2.927 | 2.607 | 2.367 | 2.125 | 1.874 |
| 15 | 23.097 | 3.052 | 2.823 | 2.526 | 2.301 | 2.074 | 1.837 |
| 16 | 24.301 | 2.945 | 2.732 | 2.454 | 2.243 | 2.029 | 1.804 |
| 17 | 25.499 | 2.851 | 2.652 | 2.390 | 2.192 | 1.989 | 1.775 |
| 18 | 26.692 | 2.768 | 2.580 | 2.333 | 2.145 | 1.952 | 1.748 |
| 19 | 27.879 | 2.693 | 2.516 | 2.282 | 2.103 | 1.919 | 1.724 |
| 20 | 29.062 | 2.625 | 2.458 | 2.236 | 2.065 | 1.889 | 1.702 |
| 21 | 30.240 | 2.565 | 2.405 | 2.193 | 2.030 | 1.862 | 1.681 |
| 22 | 31.415 | 2.509 | 2.357 | 2.155 | 1.998 | 1.836 | 1.662 |
| 23 | 32.585 | 2.458 | 2.313 | 2.119 | 1.969 | 1.813 | 1.645 |
| 24 | 33.752 | 2.412 | 2.272 | 2.086 | 1.942 | 1.791 | 1.629 |
| 25 | 34.916 | 2.369 | 2.235 | 2.056 | 1.917 | 1.771 | 1.613 |
| 26 | 36.077 | 2.329 | 2.200 | 2.028 | 1.893 | 1.752 | 1.599 |
| 27 | 37.234 | 2.292 | 2.168 | 2.001 | 1.871 | 1.734 | 1.586 |
| 28 | 38.389 | 2.258 | 2.138 | 1.977 | 1.850 | 1.718 | 1.573 |
| 29 | 39.541 | 2.225 | 2.110 | 1.954 | 1.831 | 1.702 | 1.562 |
| 30 | 40.691 | 2.195 | 2.083 | 1.932 | 1.813 | 1.688 | 1.550 |
| 31 | 41.838 | 2.167 | 2.059 | 1.911 | 1.796 | 1.674 | 1.540 |
| 32 | 42.982 | 2.141 | 2.035 | 1.892 | 1.780 | 1.661 | 1.530 |
| 33 | 44.125 | 2.116 | 2.013 | 1.874 | 1.764 | 1.648 | 1.520 |
| 34 | 45.266 | 2.092 | 1.992 | 1.857 | 1.750 | 1.636 | 1.511 |
| 35 | 46.404 | 2.070 | 1.973 | 1.840 | 1.736 | 1.625 | 1.503 |
| 36 | 47.541 | 2.048 | 1.954 | 1.825 | 1.723 | 1.614 | 1.495 |
| 37 | 48.676 | 2.028 | 1.936 | 1.810 | 1.710 | 1.604 | 1.487 |
| 38 | 49.808 | 2.009 | 1.919 | 1.796 | 1.698 | 1.594 | 1.479 |
| 39 | 50.940 | 1.991 | 1.903 | 1.783 | 1.687 | 1.585 | 1.472 |

Table 13. **VALUES OF S AND ċ FOR DETERMINING SAMPLE SIZE FOR ACCEPTANCE SAMPLING; $p_1$ AND $p_2 \leqslant .25$** (*Continued*)

| ċ | m̄ | α | | | | | |
|---|---|---|---|---|---|---|---|
| | | 0.5% | 1% | 2.5% | 5% | 10% | 20% |
| | | | | S | | | |
| 40 | 52.070 | 1.974 | 1.887 | 1.770 | 1.676 | 1.576 | 1.465 |
| 41 | 53.198 | 1.957 | 1.873 | 1.757 | 1.666 | 1.568 | 1.459 |
| 42 | 54.324 | 1.941 | 1.859 | 1.746 | 1.656 | 1.559 | 1.452 |
| 43 | 55.449 | 1.926 | 1.845 | 1.734 | 1.646 | 1.551 | 1.446 |
| 44 | 56.573 | 1.911 | 1.832 | 1.724 | 1.637 | 1.544 | 1.440 |
| 45 | 57.695 | 1.897 | 1.820 | 1.713 | 1.628 | 1.536 | 1.435 |
| 46 | 58.816 | 1.884 | 1.808 | 1.703 | 1.619 | 1.529 | 1.429 |
| 47 | 59.936 | 1.871 | 1.796 | 1.694 | 1.611 | 1.523 | 1.424 |
| 48 | 61.054 | 1.859 | 1.785 | 1.684 | 1.603 | 1.516 | 1.419 |
| 49 | 62.171 | 1.847 | 1.775 | 1.675 | 1.596 | 1.510 | 1.414 |
| 50 | 63.287 | 1.835 | 1.764 | 1.667 | 1.588 | 1.504 | 1.409 |
| 55 | 68.851 | 1.784 | 1.718 | 1.628 | 1.555 | 1.476 | 1.387 |
| 60 | 74.390 | 1.740 | 1.679 | 1.594 | 1.526 | 1.452 | 1.369 |
| 65 | 79.907 | 1.702 | 1.645 | 1.565 | 1.501 | 1.431 | 1.352 |
| 70 | 85.405 | 1.669 | 1.615 | 1.540 | 1.479 | 1.413 | 1.338 |
| 75 | 90.885 | 1.640 | 1.589 | 1.518 | 1.460 | 1.397 | 1.325 |
| 80 | 96.350 | 1.614 | 1.566 | 1.498 | 1.443 | 1.382 | 1.314 |
| 90 | 107.24 | 1.570 | 1.526 | 1.464 | 1.413 | 1.357 | 1.294 |
| 100 | 118.08 | 1.534 | 1.493 | 1.435 | 1.388 | 1.336 | 1.277 |
| 110 | 128.88 | 1.503 | 1.465 | 1.411 | 1.367 | 1.319 | 1.263 |
| 120 | 139.64 | 1.477 | 1.441 | 1.391 | 1.349 | 1.303 | 1.251 |
| 130 | 150.38 | 1.454 | 1.421 | 1.373 | 1.334 | 1.290 | 1.240 |
| 150 | 171.76 | 1.417 | 1.386 | 1.343 | 1.307 | 1.268 | 1.222 |
| 170 | 193.08 | 1.387 | 1.359 | 1.320 | 1.287 | 1.250 | 1.208 |
| 200 | 224.89 | 1.352 | 1.327 | 1.291 | 1.262 | 1.229 | 1.190 |
| 250 | 277.63 | 1.309 | 1.288 | 1.257 | 1.231 | 1.202 | 1.169 |
| 300 | 330.11 | 1.279 | 1.259 | 1.232 | 1.209 | 1.183 | 1.153 |
| 350 | 382.39 | 1.255 | 1.238 | 1.213 | 1.192 | 1.169 | 1.141 |
| 400 | 434.51 | 1.237 | 1.221 | 1.198 | 1.179 | 1.157 | 1.131 |
| 450 | 486.50 | 1.222 | 1.207 | 1.186 | 1.168 | 1.147 | 1.123 |
| 500 | 538.39 | 1.209 | 1.195 | 1.175 | 1.158 | 1.139 | 1.117 |
| 600 | 641.89 | 1.189 | 1.177 | 1.159 | 1.144 | 1.127 | 1.106 |
| 700 | 745.12 | 1.174 | 1.163 | 1.146 | 1.132 | 1.117 | 1.0980 |
| 800 | 848.12 | 1.162 | 1.151 | 1.136 | 1.123 | 1.109 | 1.0914 |
| 900 | 950.94 | 1.152 | 1.142 | 1.128 | 1.116 | 1.102 | 1.0860 |
| 1000 | 1053.6 | 1.144 | 1.134 | 1.121 | 1.110 | 1.0968 | 1.0814 |
| 1100 | 1156.1 | 1.136 | 1.128 | 1.115 | 1.104 | 1.0921 | 1.0775 |
| 1200 | 1258.6 | 1.130 | 1.122 | 1.110 | 1.0996 | 1.0880 | 1.0741 |
| 1500 | 1565.3 | 1.116 | 1.108 | 1.0977 | 1.0886 | 1.0784 | 1.0661 |
| 2000 | 2075.1 | 1.0993 | 1.0931 | 1.0840 | 1.0763 | 1.0675 | 1.0570 |

| ċ | $\overline{m}$ | $\alpha$ | | | | | |
|---|---|---|---|---|---|---|---|
| | | 0.5% | 1% | 2.5% | 5% | 10% | 20% |
| | | | | S | | | |
| 2500 | 2583.8 | 1.0884 | 1.0829 | 1.0748 | 1.0680 | 1.0602 | 1.0508 |
| 3000 | 3091.7 | 1.0803 | 1.0754 | 1.0681 | 1.0619 | 1.0548 | 1.0463 |
| 3500 | 3598.9 | 1.0741 | 1.0696 | 1.0629 | 1.0572 | 1.0507 | 1.0428 |
| 4000 | 4105.6 | 1.0692 | 1.0649 | 1.0587 | 1.0534 | 1.0473 | 1.0400 |
| 4500 | 4611.9 | 1.0651 | 1.0611 | 1.0553 | 1.0503 | 1.0445 | 1.0377 |
| 5000 | 5117.9 | 1.0616 | 1.0579 | 1.0523 | 1.0476 | 1.0422 | 1.0357 |
| 6000 | 6129.0 | 1.0561 | 1.0527 | 1.0477 | 1.0434 | 1.0385 | 1.0326 |
| 7000 | 7139.2 | 1.0518 | 1.0487 | 1.0441 | 1.0401 | 1.0356 | 1.0301 |
| 8000 | 8148.7 | 1.0484 | 1.0455 | 1.0412 | 1.0375 | 1.0332 | 1.0281 |
| 9000 | 9157.6 | 1.0456 | 1.0428 | 1.0388 | 1.0353 | 1.0313 | 1.0265 |
| 10000 | 10166.1 | 1.0432 | 1.0406 | 1.0367 | 1.0334 | 1.0297 | 1.0251 |

# Table 13. VALUES OF S AND ċ FOR DETERMINING SAMPLE SIZE FOR ACCEPTANCE SAMPLING; $p_1$ AND $p_2$ ⩽ .25 (*Continued*)

| | | $\alpha$ | | | | | |
|---|---|---|---|---|---|---|---|
| | | 0.5% | 1% | 2.5% | 5% | 10% | 20% |
| ċ | $\overline{m}$ | | | | S | | |
| 0 | 2.3026 | 459.359 | 229.090 | 90.947 | 44.890 | 21.854 | 10.319 |
| 1 | 3.8897 | 37.584 | 26.184 | 16.059 | 10.946 | 7.314 | 4.718 |
| 2 | 5.3223 | 15.753 | 12.206 | 8.603 | 6.509 | 4.829 | 3.467 |
| 3 | 6.6808 | 9.939 | 8.115 | 6.130 | 4.890 | 3.829 | 2.909 |
| 4 | 7.9936 | 7.416 | 6.249 | 4.924 | 4.057 | 3.286 | 2.587 |
| 5 | 9.2747 | 6.035 | 5.195 | 4.212 | 3.549 | 2.943 | 2.376 |
| 6 | 10.532 | 5.170 | 4.520 | 3.742 | 3.206 | 2.704 | 2.225 |
| 7 | 11.771 | 4.578 | 4.050 | 3.408 | 2.957 | 2.528 | 2.111 |
| 8 | 12.995 | 4.148 | 3.705 | 3.158 | 2.768 | 2.392 | 2.021 |
| 9 | 14.206 | 3.822 | 3.440 | 2.962 | 2.618 | 2.283 | 1.949 |
| 10 | 15.407 | 3.565 | 3.229 | 2.806 | 2.497 | 2.194 | 1.889 |
| 11 | 16.598 | 3.358 | 3.058 | 2.677 | 2.397 | 2.120 | 1.838 |
| 12 | 17.782 | 3.187 | 2.915 | 2.569 | 2.312 | 2.057 | 1.794 |
| 13 | 18.958 | 3.043 | 2.795 | 2.477 | 2.240 | 2.002 | 1.756 |
| 14 | 20.128 | 2.920 | 2.692 | 2.398 | 2.177 | 1.954 | 1.723 |
| 15 | 21.292 | 2.814 | 2.603 | 2.328 | 2.122 | 1.912 | 1.693 |
| 16 | 22.452 | 2.721 | 2.524 | 2.267 | 2.073 | 1.875 | 1.667 |
| 17 | 23.606 | 2.640 | 2.455 | 2.213 | 2.029 | 1.841 | 1.643 |
| 18 | 24.756 | 2.567 | 2.393 | 2.164 | 1.990 | 1.811 | 1.621 |
| 19 | 25.903 | 2.502 | 2.337 | 2.120 | 1.954 | 1.783 | 1.602 |
| 20 | 27.045 | 2.443 | 2.287 | 2.080 | 1.922 | 1.758 | 1.584 |
| 21 | 28.184 | 2.390 | 2.241 | 2.044 | 1.892 | 1.735 | 1.567 |
| 22 | 29.320 | 2.342 | 2.200 | 2.011 | 1.865 | 1.714 | 1.552 |
| 23 | 30.453 | 2.297 | 2.162 | 1.980 | 1.840 | 1.694 | 1.537 |
| 24 | 31.584 | 2.257 | 2.126 | 1.952 | 1.817 | 1.676 | 1.524 |
| 25 | 32.711 | 2.219 | 2.094 | 1.926 | 1.795 | 1.659 | 1.512 |
| 26 | 33.836 | 2.184 | 2.064 | 1.902 | 1.775 | 1.643 | 1.500 |
| 27 | 34.959 | 2.152 | 2.036 | 1.879 | 1.757 | 1.628 | 1.489 |
| 28 | 36.080 | 2.122 | 2.009 | 1.858 | 1.739 | 1.614 | 1.479 |
| 29 | 37.199 | 2.094 | 1.985 | 1.838 | 1.723 | 1.601 | 1.469 |
| 30 | 38.315 | 2.067 | 1.962 | 1.819 | 1.707 | 1.589 | 1.460 |
| 31 | 39.430 | 2.042 | 1.940 | 1.801 | 1.692 | 1.577 | 1.451 |
| 32 | 40.543 | 2.019 | 1.920 | 1.785 | 1.679 | 1.566 | 1.443 |
| 33 | 41.654 | 1.997 | 1.900 | 1.769 | 1.665 | 1.556 | 1.435 |
| 34 | 42.764 | 1.976 | 1.882 | 1.754 | 1.653 | 1.546 | 1.428 |
| 35 | 43.872 | 1.957 | 1.865 | 1.740 | 1.641 | 1.536 | 1.421 |
| 36 | 44.978 | 1.938 | 1.848 | 1.727 | 1.630 | 1.527 | 1.414 |
| 37 | 46.083 | 1.920 | 1.833 | 1.714 | 1.619 | 1.519 | 1.408 |
| 38 | 47.187 | 1.903 | 1.818 | 1.701 | 1.609 | 1.510 | 1.401 |
| 39 | 48.289 | 1.887 | 1.804 | 1.690 | 1.599 | 1.503 | 1.396 |

Table 13.  VALUES OF S AND ċ FOR DETERMINING SAMPLE SIZE FOR
ACCEPTANCE SAMPLING; $p_1$ AND $p_2$ ≤ .25 (*Continued*)

| ċ | $\overline{m}$ | α | | | | | |
|---|---|---|---|---|---|---|---|
| | | 0.5% | 1% | 2.5% | 5% | 10% | 20% |
| | | | | | S | | |
| 40 | 49.390 | 1.872 | 1.790 | 1.679 | 1.590 | 1.495 | 1.390 |
| 41 | 50.490 | 1.857 | 1.777 | 1.668 | 1.581 | 1.488 | 1.384 |
| 42 | 51.589 | 1.843 | 1.765 | 1.658 | 1.572 | 1.481 | 1.379 |
| 43 | 52.686 | 1.830 | 1.753 | 1.648 | 1.564 | 1.474 | 1.374 |
| 44 | 53.783 | 1.817 | 1.742 | 1.639 | 1.556 | 1.468 | 1.369 |
| 45 | 54.878 | 1.805 | 1.731 | 1.630 | 1.548 | 1.461 | 1.365 |
| 46 | 55.972 | 1.793 | 1.720 | 1.621 | 1.541 | 1.455 | 1.360 |
| 47 | 57.066 | 1.782 | 1.710 | 1.612 | 1.534 | 1.450 | 1.356 |
| 48 | 58.158 | 1.771 | 1.701 | 1.604 | 1.527 | 1.444 | 1.351 |
| 49 | 59.249 | 1.760 | 1.691 | 1.597 | 1.521 | 1.439 | 1.347 |
| 50 | 60.340 | 1.750 | 1.682 | 1.589 | 1.514 | 1.434 | 1.343 |
| 55 | 65.779 | 1.704 | 1.641 | 1.555 | 1.485 | 1.410 | 1.326 |
| 60 | 71.199 | 1.665 | 1.607 | 1.526 | 1.461 | 1.390 | 1.310 |
| 65 | 76.602 | 1.631 | 1.577 | 1.501 | 1.439 | 1.372 | 1.297 |
| 70 | 81.990 | 1.602 | 1.551 | 1.479 | 1.420 | 1.357 | 1.285 |
| 75 | 87.365 | 1.576 | 1.527 | 1.459 | 1.403 | 1.343 | 1.274 |
| 80 | 92.727 | 1.553 | 1.507 | 1.442 | 1.388 | 1.330 | 1.264 |
| 90 | 103.42 | 1.514 | 1.471 | 1.412 | 1.363 | 1.309 | 1.248 |
| 100 | 114.07 | 1.482 | 1.442 | 1.387 | 1.341 | 1.291 | 1.234 |
| 110 | 124.70 | 1.454 | 1.418 | 1.366 | 1.323 | 1.276 | 1.222 |
| 120 | 135.29 | 1.431 | 1.396 | 1.348 | 1.307 | 1.263 | 1.212 |
| 130 | 145.87 | 1.411 | 1.378 | 1.332 | 1.294 | 1.251 | 1.203 |
| 150 | 166.95 | 1.377 | 1.348 | 1.306 | 1.271 | 1.232 | 1.188 |
| 170 | 187.97 | 1.350 | 1.323 | 1.285 | 1.253 | 1.217 | 1.176 |
| 200 | 219.38 | 1.319 | 1.294 | 1.260 | 1.231 | 1.199 | 1.161 |
| 250 | 271.52 | 1.280 | 1.259 | 1.229 | 1.204 | 1.176 | 1.143 |
| 300 | 323.45 | 1.253 | 1.234 | 1.207 | 1.185 | 1.159 | 1.130 |
| 350 | 375.22 | 1.232 | 1.215 | 1.190 | 1.170 | 1.147 | 1.120 |
| 400 | 426.88 | 1.215 | 1.199 | 1.177 | 1.158 | 1.137 | 1.112 |
| 450 | 478.43 | 1.202 | 1.187 | 1.166 | 1.148 | 1.128 | 1.105 |
| 500 | 529.90 | 1.190 | 1.177 | 1.157 | 1.140 | 1.121 | 1.0992 |
| 600 | 632.63 | 1.172 | 1.160 | 1.142 | 1.127 | 1.110 | 1.0902 |
| 700 | 735.15 | 1.158 | 1.147 | 1.131 | 1.117 | 1.102 | 1.0833 |
| 800 | 837.49 | 1.147 | 1.137 | 1.122 | 1.109 | 1.0948 | 1.0777 |
| 900 | 939.68 | 1.138 | 1.129 | 1.115 | 1.103 | 1.0892 | 1.0731 |
| 1000 | 1041.8 | 1.131 | 1.122 | 1.108 | 1.0971 | 1.0844 | 1.0693 |
| 1100 | 1143.7 | 1.124 | 1.116 | 1.103 | 1.0924 | 1.0803 | 1.0659 |
| 1200 | 1245.6 | 1.119 | 1.110 | 1.0984 | 1.0883 | 1.0768 | 1.0631 |
| 1500 | 1550.9 | 1.105 | 1.0981 | 1.0876 | 1.0786 | 1.0684 | 1.0562 |
| 2000 | 2058.5 | 1.0905 | 1.0844 | 1.0754 | 1.0677 | 1.0590 | 1.0485 |

# Table 13. VALUES OF S AND ċ FOR DETERMINING SAMPLE SIZE FOR ACCEPTANCE SAMPLING; $p_1$ AND $p_2$ ≤ .25 (*Continued*)

| | | | | | $\alpha$ | | |
|---|---|---|---|---|---|---|---|
| | | 0.5% | 1% | 2.5% | 5% | 10% | 20% |
| ċ | $\overline{m}$ | | | | S | | |
| 2500 | 2565.3 | 1.0806 | 1.0751 | 1.0671 | 1.0603 | 1.0526 | 1.0433 |
| 3000 | 3071.4 | 1.0733 | 1.0683 | 1.0611 | 1.0549 | 1.0479 | 1.0395 |
| 3500 | 3577.0 | 1.0676 | 1.0631 | 1.0564 | 1.0508 | 1.0443 | 1.0365 |
| 4000 | 4082.3 | 1.0631 | 1.0589 | 1.0527 | 1.0474 | 1.0414 | 1.0341 |
| 4500 | 4587.2 | 1.0594 | 1.0554 | 1.0496 | 1.0446 | 1.0389 | 1.0321 |
| 5000 | 5091.8 | 1.0562 | 1.0525 | 1.0470 | 1.0423 | 1.0369 | 1.0304 |
| 6000 | 6100.5 | 1.0512 | 1.0478 | 1.0428 | 1.0385 | 1.0336 | 1.0278 |
| 7000 | 7108.4 | 1.0473 | 1.0442 | 1.0396 | 1.0356 | 1.0311 | 1.0257 |
| 8000 | 8115.9 | 1.0442 | 1.0412 | 1.0370 | 1.0333 | 1.0291 | 1.0240 |
| 9000 | 9122.8 | 1.0416 | 1.0388 | 1.0348 | 1.0313 | 1.0274 | 1.0226 |
| 10000 | 10129.4 | 1.0394 | 1.0368 | 1.0330 | 1.0297 | 1.0260 | 1.0214 |

**Table 13. VALUES OF S AND ċ FOR DETERMINING SAMPLE SIZE FOR ACCEPTANCE SAMPLING; $p_1$ AND $p_2$ ≤ .25 (*Continued*)**

| ċ | m̄ | 0.5% | 1% | 2.5% | 5% | 10% | 20% |
|---|---|---|---|---|---|---|---|
| | | | | | S | | |
| 0 | 1.6094 | 321.079 | 160.127 | 63.569 | 31.377 | 15.275 | 7.213 |
| 1 | 2.9943 | 28.932 | 20.156 | 12.363 | 8.426 | 5.630 | 3.632 |
| 2 | 4.2790 | 12.665 | 9.813 | 6.916 | 5.233 | 3.883 | 2.788 |
| 3 | 5.5151 | 8.204 | 6.699 | 5.060 | 4.036 | 3.161 | 2.401 |
| 4 | 6.7210 | 6.235 | 5.254 | 4.140 | 3.411 | 2.763 | 2.175 |
| 5 | 7.9060 | 5.146 | 4.428 | 3.591 | 3.026 | 2.508 | 2.025 |
| 6 | 9.0754 | 4.455 | 3.895 | 3.225 | 2.762 | 2.330 | 1.917 |
| 7 | 10.233 | 3.980 | 3.521 | 2.963 | 2.570 | 2.198 | 1.835 |
| 8 | 11.380 | 3.633 | 3.244 | 2.765 | 2.424 | 2.095 | 1.770 |
| 9 | 12.519 | 3.368 | 3.031 | 2.611 | 2.307 | 2.012 | 1.717 |
| 10 | 13.651 | 3.159 | 2.861 | 2.486 | 2.213 | 1.944 | 1.674 |
| 11 | 14.777 | 2.989 | 2.722 | 2.383 | 2.134 | 1.887 | 1.636 |
| 12 | 15.897 | 2.849 | 2.607 | 2.297 | 2.067 | 1.839 | 1.604 |
| 13 | 17.013 | 2.731 | 2.508 | 2.223 | 2.010 | 1.797 | 1.576 |
| 14 | 18.125 | 2.629 | 2.424 | 2.159 | 1.960 | 1.760 | 1.552 |
| 15 | 19.233 | 2.542 | 2.351 | 2.103 | 1.916 | 1.727 | 1.530 |
| 16 | 20.338 | 2.465 | 2.287 | 2.054 | 1.878 | 1.698 | 1.510 |
| 17 | 21.439 | 2.397 | 2.229 | 2.010 | 1.843 | 1.672 | 1.492 |
| 18 | 22.538 | 2.337 | 2.179 | 1.970 | 1.811 | 1.649 | 1.476 |
| 19 | 23.634 | 2.283 | 2.133 | 1.935 | 1.783 | 1.627 | 1.461 |
| 20 | 24.728 | 2.234 | 2.091 | 1.902 | 1.757 | 1.608 | 1.448 |
| 21 | 25.819 | 2.190 | 2.053 | 1.873 | 1.734 | 1.590 | 1.435 |
| 22 | 26.909 | 2.149 | 2.019 | 1.846 | 1.712 | 1.573 | 1.424 |
| 23 | 27.996 | 2.112 | 1.987 | 1.821 | 1.692 | 1.558 | 1.413 |
| 24 | 29.082 | 2.078 | 1.958 | 1.798 | 1.673 | 1.543 | 1.403 |
| 25 | 30.166 | 2.046 | 1.931 | 1.776 | 1.656 | 1.530 | 1.394 |
| 26 | 31.248 | 2.017 | 1.906 | 1.756 | 1.640 | 1.518 | 1.385 |
| 27 | 32.329 | 1.990 | 1.882 | 1.738 | 1.625 | 1.506 | 1.377 |
| 28 | 33.408 | 1.965 | 1.860 | 1.720 | 1.610 | 1.495 | 1.369 |
| 29 | 34.486 | 1.941 | 1.840 | 1.704 | 1.597 | 1.485 | 1.362 |
| 30 | 35.563 | 1.919 | 1.821 | 1.688 | 1.584 | 1.475 | 1.355 |
| 31 | 36.638 | 1.898 | 1.803 | 1.674 | 1.573 | 1.466 | 1.349 |
| 32 | 37.712 | 1.878 | 1.786 | 1.660 | 1.561 | 1.457 | 1.342 |
| 33 | 38.785 | 1.860 | 1.769 | 1.647 | 1.551 | 1.449 | 1.336 |
| 34 | 39.857 | 1.842 | 1.754 | 1.635 | 1.541 | 1.441 | 1.331 |
| 35 | 40.928 | 1.825 | 1.740 | 1.623 | 1.531 | 1.433 | 1.325 |
| 36 | 41.998 | 1.810 | 1.726 | 1.612 | 1.522 | 1.426 | 1.320 |
| 37 | 43.067 | 1.795 | 1.713 | 1.602 | 1.513 | 1.419 | 1.315 |
| 38 | 44.135 | 1.780 | 1.700 | 1.591 | 1.505 | 1.413 | 1.311 |
| 39 | 45.203 | 1.767 | 1.689 | 1.582 | 1.497 | 1.406 | 1.306 |

Table 13. **VALUES OF S AND $\dot{c}$ FOR DETERMINING SAMPLE SIZE FOR ACCEPTANCE SAMPLING; $p_1$ AND $p_2 \leqslant .25$** (*Continued*)

| $\dot{c}$ | $\overline{m}$ | 0.5% | 1% | 2.5% | 5% | 10% | 20% |
|---|---|---|---|---|---|---|---|
| | | | | $\alpha$ — S | | | |
| 40 | 46.269 | 1.754 | 1.677 | 1.573 | 1.489 | 1.400 | 1.302 |
| 41 | 47.335 | 1.741 | 1.666 | 1.564 | 1.482 | 1.395 | 1.298 |
| 42 | 48.400 | 1.729 | 1.656 | 1.555 | 1.475 | 1.389 | 1.294 |
| 43 | 49.464 | 1.718 | 1.646 | 1.547 | 1.468 | 1.384 | 1.290 |
| 44 | 50.527 | 1.707 | 1.636 | 1.539 | 1.462 | 1.379 | 1.286 |
| 45 | 51.590 | 1.697 | 1.627 | 1.532 | 1.456 | 1.374 | 1.283 |
| 46 | 52.652 | 1.687 | 1.618 | 1.525 | 1.450 | 1.369 | 1.279 |
| 47 | 53.713 | 1.677 | 1.610 | 1.518 | 1.444 | 1.365 | 1.276 |
| 48 | 54.774 | 1.668 | 1.602 | 1.511 | 1.438 | 1.360 | 1.273 |
| 49 | 55.834 | 1.659 | 1.594 | 1.505 | 1.433 | 1.356 | 1.270 |
| 50 | 56.893 | 1.650 | 1.586 | 1.498 | 1.428 | 1.352 | 1.267 |
| 55 | 62.182 | 1.611 | 1.552 | 1.470 | 1.404 | 1.333 | 1.253 |
| 60 | 67.458 | 1.578 | 1.522 | 1.446 | 1.384 | 1.317 | 1.241 |
| 65 | 72.722 | 1.549 | 1.497 | 1.425 | 1.366 | 1.303 | 1.231 |
| 70 | 77.977 | 1.524 | 1.475 | 1.406 | 1.351 | 1.290 | 1.222 |
| 75 | 83.223 | 1.502 | 1.455 | 1.390 | 1.337 | 1.279 | 1.213 |
| 80 | 88.461 | 1.482 | 1.437 | 1.375 | 1.325 | 1.269 | 1.206 |
| 90 | 98.916 | 1.448 | 1.407 | 1.350 | 1.303 | 1.252 | 1.193 |
| 100 | 109.35 | 1.420 | 1.382 | 1.329 | 1.286 | 1.238 | 1.183 |
| 110 | 119.76 | 1.397 | 1.361 | 1.311 | 1.271 | 1.225 | 1.174 |
| 120 | 130.15 | 1.377 | 1.343 | 1.296 | 1.258 | 1.215 | 1.166 |
| 130 | 140.52 | 1.359 | 1.328 | 1.283 | 1.246 | 1.206 | 1.159 |
| 150 | 161.23 | 1.330 | 1.301 | 1.261 | 1.227 | 1.190 | 1.147 |
| 170 | 181.91 | 1.307 | 1.281 | 1.243 | 1.212 | 1.178 | 1.138 |
| 200 | 212.83 | 1.279 | 1.256 | 1.222 | 1.194 | 1.163 | 1.126 |
| 250 | 264.24 | 1.246 | 1.226 | 1.196 | 1.172 | 1.144 | 1.112 |
| 300 | 315.50 | 1.222 | 1.204 | 1.178 | 1.156 | 1.131 | 1.102 |
| 350 | 366.67 | 1.204 | 1.187 | 1.163 | 1.143 | 1.121 | 1.0941 |
| 400 | 417.76 | 1.189 | 1.174 | 1.152 | 1.133 | 1.112 | 1.0878 |
| 450 | 468.78 | 1.177 | 1.163 | 1.142 | 1.125 | 1.106 | 1.0825 |
| 500 | 519.74 | 1.167 | 1.154 | 1.135 | 1.118 | 1.0999 | 1.0782 |
| 600 | 621.54 | 1.152 | 1.140 | 1.122 | 1.107 | 1.0908 | 1.0711 |
| 700 | 723.19 | 1.139 | 1.128 | 1.113 | 1.0990 | 1.0838 | 1.0657 |
| 800 | 824.72 | 1.130 | 1.120 | 1.105 | 1.0923 | 1.0781 | 1.0613 |
| 900 | 926.17 | 1.122 | 1.112 | 1.0985 | 1.0868 | 1.0735 | 1.0577 |
| 1000 | 1027.5 | 1.115 | 1.106 | 1.0932 | 1.0822 | 1.0696 | 1.0547 |
| 1100 | 1128.8 | 1.110 | 1.101 | 1.0886 | 1.0782 | 1.0663 | 1.0521 |
| 1200 | 1230.1 | 1.105 | 1.0965 | 1.0847 | 1.0747 | 1.0633 | 1.0498 |
| 1500 | 1533.5 | 1.0930 | 1.0858 | 1.0754 | 1.0665 | 1.0565 | 1.0444 |
| 2000 | 2038.6 | 1.0799 | 1.0738 | 1.0649 | 1.0573 | 1.0487 | 1.0384 |

**Table 13. VALUES OF S AND ċ FOR DETERMINING SAMPLE SIZE FOR ACCEPTANCE SAMPLING; $p_1$ AND $p_2 \leqslant .25$ (*Continued*)**

| ċ | $\overline{m}$ | $\alpha$ 0.5% | 1% | 2.5% | 5% | 10% | 20% |
|---|---|---|---|---|---|---|---|
| | | | | S | | | |
| 2500 | 2543.0 | 1.0712 | 1.0657 | 1.0578 | 1.0511 | 1.0434 | 1.0342 |
| 3000 | 3047.0 | 1.0647 | 1.0598 | 1.0527 | 1.0466 | 1.0396 | 1.0312 |
| 3500 | 3550.7 | 1.0598 | 1.0552 | 1.0486 | 1.0430 | 1.0366 | 1.0289 |
| 4000 | 4054.1 | 1.0558 | 1.0516 | 1.0454 | 1.0402 | 1.0342 | 1.0270 |
| 4500 | 4557.4 | 1.0525 | 1.0485 | 1.0428 | 1.0378 | 1.0322 | 1.0254 |
| 5000 | 5060.4 | 1.0497 | 1.0460 | 1.0405 | 1.0359 | 1.0305 | 1.0241 |
| 6000 | 6066.1 | 1.0453 | 1.0419 | 1.0369 | 1.0327 | 1.0278 | 1.0220 |
| 7000 | 7071.3 | 1.0418 | 1.0387 | 1.0341 | 1.0302 | 1.0257 | 1.0203 |
| 8000 | 8076.2 | 1.0391 | 1.0362 | 1.0319 | 1.0282 | 1.0240 | 1.0190 |
| 9000 | 9080.8 | 1.0368 | 1.0340 | 1.0300 | 1.0266 | 1.0227 | 1.0179 |
| 10000 | 10085.1 | 1.0349 | 1.0323 | 1.0285 | 1.0252 | 1.0215 | 1.0170 |

# Table 14

**SQUARE ROOTS OF FRACTIONAL NUMBERS FROM .001 TO .999**

## Table 14. SQUARE ROOTS OF FRACTIONAL NUMBERS FROM .001 TO .999

| Number | Square Root | Number | Square Root | Number | Square Root |
|--------|-------------|--------|-------------|--------|-------------|
| 0.001 | 0.0316 | 0.046 | 0.2145 | 0.091 | 0.3017 |
| 0.002 | 0.0447 | 0.047 | 0.2168 | 0.092 | 0.3033 |
| 0.003 | 0.0548 | 0.048 | 0.2191 | 0.093 | 0.3050 |
| 0.004 | 0.0632 | 0.049 | 0.2214 | 0.094 | 0.3066 |
| 0.005 | 0.0707 | 0.050 | 0.2236 | 0.095 | 0.3082 |
| 0.006 | 0.0775 | 0.051 | 0.2258 | 0.096 | 0.3098 |
| 0.007 | 0.0837 | 0.052 | 0.2280 | 0.097 | 0.3114 |
| 0.008 | 0.0894 | 0.053 | 0.2302 | 0.098 | 0.3130 |
| 0.009 | 0.0949 | 0.054 | 0.2324 | 0.099 | 0.3146 |
| 0.010 | 0.1000 | 0.055 | 0.2345 | 0.100 | 0.3162 |
| 0.011 | 0.1049 | 0.056 | 0.2366 | 0.101 | 0.3178 |
| 0.012 | 0.1095 | 0.057 | 0.2387 | 0.102 | 0.3194 |
| 0.013 | 0.1140 | 0.058 | 0.2408 | 0.103 | 0.3209 |
| 0.014 | 0.1183 | 0.059 | 0.2429 | 0.104 | 0.3225 |
| 0.015 | 0.1225 | 0.060 | 0.2449 | 0.105 | 0.3240 |
| 0.016 | 0.1265 | 0.061 | 0.2470 | 0.106 | 0.3256 |
| 0.017 | 0.1304 | 0.062 | 0.2490 | 0.107 | 0.3271 |
| 0.018 | 0.1342 | 0.063 | 0.2510 | 0.108 | 0.3286 |
| 0.019 | 0.1378 | 0.064 | 0.2530 | 0.109 | 0.3302 |
| 0.020 | 0.1414 | 0.065 | 0.2550 | 0.110 | 0.3317 |
| 0.021 | 0.1449 | 0.066 | 0.2569 | 0.111 | 0.3332 |
| 0.022 | 0.1483 | 0.067 | 0.2588 | 0.112 | 0.3347 |
| 0.023 | 0.1517 | 0.068 | 0.2608 | 0.113 | 0.3362 |
| 0.024 | 0.1549 | 0.069 | 0.2627 | 0.114 | 0.3376 |
| 0.025 | 0.1581 | 0.070 | 0.2646 | 0.115 | 0.3391 |
| 0.026 | 0.1612 | 0.071 | 0.2665 | 0.116 | 0.3406 |
| 0.027 | 0.1643 | 0.072 | 0.2683 | 0.117 | 0.3421 |
| 0.028 | 0.1673 | 0.073 | 0.2702 | 0.118 | 0.3435 |
| 0.029 | 0.1703 | 0.074 | 0.2720 | 0.119 | 0.3450 |
| 0.030 | 0.1732 | 0.075 | 0.2739 | 0.120 | 0.3464 |
| 0.031 | 0.1761 | 0.076 | 0.2757 | 0.121 | 0.3479 |
| 0.032 | 0.1789 | 0.077 | 0.2775 | 0.122 | 0.3493 |
| 0.033 | 0.1817 | 0.078 | 0.2793 | 0.123 | 0.3507 |
| 0.034 | 0.1844 | 0.079 | 0.2811 | 0.124 | 0.3521 |
| 0.035 | 0.1871 | 0.080 | 0.2828 | 0.125 | 0.3536 |
| 0.036 | 0.1897 | 0.081 | 0.2846 | 0.126 | 0.3550 |
| 0.037 | 0.1924 | 0.082 | 0.2864 | 0.127 | 0.3564 |
| 0.038 | 0.1949 | 0.083 | 0.2881 | 0.128 | 0.3578 |
| 0.039 | 0.1975 | 0.084 | 0.2898 | 0.129 | 0.3592 |
| 0.040 | 0.2000 | 0.085 | 0.2915 | 0.130 | 0.3606 |
| 0.041 | 0.2025 | 0.086 | 0.2933 | 0.131 | 0.3619 |
| 0.042 | 0.2049 | 0.087 | 0.2950 | 0.132 | 0.3633 |
| 0.043 | 0.2074 | 0.088 | 0.2966 | 0.133 | 0.3647 |
| 0.044 | 0.2098 | 0.089 | 0.2983 | 0.134 | 0.3661 |
| 0.045 | 0.2121 | 0.090 | 0.3000 | 0.135 | 0.3674 |

**Table 14. SQUARE ROOTS OF FRACTIONAL NUMBERS FROM** .001 **TO** .999
(*Continued*)

| Number | Square Root | Number | Square Root | Number | Square Root |
|--------|-------------|--------|-------------|--------|-------------|
| 0.136 | 0.3688 | 0.181 | 0.4254 | 0.226 | 0.4754 |
| 0.137 | 0.3701 | 0.182 | 0.4266 | 0.227 | 0.4764 |
| 0.138 | 0.3715 | 0.183 | 0.4278 | 0.228 | 0.4775 |
| 0.139 | 0.3728 | 0.184 | 0.4290 | 0.229 | 0.4785 |
| 0.140 | 0.3742 | 0.185 | 0.4301 | 0.230 | 0.4796 |
| 0.141 | 0.3755 | 0.186 | 0.4313 | 0.231 | 0.4806 |
| 0.142 | 0.3768 | 0.187 | 0.4324 | 0.232 | 0.4817 |
| 0.143 | 0.3782 | 0.188 | 0.4336 | 0.233 | 0.4827 |
| 0.144 | 0.3795 | 0.189 | 0.4347 | 0.234 | 0.4837 |
| 0.145 | 0.3808 | 0.190 | 0.4359 | 0.235 | 0.4848 |
| 0.146 | 0.3821 | 0.191 | 0.4370 | 0.236 | 0.4858 |
| 0.147 | 0.3834 | 0.192 | 0.4382 | 0.237 | 0.4868 |
| 0.148 | 0.3847 | 0.193 | 0.4393 | 0.238 | 0.4879 |
| 0.149 | 0.3860 | 0.194 | 0.4405 | 0.239 | 0.4889 |
| 0.150 | 0.3873 | 0.195 | 0.4416 | 0.240 | 0.4899 |
| 0.151 | 0.3886 | 0.196 | 0.4427 | 0.241 | 0.4909 |
| 0.152 | 0.3899 | 0.197 | 0.4438 | 0.242 | 0.4919 |
| 0.153 | 0.3912 | 0.198 | 0.4450 | 0.243 | 0.4930 |
| 0.154 | 0.3924 | 0.199 | 0.4461 | 0.244 | 0.4940 |
| 0.155 | 0.3937 | 0.200 | 0.4472 | 0.245 | 0.4950 |
| 0.156 | 0.3950 | 0.201 | 0.4483 | 0.246 | 0.4960 |
| 0.157 | 0.3962 | 0.202 | 0.4494 | 0.247 | 0.4970 |
| 0.158 | 0.3975 | 0.203 | 0.4506 | 0.248 | 0.4980 |
| 0.159 | 0.3987 | 0.204 | 0.4517 | 0.249 | 0.4990 |
| 0.160 | 0.4000 | 0.205 | 0.4528 | 0.250 | 0.5000 |
| 0.161 | 0.4012 | 0.206 | 0.4539 | 0.251 | 0.5010 |
| 0.162 | 0.4025 | 0.207 | 0.4550 | 0.252 | 0.5020 |
| 0.163 | 0.4037 | 0.208 | 0.4561 | 0.253 | 0.5030 |
| 0.164 | 0.4050 | 0.209 | 0.4572 | 0.254 | 0.5040 |
| 0.165 | 0.4062 | 0.210 | 0.4583 | 0.255 | 0.5050 |
| 0.166 | 0.4074 | 0.211 | 0.4593 | 0.256 | 0.5060 |
| 0.167 | 0.4087 | 0.212 | 0.4604 | 0.257 | 0.5070 |
| 0.168 | 0.4099 | 0.213 | 0.4615 | 0.258 | 0.5079 |
| 0.169 | 0.4111 | 0.214 | 0.4626 | 0.259 | 0.5089 |
| 0.170 | 0.4123 | 0.215 | 0.4637 | 0.260 | 0.5099 |
| 0.171 | 0.4135 | 0.216 | 0.4648 | 0.261 | 0.5109 |
| 0.172 | 0.4147 | 0.217 | 0.4658 | 0.262 | 0.5119 |
| 0.173 | 0.4159 | 0.218 | 0.4669 | 0.263 | 0.5128 |
| 0.174 | 0.4171 | 0.219 | 0.4680 | 0.264 | 0.5138 |
| 0.175 | 0.4183 | 0.220 | 0.4690 | 0.265 | 0.5148 |
| 0.176 | 0.4195 | 0.221 | 0.4701 | 0.266 | 0.5158 |
| 0.177 | 0.4207 | 0.222 | 0.4712 | 0.267 | 0.5167 |
| 0.178 | 0.4219 | 0.223 | 0.4722 | 0.268 | 0.5177 |
| 0.179 | 0.4231 | 0.224 | 0.4733 | 0.269 | 0.5187 |
| 0.180 | 0.4243 | 0.225 | 0.4743 | 0.270 | 0.5196 |

| Number | Square Root | Number | Square Root | Number | Square Root |
|--------|-------------|--------|-------------|--------|-------------|
| 0.271 | 0.5206 | 0.316 | 0.5621 | 0.361 | 0.6008 |
| 0.272 | 0.5215 | 0.317 | 0.5630 | 0.362 | 0.6017 |
| 0.273 | 0.5225 | 0.318 | 0.5639 | 0.363 | 0.6025 |
| 0.274 | 0.5235 | 0.319 | 0.5648 | 0.364 | 0.6033 |
| 0.275 | 0.5244 | 0.320 | 0.5657 | 0.365 | 0.6042 |
| 0.276 | 0.5254 | 0.321 | 0.5666 | 0.366 | 0.6050 |
| 0.277 | 0.5263 | 0.322 | 0.5675 | 0.367 | 0.6058 |
| 0.278 | 0.5273 | 0.323 | 0.5683 | 0.368 | 0.6066 |
| 0.279 | 0.5282 | 0.324 | 0.5692 | 0.369 | 0.6075 |
| 0.280 | 0.5292 | 0.325 | 0.5701 | 0.370 | 0.6083 |
| 0.281 | 0.5301 | 0.326 | 0.5710 | 0.371 | 0.6091 |
| 0.282 | 0.5310 | 0.327 | 0.5718 | 0.372 | 0.6099 |
| 0.283 | 0.5320 | 0.328 | 0.5727 | 0.373 | 0.6107 |
| 0.284 | 0.5329 | 0.329 | 0.5736 | 0.374 | 0.6116 |
| 0.285 | 0.5339 | 0.330 | 0.5745 | 0.375 | 0.6124 |
| 0.286 | 0.5348 | 0.331 | 0.5753 | 0.376 | 0.6132 |
| 0.287 | 0.5357 | 0.332 | 0.5762 | 0.377 | 0.6140 |
| 0.288 | 0.5367 | 0.333 | 0.5771 | 0.378 | 0.6148 |
| 0.289 | 0.5376 | 0.334 | 0.5779 | 0.379 | 0.6156 |
| 0.290 | 0.5385 | 0.335 | 0.5788 | 0.380 | 0.6164 |
| 0.291 | 0.5394 | 0.336 | 0.5797 | 0.381 | 0.6173 |
| 0.292 | 0.5404 | 0.337 | 0.5805 | 0.382 | 0.6181 |
| 0.293 | 0.5413 | 0.338 | 0.5814 | 0.383 | 0.6189 |
| 0.294 | 0.5422 | 0.339 | 0.5822 | 0.384 | 0.6197 |
| 0.295 | 0.5431 | 0.340 | 0.5831 | 0.385 | 0.6205 |
| 0.296 | 0.5441 | 0.341 | 0.5840 | 0.386 | 0.6213 |
| 0.297 | 0.5450 | 0.342 | 0.5848 | 0.387 | 0.6221 |
| 0.298 | 0.5459 | 0.343 | 0.5857 | 0.388 | 0.6229 |
| 0.299 | 0.5468 | 0.344 | 0.5865 | 0.389 | 0.6237 |
| 0.300 | 0.5477 | 0.345 | 0.5874 | 0.390 | 0.6245 |
| 0.301 | 0.5486 | 0.346 | 0.5882 | 0.391 | 0.6253 |
| 0.302 | 0.5495 | 0.347 | 0.5891 | 0.392 | 0.6261 |
| 0.303 | 0.5505 | 0.348 | 0.5899 | 0.393 | 0.6269 |
| 0.304 | 0.5514 | 0.349 | 0.5908 | 0.394 | 0.6277 |
| 0.305 | 0.5523 | 0.350 | 0.5916 | 0.395 | 0.6285 |
| 0.306 | 0.5532 | 0.351 | 0.5925 | 0.396 | 0.6293 |
| 0.307 | 0.5541 | 0.352 | 0.5933 | 0.397 | 0.6301 |
| 0.308 | 0.5550 | 0.353 | 0.5941 | 0.398 | 0.6309 |
| 0.309 | 0.5559 | 0.354 | 0.5950 | 0.399 | 0.6317 |
| 0.310 | 0.5568 | 0.355 | 0.5958 | 0.400 | 0.6325 |
| 0.311 | 0.5577 | 0.356 | 0.5967 | 0.401 | 0.6332 |
| 0.312 | 0.5586 | 0.357 | 0.5975 | 0.402 | 0.6340 |
| 0.313 | 0.5595 | 0.358 | 0.5983 | 0.403 | 0.6348 |
| 0.314 | 0.5604 | 0.359 | 0.5992 | 0.404 | 0.6356 |
| 0.315 | 0.5612 | 0.360 | 0.6000 | 0.405 | 0.6364 |

Table 14. **SQUARE ROOTS OF FRACTIONAL NUMBERS FROM** .001 **TO** .999
(*Continued*)

| Number | Square Root | Number | Square Root | Number | Square Root |
|--------|-------------|--------|-------------|--------|-------------|
| 0.406 | 0.6372 | 0.451 | 0.6716 | 0.496 | 0.7043 |
| 0.407 | 0.6380 | 0.452 | 0.6723 | 0.497 | 0.7050 |
| 0.408 | 0.6387 | 0.453 | 0.6731 | 0.498 | 0.7057 |
| 0.409 | 0.6395 | 0.454 | 0.6738 | 0.499 | 0.7064 |
| 0.410 | 0.6403 | 0.455 | 0.6745 | 0.500 | 0.7071 |
| 0.411 | 0.6411 | 0.456 | 0.6753 | 0.501 | 0.7078 |
| 0.412 | 0.6419 | 0.457 | 0.6760 | 0.502 | 0.7085 |
| 0.413 | 0.6427 | 0.458 | 0.6768 | 0.503 | 0.7092 |
| 0.414 | 0.6434 | 0.459 | 0.6775 | 0.504 | 0.7099 |
| 0.415 | 0.6442 | 0.460 | 0.6782 | 0.505 | 0.7106 |
| 0.416 | 0.6450 | 0.461 | 0.6790 | 0.506 | 0.7113 |
| 0.417 | 0.6458 | 0.462 | 0.6797 | 0.507 | 0.7120 |
| 0.418 | 0.6465 | 0.463 | 0.6804 | 0.508 | 0.7127 |
| 0.419 | 0.6473 | 0.464 | 0.6812 | 0.509 | 0.7134 |
| 0.420 | 0.6481 | 0.465 | 0.6819 | 0.510 | 0.7141 |
| 0.421 | 0.6488 | 0.466 | 0.6826 | 0.511 | 0.7148 |
| 0.422 | 0.6496 | 0.467 | 0.6834 | 0.512 | 0.7155 |
| 0.423 | 0.6504 | 0.468 | 0.6841 | 0.513 | 0.7162 |
| 0.424 | 0.6512 | 0.469 | 0.6848 | 0.514 | 0.7169 |
| 0.425 | 0.6519 | 0.470 | 0.6856 | 0.515 | 0.7176 |
| 0.426 | 0.6527 | 0.471 | 0.6863 | 0.516 | 0.7183 |
| 0.427 | 0.6535 | 0.472 | 0.6870 | 0.517 | 0.7190 |
| 0.428 | 0.6542 | 0.473 | 0.6877 | 0.518 | 0.7197 |
| 0.429 | 0.6550 | 0.474 | 0.6885 | 0.519 | 0.7204 |
| 0.430 | 0.6557 | 0.475 | 0.6892 | 0.520 | 0.7211 |
| 0.431 | 0.6565 | 0.476 | 0.6899 | 0.521 | 0.7218 |
| 0.432 | 0.6573 | 0.477 | 0.6907 | 0.522 | 0.7225 |
| 0.433 | 0.6580 | 0.478 | 0.6914 | 0.523 | 0.7232 |
| 0.434 | 0.6588 | 0.479 | 0.6921 | 0.524 | 0.7239 |
| 0.435 | 0.6595 | 0.480 | 0.6928 | 0.525 | 0.7246 |
| 0.436 | 0.6603 | 0.481 | 0.6935 | 0.526 | 0.7253 |
| 0.437 | 0.6611 | 0.482 | 0.6943 | 0.527 | 0.7259 |
| 0.438 | 0.6618 | 0.483 | 0.6950 | 0.528 | 0.7266 |
| 0.439 | 0.6626 | 0.484 | 0.6957 | 0.529 | 0.7273 |
| 0.440 | 0.6633 | 0.485 | 0.6964 | 0.530 | 0.7280 |
| 0.441 | 0.6641 | 0.486 | 0.6971 | 0.531 | 0.7287 |
| 0.442 | 0.6648 | 0.487 | 0.6979 | 0.532 | 0.7294 |
| 0.443 | 0.6656 | 0.488 | 0.6986 | 0.533 | 0.7301 |
| 0.444 | 0.6663 | 0.489 | 0.6993 | 0.534 | 0.7308 |
| 0.445 | 0.6671 | 0.490 | 0.7000 | 0.535 | 0.7314 |
| 0.446 | 0.6678 | 0.491 | 0.7007 | 0.536 | 0.7321 |
| 0.447 | 0.6686 | 0.492 | 0.7014 | 0.537 | 0.7328 |
| 0.448 | 0.6693 | 0.493 | 0.7021 | 0.538 | 0.7335 |
| 0.449 | 0.6701 | 0.494 | 0.7029 | 0.539 | 0.7342 |
| 0.450 | 0.6708 | 0.495 | 0.7036 | 0.540 | 0.7348 |

## Table 14. SQUARE ROOTS OF FRACTIONAL NUMBERS FROM .001 TO .999
### (Continued)

| Number | Square Root | Number | Square Root | Number | Square Root |
|--------|-------------|--------|-------------|--------|-------------|
| 0.541 | 0.7355 | 0.586 | 0.7655 | 0.631 | 0.7944 |
| 0.542 | 0.7362 | 0.587 | 0.7662 | 0.632 | 0.7950 |
| 0.543 | 0.7369 | 0.588 | 0.7668 | 0.633 | 0.7956 |
| 0.544 | 0.7376 | 0.589 | 0.7675 | 0.634 | 0.7962 |
| 0.545 | 0.7382 | 0.590 | 0.7681 | 0.635 | 0.7969 |
| 0.546 | 0.7389 | 0.591 | 0.7688 | 0.636 | 0.7975 |
| 0.547 | 0.7396 | 0.592 | 0.7694 | 0.637 | 0.7981 |
| 0.548 | 0.7403 | 0.593 | 0.7701 | 0.638 | 0.7987 |
| 0.549 | 0.7409 | 0.594 | 0.7707 | 0.639 | 0.7994 |
| 0.550 | 0.7416 | 0.595 | 0.7714 | 0.640 | 0.8000 |
| 0.551 | 0.7423 | 0.596 | 0.7720 | 0.641 | 0.8006 |
| 0.552 | 0.7430 | 0.597 | 0.7727 | 0.642 | 0.8012 |
| 0.553 | 0.7436 | 0.598 | 0.7733 | 0.643 | 0.8019 |
| 0.554 | 0.7443 | 0.599 | 0.7740 | 0.644 | 0.8025 |
| 0.555 | 0.7450 | 0.600 | 0.7746 | 0.645 | 0.8031 |
| 0.556 | 0.7457 | 0.601 | 0.7752 | 0.646 | 0.8037 |
| 0.557 | 0.7463 | 0.602 | 0.7759 | 0.647 | 0.8044 |
| 0.558 | 0.7470 | 0.603 | 0.7765 | 0.648 | 0.8050 |
| 0.559 | 0.7477 | 0.604 | 0.7772 | 0.649 | 0.8056 |
| 0.560 | 0.7483 | 0.605 | 0.7778 | 0.650 | 0.8062 |
| 0.561 | 0.7490 | 0.606 | 0.7785 | 0.651 | 0.8068 |
| 0.562 | 0.7497 | 0.607 | 0.7791 | 0.652 | 0.8075 |
| 0.563 | 0.7503 | 0.608 | 0.7797 | 0.653 | 0.8081 |
| 0.564 | 0.7510 | 0.609 | 0.7804 | 0.654 | 0.8087 |
| 0.565 | 0.7517 | 0.610 | 0.7810 | 0.655 | 0.8093 |
| 0.566 | 0.7523 | 0.611 | 0.7817 | 0.656 | 0.8099 |
| 0.567 | 0.7530 | 0.612 | 0.7823 | 0.657 | 0.8106 |
| 0.568 | 0.7537 | 0.613 | 0.7829 | 0.658 | 0.8112 |
| 0.569 | 0.7543 | 0.614 | 0.7836 | 0.659 | 0.8118 |
| 0.570 | 0.7550 | 0.615 | 0.7842 | 0.660 | 0.8124 |
| 0.571 | 0.7556 | 0.616 | 0.7849 | 0.661 | 0.8130 |
| 0.572 | 0.7563 | 0.617 | 0.7855 | 0.662 | 0.8136 |
| 0.573 | 0.7570 | 0.618 | 0.7861 | 0.663 | 0.8142 |
| 0.574 | 0.7576 | 0.619 | 0.7868 | 0.664 | 0.8149 |
| 0.575 | 0.7583 | 0.620 | 0.7874 | 0.665 | 0.8155 |
| 0.576 | 0.7589 | 0.621 | 0.7880 | 0.666 | 0.8161 |
| 0.577 | 0.7596 | 0.622 | 0.7887 | 0.667 | 0.8167 |
| 0.578 | 0.7603 | 0.623 | 0.7893 | 0.668 | 0.8173 |
| 0.579 | 0.7609 | 0.624 | 0.7899 | 0.669 | 0.8179 |
| 0.580 | 0.7616 | 0.625 | 0.7906 | 0.670 | 0.8185 |
| 0.581 | 0.7622 | 0.626 | 0.7912 | 0.671 | 0.8191 |
| 0.582 | 0.7629 | 0.627 | 0.7918 | 0.672 | 0.8198 |
| 0.583 | 0.7635 | 0.628 | 0.7925 | 0.673 | 0.8204 |
| 0.584 | 0.7642 | 0.629 | 0.7931 | 0.674 | 0.8210 |
| 0.585 | 0.7649 | 0.630 | 0.7937 | 0.675 | 0.8216 |

## Table 14. SQUARE ROOTS OF FRACTIONAL NUMBERS FROM .001 TO .999
### (*Continued*)

| Number | Square Root | Number | Square Root | Number | Square Root |
|---|---|---|---|---|---|
| 0.676 | 0.8222 | 0.721 | 0.8491 | 0.766 | 0.8752 |
| 0.677 | 0.8228 | 0.722 | 0.8497 | 0.767 | 0.8758 |
| 0.678 | 0.8234 | 0.723 | 0.8503 | 0.768 | 0.8764 |
| 0.679 | 0.8240 | 0.724 | 0.8509 | 0.769 | 0.8769 |
| 0.680 | 0.8246 | 0.725 | 0.8515 | 0.770 | 0.8775 |
| 0.681 | 0.8252 | 0.726 | 0.8521 | 0.771 | 0.8781 |
| 0.682 | 0.8258 | 0.727 | 0.8526 | 0.772 | 0.8786 |
| 0.683 | 0.8264 | 0.728 | 0.8532 | 0.773 | 0.8792 |
| 0.684 | 0.8270 | 0.729 | 0.8538 | 0.774 | 0.8798 |
| 0.685 | 0.8276 | 0.730 | 0.8544 | 0.775 | 0.8803 |
| 0.686 | 0.8283 | 0.731 | 0.8550 | 0.776 | 0.8809 |
| 0.687 | 0.8289 | 0.732 | 0.8556 | 0.777 | 0.8815 |
| 0.688 | 0.8295 | 0.733 | 0.8562 | 0.778 | 0.8820 |
| 0.689 | 0.8301 | 0.734 | 0.8567 | 0.779 | 0.8826 |
| 0.690 | 0.8307 | 0.735 | 0.8573 | 0.780 | 0.8832 |
| 0.691 | 0.8313 | 0.736 | 0.8579 | 0.781 | 0.8837 |
| 0.692 | 0.8319 | 0.737 | 0.8585 | 0.782 | 0.8843 |
| 0.693 | 0.8325 | 0.738 | 0.8591 | 0.783 | 0.8849 |
| 0.694 | 0.8331 | 0.739 | 0.8597 | 0.784 | 0.8854 |
| 0.695 | 0.8337 | 0.740 | 0.8602 | 0.785 | 0.8860 |
| 0.696 | 0.8343 | 0.741 | 0.8608 | 0.786 | 0.8866 |
| 0.697 | 0.8349 | 0.742 | 0.8614 | 0.787 | 0.8871 |
| 0.698 | 0.8355 | 0.743 | 0.8620 | 0.788 | 0.8877 |
| 0.699 | 0.8361 | 0.744 | 0.8626 | 0.789 | 0.8883 |
| 0.700 | 0.8367 | 0.745 | 0.8631 | 0.790 | 0.8888 |
| 0.701 | 0.8373 | 0.746 | 0.8637 | 0.791 | 0.8894 |
| 0.702 | 0.8379 | 0.747 | 0.8643 | 0.792 | 0.8899 |
| 0.703 | 0.8385 | 0.748 | 0.8649 | 0.793 | 0.8905 |
| 0.704 | 0.8390 | 0.749 | 0.8654 | 0.794 | 0.8911 |
| 0.705 | 0.8396 | 0.750 | 0.8660 | 0.795 | 0.8916 |
| 0.706 | 0.8402 | 0.751 | 0.8666 | 0.796 | 0.8922 |
| 0.707 | 0.8408 | 0.752 | 0.8672 | 0.797 | 0.8927 |
| 0.708 | 0.8414 | 0.753 | 0.8678 | 0.798 | 0.8933 |
| 0.709 | 0.8420 | 0.754 | 0.8683 | 0.799 | 0.8939 |
| 0.710 | 0.8426 | 0.755 | 0.8689 | 0.800 | 0.8944 |
| 0.711 | 0.8432 | 0.756 | 0.8695 | 0.801 | 0.8950 |
| 0.712 | 0.8438 | 0.757 | 0.8701 | 0.802 | 0.8955 |
| 0.713 | 0.8444 | 0.758 | 0.8706 | 0.803 | 0.8961 |
| 0.714 | 0.8450 | 0.759 | 0.8712 | 0.804 | 0.8967 |
| 0.715 | 0.8456 | 0.760 | 0.8718 | 0.805 | 0.8972 |
| 0.716 | 0.8462 | 0.761 | 0.8724 | 0.806 | 0.8978 |
| 0.717 | 0.8468 | 0.762 | 0.8729 | 0.807 | 0.8983 |
| 0.718 | 0.8473 | 0.763 | 0.8735 | 0.808 | 0.8989 |
| 0.719 | 0.8479 | 0.764 | 0.8741 | 0.809 | 0.8994 |
| 0.720 | 0.8485 | 0.765 | 0.8746 | 0.810 | 0.9000 |

**Table 14. SQUARE ROOTS OF FRACTIONAL NUMBERS FROM** .001 **TO** .999
(*Continued*)

| Number | Square Root | Number | Square Root | Number | Square Root |
|--------|-------------|--------|-------------|--------|-------------|
| 0.811 | 0.9006 | 0.856 | 0.9252 | 0.901 | 0.9492 |
| 0.812 | 0.9011 | 0.857 | 0.9257 | 0.902 | 0.9497 |
| 0.813 | 0.9017 | 0.858 | 0.9263 | 0.903 | 0.9503 |
| 0.814 | 0.9022 | 0.859 | 0.9268 | 0.904 | 0.9508 |
| 0.815 | 0.9028 | 0.860 | 0.9274 | 0.905 | 0.9513 |
| 0.816 | 0.9033 | 0.861 | 0.9279 | 0.906 | 0.9518 |
| 0.817 | 0.9039 | 0.862 | 0.9284 | 0.907 | 0.9524 |
| 0.818 | 0.9044 | 0.863 | 0.9290 | 0.908 | 0.9529 |
| 0.819 | 0.9050 | 0.864 | 0.9295 | 0.909 | 0.9534 |
| 0.820 | 0.9055 | 0.865 | 0.9301 | 0.910 | 0.9539 |
| 0.821 | 0.9061 | 0.866 | 0.9306 | 0.911 | 0.9545 |
| 0.822 | 0.9066 | 0.867 | 0.9311 | 0.912 | 0.9550 |
| 0.823 | 0.9072 | 0.868 | 0.9317 | 0.913 | 0.9555 |
| 0.824 | 0.9077 | 0.869 | 0.9322 | 0.914 | 0.9560 |
| 0.825 | 0.9083 | 0.870 | 0.9327 | 0.915 | 0.9566 |
| 0.826 | 0.9088 | 0.871 | 0.9333 | 0.916 | 0.9571 |
| 0.827 | 0.9094 | 0.872 | 0.9338 | 0.917 | 0.9576 |
| 0.828 | 0.9099 | 0.873 | 0.9343 | 0.918 | 0.9581 |
| 0.829 | 0.9105 | 0.874 | 0.9349 | 0.919 | 0.9586 |
| 0.830 | 0.9110 | 0.875 | 0.9354 | 0.920 | 0.9592 |
| 0.831 | 0.9116 | 0.876 | 0.9359 | 0.921 | 0.9597 |
| 0.832 | 0.9121 | 0.877 | 0.9365 | 0.922 | 0.9602 |
| 0.833 | 0.9127 | 0.878 | 0.9370 | 0.923 | 0.9607 |
| 0.834 | 0.9132 | 0.879 | 0.9375 | 0.924 | 0.9612 |
| 0.835 | 0.9138 | 0.880 | 0.9381 | 0.925 | 0.9618 |
| 0.836 | 0.9143 | 0.881 | 0.9386 | 0.926 | 0.9623 |
| 0.837 | 0.9149 | 0.882 | 0.9391 | 0.927 | 0.9628 |
| 0.838 | 0.9154 | 0.883 | 0.9397 | 0.928 | 0.9633 |
| 0.839 | 0.9160 | 0.884 | 0.9402 | 0.929 | 0.9638 |
| 0.840 | 0.9165 | 0.885 | 0.9407 | 0.930 | 0.9644 |
| 0.841 | 0.9171 | 0.886 | 0.9413 | 0.931 | 0.9649 |
| 0.842 | 0.9176 | 0.887 | 0.9418 | 0.932 | 0.9654 |
| 0.843 | 0.9182 | 0.888 | 0.9423 | 0.933 | 0.9659 |
| 0.844 | 0.9187 | 0.889 | 0.9429 | 0.934 | 0.9664 |
| 0.845 | 0.9192 | 0.890 | 0.9434 | 0.935 | 0.9670 |
| 0.846 | 0.9198 | 0.891 | 0.9439 | 0.936 | 0.9675 |
| 0.847 | 0.9203 | 0.892 | 0.9445 | 0.937 | 0.9680 |
| 0.848 | 0.9209 | 0.893 | 0.9450 | 0.938 | 0.9685 |
| 0.849 | 0.9214 | 0.894 | 0.9455 | 0.939 | 0.9690 |
| 0.850 | 0.9220 | 0.895 | 0.9460 | 0.940 | 0.9695 |
| 0.851 | 0.9225 | 0.896 | 0.9466 | 0.941 | 0.9701 |
| 0.852 | 0.9230 | 0.897 | 0.9471 | 0.942 | 0.9706 |
| 0.853 | 0.9236 | 0.898 | 0.9476 | 0.943 | 0.9711 |
| 0.854 | 0.9241 | 0.899 | 0.9482 | 0.944 | 0.9716 |
| 0.855 | 0.9247 | 0.900 | 0.9487 | 0.945 | 0.9721 |

## Table 14. SQUARE ROOTS OF FRACTIONAL NUMBERS FROM .001 TO .999
*(Continued)*

| Number | Square Root | Number | Square Root | Number | Square Root |
|---|---|---|---|---|---|
| 0.946 | 0.9726 | 0.966 | 0.9829 | 0.986 | 0.9930 |
| 0.947 | 0.9731 | 0.967 | 0.9834 | 0.987 | 0.9935 |
| 0.948 | 0.9737 | 0.968 | 0.9839 | 0.988 | 0.9940 |
| 0.949 | 0.9742 | 0.969 | 0.9844 | 0.989 | 0.9945 |
| 0.950 | 0.9747 | 0.970 | 0.9849 | 0.990 | 0.9950 |
| 0.951 | 0.9752 | 0.971 | 0.9854 | 0.991 | 0.9955 |
| 0.952 | 0.9757 | 0.972 | 0.9859 | 0.992 | 0.9960 |
| 0.953 | 0.9762 | 0.973 | 0.9864 | 0.993 | 0.9965 |
| 0.954 | 0.9767 | 0.974 | 0.9869 | 0.994 | 0.9970 |
| 0.955 | 0.9772 | 0.975 | 0.9874 | 0.995 | 0.9975 |
| 0.956 | 0.9778 | 0.976 | 0.9879 | 0.996 | 0.9980 |
| 0.957 | 0.9783 | 0.977 | 0.9884 | 0.997 | 0.9985 |
| 0.958 | 0.9788 | 0.978 | 0.9889 | 0.998 | 0.9990 |
| 0.959 | 0.9793 | 0.979 | 0.9894 | 0.999 | 0.9995 |
| 0.960 | 0.9798 | 0.980 | 0.9899 | | |
| 0.961 | 0.9803 | 0.981 | 0.9905 | | |
| 0.962 | 0.9808 | 0.982 | 0.9910 | | |
| 0.963 | 0.9813 | 0.983 | 0.9915 | | |
| 0.964 | 0.9818 | 0.984 | 0.9920 | | |
| 0.965 | 0.9823 | 0.985 | 0.9925 | | |

# PART FOUR

# Derivation and Accuracy of Tables and Formulas for Confidence Limits[1]

Suppose a sample of size n is drawn randomly from an infinite population in which the probability of an event—an item having a designated characteristic—is p. Based on the observed number of events c in the sample, upper and lower confidence limits $\bar{p}$ and $\underline{p}$ for the parameter p may be obtained from the cumulative binomial distribution. The confidence level may be designated in general form as $\gamma$, as $\gamma_2$ where it applies to two-sided confidence limits of both $\bar{p}$ and $\underline{p}$, and as $\gamma_1$ where it applies to a one-sided confidence limit of only $\bar{p}$ or only $\underline{p}$.

Tables 1 and 2, respectively for approximation of $\bar{p}$ and $\underline{p}$, are primary tables inasmuch as they apply to $c/n \leqslant .5$. Tables 3 and 4, also respectively for approximation of $\bar{p}$ and $\underline{p}$, are secondary in that they are limited to $c/n \leqslant .1$ for $\bar{p}$ and to $c/n \leqslant .125$ for $\underline{p}$. However, the approximation formulas based on Tables 3 and 4 are simpler than those based on Tables 1 and 2; the simplification is considered worthwhile because many sampling situations deal with $c/n \leqslant .1$ or .125.

---

[1] References cited and symbols employed in this appendix appear in Appendixes E and F, respectively.

Tables 1 and 2, together with extension formulas in Table 7, apply to n $\geqslant$ 20. To deal with n $<$ 20, Tables 5 and 6, respectively provide exact values of $\overline{p}$ and $\underline{p}$ for such n and for c/n $\leqslant$ .5. Tables 3 and 4, together with extension formulas in Table 7, respectively apply to n $\geqslant$ 40 and n $\geqslant$ 16. Tables 1–7 cover not only all values of n but also all values of c/n. When c/n $>$ .5, r = n $-$ c is substituted for c to bring the sample proportion within the scope of the tables. This procedure is discussed in more detail later in this appendix.

Altogether, Tables 1–4 and 7 permit $\overline{p}$ and $\underline{p}$ to be approximated by three different methods. All such approximations are termed $\overline{p}'$ and $\underline{p}'$, denoting that the approximations have relative accuracy of at least .999. To distinguish among the methods, we may refer to approximations based on the *Long Formulas* (Tables 1 and 2), approximations based on the *Short Formulas* (Tables 3 and 4), and approximations based on *calculated constants* (Table 7).

Tables 1–4 provide approximation constants for calculating $\overline{p}'$ and $\underline{p}'$ when c $\leqslant$ 1,000. Table 7 provides extension formulas for calculating these constants when c $>$ 1,000. It may be desirable to have similar formulas for c $\leqslant$ 1,000, for example, if it is decided to write a simple computer program to produce approximate confidence limits. Therefore, Table 7 provides formulas and other data for calculating approximation constants applicable to c $\leqslant$ 1,000. However, for c $\leqslant$ 100, confidence limits based on Tables 1–3 generally tend to be somewhat more accurate than those based on Table 7, although the latter nevertheless maintain at least .999 relative accuracy. Confidence limits based on Table 4 are the same as those based on Table 7.

Development of the tables and approximation formulas began with secondary Tables 3 and 4, which led to Tables 1, 2, and 7. Therefore to facilitate explanation it is necessary to discuss the tables out of numerical order.

## UPPER CONFIDENCE LIMIT (TABLES 1, 3, AND 5)

The binomial cumulative distribution is approximated by the Poisson cumulative distribution. Therefore the upper binomial confidence limit $\overline{p}$ is approximated by $\overline{m}/n$, where $\overline{m}$ is the upper confidence limit for the parameter m of a Poisson distribution based on an observed number of items c having a designated characteristic; m $\simeq$ np, where n is sample size and p is the proportion of items in an infinite population having a specified characteristic. The formulas for approximating $\bar{p}$ all take the basic form of $\bar{m}/n$ times an adjustment factor.

### Upper Limit, Short Formula (Table 3)

In the range of values used for confidence levels, $\bar{m}/n > \bar{p}$ for any c [1]; that is, the Poisson approximation of an upper confidence limit exceeds the binomial confidence limit. For a given c, the ratio of the approximate to the exact confidence limit decreases as n increases. The systematic nature of the error of approximation suggests that a simple adjustment would give a greatly improved approximation. Such an adjustment may take the form of adding a constant $\bar{K}$ to the denominator of the Poisson approximation so that $\bar{p} \simeq \bar{m}(n + \bar{K})$. The statistical literature [1] gives as an approximation of $\bar{p}$:

$$\bar{p} \simeq \frac{\bar{m}}{n + (\bar{m} - c)/2} \tag{1}$$

Thus $\bar{K}$ may be expressed as $(\bar{m} - c)/2$, as in Formula (1).

With the objective that $\bar{p}'$, the approximation of $\bar{p}$, should have relative accuracy of at least .999 at confidence levels $\gamma_1$ (pertaining only to $\bar{p}'$) of 99.5% or less, it was empirically found [1] that Formula (1) is satisfactory for $c > 50$ and $c/n \leqslant .1$. That is,

$$\bar{p}' = \frac{\bar{m}}{n + (\bar{m} - c)/2} \qquad \begin{array}{c} c > 50 \\ c/n \leqslant .1 \end{array} \tag{2}$$

For $c \leqslant 50$, it is necessary to replace $\bar{K}$ with a slightly different value, $\bar{k}$, to maintain at least .999 relative accuracy for $c/n \leqslant .1$ and $n \geqslant 40$:

$$\bar{p}' = \frac{\bar{m}}{n + \bar{k}} \qquad \begin{array}{c} c/n \leqslant .1 \\ n \geqslant 40 \end{array} \tag{3}$$

A separate $\bar{k}$ is required for each c and each confidence level. It was empirically found that for $c/n \leqslant .1$ and $n \geqslant 40$, a satisfactory $\bar{k}$ can be obtained by solving for $\bar{k}$ in the expression $\bar{p} = \bar{m}/(n^* + \bar{k})$. For a given c, $n^* = 5/4$ of the minimum n (or the largest integer in $5/4$ of minimum n); minimum $n = 10c$ but not less than 40; $\bar{p}$ is the exact upper binomial confidence limit at a specified confidence level for the c and n* in question [6],[1] $\overline{m}$ is the Poisson limit at a specified confidence level for the c in question [8]. The solutions for $\overline{k}$ for c ⩽ 50 are given in Table 3. The values of $\overline{k}$ in Table 3 for c > 50 were calculated from

$$\overline{k} = \frac{\overline{m} - c}{2} \qquad c > 50 \qquad (4)$$

Because of the differences in method of calculating $\overline{k}$ for c ⩽ 50 and c > 50, there is a discontinuity between c = 50 and c = 51 in the steadily rising value of $\overline{k}$.

To construct and test Table 3, $\overline{m}$ for c ⩽ 160 was obtained from [8] by converting percentage points of the chi-square distribution into Poisson limits. For c > 160 an approximation was used [1]:

$$\overline{m}' = c + A\sqrt{c + 1} + B \qquad c ⩾ 50 \qquad (5)$$

where A is the number of standard deviations for a given confidence level based on the normal distribution and B = $(A^2 + 2)/3$. At $\gamma_1$ = 90% (confidence level pertaining only to $\overline{p}$) and for c = 160, the ratio $\overline{m}'/\overline{m}$ is about 1.0001. For c > 160 and for other confidence levels, this ratio is still closer to 1. Therefore the approximation $\overline{m}'$ is not a limiting factor in obtaining relative accuracy of at least .999 for $\overline{p}'$.

To construct and test Table 3, $\overline{p}$ for n ⩽ 500 was obtained from [6] by converting lower limits for r/n, where r = n − c, into upper limits for c/n. Also used for testing were a limited number of good approximations of $\overline{p}$ for n > 500 from [7].

Selected values of $\overline{p}'$ at each confidence level were calculated on the basis of Table 3 and compared with $\overline{p}$ from [6] and [7]. Systematic empirical analysis indicated that, with trivial exceptions, $\overline{p}'$ calculated on the basis of Table 3 (and its extension formulas) has relative accuracy of at least .999 for c/n ⩽ .1 and n ⩾ 40. The trivial exceptions to the relative accuracy of .999 for Table 3 are: For c = 4 and $\gamma_1$ = 99.5% (or $\gamma_2$ = 99%), the ratio $\overline{p}'/\overline{p}$ is 1.00124 for n = 40; 1.00106 for n = 41; and .99895 to .99899 for n = 91 to 95. For c = 4 and $\gamma_1$ = 99% (or $\gamma_2$ = 98%), $\overline{p}'/\overline{p}$ is 1.00102 for n = 40.

---

[1] For c > 40, n* exceeds 500, whereas [6] provides $\overline{p}$ only for n ⩽ 500. Therefore for c > 40, n of 500 instead of n* was used to calculate $\overline{k}$; this has insignificant effect on the accuracy of $\overline{p}'$.

If relative accuracy of at least .999 is desired for these few combinations of c, n, and $\gamma$, $\overline{p}'$ should be calculated on the basis of Table 1 instead of Table 3.

**Upper Limit, Long Formula (Table 1)**

The right-hand side of formula (3) may be written $(\overline{m}/n) \times n/(n + \overline{k})$. The second term is an adjustment factor applied to the crude approximation $\overline{m}/n$. This suggests that the desired relative accuracy of .999 for $\overline{p}'$ may be obtained for a greater range of c/n and of n by introducing an additional constant in the adjustment factor. Such a revised factor is $(n - \overline{a})/(n - \overline{b})$, leading to [1]:

$$\overline{p}' = \frac{\overline{m}}{n} \times \frac{n - \overline{a}}{n - \overline{b}} \qquad \begin{array}{l} c/n \leqslant .5 \\ n \geqslant 20 \end{array} \qquad (6)$$

When the adjustment factor is $(n - \overline{a})/(n - \overline{b})$, the maximum of the absolute value of the relative error of $\overline{p}'$ can be approximately minimized for a given c by solving a pair of simultaneous equations in $\overline{a}$ and $\overline{b}$ so that $\overline{p}' = \overline{p}$ for n = 2c and for n = 4c [1]. Using $\overline{m}$ from [8] and $\overline{p}$ from [6], this was done at each confidence level to produce the values of $\overline{a}$ and $\overline{b}$ in Table 1 corresponding to c $\leqslant$ 100. It was observed for c $\leqslant$ 100 that [1]

$$\overline{a} \simeq \frac{\overline{m}}{4} \qquad (7)$$

$$\overline{b} \simeq \overline{a} - \frac{\overline{m} - c}{2} \qquad (8)$$

Based on Formulas (7) and (8), the approximation of $\overline{b}$ may be written

$$\overline{b} \simeq \frac{2c - \overline{m}}{4} \qquad (9)$$

For c > 100, $\overline{a}$ and $\overline{b}$ in Table 1 were calculated from extension Formulas (7) and (9). Values of $\overline{m}$ in Table 1 were derived from [8] for c $\leqslant$ 160 and from Formula (5) for c > 160. Because of the differences in method of calculating $\overline{a}$ and $\overline{b}$ for c $\leqslant$ 100 and for c > 100, there is a discontinuity between c = 100 and c = 101 in the steadily rising values of $\overline{a}$ and $\overline{b}$.

Selected values of $\overline{p}'$ at each confidence level were calculated on the basis of Table 1 and were compared with $\overline{p}$ from [6] and good approximations of $\overline{p}$ from [7]. Systematic empirical analysis indicated that $\overline{p}'$ calculated on the basis of Table 1 (and its extension formulas) has relative accuracy of at least .999 for c/n ≤ .5 and n ≥ 20.

### Exact Upper Limit (Table 5)

It is difficult for a simple approximation to maintain .999 relative accuracy for n < 20. Therefore, Table 5 provides exact values of $\overline{p}$ for c/n ≤ .5 and n ≤ 20, obtained for n ≥ 4 from [6], and for n ≤ 3 from [8] by converting percentage points of the beta distribution into $\overline{p}$.

### LOWER CONFIDENCE LIMIT (TABLES 2, 4, AND 6)

Tables and formulas for approximating the lower binomial confidence limit were developed and tested along lines generally similar to those for the upper limit, so that the present discussion may be shortened by reference to the preceding discussion of the upper limit.

Development and testing of formulas and tables for the upper limit were greatly facilitated by [6], which gives exact $\overline{p}$ for n = 4 to 500 and for c/n ≤ .5. However, the values of $\underline{p}$ in [8] and [9] are much less extensive, and [7] only provides (good) approximations. Therefore a more difficult problem arose in dealing with the lower limit. Fortunately this problem was eased by the tendency of the approximation formulas to be more successful for the lower limit than for the upper limit. This appears true because accuracy of approximation increases as the interval between c/n and the confidence limit diminishes and because, binomial confidence limits being asymmetrical for c/n < .5, the interval between c/n and the lower limit is smaller than the interval between c/n and the upper limit.

### Lower Limit, Short Formula (Table 4)

For the lower binomial confidence limit with c/n less than about .1, the counterpart of Formula (2) would be $\underline{p}' = \underline{m}/[n - (c - \underline{m})/2]$, where $\underline{m}$ is the lower confidence limit for the parameter m of a Poisson distribution based on an observed number of items c having a designated characteristic. However, it was found that substantially greater accuracy is provided by a simple modification [2]:

$$\underline{p}' = \frac{m}{n - (c - \underline{m} - 1)/2} \qquad \begin{matrix} c/n \leqslant .125 \\ n \geqslant 16 \end{matrix} \qquad (10)$$

Systematic testing against $\underline{p}$ or good approximations of $\underline{p}$ from [7-9] showed that, at confidence levels $\gamma_1 = 99.5\%$ or less (or $\gamma_2 = 99\%$ or less), Formula (10) has relative accuracy of at least .999 for $c/n \leqslant .125$ and $n \geqslant 16$. Accordingly, the approximation formula for Table 4 is

$$\underline{p}' = \frac{m}{n - \underline{k}} \qquad \begin{matrix} c/n \leqslant .125 \\ n \geqslant 16 \end{matrix} \qquad (11)$$

and the values of $\underline{k}$ in Table 4 are calculated for a given c and $\gamma$ from

$$\underline{k} = \frac{c - \underline{m} - 1}{2} \qquad (12)$$

Values of $\underline{m}$ for $c \leqslant 160$ were obtained from [8]. For $c > 160$, an approximation was used [2]:

$$\underline{m}' = c - 1 - A\sqrt{c} + B \qquad c \geqslant 50 \qquad (13)$$

where A is the number of standard deviations for a given confidence level based on the normal distribution and $B = (A^2 + 2)/3$. At $\gamma_1 = 95\%$ (confidence level pertaining only to $\underline{p}$) and for $c = 160$, the ratio $\underline{m}'/\underline{m}$ is about .99985. For $c > 160$ and for other confidence levels, this ratio is still closer to 1. Therefore the approximation $\underline{m}'$ is not a limiting factor in achieving relative accuracy of at least .999 for $\underline{p}'$.

**Lower Limit, Long Formula (Table 2)**
For the lower binomial confidence limit and $c/n \leqslant .5$, the counterpart of Formula (6) is

$$\underline{p}' = \frac{m}{n} \times \frac{n - \underline{a}}{n - \underline{b}} \qquad \begin{matrix} c/n \leqslant .5 \\ n \geqslant 20 \end{matrix} \qquad (14)$$

Then the maximum of the absolute value of the relative error of $\underline{p}'$ can be approximately minimized for a given c by solving a pair of simultaneous equations in $\underline{a}$ and $\underline{b}$ so that $\underline{p}' = \underline{p}$ for $n = 2c$ and for n = 4c. However, this can be done for only a few values of c; while p for n = 2c $\leqslant$ 500 is readily available [6], p for n = 4c is less so [8], [9]. For these few values of c it was observed that [2]

$$\underline{a} \simeq \frac{\underline{m}}{3} \tag{15}$$

$$\underline{b} \simeq \underline{a} + \frac{c - 1 - \underline{m}}{2} \tag{16}$$

Based on Formulas (15) and (16), the approximation of $\underline{b}$ may be written

$$\underline{b} \simeq \frac{3c - 3 - \underline{m}}{6} \tag{17}$$

The values of $\underline{a}$ and $\underline{b}$ in Table 2 for c $\leqslant$ 100 were obtained by solving a pair of simultaneous equations in $\underline{a}$ and $\underline{b}$ so that, using Formulas (14) and (16), $p' = p$ for n = 2c and $\underline{b} = \underline{a} + (c - 1 - \underline{m})/2$. For c > 100, $\underline{a}$ and $\underline{b}$ were calculated from extension Formulas (15) and (17). Because of the differences in method of calculating $\underline{a}$ and $\underline{b}$ for c $\leqslant$ 100 and c > 100, there is a discontinuity between c = 100 and c = 101 in the steadily rising values of $\underline{a}$ and $\underline{b}$; however, the discontinuity is slight and is obvious only for $\gamma_1 = 99.5\%$ ($\gamma_2 = 99\%$).

Systematic testing against p or good approximations of p from [7–9] showed that Table 2 (and its extension formulas) permit calculation of p' with relative accuracy of at least .999 for c/n $\leqslant$ .5 and n $\geqslant$ 20.

### Exact Lower Limit (Table 6)

Table 6 provides exact values of p for c/n $\leqslant$ .5 and n $\leqslant$ 20, obtained from [8].

### FORMULAS FOR APPROXIMATION CONSTANTS (TABLE 7)

For c > 57, in the case of Tables 1 and 2, and for c > 50, in the case of Table 3, Formulas (4), (7), (9), (15), and (17), respectively, permit calculation of the approximation constants $\bar{k}$, $\bar{a}$, $\bar{b}$, $\underline{a}$, and $\underline{b}$

with sufficient accuracy to achieve at least .999 relative accuracy for $\overline{p}'$ or $\underline{p}'$. In the case of Table 4, Formula (12) permits sufficiently accurate calculation of $\underline{k}$ for any value of c. For $c \geqslant 50$, Formulas (5) and (13) permit sufficiently accurate approximation of $\overline{m}$ and $\underline{m}$.

Table 7 provides modifications of Formulas (4), (7), (9), (15), and (17) to permit relative accuracy of at least .999 for $\overline{p}'$ or $\underline{p}'$ when $c \leqslant 57$. No such modification is required for Formula (12). The modified formulas entail auxiliary terms f, g, and d, which appear in Tables 7—A to 7—C. Table 7—D gives the formula for $\underline{k}$ (or $\underline{k}'$), and Table 7—E gives the formulas for $\overline{m}'$ and $\underline{m}'$.

**Upper Limit, Long Formula (6)**

Investigation [1] revealed that for $c/n \leqslant .5$, $n \geqslant 20$, and $c > 38$, Formulas (7) and (9) produce approximation constants $\overline{a}$ and $\overline{b}$ providing at least .999 relative accuracy for $\overline{p}'$, and that for $c \leqslant 38$ it is necessary to introduce an auxiliary term f as in Formulas (18) and (19):

$$\overline{a}' = \frac{\overline{m}}{4} - f \tag{18}$$

$$\overline{b}' = \frac{2c - \overline{m}}{4} - f \tag{19}$$

Empirically determined values of f for each confidence level and for $c \leqslant 38$ appear in Table 7—A.

**Lower Limit, Long Formula (14)**

Investigation [2] revealed that for $c/n \leqslant .5$, $n \geqslant 20$, and $c > 57$, Formulas (15) and (17) produce approximation constants $\underline{a}$ and $\underline{b}$ providing at least .999 relative accuracy for $\underline{p}'$, and that for $c \leqslant 57$ it is necessary to introduce an auxiliary term g as in Formulas (20) and (21):

$$\underline{a}' = \frac{\underline{m}}{3} + g \tag{20}$$

$$\underline{b}' = \frac{3c - 3 - \underline{m}}{6} + g \tag{21}$$

Empirically determined values of g for each confidence level and for $c \leqslant 57$ appear in Table 7—B.

**Upper Limit, Short Formula (3)**

Investigation [1] revealed that for c/n ⩽ .1, n ⩾ 40, and c > 50 Formula (4) produces approximation constant $\bar{k}$ providing at least .999 relative accuracy for $\bar{p}'$, and that for c ⩽ 50 it is necessary to introduce an auxiliary term d as in Formula (22):

$$\bar{k}' = \frac{\bar{m} - c}{2} + d \tag{22}$$

Empirically determined values of d for each confidence level and for c ⩽ 50 appear in Table 7–C.

**Values of $\bar{m}$ and $\underline{m}$**

For given c and $\gamma$, the formulas for $\bar{p}'$ and $\underline{p}'$ require $\bar{m}$ and $\underline{m}$. Tables 1-4 give exact $\bar{m}$ and $\underline{m}$ for c ⩽ 160. For c > 160 the tables give approximations $\bar{m}'$ and $\underline{m}'$ based on Formulas (5) and (13), respectively.[1] These formulas have great accuracy, as previously noted, and therefore are not limiting factors in obtaining relative accuracy of at least .999 for $\bar{p}'$ and $\underline{p}'$ when used in conjunction with Formulas (18) to (22). In fact, as earlier noted, Formulas (5) and (13) may be used for c ⩾ 50.

**CONFIDENCE LIMITS FOR c/n > .5**

Tables 1-7 permit calculation of confidence limits for c/n > .5 [2]. In the case of the upper limit the basic procedure is to calculate r = n − c, obtain the lower limit $\underline{p}_r$ for r/n, and calculate

$$\bar{p} = 1 - \underline{p}_r \tag{23}$$

---

[1] Approximations for $\bar{m}$ and $\underline{m}$ slightly more accurate than those given by Formulas (5) and (13) are

$$\bar{m}' = (c + 1) \left[ 1 - \frac{1}{9(c + 1)} + A \sqrt{\frac{1}{9(c + 1)}} \right]^3 \tag{5A}$$

$$\underline{m}' = c \left( 1 - \frac{1}{9c} - A \sqrt{\frac{1}{9c}} \right)^3 \tag{13A}$$

Formulas (5) and (13) were derived [1], [2] by expanding Formulas (5A) and (13A), collecting terms, and dropping minor terms. Formulas (5A) and (13A) are suggested in F. Garwood, Fiducial Limits for the Poisson Distribution, *Biometrika*, vol. 28, p. 441, 1936; they are attributed there to Wilson and Hilferty (*Natl. Acad. Sci.*, vol. 17, no. 12, p. 684, 1931).

Similarly in the case of the lower limit the procedure is to calculate $r = n - c$, obtain the upper limit $\overline{p}_r$ for $r/n$, and calculate

$$\underline{p} = 1 - \overline{p}_r \tag{24}$$

The above procedures, employing approximation constants from Tables 1-4, maintain relative accuracy of at least .999 for $\overline{p}'$ and $\underline{p}'$, within the same bounds for $r$ and $r/n$ as previously noted for $c$ and $c/n$.

Using approximation constants derived from Table 7, relative accuracy of at least .999 is generally maintained, except for the following combination of circumstances: $\underline{p}'$ is desired, $c < 100$, $c/n > .5$, and $\underline{p} < .5$. Empirical analysis indicates that relative accuracy then approaches a minimum of .998 rather than .999 [2].

# Derivation and Accuracy of Tables and Formulas for Sample Size for an Estimate of a Proportion[1]

Tables 8–13 supply formulas and constants for determining sample size either to estimate a proportion or to engage in acceptance sampling. Tables 8–12 are primarily intended for proportion sampling; of these, Tables 8 and 9 also have use for quasi-forms of acceptance sampling where only risk $\beta$ (of accepting an unsatisfactory population rate) or only risk $\alpha$ (of rejecting a satisfactory population rate) is specified. Table 13 is exclusively for acceptance sampling, where both risks $\alpha$ and $\beta$ are specified. This appendix deals only with proportion sampling; Appendix C deals with acceptance sampling.

We assume that a sample of size n is to be drawn randomly from an infinite population in which the probability of an event (item having a specified characteristic) is p; based on the number of such events c in the sample, c/n is to be used as an estimate of p, and confidence limits $\overline{p}$ and/or $\underline{p}$ are to be calculated at confidence level $\gamma$ on the basis of c/n. We further assume that c/n $\leqslant$ .5; if in fact c/n $>$ .5, we substitute r/n for c/n, where r = n − c. (The upper and lower

---

[1]References cited and symbols employed in this appendix appear in Appendixes E and F, respectively.

confidence limits based on r/n are respectively the complements of the lower and upper limits based on c/n.)

Initially we assume that both the upper and lower confidence limits $\overline{p}$ and $\underline{p}$ are desired; later we deal with situations where only one limit is desired. We designate the confidence level as $\gamma_2$ when both limits are desired, and as $\gamma_1$ when only one limit is desired.

We assume that the statistician wants to determine a value of n that will provide a specified maximum error margin $\hat{e}$ at a specified confidence level $\gamma$. The length of a confidence interval $\overline{p} - \underline{p}$ is the sum of two error margins: $\overline{e} = \overline{p} - c/n$ and $\underline{e} = c/n - \underline{p}$. In the usual symmetrical approach to a two-sided confidence interval (equal confidence levels for $\overline{p}$ and $\underline{p}$), $\overline{e}$ is the larger error margin when the binomial distribution is used and c/n $<$ .5. We assume that the statistician in specifying a maximum error margin $\hat{e}$ has in mind the larger error margin $\overline{e}$ when both $\overline{p}$ and $\underline{p}$ are desired. Of course $\hat{e}$ pertains to $\overline{e}$ if only $\overline{p}$ is desired, or to $\underline{e}$ if only $\underline{p}$ is desired.

The binomial error margin e depends on the observed sample proportion c/n. For given n and $\gamma$, e generally increases with c/n, in most cases until c/n $=$ .5. As n grows large, it becomes increasingly true that e reaches a maximum at c/n $=$ .5; this is nearly true once n reaches about 100. Therefore, for practical purposes, a reasonably conservative approach is to determine n so that if c/n $=$ .5, e will not exceed $\hat{e}$.

However, information frequently exists about the proportion of events that may be anticipated in the sample. This information may come from knowledge about the population, experience with similar populations, a pilot sample, etc. If the anticipated proportion is appreciably below .5, then for specified $\hat{e}$ and $\gamma$ the sample size can be reduced. We denote the largest anticipated sample proportion by $\hat{p}$ (which may be .5).

We now introduce the following criterion of Appropriate Sample Size: If c/n $=$ $\hat{p}$, then e $=$ $\hat{e}$ (as nearly as integers for c and n permit).

The following discussion develops two sets of procedures that overall meet the criterion of Appropriate Sample Size with fairly good accuracy and at the same time are reasonably simple: a Poisson Procedure for $\hat{p} \leqslant$ .25, and a Modified Normal Procedure for $\hat{p} >$ .25. Furthermore, through a simple correction factor, a refined degree of accuracy can be obtained for n so that the criterion of Appropriate Sample Size is met exactly or very nearly. It was empirically found that use of different procedures for $\hat{p} \leqslant$ .25 and $\hat{p} >$ .25 serves to maximize the accuracy of n before applying the correction factor. The greater this preliminary accuracy, the more dispensable or else the more effective is the correction factor.

## POISSON PROCEDURE (FOR $\hat{p} \leqslant .25$) WHEN BOTH $\bar{p}$ AND p OR ONLY $\bar{p}$ ARE DESIRED (TABLE 8)

For specified $\hat{p}$, $\hat{e}$, and $\gamma_1$ or $\gamma_2$ and defining $\hat{\bar{p}}$ (anticipated value of $\bar{p}$) = $\hat{p} + \hat{e}$, a procedure for determining sample size [3] may be developed on the basis of a fairly good approximation for $\bar{p}$ given in Appendix A:

$$\bar{p} \simeq \frac{\bar{m}}{n + (\bar{m} - c)/2} \tag{1}$$

where $\bar{m}$ is the upper confidence limit for the parameter m of a Poisson distribution based on an observed number of events c. Rearrangement of terms leads to

$$\frac{\bar{m}}{c} \simeq \frac{\bar{p}(2 - c/n)}{(c/n)(2 - \bar{p})} \tag{25}$$

Assuming c/n = $\hat{p}$ and $\bar{p}$ = $\hat{\bar{p}}$, Formula (25) becomes

$$\frac{\bar{m}}{c} \simeq \frac{\hat{\bar{p}}(2 - \hat{p})}{\hat{p}(2 - \hat{\bar{p}})} \tag{26}$$

We may define

$$Q = \frac{\bar{m}}{c} \tag{27}$$

$$\hat{Q} = \frac{\hat{\bar{p}}(2 - \hat{p})}{\hat{p}(2 - \hat{\bar{p}})} \tag{28}$$

Table 8 shows for various values of Q and $\gamma_1$ or $\gamma_2$ the corresponding values of c. Assuming c/n = $\hat{p}$,

$$n = \frac{c}{\hat{p}} \qquad \hat{p} > 0 \tag{29}$$

Since $\hat{Q} \simeq Q$, the sample-size procedure is as follows: (1) Calculate $\hat{Q}$ from Formula (28). (2) In Table 8 for the specified $\gamma_1$ or $\gamma_2$ find Q nearest to $\hat{Q}$. Find c corresponding to Q. If necessary, interpolate linearly between values of Q bounding $\hat{Q}$ to find c; c is an integer. (3) Calculate n from Formula (29). Inasmuch as n is based on the Poisson distribution, this may be referred to as the Poisson Procedure for determining sample size.

The fact that c and n are integers may make it impossible for the sample to exactly meet the specifications for $\hat{p}$ and/or $\hat{e}$. Inability to meet specifications tends to increase as n and c become small, for rounding to an integer then has a relatively large effect. If desired, one may empirically adjust the sample size to meet the specification $\bar{e}$ as nearly as possible, to meet the specification $\hat{p}$ as nearly as possible, or to achieve a compromise that comes quite close to both $\hat{e}$ and $\hat{p}$.

Because c is an integer, Q in Table 8 can take only a limited number of values. In going from $\hat{Q}$ to the nearest tabular value of Q, the effect of rounding upward must be taken into account. The result of using Q higher than $\hat{Q}$ is to reduce c and thereby reduce sample size, with the possibility that $\bar{e} > \hat{e}$ and the criterion of appropriate sample is therefore not met. Hence it may be desirable to round *down* from $\hat{Q}$ to the nearest lower value of Q even though a higher value of Q is closer to $\hat{Q}$. This situation is particularly apt to occur when c is small. In the area of Table 8 where interpolation may have to be used to find c (i.e., where c is relatively large), and when c is adjusted to the nearest integer, it must similarly be recognized that sample size may become too small as the result of adjusting c downward. However, since c is relatively large, rounding c has relatively small effect on sample size.

The Poisson Procedure based on Table 8 has good accuracy overall. If we designate the Appropriate Sample Size as $n_a$, the relative error is $(n - n_a)/n_a$. The relative error (disregarding the effect of integers) is always in the conservative direction (positive); that is, Formulas (28) and (29) tend to overstate the sample size. Empirical analysis indicates that for $\gamma_2$ from 99 to 60% and for $n_a$ between 100 and 100,000, the maximum relative error is about 4% (at $n_a = 100$, $\hat{p} = .25$, and $\gamma_2 = 99\%$), and that for $n_a$ between 20 and 100 it is about 10% (at $n_a = 20$, $\hat{p} = .25$, and $\gamma_2 = 99\%$). The relative error approaches zero as $n_a$ increases, as $\hat{p}$ decreases, and as $\gamma$ decreases.

The accuracy of n may be tested by calculating $\bar{p}$ on the basis of c, n, and $\gamma$ and calculating $\bar{e} = \bar{p} - c/n$. If $\bar{e} = \hat{e}$, the criterion of

Appropriate Sample Size is met. Otherwise one may use Formula (31) below to obtain a corrected sample size n'.

Based on the normal distribution as an approximation of the binomial distribution, a formula frequently given for determining sample size is

$$n = \frac{A^2 \hat{p}\hat{q}}{\hat{e}^2} \qquad \hat{p} > 0 \qquad (30)$$

where A is the number of standard deviations for a given confidence level and $\hat{q} = 1 - \hat{p}$. Formula (30) suggests that sample size varies inversely with the square of the error margin. Given the error margin $\overline{e}$ for n and given the specified error margin $\hat{e}$ for a corrected sample size n', we may write $n'/n = \overline{e}^2/\hat{e}^2$, or

$$n' = n\left(\frac{\overline{e}}{\hat{e}}\right)^2 \qquad (31)$$

The preceding discussion assumes that $\gamma_2$ is used, namely, that after the sample is taken a confidence statement will be made about both the upper and lower confidence limits. If a statement is to be made only about the upper limit, reference is made to $\gamma_1$ rather than $\gamma_2$ in Table 8.

Values of $\overline{m}$ for construction of Table 8 are derived from [8] for $c \leqslant 160$ and from Formula (5) for $c > 160$. Values of $\overline{m}$ underlying Q in Table 8 contain more significant figures than $\overline{m}$ in Tables 1 and 3.

## POISSON PROCEDURE (FOR $\hat{p} \leqslant .25$) WHEN ONLY $\underline{p}$ IS DESIRED (TABLE 9)

The formulas and procedures relating to Table 9 are essentially analogous to those relating to Table 8. The anticipated proportion of events in the sample is $\hat{p}$; the specified confidence level is $\gamma_1$; the lower confidence limit is $\underline{p}$; the error margin is $\underline{e} = c/n - \underline{p}$; the anticipated (specified) error margin is $\hat{e}$; and the anticipated lower confidence limit is $\underline{\hat{p}} = \hat{p} - \hat{e}$. By the criterion of Appropriate Sample Size, $\underline{e} = \hat{e}$ if $c/n = \hat{p}$.

The sample-size procedure is based on Formula (32), which is a less accurate but more easily manipulated version of Formula (10):

$$\underline{p} \simeq \frac{\underline{m}}{n - (c - \underline{m})/2} \qquad (32)$$

Assuming $c/n = \hat{p}$ and $\underline{p} = \hat{p}$, this leads to

$$\frac{m}{c} \simeq \frac{\hat{p}(2 - \hat{p})}{\underline{\hat{p}}(2 - \underline{\hat{p}})} \tag{33}$$

We may define

$$R = \frac{m}{c} \tag{34}$$

$$\hat{R} = \frac{\hat{p}(2 - \hat{p})}{\underline{\hat{p}}(2 - \underline{\hat{p}})} \tag{35}$$

Table 9 shows for various values of R and $\gamma_1$ the corresponding values of c. Assuming $c/n = \hat{p}$,

$$n = \frac{c}{\hat{p}} \qquad \hat{p} > 0 \tag{29}$$

Since $R \simeq \hat{R}$, the sample-size procedure is as follows: (1) Calculate $\hat{R}$ from Formula (35). (2) In Table 9 for the specified $\gamma_1$ find R nearest to $\hat{R}$. Find c corresponding to R. If necessary, interpolate linearly between values of R bounding $\hat{R}$ to find c; c is an integer. (3) Calculate n from Formula (29).

The Poisson Procedure based on Table 9 has good accuracy overall. The relative error (disregarding the effect of integers) is always in the conservative direction; that is, Formulas (35) and (29) tend to overstate the sample size. Empirical analysis indicates that for $\gamma_1$ from 99.5% to 80% and for $n_a$ (Appropriate Sample Size) between 100 and 100,000 the maximum relative error is about 3% (at $n_a = 100,000$, $\hat{p} = .25$, and $\gamma_1 = 99.5\%$) and that for $n_a$ between 20 and 100 it is substantially zero. The relative error approaches zero as $n_a$, $\hat{p}$, and $\gamma_1$ decrease.

The accuracy of n may be tested by calculating $\underline{p}$ on the basis of c, n, and $\gamma$ and calculating $\underline{e} = c/n - \underline{p}$. If $\underline{e} = \hat{e}$, the criterion of Appropriate Sample Size is met. Otherwise one may use Formula 36—obtained in the same manner as Formula (31) on the basis of Formula (30)—to produce a corrected sample size n':

$$n' = n\left(\frac{\underline{e}}{\hat{e}}\right)^2 \tag{36}$$

Problems arising from the use of integers for c and n are analogous to those already discussed in connection with Table 8. Values of $\underline{m}$ for construction of Table 9 are derived from [8] for $c \leqslant 160$ and from Formula (13) for $c > 160$. Values of $\underline{m}$ underlying R in Table 9 contain more significant figures than $\underline{m}$ in Tables 2 and 4.

## ALTERNATIVE POISSON PROCEDURES (FOR $\hat{p} \leqslant .25$, TABLE 11)

In place of Table 8, Formula (38) below may be used to determine integer c [3]. Otherwise the Poisson Procedure is the same as already described in connection with Table 8.

Formula (5) in Appendix A gives as an approximation $\overline{m} = c + A\sqrt{c + 1} + B$, where A is the number of standard deviations for a given confidence level based on the normal distribution and $B = (A^2 + 2)/3$. Thus we may write as an approximation $\overline{m}/c = Q = (c + A\sqrt{c + 1} + B)/c$, which transforms into $(Q - 1)(c + 1) - A\sqrt{c + 1} - (Q + B - 1) = 0$. Quadratic solution for c yields

$$c = \left[ \frac{A + \sqrt{A^2 + 4(Q - 1)(Q + B - 1)}}{2(Q - 1)} \right]^2 - 1 \qquad p > 0 \ (37)$$

Assuming $c/n = \hat{p}$ and $\overline{p} = \hat{\overline{p}}$, Formula (37) may be written

$$c = \left[ \frac{A + \sqrt{A^2 + 4(\hat{Q} - 1)(\hat{Q} + B - 1)}}{2(\hat{Q} - 1)} \right]^2 - 1 \qquad p > 0 \ (38)$$

When the confidence level is $\gamma_2$ (both $\overline{p}$ and $\underline{p}$ desired), A in Formula (38) designates the number of standard deviations based on two tails of the normal distribution. When the confidence level is $\gamma_1$ (only $\overline{p}$ desired), A designates the number of standard deviations based on one tail of the normal distribution. Values of $A, A^2$, and B for $\gamma_1$ and $\gamma_2$ are given in Table 11. Formula (38) has high accuracy in the sense that, for specified $\hat{p}$, $\hat{e}$, and $\gamma$, it yields substantially the same values of c as does Table 8.

In place of Table 9, Formula (39) may be used to determine integer c. Otherwise the Poisson Procedure is the same as already described in connection with Table 9.

$$c = \left[ \frac{A + \sqrt{A^2 - 4(B - 1)(1 - \hat{R})}}{2(1 - \hat{R})} \right]^2 \qquad \hat{p} > 0 \qquad (39)$$

Formula (39) is based on the approximation for $\overline{m}$ given by Formula (13) and is derived in essentially the same way as Formula (38). Inasmuch as Formula (39) pertains only to the lower confidence limit, A is the number of standard deviations for a given confidence level based on one tail of the normal distribution. Formula (39) has high accuracy in the sense that, for specified $\hat{p}$, $\hat{e}$, and $\gamma$, it yields substantially the same values of c as does Table 9.

The term inside the radical in Formula (39) may be negative in a very few cases: when $\gamma_1 = 99.5\%$ and c should be 1, 2, or 3, and when $\gamma_1 = 99\%$ and c should be 1 or 2. Hence c should be obtained from Table 9 when the term inside the radical is negative.

### POISSON PROCEDURE FOR $\hat{p} = 0$ (TABLE 12)

Formulas (29) and (38) presume that $\hat{p} > 0$. If $\hat{p} = 0$, only the upper confidence limit is in point and Formula (40) below is used. Assuming $c/n = \hat{p} = 0$ and $\overline{p} = \hat{e}$, rearrangement of terms in Formula (1) leads to [3]

$$n = \frac{\overline{m}}{\hat{e}} - \frac{\overline{m}}{2} \qquad \hat{p} = 0 \qquad (40)$$

Table 12 gives values of $\overline{m}$ for $c = 0$ (corresponding to $\hat{p} = 0$) at several confidence levels.

### MODIFIED NORMAL PROCEDURE (FOR $\hat{p} > .25$, TABLE 10)

If we presume that the value of the parameter p is $\overline{p}$, we can write as an approximation, based on the normal distribution with a continuity correction [3], [5],

$$\overline{p} = \frac{c}{n} + A\sqrt{\frac{\overline{p}\,\overline{q}}{n}} + \frac{1}{2n} \qquad (41)$$

where $\overline{q} = 1 - \overline{p}$. Letting $\overline{e} = \overline{p} - c/n$, Formula (41) becomes

$$\overline{e} = A\sqrt{\frac{\overline{p}\,\overline{q}}{n}} + \frac{1}{2n} \qquad (42)$$

Transferring the term $1/2n$ to the left side of Formula (42), squaring both sides, and rearranging terms leads to

$$n = \frac{A^2 \overline{p}\,\overline{q} + \overline{e}}{\overline{e}^2} - \frac{1}{4n\overline{e}^2} \tag{43}$$

The relative effect of $1/4n\overline{e}^2$ on sample size is measured by $(1/4n\overline{e}^2)/n = 1/4n^2\,\overline{e}^2$. Inspection indicates that $1/4n^2\,\overline{e}^2$ is largest when $\gamma_2$, $c/n$, and $n$ are minimal. Within the intended scope of the Modified Normal Procedure, the minimal value is 60% for $\gamma_2$ and .25 for $c/n$, and we may assume a practical lower limit of 20 for $n$. Accordingly in Formula (41) we insert .84162 for A, .25 for $c/n$, and 20 for n, yielding $1.03541\overline{p}^2 - .58542\overline{p} + .07563 = 0$. Quadratic solution yields $\overline{p} = .36563$; thus $\overline{e} = .36563 - .25 = .11563$. With $n = 20$ and $\overline{e} = .11563, 1/4n\overline{e}^2$ is about 1, and $1/4n^2\,\overline{e}^2$ is about 5%. Solution for other values of n (including n below 20) shows that $1/4n\overline{e}^2$ ranges from a low of about .5 to a maximum of about 2 for $\gamma_2 = 60\%$ and $c/n = .25$. Hence $1/4n^2\,\overline{e}^2$ diminishes almost inversely in proportion to n; by the time n is 100, $1/4n^2\,\overline{e}^2$ is only about 1% for $\gamma_2 = 60\%$ and $c/n = .25$. At higher confidence levels and higher values of $c/n$, $1/4n^2\,\overline{e}^2$ is still smaller than indicated.

In view of the small effect of $1/4n\overline{e}^2$ on sample size, this term may be dropped from Formula (43). If at the same time we substitute the anticipated values $\hat{p}, \hat{q}$, and $\hat{e}$ for $\overline{p}, \overline{q}$, and $\overline{e}$, Formula (43) becomes [3]

$$n = \frac{A^2 \hat{p}\,\hat{q} + \hat{e}}{\hat{e}^2} \tag{44}$$

When the confidence level is $\gamma_2$ (both $\overline{p}$ and $\underline{p}$ desired), A in Formula (44) designates the number of standard deviations based on two tails of the normal distribution. When the confidence level is $\gamma_1$ (only $\overline{p}$ desired), A designates the number of standard deviations based on one tail of the normal distribution. Values of A for $\gamma_1$ and $\gamma_2$ are given in Table 10.

In similar fashion, if we presume that the value of the parameter p is $\underline{p}$, we obtain

$$n = \frac{A^2 \hat{p}\,\hat{q} + \hat{e}}{\hat{e}^2} \tag{45}$$

where $\hat{q} = 1 - \hat{p}$. Inasmuch as Formula (45) pertains only to the lower confidence limit, A is the number of standard deviations for a given confidence level based on one tail of the normal distribution.

Formula (44) has good accuracy overall (for $\hat{p} > .25$). The relative error—$(n - n_a)/n_a$—may be positive or negative, representing over-statement or understatement of sample size, respectively. Systematic empirical analysis indicates that for $\gamma_1$ between 99.5% and 80% (or $\gamma_2$ between 99% and 60%) and for $n_a$ between 100 and 100,000 the relative error is between approximately −3% and +5% and that for $n_a$ between 20 and 100 it is between approximately −11% and +5%.

Formula (45) is somewhat less accurate than Formula (44) but is nevertheless quite serviceable for the great bulk of practical sample sizes and confidence levels most frequently used. The relative error may be positive or negative. Systematic empirical analysis indicates that for $\gamma_1$ between 99.5% and 80% and for $n_a$ between 100 and 100,000 the relative error is between approximately −10% and +3%, and that for $n_a$ between 20 and 100 it is between approximately −25% and +11%.

# Derivation and Accuracy of Tables and Formulas for Sample Size for Acceptance Sampling[1]

Tables 8, 9, and 13, supplemented by Tables 11 and 12, provide formulas and other data for acceptance sampling. In addition, required values of $\overline{m}$ and $\underline{m}$ are provided by Tables 1–4 and 7–E. Tables 8 and 9 are primarily intended for proportion sampling but also serve for quasi-forms of acceptance sampling, where only risk $\beta$ (accepting an unsatisfactory population rate $p_2$) or only risk $\alpha$ (rejecting a satisfactory population rate $p_1$) is specified. Table 13 is exclusively for acceptance sampling proper, where both risk $\alpha$ and risk $\beta$ are specified.

## ACCEPTANCE SAMPLING WITH RISKS $\alpha$ AND $\beta$ SPECIFIED (TABLE 13)

Acceptance sampling deals with the proportion p of error items (containing a designated error or defect) in the population. Based on the number of error items c in the sample, the population is either

[1] References cited and symbols employed in this appendix appear in Appendixes E and F, respectively.

accepted as satisfactory or rejected as unsatisfactory. One specifies $p_2$ as an unsatisfactory rate and $\beta$ as the risk (probability) of accepting a population with an error rate $p_2$; and one specifies another error rate $p_1$ as a satisfactory rate and $\alpha$ as the risk of rejecting a population with an error rate $p_1$. Sample size n and acceptance number $\dot{c}$ are chosen to meet these specifications. If the sample contains $\dot{c}$ or fewer error items, the population is accepted as satisfactory. Defining $\dot{c}_1 = \dot{c} + 1$, the population is rejected if the sample contains $\dot{c}_1$ or more error items; $\dot{c}_1$ is the rejection number.

For a sample of suitable size n drawn from a population with error rate p, there is probability $\beta$ of getting $\dot{c}$ or fewer error items. From the viewpoint of acceptance sampling, if $p = p_2$, there is $\beta$ risk the sample will contain $\dot{c}$ or fewer error items. From the viewpoint of proportion sampling, if there are $\dot{c}$ error items in a sample of n, it is inferred that $\overline{p}$ is the upper confidence limit at $\gamma_1 = 1 - \beta$. Thus the relationship between $\dot{c}$ and $\overline{p}$ corresponds to that between $\dot{c}$ and $p_2$. If n contains $\dot{c}$ error items and if the confidence level $\gamma_1$ is the complement of risk $\beta$, the upper confidence limit $\overline{p}$ has the same value as the unsatisfactory error rate $p_2$. Similarly, if n contains $\dot{c}_1$ error items and if the confidence level $\gamma_1$ is the complement of risk $\alpha$, the lower confidence limit has the same value as the satisfactory error rate $p_1$. The lower confidence limit based on $\dot{c}_1/n$ may be designated $\underline{p}_1$. And we may designate $\underline{m}_1$ as pertaining to $\dot{c}_1$.

Previous formulas have used the general term c, which includes the specific value $\dot{c}$. In the following development, c and $\dot{c}$ are considered equivalent.

Based on Formula (10),

$$\underline{p}_1 \simeq \frac{\underline{m}_1}{n - (\dot{c}_1 - \underline{m}_1 - 1)/2} \tag{46}$$

Substituting $\dot{c}$ for $\dot{c}_1 - 1$, this becomes

$$\underline{p}_1 \simeq \frac{\underline{m}_1}{n - (\dot{c} - \underline{m}_1)/2} \tag{47}$$

Rearrangement of terms leads to

$$\frac{\underline{m}_1}{\dot{c}} \simeq \frac{\underline{p}_1 (2 - \dot{c}/n)}{(\dot{c}/n)(2 - \underline{p}_1)} \tag{48}$$

Dividing Formula (25) by Formula (48) yields

$$\frac{\overline{m}}{\underline{m}_1} \simeq \frac{\overline{p}(2 - \underline{p}_1)}{\underline{p}_1(2 - \overline{p})} \tag{49}$$

In view of the equivalence between $\overline{p}$ and $p_2$ and between $\underline{p}_1$ and $p_1$, Formula (49) may be written

$$\frac{\overline{m}}{\underline{m}_1} \simeq \frac{p_2(2 - p_1)}{p_1(2 - p_2)} \tag{50}$$

We may define

$$S = \frac{\overline{m}}{\underline{m}_1} \tag{51}$$

$$\hat{S} = \frac{p_2(2 - p_1)}{p_1(2 - p_2)} \tag{52}$$

Table 13 shows for various values of S and for various combinations of $\alpha$ for $p_1$ and $\beta$ for $p_2$, the corresponding values of acceptance number $\dot{c}$. Since $\hat{S} \simeq S$, the procedure for determining $\dot{c}$ and n is as follows: (1) Calculate $\hat{S}$ from Formula (52). (2) In Table 13 for the desired combination of $\alpha$ and $\beta$ find S nearest to $\hat{S}$. Find $\dot{c}$ corresponding to S. If necessary, interpolate linearly between values of S bounding $\hat{S}$ to find $\dot{c}$; $\dot{c}$ is an integer. (3) Find $\overline{m}$ corresponding to $\dot{c}$ in Table 13; or else in Table 1 or 3, or by the formula in Table 7–E, find the value of $\overline{m}$ corresponding to $\dot{c}$ for $\gamma_1 = 1 - \beta$. (4) Calculate n from Formula (53):

$$n = \frac{\overline{m}}{p_2} - \frac{\overline{m} - \dot{c}}{2} \tag{53}$$

Formula (53) is a rearrangement of terms in Formula (1), with $p_2$ substituted for $\overline{p}$.

The above procedure for obtaining n and $\dot{c}$ is intended for rates $p_1$ and $p_2 \leqslant .25$. Then n is substantially accurate in the sense that the probability is $\beta$ of obtaining $\dot{c}$ or fewer error items if $p = p_2$, and the probability is $\alpha$ of obtaining $\dot{c}_1$ or more error items if $p = p_1$. Formula (53) tends to favor accuracy with respect to $p_2$ and risk $\beta$

rather than with respect to $p_1$ and risk $\alpha$. This is equivalent to stating that Formula (53) seeks to meet the following criterion: If the number of error items c in n equals the acceptance number $\dot{c}$, then $\overline{p} = p_2$ at confidence level $\gamma_1 = 1 - \beta$. Tests for sample sizes to 1,000, based on values of $p_2$ and $p_1$ typically employed, indicate that n obtained from Formula (53) generally tends to be within one or two items of meeting this criterion.

If it is desired to have sample size favor accuracy with respect to $p_1$ and risk $\alpha$, sample size may be calculated from Formula (54), which is a rearrangement of terms in Formula (10), with $\dot{c}_1$ substituted for c, $\underline{m}_1$ for $\underline{m}$, and $p_1$ for $\underline{p}$:

$$n = \frac{\underline{m}_1}{p_1} + \frac{\dot{c}_1 - \underline{m}_1 - 1}{2} \tag{54}$$

where $\underline{m}_1$ is the value of $\underline{m}$ for $\dot{c}_1$. Values of $\underline{m}$ and c for $\gamma_1 = 1 - \alpha$ may be obtained from Table 2, 4, or 7—E; c is read as $\dot{c}_1$, and $\underline{m}$ as $\underline{m}_1$. Formula (54) has high accuracy, comparable with that of Formula (53), and seeks to meet the following criterion: If the number of error items in n equals the rejection number $\dot{c}_1$, then $\underline{p}_1 = p_1$ at confidence level $\gamma_1 = 1 - \alpha$.

To test the value of n obtained from Formula (53), one may calculate $\overline{p}$ based on acceptance number $\dot{c}$, n, and $\gamma_1 = 1 - \beta$. If $\overline{p} = p_2$ (as nearly as integers for $\dot{c}$ and n allow), sample size is correct. Otherwise one may empirically adjust n; in view of the accuracy of Formula (53), the adjustment will typically be on the order of one or two sample items. Similarly, to test the value of n obtained from Formula (54), one may calculate $\underline{p}_1$ based on rejection number $\dot{c}_1$, n, and $\gamma_1 = 1 - \alpha$. If $\underline{p}_1 = p_1$ (as nearly as integers for $\dot{c}_1$ and n allow), sample size is correct. Otherwise one may empirically adjust n, with the adjustment typically being on the order of one or two items.

Formula (53) has a slight tendency to produce smaller sample sizes than Formula (54). It should be borne in mind that upward adjustment of n, with acceptance number $\dot{c}$ (and rejection number $\dot{c}_1$) remaining the same, serves to decrease the risk of accepting a population with error rate $p_2$ but to increase the risk of rejecting a population with error rate $p_1$. This suggests the desirability of calculating sample size on the basis of Formula (53), thereby avoiding an unwanted increase in the risk of rejection.

Table 13 is an adaptation of Cameron's table [4] but is broader in scope and accompanied by refinements and changes in use. Table 13 contains more risk probabilities $\alpha$ and $\beta$, together with combinations of these probabilities, and it supplies many more values of $S = \overline{m}/\underline{m}_1$ with corresponding values of $\dot{c}$ and $\overline{m}$. Cameron's table is entered on the basis of $p_2/p_1$; for more accurate sample size, Table 13 is entered on the basis of $\hat{S} = (p_2/p_1) \times (2 - p_1)/(2 - p_2)$. Cameron's table suggests calculating $n = \underline{m}_1/p_1$; use of Formula (53) or (54) yields a more accurate sample size. (It is appropriate at this point to acknowledge that Tables 8 and 9 and their accompanying procedures for determining sample size are patterned after Cameron's table.)

Values of $\overline{m}$ and $\underline{m}_1$ for construction of Table 13 are derived from [8] for $c \leqslant 160$ and from Formulas (5) and (13) for $c > 160$. Values of $\overline{m}$ and $\underline{m}_1$ underlying $S$ in Table 13 contain more significant figures than $\overline{m}$ and $\underline{m}$ (read as $\underline{m}_1$) in Tables 1–4 and than $\overline{m}$ in Table 13.

### Alternative Procedure for Obtaining $\dot{c}$

In place of Table 13, Formula (55) below may be used to determine $\dot{c}$. Otherwise the procedure for obtaining $\dot{c}$ is the same as already described in connection with Table 13. Based on Formulas (5) and (13) in Appendix A, we may write as approximations $\overline{m} = c + A_\beta \sqrt{c + 1} + B_\beta$ and $\underline{m}_1 = c - A_\alpha \sqrt{c + 1} + B_\alpha$; A is the number of standard deviations for a given risk level based on the normal distribution, and $B = (A^2 + 2)/3$; the subscripts $\alpha$ and $\beta$ denote that A and B pertain to risk $\alpha$ or $\beta$. Thus we may write as an approximation $\overline{m}/\underline{m}_1 = S = \hat{S} = (c + A_\beta \sqrt{c+1} + B_\beta)/(c - A_\alpha \sqrt{c+1} + B_\alpha)$.    This transforms into $(\hat{S} - 1)(c + 1) - (\hat{S}A_\alpha + A_\beta)\sqrt{c+1} + (\hat{S}B_\alpha - B_\beta - \hat{S} + 1) = 0$. Substituting $\dot{c}$ for c, quadratic solution for $\dot{c}$ yields

$$\dot{c} = \left[ \frac{(\hat{S}A_\alpha + A_\beta) + \sqrt{(\hat{S}A_\alpha + A_\beta)^2 - 4(\hat{S} - 1)(\hat{S}B_\alpha - B_\beta - \hat{S} + 1)}}{2(\hat{S} - 1)} \right]^2 - 1$$

(55)

Values of A and B for $\alpha$ and $\beta$ are given in Table 11. (However, because of its complexity, the above formula is not presented and an example is not given in Table 11 and Chapter Three.)

### QUASI-FORMS OF ACCEPTANCE SAMPLING
### (TABLES 8, 9, 11, AND 12)

In some special situations, such as auditing, it may be desired to follow an acceptance-sampling approach only with respect to risk $\beta$.

To illustrate, an auditor checking a population of purchase vouchers for a specified type of error may seek assurance through a sample that the proportion of vouchers containing such error does not exceed a rate considered consistent with audit objectives. In the language of proportion sampling, he might wish to be 95% confident that the error rate does not exceed, say, .04. In terms of acceptance sampling, he would define $p_2$ as .04 and $\beta$ as 5%. Provided he can further specify $\hat{p}$, the anticipated error rate in the sample, he can determine sample size and acceptance number. Based on the equivalence between $\overline{p}$ and $p_2$, Formula (28) becomes

$$\hat{Q} = \frac{p_2 (2 - \hat{p})}{\hat{p}(2 - p_2)} \tag{56}$$

Then Table 8 and the accompanying procedure previously described are used to determine $\dot{c}$, the acceptance number, and n is determined from Formula (53), using $\overline{m}$ from Table 1, 3, or 7—E. Alternatively, $\dot{c}$ may be obtained from Formula (38), which appears in Table 11, instead of from Table 8, and Formula (53) is used as before to determine n.

When $\hat{p} = 0$, the acceptance number is zero and Formula (53) is again employed to determine n. However, $\dot{c}$ may then be omitted from the formula, resulting in

$$n = \frac{\overline{m}}{p_2} - \frac{\overline{m}}{2} \qquad \hat{p} = 0 \tag{57}$$

Similarly there may be situations where it is desired to follow an acceptance-sampling approach only with respect to risk $\alpha$. For example, one might seek assurance through a sample that the error rate is at a level high enough to warrant some course of action. Based on the equivalence between $\underline{p}_1$ and $p_1$, Formula (35) becomes

$$\hat{R} = \frac{p_1 (2 - \hat{p})}{\hat{p}(2 - p_1)} \tag{58}$$

Then Table 9 and the accompanying procedure previously described are used to determine $\dot{c}_1$, the rejection number; the values of c in Table 9 are read as $\dot{c}_1$. n is determined from Formula (54), using $\underline{m}_1$ from Table 2, 4, or 7—E; the values of c and $\underline{m}$ in these tables are read as $\dot{c}_1$ and $\underline{m}_1$. Alternatively, $\dot{c}_1$ may be obtained from Formula (39), which appears in Table 11, instead of from Table 9 (again reading c as $\dot{c}_1$ in the formula); and Formula (54) is used as before to determine n.

It should be noted that although $\hat{p}$ denotes the anticipated error rate in both Formulas (56) and (58), it refers to the highest anticipated rate in Formula (56) and to the lowest anticipated rate in Formula (58).

# Derivation of Formulas for the Finite Population Correction[1]

The tables, formulas, and procedures for obtaining confidence limits and sample sizes discussed in Appendixes A–C assume an infinite population. This assumption is now removed.

If an attribute sample of size n is drawn without replacement from a finite population of size N, the hypergeometric distribution correctly provides the probabilities on which confidence limits are based. The difficulties of working with the hypergeometric distribution may be avoided by applying a finite population correction (FPC) to the binomial confidence limit, yielding a good approximation of the exact confidence limit. Thus, in the case of the upper confidence limit [1], [5],

$$\overline{p}_F = \frac{c_F}{n_F} + \left(\overline{p} - \frac{c_F}{n_F}\right)\sqrt{\frac{N - n_F}{N - 1}} \tag{59}$$

where $\overline{p}_F$ designates the upper confidence limit incorporating the

[1]References cited and symbols employed in this appendix appear in Appendixes E and F, respectively.

FPC, $n_F$ is the sample size incorporating the FPC, $c_F$ is the number of events in $n_F$, and $\bar{p}$ is based on $c_F/n_F$.

Designating $\bar{p}_F - c_F/n_F$ as $\bar{e}_F$ and $\bar{p} - c_F/n_F$ as $\bar{e}$, Formula (59) becomes

$$\bar{e}_F = \bar{e}\sqrt{\frac{N - n_F}{N - 1}} \qquad (60)$$

If n (without the FPC) results in an error margin equal to $\bar{e}_F$, Formula (31) in Appendix B suggests that $\bar{e}_F/\bar{e} = \sqrt{n_F/n}$. Thus Formula (60) becomes $\sqrt{n_F/n} = \sqrt{(N - n_F)/(N - 1)}$. Squaring both sides and rearranging terms, we obtain the well-known FPC formula

$$n_F = \frac{n \times N}{n + N - 1} \qquad (61)$$

If $n_F$ in Formula (60) does not result in $\bar{e}_F$ being equal to specified $\hat{e}$, we may state that $n_F'$ results in $\hat{e}$; that is, $\hat{e} = \bar{e}'\sqrt{(N - n_F')/(N - 1)}$. Based on Formula (31), $\bar{e}/\bar{e}' = \sqrt{n_F'/n_F}$. Thus $\bar{e}_F/\hat{e} = \sqrt{n_F'/n_F} \times \sqrt{(N - n_F)/(N - n_F')}$. Squaring both sides and rearranging terms leads to

$$n_F' = \frac{n_F \times N}{n_F + (N - n_F)(\hat{e}/\bar{e}_F)^2} \qquad (62)$$

Formula (59) may be stated in general form so as to apply to any n by substituting c for $c_F$ and n for $n_F$:

$$\bar{p}_F = \frac{c}{n} + \left(\bar{p} - \frac{c}{n}\right)\sqrt{\frac{N - n}{N - 1}} \qquad (63)$$

Similarly, a good approximation to the lower confidence limit for a finite population is provided by [2], [5]

$$\underline{p}_F = \frac{c}{n} - \left(\frac{c}{n} - \underline{p}\right)\sqrt{\frac{N - n}{N - 1}} \qquad (64)$$

Formula (64), with $c_F$ and $n_F$ in place of c and n, leads in the same manner as Formula (59) to Formula (61) and to

$$\underline{e}_F = \underline{e}\sqrt{\frac{N - n_F}{N - 1}} \tag{65}$$

$$n_F' = \frac{n_F \times N}{n_F + (N - n_F)(\hat{e}/\underline{e}_F)^2} \tag{66}$$

Since $\sqrt{(N - n)/(N - 1)} \simeq \sqrt{1 - n/N}$, and since $\sqrt{1 - n/N} \simeq 1 - n/2N$ when $n/N$ is substantially smaller than 1, we may replace $\sqrt{(N - n)/(N - 1)}$ by $1 - n/2N$ in Formula (63). Hence Formula (63) becomes

$$\overline{P}_F \simeq \frac{c}{n} + \left(\overline{p} - \frac{c}{n}\right)\left(1 - \frac{n}{2N}\right)$$

Rearrangement of terms leads to

$$\overline{P}_F \simeq \overline{p} - \frac{n\overline{p} - c}{2N} \tag{67}$$

Similarly, Formula (64) leads to

$$\underline{P}_F \simeq \underline{p} + \frac{c - n\underline{p}}{2N} \tag{68}$$

## APPENDIX E

# References Cited
# in Appendixes A–D

[1]  Anderson, T. W., and Herman Burstein: Approximating the Upper Binomial Confidence Limit, *J. Am. Statist. Assoc.*, pp. 857–861, September, 1967.

[2]  —— and ——: Approximating the Lower Binomial Confidence Limit, *J. Am. Statist. Assoc.*, pp. 1413–1415, December, 1968.

[3]  —— and ——: "Determining the Appropriate Sample Size for Confidence Limits for a Proportion," Technical Report No. 3 prepared under the auspices of the Office of Naval Research, Department of Statistics, Stanford University, Stanford, Calif.

[4]  Cameron, J. M.: Tables for Constructing and for Computing the Operating Characteristics of Single Sampling Plans, *Ind. Quality Contr.*, pp. 37–39, July, 1952.

[5]  Cochran, William G.: "Sampling Techniques," 2d ed., John Wiley & Sons, Inc., New York, 1963, pp. 58, 60.

[6]  Cooke, J. R., M. T. Lee, and J. P. Vanderbeck: "Binomial Reliability Table (Lower Confidence Limits for the Binomial Distribution)," U.S. Naval Ordnance Test Station, China Lake, Calif., 1964. (Available from National Bureau of Standards, Clearinghouse for Federal Scientific and Technical Information, Springfield, Va.)

[7]  Hald, A., and E. Kousgaard: A Table for Solving the Equation B(c, n, p) = P for c = 0 (1) 100 and 15 values of P, Institute of Mathematical Statistics, Univ. of Copenhagen, November, 1966.

454

[8]  Harter, H. Leon. "New Tables of the Incomplete Gamma-Function Ratio and of Percentage Points of the Chi-Square and Beta Distributions," Aerospace Res. Labs., U.S. Air Force, 1964.

[9]  Leone, F. C., G. E. Haynam, J. T. Chu, and C. W. Topp: "Percentiles of the Binomial Distribution," Case Institute of Technology, Cleveland, Ohio, June, 1960.

# APPENDIX F

# Glossary of Symbols[1]

| A | Number of standard deviations for a given confidence level $\gamma$, based on the normal distribution, and also for a given risk level $\alpha$ or $\beta$ ($\gamma_1 = 1 - \alpha$ and $\gamma_1 = 1 - \beta$). Values of A are given in Tables 7—E, 10, and 11. |
|---|---|
| $(A_\alpha)$ | A pertaining to $\alpha$. |
| $(A_\beta)$ | A pertaining to $\beta$. |
| $\overline{a}$ | Constant in Table 1, for given $\gamma$ and c, for calculating $\overline{p}'$. |
| $\overline{a}'$ | Value of $\overline{a}$ obtained from formula in Table 7—A; $a' = \overline{m}/4 - f$. |
| $\underline{a}$ | Constant in Table 2, for given $\gamma$ and c, for calculating $\underline{p}'$. |
| $\underline{a}'$ | Value of $\underline{a}$ obtained from formula in Table 7—B; $a' = \underline{m}/3 + g$. |
| B | Term in formulas for $\overline{m}'$ and $\underline{m}'$ and for c, $\dot{c}$, and $\dot{c}_1$; $B = (A^2 + 2)/3$. Values of B are given in Tables 7—E and 11. |
| $(B_\alpha)$ | B pertaining to $\alpha$. |
| $(B_\beta)$ | B pertaining to $\beta$. |
| $\overline{b}$ | Constant in Table 1, for given $\gamma$ and c, for calculating $\overline{p}'$. |
| $\overline{b}'$ | Value of $\overline{b}$ obtained from formula in Table 7—A; $\overline{b}' = (2c - \overline{m})/4 - f$. |
| $\underline{b}$ | Constant in Table 2, for given $\gamma$ and c, for calculating $\underline{p}'$. |

[1] ( ) denotes symbols appearing only in the appendixes.

$\underline{b}'$     Value of $\underline{b}$ obtained from formula in Table 7–B; $\underline{b}' = (3c - 3 - \underline{m})/6 + g$.

c     Number of events (items having a designated characteristic) in the sample; c may be actual or anticipated. Formulas for anticipated c are given in Table 11.

$c_F$     c (anticipated), reduced by the finite population correction; $c_F = n_F \hat{p}$.

c/n     Proportion of events in the sample.

$\dot{c}$     Acceptance number—maximum number of error (defective) items in the sample permitting the lot (population) to be accepted as satisfactory. Values of $\dot{c}$ are given in Table 13 when rates $p_1$ and $p_2$ and risks $\alpha$ and $\beta$ are specified (alternatively, formula for $\dot{c}$ is given in Appendix C, page 448) and given in Table 8 when only rate $p_2$ and risk $\beta$ are specified (alternatively, formula for $\dot{c}$ is given in Table 11).

$\dot{c}_F$     $\dot{c}$ reduced by the finite population correction; $\dot{c}_F = (n_F \times \dot{c})/n$.

$\dot{c}_1$     Rejection number—minimum number of error (defective) items in the sample permitting the lot (population) to be rejected as unsatisfactory; $\dot{c}_1 = \dot{c} + 1$. $\dot{c}$ is given in Table 13 when rates $p_1$ and $p_2$ and risks $\alpha$ and $\beta$ are specified; $\dot{c}_1$ is given in Table 9 when only rate $p_1$ and risk $\alpha$ are specified (alternatively, formula for $\dot{c}_1$ is given in Table 11).

$\dot{c}_{1F}$     $\dot{c}_1$ reduced by the finite population correction; $\dot{c}_{1F} = (n_F \times \dot{c}_1)/n$.

d     Auxiliary term in Table 7–C for calculating $\overline{k}'$.

e     Error margin; interval between c/n and $\overline{p}$ or between c/n and $\underline{p}$ .

$\overline{e}$     Upper error margin; $\overline{e} = \overline{p} - c/n$.

$\underline{e}$     Lower error margin; $\underline{e} = c/n - \underline{p}$.

$\hat{e}$     Anticipated value of e. $\hat{e}$ refers to $\hat{\overline{e}}$ when both $\overline{p}$ and $\underline{p}$, or only $\overline{p}$, are desired; $\hat{e}$ refers to $\hat{\underline{e}}$ when only $\underline{p}$ is desired.

$\hat{\overline{e}}$     Anticipated value of $\overline{e}$.

$\hat{\underline{e}}$     Anticipated value of $\underline{e}$.

$\overline{e}_F$     $\overline{e}$ reduced by the finite population correction;

$$\overline{e}_F = \overline{e}\sqrt{(N - n_F)/(N - 1)}.$$

$\underline{e}_F$     $\underline{e}$ reduced by the finite population correction;

$$\underline{e}_F = \underline{e}\sqrt{(N - n_F)/(N - 1)}.$$

f     Auxiliary term in Table 7–A for calculating $\overline{a}'$ and $\overline{b}'$.

g     Auxiliary term in Table 7–B for calculating $\underline{a}'$ and $\underline{b}'$.

(K̄)     Constant for calculating $\overline{p}' = \overline{m}/(n + \overline{K})$ when c/n $\leqslant$ .1 and $\dot{c} > 50$; for $c > 50$, $\overline{K} = (\overline{m} - c)/2$; for $c \leqslant 50$, $\overline{K}$ is replaced by $\overline{k}$ .

$\overline{k}$     Constant in Table 3, for given $\gamma$ and c, for calculating $\overline{p}'$.

$\overline{k}'$     Value of $\overline{k}$ obtained from formula in Table 7–C; $\overline{k}' = (\overline{m} - c)/2 + d$.

$\underline{k}$     Constant in Table 4, for given $\gamma$ and c, for calculating $\underline{p}'$.

$\underline{k}'$     Value of $\underline{k}$ obtained from formula in Table 7–D; $\underline{k}' = (c - 1 - \underline{m})/2$.

m     Parameter of a Poisson distribution.

$\overline{m}$     Upper confidence limit for m, based on c (or ċ) and $\gamma$. Values are given in Tables 1, 3, and 13.

$\overline{m}'$     Value of $\overline{m}$ obtained from formula in Table 7–E; $\overline{m}' = c + A\sqrt{c + 1} + B$; $c \geqslant 50$. ($\overline{m}'$ may be based on ċ instead of c.)

$\underline{m}$     Lower confidence limit for m, based on c and $\gamma$. Values are given in Tables 2 and 4.

$\underline{m}'$     Value of $\underline{m}$ obtained from formula in Table 7–E; $\underline{m}' = c - 1 - A\sqrt{c} + B$; $c \geqslant 50$.

$\underline{m}_1$     Lower confidence limit for m, based on $\dot{c}_1$ and $\gamma_1$. Values are given in Tables 2 and 4 (by reading c as $\dot{c}_1$ and $\underline{m}$ as $\underline{m}_1$).

$\underline{m}'_1$     Value of $\underline{m}_1$ obtained from formula in Table 7–E; $\underline{m}'_1 = \dot{c}_1 - 1 - A\sqrt{c_1} + B$.

N     Population size.

n     Sample size.

n′     n adjusted for appropriate sample size (so that $e = \hat{e}$ if $c/n = \hat{p}$); $n' = n \times (e/\hat{e})^2$.

$n_F$     n reduced by the finite population correction; $n_F = (n \times N)/(n + N - 1)$.

$n'_F$     $n_F$ adjusted for Appropriate Sample Size (so that $e_F = \hat{e}$ if $c/n = \hat{p}$, in the case of proportion sampling, and so that $\overline{p}_F = p_2$ if $c = \dot{c}_F$, or $\underline{p}_F = p_1$ if $c = \dot{c}_{1F}$, in the case of acceptance sampling). In the case of proportion sampling, $n'_F = (n_F \times N)/[n_F + (N - n_F)(\hat{e}/e_F)^2]$; in the case of acceptance sampling $n'_F$ is found by trial and error, using linear interpolation based on trial values of $n_F$.

(n*)     Largest integer in $12.5 \times c$ but not less than 50.

p     Proportion of events in the population.

$\hat{p}$     Anticipated value of c/n.

$\overline{p}$     Upper confidence limit for p based on c/n. Exact values of $\overline{p}$ for $c/n \leqslant .5$ (approximately) and $n \leqslant 20$ are given in Table 5.

$\overline{p}'$     Approximation of $\overline{p}$:
Table 1, Long Formula: for $c/n \leqslant .5$ and $n \geqslant 20$,

$$\overline{p}' = \frac{\overline{m}}{n} \times \frac{n - \overline{a}}{n - b}$$

Table 3, Short Formula: for $c/n \leqslant .1$ and $n \geqslant 40$,

$$\overline{p}' = \frac{\overline{m}}{n + \overline{k}}$$

$\hat{\overline{p}}$     Anticipated upper confidence limit; $\hat{\overline{p}} = \hat{p} + \hat{\overline{e}}$.

$\overline{p}_F$    $\overline{p}$ reduced by the finite population correction;
$\overline{p}_F = c/n + (\overline{p} - c/n)\sqrt{(N-n)/(N-1)}$.

$\underline{p}$    Lower confidence limit for p based on c/n. Exact values of $\underline{p}$ for $c/n \leqslant$ .5 (approximately) and $n \leqslant 20$ are given in Table 6.

$\underline{p}'$    Approximation of $\underline{p}$:
Table 2, Long Formula: for $c/n \leqslant .5$ and $n \geqslant 20$,

$$\underline{p}' = \frac{m}{n} \times \frac{n-a}{n-b}$$

Table 4, Short Formula: for $c/n \leqslant .125$ and $n \geqslant 16$,

$$\underline{p}' = \frac{m}{n-k}$$

$\hat{\underline{p}}$    Anticipated lower confidence limit; $\hat{\underline{p}} = \hat{p} - \hat{\underline{e}}$.

$\underline{p}_F$    $\underline{p}$ increased by the finite population correction;
$\underline{p}_F = c/n - (c/n - \underline{p})\sqrt{(N-n)/(N-1)}$.

$(\underline{p}_1)$    Lower confidence limit for p based on $\dot{c}_1/n$.

$\overline{p}_r$    Upper confidence limit based on r/n; based on c/n, $\underline{p} = 1 - \overline{p}_r$.

$\underline{p}_r$    Lower confidence limit based on r/n; based on c/n, $\overline{p} = 1 - \underline{p}_r$.

$p_1$    Satisfactory error (defective) rate.

$p_2$    Unsatisfactory error (defective) rate.

$Q$    $\overline{m}/c$; values of Q are given in Table 8.

$\hat{Q}$    Approximation of Q; $\hat{Q} = \hat{\overline{p}}/\hat{p} \times (2 - \hat{p})/(2 - \hat{\overline{p}})$ or
$\hat{Q} = p_2/\hat{p} \times (2 - \hat{p})/(2 - p_2)$.

$\hat{q}$    $1 - \hat{p}$.

$\hat{\overline{q}}$    $1 - \hat{\overline{p}}$.

$\hat{\underline{q}}$    $1 - \hat{\underline{p}}$.

$R$    $\underline{m}/c$; values of R are given in Table 9.

$\hat{R}$    Approximation of R; $\hat{R} = \hat{\underline{p}}/\hat{p} \times (2 - \hat{p})/(2 - \hat{\underline{p}})$ or
$\hat{R} = p_1/\hat{p} \times (2 - \hat{p})/(2 - p_1)$.

$r$    $n - c$.

$r/n$    $(n - c)/n$.

$S$    $\overline{m}/\underline{m}_1$ based on $\dot{c}$ and $\dot{c}_1$. Values of S are given in Table 13.

$\hat{S}$    Approximation of S; $\hat{S} = p_2/p_1 \times (2 - p_1)/(2 - p_2)$.

$\alpha$    Risk of rejecting a lot (population) with rate $p_1$.

$\beta$    Risk of accepting a lot (population) with rate $p_2$.

$\gamma$    Confidence level, general term denoting either $\gamma_1$ or $\gamma_2$.

$\gamma_1$   Confidence level for statement about only $\bar{p}$ or only $\underline{p}$.

$\gamma_2$   Confidence level for statement about both $\bar{p}$ and $\underline{p}$.

# Index